Advances in Intelligent Systems and Computing

Volume 1137

The series "Advances in Intelligent Systems and Computing" contains publications on theory, applications, and design methods of Intelligent Systems and Intelligent Computing. Virtually all disciplines such as engineering, natural sciences, computer and information science, ICT, economics, business, e-commerce, environment, healthcare, life science are covered. The list of topics spans all the areas of modern intelligent systems and computing such as: computational intelligence, soft computing including neural networks, fuzzy systems, evolutionary computing and the fusion of these paradigms, social intelligence, ambient intelligence, computational neuroscience, artificial life, virtual worlds and society, cognitive science and systems, Perception and Vision, DNA and immune based systems, self-organizing and adaptive systems, e-Learning and teaching, human-centered and human-centric computing, recommender systems, intelligent control, robotics and mechatronics including human-machine teaming, knowledge-based paradigms, learning paradigms, machine ethics, intelligent data analysis, knowledge management, intelligent agents, intelligent decision making and support, intelligent network security, trust management, interactive entertainment, Web intelligence and multimedia.

The publications within "Advances in Intelligent Systems and Computing" are primarily proceedings of important conferences, symposia and congresses. They cover significant recent developments in the field, both of a foundational and applicable character. An important characteristic feature of the series is the short publication time and world-wide distribution. This permits a rapid and broad dissemination of research results.

** Indexing: The books of this series are submitted to ISI Proceedings, EI-Compendex, DBLP, SCOPUS, Google Scholar and Springerlink **

More information about this series at http://www.springer.com/series/11156

Álvaro Rocha · Carlos Ferrás ·
Carlos Enrique Montenegro Marin ·
Víctor Hugo Medina García
Editors

Information Technology and Systems

Proceedings of ICITS 2020

 Springer

Editors
Álvaro Rocha
DEI/FCT
University of Coimbra
Coimbra, Portugal

Carlos Ferrás
Facultad de Geografia
University of Santiago de Compostela
Santiago de Compostela, Spain

Carlos Enrique Montenegro Marin
Facultad de Ingeniería
Universidad Distrital Francisco José de
Bogota, Colombia

Víctor Hugo Medina García
Facultad de Ingeniería
Universidad Distrital Francisco José de
Bogota, Colombia

ISSN 2194-5357   ISSN 2194-5365 (electronic)
Advances in Intelligent Systems and Computing
ISBN 978-3-030-40689-9   ISBN 978-3-030-40690-5 (eBook)
https://doi.org/10.1007/978-3-030-40690-5

This Springer imprint is published by the registered company Springer Nature Switzerland AG
The registered company address is: Gewerbestrasse 11, 6330 Cham, Switzerland

Preface

This book is composed of the papers written in English and accepted for presentation and discussion at The 2020 International Conference on Information Technology & Systems (ICITS'20). This conference had the support of the University Distrital Francisco José de Caldas, IEEE Systems, Man, and Cybernetics Society, and AISTI (Iberian Association for Information Systems and Technologies). It took place in Bogotá, Colombia, February 5–7, 2020.

The 2020 International Conference on Information Technology & Systems (ICITS'20) is an international forum for researchers and practitioners to present and discuss the most recent innovations, trends, results, experiences, and concerns in the several perspectives of information technology and systems.

The Program Committee of ICITS'20 was composed of a multidisciplinary group of 152 experts and those who are intimately concerned with information systems and technologies. They have had the responsibility for evaluating, in a 'double-blind review' process, the papers received for each of the main themes proposed for the conference: (A) information and knowledge management; (B) organizational models and information systems; (C) software and systems modeling; (D) software systems, architectures, applications, and tools; (E) multimedia systems and applications; (F) computer networks, mobility, and pervasive systems; (G) intelligent and decision support systems; (H) big data analytics and applications; (I) human–computer interaction; (J) ethics, computers, and Security; (K) health informatics; (L) information technologies in education; (M) antenna systems and technologies.

ICITS'20 also included several workshop sessions taking place in parallel with the conference ones. They were sessions of WMETACOM 2020 – 3rd Workshop on Media, Applied Technology and Communication, and of IKIT 2020 – 1st Workshop on Information and Knowledge in the Internet of Things.

ICITS'20 received more than 300 contributions from 32 countries around the world. The papers accepted for presentation and discussion at the conference are published by Springer (this book) and by AISTI and will be submitted for indexing by ISI, EI-Compendex, SCOPUS, DBLP, and/or Google Scholar, among others.

We acknowledge all of those that contributed to the staging of ICITS'20 (authors, committees, workshop organizers, and sponsors). We deeply appreciate their involvement and support that was crucial for the success of ICITS'20.

February 2020   Álvaro Rocha
                                                                Carlos Ferrás
                                          Carlos Enrique Montenegro Marin
                                            Víctor Hugo Medina García

Organization

Conference

Honorary Chair

Álvaro Rocha   University of Coimbra, Portugal

Scientific Committee Chair

Carlos Ferrás Sexto   University of Santiago de Compostela, Spain

Local Organizing Chair

Julio Baron Velandia   Universidad Distrital Francisco José de Caldas, Colombia

Local Organizing Committee

Carlos Enrique Montenegro Marín   Universidad Distrital Francisco José de Caldas, Colombia

Victor Hugo Medina García   Universidad Distrital Francisco José de Caldas, Colombia

Alvaro Espinel Ortega   Universidad Distrital Francisco José de Caldas, Colombia

Giovanny Mauricio Tarazona Bermudez   Universidad Distrital Francisco José de Caldas, Colombia

Julio Baron Velandia   Universidad Distrital Francisco José de Caldas, Colombia

Roberto Ferro Escobar   Universidad Distrital Francisco José de Caldas, Colombia

José Ignacio Rodríguez   Universidad Distrital Francisco José de Caldas,
  Molano   Colombia
Paulo Alonso Gaona-Garcia   Universidad Distrital Francisco José de Caldas,
                                  Colombia

Scientific Committee

Aashish Bardekar   Sipna College of Engineering and Technology,
                                  Amravati, India
Alessandro Kraemer   UTFPR, Brazil
Alexandra González   Universidad Técnica Particular de Loja, Ecuador
Alexandru Vulpe   University Politehnica of Bucharest, Romania
Amal Al Ali   University of Sharjah, United Arab Emirates
André Marcos Silva   University Adventist of São Paulo, Brazil
André Kawamoto   Federal University of Technology, Brazil
Angeles Quezada   Universidad Autonoma de Baja California,
                                  Mexico
Angelica Caro   Universidad del Bío-Bío, Chile
Ania Cravero   University de La Frontera, Chile
Ankur Bist   KIET Ghaziabad, India
António Augusto Gonçalves   Universidade Estacio de Sá, Brazil
Antonios Andreatos   Hellenic Air Force Academy, Greece
Boris Shishkov   ULSIT/IMI - BAS/IICREST, Bulgaria
Borja Bordel   Universidad Politécnica de Madrid, Spain
Bruno da Silva Rodrigues   Universidade Presbiteriana Mackenzie, Brazil
Carlos Cares   Universidad de La Frontera, Chile
Carlos Carreto   Polytechnic of Guarda, Portugal
Carlos Grilo   Polytechnic of Leiria, Portugal
Carlos Hernan Fajardo Toro   Universidad Ean, Colombia
Claudio Cuevas   Federal University of Pernambuco, Brazil
Dália Filipa Liberato   ESHT/IPP, Portugal
Dante Carrizo   Universidad de Atacama, Chile
Diego Ordóñez-Camacho   Universidad Tecnológica Equinoccial, Ecuador
Dusan Petkovic   TH Rosenheim, Germany
Eddie Galarza   Universidad de las Fuerzas Armadas, Ecuador
Edgar Serna   Universidad Autónoma Latinoamericana,
                                  Colombia
Efraín R. Fonseca C.   Universidad de las Fuerzas Armadas ESPE,
                                  Ecuador
Enrique Carrera   Universidad de las Fuerzas Armadas ESPE,
                                  Ecuador
Erika Upegui   Universidad Distrital Francisco José de Caldas,
                                  Colombia
Ewaryst Tkacz   Silesian University of Technology, Poland
Felix Blazquez Lozano   University of A Coruña, Spain

Filipa Ferraz	University of Minho, Portugal
Felipe Machorro-Ramos	Universidad de las Américas Puebla, Mexico
Francisco Brito Filho	UFERSA, Brazil
Francisco Javier Valencia Duque	Universidad Nacional de Colombia, Colombia
Francisco Valverde	Universidad Central del Ecuador, Ecuador
Franklim Silva	Universidad de las Fuerzas Armadas, Ecuador
Gabriel Elías Chanchí Golondrino	Institución Universitaria Colegio Mayor del Cauca, Colombia
Gabriel Elías Chanchí Golondrino	Universidad de Cartagena, Colombia
George Suciu	BEIA, Romania
Hugo Peixoto	University of Minho, Portugal
Igor Aguilar Alonso	Universidad Nacional Tecnológica de Lima Sur, Peru
Ildeberto Rodello	University of São Paulo, Brazil
Irene Rivera-Trigueros	University of Granada, Spain
Isabel Pedrosa	Coimbra Business School ISCAC, Portugal
Ismael Gutierrez Garcia	Universidad del Norte, Colombia
Jan Kubicek	Faculty of Electrical Engineering and Computer Science VŠB-TUO, Czech Republic
Javier Criado	University of Almería, Spain
Javier Guaña-Moya	Pontificia Universidad Católica del Ecuador, Ecuador
Javier Medina	Universidad Distrital Francisco José de Caldas, Colombia
Jhony Alexander Villa-Ochoa	Universidad de Antioquia, Colombia
Joan-Francesc Fondevila-Gascón	EAE Business School, CECABLE, Spain
João Carlos Souza	University of Brasilia, Brazil
João Paulo Pereira	Polytechnic of Bragança, Portugal
João Vidal de Carvalho	ISCAP/IPP, Portugal
John Waterworth	Umeå University, Sweden
Jorge Buele	Universidad Técnica de Ambato, Ecuador
Jorge Luis Pérez	Universidad de Las Américas, Ecuador
Jose Aguilar	Universidad de los Andes, Venezuela
José Luís Reis	ISMAI – Instituto Universitário da Maia, Portugal
José Luís Silva	ISCTE-IUL and Madeira-ITI, Portugal
José Varela-Aldás	Universidad Indoamérica, Ecuador
Juan Pablo D'Amato	UNCPBA/CONICET, Argentina
Júlio Menezes Jr.	Federal University of Pernambuco, Brazil
Juncal Gutiérrez-Artacho	University of Granada, Spain
Jussi Okkonen	University of Tampere, Finland

Leandro Flórez Aristizábal	Antonio Jose Camacho University Institute, Colombia
Leonardo Botega	UNIVEM, Brazil
Leidy Yohana Flórez Gómez	Universidad Autónoma de Bucaramanga, Colombia
Lorena Siguenza-Guzman	Universidad de Cuenca, Ecuador
Lornel Antonio Rivas Mago	Universidad Andina del Cusco, Peru
Lukasz Tomczyk	Pedagogical University of Cracow, Poland
Luis Camacho	SUNY Empire State College, USA
Mafalda Teles Roxo	INESC TEC, Portugal
Manuel Au-Yong-Oliveira	University of Aveiro, Portugal
Marc Polo	Ramon Llull University, Spain
Marcelo de Paiva Guimarães	Federal University of São Paulo, Brazil
Marcelo V. Garcia	University of the Basque Country, Spain
Marciele Berger	University of Minho, Portugal
Marc Gonzalez Capdevila	FACENS, Brazil
Marco Quintana	UIDE, Ecuador
Maria Amelia Eliseo	Universidade Presbiteriana Mackenzie, Brazil
María de la Cruz del Río-Rama	University of Vigo, Spain
Maria de las Mercedes Canavesio	UTN - Facultad Regional Santa Fe, Argentina
Maria José Sousa	University of Coimbra, Portugal
María Teresa García-Álvarez	University of A Coruna, Spain
Mario Ron	Universidad de las Fuerzas Armadas ESPE, Ecuador
Maristela Holanda	University of Brasilia, Brazil
Michail-Alexandros Kourtis	NCSR Demokritos, Greece
Millard Escalona	Universidad Israel, Ecuador
Monica Leba	University of Petrosani, Romania
Neeraj Gupta	KIET Ghaziabad, India
Nelson Rocha	University of Aveiro, Portugal
Nikolai Prokopyev	Russian Academy of Sciences, Russia
Niranjan S. K.	JSS Science and Technology University, India
Norka Bedregal	Universidad Nacional de San Agustín de Arequipa, Peru
Oscar Medina	UTN-FRC, Argentina
Osvaldo Oliveira	UNIFACCAMP, Brazil
Pablo Alejandro Quezada Sarmiento	Universidad Internacional del Ecuador, Ecuador
Pablo Pico-Valencia	Pontifical Catholic University of Ecuador (Esmeraldas), Ecuador
Patricia Acosta	Universidad de Las Américas, Ecuador
Patricia Alexandra Quiroz Palma	Universidad Politécnica de Valencia, Spain

Paulo Batista	University of Évora, Portugal
Parama Bhaumik	Jadavpur University, India
Paulus Isap Santosa	Gadjah Mada University, Indonesia
Pedro Liberato	ESHT/IPP, Portugal
Piotr Kulczycki	Systems Research Institute, Polish Academy of Sciences, Poland
Ramayah T.	Universiti Sains Malaysia, Malaysia
Ramiro Delgado	Universidad de las Fuerzas Armadas ESPE, Ecuador
Renato Jose Sassi	Universidade Nove de Julho, Brazil
Rilwan Sabo Muhammad	Abubakar Tafawa Balewa University, Bauchi, Nigeria
Rodrigo Campos Bortoletto	São Paulo Federal Institute, Brazil
Rodrigo Hübner	UTFPR, Brazil
Rosa Galleguillos-Pozo	Universidad Politecnica de Cataluña, Spain
Ruth Reátegui Rojas	Universidad Técnica Particular de Loja, Ecuador
Samanta Patricia Cueva Carrión	Universidad Técnica Particular de Loja, Ecuador
Sampsa Rauti	University of Turku, Finland
Samuel Sepúlveda	Universidad de La Frontera, Chile
Sanaz Kavianpour	University of Technology, Malaysia
Sandra Costanzo	University of Calabria, Italy
Santoso Wibowo	CQUniversity, Australia
Saulo Barbará Oliveira	Universidade Federal Rural do Rio de Janeiro, Brazil
Sylvie Ratté	École de Technologie Supérieure, Canada
Sussy Bayona Ore	Universidad Nacional Mayor de San Marcos, Peru
Teresa Guarda	State University of Santa Elena Peninsula, Ecuador
Valeria Farinazzo Martins	Mackenzie Presbyterian University, Brazil
Victor Manuel Cornejo Aparicio	Universidad Nacional de San Agustín de Arequipa, Peru
Vitor Santos	NOVA IMS, Portugal
Wilmar Yesid Campo Muñoz	Universidad del Quindío, Colombia

Contents

Organizational Models and Information Systems

Review of Information Systems with Technological Development for Tourism Planning with an Emphasis on Host Communities

Marcia Ivonne Lara Silva[✉] [iD], Luz Andrea Rodríguez Rojas[✉] [iD], and Edgar Jacinto Rincón[✉] [iD]

Universidad Distrital Francisco José de Caldas,
Cra. 7 #40b-53, Bogotá, Colombia
ing.marcialara@gmail.com, ing.rodriguezla@gmail.com,
edgar.rincon.rojas@gmail.com

Abstract. Tourism planning focuses on the traction of tourists to generate economic growth; today, it is also necessary to understand how host communities are affected by this activity, since negative impacts have been evidenced by displacement, deculturization, affectation to their dignity, among others, in contravention of the main objectives of tourism. It has been observed that this planning has not been accompanied, to date, by technological developments that facilitate the analysis of data, causing a large amount of information to be lost. A documentary review was carried out aimed at identifying information systems that make use of information technologies, conditioned to the tourist planning of a territory. Findings: the phases of commercialization, execution and post-sale of the tourism, have technological tools designed for this purpose, about 6:1 to the planning, design, and execution, evidencing disadvantage in the management of the information in early ages of the tourist development. The bibliography does not include an information system for tourism planning. There is technological development for commercialization and sale in America and Europe consistent with the contribution made by the tourist activity to the GDP in each continent, 8.6%, and 11% respectively. However, the use of technology is not proportional to the planning, since it does not facilitate the observation and monitoring of the plans or their impacts on the destinations, resulting in the rejection of the communities.

Keywords: Early warning · Information system · Tourism planning · Host communities

1 Tourism

1.1 Negative Impacts of Tourism

Since the decade of the 70's the topic of early warnings has had the attention of researchers around the world, topics such as the prediction of natural disasters, food security and health projections, school desertion, migration and the human sense, have

© Springer Nature Switzerland AG 2020
Á. Rocha et al. (Eds.): ICITS 2020, AISC 1137, pp. 3–13, 2020.
https://doi.org/10.1007/978-3-030-40690-5_1

been studied with emphasis in the last decade, while the financial and economic fluctuations have been permanently studied during the last 4 decades.

It is interesting to observe how the study of early warnings has not been given in detail in terms of specific economic activities, but rather as the study of the behavior of some phenomena (environmental, social or economic), in order to be able to anticipate and mitigate their possible negative impacts, that is, from a causal point of view always using the tools of predictive analysis, as has been proposed by the whole positivist movement in its argumentation [1].

Perhaps this is the reason why until now tourism has not had an information system in terms of prevention, before degradation. However, if tourism is understood as a social phenomenon that studies the displacement of individuals to develop activities inherent to their daily life in terms of work, health, education and/or enjoyment, it allows its application as a phenomenon, but also as an economic activity. In this sense, the vision of explaining and understanding can complement the causal position of impacts, and allow understanding to make sense of what is observed [1].

Early warnings in tourism have been investigated since 1980, through the phases of the life cycle that a tourist destination may have, the subject has been related to the methods under the acronym TALC - Tourist Area Life Cycle, which speaks of the stagnation of the tourist destination as a phase of its life cycle, after which comes its decline or rejuvenation [2, 3]; In this sense, it is observed how the study of the tourist destination is carried out as if it were one more consumer product in the market and not from the prevention towards the territory.

The absence of an information system applied to tourism planning does not facilitate the identification of elements that could mitigate the impacts on the environment and on the host communities. Such is the case of the Marietas Islands in Mexico [4], the Caño Cristales River in Colombia [5], Venice in Italy [6], Amsterdam in Holland [7], Barcelona in Spain [8], Machupicchu in Peru [9], the Great Wall in China [10], to mention only some of the most recognized at the level of environment; all previous tourist destinations were negatively impacted by the large influx of tourists on which no preventive action was carried out as a result of tourism planning.

Other exposures besides environmental ones are observed in host communities composed of ethnic peoples, among others, affected for example, when tourist demand falls and they become massively unemployed, counterbalancing the proposed strategies for poverty reduction, as has been evidenced in Kenya when the influx of tourists to safaris decreases, where people belonging to minorities are employed and especially those with low economic possibilities [11–13]. Also, through the staging of the culture, traditions, and habitat of the destination receiving tourism, exposing ethnic elements in an unsustainable way [14, 15].

In this sense, authenticity, identity, the transformation of rituals, disrespect for individuals as persons being treated as tourist attractions and the environment itself have ended up with important alterations, as can be seen in the main Latin American tourist destination, Cancun, Quintana Roo - Mexico, where the vertical integration of tourism has motivated the massive arrival of tourists in a destination with an ethnic population, which has resulted in the displacement of 65% of the Mayan population, together with the loss of their language, abrupt changes in their way of dressing and transformations in their space [14].

In territories where tourism is used and ethnic host communities are present, the sense of identity of the population groups that inhabit it is affected, since their culture, tradition, language and even their way of life are assumed as tourist attractions, thus both their cultural identity and ethnic pride can be vulnerable.

1.2 Colombia - Destinations

In Latin America, Colombia is not the exception either. Promotion strategies have increased for destinations already identified as successful, but also for those that until now have not been fully exploited. Thanks to the open data portal www.datos.gov.co there is access to some data on international arrivals to the country. For data processing, the last 3 years are taken: 2016, 2017 and 2018. International arrivals have been plotted to facilitate understanding and to measure the importance of constructing a model for rural tourism planning (Figs. 1 and 2).

Countries that contribute more than 10,000 tourists per year to Colombia.

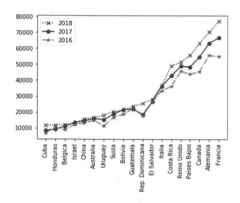

Fig. 1. >10.000 y <100.000 tourists by country. Source: Author. Adapted www. datos.gov.co

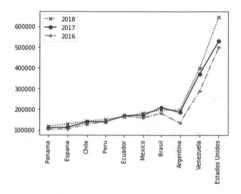

Fig. 2. >100.000 tourists by country. Source: Author. Adapted www.datos.gov.co

Reception of Tourists by Department. For their part, the departments of Colombia have the following reception of tourists.

It can be seen that the largest influx of tourists goes to the departments where the main cities Figs. 5 and 6 are located, but with the commercial effort of the National Government, it is the regions away from natural environments that are having the greatest visibility Figs. 3 and 4, to positively impact their economy. It is important to understand the importance of being able to plan an economic activity to be developed in a territory, to prevent undesirable influences in the host communities as well as in the environment.

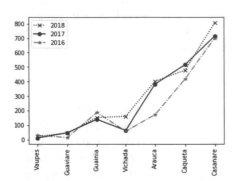

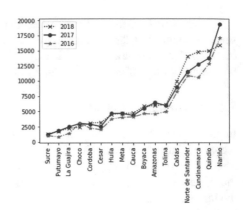

Fig. 3. <1.000 tourists by depart. Source: Author. Adapted www.datos.gov.co.

Fig. 4. >1.000 y <20.000 tourists by depart. Source: Author. Adapted www.datos.gov.co.

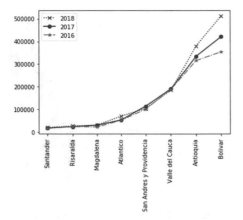

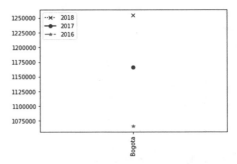

Fig. 5. >20.000 y <1.000.000 tourists by depart. Source: Author. Adapted www.datos. gov.co.

Fig. 6. >1.000.000 tourists by depart. Source: Author. Adapted www.datos.gov.co.

The absence of an expert information system to support tourism planning will make it difficult to monitor and review, to avoid detours or to adjust strategies in each region, thereby endangering not only Colombia's great biodiversity in its natural territory [16], but also the host ethnic communities: "87 indigenous ethnic groups, 3 differentiated groups of Afro-Colombian population and the ROM or gypsy people; 64 Amerindian languages are spoken, in addition to the bandé… where there are also 710 titled resguardos located in 27 departments" [17].

2 Information Systems Adopted in Tourism

Sustainable tourism is an activity that involves - society, economy, and environment [18], it is to be expected that information systems will develop and accompany the activity, just as they have strengthened processes in other fields of knowledge, at the same rate of growth that is evident in the world [19]. The application of technology based on information is observed in tourism in marketing activities and development of dissemination channels [20], taking into account suggestions based on the interests of travelers [21–24], the creation of creative tourist packages, the power to facilitate the tourist's early visualization of the destination he will visit, either through 3D systems or sensory systems [19, 26], as well as the tourism logistics network [27].

There have also been applications to strengthen sales processes, the purchase of lodging services through market segmentation [28], to bring the tourist experience closer to the emotion of the host at the destination [29, 30], to train communication skills in the English language while doing tourism [31], sale of services in the sector [32–35], to qualify the tourist service received [20, 36–41], minimize the risk of the tourist by natural disasters in the destination he visits [42]. At the same time, tourism has been used as a brand positioning tool for travel destinations [37].

The environment and society in the development of the tourist activity have not been left without attention, through geographic information systems [43] it has been possible to identify the effects that are generated on the destinations, locating the areas where the territory has suffered modifications due to the presence of tourists [44, 45]. It has also been necessary to optimize public energy service networks due to the high presence of tourists, such as Galapagos, using CYMDIST software [46]. Information systems have been applied to analyze the levels of irritation of the hosts of destinations located in cross-border destinations in Mainland China [47] and the phase in which the destination is located according to Butler's TALC methodology [3].

In terms of tourism planning, understood as the prior organization of the destination to be intervened [48], it should be seen as a strategy that facilitates the recognition for the development of a territory, through the organization, evaluation, and integration of all its elements [43, 49]; in this sense, geographic information systems have supported the establishment of tourism inventories [43], and techniques have also been developed to determine tourism demand [50], based on the tourism supply already established in the territory [37, 51]. STELLA software has even been used to create simulations of tourism behavior, which can be shared at any time and with anyone [52]. In Finland, PPGIS [53] has been brought closer to tourism to determine those destinations that are most attractive to visitors and those that have not been taken into account [54]. Scenario analyses have also been carried out to determine the sustainability of the destination with tourist activity [55] See Table 1 Information systems by continent.

However, although there is a variety of uses in different tourism activities, the development, and application of information-based technology in tourism are still considered very poor. There is no information system in the document review that would allow the follow-up or monitoring of tourism activity, from territorial planning, all the bibliography shows analyses after the activity, without these being linked to previous planning.

Table 1. Information systems by country (Source: Author)

Information System	Application	Year	Country	Phase
Abstract Remote Sensing (RS) and Geography Information System (GIS) technologies	Ecological analysis and territorial planning	2018	China	Planning
Creative Travel Management System Based on Software Reuse and Abstraction Techniques	Creative design of tourist packages	2017	China	Design
Intelligent tourism system and a practical design artefact	Exploration of the sustainable values of tourist experiences and the behaviour of local staff	2017	Japón	
Preference Ranking Organization Method for Enrichment Evaluation – PROMETHEE	Market segmentation	2002	Brasil	
Independent travel recommendation algorithm for professional travelers implemented on Android, using Java	Behavioral Suggestions	2017	China	Commercialization
Augmented reality system	Provide information about the environment and improve perception.	2017	Russia	
"Itchy feet", a prototype implementing this 3D e-Tourism environment	Show benefits of an environment	2007	Austria	
Geographic Information System to manage tourist information in Gijón	Make it easier for tourists to find places	2012	Austria	
Google Travel	Suggestions by behavior	2018	Chipre	
System Using Social Pertinent Trust Walker Algorithm - Travel Recommendation System	Suggestions by behavior	2016	India	

(continued)

Table 1. (*continued*)

Information System	Application	Year	Country	Phase
Airbnb	To offer lodging to the hosts of the city	2018	EEUU	Sales
Sabre Holding (Semi Automated Business-Related Enterprise) - Sistemas de Distribución Global (A.A. e IBM)	Reservations air transport, lodging and travel agencies	2018	EEUU	
Travelport		2018	EEUU	
Travelsky	Control of sales and operation of civil air operations	2018	China	Sales
Amadeus - Sistemas de Distribución Global (Air France, Iberia, Lufthansa y SAS)	Management of hotels, airlines, travel agencies, airports, ground transportation, railway companies, centralized	1987	España	
Immersive VR applications dedicated to training English for Tourism Purpose (ETP) skills	Practice languages while travelling	2017	Japón	Ejecution
TripAdvisor	To qualify the experience of the client in the reception of the service	2016	EEUU	Rating
Hotels.com		2016	EEUU	
Ctrip - Ahora Trip		2016	China	
Booking Holdings		2016	Holanda	
Trivago		2018	Alemania	
Geographic Information Systems	Confirm the carbon footprint left by tourism in the territory	2017	Thailandia	Monitoring

As a result of Table 1, the phases of commercialization, execution and post-sale of the tourism with greater development of technological tools are observed. Planning, design, and execution have less technological support, which makes it difficult to manage information at the early stages of tourism development and monitoring. It is observed that the development of information systems for the use of the tourism sector has had greater relevance for the marketing and sales phases in Europe and America.

2.1 Description of the SIPLAT Model

SIPLAT stands for Tourism Planning Information System. It is an expert system that records the fundamental information to support tourism planning in a territory with an important emphasis on the care and protection of host communities.

Its main function is diagrammed as follows (Figs. 7 and 8):

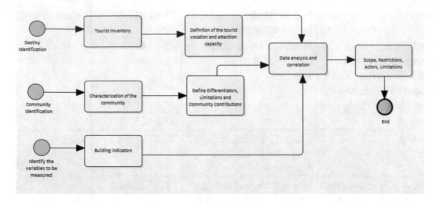

Fig. 7. Function of the tourism planning system. Source: Author

Modular design of SIPLAT:

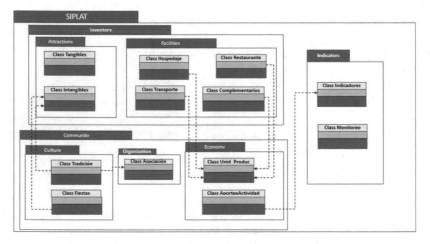

Fig. 8. Modularity SIPLAT. Source: Author

3   Conclusions

It is necessary to guarantee the reliability and accuracy of the data collected from the tourist inventor and the community that inhabits the destination, to guarantee the learning process.

It is necessary to homologate the existing information to carry out the first learning exercises of the expert system to be designed.

The expert system will provide an analysis of the most representative variables to identify the greatest amount of risks to the community during tourism planning during the execution of the tourism exercise.

References

1. Martínez, J.: La unidad del método científico. In: COntextox XVIII, pp. 43–76 (1990)
2. E. Comission: The tourism area 'life cycle': a clarification. Ann. Tour. Res. **38**(3), 1185–1187 (2011)
3. Diedrich, A., García-Buades, E.: Local perceptions of tourism as indicators of destination decline. Tour. Manag. **30**(4), 512–521 (2009)
4. Sermanat, La conanp anuncia que la playa del amor cerrará a partir del 9 de mayo 2016
5. Semana Sostenible, Cierran Caño Cristales a los turistas debido a la sequía (2017)
6. Hosteltur, Venecia, un destino en jaque por el turismo de masas y la corrupción (2014)
7. El Confidencial, En Ámsterdam no caben más turistas: 'Vivir aquí se ha convertido en una pesadilla' (2017)
8. Euronews, Barcelona: El problema del turismo masivo (2017)
9. La República – Perú: Turismo en Machupicchu sobrepasa límites en agosto (2013)
10. National Geographic: El deterioro de la Muralla China (2015)
11. UNNWTO: El turismo y la atenuación de la pobreza (2010). https://step.unwto.org/es/content/el-turismo-y-la-atenuacion-de-la-pobreza
12. Bolwell, W.D.: Reducir la pobreza a través del turismo (2009)
13. Akama, J.: Western environmental values and nature-based tourism in Kenya. Tour. Manag. **17**, 567–574 (1996)
14. Pereiro, X.: Los efectos del turismo en las culturas indígenas de América Latina. Rev. Esp. Antropol. Am. **43**(1), 155–174 (2013)
15. Ávila Romero, A.: Análisis del Turismo alternativo en comunidades indígenas de Chiapas, México. Etudes cCribeennes, vol. 31–32 (2015)
16. Colciencias, Colombia, el segundo país más biodiverso del mundo (2016)
17. PNUD: Colombia (2019). http://www.co.undp.org/content/colombia/es/home/countryinfo.html
18. Organización Mundial del Turismo: Medición creíble del turismo sostenible para mejorar la toma de decisiones (2018)
19. Bogomazova, I.V., Stenyushkina, S.G.: Excursion tours and the possibility of using augmented reality technologies for improving the local tourist attractiveness. J. Environ. Manag. Tour. **8**(4), 943–951 (2017)
20. Abrahams, B.: Tourism information systems integration and utilization within the semantic WEB, October 2006
21. Ravi, L., Vairavasundaram, S.: A collaborative location based travel recommendation system through enhanced rating prediction for the group of users. Comput. Intell. Neurosci. **2016** (2016)
22. Pan, Q., Wang, X.: Independent travel recommendation algorithm based on analytical hierarchy process and simulated annealing for professional tourist. Appl. Intell. **48**, 1–17 (2017)
23. Dergiades, T., Mavragani, E., Pan, B.: Google trends and tourists' arrivals: emerging biases and proposed corrections. Tour. Manag. **66**, 108–120 (2018)
24. Parreño, A.: Sistema De Informac Ión Geográfica Para Gestionar La Información Turistica De La Ciudad De Gijó N, Asturias (2012)

25. Ma, S., Yang, H., Shi, M.: Developing a creative travel management system based on software reuse and abstraction techniques. In: Proceedings - International Computer Software and Applications Conference, vol. 2, pp. 419–424 (2017)
26. Berger, H., Dittenbach, M.: Opening new dimensions for e-Tourism. Virtual Real. **11**, 75–87 (2007)
27. Torres, J.: Criterios básicos para la Caracterización de Redes Logísticas en la RAPE central (2017)
28. Souza, A., Silva, A., de Barbosa, L.M.: Understanding consumers' reluctance to purchase hotel services online: what makes it so risky? Pasos-Revista Tur. Y Patrim. Cult. **14**(5), 1253–1266 (2016)
29. Lim, C., Mostafa, N., Park, J.: Digital Omotenashi: toward a smart tourism design systems. Sustainability **9**(12) (2017)
30. Airbnb: Airbnb   (2018).   https://www.airbnb.com.co/?af=43720035&c=.pi0. pk57951510207_264142931501_c_21910762233&gclid= CjwKCAjw9qfZBRA5EiwAiq0AbeSaywAQAZldtNo3eiA_Gf4vmO8GnDkGeL_ eEF2l6Mpy2RpJQ5UVRhoChb0QAvD_BwE
31. Lee, B., Shih, H., Chou, Y., Chen, Y.: Educational Virtual Reality implementation on English for Tourism Purpose using knowledge-based engineering. In: IEEE International Conference on Applied System Innovation, pp. 792–795 (2017)
32. Amadeus: Amadeus (2018). http://www.amadeus.com
33. Sabre: Sabre (2018). https://www.sabre.com/
34. Travelsky: Travelsky (2018). http://www.travelsky.net
35. Travelport: Travelport (2018). https://www.travelport.com/sites/default/files/timline_master_ v_3-2016_usl-es-es.pdf
36. Leung, R., Au, N., Liu, J., Law, R.: Do customers share the same perspective? A study on online OTAs ratings versus user ratings of Hong Kong hotels. J. Vacat. Mark. **24**(2), 103–117 (2016)
37. Lin, C.H., Kuo, B.Z.L.: The moderating effects of travel arrangement types on tourists' formation of Taiwan's unique image. Tour. Manag. **66**, 233–243 (2018)
38. Vincent, C.T.P.: Amateur versus professional online reviews: impact on tourists' intention to visit a destination. Tourism **66**(1), 35–51 (2018)
39. Booking: Booking (2018). www.booking.com
40. Tripadvisor: Tripadvisor (2018). https://tripadvisor.mediaroom.com
41. Trivago: Trivago (2018). www.trivago.com.co
42. Leelawat, N., Suppasri, A., Latcharote, P., Imamura, F., Abe, Y., Sugiyasu, K.: Increasing tsunami risk awareness via mobile application. In: IOP Conference Series: Earth and Environmental Science, vol. 56, no. 1 (2017)
43. Niño, S., Danna, J.: Los Sistemas de Información Geográfica (SIG) en turismo como herramienta de desarrollo y planificación territorial en las regiones periféricas. CIDADES, Comunidades e Territórios **32**, 18–39 (2016)
44. Prueksakorn, K., Gonzalez, J.C., Keson, J., Wongsai, S., Wongsai, N., Akkajit, P.: A GIS-based tool to estimate carbon stock related to changes in land use due to tourism in Phuket Island, Thailand. Clean Technol. Environ. Policy **20**, 561–571 (2018)
45. Zhang, T., Ji, H., Hu, Y., Ye, Q., Lin, Y.: Application of classification algorithm of machine learning and buffer analysis in torism regional planning. In: International Archives of the Photogrammetry, Remote Sensing and Spatial Information Sciences - ISPRS Archives, vol. 42, no. 3, pp. 2297–2302 (2018)
46. Morales, D., Besanger, Y., Medina, R.: Complex distribution networks: case study Galapagos Islands. Stud. Syst. Decis. Control **145**, 251–281 (2018)

47. Zhang, J., Wong, P., Lai, P.: A geographic analysis of hosts' irritation levels towards mainland Chinese cross-border day-trippers. Tour. Manag. **68**, 367–374 (2018)
48. Osorio, M.: La planificación turística. Enfoque y modelos. Mintzberg **8**(1), 291–314 (2006)
49. Cornejo, L.E.: Planificación turística sustentable en la Región de Coquimbo. Evaluación y aporte metodológicos. Boletín CF + S **42/43**, 27–37 (2010)
50. Xinyan, Z., Haiyan, S.: An integrative framework for collaborative forecasting in tourism supply chains. Int. J. Tour. Res. **20**(2), 158–171 (2017)
51. González-ramiro, A., Gonçalves, G., Sánchez-ríos, A., Jeong, J.S.: Using a VGI and GIS-based multicriteria approach for assessing the potential of rural tourism in extremadura (Spain). Sustainability **8**, 1–15 (2016)
52. Tan, W.J., Yang, C.F., Château, P.A., Lee, M.T., Chang, Y.C.: Integrated coastal-zone management for sustainable tourism using a decision support system based on system dynamics: a case study of Cijin, Kaohsiung, Taiwan. Ocean Coast. Manag. **153**, 131–139 (2018)
53. Sieber, R.: Public participation geographic information systems: a literature review and framework. Ann. Assoc. Am. Geograph. **96**, 491–507 (2006)
54. Kantola, S., Uusitalo, M., Nivala, V., Tuunlentie, S.: Tourism resort users' participation in planning: testing the public participation geographic information system method in Levi, Finnish Lapland. Tour. Manag. Perspect. **27**, 22–32 (2018)
55. Mai, T., Smith, C.: Scenario-based planning for tourism development using system dynamic modelling: a case study of Cat Ba Island, Vietnam. Tour. Manag. **68**, 336–354 (2018)

A Systematic Review on IoT and E-Services Co-production

Janine Ulrich$^{(\boxtimes)}$ ⓘ, Luana Gattass e Silva ⓘ, and Cristiano Maciel ⓘ

Federal University of Mato Grosso – UFMT, Cuiabá, Brazil
janineulrichl@gmail.com

Abstract. The availability of different types of data collected by Internet of Things-IoT can be exploited to enhance transparency and promote government action for citizens. However, in this field, how has the approximation between government and citizens been achieved for the co-production of digital public services? Thus, this research aims to identify the applicability of IOT in the coproduction of public services and what the characteristics of related models applicable to public administration. In order to do so, this article carried out a systematic review of the literature, in the period from 2012 to 2018. For this purpose, the database of advanced search Scopus. In general, many authors see IoT as facilitating the concept of smart cities, as well as a new way of managing communities based on information sharing and collaboration. Furthermore, the data collected through IoT can improve the capacity to predict problems in the development of a collective intelligence. The review has shown that there are advantages in the co-production of public services from the use of IoT as an instrument for reaching public value for public administration. However, the exploration of quality aspects of e-government service offerings and the understanding of user preferences are research questions that need to move forward in the face of relevance to society.

Keywords: Internet of Things · e-Government · Information and Communication Technologies · Co-production · Smart Government

1 Introduction

Co-production comes as a new role and as a collaborative trend in both public and private organizations. According to Ramaswamy and Gouillart [1], the term co-production is used to address how social and technological changes allow organizations, groups, and individuals to interact, collaborate and solve problems, creating value together. In the Government, the citizen is changing his profile from only passive consumers of products and services to a more active role. Thus, the growing evolution of the New Information and Communication Technologies (NICTs) provide support to these processes in public and private organizations, creating a new field of opportunities for organizations to meet their customers' needs [2]. Governments have been adopting strategies for the use of Information and Communication Technologies (ICTs) as a tool for providing services, information and providing channels for the participation of the population, thus building a differentiated model of public management called e-government.

© Springer Nature Switzerland AG 2020
Á. Rocha et al. (Eds.): ICITS 2020, AISC 1137, pp. 14–23, 2020.
https://doi.org/10.1007/978-3-030-40690-5_2

In light of this, smart government can be conceptualized as an adaptive evolution of government, in response to rapid digital technology change, to innovate the ways in which government can achieve better citizen engagement, accountability, and inter-operability [3].

The popularization of the Internet and its access starting from mobile devices can be pointed as the most visible part of the change in communication media. Such current Internet scenery is promising and is favorable to the construe of ideas, to the out-break of concepts and theories, as well as to the proliferation of new technologies, among them, the Internet of Things. IoT [4] links to the capacity of the objects, online making information available regarding their operation [5]. Lacerda and Lima-Marques [6], sustain that, online, objects are capable to accomplish actions in an independent way and to generate data in exponential quantity and variety, as product of interactions.

Thus, IoT provides easy access and interaction with a wide range of devices such as surveillance cameras, monitoring sensors, displays and vehicles. All such infrastructure is useful to the electronic government. The increasing use of IoT will foster the development of several applications making use of the enormous amount and data variety generated by those objects, to provide new services to citizens, businesses and public service administration. Such paradigm really finds application in many different domains for public administration, such as smart power management and smart net-works, automotive, traffic management and many others [7].

The National Intelligence Consulate – NIC [8], a medium-and long-term strategic thinking center within the United States Intelligence Community (IC), predicts that by 2025, ubiquitous computing will have technological advances through IoT and will be present in Food packages, furniture, environment sensors, paper documents, and more. Furthermore, it highlights the future emerging opportunities, starting from the idea that popular demand combined with technological advances could boost the widespread dissemination of IoT, which would allow, within the next Internet generation, ines-timably contribution for economic development [8].

Urban IoT provides the collection of various types of data. Such availability allows the increase of public transparency, and enables various governmental actions for citizens such as transportation and parking, lighting, surveillance and maintenance of public areas, cultural heritage preservation, garbage collection, salubriousness in hos-pitals and school, with a number of benefits in the management and optimization of public services [9].

According to Szkuta, Pizzicanella and Osimo [10], the concept of co-production emerged in the 70's to define services delivery with a high degree of user involvement, reappearing in the research agenda, with the advent of digital services. The authors also say that co-production introduces new services, potentially making the public sector more responsive to the user needs, which allow co-production to replace the provision of public services paid with volunteer resources, as a potential source of economics.

In the present research, given the widespread dissemination of IoT and co-production, we are looking into: how effective has the approximation between gov-ernment and citizens to co-production of digital public services have been? To such end, a systematic of the literature was carried out with the aim of identifying the applicability of IoT in the co-production of public services and the characteristics of the related models applicable to the public administration. The results of the research are discussed in this article, indirectly, seeking to boost public administration to adopt co-production strategies and solutions of Information and Communication Technology-ICT in the management of public issues, increasing the quality of services offered to citizens, reducing the operational costs of public administration, aiming at better use of public resources. This article is structured in sections. In Addition to this introductory section, the next one defines the base methods for the research, Sect. 3 shows result analysis, then Sect. 4 with final comments and future discussions.

2　Methodology

The carried out methodology within the present work was a systematic review that, according to Castro [11], it is a method used in a planned review to answer a specific question and uses explicit and systematic methods to identify, select, and critically evaluate the studies, and to collect and assess data from the studies included in the review. The work is set into 3 steps: 1. Define the objective and research questions; 2. Select the keywords and databases, for localization and literature selection; 3. Identify and analyze relevant articles for data extraction, evaluation of collected data and interpretation and synthesis of results.

The work of Almeida et al. [12], for having carried out a systematic review on the co-production of digital services, it was used during the first phase to elucidate the concepts of the area. After that, the objective and the research questions were set, synthesized in the following question: How has the approximation between government and citizens to co-production of digital public services have been effective? The dismemberments of this problem will result in the analysis of the necessary points for the theme investigation, such as: how many and where from are the publications made on this subject; What is the applicability of IoT in the co-production of public services and what are the characteristics of the related models applicable to public administration?

After the formulation of the questions, we proceeded to keywords definition, for the publication selection concerning the co-production of public services using IoT for the progress of E-government (E-gov). Once set the String ("IoT "OR" Internet of Things "OR" Internet of Everything "or" Web 3.0 ") AND ("Digital Government "OR" E-government "OR" E-gov "OR" egovernment") AND ("Co-production "or" Co-production "OR" Co-participation "or" Co participation "OR" Citizen sourcing "or" Collaborative Government").

For the publication identification and selection, a search was carried out in the Scopus database. Since 2004, Scopus proposes a comprehensive view of global scientific production in all areas through tools to track, analyze, and visualize scientific publications. The query base was set for advanced search, with the insertion of the String search defined, using the search operator "ALL", in order to cover all fields. The consultation was held on December 02, 2018.

As for the inclusion criteria, we considered the studies regarding the applicability of IoT in the co-production of public services and the characteristics of the related models applicable to public administration. Publications must contain an International Standard

Serial Number – ISSN or International Standard Book Number – ISBN, to be in the English language and must have been published until the consulting date.

The content of the publications was identified and analyzed in order to guarantee the analysis of the studied texts. As an exclusion criterion, the strategy for identifying keywords and their synonyms was chosen to refuse documents. Cumulatively, we did not consider the publications only bringing the terms in the bibliographical references and then those they treated co-production, IoT and e-government in a tangential way. Then the studies were organized by publication year, from the most recent to the oldest, in order to enable the descriptive and, graphic presentation of the results.

3 Result Analysis

This section sets the result into quantitative and qualitative analysis.

3.1 Quantitative Analyze

From the first step application on the systematic review of literature, 18 results were mapped on the Scopus database[1], considering the following types: 6 Articles, 5 Conferences Papers, 3 Books, 3 Books Chapters, and 1 review, in order to identify the applicability of IoT in the coproduction of public services, worldwide. In applying the exclusion criteria and analyzing the scientific works under the same objective, there were 06 scientific publications, considering: 3 congress articles, 2 book chapters, and 1 article, from different countries. Find Table 1 with the identified papers, considering publication year, after applying the exclusion criteria.

Table 1. Scientific publications. (Source: Elaborated by the researcher, Scopus, 2018).

N	Year	DocType	Author	Country	Document title
1	2018	Article	Webster, William, Leleux	UK	Smart governance: Opportunities for technologically-mediated citizen co-production
2	2017	Conf. Papers	Simonofski et al.	UK, Belgium	Citizen participation in smart cities: Evaluation framework proposal
3	2017	Book Chapter	Anthopolos	Greece	Smart government: A new adjective to government transformation or a trick?
4	2016	Book Chapter	Paletti	UK	Co-production through ICT in the public sector: When citizens reframe the production of public services
5	2016	Conf. Papers	Fugini, Teimourikia	Italy	The role of ICT in co-production of e-government public services
6	2012	Conf. Papers	Molinari	Italy	Innovative business models for smart cities: Overview of recent trends

---

[1] https://www.scopus.com/.

Answering the question: How many and where are the publications made on this subject?

A concentration of scientific production in Europe was found, the 06 identified publications refer to European productions, although concomitantly in more than one country, being the largest concentration in United Kingdom (3), followed by Italy (2), Belgium, and Greece (1). The publications start in 2012, and are concentrated in the period from 2016 to 2018, according to the graph below, resulting from the bibliometric analysis provided by the Scopus platform. It is also observed that the 6 identified countries are classified as Top Leading Countries in E-Government Development The United Nations E-Government Survey presents a systematic assessment of the use and potential of information and communication technologies to transform the public sector by enhancing efficiency, effectiveness, transparency, accountability, access to public services and citizen participation in the 193 Member States of the United Nations, and at all levels of development [13].

We begin with a survey of the study areas of the selected documents, where each document can be classified in several areas. We observed that 5 of them are from the Computer Science area, thus highlighting the importance of the area related to the object of this article. Still 4 of them are classified as Business studies, Management and Accounting, 3 as decision sciences; 2 Mathematics; 2 Social sciences, in addition to Biochemistry, Genetics and Molecular Biology, Engineering, Chemical Engineering, Energy, Materials Science that pointed 1 reference each. The papers in the systematic review are recent and have shown how public administration is being pushed to adopt coproduction strategies and ICT solutions in public management, increasing the quality of services offered to citizens, reducing the operational administration costs with a view to better use of public resources.

The selected publications in this research also pointed out that the concept of "IoT" is characterized by initiatives capable of connecting people, sensors and sharing data between communities. The definition and common vision is that IoT connects objects such as sensors, tags, smartphones, machines, buildings to the virtual world, generating a network of multiple human and non-human agents. Many authors see IoT as a facilitator of the Smart Government concept as well as a new way to manage communities based on information sharing and collaboration. In fact, the collected data can improve the ability to predict problems in the development of a collective intelligence. The new technologies and paradigms (Internet of things, Cloud Computing or the embeddedness of sensors) will play an important role in the transformation of cities, but it is the way they are applied to that makes Smart Cities [14].

3.2   Qualitative Analyze

Therefore, we will analyze it qualitatively, in chronological order, from the most current to the oldest publications. Answering the following question: What is the applicability of IoT in co-producing public services and what are the characteristics of related models applicable to public administration?

According to Webster, William, Leleux [15], a series of new innovative practices of citizen participation is presented, facilitated by new ICT: as living laboratories; Hackathons Fablabs and creative spaces; Smart Urban Laboratories; Citizens' Panel; Gamification; "open" Data Sets, and Crowdsourcing. The authors also propose a framework named "Technologically mediated municipal reciprocity model" to be used by public administrators to meet the principles of smart governance. This model highlights the necessary components to facilitate reciprocal relationships in which municipalities and citizens engage, interact, and co-produce using ICTs and social media. Moreover, in which there are incentives and rewards identified for both parties, in Creation of the shared governance of the smart city and thus enhances a better understanding of the necessary ingredients for smart government [15].

Simonofski et al. [14], investigated and proposed the way that citizens can transform a city into a smart city. By engaging in the (1) democratic process, (2) co-creating, acting as suppliers of experience and competencies to propose better Solutions and (3) using the city's ICT infrastructure proactively, including the IoT, transforming the citizen into data collectors, using mobile devices or other technologies, thus, providing the citizen with an integral part of the smart city. The New technologies and paradigms (IoT, Cloud Computing or the embeddedness of sensors) will play an important role in transforming cities, but it is the way they are applied, which makes them Smart Cities.

The authors also design a framework to organize the "Citizen Participation Evaluation" that can be useful in different ways: (1) as a tool for evaluating smarts cities strategies; (2) as a governance tool for government officials willing to invest in a citizen-oriented smart city strategy; (3) as a tool of creativity, enabling the comparative analysis of best practices for a criterion or category of criteria in different smart cities. From this comparative analysis, new means of citizen participation could be conceived and implemented.

Anthopoulos [16], created a Unified Smart Government framework that combines approaches given by several scholars. This framework shows that Smart Gov can be considered the evolution of the Digital Gov and the Open Gov and Smart City at the local level. It also describes the 3 basic components for the implementation of Smart Gov: Data, innovation and emerging technologies. IoT is highlighted in the article dealing with the Smart Government. It describes that technology provides Governments with the tools they have never had before, such as the IoT, which changes the information collection and the processes flow, while allowing a straight-forward and continuous connection with the community. The co-produce with citizens theme also gets prominence in the article in presenting the types of coproduction: Citizen-to-Government; Government to Citizen and Citizen to Citizen [17].

Paletti [18], describes how the co-production practices through ICTs are changing the form of the provision of public services. Inducing structural changes in public administration in order to enable future citizens to actively take part in the co-production of services. There are many apps helping citizens to report problems or be part of the public service. Such applications can be divided into 3 categories: (1) report problems, send location, image; (2) Crowdsourcing data, through sensors and (3) involving citizens in the provision of public services. A strategy for co-production engagement is the use of gamification and a dedicated incentive system rewarding co-producers. In This Way, the State role is to work on the governance, improve its competencies, thus becoming the Coordinator and facilitator of the process, seeking to generate Public Value through public services. The State proposes an open source electronic government platform, with new forms of co-production, which allows a process of agile development, ascendant and responsive concerning the society, promoting innovation, reducing costs and satisfying the needs of citizens.

Fugini and Teimourikia [19], analyzed the advances of ICT in the improvement and incorporation of co-production elements. According to the authors, co-production is being considered as the paradigm representing the involvement and collaboration of end users in improving the services of electronic government applications. Having a large involvement of managers, private and public organizations, individuals or communities not only as consumers, but also as decision makers in the provision of services and throughout the life cycle of such services. In General, that ICT alone is not sufficient to provide better services or involve users in co-production work. An early and complete involvement of the user, together with political and organizational support, is necessary so that co-production is considered.

The authors propose 5 dimensions of service delivery to assess the quality of the offered services by e-government: (1) Strategic dimension; (2) Economic and financial dimension; (3) Organizational and Technological Dimension; (4) The social dimension; (5) The political and democratic dimension.

They also describe the elements that are believed to facilitate co-production: (1) Involvement of stakeholders and users from the outset; (2) Rewarding co-producers; (3) Cost Reduction through ICT technology; (4) Easy and accessible means of communication; (5) Considering diversity; (6) The ICT platform can be easily used by non-technical users; (7) Built on existing platforms; (8) Platforms to collect feedback, report, and user reviews; (9) Offering training for co-producers; (10) Organization of co-production activities; and (11) Privacy and security. It follows to basic and innovative ICT tools and paradigms, which can play a role in facilitating co-production in electronic government services: (1) ICT Hardware and Communication Infrastructures: (a) IoT and Smart Cities; b) Mobility; (2) The World Wide Web: (a) Semantic Web Services and (3) Information, Data Management and Innovation: (a) Big Data systems and Knowledge Management; (b) Crowdsourcing and Open Innovation; (c) Social Network and Gamification.

According Molinari [20], well-known concepts such as sustainability, quality and impact of the public service are driven by the relationship between the personal/collective experience of a co-produced electronic service and the transformation of individual behaviors/mass taking place at the social level, due to the interaction between the Internet of People (IoP); The Internet of Services (IoS) and The Internet of Things (IoT).

For the mentioned author, the economic crisis results in low public budgets and increased users' expectations require policies that fit the local goals. Such challenges require that (local) Governments: deliver services that are transformational rather than merely responsive, and focus on transforming the relationship between the public sector and the users. In a large number of cases, what happens when a new (traditional) e-Government service is deployed is that the cumulative costs per user served largely exceed the benefits and this lasts for a long period until the relationship is reversed, assuming it is reversed.

The author presents a new business environment, in which governments are, not more service providers, taking the role of "service facilitators". From a governmental standpoint, most efforts still have infrastructure implementations focus. The proposed scheme enables the risk transfers and uncertainties of the necessary investment, to specific commercial actors that must be held accountable for the risks and results inherent to their ventures.

4 Conclusion

This article sought to review the literature in order to identify the applicability of IoT in the co-production of public services, as well as to identify the characteristics of the related models applicable to government. Despite the relevance of the theme, studies are still recent, within 06 publications identified in Europe, so we can point out not to be a very widespread practice worldwide. Many of the authors see IoT as a facilitator of the innovative practices of co-production of digital public service, citizens engage and as a new way to manage communities based on information sharing and collaboration. In addition, data collected via IoT can improve the ability to predict problems in the development of a collective intelligence. The New ICTs and paradigms (Internet of Things, Cloud Computing or the embeddedness of sensors) will play an important role in transforming the smart government.

ICTs, including IoT, are channels that mediate the inputs of public needs with the needs of citizens, playing an important role in the smart government process. However, ICTs alone are not sufficient to provide better services or involve users at work for co-production. Said that, it is up to the State to be the coordinator and facilitator, as to the citizen to be a supplier of experiences and competencies to propose or create better solutions, or even acting as a data collector, consciously or not, being an integral part of the process. The New technologies and paradigms will play an important role in transforming cities. However, it is the way technologies are applied that enables Smart Cities [14].

According to the assessed literature, co-production through ICTs has contributed to a smart government, providing significant increases in quality, transparency, and cost reduction on public e-services, as well as giving opportunity its sustainability, and also, generating a greater perception of the public value of the State. Such process influences several aspects of daily life and the behavior of the citizen, causing a greater civic engagement of society.

A strategy for engagement and effective has the approximation between government and citizens to co-production of digital public services are the use of gamification and an incentive rewarding co-producers.

In an indirect way, this article mapped out the frameworks that the selected scientific papers presented, in order to contribute to the transformation of government Indirectly, this article mapped the milestones presented by the selected scientific papers in order to contribute to the transformation of government by enhancing smart governance, as: Technologically Mediated Municipal Reciprocity Model, bringing principles of smart governance; Unified Framework of Smart Governance approaches showing that smart Government can be considered the evolution of e-government;

Evaluation Method of Citizen Participation, as an evaluation tool, which can be used for governance and creativity of smart cities strategies; Indicators for Evaluation of e-Services; Proposal of an electronic government platform outsourcing the creation of new modules for the production of public e-services for an ecosystem of citizens and companies; ideal path to the sustainability of the public service.

The Exploration of quality aspects of the offered services by the electronic government and the comprehension of the user's preferences are research issues not prioritized by the consulted bibliography, despite its relevance to society.

From the literature review and analysis of the articles found, this research provides a greater understanding of the subject, also, pointing out some identified trends that may motivate future research, such as privacy, information security, open data, interoperability, collaborative culture, confidence of those involved in co-production and public gain through public value in smarts cities. Also, that these issues are directly subject to an IoT and/or co-production of services and deserve attention on the part of those who are clinging to these technologies.

The present research enabled a greater understanding of the subject, as well as pointing out some future trendy, such as privacy, data security, open data, interoperation, collaboration culture, and trust on the people involved.

Moreover, it is important to reinforce that the improvement and updating of this review, as well as its possible publication, considered as the final objective of this study, deserves to take place regarding the emergence of new studies on the subject. In this way, the government's stakeholders can offer a broader and, at the same time, summarized the view of the issue.

References

1. Ramaswamy, V., Gouillart, F.: Building the co-creative enterprise. Harvard Bus. Rev. **88**, 100–109 (2010)
2. Rasool, G., Pathania, A.: Revisiting marketing mix study of evidences for investigating innovative role of technology in co-creation. J. General Manag. Res. **1**(1), 37–50 (2014)
3. Gil-Garcia, J., Zhang, J., Puron-Cid, G.: Conceptualizing smartness in government: an integrative and multi-dimensional view. Gov. Inf. Q. **33**(3), 524–534 (2016). https://doi.org/10.1016/j.giq.2016.03.002
4. Sato, K.: Mobilidade, Comunicação e Consumo: Expressões da telefonia celular em Angola, Brasil e Portugal, 366 p. Doctoral thesis – Graduação em Ciências da Comunicação, Academia de Comunicação e Artes, Universidade de São Pauo, São Paulo (2015)
5. Sobrinho, G.O.: Serviço de resolução e descoberta de informações sobre objetos em sistemas baseados em RFID. Tese de Doutorado em Engenharia, Universidade de São Paulo, São Paulo (2013). http://www.teses.usp.br/teses/disponiveis/3/3141/tde-16102013-162918/pt-br.php. Accessed Nov 2018
6. Lacerda, F., Lima-Marques, M.: Da necessidade de princípios de Arquitetura da Informação para a Internet das Coisas. Perspectivas em Ciência da Informação **20**(2), 158–171 (2015). https://doi.org/10.1590/1981-5344/2356. FapUNIFESP (SciELO)
7. Bellavista, P., et al.: Convergence of MANET and WSN in IoT urban scenarios. IEEE Sens. J. **13**(10), 3558–3567 (2013). https://doi.org/10.1109/jsen.2013.2272099. Institute of Electrical and Electronics Engineers (IEEE)

8. U.S. Government. NIC, National Intelligence Consulate - (Org.). Global trends 2025: a transformed world. Reports & Publications, Washington, 99 p. (2008). https://www.dni.gov/files/documents/Newsroom/Reports%20and%0Pubs/2025_Global_Trends_Final_Report.pd. Accessed Nov 2018. (978-0-16-0)
9. Zanella, A., et al.: Internet of things for smart cities. IEEE Internet Things J. [s.l.] **1**(1), 22–32 (2014). https://doi.org/10.1109/jiot.2014.2306328
10. Szkuta, K., Pizzicannella, R., Osimo, D.: Collaborative approaches to public sector innovation: a scoping study. Telecommun. Policy **38**(5–6), 558–567 (2014). https://doi.org/10.1016/j.telpol.2014.04.002. Accessed Nov 2018
11. Castro, A.: Curso de revisão sistemática e metanálise. 2011 São Paulo: Led-dis/unifesp, 11 p. (2015) http://www.virtual.epm.br/cursos/metanalise. Accessed Nov 2018
12. Almeida, G., et al.: Co-production of digital services: definitions, frameworks, cases and evaluation initiatives - findings from a systematic literature review. In: Electronic Government and the Information Systems Perspective, Brazil, pp. 3–19. Springer (2018). https://doi.org/10.1007/978-3-319-98349-3_1
13. United Nations. Top loading countries in E-government development. E-Government Survey 2018. https://publicadministration.un.org/egovkb/en-us/Resources/Infographics. Accessed Jul 2019
14. Simonofski, A., et al.: Citizen participation in smart cities: evaluation framework proposal. In: 19th Conference on Business Informatics (CBI), pp. 227–236. IEEE, July (2017). https://doi.org/10.1109/cbi.2017.21
15. Webster, C., William, R., Leleux, C.: Smart governance: opportunities for technologically-mediated citizen co-production. Inf. Polity **23**(1), 95–110 (2018). https://doi.org/10.3233/ip-170065
16. Anthopoulos, G.: Smart government: a new adjective to government transformation or a trick? In: Public Administration and Information Technology, Greece, pp. 263–293. Springer (2017). https://doi.org/10.1007/978-3-319-57015-0_6
17. Linders, D.: From e-government to we-government: defining a typology for citizen co-production in the age of social media. Gov. Inf. Q. **29**(4), 446–454 (2012)
18. Paletti, A.: Co-production through ICT in the public sector: when citizens reframe the production of public services. LNISO, pp. 141–152 (2016). Springer (2016). https://doi.org/10.1007/978-3-319-40265-9_10
19. Fugini, M., Teimourikia, M.: The role of ICT in co-production of e-government public services, pp. 119-139. Springer (2016). https://doi.org/10.1007/978-3-319-30558-5_8
20. Molinari, F.: Innovative business models for smart cities: overview of recent trends. In: Proceedings of the 12th European Conference on Egovernment, Milan, Italy, vol. 1, pp. 483–492. Institute of Public Governance and Management, ESADE Barcelona, Campus Sant Cugat, Barcelona, Spain 14 June 2012 (2012)

Conceptual Framework for Social Media Usage in Public Services – An Indian Perspective

D. V. R. Subrahmanyam[1]([⊠]), M. V. Rama Prasad[2],
and D. V. Sahrudh[3]

[1] GITAM Institute of Management, GITAM University, Visakhapatnam, India
`dvrsubra@gmail.com`
[2] GITAM School of Business, GITAM University, Bengaluru, India
`ramprasad.musunuri@gitam.edu`
[3] GITAM Institute of Technology, GITAM University, Visakhapatnam, India
`sahrudh@gmail.com`

Abstract. Individuals are becoming more social with the penetration of social media into their day to day life, whether anyone accepts or not, everyone is gearing towards consumption shift with digital content in their life as they spend more and more time online. Every organization focused on the utilizing the power of social media and devising own strategy for marketing of products and services. In general, service sector is one area where these services are greatly influenced by the services offered by social media. This paper begins with introduction to the social media, classifying the service sector and explores further on how Public service sector in India is utilizing the social media for expanding the quality and reach of their services. Government as well as private sector organizations have their own practices to make use of Information Management and to be specific, social media services. This study also depicts the role of social media in influencing the services offered to the public and an approach towards the expected plan of action.

Keywords: Social media · Decision making · Products and services · Social marketing

1 Introduction

1.1 Background

As mobile devices slowly became un-detachable body parts of individuals and social networks became a medium to engage with friends, families and even organizations are trying to explore ways to reach customer through social media. It became a way of tapping the power of the social media to promote new products, services and breakthrough ideas. Every organization is trying to leverage their social media presence and apply strategies for selling their products and services, including them in their business applications. Efforts are intended to engage the customer, understanding customer online behavior, quickly provide relevant information, analyze customer buying patterns to improve cross selling and upselling, improve the customer reach, building their brand and continuous improvement of marketing strategies through social media. With the help of the study on public service sector, different services promoted on social media we have lined up a case study of typical government services to explain the usage and recommendations.

Power of social media: Some people consider it as best utilization of time, others as waste of time where as some consider it as time pass. In all these cases, people spend sufficient time on social media like blogs, videos, music, feedback, photos, news etc. "GeoCities" was named as one of the earliest social networking websites on the internet, appearing in November 1994, then by "Classmates" in December 1995 and followed by "Six Degrees" in May 1997. The corporate job-oriented site "LinkedIn" appeared in May 2003; "Orkut" entered in January 2004; "Facebook" in February 2004; the text-based service "Twitter" with posts called "tweets", having limitation of 140 characters, entered in July 2006; and "Google+" in July 2011 [1].

1.2 Definitions

Social media are interactive computer-facilitated technologies that enable the creation and sharing of info, career interests, ideas, and other forms of expression via virtual societies and networks [2]. The variety of stand-alone and built-in social media services currently available introduces definition challenges; however, there are some common features [3] like interactive web 2.0 applications, connecting users with one another, using desktop, laptop, mobile phones or tablets, user generated content such as texts, photos, comments, voice, lip sync, videos and data exchanged through online interactions.

Social Networks formed through social media change the way groups of people interact and communicate or stand with the votes. They introduce considerable and pervasive changes to messages between organizations, groups, and people [2]. These changes are the focus of the emerging fields of techno self-studies. Some of the most popular social media websites, with over 100 million registered users, include Facebook, Facebook Messenger, YouTube, Instagram, WhatsApp, Weibo, Twitter, Tumblr, Telegram, Baidu Tieba, WeChat, LinkedIn, Snapchat, Pinterest, Viber, LINE, Tiktok, Smule.

The widest definition of the Service Sector covers all industries except goods-producing sector - agriculture, construction, mining and manufacturing. Under this definition, services include transportation, communications, public utilities, wholesale and retail trade, finance, insurance, real estate, other personal and business services, and government Variations in definition excludes all government services at all levels to have only private personal and business services even excluding transportation, communications, wholesale and retail trade, finance, insurance and real estate from it [4]. For our study we tried to cover all the services as shown below (Fig. 1).

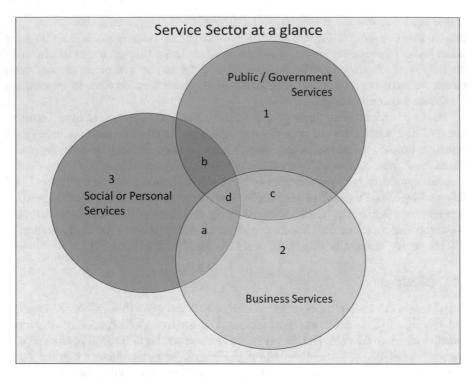

Fig. 1. Service sector representation (Source: own interpretation)

The three regions in the above diagram briefly represent classification of service sector into personal, business and government services, classification includes:

1. Public or Government Services- Public administration, police, defense and other services, social security, education, health, legal and information services.
2. Business Services – Banking, insurance, consultancy, transport, trade - retail, repair & maintenance.
3. Personal or Private Services – Hotel, Charitable trusts, research foundations and societies, beauty parlor, Travel, marriage lines, advertising.

This list is not exhaustive, but it helps to establish the diverse and intricate nature of the service sector.

We tried to explain the intersection of these three sectors in below points:

(a) Business + Personal – Services offered by organization and consumed by individuals like booking Hotel, travel tickets, entertainment – movies, theatre, amusement parks.
(b) Personal + Government – Services offered by government for public consumption like RTI, registration services – slot booking, traffic violation tickets, Ration card, fines and penalties, Aadhar card, Pan card, Voter card etc.

(c) Business + Government – Services offered by government for the benefit of businesses like Company/SEZ registrations, license renewals, tenders and notices.

(d) Government + Business + Personal – Offered by business, supported by government and consumed by Individuals – Public sector services which are business nature offered to individuals, like gas cylinders booking, Ration status, IRCTC ticket booking.

2 Literature Review

Social Media. In a more related and simple description, Miranda et al. [5] explained social media as "mainly conceived of as a medium wherein ordinary people in ordinary social networks (as opposed to professional journalists) can create user-generated news". Bharati et al. [6] refer to social media as a technology "not focused on transactions but on collaboration and communication across groups both inside and outside the firm.". Few other authors Spagnoletti et al. in 2015 [7] and Xu and Zhang et al. in 2013 [8] referred it to social media as a set of internet-based applications, which are aimed at helping the make, update and exchange of user-generated content, whereas establishing new associations between the content makers themselves. Lastly, Tang et al. [9]. In 2012 also identify social media as user-generated media, which is a source of "online information created, initiated, circulated, and used by consumers intent on educating each other about products, services, brands, traits and issues".

As explained by Greenwood in Pew Research Center documentation on the variety of ways social media used by Americans to get information and interact with others. Majority of Americans agreed that they follow news via social media, and half of the public has started to use these sites to follow the 2016 presidential election. It was also observed that Americans are using social media to take mental breaks while on the job or look out for job opportunities, in context of the work [10]. They also engage their discussions on complex privacy issues in social media sites, in the context of security and privacy. Since 2012, Government has started tracking specific websites and social media platforms that users connect in their social networking [10]. Sung et al. proved using Facebook data of colleges and universities across the US to show that people in the same class or same year or same major tend to form denser groups/networks [11].

Service Sector. As explained by Lakhe and Mohanty, Service means a functional component that is deeply involved with design planning, procurement, engineering, repair, maintenance, warehousing, distribution and marketing. Characteristics of services include intangibility, Heterogeneity, Inseparability, Perishability. Services are intangible, making it hard for likely customers to know what they will obtain and what worth it will hold for them. Indeed, some consultants or providers of investment services, offer without guarantees of the value for money paid. Since the quality of these services depends mostly on the quality of the people offering the services, "people costs" are normally a high fraction of service costs. Whereas a producer may use expertise, simplification, and other techniques to reduce the cost of products sold, the service provider often faces a relentless form of increasing costs [12].

According to Planning commission of India, Chief Economic Adviser KV Subramanian's maiden Economic Survey 2018-19,

The services sector that accounts for 54% of India's Gross Value Added (GVA), saw its growth moderating to 7.5% in 2018-19 from 8.1% in 2017-18 [13].

Social media influence on service sector: In a study, Zhang suggested that product/service info on vendor's websites are often partial and propose a *Latent Dirichlet Allocation model* to reveal the useful balancing hidden information in customer assessments [14].

In another study, Yan et al. [15] evaluate revisit intents for restaurants, and find that food and service quality, price and value for money, and the hygienic atmosphere govern such intentions. In a parallel discussion, Wu et al. [16] suggest that consumers integrate product value with seller reviews in configuring their inclination to pay. Aggarwal et al. interestingly reveal positive effects of employee negative posts on company reputation, given that such posts creates interest to larger audience [17].

Above studies reveal that customer focus is on evaluating the legitimacy of product/service feedbacks/recommendations published online and review volume is often qualified by purchasers' attitude to bearing risk. Most studies identify that the mixture of customer's interest and available reviews help them to compare and choose products/services that offer best value.

Social Media in Marketing Strategies. Social media sites are now a huge part of marketing strategies, and the familiar studies showcase of the extent to which social media is being combined in marketing strategies. Rishika et al. validate how higher social media activity directly associates with higher involvement and customer patronage [18]. García-Crespo et al. study the continuous interaction between customers and establishments, as it impacts the social network environment with implications on marketing and new product advancement [19]. Dou et al. research enhancing the strength of a system by regulating the embedded social media features with the right market penetration and pricing tactics [20].

In a slightly different study, Oh et al. [21] examine the pricing models for newspaper online and fount that charging for formerly free online content has a uneven impact on word of mouth for niche and widespread topics/articles. Lee et al. [22] study the reputation of social commerce in marketplace to find that Facebook likes improved the surge in sales, drives traffic, and introduces social presence in the shopping experience. Chen et al. [23] study music sales on "MySpace" to find that timing, propagation, and content of the personalized message has substantial effect on sales. Qiu et al. [24] analyzed "YouTube" information to find that learning and network mechanisms economically and statistically influence video views. Susarla et al. [25] inspected video and user info datasets of YouTube and found that the success of a video largely depends on social exchanges to determine its impact. Oestreicher-Singer and Zalmanson [26] reveal that the more viability of firms when they join social media in purchasing and consumption practices, rather than using it as an ancillary for soft marketing online.

All these studies on social media and service offering reveals the significance of inclusion of social media in marketing strategies, the importance of community structure and patterns in using social media for marketing tenacities. Also, classifying specific customer sections across social sites, for example, members of a group or forum, helps e-marketers to focus specific customers based on their demographic patterns and comparable interests.

2.1 Objectives of Study

The primary objective of the study is to understand the overall classifications of the service sector and their usage of social media with respect to the growth opportunities and challenges to the government sector.

Objectives are listed as follows:

- To study the current usage of social media in government sector.
- To analyze the service offerings of government sector against certain social media attributes.
- To observe the extent of usage and hurdles in the process.

The opportunity at hand is to recognize a mechanism that would help us leveraging the power of social media and its influence into service sector, have a clear focus to address the advancements and challenges.

2.2 Research Methodology

The paper searches the various dimensions attached to the concept of social media influence on service sector. It is a descriptive study, where examples are mentioned based on the situations from some secondary material obtained and some from the personal experiences of the authors and friends.

2.3 Key Attributes for Conceptual Framework

Following are the attributes used to examine the services and their interactions shown below in Fig. 2.

- Engagement
- Promotional activities
- Impact of user feedback
- Real time conversation/monitoring

Fig. 2. Conceptual framework (Source: own interpretation)

2.4   Key Benefits to Service Sector

This study suggests, the actions that can help government to develop or improve their social marketing strategies. It is very critical for any sector to involve the customer/public, invite them into their social presence and cut down the time to market in their promotion of products/services.

(a) to develop brand awareness – Builds followers
(b) to build relationships – Customer engagement
(c) to increase website traffic – Improve web presence
(d) to get rewards through social media – Retain customers

3   Government Services - Analysis

To understand the services offered by the government sector shown in Table 1 below, against various attributes like engagement, promotional activities, impact of user feedback, and real time conversations.

Table 1. Usage of social media in government services

Attribute	Engagement	Promotional activities	Impact of user feedback	Real time conversation (or) monitoring
Brief description	Maintain relationship, retention, two-way communications	Building awareness, conduct campaigns, reach public quicker and a cheaper way	To minimize importance of middlemen, to understand market, to get feedback, complaints, to bring in transparency	Introducing Chat, virtual assistant, chatbot services, provide quick responses, bring IVR system
Current usage	* Every Ministry (e.g. HRD) has social media presence * Citizens are engaged/follow to ministers thru social media * SMS for passport, gas booking, RTO tickets etc. are in use * Local level services are using it (e.g. Kanpur city portal)	* Several government services are informed thru social media * Tenders notifications, government auctions are made available *Services of Health ministry, Save a girl child, swacchbharat, save water etc. announcements	* Created digital revolution with complaints on IRCTC services * Karnataka Govt has dept services through social media * Discussion boards to analyze the social media	*Used to get grievances, complaints, to track and resolve civic issues * Sending SMS on the statuses, nearby toilets in the vicinity of location, IVR implementations
Hurdles	* Reaching few people * Lack of awareness of government programs * Unpredictable conversations may become abuse * Repetition of similar posts/conversations	* Lack of Education and accessibility to the social media * Rural areas still do not have network facility	* Misuse of direct communication * False alarms * False impression at international community * Negative messages of Rs.2000 note ban	* Resistance from departments/ministries, local bodies * Bandwidth in Tier-2/3 or Rural areas is very limited
Recommendation	* Enable dedicated and centralized social media services to engage public * Including Facebook and WhatsApp in engaging consumers * Limit the scope by constant monitoring * Propose earning of rewards for every service utilized	* Inclusions of social media campaigns * Campaigns like Donate blood with calendar on twitter, WhatsApp * Develop Ad and Content Calendars * introducing services like budget, tourism, Industrial policy etc.	* Govern social media to avoid mis-representations * Regulating the social media channels * Imposing fines on falsifications * Develop a Social Media Strategy to the respective service	* WhatsApp call inclusion is cheaper and video call possibilities in rescue operations * Proper maintenance and Optimization plan for growth * Develop Reports and end user Communication

(*continued*)

Table 1. (*continued*)

Attribute	Engagement	Promotional activities	Impact of user feedback	Real time conversation (or) monitoring
Success stories	* Kanpur city implemented * She team campaign in Hyderabad * RTO sends receipts * Gas booking status * Passport status updates	* Rupee logo and other department logo creation contests * Twitter hashtags for #election reminder, #election results * Announcements for peace on Ayodhya verdict, Demonetization	* Swachh survekshan * News articles * Sting operation on corruption * Tweets by politicians in response to public	* Public governance and redressal (PGR) called Puraseva in state of AP * Education department announces results through social media and internet

3.1   Benefits of Using Social Media Services

- Leverage the community's knowledge, skills, creativity/ideas to establish a two-way communication channel to improve services.
- Enable increased interaction with public in a timely manner.
- Faster time to spread services-related information at a cheaper cost.
- Continuous quality improvement with feedback collected.

3.2   Challenges in Using Social Media Services

- Where public is in remote locations without internet and social media
- When intention is to reach public slower than the other mediums
- When it is difficult to understand the service need

4   Recommendations

Looking at the analysis of the government sector services and following are the recommendations consolidated as well as explained a decent plan to include social media in the service sector:

- Need to build dedicated, centralized and integrated government service platform with social media.
- Develop a social media strategy to their respective services
- Introduce earning of rewards to public for every service utilized
- Develop advertisement and content calendars
- Develop reports and end user communication
- Prepare maintenance and optimization plan for growth

Further research may be conducted by building on hypothesis to include the above plan of action and recommendations to validate results through statistical analysis.

Acknowledgements. We would like to thank everyone who has provided their inputs and personal experiences with the social media and special thanks to all the authors and researchers whom we have cited or referred and the reviewers who took their time to thoroughly review our submissions to provide valuable feedback.

References

1. Hudson, G., Léger, A., Niss, B., Sebestyén, I., Vaaben, J.: JPEG-1 standard 25 years: past, present, and future reasons for a success. J. Electron. Imaging **27**(04), 040901 (2018)
2. Kietzmann, J.H., Hermkens, K.: Social media? Get serious! Understanding the functional building blocks of social media. Bus. Horiz. **54**(3), 241–251 (2011)
3. Obar, J.A., Wildman, S.S.: Social media definition and the governance challenge: an introduction to the special issue. Telecommun. Policy SSRN **54**(3), 241–251 (2015)
4. Nayyar, G.: The Service Sector in India's Development. Cambridge University Press, New York (2012)
5. Miranda, S.M., Young, A., Yetgin, E.: Are social media emancipatory or hegemonic? Societal effects of mass media digitization. MIS Q. **40**, 303–329 (2016)
6. Bharati, P., Zhang, C., Chaudhury, A.: Social media assimilation in firms: investigating the roles of absorptive capacity and institutional pressures. Inf. Syst. Front. **16**, 257–272 (2014)
7. Spagnoletti, P., Resca, A., Sæbø, Ø.: Design for social media engagement: insights from elderly care assistance. J. Strategic Inf. Syst. **24**, 128–145 (2015)
8. Xu, S.X., Zhang, X.M.: Impact of Wikipedia on market information environment: evidence on management disclosure and investor reaction. MIS Q. **37**(4), 1043–1068 (2013)
9. Tang, Q., Gu, B., Whinston, A.B.: Content contribution for revenue sharing and reputation in social media: a dynamic structural model. J. Manag. Inf. Syst. **29**, 41–76 (2012)
10. Greenwood, S., Perrin, A., Duggan, M.: Social media update. Pew Res. Center Internet Technol. **11**, 1–19 (2016). https://www.pewresearch.org/internet/2016/11/11/social-media-update-2016/
11. Sung, Y.S., Wang, D., Kumara, S.: Uncovering the effect of dominant attributes on community topology: a case of Facebook networks. Inf. Syst. Front. **20**, 1041–1052 (2016)
12. Mohanty, R.P.: Lakhe, R.R.: TQM in the Service Sector. Jaico Publishing House, Mumbai (2001)
13. Subramanian, K.: Economic Survey of India. Economic Survey, India (2019)
14. Piramuthu, S., Zhang, J.: Product recommendation with latent review topics. Inf. Syst. Front. **20**, 617–625 (2018)
15. Yan, L., Peng, J., Tan, Y.: Network dynamics: how can we find patients like us? Inf. Syst. Res. **26**(3), 496–512 (2015)
16. Wu, J., Gaytán, E.A.A.: The role of online seller reviews and product price on buyers' willingness-to-pay: a risk perspective. Eur. J. Inf. Syst. **22**, 416–433 (2013)
17. Aggarwal, R., Gopal, R., Sankaranarayanan, R., Singh, P.V.: Blog, blogger, and the firm: can negative employee posts lead to positive outcomes? Inf. Syst. Res. **23**, 306–322 (2012)
18. Rishika, R., Kumar, A., Janakiraman, R., Bezawada, R.: The effect of customers' social media participation on customer visit frequency and profitability: an empirical investigation. Inf. Syst. Res. **24**, 108–127 (2013)
19. Garcia-Crespo, A., Colomo-Palacios, R., Gomez-Berbis, J.M., Ruiz-Mezcua, B.: SEMO: a framework for customer social networks analysis based on semantics. J. Inf. Technol. **25**, 178–188 (2010)

20. Dou, Y., Niculescu, M.F., Wu, D.J.: Engineering optimal network effects via social media features and seeding in markets for digital goods and services. Inf. Syst. Res. **24**, 164–185 (2013)
21. Oh, W., Moon, J.Y., Hahn, J., Kim, T.: Research note—leader influence on sustained participation in online collaborative work communities: a simulation-based approach. Inf. Syst. Res. **27**, 383–402 (2016)
22. Lee, K., Lee, B., Oh, W.: Thumbs up, sales up? The contingent effect of Facebook likes on sales performance in social commerce. J. Manag. Inf. Syst. **32**, 109–143 (2015)
23. Chen, H., De, P., Hu, Y.J.: IT-enabled broadcasting in social media: an empirical study of artists' activities and music sales. Inf. Syst. Res. **26**, 513–531 (2015)
24. Qiu, L., Tang, Q., Whinston, A.B.: Two formulas for success in social media: learning and network effects. J. Manag. Inf. Syst. **32**, 78–108 (2015)
25. Susarla, A., Oh, J.H., Tan, Y.: Social networks and the diffusion of user-generated content: evidence from YouTube. Inf. Syst. Res. **23**, 23–41 (2012)
26. Oestreicher-Singer, G., Zalmanson, L.: Content or community? A digital business strategy for content providers in the social age. MIS Q. **37**(2), 591–616 (2013)

Extending Persuasive System Design Frameworks: An Exploratory Study

Geovanna Evelyn Espinoza Taype[(⊠)]
and Maria Cecilia Baranauskas Calani

University of Campinas, UNICAMP, Campinas-São Paulo, Brazil
evelynespinozataype@gmail.com,
c.baranauskas@gmail.com

Abstract. Many designers, developers, and stakeholders in information systems design attempt to create persuasive technologies, although many times without a formal support from persuasive design approaches. In addition research on design methodology is one of the most challenging issues in the field of persuasive technology. This paper intends to discuss and expand the persuasive design frameworks regarding four key aspects: considering the socially-awareness in the process of designing persuasive systems; inclusion of persuasive principles not only for web and mobile applications but also for pervasive applications; the extension of ethical aspects in the design of persuasive information systems; and the inclusion stakeholder behavior to influence achieve a target behavior. We conclude this paper with prospections for frameworks to persuasive systems design and future lines of works.

Keywords: Persuasive system design · Pervasive technology · Human computer interaction · Information system design

1 Introduction

Persuasive design is an approach to design which intends to influence human behavior through product or services features. According to a definition provided by Fogg [5, 6], persuasive design focuses on increasing motivation, abilities and triggering behavior of consumers. The aim of this paper is to review literature frameworks to support designers in developing a persuasive system, discussing and extending some aspects of them, in accordance to the challenges of the contemporary technologies and their raised necessities.

In a research conducted by Oinas-Kukkonen and Harjumaa [10], they show that many information systems have few persuasive aspects to attract people to use different applications; as a consequence, many researchers state the importance of including persuasive design in information systems, especially those that aim at the sustainable transformations of our lives toward wellbeing.

Currently, there are several attempts to create persuasive technologies and implementing information systems, especially in the field of sustainability. However, literature has shown they often fail, as pointed out by Fogg [5, 6]; one problem is that a lot of projects are overambitious, for designers that have never had created a persuasive system before. Selecting a challenging behavior such as eating healthy food as a target,

© Springer Nature Switzerland AG 2020
Á. Rocha et al. (Eds.): ICITS 2020, AISC 1137, pp. 35–45, 2020.
https://doi.org/10.1007/978-3-030-40690-5_4

especially for some users that love fast food, it is a challenge because it involves difficult change of behavior, as other endeavors in changing behavior through the use of technology.

Several researchers on persuasive design have introduced models and frameworks for persuasive design: Fogg [5, 6] proposed eight steps to design persuasive technologies, while Oinas [11] also proposed a model and principles for persuasive system design. Mustaquim and Nyström [9] proposed a persuasive system design for sustainability considering a cognitive dissonance model, and Stibe [14] proposed a conceptual framework for the design and evaluation of persuasive systems focusing on the ecosystems. Although with unquestionable contributions, these models and frameworks do not address, or make explicit, methodological and ethical aspects of the process of designing applications with persuasive considerations. In this paper we argue that there is room for research in the field considering those aspects and the participation of stakeholders to address the subject.

The paper is organized in four sections: the next section presents background and related work on persuasive design frameworks. The third section presents a case study to clarifying persuasion in a socially aware perspective, underling key points in the proposed framework, the fourth section briefly present a preliminary practical experimentation of the framework with a case study, and lastly, concludes, pointing out further possibilities of investigation on the theme.

2   Background and Related Work

Research on persuasive system design has mostly taken as foundation the works of Fogg [4] and Oinas [11]. Table 1 summarizes a chronology of some of the main contributions in literature on persuasive systems design frameworks and platform of applications respectively.

Table 1. Frameworks for persuasive design.

Years	Researches	Type of application	Reference
2003	Persuasive Technology: Using Computers to Change What We Think and Do	Web	Fogg [6]
2009	Persuasive Systems Design-Key Issues Model and System Features	Web/Mobile	Oinas [15]
2014	Designing Persuasive Systems for Sustainability – A Cognitive Dissonance Model	Web/Mobile	Mustaquim and Nyström [13]
2015	Towards a Framework for Socially Influencing Systems	Web/Mobile	Stibe [19]
2018	Transforming Sociotech Design	Web/Mobile	Stibe, Christensen and Nyström [21]
2018	A Framework for Design and Development of Persuasive Mobile Systems	Web/Mobile	Murillo et al. [11]

An analysis of the mainstream frameworks in literature was conducted. From that analysis we identified that the target systems usually address web and mobile applications but not pervasive and ubiquitous scenarios of applications. Additionally, all these researches have a common base of principles for designing persuasive information systems. A set of principles usually adopted is summarized in the next section.

2.1 Analyzing Principles

Frameworks regarding persuasive system design have common aspects that shape these frameworks as shown in Table 2. The first framework proposed by Fogg in [4], is composed by 5 principles. Oinas [11] presents 28 principles (five of these principles came from Fogg's principles). The framework focusing in persuasive design for sustainability and ecosystems proposed by Mustaquim and Nyström [9] considers 5 principles. These principles are different from the proposed by Fogg and Oinas. Stibe [12, 13] present 7 principles in his model for leveraging powers of social influence, in his framework the highlighted seven principles are inspired by those seven principles categorized as social support proposed by Oinas. Murillo [8], in his proposed framework, considers the 28 principles proposed by Oinas as part of his framework for design and development of persuasive mobile systems. Table 2 shows a synthesis of the different principles used by the frameworks.

Table 2. Analyzing persuasive design principles from the literature.

N	Principles	2003 Fogg [6]	2009 Oinas [15]	2014 Mustaquim and Nyström [13]	2015 Stibe [19]	2018 Murillo [11]
1	Reduction	X	X			X
2	Tunneling	X	X			X
3	Tailoring	X	X			X
4	Personalization		X			X
5	Self-monitoring	X	X			X
6	Simulation		X			X
7	Rehearsal		X			X
8	Praise		X			X
9	Rewards		X			X
10	Reminders		X			X
11	Suggestion	X	X			X
12	Similarity		X			X
13	Liking		X			X
14	Social role		X			X
15	Trustworthiness		X			X
16	Expertise		X			X

(*continued*)

Table 2. (*continued*)

N	Principles	2003 Fogg [6]	2009 Oinas [15]	2014 Mustaquim and Nyström [13]	2015 Stibe [19]	2018 Murillo [11]
17	Surface credibility		X			X
18	Real world feel		X			X
19	Authority		X			X
20	Third party endorsements		X			X
21	Verifiability		X			X
22	Social Learning		X		X	X
23	Social Comparison		X		X	X
24	Normative Influence		X		X	X
25	Social facilitation		X		X	X
26	Cooperation		X		X	X
27	Competition		X		X	X
28	Recognition		X		X	X
29	Equitability			X		
30	Inclusiveness			X		
31	Optimality			X		
32	Privacy			X		
33	Transparency			X		

Some literature research work applied some of the 33 principles in different domains. For example: Stibe [12, 13] uses seven principles in his model for leveraging powers of social influence to change behaviors and attitudes of users. In his framework the highlighted seven principles are inspired by those seven principles categorized as social support proposed by Oinas. Murillo [8], in his proposed framework, considers the 28 principles proposed by Oinas as part of his framework for design and development of persuasive mobile systems. Table 2 shows a synthesis of the different principles used by the frameworks.

According to the information shown in Table 2; 12 principles identified are the most widely used to design persuasive systems. Seven of these principles (Social Learning, Social Comparison, Normative Influence, Social Facilitation, Cooperation, Competition, Recognition) broadly used are focused on social influence, because as a substantive phenomenon these have the potential to influence behavior or attitudes through the influences by other people. Five principles (Reduction, Tunneling, Tailoring, Self-monitoring and Suggestion) are used for helping the user with tasks. Principles less used by the researchers are those focused on sustainability, proposed by

Mustaquim and Nyström (Equitability, Inclusiveness, Optimality, Privacy and Transparency).

2.2 Analyzing Persuasive Principles in Terms of Design Principles

Regarding the persuasive principles found in literature and taking into account what already exist in terms of principles in systems design such as those from Usability (proposed by Nielsen in [20]), Universal Design (Mace [19]) and the Laws of Simplicity (proposed by Maeda in [18]), a brief analysis identified commonalities and differences among them. Some of the principles coming from the persuasive design literature are present in the universal design principles and in the laws of simplicity. Illustrating: the reduction principle is already part of the laws of simplicity and the equitability principle is also part of the universal design principles.

2.3 Synthesizing and Clustering Principles for Persuasive Design

After analyzing the principles found in literature, including those from the design frameworks mentioned, and studying different aspects in pervasive applications, they are organized in groups as shown in Table 3. *Simplifying tasks* group is constituted by those principles aimed at persuading people by helping with tasks in the system in order to achieve the desired behavior [4, 11], For example: an application that shows me information about the quantity of calories consumed by day, can help me to monitor and control my behavior to consume less calories. The *Usability* group principles address persuasion through the system usability to achieve specific goals about the desired behavior. The *Social Influence* group principles address the influence on other people. For example: people can influence to do a behavior, contributing with information to the person aiming at achieving the desired behavior [11–13]. The *Credibility group* support persuasion through the veracity of information and authority that are shown to the user. For example: the information showed in the web site regarding the authority of office government [11].

Table 3. Clusters of principles for (Persuasive) design.

Simplifying tasks	Usability	Credibility	Social influence
Reduction	Reminders	Trustworthiness	Social Learning
Tunneling	Suggestion	Expertise	Social Comparison
Tailoring	User control and freedom	Surface Credibility	Normative Influence
Personalization	Flexibility	Real world feel	Social facilitation
Self-monitoring	Emotion	Authority	Cooperation
Simulation	Simplicity	Third party endorsements	Competition
Rehearsal	Perceptible information	Verifiability	Recognition
Similarity		Equitability	Inclusiveness
Optimality		Transparency	Praise
		Privacy	Rewards
			Liking
			Social role

We also identified a gap with relation to ubiquitous and enactive applications. The next section explores this subject based on the Socially Aware Design approach [3].

3   Clarifying Persuasion in a Socially-Aware Perspective: A Case Study

The 'problem' of persuasive system design was analyzed with the Socially Aware Design (SAwD) tool [1], a collaborative web system for socially aware clarification of design problems. This tool introduces a socially aware design approach to support early design activities, when a problem is being understood and a solution is being proposed. This tool helps to articulate ideas coming from Organizational Semiotics [7] and Participatory Design [1] towards the problem solutions.

For the clarification of the problem using the SawD tool, meetings and activities were developed collaboratively by professors, students and members of the InterHAD research team. InterHAD is a Laboratory on Human-Digital Artifact Interaction in which graduate and undergraduate students of the Institute of Computing from UNICAMP participate. Meetings were scheduled and brainstorming activities were carried out with the aforementioned research team, using the SAwD tool to identify the stakeholders direct and indirectly interested in the 'problem' of persuasive design. After identified stakeholders, persuasion in design and frameworks for persuasive design were discussed. Despite the fact that Fogg [4] defines persuasion broadly as an attempt to shape, reinforce or change attitudes, behaviors, feeling or thoughts about an issue, object or action without using coercion or deception, the issue of manipulation was raised, i.e. persuasion could be for good as well as for bad intentions. Persuasion in design brings a great responsibility. For this reason the ethical aspects take a great importance and should be part of the guidelines for the design of information systems, especially those with persuasive aspects. Additionally, the team discussed and identified the necessity to seek equilibrium between all the forces stakeholders bring to the process of design. The stakeholders should be part of the process of design, to guarantee ethical aspects, to not incur in deception or coercion. This participatory design could guarantee the equilibrium in the design and inclusion of persuasive aspects in information system.

3.1   Key Points to Consider in Persuasive Design Frameworks

After having clarified persuasion identifying stakeholders with the SAwD tool, and analyzed the issues raised by the team, some demands and opportunities to the research agenda in persuasive design were identified. Basically, four main aspects are proposed:

A. Including the **socially aware design as a part of the process** of persuasive design of information systems, to better characterize the involved stakeholders, raising issues the interested parties might have and proposing ideas of solutions to them.

B. Extending the **principles of persuasive** design to include principles that address specificities of pervasive ubiquitous and enactive applications; for instance, the involvement of the body and personal data in pervasive scenarios (e.g. IoT scenarios). The set of principles should be the result of practices in A.

C. Including **ethical c**odes and values as part of guidelines for the persuasive system design frameworks. Those codes should be raised and discussed by those involved in the socially-aware design process of A., and as part of B.

D. Including **stakeholders' behavior** in the process to design persuasive information system.

Each of these proposed topics is summarized in the following sections.

3.1.1   Socially-Aware Design in Persuasive Information Systems

The Socially-Aware Design argues for a design process that starts with the identification of the main stakeholders, their main abilities and needs. Each stakeholder will play a role as design partner in the process of designing persuasive applications. In participatory practices they will help to achieve a better equilibrium and awareness in the inclusion of persuasive aspects in the application design. The social aware design brings to participatory discussions a structured and systematic view of the problems. In order to propose solutions for the design of complex system with persuasive aspect, the social aware design enables the problem articulation method to understand problems from different perspectives, raising ideas of solutions among and for different stakeholders in a participatory way [1].

To the best of our knowledge, we did not find in the literature frameworks that include the main stakeholders as design partners in the design process of system with persuasive intents. The Socially-Aware Design Model [1] might help in achieving equilibrium between all main stakeholders' interests during design to include persuasive aspects in the information systems, while respecting ethical principles of the social group.

3.1.2   Stakeholder Behavior

The stakeholders' behavior in the process of design is a key point to consider at the design of persuasive information systems. Stakeholders can be highly involved with the target behavior or at reverse could be less involved with the target behavior, rationally or emotionally. The stakeholders that are highly involved with the target behavior can influence others less involved with the target behavior. In the literature, we did not find this perspective in frameworks to design persuasive systems.

3.1.3   Principles Towards Pervasive Applications

Analyzing and observing the identified persuasive principles, some aspects should be considered as part of the set, when we think of pervasive and ubiquitous applications; two groups of principles are suggested: a hedonic and an enactive.

The *hedonic* group is related to principles to persuade people through the fun, humor or sensory and emotional experiences.

- **Fun:** is a special type of motivation supported by humor; it can be used in conversations and can contribute to develop relationships. The idea is infusing a bit of appropriate, contextual and humor to influence a positive behavior. The system should support our overall mood and hedonic motivation, which is defined as the fun or pleasure [17] in using systems.

- **Sensory:** the system should support interaction that is physical, reflexive, palpable, material, close to the body [15].
- **Emotional:** the system should support emotional experiences to achieve a behavior, stimulating our mind and our sense of achievement.

The *enactive* group is constituted of principles to persuade people through technologies as continuous, ubiquitous and intelligent accompaniment of the human actor or as a direct extension of the person's perceptual and cognitive apparatus involved in participation in the system-living and acting with the system instead of just using it [22]. This principle is featured for supporting autonomy and proactivity in taking decisions.

- **Proactivity** refers to identifying and making decisions that anticipate people's needs; for proactivity to be effective, it is crucial that a pervasive computing system tracks the user intent. Nevertheless, proaction should be carefully designed, because it can annoy a user and thus defeat the original goal [16], or be in conflict with ethical principles that the group decides to follow.
- **Autonomy** is the capacity to manage internal self-constructive processes to control, change and regulate the exchange with the ambience [21, 24]. This principle should be in balance to the system's proactivity.

3.1.4   Ethical Principles

Towards a practical approach to ethics, we suggest to consider it in reflections along the process of designing persuasive systems. We as designers and the stakeholders should create technologies ethically acceptable in that social group. Some ethical values identified from the mainstream literature to be considered as a start point in the design process are:

- **Privacy:** The system should keep user's behavior private [9].
- **Design Transparency:** The system should maintain transparency in design for dealing with different outcomes. The system must not misinform in order to achieve their persuasive end [9].
- **Benevolence:** The system should take more responsibility of their user's wellbeing, to give positive benefit to the person. The systems should consider the person's mental and physical health. This principle was identified with the case study.
- **Responsibility:** The systems should assume responsibility for all reasonably predictive outcome of their use [23].
- **Design Motivation:** The creator of a persuasive technology should disclose their intention, motivation and intended outcomes [23].

3.2   A Preliminary Practical Experimentation of the Framework

The practical use of the framework involved the design of a socially-aware persuasive information system to motivate people to eat healthy food. Taking into account the key points presented in this framework, a process was carried out to design the system. The main steps involved: stakeholders' identification, stakeholders' behavior identification, persuasion frame (to articulate stakeholders' problems and ideate solutions with the principles proposed in this work), definition of requirements, and prototypes design. As a result of the design, the information system was developed, which includes persuasive features at the social and physical ambience surrounding the person. This means, the person is immersed in the ambience with computational pervasive applications. A preliminary pilot experiment was carried out with the system. The results of this experiment surprised us, as a person that did not like food with vegetables, liked the experience and even was interested in knowing the drink recipe!.

4 Conclusion

Although there are several models and frameworks to persuasive design, literature hardly presents them in support of the design of pervasive, ubiquitous applications and their contexts. In this paper we investigate the issue by analyzing different frameworks and principles for persuasive design found in the mainstream literature, organizing them in a clustering of four groups: simplifying tasks, usability, credibility and social influence.

New considerations are suggested for persuasive system design frameworks especially considering pervasive, ubiquitous applications. Among the suggested aspects to consider we include using the socially-aware design approach to raise and discuss ethical aspects of the prospective system, through the analysis and participation of stakeholders, taking a count stakeholder behavior anticipating their issues in using the system and proposing solutions to them. Moreover we suggest two other groups of principles: a hedonic and an enactive, to add to a set of initial ethical values.

Ongoing work involves to evaluate the four points proposed: socially aware design as a part of the process of persuasive design of information systems, stakeholders behavior focused to target behavior as a input to design persuasive system, persuasive principles to pervasive, ubiquitous, enactive applications, and the ethical codes as part of a proposed framework. Ongoing work involves present a formal framework with process, steps, artifacts to guide the design of persuasive information systems. Integrating results of this study will be available and free in an OpenDesign platform (https://opendesign.ml/), to apply the proposed framework within the practical context of development pervasive systems [2].

Acknowledgement. This research is sponsored by the Council for Scientific and Technological Development (CNPQ) supported by the Computing Institute from Campinas University (UNICAMP), greeted by the INTERHAD research laboratory group in Brazil, and the Project Management Professional (PMPPeru) to allow the international experience with Peru.

References

1. Valderlei, S.J., Pereira, R., Bastos, B.S., Duarte, E.F., Baranauskas, M.C.C.: SAwD – socially aware design: an organizational semiotics-based CASE tool to support early design activities. In: Socially Aware Organisations and Technologies, Impact and Challenges: 17th IFIP WG 8.1 International Conference on Informatics and Semiotics in Organisations, ICISO 2016, Proceedings, no. 477, pp. 60–69. Springer, Campinas (2016)
2. Baranauskas, M.C.C.: OpenDesign: techniques and artifacts for socially aware design of computer system. FAPESP Project #15/24300-9 (2015)

3. Baranauskas, M.C.C.: Social awareness in HCI. Interactions **21**(4), 66–69 (2014)
4. Fogg, B.: Persuasive Technology: Using Computers to Change What We Think and Do. Morgan Kaufmann, San Francisco (2003)
5. Fogg, B.: A behavior model for persuasive design. In: Proceedings of the 4th International Conference on Persuasive Technology, pp. 40–43. ACM (2009)
6. Fogg, B.: Creating persuasive technologies: an eight-step design process. In: Proceedings of the 4th International Conference on Persuasive Technology, pp. 44–48. ACM (2009)
7. Liu, K.: Semiotics in Information Systems Engineering. Cambridge University Press, Cambridge (2000)
8. Murillo, M.M.F., Vazquez, B.M., Cota, C.X.N., Nieto, H.J.I.: A framework for design and development of persuasive mobile systems. In: 2018 International Conference on Electronics, Communications and Computers (CONIELECOMP), Cholula, Mexico, pp. 59–66. IEEE (2018)
9. Mustaquim, M., Nyström, T.: Designing persuasive systems for sustainability – a cognitive dissonance model. In: Proceedings of the 22nd European Conference on Information Systems (ECIS) 2014, pp. 9–11. AIS (2014)
10. Oinas-Kukkonen, H., Harjumaa, M.: A systematic framework for designing and evaluating persuasive systems. In: Oinas-Kukkonen, H., Hasle, P., Harjumaa, M., Segerståhl, K., Øhrstrøm, P. (eds.) International Conference on Persuasive Technology, vol. 5033, pp. 164–176. Springer, Heidelberg (2008)
11. Oinas-Kukkonen, H., Harjumaa, M.: Persuasive systems design: key issues, process model, and system features. Commun. Assoc. Inf. Syst. **24**(1), Article 28 (2009)
12. Stibe, A.: Advancing typology of computer-supported influence: moderation effects in socially influencing systems. In: MacTavish, T., Basapur, S. (eds.) Persuasive Technology. Lecture Notes in Computer Science, vol. 9072, pp. 253–264. Springer, Cham (2015)
13. Stibe, A.: Towards a framework for socially influencing systems: meta-analysis of four PLS-SEM based studies. In: MacTavish, T., Basapur, S. (eds.) Persuasive Technology, vol. 9072, pp. 172–183. Springer, Heidelberg (2015)
14. Stibe, A., Kjær Christensen, A.K., Nyström, T.: Transforming sociotech design (TSD). In: 13th International Conference on Persuasive Technology, Persuasive 2018, Waterloo, Canada (2018)
15. Gulliksson, H.: Pervasive Design, 4th edn. Videoiterna, Umea (2015)
16. Satyanarayanan, M.: Pervasive computing: vision and challenges. IEEE Pers. Commun. **8**, 10–17 (2001)
17. Venkatesh, V., Thong, J.Y., Xu, X.: Consumer acceptance and use of information technology: extending the unified theory of acceptance and use of technology. MIS Q. **36**, 157–178 (2012)
18. Maeda, J.: The Laws of Simplicity. MIT Press, London (2006)
19. Mace, R.: Universal Design. http://universaldesign.ie/What-is-Universal-Design/The-7-Principles/. Accessed 08 Nov 2018
20. Nielsen, J.: Usability Engineering. https://www.nngroup.com/articles/ten-usability-heuristics/. Accessed 08 Nov 2018
21. Thompson, E., Stapleton, M.: Making sense of sense-making: reflections on enactive and extended mind theories. Topoi **28**, 23–30 (2009)
22. Kaipainen, M., Ravaja, N., Tikka, P., Vuori, R., Pugliese, R., Rapino, M., Takala, T.: Enactive systems and enactive media: embodied human-machine coupling beyond interfaces. Leonardo **44**(5), 433–438 (2011)

23. Borgefalk, G., Leon, N.: The ethics of persuasive technologies in pervasive industry platforms: the need for a robust management and governance framework. In: 14th International Conference, PERSUASIVE 2019, Limassol, Cyprus (2019)
24. Thompson, E., Stapleton, M.: Making sense of sense-making: reflections on enactive and extended mind theories. Topoi **28**(1), 23–30 (2009)

A New Mathematical Model
for the Vehicle Routing Problem
with Backhauls and Time Windows

Daniela Quila[1], Daniel Morillo[2], Guillermo Cabrera[3], Rodrigo Linfati[4],
and Gustavo Gatica[1(✉)]

[1] Universidad Andres Bello, Santiago, Chile
dp.quila@gmail.com, ggatica@unab.cl
[2] Pontificia Universidad Javeriana - Cali, Cali, Colombia
daniel.morillo@javerianacali.edu.co
[3] Pontificia Universidad Católica de Valparaíso, Valparaíso, Chile
guillermo.cabrera@pucv.cl
[4] Universidad del Bío-Bío, Concepción, Chile
rlinfati@ubiobio.cl

Abstract. This investigation presents a new mathematical model to solve the vehicle routing problem with backhauls and time windows (VRPBTW). In this problem, customers are divided into two subsets, for delivery and collection. Each vehicle leaves the warehouse to deliver merchandise to linehauls customers. Subsequently, it makes a collection of merchandise to backhauls customers and returns to the departure warehouse. In this proposal, the objective is to minimize the total distance, satisfying all restrictions. In addition, the number of vehicles to make the route is minimized. The model has been evaluated based on artificial data adapted from the literature, which includes demand for 10, 15, 20 and 30 nodes. The computational results contribute to validate the approach and scale the problem for future work.

Keywords: Combinatorial optimization · VRP · Backhauls and linehauls · Time windows · Integer programming

1 Introduction

The vehicle routing problem (VRP) is one of the most important logistical problems in the field of combinatorial optimization. The objective function of VRP is to determine a set of vehicle routes from a central warehouse in order to carry out transportation requests for delivery or collection at a minimum cost [13]. The problems present in real-life scenarios tend to be much more complex. In this investigation we consider two generalizations from the base problem:

- In practice, not only must one bring commodities from the warehouse to clients, but one should also pick up a number of clients (or their commodities) and return them to the warehouse. In real life, this problem is solved resolving two routes for vehicles with capacity, one for the delivery clients (linehauls)

© Springer Nature Switzerland AG 2020
Á. Rocha et al. (Eds.): ICITS 2020, AISC 1137, pp. 46–53, 2020.
https://doi.org/10.1007/978-3-030-40690-5_5

and another for the collection clients (backhauls). However, this approach does not create highly efficient solutions. It is more profitable to serve – on the same route – the delivery and collection clients using the same vehicles that carry out the delivery route. This problem is known and modeled in literature as the "Vehicle routing problema with backhauls" [12]. One application of VRPB is the way in which returnable/reusable bottles are handled. The full bottles are delivered to clients and the empty bottles are collected to be recycled [9].

– There are a growing number of cities that are incorporating electric buses in public transportation [4]. For route planning, one of the principle characteristics to consider is the limited capacity of the battery and the long charging time. This additional restriction is known as a window of time [1]. In literature, a client is associated with an interval of time that establishes the start and the end of service, as well as its duration [10]. This restriction can be generalized for any type of system that requires an interval of time to satisfy a client.

The vehicle routing problem with backhauls and time windows (VRPBTW) was introduced by [5]. Each client (linehaul and backhaul) should be visited within a specific interval of time. The lower and upper limits on the window of time define the earliest and the latest time to begin a service for the client. Moreover, each client has a specific service time, which is the time that vehicle dedicates to the loading or unloading of commodities. Accordingly, the total time for the route is the sum of the travel time, the wait time, and the service time. Gélinas et al. demonstrate that finding a viable solution to VRPBTW with a fixed number of vehicles is a problem with NP-Hard complexity [5]. The applications of VRPBTW arise in both in public and private sectors, like the programming of airlines and railway fleets.

2 Literature Review

VRP is the foundation of planning and distribution processes. Since the first publication on the topic in the 1950s [2], there have been an important number of studies developed to address versions of this problem [3,13]. This problem and versions of it have continually piqued the interest of the academic community due to its practical relevance and inherent difficulty.

Although there are some heuristic, metaheuristic and precise algorithms in literature to resolve VRP and its variants (CVPR, VRPB, VRPTW) only a few recent studies have been dedicated to VRPBTW. Gélinas et al. proposed a new branching strategy for the approaches to branch and bound based on column generation [5]. Later, Taillard et al. designed a heuristic search in Tabu for the VRPBTW with priority given to the client [11]. Pradenas et al. considered the environmental impact of vehicles and sought to minimize the greenhouse gas emissions [7]. Additionally, metaheuristic hybrids have been developed like those in the work of Küçükoğlu and Öztürk [6], who proposed an algorithm that combined Simulated Annealing and Tabu Search to achieve competitive results.

Recently, Reil et al. designed a heuristic for VRPBTW with three-dimensional load limitation [8].

To date, heuristic and metaheuristic algorithms have been proposed to resolve VRPBTW, which has the objective function of minimizing the distance covered. The principle contribution of this investigation is to introduce a new mathematical model, the objective function of which allows it to minimize the distance covered and at the same time reduce the number of vehicles used to execute the route.

3   Proposed Mathematical Model

The VRPBTW is formulated based on İlker Küçükoğlu and Nursel Öztürk existing mathematical model [6]. $G(V, \bar{A})$ is an undirected graph with a set of vertices $V = \{0\} \cup L \cup B$ where L and B are subsets that represent the Linehaul and Backhaul clients, respectively. The set of arcs $\bar{A}$ represent all of the possible connections between the nodes, $\bar{A} = A_1 \cup A_2 \cup A_3$ where:

$$A_1 = \{(i,j) \in \bar{A} : i \in L \cup \{0\}, j \in L\}$$
$$A_2 = \{(i,j) \in \bar{A} : i \in L, j \in B \cup \{0\}\}$$
$$A_3 = \{(i,j) \in \bar{A} : i \in B, j \in B \cup \{0\}\}$$

In this way, the set of arcs, $\bar{A}$, is partitioned in three subsets: A_1 has all of the arcs from delivery to one linehaul vertex and from the linehaul vertex to the other linehaul vertex. A_2 contains all of the arcs from one linehaul vertex to a backhaul vertex and from that final one to the initial delivery. A_3 contains all of the arcs from one backhaul vertex to another backhaul vertex and then, to the delivery. Accordingly, $\bar{A}$ does not contain any arc that makes the problem unfeasible. Additionally, for each $i \in V$ there are two sets: $A_i^+ = \{j : (i,j) \in \bar{A}\}$ and $A_i^- = \{j : (j,i) \in \bar{A}\}$. The first set defines the nodes that can be directly reached from node i. The second set defines the nodes that can directly reach node i.

This investigation aims to minimize the distances for each route and, simultaneously, minimize the quantity of vehicles that are needed in order to carry out the routes, with a penalty if all of the vehicles are used. With these parameters, the VRPBTW can be developed using the following model proposed by Mixed-Integer Programming (MIP).

3.1   Problem Notation

- K: set of vehicles.
- C_{ij}: distance from node i to node j, $(i,j) \in \bar{A}$.
- t_{ij}: time to travel between nodes i and j, $(i,j) \in \bar{A}$.
- a_i: the earliest arrival time of client i, $\forall i \in \bar{V} \backslash \{0\}$.
- b_i: the latest arrival time of client i, $\forall i \in \bar{V} \backslash \{0\}$.
- s_i: the service time of client i, $\forall i \in \bar{V} \backslash \{0\}$.
- d_j: demand associated with the node j (line o back).

- U_k: capacity of vehicle k, $\forall k \in K$.
- T_{max}: maximum time allowed for the vehicle routes.
- M: large value for penalty.

3.2 Decision Variables

- $x_{ijk} : \begin{cases} 1 & \text{if the vehicle } k \text{ travels fromn client } i \text{ to } j. \\ 0 & \text{otherwise.} \end{cases}$
- $W_{ik} :$ time from the start of the service of vehicle k to client i.

3.3 Objective Function

$$\text{Min } z = \sum_{(i,j) \in A_1} \sum_{k \in K} U_{ij} x_{ijk} + 100 \sum_{j \in A_0^+} \sum_{k \in K} x_{0jk} \tag{1}$$

3.4 Constraints

$$\sum_{(i,j) \in A_1} d_j x_{ijk} \leq U_k \qquad \forall k \in K \tag{2}$$

$$\sum_{(i,j) \in A_2 \cup A_3; j \in B} d_j x_{ijk} \leq U_k \qquad \forall k \in K \tag{3}$$

$$\sum_{j \in A_0^+} x_{0jk} \leq 1 \qquad \forall k \in K \tag{4}$$

$$\sum_{i \in A_0^-} x_{i0k} \leq 1 \qquad \forall k \in K \tag{5}$$

$$\sum_{i \in A_j^-} \sum_{k \in K} x_{ijk} = 1 \qquad \forall j \in V \backslash \{0\} \tag{6}$$

$$\sum_{j \in A_i^+} \sum_{k \in K} x_{ijk} = 1 \qquad \forall i \in V \backslash \{0\} \tag{7}$$

$$\sum_{i \in A_j^-} x_{ijk} = \sum_{i \in A_j^+} x_{jik} \qquad \forall j \in V \backslash \{0\}, \forall k \in K \tag{8}$$

$$W_{jk} \geq t_{0j} - M(1 - x_{0jk}) \qquad \forall j \in V \backslash \{0\}, \forall k \in K \tag{9}$$

$$W_{0k} \geq W_{ik} + s_i + t_{i0} - M(1 - x_{i0k}) \qquad \forall i \in V \backslash \{0\}, \forall k \in K \tag{10}$$

$$W_{jk} \geq W_{ik} + s_i + t_{ij} - M(1 - x_{ijk}) \qquad \forall i \in V \backslash \{0\}, \forall j \in V \backslash \{0\}, \forall k \in K \tag{11}$$

$$a_i \leq W_{ik} \leq b_i \qquad \forall i \in V \backslash \{0\}, \forall k \in K \tag{12}$$

$$0 \leq W_{0k} \leq T_{max} \qquad \forall k \in K \tag{13}$$

$$x_{ijk} \in \{0,1\} \qquad \forall i \in V, \forall j \in V, \forall k \in K \tag{14}$$

$$W_{ik} \geq 0 \qquad \forall i \in V, \forall k \in K \tag{15}$$

The objective function (1) minimizes the distances and the number of vehicles used in the model. Constrains (2) to (3) ensure that the demand does not surpass the vehicle capacity. Constrains (4) to (5) ensure that all of the routes carried out by the vehicle drop off the delivery and return to the same place. Constraints (6) to (7) guarantee that exactly one arc enters and leaves each client vertex. Constraint (8) guarantee that a vehicle leaves the same client vertex in which it entered (balance constraints). Constraints (9) to (11) guarantee the viability of the schedule according to the time considerations and define the variables associated with the service start times in all of the nodes. Constraints (12) to (13) guarantee that each client is visited within the specific window of time. Finally, constraints (14) to (15) define the nature of the decision variables.

4   Computational Results

The proposed model has been implemented using AMPL (A Mathematical Programming Language) IDE and solved with IBM CPLEX 12.8.0.0. The model was carried out on a PC with an Intel i5 7th generation processor running at a speed of 2.50 GHz and with 4 GB of RAM memory. There were optimal solutions obtained for a considerable number of small and medium (up to 30 nodes) instances generated with a 90-minute limit on computing time. Although the results are highly competitive, considering the complexity of the problem, for instances of a larger size it would be feasible to explore using metaheuristic methodologies for a solution. The results achieved in the computational experiments validate the proposed model and are summarized in Table 1.

Table 1. Artificial instances generated for VRPBTW.

Instances	Nodes	Linehauls	Backhauls	Vehicles	Optimal distance achieved
P01	10	4	5	4	558
P02	10	6	3	5	613
P03	10	4	5	4	1002
P04	15	9	6	5	760
P05	15	10	5	5	728
P06	15	12	3	5	767
P07	20	9	11	6	941
P08	20	8	12	8	1152
P09	30	16	14	12	1250
P10	30	16	14	12	1165

The instance shown next corresponds to $P03$ from the previous table. Table 2 shows the result of the decision variable x_{ijk}.

Table 2. Results of the decision variable x_{ijk} for the $P03$ instance.

Value of the basic decision variables		
$x_{014} = 1$	$x_{251} = 1$	$x_{601} = 1$
$x_{021} = 1$	$x_{394} = 1$	$x_{702} = 1$
$x_{042} = 1$	$x_{472} = 1$	$x_{804} = 1$
$x_{134} = 1$	$x_{561} = 1$	$x_{984} = 1$

Figure 1 demonstrates graphically the solution represented by the decision variables (Table 2) according to the data generated in the P03 instance, where there are four vehicles available. However, the results demonstrate that for this instance only three vehicles were used. This result validates the contribution of the mathematic model proposed in this investigation.

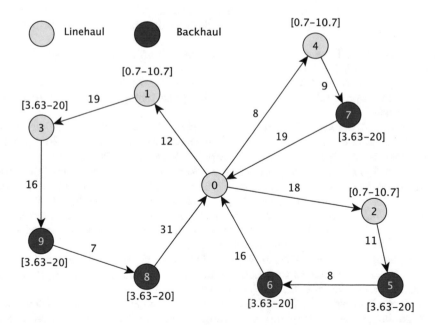

Fig. 1. Result graph for the $P03$ instance.

5   Conclusions and Future Investigation

This article has addressed the vehicle routing problem with backhauls and time windows (VRPBTW). It has proposed a new mathematical model that includes a reduction in the number of vehicles used and, simultaneously, reduces the cost per route. The results show that the optimal solution is achieved for problems up with up to 30 nodes.

Based on the computational results, the mathematical model proposed is scalable considering its complexity. Moreover, in the model one can easily include a multi-objective approach, complementing the proposal. Given the nature of the problem, one could explore metaheuristic and simheuristic techniques in order to generate possible solutions in less computational time.

Currently, this work has added the variant of variable demand. This is resolved using the Sample Average Approximation Method (SAA) and validation using Monte Carlo simulation with artificially generated data.

References

1. Cortés-Murcia, D.L., Prodhon, C., Afsar, H.M.: The electric vehicle routing problem with time windows, partial recharges and satellite customers. Transp. Res. Part E: Logist. Transp. Rev. **130**, 184–206 (2019). https://doi.org/10.1016/j.tre.2019.08.015
2. Dantzig, G.B., Ramser, J.H.: The truck dispatching problem. Manag. Sci. **6**(1), 80–91 (1959). https://doi.org/10.1287/mnsc.6.1.80
3. Eksioglu, B., Vural, A.V., Reisman, A.: The vehicle routing problem: a taxonomic review. Comput. Ind. Eng. **57**(4), 1472–1483 (2009). https://doi.org/10.1016/j.cie.2009.05.009
4. Gatica, G., Ahumada, G., Escobar, J., Linfati, R.: Efficient heuristic algorithms for location of charging stations in electric vehicle routing problems. Stud. Inform. Control **27**, 73–82 (2018). https://doi.org/10.24846/v27i1y201808
5. Gélinas, S., Desrochers, M., Desrosiers, J., Solomon, M.M.: A new branching strategy for time constrained routing problems with application to backhauling. Ann. Oper. Res. **61**(1), 91–109 (1995). https://doi.org/10.1007/bf02098283
6. Küçükoğlu, İ., Öztürk, N.: An advanced hybrid meta-heuristic algorithm for the vehicle routing problem with backhauls and time windows. Comput. Ind. Eng. **86**, 60–68 (2015). https://doi.org/10.1016/j.cie.2014.10.014
7. Pradenas, L., Oportus, B., Parada, V.: Mitigation of greenhouse gas emissions in vehicle routing problems with backhauling. Expert Syst. Appl. **40**(8), 2985–2991 (2013). https://doi.org/10.1016/j.eswa.2012.12.014
8. Reil, S., Bortfeldt, A., Mönch, L.: Heuristics for vehicle routing problems with backhauls, time windows, and 3D loading constraints. Eur. J. Oper. Res. **266**(3), 877–894 (2018). https://doi.org/10.1016/j.ejor.2017.10.029
9. Ropke, S., Pisinger, D.: A unified heuristic for a large class of vehicle routing problems with backhauls. Eur. J. Oper. Res. **171**(3), 750–775 (2006). https://doi.org/10.1016/j.ejor.2004.09.004
10. Solomon, M.M.: Algorithms for the vehicle routing and scheduling problems with time window constraints. Oper. Res. **35**(2), 254–265 (1987). http://www.jstor.org/stable/170697

11. Taillard, É., Badeau, P., Gendreau, M., Guertin, F., Potvin, J.Y.: A tabu search heuristic for the vehicle routing problem with soft time windows. Transp. Sci. **31**(2), 170–186 (1997). https://doi.org/10.1287/trsc.31.2.170
12. Toth, P., Vigo, D.: An exact algorithm for the vehicle routing problem with back-hauls. Transp. Sci. **31**, 372–385 (1997). https://doi.org/10.1287/trsc.31.4.372
13. Toth, P., Vigo, D.: Vehicle Routing: Problems, Methods, and Applications, 2nd edn. SIAM - Society for Industrial and Applied Mathematics, Philadelphia (2014). https://www.xarg.org/ref/a/1611973589/

Using Architecture Patterns in the Conceptual Model of an eGov Software

Oscar Carlos Medina(✉), María Soledad Romero(✉),
Rubén Aníbal Romero(✉), Siban Mariano Martin(✉),
and Marcelo Martín Marciszack(✉)

CIDS - Centro de Investigación, Desarrollo y Transferencia de Sistemas de
Información, Universidad Tecnológica Nacional – Facultad Regional Córdoba,
Cruz Roja Argentina y Maestro López s/n, Ciudad Universitaria,
Córdoba, Argentina
{omedina, marciszack}@frc.utn.edu.ar,
romeroma.soledad@gmail.com, romeroa.ruben@gmail.com,
smarianomartin@gmail.com

Abstract. *Context:* A model to reuse of a successful solution for the same problem, considering different contexts, is called a Pattern. Patterns can be classified in different types from Software Engineering point of view. *Objective:* This work focuses on the use of architecture patterns for Conceptual Modelling of a software product. *Methods:* A study case on a process that supports an occupational safety monitoring application in the Public Sector was analyzed. This experience can be used as a baseline to be used in similar Electronic Government systems. *Results:* The application of Architecture Patterns allows to define Software Architecture planning the system structure in a middle layer of the Conceptual Modelling. *Conclusion:* It is feasible to incorporate architecture patterns to the eGov system modelling phase and to increase the software quality level that implements them. The patterns generated from this experience will be used to define a model to analyze the application of patterns in Conceptual Modelling of Electronic Government systems.

Keywords: Patterns · Architecture · Electronic Government · Occupational safety · eGov

1 Introduction

Electronic Government is *"the application of Information and Communication Technologies to Government processes"* [1]. Usually denoted as e-Government or eGov ("e" prefix means electronic), is a holistic vision that includes information systems supporting processes in the Public Sector.

In a similar way as some information systems design is built upon a set of business best practices, such as management and accounting systems, it would be beneficial to have Electronic Government Good Practices as a reference in the construction of public software. Government Good Practices are, in this context, *"every initiative and*

Á. Rocha et al. (Eds.): ICITS 2020, AISC 1137, pp. 54–63, 2020.
https://doi.org/10.1007/978-3-030-40690-5_6

experience that help to improve the effectiveness of government actions and affect positively in citizen life conditions, achieving a measurable impact in communities" [2].

Processes supporting E-Gov applications are related to Good Government Practices. The standardized and accurate description of successful experiences is a significant tool to take advantage of previous knowledge when developing a new E-Government system for the same process. Electronic Government debate has regained strength due to new realities, such as Smart Cities, where *"governments open to participation to new policy work, education, social, health and public safety-oriented citizenship"* [3].

1.1 Architecture Patterns

A "pattern" is a model that contains the generic description of a process, according to Software Engineering. The pattern concept was used for the first time by the architect Christopher Alexander, and it can be read in his books "A pattern language" [4] and "The Timeless Way of Building" [5]. In them, he defines *"each pattern describes a problem which occurs over and over again in our environment, and then describes the core of the solution to that problem"* [5].

The context-problem-solution triad is used in software design by Gamma, Helm, Johnson and Vissides, who publishes the most important work about this subject: "Design Patterns: Elements of Reusable Object-Oriented Software" [6], in accordance to object-oriented programming paradigm. Patterns for the different stages of software development were introduced in this book, with specific types for each one of them, such as patterns that model business processes or user interfaces.

Using a different approach, Buschmann [7] defines another pattern type, the Architecture Pattern. These patterns are a predefined set of subsystems with their responsibilities and a series of suggestions to organize the different components.

The goal of Architecture Patterns is to represent the different elements that shape a software product, and their interrelation in the Conceptual Model of the system. Unlike software design patters, which are based in object-oriented development, Architecture Patterns are defined at a higher level of abstraction.

This Software Architecture includes, according to Pérez-Campanero Atanasio [8]:

- The highest-level design of the system structure.
- Required patterns and abstractions to guide the software construction of a system.
- The foundation for software professionals to work on a common line that allows to cover restrictions and achieve the goals of the system.
- The goals of the system, not only the functional ones, but also those related with maintenance, auditing, flexibility and interaction with other systems.
- The restrictions that limit the construction of the system in relation to the available technologies to implement it.

Software Architecture defines, in an abstract manner, the components that carry out any task of the information process, their interfaces and the communication between them.

There are many Architecture Patterns, and new ones are emerging in concordance to the needs of the development technologies. A commonly used classification is the one proposed by Richards [9]:

- Layer Architecture.
- Event-Oriented Architecture.
- Asynchronous systems with asynchronous data flow.
- Microkernel Architecture.
- Microservices Architecture.
- Space based architecture.

Layer Architecture and Microservices Architecture will be briefly explained, given that they will be used in the selected study case.

Layer Architecture is the most used model since it is supported by data bases and an important quantity of applications store their information in tables. The software is organized so that data are entered in the top layer and keep descending through every layer until they reach the bottom layer, which usually is a database. Each layer has a specific task, such as verifying data coherence or value reformatting to maintain consistence.

Model View Controller (MVC) structure is the standard software development focus suggested for web development. The top layer of this stratified architecture is the View layer, frequently containing CSS, JavaScript and HTML with embedded dynamic code. The middle layer is the Controller, containing several rules and methods to transform the data that travels between the view and the Model.

The advantage of a Layer Architecture is splitting the individual implementation difficulties, which means that every layer can focus in their own function. This architecture can also have additional open layers, such as a service layer.

Microservices Architecture has the goal to create a series of small different programs, and then to create a new small program for each function that gets added. This focus is similar to the Event-Oriented and Microkernel ones, but is mainly used when the different tasks are easily split into services. In many cases, a task can require different quantities of processing and may vary in its use. Upon implementing these services in a separate manner, each one can scale up and down independently according to the demand it has.

The service orientation is an agnostic implementation paradigm that may be carried out with any adequate technological platform, even though it is used on web services nowadays. In the same sense that an agnostic person declares that any divine knowledge is inaccessible to human understanding, an agnostic software architecture service does not depend of a proprietary technology nor of a specific business process.

1.2   The Application of Patterns in Conceptual Modelling

The aim of the Conceptual Model of a software product is to detect and characterize significant concepts of a problem domain, recognizing the features and their inherent relationships. According to Sommerville [10], Conceptual Model of a system is carried out during the first activities of "Rational Unified Process" of software development: business modelling, requirements elicitation, analysis and design. Recent empiric studies, such as the one by Pacheco et al. [11], search to find mature and efficient techniques for requirement elicitation. A pattern that has proven to be successful can be selected once the problem has been correctly described through these techniques. In order to reuse knowledge from previous experiences, a pattern is applied to Conceptual Modelling. These experiences are encapsulated in concrete analysis and design solutions, allowing the verification and validation of functional requirements.

Several benefits would be obtained with the inclusion of the Conceptual Model of public software. According to Beck et al. [12], the use of patterns offers the following advantages:

- Help to diminish development times
- Allow a better communication
- Help to reduce design errors, showing the essential parts of design
- Are reusable and implement the use of "good practices".

The proposed introduction of Patterns in Conceptual Modelling includes two levels covering different aspects (see Fig. 1).

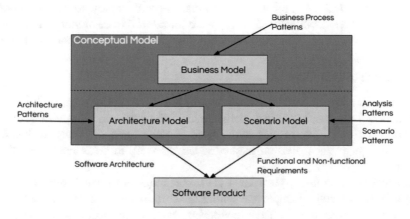

Fig. 1. Introduction of patterns in Conceptual Modelling

Business Processes Patterns may be used to define the Business Model. In the next level of abstraction, Analysis and Scenario Patterns may be used in the Scenario Model to specify functional and non-functional requirements and Architecture Patterns may be used to define the Architecture Model.

Note that, even though Architecture Patterns are a part of the logic architecture of a system, thus making it necessary to define them in the construction stage, it is feasible and convenient to include them in the Conceptual Model. The Architecture Model in this instance determines the development strategy that will be used in a level of abstraction independent of the platform used.

It is desirable to obtain these advantages by including patterns in Conceptual Modelling and in Electronic Government processes, proposing a study case as an example in the present work. Thus, a problem was identified and a successful solution was effectively implemented in production within a Government context, and it is described by using patterns.

It is important to mention that reusability is one of software quality dimensions. A software quality measuring model is needed to validate the effectiveness of a tool that facilitates reuse. There are specific measuring models for Electronic Government, explained in [1, 13–16]. They propose frameworks to assess quality in electronic services and government websites.

Quality optimization due to the inclusion of patterns in Conceptual Modelling will be validated by means of a specific quality measuring model for e-Government systems, such as MoQGEL Model [16]. Therefore, this measure can be considered as an objective and quantifiable indicator.

2   Materials and Methods

We will approach the monitoring process of occupational safety that supports a mobile application called "Ubicuo" [17]. This software has been successfully implemented in a Public Entity, as well as in other organizations. We will employ Architecture Patterns in this paper. Three Architecture Patterns were introduced in de architecture modelling of the actual version of "Ubicuo", evaluating their impact and comparing the results.

2.1   Selected Study Case

EPEC (Córdoba Province Energy Corporation) is a public self-governed entity, whose main activity is to supply electric energy to Córdoba province, Argentina [18].

The provision of this service is managed in geographic divisions, with different workers assigned. Each division is subdivided, as well, in areas related to different branches.

In accordance to ISO 14.001:2004 [19], that requires to watch over the occupational safety and health of employees within the corporation, a monitoring process was developed and carried out. This monitoring process of occupational safety was implemented with the "Ubicuo" mobile application. It is important to point out that old practices were improved during the process development. One of such improvements was the inclusion of technologies to a process that was previously manual and supported by paper. In this sense, the publication by OIT (International Work Organization): "Something to point out has been the use of technological solutions, not only to improve employee registration by their employer, but also to facilitate the work of inspectors on site. In a time where the use of technology is crucial, the creation of mobile application has facilitated the supervision of occupational obligations compliance" [20].

Monitoring activity consists in the execution of inspections to gather risk warnings present in work areas and in employee's actions. The employees the Public Entity make observations which are registered through this process, allowing the generation of reports that notify and clearly show the risks found. Thus, is achieved the main goal of the process is achieved.

First, it is worth noting that the process starts with the inspection activity. In this context, an inspection is the action, planned or unplanned, of an observer registering an observation, be it a "risk" event, or any observable data. This broad scope of inspection would also allow generating not only a risks knowledge base, but also a learnt lesson one. In addition, the concept of risk is the one adopted by the OIT, referred to: "a physical situation with potential to cause personal harm, property or environmental damage or a combination of these" [21].

The roles of "inspector" and "monitor" were established by the organization to carry out the inspection activity. The inspector has the responsibility to regularly visit the facilities and register every possible observation. With that in mind, they are previously trained in "Ubicuo" software. It consists in a mobile application that allows systematizing the observations in each inspection. This system works by capturing georeferenced photography, adding the intervention level and a categorization along a description and an action proposal for its resolution. Captured information is sent to servers through internet to "Ubicuo" database, making it immediately available for consulting and processing. In a similar way, the monitor uses the tool to load observations, but in a spontaneous manner, or to confirm solutions to already registered problems. These functionalities are a good practice that responds to a determined context providing every documentary element necessary and evidentiary to solve the central problem of controlling occupational safety and health.

All collected data is stored in the cloud. This feature permits drafting adequate criteria to systematize the information, and to generate indicators automatically. This way, information management enables the generation of statistics that reduce the uncertainty in decision taking. The main difficulties that need to be dealt by this stage is the processing of large volume of historic data and a limited storing capacity.

2.2 Using Architecture Patterns in the Conceptual Model

According to the classification proposed by Richards and explained previously, the architecture modelling of "Ubicuo" combines a layer architecture strategy with a microservices architecture. Model View Controller, Generic Repository and Unit of Work patterns were introduced. The Fig. 2 shows the implementation of these patterns in the architecture modelling of "Ubicuo". The way of working for each pattern will now be detailed.

The MVC Pattern is an architectonic pattern used to classify the information, the system logic and the user interface. In this type of architecture there is a central system or controller that manages the inputs and outputs of the system, one or many models that search for the needed data and information and an interface that shows the results to the end user. It is frequently used in web development given that the need to interact with several languages to create a web site facilitates the confusion between each component if they are not split in an adequate manner. This pattern allows the modification of each one of its components without affecting the others. It was introduced in the mobile application architecture of "Ubicuo" in the business and presentation layer for the user login.

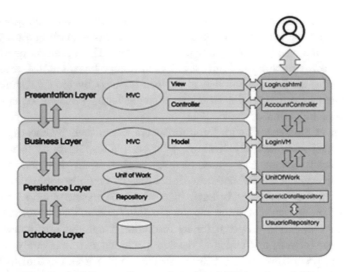

Fig. 2. Architecture Patterns introduced "Ubicuo" modelling.

The Generic Repository Pattern helps to structure applications whose functionality and collaboration is purely data-driven. A Generic Repository keeps the common data where the application components work, that access and modify the data in a shared repository. The state of these data in the shared repository urges to control the specific components flow.

The direct access to the data may imply: duplicated code, programming errors potential risk, difficulties to centralize rules related to the data, such as cache storing. A Generic Repository intercedes between the domain and the data forwarding layers, acting as set of domain objects in memory. The client objects build request specifications in a declarative manner and send them to the repository to be satisfied. The objects may be added and deleted from the Generic Repository, as they could be from a simple object set, and the forwarding code encapsulated by the repository carries out the appropriate operations behind the scenes. The use of the Generic Repository pattern allows to establish standard method to access the database, such as the implementation of the generic method GetAll() in the software persistence layer of "Ubicuo" software, communicating to the SQL Server database engine.

Object-oriented applications and relational databases use different mechanisms to structure data. The object-oriented model domain knows the relational database scheme and vice versa.

Forwarding between an object-oriented model and a relational scheme introduces additional complexity in an application. For example, the collections and inheritance are not present in the database relations, while the relational constructions such as SQL queries are not primitives of the behaviour of conventional object-oriented languages. It is easy to add access code to the database to an object of the application, but it makes the implementation of the object unnecessarily fragile and complex if the scheme or the database change. A solution is needed that is stable and lightly coupled.

Finally, the Unit of Work Pattern introduces a mapper to persist each type of object of the application with the responsibility to transfer data from the object to the relational database and vice versa. The Unit of Work operates as a data forwarding unit that carries out every necessary data transformation and guarantees coherence between two representations. At the same time, it allows the evolution of the object-oriented domain and the relational database scheme, as it can be seen in the Fig. 2 of the Persistence Layer of "Ubicuo". This design also simplifies the unit test. Upon inserting and updating data in the database, the Unit of Work Pattern helps the data forwarding unit to understand which objects have been changed, created or destructed.

3   Results

In order to evaluate the effectiveness of the applicability of the Unit of Work, a previous version of "Ubicuo" which did not have it implemented was analyzed. Benefits were identified in three different features: instantiated variables for read-write access to the database, number of write accesses to the databases and "rollback" execution.

Note that the actual version of "Ubicuo" includes the Unit of Work Pattern in every module. The duo of MVC and Generic Repository Patterns is applied in 14 of 15 modules, that is in the 93,33% of the development, while it is applied in 47 of 55 modules of the views, that is an 85,45% of the total.

The comparative table of the features evaluated with and without the Architecture Patterns can be seen in Table 1.

In relation to the database read access, the gain is from 1 to 7 variables defined at module level. In the same sense, write operations to the database is optimized from 1 to 5 instantiated variables and from 1 to 5 write accesses to databases that are carried out in a single call.

Table 1. Comparative results with and without Architecture Patterns.

Heading level	Without Architecture Patterns	With Architecture Patterns
Instanced variables for reading	5 variables for 5 reading operations	1 variables for 7 reading operations
Instanced variables for writing	3 variables for 3 writing operations	1 variables for 5 writing operations
Write access to database	3 accesses for 3 writing operations	1 access for 5 writing operations
Rollback execution	Not supported	Supported

The rollback operation allows discarding a change to the database upon detection of an error in the operations, leaving the database without inconsistencies and in the same previous state to the execution of change sentences. When Architecture Patterns were not used, it was necessary to guarantee this functionality with other type of mechanism.

From a software quality focus, the advantages obtained from the application of Architecture Patterns affect the following dimensions of the software product quality:

- Maintenance: resulting from modularity, reusability and capacity to introduce changes in the models.
- Security: by collaborating to the integrity of the database.
- Performance efficiency: by reducing the quantity of instantiated variables and database accesses.

4   Conclusions

In light of the results obtained in the previous sections, we can conclude that a pattern catalog can be built from this experience [13].

The reuse property of patterns makes them ideal candidates to introduce them in the Conceptual Modelling activity, which will be used to formalize the initial definition of a new system.

The use of standardized patterns to characterize eGov software is feasible and promotes reusability. This is especially relevant in the particular case public software, where these processes support Electronic Government applications.

In addition, the graphical representation of patterns eases the communication of the fundamental parts between different actors of a process, promoting a better comprehension of the system.

In this sense, the consolidation of distributed systems and the rise of web and mobile application required that more attention is payed to middleware, which is the software that exchanges data between applications. The architectonic patterns are focused on the software architecture predicament and describe broad design problems that are solved by a structural approach for a specific context, thus the solution has limitations and restrictions.

Therefore, it can be concluded that the application of patterns allows to define Software Architecture planning the system structure in a middle layer of the Conceptual Modelling. It was identified that the generation of this type of patterns promotes the reuse of successful architectonic configurations, helping to describe the essential to the process.

The use of patterns in the Conceptual Model increases the level of quality of the software that implements them.

Finally, we conclude that, given that the effectiveness of applying these patterns has been proven in concrete results, they will be used to define an analysis model for the application of patterns in Conceptual Modelling of Electronic Government systems. Furthermore, considering the findings of the study case approached in this work, it can be inferred that, based on a problem specification and its particular context, it is possible to define a pattern selection method from an eGov pattern catalog.

References

1. Aproximación descriptiva a las Buenas Prácticas de Gobierno Electrónico y a su incorporación en el Modelado Conceptual de Sitios Web Públicos de Argentina. http://rtyc.utn.edu.ar/index.php/rtyc/article/view/107. Accessed 11 Sept 2019
2. Varela Rey, A.: Beneficios del intercambio de buenas prácticas municipals. INNOTEC Gestión **7**, 55–59 (2016). Laboratorio Tecnológico del Uruguay
3. Estrada, E., Maciel, R., Peña Pérez Negrón, A., López Lara, G., Larios, V., Ochoa, A.: Framework for the analysis of smart cities models. In: Mejia, J., et al. (ed.): Trends and Applications in Software Engineering. Proceedings of the 7th International Conference on Software Process Improvement (CIMPS 2018), pp. 261–269. Springer (2019)
4. Alexander, C.: A Pattern Language. Oxford University Press, New York (1977)
5. Alexander, C.: The Timeless Way of Building. Oxford University Press, New York (1979)
6. Gamma, E., Helm, R., Johnson, R., Vissides, J.: Design Patterns: Elements of Reusable Object-Oriented Software. Addison-Wesley, Boston (1994)
7. Buschmann, F., Sommerlad, P., Stal, M., Meunier, R., Rohnert, H.: Pattern Oriented Software Architecture: A System of Patterns, vol. 1. Wiley, Hoboken (1996)
8. Desarrollo de Software y de las Arquitecturas de Software. http://www.issi.uned.es/doctorado/softwarch/DEA/Trabajos%2010/Trabajo%20de%20Investigacion%20-%20Juan%20Antonio%20Perez-Campanero.pdf. Accessed 11 Sept 2019
9. Richards, M.: Software Architecture Patterns. O'Reilly, Newton (2015)
10. Sommerville, I.: Ingeniería de Software, 9a Edición en español. Pearson (2011)
11. Pacheco, C., García, I., Reyes, M.: Requirements elicitation techniques: a systematic literature review based on the maturity of the techniques. IET Softw. J. **12**(4), 365–378 (2018)
12. Beck, D., Coplien, J., Crocker, R., Dominick, L., Meszaros, G., Paulisch, F.: Industrial experience with design patterns. In: ICSE-18 International Conference on Software Engineering, pp. 103–113. Technical University of Berlin, Germany (1996)
13. Medina, O.C, Marciszack, M.M., Groppo, M.A.: Proposal for the patterns definition based on good practices for the electronic government systems development. In: 13th Iberian Conference on Information Systems and Technologies (CISTI), Cáceres, Spain (2018)
14. Rodríguez, R.A.: Marco de Medición de calidad para gobierno electrónico. Ph.D. thesis, Universidad Nacional de la Plata (2011)
15. Sá, F.A., Rocha, A., Pérez Cota, M.: From the quality of traditional services to the quality of local e-Government online services: A literature review. Gov. Inf. Q. **33**, 149–160 (2016)
16. Sá, F.A., Rocha, A.: Qualidade do GovernoEletrónico. Modelo MoQGE. Ed. Sílabo (2017)
17. Ubicuo - Nuestros productos. http://www.ubicuo.com.ar. Accessed 11 Sept 2019
18. EPEC - ¿Quiénes somos?. https://www.epec.com.ar/institucional/quienes-somos. Accessed 11 Sept 2019
19. ISO 14000 family - Environmental management. https://www.iso.org/iso-14001-environmental-management.html. Accessed 11 Sept 2019
20. Organización Internacional del Trabajo: Notas sobre tendencias de la inspección del trabajo. OIT, FORLAC (2015)
21. Organización Internacional del Trabajo: Inspección de seguridad y salud en el trabajo – Módulo de formación para inspectores, OIT (2017)

Information and Knowledge
Management

Quality in Documentation: Key Factor for the Retrieval Process

Diana Suárez López[(⊠)] ⓘ and José Maria Alvarez Rodriguez ⓘ

Universidad Carlos III, Madrid, Spain
100394339@alumnos.uc3m.es

Abstract. This work aims to identify the latest progress, techniques and/or models implemented to improve the information quality in the information retrieval process, as well as the defined metrics to evaluate the quality of the content; therefore, a review of research work carried out in the last five years is presented, in the analysis carried out it was shown that 44% of the retrieval systems focus on improving the indexation of the documents allowing better accessibility, which leads to optimal growth at the time of extraction, the other 66% points to the quality of the response of the queries; as a result, it was found that the common factors that affect quality when searching, retrieving and representing the information available to the user, among which are user satisfaction, the quality of the classification and the quality of the text itself.

Keywords: Information retrieval · Quality · Scientific documentation

1 Introduction

Quality is understood as the set of inherent characteristics associated with an object, product or service, in this case a document or a web site, becoming an indispensable requirement, closely related to the satisfaction that it can provide for the user; In this sense, the quality of the information must be relevant, constantly updated, in such a way that it can generate value and usefulness to those who need it; To guarantee compliance with these characteristics, technical, financial and human efforts are required.

The quality of the information is one of the main problems presented by the web [1, 2] because of the large volume that is generated permanently [3], coming from different sources of information, especially those resulting from scientific activity; in such a way that they hinder their search and access due to their heterogeneity, quality and visibility, through the implementation of the different techniques and models that exist to retrieve said information, from public information hosted on the Internet to scientific documentation stored in prestigious databases and repositories, in this sense the low quality of the information increases the extraction costs [4].

Taking into account the above, an analysis was carried out that allowed identifying the main problems associated with the quality of scientific documentation as a key element to implement an information retrieval system (SIR), for which scientific publications of the last ones were consulted. five years, in which deficiencies were found with respect to: the quality in the documentary classification, the satisfaction of

© Springer Nature Switzerland AG 2020
Á. Rocha et al. (Eds.): ICITS 2020, AISC 1137, pp. 67–74, 2020.
https://doi.org/10.1007/978-3-030-40690-5_7

the need for information, this with respect to the quality of the results of the consultation, the quality detected by both humans and automatic way, the metrics to measure the quality of the text, as well as specific metrics for scientific documentation, models to predict the quality and finally the accessibility and usability.

2 Literature Review

2.1 The Need of Information

The need for information is the first step to start the search process, this information can be retrievable by combining different technologies and/or tools such as databases, search engines, thesauri, among others, being in this sense the retrieval of information the second step to determine the need for information, according to [5] the proper management of these tools can contribute to achieve a quality retrieval. Although the handling of information is an issue that has been addressed for a long time, it still presents some difficulties to manage it. When implementing the search tools, retrieve a large number of documents, only a small part of which is relevant to the user's query [6] however, they do not always appear first in the query output.

According to [7] the correct representation of the need for information and the time the user takes to execute the designed strategy with a view to retrieve the appropriate information, are key factors for the information retrieval system be successful; in that sense, this need must be interpreted and carried to terms that the SIR can understand.

Similarly [8] states that the process of information retrieval covers some cognitive problems when representing both the information needs and the knowledge contained in the documents, which makes it difficult to select the relevant documents to be shown to the user.

For [9], the lack of a semantic web makes it impossible to find precisely what information is needed; which leads to the user cannot meet their need for information, for it the authors propose to organize and structure the data and attributes in order to give them more context and meaning, this is achieved using metadata.

The metadata appear as an alternative to the internal data record from the resources, to improve the results of searches on the web in terms of ease, speed and accuracy [10], as well as helping to organize the contents; therefore it is essential to standardize them to improve the quality of the results. Among the most used metadata is Dublin Core, in charge of cataloging digital resources in a general way, in accordance with [11] this model has become an operational infrastructure for the development of the Semantic Web, adaptable and applicable to any digital information need.

In the sense [12] states that the first step to perform a search and retrieval is to ensure that the metadata to which they are accessed have a certain level of quality, since the main objective of the metadata is to facilitate its search, retrieval and evaluation, helping to satisfy the need for information. In the work presented by the authors, three metrics were defined to evaluate the quality of the metadata: completeness, consistency and coherence metrics.

In accordance with [13] user satisfaction with the complexity of the semantic web cannot be measured with typical metrics of information retrieval, it is necessary to use ontologies to determine the power and complexity to carry out the query, since the same concept can be expressed in different vocabulary or metadata in different terms and in different languages. In some occasions the result will depend on the meaning of the concept defined by experts and the interpretations generated automatically by advanced algorithms, in the investigation of [14] an automatic reasoning is presented where the automatically collected data is verified. They are contributed by humans.

2.2   The Documentary Classification

The classification of documents is nothing more than the division of documents into groups with similar characteristics, which allows an optimal search on the Internet. The grouping of documents is applicable in several fields of knowledge such as segmentation of markets, customers, monitoring and analysis of content, in the classification of documents as part of the process of information retrieval, among others. To improve the effectiveness of the tasks performed on documents and collections [15], information requests should be better understood using natural language processing (PLN), in order to provide approaches to improve quality focused on precision.

The organization of documentation and understanding of natural language, its construction, taxonomies and ontologies is a complex process that requires time and effort [16]. In that sense, the classification of documents can be improved by balancing the relevance of the document for the user's query and the estimated quality of the document [17]; taking into account that users sometimes require the system to return as a result of their searches, parts of documents and not complete documents, as is usually the case in traditional SIR [18].

3   Methodology

The methodology implemented is based on a systematic review protocol, in which the following steps have been defined.

1. Research question: the objective of this paper is to provide an answer to the following question:
   ¿What are the advances, techniques, models implemented to improve the quality of information in the information retrieval process, as well as the metrics defined to evaluate the quality of the content that is intended to be retrieved?
2. Search string: two search strings have been formulated to improve the result of the query.

   - (Information retrieval Y (natural language programming) AND (semantic web or metadata)).
   - (Scientific documentation Y (quality metrics) Y (content metrics)).

3. Bibliographic data bases: large databases were selected in the field of information technology, in addition Google Scholar was used to obtain results from other sources.

   - ACM Digital Library: is the world's most comprehensive database of full-text articles and bibliography in the field of information technology and computers.
   - IEEE Xplore: digital research library with access to magazines and congress proceedings, conferences, standards and educational courses of IEEE, related to computer science, electrical and electronic engineering.
   - Sciencie Direct (Elsevier): is the largest publisher of books and scientific literature in the world, offers services to more than 30 million scientists, students and health professionals and information.
   - SpringerLink: is a comprehensive online search platform that allows access to more than 5 million resources and content that make up the most complete online collection of books, magazines, reference works, protocols and databases of science, technology and medicine.

4. Selection: articles published in both English and Spanish were taken into account in the last years (2012–2018) relevant in terms of their content, updating, field of application in various areas including computer science and medicine, these being the most relevant in terms of the implementation of information retrieval systems. A taxonomic analysis is made in a spreadsheet, in which metadata are extracted as title, year of publication, techniques, technologies, methods used, results and evaluation, in order to determine those jobs that meet the requirements.

5. Data analysis: a qualitative evaluation consisting of grouping the documents according to their purpose and relevance for this study was carried out; in this way we select taking into account the contributions in terms of obtaining a better quality and the evaluation metrics.

4   Analysis of Result

Below is a taxonomic analysis of research found in different sectors and topics, such as health, computer science, librarianship, among others, it should be noted that the sectors were selected randomly, the results were grouped taking into account the SIR and metrics quality (Table 1).

Table 1. Contributions to the improvement of the SIR.

Work	Purpose		Evaluation
	Indexing	Improvement	
[18]	x		Metrics of grouping quality and user need satisfaction
[19]		x	Compare the feature sets obtained in the database using the cosine similarity function
[20]		x	Precision, performance and efficiency metrics
[21]		x	Accuracy and completeness metrics in the comparison documents
[22]		x	Source metrics: IF, SJR and ranking core Author metrics: index H and G Article metrics: AR index and number citation
[23]	x		Metrics of distance and decrease in the number of operations
[24]		x	Metrics of similarity and accuracy Comparative analysis
[25]	x		Introductions of the new concepts: meta-terms
[26]		x	Precision and coverage metrics Relevance evaluation
[27]	x	x	Metrics of the thesaurus quality, according to the weight for the collection of the document
[28]	x		Ranking and impact factor web of Science
[29]		x	Metrics to assess the speed in obtaining relevant documents
[30]	x	x	Precision Metrics and Average Reciprocal Ranking
[31]		x	Article metrics: source of publication, author quality, seniority time
[32]		x	Accuracy and completeness metrics that allowed demonstrating the quality, relevance of information retrieval with semantic annotation
[33]	x		Retrieval of repositories in scientific documents the information about them to index them in the system: bibliographic data and the topics of each document
[34]	x		Metrics of time consumed in the process
[35]		x	Accuracy and coverage metrics

The previous taxonomy shows that 66.6% of the researches consulted in recent years focus on improving the quality of the results of the consultation through the combination of various techniques, methods and algorithms, integrating both traditional and modern models, the other 44.4% seeks to index documents and contents in the best way, this being a key element to improve retrieval given its relevance; it can be said that among the indicators most used are the ranking of publication (27.7%), accuracy (33.3%), completeness (33.3%), coverage (5.5%) and decrease of the time (16.6%) of the SIR response; Each of these elements are validated by means of metrics and the evaluation of user satisfaction.

With respect to the publication ranking indicators applied to the information retrieval process, it was found that the quality of the document is based on its source of publication (either indexed journals or scientific events), the quality of the author and the quality of the article in yes, including the time of seniority; for which impact factor (IF) metrics, SJR index, CORE ranking, H index, G index, AR index have been defined, these indicators being the common denominator of the analyzed works; in this way most of these indicators are based on the appointment, despite the absence of an unequivocal relationship between citations and merit or scientific quality [36].

5 Conclusions

In spite of all the efforts made so that the information retrieval systems are more efficient in terms of the results of the consultation and thus achieve a greater precision that satisfies the user's need; there are still some challenges and problems to solve:

- Must define metrics that allow to evaluate if the results of the query are relevant for the user, for it is proposed that the SIR can return only parts of the document and not the complete document, according to the client's need.
- Define metrics to evaluate the quality of the document in terms of its content and not the publication indicators, taking into account that on the Internet we find a significant amount of articles that are not housed in scientific databases or indexed journals; this will allow to expand the search in different sources of information.
- Structure the metadata in such a way that the concepts and/or definitions of the contents can be understood to achieve a quality retrieval.
- The development of an SIR is required to reduce processing times, in such a way that it minimizes operational costs; for this it is important that researchers and developers consolidate their knowledge in techniques and/or methods that allow optimizing the retrieval processes.

There are still many challenges to be solved in terms of information retrieval systems, there are few areas in which research has been conducted such as health, computer science, education, librarianship and marketing among others. Researchers focus their work on improving the quality of information retrieval, combining all kinds of techniques, models and technologies that allow to achieve more accurately and exhaustively the results thrown in the consultations, however you need to delve a bit more into the quality of the documents, content or the object that is required to retrieve, the higher the quality of the contents, the better their retrieval will be.

As a work, we intend to design a quality model allows us to evaluate documents in the computer sciences area, for this purpose, the elements can be recoverable in the field of content and accessibility will be defined, through the experiments you must ensure and verify that there is a link between quality and information retrieval and its impact.

References

1. Rosell León, Y., Senso Ruiz, J., Leiva Mederos, A.: Design of an ontology for the management of heterogeneous data at the universities: methodological framework. Cuban J. Inf. Health Sci. **27**(4), 545–567 (2016)
2. Martín Fombellida, A.B., Sáez Lorenzo, M., Iglesias de Sena, H., Alonso Sardón, M., Arévalo, J.A., Mirón Canelo, J.: Does the information about self-medication available in the Internet meet standards of quality? Cuban J. Inf. Health Sci. **27**(1), 19–34 (2016)
3. Hernández-Leal, E.J., Duque-Méndez, N.D., Moreno-Cadavid, J.: Big Data: an exploration of research, technologies and application cases. Technological **20**(39), 17 24 (2017)
4. Feilmayr, C.: Optimizing selection of assessment solutions for completing information extraction results. Comput. Syst. **17**(2), 169–178 (2013)
5. Márquez, D., Hernández, Y., Ochoa-Ortiz, A.: Evaluation of information retrieval algorithms within an energy documents repository. In: Castro, F., Miranda-Jiménez, S., González-Mendoza, M. (eds.) Advances in Computational Intelligence, MICAI 2017. Lecture Notes in Computer Science, vol. 10633. Springer, Cham (2018)
6. Iqbal, M., Abid, M., Khursheed, F.: Information retrieval process on the web: a survey on web crawler types & algorithms. IJICTT **2**(1), 15 (2015)
7. Cuba, Y., Olivera, D.: Los metadatos, la búsqueda y recuperación de información desde las Ciencias de la Información. e-Ciencias de la Información **8**(2) (2018). https://doi.org/10. 15517/eci.v8i2.30085
8. Fonseca Reyna, Y.: Recuperación de la información: taxonomía de sus modelos. Rev. Cubana de Cienc. Informáticas **6**(2), 1–8 (2012)
9. Martínez Arellano, F., Amaya Ramírez, M.: El papel de los metadatos en la Web Semántica. Biblioteca Universitaria **20**(1), 3–10 (2017)
10. Castro-Romero, A., González-Sanabria, J., Ballesteros-Ricaurte, J.: Technologies for metadata management in scientific articles. Eng. Compet. **17**(2), 123–134 (2015)
11. Piedra, J., Chicaiza, P., Quichimbo, V., Saquicela, E., Cadme, J., Lopez, M., Espinoza, E.: Marco de Trabajo para la Integración de Recursos Digitales Basado en un enfoque de Web Semántica. Rev. ibérica de Sistemas y Tecnologías de Información **3**(3), 55–70 (2015)
12. Tabares, V., Duke, N., Moreno, J., Ovalle, D., Vicari, R.: Evaluación de la calidad de metadatos en repositorios digitales de objetos de aprendizaje. Rev. Interamericana dBibliotecología **36**(3), 183–195 (2013)
13. Morato, J., Sánchez-Cuadrado, S., Ruiz-Robles, A., Moreiro-González, J.A.: Visualización y recuperación de información en la web semántica. El Profesional de la información **23**(3), 319–329 (2014)
14. Sabou, M., Aroyo, L., Bontcheva, K., Bozzon, A., Qarout, R.K.: Semantic web and human computation: the status of an emerging field. Semant. Web, 1–12 (2015). Preprint
15. Vallejo, D.: Clustering of documents with size restrictions. Master thesis, University of Valencia, Spain (2016)
16. Barros-Justo, J.: Mining unstructured data to support requirements elicitation by using controlled vocabularies: a systematic mapping study. DYNA **82**(193), 165–169 (2015)
17. Louis, A., Nenkova, A.: A corpus of science journalism for analyzing writing quality. Dialogue Discourse **4**(2), 87–117 (2013)
18. Magdaleno, D., Fuentes, I.E., Cabezas, M., Garcia, M.M.: Recuperación de información para artículos científicos soportada en el agrupamiento de documentos XML. Rev. Cubana de Ciencias Informáticas **10**(2), 57–72 (2016)
19. Deo, A., Gangrade, J., Gangrade, S.: A survey paper on information retrieval system. Int. J. Adv. Res. Comput. Sci. **9**(1), 778–781 (2018)

20. Bhatia, P.K., Mathur, T., Gupta, T.: Survey paper on information retrieval algorithms and personalized information retrieval concept. Int. J. Comput. Appl. **66**(6), 14–18 (2013)
21. Solarte-Pabón, O., Millán-González, M.E.: Propuesta para extender semánticamente el proceso de recuperación de información. Rev. EIA **11**(22), 51–65 (2014)
22. Kuna, H., Martin, R., Martini, E., Solonezen, L.: Desarrollo de un Sistema de recuperación de información para publicaciones científicas del area de ciencias de la Computación. Rev. Latinoamericana de Ingeniería de Softw. **2**(2), 107–114 (2014)
23. Britos, L., Kasian, F., Luduena, V., Merenda, F., Printista, A.M., Reyes, N.S., Deco, C.: Recuperación de datos e información en bases de datos masivas. In: XX Workshop de Investigadores en Ciencias de la Computación (WICC, Universidad Nacional del Nordeste) (2018)
24. Noblejas, C.J., Rodríguez, A.P.: Recuperación y visualización de información en Web of Science y Scopus: una aproximación práctica. Investigación Bibliotecológica: archivonomía, bibliotecología e información **28**(64), 15–31 (2014)
25. Morán, A.A., Naumis, C.: Métodos y tendencias de recuperación de información biomédica y genómica basados en las relaciones semánticas de los tesauros y los MeSH. Investigación bibliotecológica **30**(68), 109–123 (2016)
26. Osorio-Zuluaga, G.A., Méndez, N.D.D.: Collaborative construction of metadata and full-text dataset. In: IEEE Latin American Conference on Learning Objects and Technology (LACLO), pp. 1–6 (2016)
27. Urdiciain, B.G., Sánchez, R.: Técnicas de recuperación de información aplicadas a la construcción de tesauros. TransInformaçao **26**(1), 19–26 (2014)
28. Calle-Velasco, G.D.L.: Modelo basado en técnicas de procesamiento de lenguaje natural para extraer y anotar información de publicaciones científicas. Doctoral dissertation, ETSI_Informatica (2014)
29. Ryckeboer, H.E., Spositto, O.M., Bossero, J.C., Barone, M.: Recuperación de la información. In: XX Workshop de Investigadores en Ciencias de la Computación WICC 2018, Universidad Nacional del Nordeste (2018)
30. Gracía, P., Ferney, Y.: Análisis Comparativo del Desempeño y Costo Computacional de una Infraestructura de Almacenamiento y Procesamiento Distribuido para el Procesamiento de Colecciones de Texto (2017)
31. Kuna, H., Martin, R., Martini, E., Solonezen, L.: Development of an information retrieval system for scientific publications in the area of computer science. Lat. Am. J. Softw. Eng. **2** (2), 107–114 (2014)
32. Viltres Sala, H., Rodríguez Leyva, P., Febles, J., Estrada Sentí, V.: Procesamiento Semántico de información en Sistemas de Recuperación de Información. Rev. Cubana de Ciencias Informáticas **12**(1), 102–116 (2018)
33. Santos Gago, J.M., Álvarez Sabucedo, L.M., Fernández Iglesias, M.J., Míguez Pérez, R., Alonso Roris, V.M., Mikic Fonte, F.: Diseño de un marco semántico para la recuperación contextualizada de documentos científicos en el ámbito sanitario. Nutrición Hosp. **27**, 59–66 (2012)
34. Singh, V., Saini, B.: An effective tokenization algorithm for information retrieval systems. In: Computer Science & Information Technology (CS & IT), pp. 109–119 (2014)
35. Gamallo, P., García González, M.: Técnicas de Procesamiento del Lenguaje Natural en la Recuperación de Información. NovATIca **219**, 42–47 (2012)
36. Nassi-Caló, L.: Evaluation metrics in science: current status and prospects. Rev. Latino-Am. Enfermagem. **25** (2017)

Artificial Neural Networks for Discovering Characteristics of Fishing Surveillance Areas

Anacleto Correia[1](✉) ⓘ, Ricardo Moura[1,2] ⓘ, Pedro Agua[1] ⓘ,
and Victor Lobo[1] ⓘ

[1] CINAV, Naval Academy, 2810-001 Almada, Portugal
anacleto.coreia@gmail.com
[2] Center for Mathematics and Applications (CMA), 2829-516 Almada, Portugal

Abstract. The demographic pressure entails over-exploitation of the coastal regions and the consumption of marine resources in a non sustainable manner, jeopardizing the species renewal. Several species are currently facing great threat of disappearing from Portuguese coastal waters, namely the *Sardina pilchardus*, due to illegal, unregulated or not reported fishing. The Portuguese Navy performs regular surveillance and monitoring of fishing activities for law enforcement. Those actions gather useful information about the fishing activity, specifically about the types of fishing gear used. Since the geo-spatial data on a regular map, by itself, was not enough to present a clear picture regarding the predominant type of fishing gears used for captured sardine in the Portuguese coastal areas, we applied an artificial neural network to georeferenced information in order to derive a new layer with the areas where the fishing gears used for *Sardina pilchardus* fishing are most likely to be found.

Keywords: Artificial neural networks · Fishing surveillance · Geo-spatial information

1 Introduction

The Earth surface is composed by around 72% of water and some 96% of it is saline water, so oceans strongly affect the global ecosystem. As a source of many resources the oceans are being subject to constant pressures by human activities, particularly due to exponential population growth across certain regions of the globe [1]. This demographic pressure entails numerous problems such as over-exploitation of coastal regions, as a particular form of the "tragedy of the commons archetype" [2] and driving the consumption of living sea resources into a non-sustainable level, jeopardizing species renewal. Several species are presently facing a great threat of disappearing, mainly due to illegal, unregulated or not reported fishing.

In the last decades, all over the world, a self-awareness that oceans governance must be approached in a strict way has emerged, fostering sustainable development by adopting preventive actions towards a balanced ecosystem approach. The United Nations established that its state members are responsible for their own coastal areas and seas, considering that it is a common mankind heritage [3]. Likewise, international principles and standards of behaviour were established to promote responsible actions,

© Springer Nature Switzerland AG 2020
Á. Rocha et al. (Eds.): ICITS 2020, AISC 1137, pp. 75–83, 2020.
https://doi.org/10.1007/978-3-030-40690-5_8

to ensure the aquatic living resources' effective conservation, and to ensure that management and development of the ecosystem and its biodiversity develops in a responsible manner [4].

Illegal fishing represents a big threat to the global sea life. It reduces the number of fish, destroys their habitats, distorting competition by putting honest fisherman in an unfair disadvantage and destroying the coastal communities' subsistence, especially in developing countries. The European Union, being the biggest world fish importer, adopted an innovative policy to fight illegal fishing all over it fishing areas, by not allowing the commercial transaction of illicit fishing products into its markets, unless they are legally certified [5].

As a European Union (EU) country, Portugal stands out for its peripheral location and for its vast sea area, as a result of the extensive continental coast shore together with the Madeira and Azores archipelagos. The maritime areas play an important role due to their dimensions and geostrategic position, where some of the busiest commercial routes take place, which translates into an intense maritime activity. It is thus a national priority to have adequate surveillance and monitoring of its waters, to ensure not only maritime safety, but also sustainable exploitation of sea life.

The Portuguese Navy is a major actor in performing surveillance and monitoring actions for the country's sovereign and jurisdiction areas, namely by applying and verifying fishing boats' compliance with applicable laws and maritime regulations. The result of surveillance activity generates useful information about the fishing activity, specifically about the fishing techniques used by vessels subject of inspections, the amount and species of captured fish, paths taken, and the geographic location of the fishing boats. The collected data are samples that, treated with adequate data analysis techniques, may contribute for a better understanding of fishing activity and the efficacy of the inspection activities in preserving natural resources.

The aim of the current study was to gather information that allows for adequate planning of sardine fishing monitoring, a species whose capture is under strong constraints and attention, due to critical stock reductions. With that in mind, supervised artificial neural networks (ANNs) together with geo-spatial mapping [6–8] was applied to historical data, to allow for a better planning of monitoring actions. Using geo-spatial positional variables, *longitude* and *latitude*, as predictor variables, an ANN model will be constructed to predict what fishing gear will most probably be used. We can then construct a predictor map which will allow an overview of the predominant areas of certain fishing gears.

This paper has 5 sections. This section presents the context and motivation for the study. Section 2 presents previous work related with modelling of fishing resources preservation. Section 3 describes the data gathered and used by the proposed model. Section 4 describes the configuration of the artificial neural networks for geo-mapping together with a visualization of the areas where different fishing gears are used for *Sardina pilchardus* capture in Portugal's coastal areas. In the last section, we make some final remarks, providing some insight for future work.

2 Literature Review

Management of fishery resources while maintaining ecosystem sustainability is a worldwide main fishing management objective. In order to respond to this demand some studies have been developed on the selective evolution of fishing resources, in order to understand the changes caused by behavioural traits, physiology and morphology of the captured species, with the purpose of avoiding the stock's point of no-return. For this reason, quantifying and predicting the effect of the evolution in the fishing sector must constitute a research priority [9].

In a recent paper by Peck *et al.* [10], the statistical evolution of maritime living resources is approached by distribution models of species (DMS). Such models incorporate spatial data, fishing data and physical factors, allowing for an understanding of the spatial and temporal changes of the species distribution. DMS are commonly used to predict the current situation and project future distributions, where the obtained results may be integrated with other models or be presented as analysis maps. The statistical models allow for the study of correlations but do not provide a cause-and-effect relationship, hence there is a need for including new biotic and abiotic patterns beyond the historical observations [10].

An improvement on previous models came from the study of Laugen *et al.* [9], which addresses the evaluation of the fishing evolution impact awareness (EvoIA). This project modelled the evolution of fishing gear which impacts the maritime ecosystem, promoting the adoption of more adequate management measures to promote sustainable fishing. To achieve such goals, previously validated evaluation models were used - demographic dynamic models, socioeconomic dynamic models and management and strategy models - together with developed models of fishing impacts by statistical methods, in order to infer about the changes in the ecosystem over time. In general, the new models allowed to understand the effect of environmental variables, excessive fishing, abiotic factors and other (resulting from the ecosystem dynamics) over the growth patterns, maturation and reproduction of the species. Therefore, project EvoIA supplies a reference framework allowing the combination of different models with the purpose of evaluating the over-fishing impact in marine resources. However, this combination of models has two constraints. The first constraint is related to the fact that the models were developed separately, for different contexts, thus making it hard to conclude about the over-fishing impacts. The other constraint is due to the fact that not all the ecosystem external and internal factors were considered, thus the need to resort to more elaborate quantitative tools taking into consideration the evolutionary rate differential of each species.

The previously mentioned works emphasize that besides biotic and abiotic factors, it is necessary to consider the human factors, namely fishing activity, and how it influences the maritime ecosystems. The spatial and temporal distribution of the fishing effort allows for the evaluation of fishing interactions with the species studied, allowing the development of "fishing pressure indicators" and the evaluation of the dynamics of fishing activity motivated by external factors, such as regulatory changes.

In contrast to statistical methods common in studies of maritime ecosystems, the present work relies on machine learning techniques with a specific objective: attain the spatial distribution of sardine fishing vessels, by type of gear, in order to support the planning of fishing surveillance activities, as well as the verification of this species regulations compliance by the inspected vessels.

3 Data Collection

The historical data collected concerns the period between January 2014 and December 2015, extracted from a Portuguese Navy database, related to information gathered from fishing monitoring records (Fig. 1) where the sardine species (Fig. 2) was present.

Fig. 1. Illustration of monitoring of traditional sardine fishing vessels in Portuguese coastal waters.

The *Sardina pilchardus* species exists in the Northeast Atlantic, Iceland (although rare), North Sea, Senegal, Mediterranean, Marmara Sea and Black Sea. It is found between 10 and 100-m depths, however during daytime it is more commonly found to be between 25 and 55 m. At night, it approaches the surface, rising to ten meters. It breeds between 20 and 25 m deep near the coast or up to 100 km offshore from September to May on Western Europe's coast and the Mediterranean. It feeds on crustaceans, plankton and even larger organisms.

The data extracted from the database included: the quantity of sardines observed in the inspection, the type and size of the inspected vessel, the date of the inspection event, the naval unit that performed the inspection, the type of gear used, and the geographical position at the moment of inspection. Regarding the type of gear used, the two main types are found to be the *seines without purse lines* and *purse seines*, while other typologies have a very low number of occurrences. Fishing with seines (Fig. 3) is a fishing technique that uses a wall made of a long and tall fishnet which is dropped in a way that surrounds the fish and reduces the escape paths.

Fig. 2. Sardine (*Sardina pilchardus*). Scientific illustration by Pedro Salgado [11]

Seines without purse lines   Purse seines

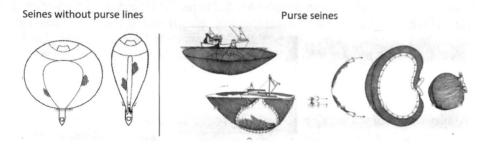

Fig. 3. Illustration of gear types: seines without purse lines and Purse seines.

This technique is used for fishing small pelagic fish, like the sardine. According to current regulation, it is forbidden to use fishnets with mesh sizes inferior to 16 mm. For the last years, seines fishing has again become a dominant fishing technique, overtaking the use of multipurpose fishing and trawl fishing, with the captured volume using seines increasing significantly, hence justifying an increased effort in checking maritime regulations and applicable law compliance.

4   Applying ANNs for Geo-mapping Visualization

This section aims to provide exploratory information about the sample of fishing surveillance historic data (from years 2014 and 2015), concerning the type of gear found in sardine inspection data.

The dataset comprised the following relevant variables for the study:

– *Year* - 2014 and 2015;
– *Month* - Numeric value [1, 12] ⊂ N is the inspection's month;
– *Quantity* - Numeric value of kilograms of observed sardine found in the inspection;
– *Gear* - Categorical value that describes the type of gear used by the inspected fishing boat;
– *Longitude* - GPS longitude coordinate of the inspection;
– *Latitude* - GPS latitude coordinate of the inspection.

For our study, we created an additional categorical variable, *Type of gear*, with three possible values: *A* - Encircling nets, *B* - Encircling nets with purse lines; *C* - Other. The C class aggregates all other fishing techniques, such as Scottish/Danish Encircling, pot gear, creels, *etc.*, which had low representativeness.

In Fig. 4, one may see the location of the fishing vessels which had sardines in their possession when the Navy inspected them, with different colours representing the *Type of gear* for each vessel.

Since it is not easy to visualize the predominance of the fishing gear in each inspected area, it would be interesting to be able to predict for each area the preferred types of fishing gears. We thus devised a classification model that receives *Longitude* and *Latitude* as input variables and produces as target class the *Type of gear*. Several Artificial Neural Networks (ANNs), following the *ScikitLearn* Multi-layer Perceptron [12] algorithm, were trained using different numbers of hidden layers. All the ANNs models used the rectified linear unit function (ReLu) as activation function, an adaptive moment estimation (Adam) solver (stochastic gradient based optimizer) for weight optimization and $\alpha = 0.01$ for the weight decay. Table 1 illustrates the results of the area under the ROC curve and the accuracy (proportion of correct classifications) of the model for the trained data and also for the test by 10-fold stratified cross-validation. After the construction of the model, one combines the model with the Orange Geo-map application that displays coloured regions for all the selected areas of interest by depicting the predicted class for every set of *longitude* and *latitude* coordinates determined by the Geo-map application [13].

It can easily be seen that an ANN model with 8 neurons in the hidden layer presents a good compromise, because more neurons will not increase accuracy significantly, and would probably only overfit the data, since the peak of accuracy in the validation set occurs with 8 neurons in the hidden layer. During the training of multiple models with 8 hidden neurons, we compared the accuracy (proportion of correct classifications per class) of each model and chose the one with highest accuracy for each of the three categories of *Type of gear*, *A*, *B* or *C*.

Figure 5 illustrates the application of the chosen model to Portuguese coastal areas, where it is possible to observe the predominance of the *Type of gear* per given area. Although it is clear from the picture that all types of gears are used in the Portuguese coast from the North tip down to *Cabo da Roca*, and there is no predominant one. From Lisbon down to *Sagres*, one may observe a dominance of the encircling nets with purse lines technique, together with some casual cases of encircling nets without the purse lines. The *Algarve* coastal region has a predominance of cases where the technique of encircling nets does not make use of purse lines.

Table 1. Results of the application of several ANNs in order to predict the classification of the class *Type of gear* from *Latitude* and *Longitude*

Hidden	Train data		Cross-validation data	
Neurons	Area under ROC	Accuracy	Area under ROC	Accuracy
1	0.712	0.608	0.612	0.529
2	0.731	0.627	0.684	0.588
4	0.725	0.608	0.769	0.569
6	0.732	0.588	0.785	0.569
8	0.752	0.647	0.785	0.627
12	0.757	0.608	0.771	0.588
16	0.754	0.686	0.749	0.569

Fig. 4. Location of the vessels at the moment of inspection, where the color represents the type of fishing gear (*A* - Encircling fishnets, *B* - Encircling nets with purse; *C* - Other)

Fig. 5. Overlay of predominance of fishing gear types (*A* - Encircling fishnets, *B* - Encircling fishnets with purse - lines; *C* - Other) for Portugal's coastal areas.

Since location and time of Portuguese Navy inspections are random, they can be a representative estimate of the quantity of sardines fished. In the future, it would be interesting to compare these monthly quantities with the monthly quantity loaded at the national ports. The purpose is thus to ascertain if there is a relation between the values of the inspected records and the port national records.

5   Conclusion

An overview of the information gathered by the Portuguese Navy during vessel inspections were presented, particularly for the *Sardina pilchardus* species. A Geo-map jointly with a classification model, an Artificial Neural Network model, allowed the production of a map that summarizes the type of fishing gear used in different areas. Portugal presents three very distinctive areas of fishing gear predominance: in the south encircling fishnets without purse lines predominate, in the southwest area (from Sagres to Setúbal) encircling fishnets with purse lines predominates, and the remaining areas do not have a main preferred fishing gear. In future developments, it would be interesting to perform a cluster analysis, by using K-means, hierarchical clustering or Self-Organizing Maps, in the same Geo-Spatial Map and construct a brief description of each cluster.

Acknowledgments. This work was funded by the Portuguese Navy.

References

1. Meadows, D., Randers, J., Meadows, D.: Limits to Growth – The 30-Year Update. Chelsea Green Publishing Co., Hartford (2004)
2. Senge, P.: The Fifth Discipline – The Art & Practice of The Learning Organization. Random House Business Books, London (2006)
3. United Nations: United Nations Convention on the Law of the Sea (1982). https://www.un.org/Depts/los/conventionagreements/texts/unclos/unclose.pdf. Accessed 15 Sept 2019
4. Code of Conduct for Responsible Fishing, Food and Agriculture Organization of the United Nations (FAO) (1993)
5. Kroodsma, D., Miller, N., Roan, A.: The Global View of Transhipment: Preliminary Findings. Global Fishing Watch and SkyTruth (2017). http://globalfishingwatch.org. Accessed 15 Sept 2019
6. Kanevski, M., Pozdnukhov, A., Timonin, V.: Machine learning algorithms for geospatial data. Applications and software tools. In: iEMSs 2008: International Congress on Environmental Modelling and Software Integrating Sciences and Information Technology for Environmental Assessment and Decision Making (2008)
7. Gopal, S.: Artificial neural networks in geospatial analysis. In: International Encyclopedia of Geography: People, the Earth, Environment and Technology: People, the Earth, Environment and Technology, pp. 1–7 (2016). https://doi.org/10.1002/9781118786352.wbieg0322. Accessed 15 Sept 2019
8. Suryanarayana, I., Braibanti, A., Rao, R.S., Ramam, V.A., Sudarsan, D., Rao, G.N.: Neural networks in fisheries research. Fish. Res. **92**(2–3), 115–139 (2008)
9. Laugen, A., Engelhard, G., Whitlock, R., Arlinghaus, R., Dankel, D., Dunlop, E., Dieckmann, U.: Evolutionary impact assessment: accounting for evolutionary consequences of fishing in an ecosystem approach to fisheries management. Fish Fish. **15**(1), 6596 (2014)
10. Peck, M., Arvanitidis, C., Butenschon, M., Canu, D., Chatzinikolaou, E., Cucco, A., Wolfshaar, K.: Projecting changes in the distribution and productivity of living marine resources: a critical review of the suite of modeling approaches used in the large European project VECTORS. Estuarine and Coastal Marine Science (2016). https://www.researchgate.net/publication/303534106. Accessed 15 Sept 2019

11. Sá, S.: Radiografia aos nossos peixes. Revista Visão Julho (2016). http://visao.sapo.pt/ambiente/agricultura/2016-07-24-Radiografia-aos-nossos-peixes. Accessed 15 Sept 2019
12. Pedregosa, F., Varoquaux, G., Gramfort, A., Michel, V., Thirion, B., Grisel, O., Blondel, M., Prettenhofer, P., Weiss, R., Dubourg, V., Vanderplas, J., Passos, A., Cournapeau, D., Brucher, M., Perrot, M., Duchesnay, E.: Scikit-learn: machine learning in Python. J. Mach. Learn. Res. **12**, 2825–2830 (2011)
13. Demšar, J., Curk, T., Erjavec, A., Gorup, Č., Hočevar, T., Milutinovič, M., Možina, M., Polajnar, M., Toplak, M., Starič, A., Štajdohar, M.: Orange: data mining toolbox in Python. J. Mach. Learn. Res. **14**(1), 2349–2353 (2013)

K-Means Clustering for Information Dissemination of Fishing Surveillance

Anacleto Correia[1(✉)] , Ricardo Moura[1,2] , Pedro Agua[1] ,
and Victor Lobo[1]

[1] CINAV, Naval Academy, 2810-001 Almada, Portugal
anacleto.coreia@gmail.com
[2] Center for Mathematics and Applications (CMA), 2829-516 Almada, Portugal

Abstract. The Portuguese Navy is responsible for monitoring the largest Exclusive Economic Zone in Europe. The most captured species in this area are *Scomber colias* and *Trachurus trachurus*, commonly called Mackerel and Horse Mackerel, respectively. One of the Navy's missions is pursuing actions of fishing surveillance to verify the compliance of proceedings with the species' fishing activity regulation. This monitoring actions originate data that represents a sample of the fishing activity in the area. The collected data, analysed with adequate data mining techniques, makes it possible to extract useful information to better understand the fishing activity related to Mackerel and Horse Mackerel, even if the full data set cannot be disclosed. With this in mind the authors used a non-supervised learning technique, the K-Means algorithm, which grouped data in clusters by its similarity and made a summarized description of each cluster with the purpose of releasing a general overview of such records. The information obtained from the clusters led the authors to deepen the study by performing a comparison of the monthly average quantity recorded per vessel for the two species in order to infer about the relation between captured quantity Mackerel and Horse Mackerel over time.

Keywords: K-Means clustering · Fishery surveillance · Geo-spatial data

1 Introduction

Oceans cover about 71% of the Earth's surface, exerting a strong influence on the global ecosystem. However, it turns out that oceans are increasingly subject to pressures resulting from human actions, in particular from the exponential growth of population in certain areas. The great demographic pressure entails several problems, including the over-exploitation of coastal regions and irrevocably increasing consumption of living marine resources in a non-sustainable manner, endangering the ability of species to renew themselves. Many species are already seriously threatened by over-fishing, mainly due to illegal, unregulated or unreported fishing.

In recent decades, there has been a growing awareness around the world that ocean management and its governance must be rigorously addressed, seeking sustainable development and taking preventive action in accordance with a balanced eco-systemic view [1]. The United Nations established that states are responsible for their coastal

Á. Rocha et al. (Eds.): ICITS 2020, AISC 1137, pp. 84–93, 2020.
https://doi.org/10.1007/978-3-030-40690-5_9

zones and the sea, which is considered a mankind common heritage [2]. Similarly, international principles and standards of behaviour for responsible practices have been established to ensure effective conservation, management and development of aquatic living resources, while respecting the ecosystem and biodiversity [3], limiting the effects of "the tragedy of the commons archetype" [4].

Illegal fishing is a major threat to global marine resources. It depletes fish stocks, destroys marine habitats, distorts competition, puts honest fishermen at an unfair disadvantage and destroys the livelihoods of coastal communities, particularly critical in developing countries. As the world's largest importer of fishery products, the European Union (EU) has adopted an innovative policy to combat illegal fishing worldwide by forbidding illicit fishery products to trade in the EU area unless they are legally certified [5].

Portugal as a member of the EU stands out for its peripheral location and the vast area of maritime spaces, which result from an extensive continental coastline and encompassing the archipelagic systems of Madeira and the Azores. Maritime spaces are of undeniable importance given their sizes and geostrategic position, at which some of the world's busiest commercial shipping routes take place, which translates into intense maritime activity such as fishing, among others. It is a demand on the country to ensure adequate surveillance and monitoring of its waters, ensuring not only maritime safety but also the sustainable exploitation of marine resources.

The Navy is a vector in the execution of surveillance and monitoring actions, namely in the field of maritime laws and regulations enforcement, for areas under the sovereignty or jurisdiction of its country, which includes fisheries surveillance. The results of inspection activities give rise to useful information set about fishing activity, in particular the fishing gear used by the inspected vessels, the quantities and species of fish observed during the inspection actions, the path and the geographical position of the vessels inspected. The data collected constitute a sample that when treated with the proper techniques, may contribute to better knowledge of fishery and inspection actions effectiveness, promoting the conservation of fish resources.

The objective of the present study was to obtain indicators that would allow the proper planning of Mackerel and Horse Mackerel fishing inspection actions, a species whose capture achieves high volumes in the Portugal. To this end, an unsupervised technique of clustering was applied, the K-Means clustering, taking the historical data from the Portuguese Navy databases, where these two species were recorded, in order to, without compromising the data confidentiality, provide an overview of such fishing activity. In order to produce robustness by effective seeding the K-Means algorithm will be applied to the data by using the k-means++ initialization/seeding by using a weighted probability distribution to determine the starting k centers [6, 7].

This article is divided into 5 sections. The current section provides the problem context and the motivation for taking the study. Section 2 refers to some studies, previously made in the preservation of fish stocks. Section 3 describes the data used in the study. The application of the unsupervised technique is described in Sect. 4, as well as monthly comparisons between the two species, while searching for an association between them. Finally, Sect. 5 summarizes the main results of the study.

2 Literature Review

The sustainable management of fishery resources is a main objective for fisheries ecosystem management on a worldwide basis. To engage this need, studies have been developed on the selective development of fish resources, in order to try to understand the observable changes in behavioural traits, physiology and morphology of the species captured in order to avoid such stocks reaching a point of no return. Therefore, quantifying and predicting the effect of developments affecting fisheries and fishing industry should be a research priority [8].

From this perspective, Peck *et al.* [9], model the changes and distribution of living marine resources with a statistical model of species distribution (MSD). This model incorporates spatial data, fisheries data and physical factors, which may allow us to understand the spatial changes or temporal distribution of the species. The MSD are commonly used to gain insight regarding the current situation and to project future distributions, where the results may be incorporated into other models or simply be presented as analysis maps. Through the use of statistical models correlations may be obtained, however these do not ensure an understanding of cause and effect structure at play behind observed data, hence the need to include new biotic and abiotic patterns, which can go beyond historical observations [9].

The improvement from the above-mentioned models, has roots in the study of Laugen *et al.* [8], which addresses the evolutionary impact assessment of fisheries (EvoIA). This project models the evolution of fishing practices in marine ecosystems, thus suggesting the adoption of the most appropriate management measures for fisheries sustainability. For this purpose, valuation models previously validated were used - the demographic dynamics model, the socioeconomic dynamics model and management strategy models-, in addition to models developed by statistical methods to determine the impact on fisheries and to infer the changes in ecosystems over time. In general, the new models obtained allow an understanding of the effects of environmental variables, excessive fishing, abiotic factors and other variables (resulting from ecosystem dynamics) on the patterns of growth, maturation and reproduction of the species. Hence, the EvoIA project provides a framework that allows different models to be combined to assess the impacts of overfishing on aquatic resources. However, the used approach, resulting from a combination of models, presents at least two difficulties. The first one has to do with the fact that the combined models have been developed in isolation, in different contexts, therefore being difficult to conclude about the impact resulting from fishing exploitation. As a second difficulty, it does not allow for the consideration of all external and internal factors of ecosystems, therefore being necessary to resort to more sophisticated quantitative tools that allow the inclusion of the evolutionary differential rate of each species. The previously mentioned work suggests that in addition to the biotic and abiotic factors, it is necessary to take into account human factors affecting such fisheries, as they end up influencing the marine ecosystems. The spatial and temporal distribution of fishing efforts allow us to evaluate the fishery interactions with the species under study, facilitating the development of fishing pressure indicators, enabling the assessment of the displacement of fishing activity motivated by external factors, such as the amendments to the regulation rules.

Instead of statistical methods of the marine ecosystem, the present work uses automatic learning techniques of clustering for a more specific purpose: to determine the spatial distribution of Mackerel and Horse Mackerel in fishing vessels, by building a description of different clusters which have patterns in common in the Portuguese sovereignty sea, such as the quantity per cluster location, the dominant type of gear and dominant captured species, with the purpose of supporting more efficient inspection planning.

3   Data Collection

The objective in using a clustering technique to obtain clusters with intra-cluster similarity and inter-cluster dissimilarity is to provide an overview of the data [10, 11] that cannot be accessed by the general public, regarding the fishing of Mackerel (Fig. 1) and Horse Mackerel (Fig. 2).

Fig. 1. Mackerel. Scientific illustration by Pedro Salgado [12]

Fig. 2. Horse Mackerel. Scientific illustration by Pedro Salgado [12]

The historical data used was related with the period from January 2014 to December 2015, regarding information of reported fishery inspections (Fig. 3) of vessels containing the two referred species. The metadata from the extracted information included: the name of the species, the weight of the species captured by the fishing vessel until the moment of inspection, the type and size of the fishing vessel, the date of the enforcement action, the naval enforcement unit, the type of fishing gear used by the vessel and the geographical position at the time of the inspection action.

4   Description of the Data Using K-Means Clustering

Taking the dataset of the fishing vessels inspection with records for *Mackerel* and *Horse Mackerel*, an overview of the relevant information was made available through the support of the K-Means clustering and visualization techniques [13], using the Python based software Orange [14]. The result was clusters of data which will contain information about patterns concerning the fishing activity.

Fig. 3. Illustration of a fishery surveillance action

The full dataset comprises the following relevant variables for the study:

- *Species* - *Mackerel* and *Horse Mackerel*
- *Year* - 2014 and 2015;
- *Month* - Number of the month [1, 12] when the inspection was made;
- *Quantity* - Numeric variable of weight of observed fish in kilograms;
- *Gear* - Categorical variable with the description of the fishing gear used by fishing vessels;
- *Longitude* - GPS longitude coordinate of the location of inspection;
- *Latitude* - GPS latitude coordinate of the location of inspection;

Due to the extensive number of different categories of variable *Gear*, one more variable was created, *Type of gear*, divided into six categories, *A* - Encircling fishnets, *B* - Encircling fishnets with purse lines; *C* - Otter Trawls; *D* - Gillnets/Bottom-set Gillnets/Trammel fishnets; *E* - Hook and line/Single line/Longline/Troll line; and *O* - 'Other'. The last category comprises numerous gear types and techniques that were found in very few records, not having enough representativeness to have a category on their own.

In Fig. 4, one may observe the fishing vessels locations at the moment of inspection, for the records of the dataset, defined by clusters' hit points.

Figure 4 provides an overview of the fishing areas. More relevant details were extracted by applying a clustering algorithm to divide the data into clusters and present the general description by cluster.

In order to apply the K-Means algorithm, the numerical variables, such as *Quantity*, *Longitude* and *Latitude*, were normalized to prevent a variable of to have more weight than the others when the technique was used.

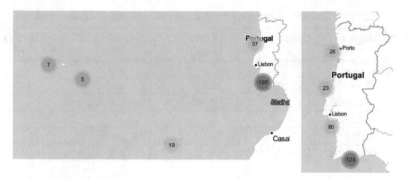

Fig. 4. Hit points of the inspected areas whose vessels possessed *Mackerel* and *Horse Mackerel*

Considering $X = \{x_1, \ldots, x_n\}$, the variable to be normalized by Min-Max normalization, the new normalized variables become:

$$X_i = \frac{x_i - min(X)}{max(X) - min(X)}.$$

The normalized variables have a minimum of 0 and a maximum of 1, and thus identical weights between them, as well as with any other binary variable. After normalizing the referred variables, the K-Means algorithm is applied, with k-means++ initialization of the centroids [6] and re-running the algorithm 100 times for robustness purposes using only the normalized variables of *Quantity*, *Longitude* and *Latitude*, together with the (binary) dummy variables generated from variables *Species* and *Type of gear*, disregarding the other attributes. This algorithm was applied considering a range of clusters from 2 to 13 and computing its average Silhouette [15], a consistency validation measure which measures the intra-cluster similarity and inter-cluster dissimilarity, simultaneously. Since the values of this measure ranges from −1 to 1, the objective is to present cluster with average Silhouette value closer to 1. Table 1 presents the results of the average Silhouettes per number of clusters.

Table 1. Silhouette scores for the application of K-Means algorithm from 2 to 13 clusters

Clusters	2	3	4	5	6	7	8	9	10	11	12	13
Silhouette	**0.936**	0.869	**0.870**	0.821	**0.821**	0.771	0.772	**0.773**	0.767	0.767	0.769	0.766

Despite the average Silhouette highest value being recorded when the number of clusters is 2, we chose several values of k, namely $k = 2, 4, 6, 9$, for the number of clusters since these represent local maximums of the average Silhouette in regards to k and thus we may perform a cluster description for each case that will enable a better overview of data, by finding patterns related to the different areas and different quantities, for both species in this study.

Table 2. General description of the information within K-Means clusters

Cluster	Species	Quantity	Type of gear	General Location (>1 record)
Two clusters (k = 2)				
1	Both	[0.5, 12000]	*All*	All areas
2	Mackerel	[16000, 40000]	*A, B, O*	Setúbal, Algarve
Four clusters (k = 4)				
1	Both	[0.5, 4008]	*All*	All areas
2	Mackerel	[20000, 30000]	*A, B, O*	Setúbal, Algarve
3	Both	[4500, 16000]	*A, B, C*	Leiria, Setúbal, Algarve
4	Mackerel	[38000, 40000]	*A, B*	Setúbal
Six clusters (k = 6)				
1	Both	[0.5, 1500]	*All*	All areas
2	Mackerel	[24000, 30000]	*A, O*	Sesimbra, Algarve
3	Both	[6000, 12000]	*A, B, C*	Leiria, Setúbal, Algarve
4	Mackerel	[38000, 40000]	*A, B*	Setúbal
5	Both	[1700, 4500]	*A, B, C*	Leiria, Setúbal, Algarve
6	Mackerel	[16000, 20000]	*B*	Setúbal
Nine clusters (k = 9)				
1	Both	[750, 2200]	*A, B, C*	Aveiro, Leiria, Setúbal, Algarve, Madeira
2	Mackerel	[38000, 40000]	*A, B*	Setubal
3	Both	[6000, 8000]	*A, B, C*	Algarve
4	Mackerel	[24000, 27000]	*A*	Sesimbra
5	Both	[0.5, 660]	All	All areas
6	Mackerel	[16000, 22000]	*B*	Setubal
7	Both	[2350, 4500]	*A, C*	Setubal, Algarve
8	Mackerel	30000	*A, O*	Setubal, Algarve
9	Mackerel	[10000, 12000]	*A*	Algarve

A 'cluster-by-cluster' general description is presented in Table 2.

From Table 2 and some data analysis it is possible to retrieve some key indicators:

- In Azores, the quantity of *Mackerel* and *Horse Mackerel* inspected were always smaller than 660 kg;
- Setubal, Sesimbra and Algarve presented the highest values for quantities of fish, in particular *Mackerel*, from vessels that used mainly the encircling fishnets with or without purse lines;
- Records for vessels with *Horse Mackerel* only showed quantities bellow 8000 kg;
- The categories *D*, *E* and *O* of *Type of gear* have only records associated to small quantities of fish;
- In the areas of the north of Setubal, the fish quantities recorded are all under 2200 kg.

From the above description it is possible to observe that the quantities of fished *Mackerel* may be much higher than the ones registered at vessels for *Horse Mackerel*, thus we think that it should be interesting to study the variability of the average quantity of these two species per month. By aggregating the quantities per month, we can see in Fig. 5 a scatter graph that depicts the relation for the monthly average quantities of both species captures.

Note that the relation between both species' monthly average capture quantities are in some way proportionally reverse, suggesting that, when large quantities of one species is registered, very small quantities of the other is registered.

Another perspective from this inverse relation can be achieved by presenting a timeline graph overlaying both quantities, as can be portrayed in Fig. 6.

We may see that the peak quantities of captured *Mackerel* coincide with very low values of *Horse Mackerel* capture (Fig. 6), especially in the winter seasons. Moreover, only in March do both species have roughly the same quantities recorded by the Navy inspections. The 'inversion of quantities' unexpectedly suggests this pattern should be further studied and, constitute a reliable indicator of this species quantity in the Portuguese Sea.

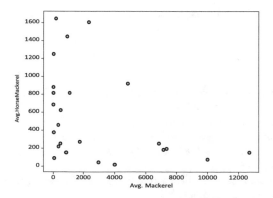

Fig. 5. Scatter graph with the monthly average quantities of *Mackerel* and *Horse Mackerel* in 2014 and 2015.

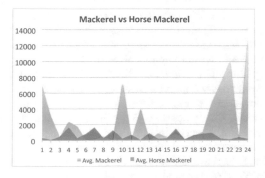

Fig. 6. Line graph of the average quantities of *Mackerel* and *Horse Mackerel* over time (January 2014–December 2015).

5   Conclusions

Even without releasing the full information set, due to data confidentiality, by applying K-Means as a 'clustering technique', it is possible to describe the patterns of *Mackerel* and *Horse Mackerel* fishing in the Portuguese sea waters, which makes the technique particularly interesting for supporting future planning of the inspection activities when conducting a fishing resources preservation strategy. From the clusters' summarized created information, it was possible to identify what are the areas with the highest recorded volume of the two species and also the dominant type of fishing gear in the respective area. It is relevant to point out the great volume of Mackerel recorded in the area of Setúbal, and consequently the use of Encircling fishnets. Moreover, it was observed an inverse proportionality between the monthly recorded quantity for *Mackerel* and *Horse Mackerel*, denoting that when a higher quantity of one species is recorded, the other species most likely will show a lower recorded quantity. This fact alongside with the recorded smaller quantities of *Horse Mackerel* should be studied in-depth taking data from other origins in order to better understand the causes of the variations, in the Portuguese waters, of both species' quantities.

Acknowledgments. This work was funded by the Portuguese Navy.

References

1. Meadows, D., Randers, J., Meadows, D.: Limits to Growth – The 30-Year Update. Chelsea Green Publishing Co., Hartford (2004)
2. United Nations: United Nations Convention on the Law of the Sea (1982). https://www.un.org/Depts/los/convention_agreements/texts/unclos/unclos_e.pdf. Accessed 15 Sept 2019
3. Code of Conduct for Responsible Fishing, Food and Agriculture Organization of the United Nations (FAO) (1993)
4. Senge, P.: The Fifth Discipline – The Art & Practice of The Learning Organization. Random House Business Books, London (2006)
5. Kroodsma, D., Miller, N., Roan, A.: The Global View of Transhipment: Preliminary Findings. Global Fishing Watch and SkyTruth (2017). http://globalfishingwatch.org. Accessed 15 Sept 2019
6. Arthur, D., Vassilvitskii, S.: k-means++: the advantages of careful seeding. In: Proceedings of the Eighteenth Annual ACM-SIAM Symposium on Discrete Algorithms, pp. 1027–1035. Society for Industrial and Applied Mathematics, Philadelphia (2007)
7. Bachem, O., Lucic, M., Hassani, H., Krause, A.: Fast and provably good seedings for k-means. In: Advances in Neural Information Processing Systems, pp. 55–63 (2016)
8. Laugen, A., Engelhard, G., Whitlock, R., Arlinghaus, R., Dankel, D., Dunlop, E., Dieckmann, U.: Evolutionary impact assessment: accounting for evolutionary consequences of fishing in an ecosystem approach to fisheries management. Fish Fish. **15**(1), 6596 (2014)
9. Peck, M., Arvanitidis, C., Butenschon, M., Canu, D., Chatzinikolaou, E., Cucco, A., Wolfshaar, K. Projecting changes in the distribution and productivity of living marine resources: a critical review of the suite of modeling approaches used in the large European project VECTORS. Estuarine and Coastal Marine Science (2016). https://www.researchgate.net/publication/303534106. Accessed 15 Sept 2019

10. Jain, A.K., Murty, M.N., Flynn, P.J.: Data clustering: a review. ACM Comput. Surv. **31**(3), 264–323 (1999)
11. Han, J., Kamber, M., Tung, A.: Spatial clustering methods in data mining. In: Geographic Data Mining and Knowledge Discovery, pp. 188–217. Taylor & Francis, London (2001)
12. Sá, S.: Radiografia aos nossos peixes. Revista Visão Julho (2016). http://visao.sapo.pt/ambiente/agricultura/2016-07-24-Radiografia-aos-nossos-peixes. Accessed 15 Sept 2019
13. Jin, X., Han, J.: K-means clustering. In: Encyclopedia of Machine Learning and Data Mining, pp. 695–697 (2017)
14. Demšar, J., Curk, T., Erjavec, A., Gorup, Č., Hočevar, T., Milutinovič, M., Možina, M., Polajnar, M., Toplak, M., Starič, A., Štajdohar, M.: Orange: data mining toolbox in Python. J. Mach. Learn. Res. **14**(1), 2349–2353 (2013)
15. Rousseeuw, P.J.: Silhouettes: a graphical aid to the interpretation and validation of cluster analysis. Comput. Appl. Math. **20**, 53–65 (1987). https://doi.org/10.1016/0377-0427(87)90125-7

Formal Cross-Domain Ontologization
of Human Knowledge

Ekaterina Isaeva[1]([⊠]) [iD], Vadim Bakhtin[2] [iD], and Andrey Tararkov[1]

[1] Perm State University, Perm, Russia
ekaterinaisae@gmail.com, tararkov_av@mail.ru
[2] Perm National Research Polytechnic University, Perm, Russia
bakhtin_94@bk.ru

Abstract. We introduce a new software solution to the problem of labour- and time-consuming linguistic crossdomain ontological modelling. TSGraph contributes to the interaction of a cognitive linguist with an automatically collected terminological database. The program builds ontology of several knowledge domains and highlights their overlapping in terms of semantic relations of ontological areas. The relations are parsed from Datamuse API, "a word-finding query engine for developers" [1]. TSGraph involves a number of technologies including Java 8, Spring Framework, Maven, and MySql. To build the graph we use the graph visualization library arbor.js which is a library of java script language.

The current version of TSGraph includes terminology of three domains, namely Computer Security, Biology, and Chemistry. The terminological database is collected by means of a supervised machine learning technology, implemented with TSBuilder, which is our previous development. For visual clustering, the terms have attraction points highlighted in different colours and labeled according to the knowledge domains they represent. Links within and between clusters are highlighted in relevant colours. Pop-up windows provide information on term's semantic relations, specifically semantically related term and the knowledge domain it belongs to.

The app will be useful to linguists to determine crossdomain synonymy. Further efforts are required to add morphological, syntactic, and other types of relations between terms and to extend the number of domains. These are the features, which are in high demand by researchers in the field of cognitive terminology.

Keywords: Ontology · Datamuse · Spring Framework · Maven · MySql · Java 8 · Cluster · Arbor.js library · Synonymy · Computer linguistics · Visualization · Crossdomain overlapping · Term · Terminology · TSGraph · TSBuilder

1 Introduction

One of the modern methods of studying and ordering special knowledge is based on the assumption that language structure reflects conceptual structure, and that it is possible to understand the essence of human thinking via the language as a mirror of this

© Springer Nature Switzerland AG 2020
Á. Rocha et al. (Eds.): ICITS 2020, AISC 1137, pp. 94–103, 2020.
https://doi.org/10.1007/978-3-030-40690-5_10

process [2]. In this perspective, when addressing the problem of special knowledge acquisition, one should start off by term system modelling. Following the cognitive approach, we focus on the structure of special concept, its multidimensionality, semantic and syntactic properties of terms designating the concept (for more detail see [3]. In this paper we seek to address the challenging area of cognitive linguistics, particularly crossfield ties of terms through joint modelling of ontologies of several scientific (knowledge) domains. Simultaneously, we focus on the crucial point of automation in cognitive linguistic research. Currently, we introduce the second step of our project aimed at optimization of human interaction with an automatically collected terminological database.

Due to interdisciplinarity of our approach, this paper includes the key terminology of different areas, namely, *term*, which can be defined as a word of natural language, that has a special meaning realizing in professional communication in a particular subject area or knowledge domain; *ontology*, which is roughly determined as a system of concepts with relations and interdependencies of different types and the formal representation of this system in a man – machine readable way; *subject area* or *knowledge domain (domain)* mean the broad area of scientific or professional communication, such as Physics, Chemistry, Biology, Computer security, Maths etc.; *cluster* is understood here as the database elements grouping according to certain classifying rules or preset features.

The paper is organized as follows: (1) *Introduction*, where the topicality, the aims, the scope, and the key terms are described; (2) *Rationale for ontological modelling* which provides the background knowledge on ontological modelling as the way of human – computer interaction; (3) *TSGraph development steps* aimed at the description of methods, technologies, and the programming specifics of our project; (4) *Results*, which demonstrate the implementation of the project; and (5) *Conclusion* on theoretical and practical value, limitations, and further development discussed in the final part of the article.

2 Rationale for Ontological Modelling

The concept of ontology first appeared in philosophy to designate a system of categories representing a certain vision of the world. Thus, ontology is a reconstructed model of the world. Gruber defines ontology in terms of terminology as a detailed description of conceptualization [4]. Problem-oriented ontology in a certain knowledge domain, fixing its concepts with their links and properties is a new environment for special knowledge engineering. Ontologies reflect the result of conceptualization of reality through a detailed definition of concepts, i.e. terms and their relationships in a particular field of knowledge. Multidomain approach makes it possible to examine overlaps of ontologies of different domains, thus allowing inferences of crossfield metaphorization of terms.

As is known, creation of ontological databases is a very labor-intensive and long-term process, which nowadays seems inconceivable without the use of information-computer technologies. Computer's function is to generate concepts, set the limits of their applicability, tag them to subdomains, label terms' relations in compliance with predefined rules. To fulfill this task there is a range of tools for formalized description of data semantics in Semantic Web, e.g.:

- the system of formal languages (RFD/OWL), the CycL language for ontological knowledge base, Standard Upperlevel Ontology (SUO));
- formalized tools for domain semantics description (e.g. Unified Modeling Language (UML), OilEd editor based on the OWL language);
- easy-to-use software tools created under the concept of Mind Map (MindMapper, Mindjet MindManage, FreeMind, Mind42.com etc.);
- software developed for modeling subject areas both on the basis of formalized languages (e.g., IBM Rational Rose and alike, used mainly in the database design and activities related to the analysis of business processes and control systems) and with the support of the semantics of the natural language (e.g., InTez, i.e., ontology editor based on a universal lexical interpretation of ontology, ResearchCyc, i.e., volumetric ontological knowledge base), and others.

Generally, the process of building ontology consists of the following three steps:

(1) On the basis of linguistic ontology, the dictionary of documents is built, and the frequency scheme of the database is generated. Each document is replaced by a vector of frequencies of dictionary words. An alternative way of building vectors is to use document vectorization algorithms, such as word2vec.
(2) The resulting array is clustered using self-organizing Kohonen maps or another suitable method of cauterization.
(3) An expert analysis obtained sets of concepts of ontology from a certain cluster and the nearest ones. The expert checks the links between these objects and, in case of their absence in ontology, defines them.

As a rule, the first stage is usually performed manually.

However, taking into account the fact that ontology is a definition of relations between some objects, it is worth while using means of artificial intelligence and machine learning to collect a database and set interterm relations. Neural networks, in comparison with the methods of mathematical statistics, cope with poorly formalized tasks, including the identification and categorization of terms from natural language texts (for more information see [5, 6]). Neural networks by their nature are universal approximators, which allow to model very complex regularities. Thus, it is reasonable to use neural networks as means of primary processing for simplification of work of the experts developing ontology.

Following this line of reasoning our research team has initiated a project on the study of ontological overlap. We started from the software for automated term system development based on supervised machine learning (for more information on the development of TSBuilder see [7]. This helped us to collect a multidomain database of categorized terms. To optimize the user experience, we conceive our work on TSGraph as means of ontological visualization, which provides multimodal representation of cross domain-relations of terms.

3 TSGraph Development Steps

The database, which was used in the process of building our ontological model, had been collected in two stages. At the first stage, linguists together with domain experts manually selected terms from open-access scientific texts in certain subject areas. Further, a software solution (TSBuilder) for automated identification and classification of terms from scientific texts was developed. In the first version, we used the snowball stemmer and the method of matching with the word lemmas from the database collected at the first stage. The neural network based on a multi-layer perceptron was used for classification, and it assessed whether the term found belonged to one of the basic fields of the ontological model. Later on, the terminology base was replenished with the help of TSBuilder, which managed to increase the efficiency of linguists' search for terminology. The initial subject area for experiments was Computer virology, then ontological models in biology and chemistry were added, and the terminology base was extended.

At that stage visualization of the term system was not available, there was only a base of terms, which could be used for manual terminology systematization. This underpinned our idea to create a software package, which would provide a graphical representation of the ontological model for one or more subject areas. TSGraph was supposed to build an undirected graph, in which the vertices were the terms of the subject area, and the links between certain terms reflect a high degree of synonymy (semantic similarity) of concepts. Links could combine two terms from the same domain, as well as terms from different domains, as the term could have different meanings when used depending on the context of the subject area. To define semantically related terms, it was decided to use the API of Datamuse [1], since "it embeds WordNet and provides a wide range of features including autocomplete on text input fields, ranking of search relevancy, assistance in writing apps, and word games" [8]. This resource allows you to get a list of words which are as close as possible in meaning to a term in question. The length of the list can be adjusted with the help of query parameters. The license allows us to use this resource to get information on various words without restrictions, but for processing more than 100,000 requests from our resource, in the future we will need to get a special token to work with their API. To obtain results Datamuse operates the following freely available data, relevant to our project:

Corpus-based data: The Google Books Ngrams data set is used to build the language model that scores candidate words by context, and also for some of the lexical relations. word2vec is used for reranking result sets by topic (the "topics" parameter). word2vec as well as the

*excellent Paraphrase Database are used to backfill the results for single-word "means-like"
constraints (the "ml" parameter); in particular, the "XXL" lexical paraphrases are used,
without modification.*

*Semantic knowledge: WordNet 3.0 is used for several of the static semantic lexical relations.
For the "means-like" ("ml") constraint, dozens of online dictionaries crawled by OneLook are
used in addition to WordNet [9].*

This proves the reliability of the Datamuse resource, since corpora provide massive
data for efficient machine learning in the area of extraction of up-to-date word
meanings in specific contexts. The fact that The Google Books Ngrams are not limited
to any particular domain but provide data on different subject areas means that there
will not be any constrains for adding new term systems into our ontology. The usage of
WordNet is conventional and reasonable for determining semantic distance of words in
the English language, especially for heterogeneous in terms of domains and morphological features data we have, because WordNet provides "Nouns, verbs, adjectives and
adverbs <...> grouped into sets of cognitive synonyms (synsets), each expressing a
distinct concept <...> interlinked by means of conceptual-semantic and lexical relations" [10].

4   Specifics of TSGraph Architecture

As mentioned earlier TSGraph provides visualization of semantic similarity of terms in
TSBuilder database. To select synonyms from the terms uploaded from the database
the linear search algorithm is used, then the search for the synonyms among other terms
is done in the same way. If there is an intersection of synonyms (or parts of a syn-
onym), the "connection" between them is formed.

For more convenient visual clustering, the terms have so-called attraction points
highlighted in different colours and labeled according to the subject area or knowledge
domain. Links within clusters (domains) and between them are also highlighted in
relevant colours.

The algorithm of the graph includes the following steps.

(1) Terms and their synonyms are uploaded according to predetermined clusters
    (domains) (other filters can also be applied).
(2) In each cluster, we loop over all the terms, comparing them in pairs with the terms
    from the rest of the subject areas, building a list of words synonymous with the
    term in question (the synonyms have been parsed to the database from Datamuse).
    The list is stored in the RAM of the server while the information is being
    processed.
(3) With the linear search in the list of potential synonyms we check whether at least
    part of a synonym is included into another term as a substring. In this case
    relationship is established.

(4) The terms are located to the vertices of the graph. In our case, the server interprets the terms as strings, thus, no additional transformations are required to store or present them. Additionally, dummy nodes are created, which are "points of attraction" of the graph nodes, which belong to the same domain.

(5) Terms' relations are visualized in the form of graph edges, using different types of lines for internal and external relations. The visualization uses the arbor.js library, which allows us to dynamically change the graph if necessary.

(6) The cluster centers are visualized with corresponding colour areas and domain names. At the moment, colours are stored in the database as one of the parameters of the subject area, in future relevant customization may be added.

In the project, we use the graph visualization library arbor.js which is a library of java script language based on another, better known library, jQuery [11]. From the server, where the necessary data for the graph was extracted and built, a list of the graph nodes and edges comes in the form of json description, which is later drawn using javascript and arbor.js in the browser.

We apply the following json structure [12]:

```
{
  "nodes": [
    {
      "term": {
        "id": <Unic term identifier>,
        "term": "<Term in the string form >",
        "clazz": {
          "project": {
            "id": <Unic domain identifier >,
            "project": "<Domain name >",
            "color": "<colour in the form of R,G,B>"
          }
        }
      }
    },
        ...
  ]
}
```

The API used for the search of synonyms, namely Datamuse, is a web resource that can respond to queries to find a certain number of semantically close words or expressions. For instance, if we want to determine semantically related words or expressions to the word *brain*, we receive the following feedback to our request (Fig. 1):

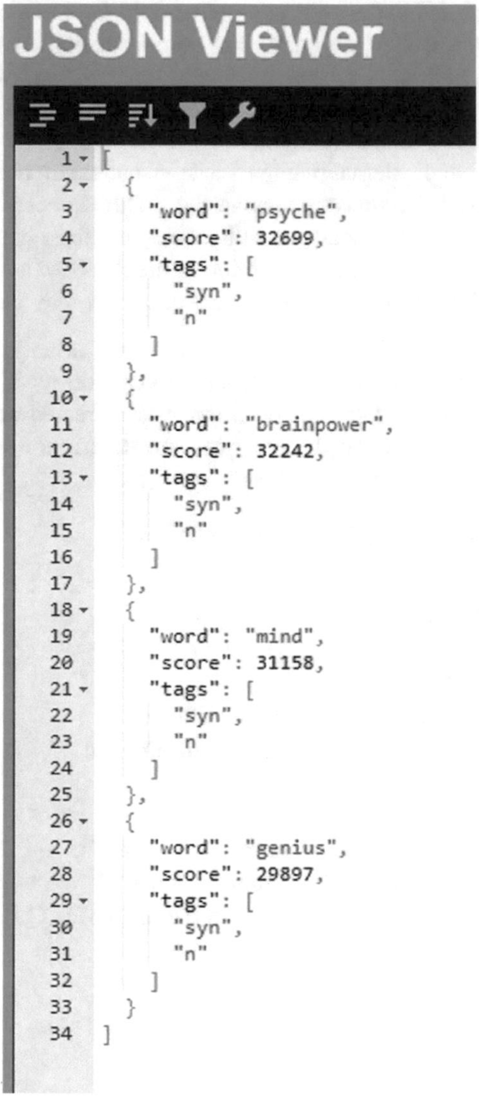

```
JSON Viewer

1 ▾ [
2 ▾    {
3         "word": "psyche",
4         "score": 32699,
5 ▾      "tags": [
6            "syn",
7            "n"
8         ]
9      },
10 ▾    {
11        "word": "brainpower",
12        "score": 32242,
13 ▾     "tags": [
14           "syn",
15           "n"
16        ]
17     },
18 ▾    {
19        "word": "mind",
20        "score": 31158,
21 ▾     "tags": [
22           "syn",
23           "n"
24        ]
25     },
26 ▾    {
27        "word": "genius",
28        "score": 29897,
29 ▾     "tags": [
30           "syn",
31           "n"
32        ]
33     }
34 ]
```

Fig. 1. An example of search for semantically nearest words in Datamuse (source https://www. datamuse.com)

TSGraph involves a number of technologies including:

- Java 8, i.e., a cross-platform programming language in which the server part of the application is implemented;
- Spring Framework, i.e., a framework for development in Java; in this case it is used to implement REST API for communication with the server;

- Maven, i.e., a means of assembling a project;
- MySql, i.e., a relational database used to store terms.

Synonyms for terms are preloaded and stored in the local TSGraph database for efficient operation. Its structure provides for the storage of terms and contexts of their use. Currently, the first version of the application runs on a remote server and is available for use. As a remote server we have chosen the Ubuntu operating system with access for SSH configuration.

5   Results

The beta version of TSGraph is available at a request. Figure 2 shows the current version of three-domain ontological model.

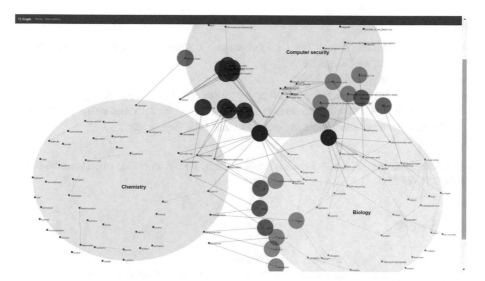

Fig. 2. Ontological model representation window for three domains.

The view window shows a non-oriented graph built for the three domains (Computer Security, Biology, and Chemistry), in which terms belonging to different areas can be distinguished by the colors of the vertices in which they are located. Lines of different colours show both the connections within and between the domains. The terms of each domain are combined into a cluster and are arranged around it on the graph. The terms highlighted with coloured circles have connections with terms in other subject areas, the more intense the colour is, the more connections the term has.

6 Conclusion

This work is the second step following the development of the program for automated term identification. The software is intended to streamline manual linguistic work for conceptual or ontological modelling in the framework of cognitive terminology. The findings add to a growing body of literature on computer linguistics, text mining, ontological modelling, as well as to our understanding of human-computer interaction.

Our work clearly has some limitations, such as lack of automated one-click extension of the database for ontological modelling or the number of domains, and restriction on the amount of terms represented on the graph. The present version of TSGraph includes only synonymic relations of terms, thus, the picture of different domain terminology overlap is still incomplete. Consequently, further efforts are required to visualize for the researchers this complex knowledge – language interdependence through the prism of computer ontological modelling.

That is why, our further work will concentrate on the following program modification:

– extend the choice of relations (links) beside the semantical distance of words;
– add new domains (subject areas);
– expand the terminology base;
– add filters for flexible information aggregation and visualization, e.g. selecting the terms with the most intercluster relations, changing the number of terms to be selected, manipulating with the colors of the links/areas, etc.

Acknowledgements. The reported study was funded by RFBR according to the research project № 18-012-00825 A.

References

1. Datamuse API. http://www.datamuse.com/api/
2. Langacker, R.W.: Foundations in Cognitive Grammar: Theoretical Prerequisites, vol. 1. Stanford University Press, Stanford (1987)
3. Isaeva, E., Burdina, O.: Transdiscursive term transformation: the evidence from cognitive discursive research of the term 'virus'. In: I Ferrando, I.N. (ed.) Current Approaches to Metaphor Analysis in Discourse, pp. 79–110. De Gruyter, Berlin (2019). https://doi.org/10. 1515/9783110629460-005
4. Gruber, T.: Ontology. In: Liu, L., Tamer Özsu, M. (eds.) Encyclopedia of Database Systems. Springer, New York (2009). http://tomgruber.org/writing/ontology-definition-2007.html
5. Bakhtin, V.V., Isaeva, E.V.: Developing an algorithm for identification and categorization of scientific terms in natural language text through the elements of artificial intelligence. In: Proceedings of 2018 14th International Scientific-Technical Conference on Actual Problems of Electronic Instrument Engineering, APEIE 2018, pp. 386–390 (2009) https://doi.org/10. 1109/apeie.2018.8545317

6. Bakhtin, V.V., Isaeva, E.V.: New TSBuilder: shifting towards cognition. In: Proceedings of the 2019 IEEE Conference of Russian Young Researchers in Electrical and Electronic Engineering, ElConRus 2019, 28 February 2019, pp. 179–181 (2019) https://doi.org/10.1109/eiconrus.2019.8656917. Art. № 8656917
7. Isaeva, E.V., Suvorova, V.A., Bakhtin, V.V.: Automatic Documentation Math. Linguistics. **50**(3), 104–111 (2016). https://doi.org/10.3103/S0005105516030031. Allerton Press
8. Gallant, M., Isah, H., Zulkernine, F., Khan, S.: Xu: an automated query expansion and optimization tool. In: 43rd IEEE Annual Computer Software and Applications Conference, COMPSAC 2019; Milwaukee, USA, vol. 1. pp. 443–452 (2019)
9. Datamuse. https://www.datamuse.com
10. WordNet. https://wordnet.princeton.edu/
11. jQuery. https://code.jquery.com/
12. Arbor.js. http://arborjs.org/

The Development of a Business Intelligence Web Application to Support the Decision-Making Process Regarding Absenteeism in the Workplace

Sara Oliveira[1] , Marisa Esteves[2(✉)] , Rui Cernadas[3] ,
António Abelha[2] , and José Machado[2]

[1] University of Minho, Campus Gualtar, 4470 Braga, Portugal
sara.oliveira.3996@gmail.com
[2] Algoritmi Research Center, University of Minho, Campus Gualtar,
4470 Braga, Portugal
{marisa,abelha,jmac}@di.uminho.pt
[3] Continental Mabor – Indústria de Pneus, Lousado,
4760 V. N. Famalicão, Portugal
rui.cernadas@conti.de

Abstract. Nowadays, one of the biggest concerns of industries all over the world is situations regarding absenteeism, since it has a great impact on the productivity and economy of companies, as well as on the health of their employees. The major causes of absenteeism appear to be work accidents and sickness leaves, which lead to the attempt by companies of understanding how the workload is related to the health of their collaborators and, consequently, to absenteeism. Thus, this paper proposes the design and development of a Web Application based on Business Intelligence indicators in order to help the health and human resources professionals of a Portuguese company analyse the relation between absenteeism and the health and lifestyle of employees, with the intention of concluding whether the work executed on the company is harming workers' health. Furthermore, it is intended to discover the principal motives for the numerous and more frequent absences in this company, so that it is possible to decrease the absenteeism rate and, hence, improve the decision-making process. This platform will also provide higher quality healthcare and the possibility to find patterns in the absence of collaborators, as well as reduce time-waste and errors.

Keywords: Information and Communication Technology · Business Intelligence · Decision-making process · Absenteeism · Web application · Health and human resources professionals

1 Introduction

Absenteeism refers to the absence of individuals to work and it can be caused by work accidents, personal illness, and family issues, among others [1]. It has a tremendous influence on industries, since it decreases their productivity and economy, which highly

Á. Rocha et al. (Eds.): ICITS 2020, AISC 1137, pp. 104–113, 2020.
https://doi.org/10.1007/978-3-030-40690-5_11

concerns companies. Consequently, it creates the necessity to analyze this issue with the aim of comprehending the main motives so that it is possible to justify and decrease high absenteeism rates, as well as understand if workload is somehow related to the absences and causing harm to the health of workers [2].

Thus, in order to help health and human resources professionals of a Portuguese company analyze absenteeism, a solution was proposed. It consisted in designing and developing a Web application based on Business Intelligence (BI) indicators. This platform will allow the identification of the causes for an increasing absenteeism rate and the analysis of the relation between the great number of absences and the health and lifestyle of their collaborators. Furthermore, it is also planned to find patterns in the absences of workers, as well as solutions to decrease such rate.

Regarding the structure of this paper, the state of art related to this project is described in Sect. 2. Thereafter, in Sect. 3, the research strategies chosen to conduct this study are briefly presented. Then, in Sect. 4, the case study is discussed with the intention of revealing the main issues faced by the company, as well as validate the pertinence of this work. Subsequently, Sect. 5 demonstrates the results achieved, which contain the architecture for the Web application and the BI indicators created. In Sect. 6, a discussion of the results is presented and, lastly, in Sect. 7, the principal conclusions are enhanced and future work is suggested.

2 State of the Art

Over the last few years, Health Information and Communication Technology (ICT) was proved to be essential to healthcare facilities. ICT consists in the use of electronic tools that allows the processing, transmission, storage, access, and use of medical data in order to provide an easier and faster manipulation of data in healthcare organizations [3]. Besides that, health ICT has the capability of increasing access, quality, and efficiency of healthcare processes, as well as reducing time-waste, clinical errors, and costs [3–5].

Business Intelligence (BI) is the most relevant ICT concept, since it enables the collection, transformation, analysis, and organization of data from multiple internal and external sources of information for a faster and more effective decision-making process [6–9]. Thus, BI transforms a large amount of raw data into valuable knowledge, which provides the possibility of making fact-based decisions [10, 11].

On the top of that, BI offers numerous tools, including the Extract, Transform, and Load (ETL) process that is capable of extracting information from one or more sources, clean-up and normalize the data and, finally, load the data into a data warehouse (DW) [12, 13]. Moreover, BI also resorts on data warehousing, which is a tool that is able to store and organize the previously treated data [14]. Additionally, this technology allows the visualization and interpretation of information through Reporting, Data Mining (DM), and Online Analytical Processing (OLAP) processes, which enable the elaboration of reports, the definition of patterns, and the study and manipulation of data stored in the DW from different perspectives, respectively [15–17].

Nevertheless, in spite of the increasing adoption of BI systems in healthcare facilities, there are still great challenges that must be overcome in order to implement this technology in healthcare organizations, such as technical issues and familiarity with ICT [3, 7]. Therefore, it is necessary to investigate this problem in order to understand how to apply successfully the BI methodology in the healthcare industry [4, 18].

3   Research Strategies

It was fundamental to select a methodology that would conduct this project within an organized path and well-defined rules and, consequently, guarantee its success. Thus, the research strategy that is being followed is Design Science Research (DSR), since it is the most adequate methodology for ICT research projects [19, 20].

DSR has the main goal of creating and evaluating innovative ICT artefacts with the aim of solving organizational problems [19, 20]. Therefore, this research strategy is consisted by the following steps: "Problem Identification and Motivation", "Definition of the Objectives for a Solution", "Design and Development", "Demonstration", "Evaluation", and "Communication" [19].

For the first three steps, it was needed to perform interviews and meetings with the target audience, i.e., health and human resources professionals. The information acquired through these steps was essential to identify the principal challenges and to establish the objectives in order to execute an adequate solution.

The fourth and fifth steps are currently being accomplished. The application is being presented to the target audience as it is being developed with the intention of understanding if the defined objectives are being encountered and if there is the necessity of modifications. Additionally, it is important to mention that a Proof of Concept (PoC) was performed in order to analyze the impact of the proposed solution in the company.

Lastly, the final step of DSR is also being executed, which consists in announcing the novelty and value of this work, being the intention of this paper.

It is important to refer that ethical issues are being taken into consideration in order to guarantee the safety and confidentiality of all of the information provided.

4   Case Study

The case study presented in this paper consists of a problem concerning absenteeism in a Portuguese company that produces various types of tires, which is being managed by a German multinational company (Continental). The issues and main challenges that motivated the development of the proposed solution will be discussed in this section.

It was necessary to reunite with the health and human resources professionals of the referred company in order to visualize the way absenteeism has been analyzed for the past few years and to comprehend their needs and the problems that must be solved. As a result of that, it is possible to list the challenges faced by these professionals:

- The company has the entire data concerning absenteeism in Excel files. This situation makes the analysis of information and the decision-making process a lot more difficult and time-consuming, since the data are disorganized and harder to interpret;
- There is the necessity of attempting to identify patterns within the absences of collaborators and this analysis can lead to errors, since the Excel file method makes the study of absenteeism and, consequently, of a vast amount of data more confusing;
- The analysis of absenteeism through Excel files is harder due to the fact that all the recorded data of the employees, i.e., personal, health, and absence information, are kept in various files.

Thereby, it was proven the necessity that these professionals have for a solution that will allow them to analyze the problem of absenteeism in a faster and more efficient way. The proposed solution is, as previously mentioned, the design and development of a Web application with BI clinical and performance indicators in order to facilitate the finding of patterns in the absences of collaborators and the analysis of the absenteeism rate and, consequently, improve the decision-making process so that it is possible to enhance the healthcare provided and reduce the absenteeism rate.

5   Results

In the present section, the results of the proposed solution are described. The first subsection, Subsect. 5.1, presents the architecture of the Web application and all of the functionalities available for its users. Thereafter, some of the BI indicators created in order to support the analysis of absenteeism and the decision-making process are shown and discussed in Subsect. 5.2.

5.1   Architecture of the Web Application

Before designing the Web application, a database had to be created in order to store all the necessary data since the company for which the solution is being developed has all the information related to absenteeism in Excel files and, consequently, does not have any database implemented. The MySQL database was selected due to the fact that it is a secure and reliable relational database. Thus, a DW was created through the ETL process. The DW contains absence, health, and personal information of the employees.

Moreover, the Web application is being developed with the ReactJS library, which is a JavaScript library written in JSX, i.e., a syntax extension to JavaScript [21]. Additionally, the sharing of information between the application and the DW is made through a REST API in NodeJS with SQL queries, more precisely in ExpressJS [22]. In Fig. 1, it is presented a schematic representation of the architecture of the application.

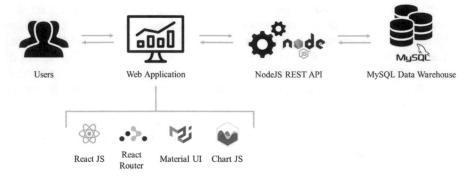

Fig. 1. Schematic representation of the architecture of the Web application.

Firstly, the user needs to sign into the Web application with his login credentials in order to use the application. It must be mentioned that the login information of all the users is stored in the DW. Furthermore, the application must be accessed using a Web browser and it is distributed to all the health and human resources professionals through the Intranet, which is the private network of the company.

Then, the user may access the modules and functionalities available in the application, namely:

- Dashboard: it displays basic data related to the absenteeism in the company, such as the absenteeism rate in the previous 12 months, the latest five absent collaborators, and the three principal motives of absence, among others;
- Management of Absence Records: it allows the visualization of the absence entries of each collaborator. Additionally, the human resources professionals can add, update or exclude absence records;
- Management of the Collaborators: it enables the visualization of the personal information of each collaborator. Moreover, the human resources professionals can add, update or exclude collaborators from the list;
- BI Indicators Page: this module presents all of the BI clinical and performance indicators related to the number of collaborators missing and the absenteeism rate so that it is possible for the user to analyze the relation between the health and lifestyle of the employees and the absenteeism in the company, as well as to identify the causes for the increasing rate and understand how to decrease such rate;
- Management of Notifications: the user may consult the notifications received and also select the type of alerts he wants to receive, such as when a collaborator is missing for 30 days or more, when a collaborator from a determined department initiates an absence, among others;
- Registration of Users: the users with administrator accounts are able to register new accounts for health and human resources professionals, as well as edit and delete, since the application is restricted to these professionals and cannot be accessed by the entire company;
- Profile: it is possible for the user to view and update his personal data;
- Sign out: it allows the user to sign out of his account.

5.2   Business Intelligence Indicators

In order to accomplish the previously stated objectives, it was necessary to create BI indicators, since it will help with the analysis of the problem, as well as with the decision-making process. Therefore, these indicators were created with the ChartJS library, a flexible and simple JavaScript charting library that can be used on ReactJS applications with the data stored in the MySQL database [23]. Moreover, a RESTful API written in ExpressJS, a NodeJS Web framework, with SQL queries is being developed with the aim of enabling the sharing of data between the frontend application and the backend DW [24].

Thus, the next sixteen BI indicators were created:

- Number of absent collaborators by month;
- Number of absent collaborators by cause of absence;
- Number of absent collaborators by academic level;
- Number of absent collaborators by city of residence;
- Number of absent collaborators by body mass index (BMI);
- Number of absent collaborators by disease;
- Number of absent collaborators by period of absence;
- Absenteeism rate by gender;
- Absenteeism rate by age group;
- Absenteeism rate by department;
- Absenteeism rate by smokers and non-smokers;
- Absenteeism rate by years of work in the company;
- Absenteeism rate by parents and non-parents;
- Absenteeism rate by who wear glasses and who does not;
- Absenteeism rate by those who exercise regularly and those who do not;
- Absenteeism rate by those who do rotating shifts and those who do not.

However, it must be mentioned that some of the previously stated BI indicators were not created yet since there is not enough data available. On the other hand, it is important to refer that these BI indicators were built considering all 2018 and 2019 until the month of august (included).

One of the BI indicators created is presented in Fig. 2.

In Fig. 2, it is demonstrated the BI indicator related to the number of absent collaborators by period of absence during 2018. It shows that there were 119, 501, and 211 collaborators absent for a period of three or less days, three to thirty days, and more than thirty days, respectively. Therefore, it is possible to conclude that the period for which the workers were commonly more absent was the mid-term absence, i.e., the period of three to thirty days. Furthermore, it can be assumed that most of the employees that missed work in 2018 did not have serious complicated health issues and were able to return to work in less than a month.

Another indicator was created in order to allow the study of the number of absent collaborators by cause of absence in 2018 and it enabled the discovery of the three

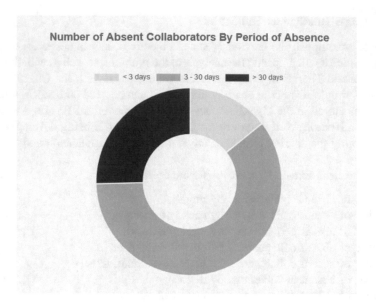

Fig. 2. Business Intelligence indicator concerning the number of absent collaborators by period of absence in the year of 2018.

principal causes of absence in the company, which were Short Term Disability (STD) – Natural Disease, Work Accident, and Parental Leave with 373, 143, and 136 absent employees, respectively. Thus, with this information, it can be concluded that the work done in this company is damaging the health of collaborators through working accidents and the development of diseases. Besides that, it is also possible to visualize that many collaborators are becoming parents.

Moreover, it was also created a BI indicator concerning the absenteeism rate by gender in the year of 2018 and it is possible to analyze that men and women had an absenteeism rate of 5.93% and 4.06%, respectively. Thereby, this data allows to determinate that men had a higher absenteeism rate than women during 2018, which can also be explained by the fact that there are more men than women working at this company.

The formula that enabled the calculation of the absenteeism rate is demonstrated in following equation:

$$\frac{\text{Number of absent days}}{\text{Total number of collaborators} \times \text{Number of worked days}} \times 100\% \qquad (1)$$

However, it is possible that the previously referred absenteeism rates are miscalculated, since the number of worked days is unknown and it was necessary to use an estimated value to calculate those percentages.

Moreover, it must be mentioned that it is planned to create more BI indicators in order to provide to the company an efficient and reliable analysis of the problem.

6 Discussion

In order to present the practicality, effectiveness, value, and quality of the Web application, a Proof of Concept (PoC) was made. Thus, a SWOT analysis was necessary in order to enhance the strengths and identify the weaknesses of the solution, which are internal factors, and to also acknowledge the opportunities and threats, which are external factors.

The main strengths of the Web application are the following:

- Enables the study of the absenteeism in the company and improves the quality of the decision-making process;
- Decreases times-loss and the number of errors occurred on the analysis of the problem comparing to the Excel file method;
- Easier access and visualization of the information and, consequently, faster interpretation of the data;
- Makes the attempt of discovering patterns in the absences of employees a simpler task;
- High usability due to the fact that it is an intuitive tool with the information displayed in an organized way;
- High scalability since new features can be easily inserted in the application.

On the other hand, one weakness of the Web application can be outlined:

- Required intranet connectivity to use the application.

Regarding the opportunities of the solution, the next points were identified:

- Implementation of a Data Mining (DM) module in order to allow the prediction of future tendencies on the absences of the collaborators with the help of past registries of absences;
- The development of new functionalities and modules with the aim of expanding the application;
- Elimination of the Excel files method, since it provides a higher risk of errors;
- Possibility of inserting new information in the DW, as well as of creating new BI clinical and performance indicators.

Finally, the threats for the proposed solution are as follow:

- Problems associated with the private network connection, i.e., Intranet;
- The potential lack of acceptance to use this solution by health and human resources professionals.

7 Conclusion and Future Work

The work described throughout this paper was created in order to help solving an absenteeism problem in a Portuguese company. So, a Web application with BI indicators was designed and developed with the aim of finding the causes to an increasing absenteeism rate and also of studying the possibility of the work done in the company be damaging the health of collaborators. Furthermore, it was also intended to analyze if the lifestyle of employees contribute to a more absent behavior.

Thus, through the analysis of some of the BI indicators created, it is possible to confirm and conclude that the workload of the company is causing harm to the health of collaborators, since two of the main causes of absence in the company are diseases and work accidents. This information also justifies the need of such application, proving that it is easier, faster, and less prone to errors to analyze absenteeism through the Web application, comparing to the Excel files method. However, there is not yet enough data in order to create more BI indicators to make more precise conclusions regarding the relation between absenteeism and the health and lifestyle of collaborators, as well as the actions needed to decrease such absenteeism rate.

So, regarding future work, it is intended to build more BI performance and clinical indicators as soon as more data are provided by the company. Additionally, it is also planned to implement a Data Mining (DM) module in order to help predicting future patterns and tendencies on absenteeism based on past records of absences. Finally, it is foreseen to test the application with the target audience so that it is possible to understand if the objectives are being accomplished and if changes are needed.

Acknowledgements. This work has been supported by FCT – *Fundação para a Ciência e Tecnologia* within the Project Scope: UID/CEC/00319/2019.

References

1. Aldana, S.G., Pronk, N.P.: Health promotion programs, modifiable health risks, and employee absenteeism. J. Occup. Environ. Med. **43**(1), 36–46 (2001). https://doi.org/10.1097/00043764-200101000-00009
2. Blau, G.J., Boal, K.B.: Conceptualizing how job involvement and organizational commitment affect turnover and absenteeism. Acad. Manag. Rev. **12**(2), 288–300 (1987). https://doi.org/10.5465/amr.1987.4307844
3. Gagnon, M.P., Desmartis, M., Labrecque, M., et al.: Systematic review of factors influencing the adoption of information and communication technologies by healthcare professionals. J. Med. Syst. **36**(1), 241–277 (2012). https://doi.org/10.1007/s10916-010-9473-4
4. Esteves, M., Abelha, A., Machado, J.: The development of a pervasive web application to alert patients based on business intelligence clinical indicators: a case study in a health institution. J. Wirel. Netw., 1–7 (2019). https://doi.org/10.1007/s11276-018-01911-6. Springer
5. Alpuim, A., Esteves, M., Pereira, S., Santos, M.F.: Monitoring time consumption in complementary diagnostic and therapeutic procedure requests. In: Health Care Delivery and Clinical Science: Concepts, Methodologies, Tools, and Applications, pp. 1553–1579. IGI Global (2018). https://doi.org/10.4018/978-1-5225-3926-1.ch078
6. Negash, S., Gray, P.: Business intelligence. In: Handbook on Decision Support Systems 2. International Handbooks Information System, pp. 175–193. Springer, Berlin (2008). https://doi.org/10.1007/978-3-540-48716-6_9
7. Foshay, N., Kuziemsky, C.: Towards an implementation framework for business intelligence in healthcare. Int. J. Inf. Manag. **34**(1), 20–27 (2014). https://doi.org/10.1016/j.ijinfomgt.2013.09.003

8. Reis, R., Mendonça, A., Ferreira, D.L.A., Peixoto, H., Machado, J.: Business intelligence for nutrition therapy. In: Healthcare Policy and Reform: Concepts, Methodologies, Tools, and Applications, pp. 459–474. IGI Global (2019). https://doi.org/10.4018/978-1-5225-6915-2. ch022

9. Esteves, M., Miranda, F., Machado, J., Abelha, A.: Mobile collaborative augmented reality and business intelligence: a system to support elderly people's self-care. In: Rocha, Á., Adeli, H., Reis, L., Costanzo, S. (eds.) Trends and Advances in Information Systems and Technologies, WorldCIST 2018, vol. 747, pp. 195–204. Springer, Cham (2018). https://doi.org/10.1007/978-3-319-77700-9_20

10. Mach, M.A., Abdel-Badeeh, M.S.: Intelligent techniques for business intelligence in healthcare. In: 2010 10th International Conference on Intelligent Systems Design and Applications, pp. 545–550. IEEE, Cairo (2010). https://doi.org/10.1109/isda.2010.5687209

11. Mettler, T., Vimarlund, V.: Understanding business intelligence in the context of healthcare. Health Inf. J. **15**(3), 254–264 (2009). https://doi.org/10.1177/1460458209337446

12. Bansal, S.K.: Towards a semantic extract-transform-load (ETL) framework for big data integration. In: 2014 IEEE International Congress on Big Data, pp. 522–529. IEEE, Anchorage (2014). https://doi.org/10.1109/bigdata.congress.2014.82

13. Esteves, M., Miranda, F., Abelha, A.: Pervasive business intelligence platform to support the decision-making process in waiting lists. In: Next-Generation Mobile and Pervasive Healthcare Solutions, pp. 186–202. IGI Global (2018). https://doi.org/10.4018/978-1-5225-2851-7.ch012

14. Golfarelli, M., Maio, D., Rizzi, S.: Conceptual design of data warehouses from E/R schemes. In: Proceedings of the Thirty-First Hawaii International Conference on System Sciences, vol. 7, no. 1, pp. 334–343. IEEE, Kohala Coast (1998). https://doi.org/10.1109/hicss.1998. 649228

15. Brandão, A., Pereira, E., Esteves, M., Portela, F., Santos, M., Abelha, A., Machado, J.: A benchmarking analysis of open-source business intelligence tools in healthcare environments. Information **7**(4), 57 (2016). https://doi.org/10.3390/info7040057

16. Chaudhuri, S., Dayal, U., Narasayya, V.: An overview of business intelligence technology. Commun. ACM **54**(8), 88–98 (2011). https://doi.org/10.1145/1978542.1978562

17. Olszak, C.M., Batko, K.: The use of business intelligence systems in healthcare organizations in Poland. In: 2012 Federated Conference on Computer Science and Information Systems (FedCSIS), pp. 969–976. IEEE, Wroclaw (2012)

18. Bonney, W.: Applicability of business intelligence in electronic health record. Procedia-Soc. Behav. Sci. **73**, 257–262 (2013). https://doi.org/10.1016/j.sbspro.2013.02.050

19. Peffers, K., Tuunanen, T., Rothenberger, M.A., Chatterjee, S.: A design science research methodology for information systems research. J. Manag. Inf. Syst. **24**(3), 45–77 (2007). https://doi.org/10.2753/mis0742-1222240302

20. Hevner, A., Chatterjee, S.: Design science research in information systems. In: Design Research in Information Systems. Integrated Series in Information Systems, vol. 22, pp. 9–22. Springer, Boston (2010). https://doi.org/10.1007/978-1-4419-5653-8_2

21. Facebook Inc. React – a JavaScript library for building user interfaces. https://reactjs.org/. Accessed 17 Aug 2019

22. Node.js. Introduction to Node.js. https://nodejs.dev/. Accessed 17 Aug 2019

23. Chart.js. Chart.js – simple yet flexible JavaScript charting for designers & developers. https://www.chartjs.org/. Accessed 18 Aug 2019

24. Express.js. Express.js – the fast, unopinionated, minimalist web framework for node. https:// github.com/expressjs. Accessed 18 Aug 2019

The Intention to Use E-Commerce Using Augmented Reality - The Case of IKEA Place

Carlos Alves[1] and José Luís Reis[1,2](✉)

[1] IPAM, Portuguese Institute of Marketing, Porto, Portugal
carlos@3bc.pt, jreis@ipam.pt
[2] Research Unit CEDTUR/CETRAD, ISMAI, Maia University Institute,
Maia, Portugal
jreis@ismai.pt

Abstract. The evolution of technologies stimulates the e-commerce market. Recently, augmented reality has been creating significant value in the retail sector. This study aims to analyse an application, IKEA Place, which allows consumers to shop online using augmented reality technology without the need for a marker to identify the surface (Markerless Augmented Reality). The model to evaluate this study was based on the constructs of Technology Acceptance Model (TAM) and other models. The sustainability of the study is exploratory and involves identifying whether this application aims to create more confidence and convenience in the acquisition of a product through the support of this technology. The study was carried out in the North of Portugal, obtaining a convenience sampling in the Porto and Braga regions. The results concluded that there are no significant differences based on sociodemographic data, the operating system used, familiarity with the organization related to the intended use. In contrast, technology changes have been found to offer consumers more confidence and convenience of purchase. Significant results were also obtained in terms of user experience, the preference of this application to make online purchases over existing channels, the future use of such technologies by users to purchase a product and the attraction of purchase regarding perceived ease of use of the application under analysis.

Keywords: Augmented reality · e-commerce · TAM · IKEA Place · Intent to use

1 Introduction

Virtual Reality (VR) is increasingly present in people's daily lives, especially in younger generations such as Gen Z and Millennials through video games. Billinghurst, Clark and Lee state that one of the greatest opportunities for AR is to use technology for product marketing, as marketing is intended to capture attention and provide motivation to learn more about the product and RA technology can be used to create memorable experiences [1]. The general objective of this investigation is to verify how the Portuguese user makes a purchase through an augmented reality e-commerce platform and feels confident and convenient when making this purchase.

© Springer Nature Switzerland AG 2020
Á. Rocha et al. (Eds.): ICITS 2020, AISC 1137, pp. 114–123, 2020.
https://doi.org/10.1007/978-3-030-40690-5_12

To achieve the objective of this work, a questionnaire was used for users of IKEA's Augmented Reality (AR) application IKEA Place, based on a questionnaire to analyze the experience of using the application.

This article is made up of a section of outlines which applications IKEA developed prior to IKEA Place and some related studies that in turn fostered the existence of this article. Another section presents which constructs are used to design the conceptual model, which authors address these constructs in their publications and the research hypotheses that were determinant for the purpose of the study. The Sect. 4 explains the methodology that was used in the study, including how data were collected and processed as well as the characterization of the sample. Section 5 presents the data analyze and the result discussion of the research hypotheses, finally, the study conclusions are presented.

2 IKEA and Computer Applications Studies

In 1960 Ingvar Kamprad invented a different concept of furniture, being the first organization in the industry to push for self-service, excellent design and affordable furniture for all people [2].

2.1 IKEA Applications

Currently IKEA has 3 applications, IKEA Place, IKEA Catalog APP and IKEA Store APP. IKEA Place is an application that aims to provide an AR experience by placing 3D models of your products in a given space and thus allowing you to purchase these products within the application itself. IKEA Catalog APP is an application capable of placing the catalog that is printed on paper on a platform such as a tablet or smartphone, thus allowing for faster product search. The IKEA Store APP is an in-store app that gives customers the chance to plan their shopping list at home, view product stock, scan items to add to their store. list, among other features beneficial to the customer who purchases within an IKEA physical surface [3].

IKEA launched its first AR-based catalog in 2013 with the aim of reducing returns from online shopping or traditional methods so that, with this new technology, the customer through their mobile phone can see in real dimensions the products they want to buy [4].

2.2 IKEA Studies

Of the various studies on IKEA, the study of its catalog using AR from Rese et al. who analyzed four AR applications in order to understand the acceptance of the technologies, two of them using markers and two without using them, to point out that the application of IKEA was in the group of marker applications. Thus, the results obtained were satisfactory due to the robustness of the technology acceptance model [5].

Recently a study was done on an application of IKEA, IKEA AR, where the main objective would be to verify the influence of AR on purchase intention. The study states that little is known about whether AR was viable in order to enrich buying behavior and thus produce favorable results. Through an experimental methodology, it was analyzed in generation Y where a possible purchase intention was compared between the IKEA AR application and the existing website at the time. It was concluded that using the IKEA AR application was more useful and enjoyable, providing higher purchase intentions than the site itself [6]. IKEA discontinued this application and launched another one about two months after the release of Apple's ARKit. The IKEA Place app, launched in September 2017 has over 3,200 products and the innovation of this app is that it does not require markers to recognize the surface, thus eliminating the need for a paper catalog and only requiring an Apple iPhone 5s or higher smartphone with iOS 11 operating system version installed. This application is so precise that the user can see the texture of the fabrics and with that IKEA wants to change the paradigm of online shopping, thus creating the ease of decision making and at the same time creating the milestone as one of the first organizations "selling furniture of the future" [7].

Once installed on an iOS device, the app is intuitive since after the app's loading screen, handouts for using it are displayed. To add a new product, press the "+" button and select the desired product. After that, the surface is mapped through the camera and the object will appear. Then select where you want to place the object by dragging it to the selected position.

In March 2018 the IKEA Place Android version was released which allows users to securely place the 3D renderings in the app. IKEA opens a world of possibilities for users to use this application on any ARCore compatible device [8].

3   Research Reference Framework and Conceptual Model

This section presents the practical and empirical frameworks, based on the literature review, as well as the conceptual model for research and data collection.

3.1   Reference Framework and Research Hypotheses

Based on the literature review, the research hypotheses are a response previously developed to support the research questions, based on studies related to the same theme [27]. Based on the literature review, the following research hypotheses were defined:

H1. Purchasing confidence is positively related to perceived utility [14, 16, 24, 25].
H1.1. Convenience of purchase is positively related to perceived utility [14, 16, 24, 25].
H1.2. Purchase confidence is positively related to intention to use [5, 14, 16, 19, 20, 22, 24, 25].
H1.3. Convenience of purchase is positively related to intention to use [5, 14–16, 19, 20, 22, 24, 25].
H2. Intention to use is positively related to age [5, 14, 15, 18–20, 22].
H2.1. Intent to use is positively related to Gender [14, 15, 19, 20, 22].
H2.2. Intention to use is positively related to educational level [5, 14, 15, 18–20, 22].
H2.3. Intent to use is positively related to place of residence [5, 14, 15, 19, 20, 22].

H3. Intent to use is positively related to the operating system [5, 14, 15, 19, 20, 22].

H4. Intention to use is positively related to familiarity [5, 14, 15, 19, 20, 22].

H5. Intention to use is positively related to attractiveness of purchase [5, 14, 15, 19, 20, 22, 23].

H6. Ease of purchase is positively related to perceived utility [5, 14, 15, 17, 20, 22, 24, 25].

H7. Perceived utility is positively related to purchase attractiveness [5, 14, 17, 18, 20, 22, 25].

H8. Perceived ease of use is positively related to attractiveness of purchase [5, 14, 16–18, 20, 22, 24].

For the construction of the conceptual model, concepts presented in Table 1 were obtained from the literature review, which support the research methodology.

Table 1. Main research preferences

Concept	Definition	Author
Augmented reality	RA, at its core, transforms volumes of data and analysis into images or animations that overlap the real world	[9–13]
Perceived utility	Degree from which an individual believes using a system improves performance	[5, 14–22]
Perceived ease of use	The degree to which an individual believes that using a system requires no effort	[5, 14–22]
Intent to use	Degree indicating the extent to which an individual is willing to use a system	[5, 14–22]
Purchase attractiveness	Factor that determines whether a user wants to make a purchase	[23]
Convenience	Shopping convenience is perceived as the ability to shop virtually anytime from a variety of locations without visiting a store	[24, 25]
Ease of purchase	Avoid physical and emotional difficulties buying from other channels	[24]
Confidence	An individual's willingness to believe in a transaction with an entity about which he or she feels confident about his or her intentions	[16, 26]

3.2 Conceptual Model for IKEA Place Evaluation

IKEA Place's evaluation model aims to analyze the experience of using the application under analysis. The model presented in Fig. 1 is based on several models such as TAM - Technology Acceptance Model [14, 24], TAM for online shopping [16], TAM 3 [20], Augmented Reality Interactive Technology TAM [17] and extension of TAM to accept augmented reality applications [5].

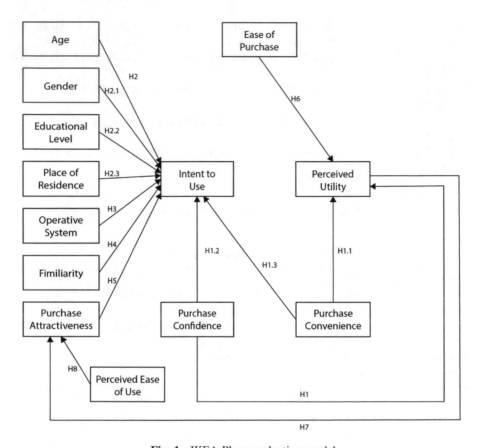

Fig. 1. IKEA Place evaluation model

This evaluation model aims to verify the existence of a positive relationship between the intention to use the application and the following factors: age, gender, educational level, place of residence, familiarity (if the individual has already bought it on an IKEA retail area), purchase attractiveness, purchase confidence, and purchase convenience. As for perceived utility, conclude whether there is a positive relationship between purchase confidence, convenience of purchase and ease of purchase. Finally, find out if there is a positive relationship between attractiveness and perceived ease of use and perceived utility factors.

The databases used to find the articles that support the Research reference framework ar Google Scholar, Science Direct, SCOPUS and ResearchGate. The keywords used to find the principal articles are augmented reality, interactive marketing, Ikea augmented reality, TAM, Technology Acceptance Model.

4   Methodology

This paper aims to analyze the use of the IKEA Place application, with the objective of verifying the intended use, perceived utility and purchase attraction, as well as analyzing the reactions of those who use it. The analysis of the use of the application was made observing the respondents to try the application. Simultaneously, an analysis was performed using a notebook, recording notes on the user experience. After using the application, respondents answered a questionnaire about their experience of using applications based on questions from various technology acceptance models.

Based on the questionnaire answered, the information was combined with the observations made to reach conclusions with the following objectives:

General objective: Check how the Portuguese user makes a purchase through an augmented reality e-commerce platform and if he feels the confidence and convenience to make that purchase.

Specific objectives:

(1) Identify whether technology changes offer the online shopper more confidence and convenience.
(2) Verify that sociodemographic factors influence a user in view of the intention to use an augmented reality e-commerce application.
(3) Check the influence of the operating system if a user wanted to purchase a product using an e-commerce application using augmented reality technology.
(4) Identify whether the user after physically knowing a commercial surface could buy a product online using an augmented reality application from that organization.
(5) Identify if the user after learning about an augmented reality ecommerce application is attracted to buy again using this technology.
(6) Check if the user finds it easier to make a purchase with the support of augmented reality technology.
(7) Identify if the user finds augmented reality to buy products useful.
(8) Identify whether the augmented reality e-commerce application that is easy to use makes the user more attracted to buying a product.

4.1   Research Methods and Data Collection Techniques

For this investigation it was used quantitative methodology and qualitative methodology, that is, a mixed methodology was made in an exploratory study.

Given the scenario presented, for the research under analysis an exploratory study was carried out that aims to recognize little studied realities and raise hypotheses for understanding these realities [27]. Exploratory research aims to provide clarifications and understandings about the object under study [28].

The data in question were quantitatively related through the data extracted from the questionnaire based on the constructs of the TAM and other models and studies obtained through the literature review.

The data collection typology used was a convenience sampling based on primary data. For data collection the IKEA Place study application was used. The respondent was told that he would have to install the app and add a chair to his environment. The chair name (Odger) was provided on a sheet if it did not appear on the application's "Highlights" screen. The respondent then proceeded to a simulation of buying the chair. Login data has been provided (by the investigator) to reduce data collection time. If the respondent's smartphone was not compatible, the investigator had two pre-prepared smartphones so that the application could be freely installed.

Data collection based on the qualitative method was based on non-participant observation. The non-participant observation took place at the time of using the IKEA Place application, in order to obtain participant's user experience comments and annotations.

Another research method that was used as a data collection technique was the questionnaire. The questionnaire was developed on the Google Forms platform and made available to respondents after testing the application.

4.2   Sample Characterization

The age group with the highest percentage of respondents was 56.1%, aged 25–34, followed by 22%, ages 18–24. With the smallest number of respondents were the ages between 35–44 years, which obtained a percentage of 9.8% followed by the age group of 45–54 years, with a percentage of 12.2%.

Regarding the gender of the respondents, 47.6% were female and 52.4% were male.

As for academic qualifications, 67.1% hold a bachelor's degree, 20.7% hold a Secondary or Vocational Education degree and 7.3% hold a master's degree. Each of the PhD, Higher Professional Technical Course and Technological Specialization Course items hold 1.2% of the sample.

Regarding the place of residence of respondents, it can be analyzed that the vast majority reside in the Braga area, with a total of 69.5% and the rest in the Porto area with a total of 30.5% of respondents.

Regarding the operating system, most respondents were iOS users with 58.5% and 41.5% were users of the Android operating system.

5   Analysis and Discussion of Results

For data analysis, we performed a descriptive analysis in order to obtain the values for the minimum, maximum, average and standard deviation - see Table 2.

Table 2. Descriptive analysis of the data obtained

	N	Minimum	Maximum	Average	Standard deviation
Ease of use	82	2,48	7,00	6,21	1,11
Perceived utility	82	2,77	7,00	5,73	1,05
Intent to use	82	2,58	7,00	5,55	1,17
Purchase attractiveness	82	2,50	7,00	6,38	0,72
Purchase convenience	82	4,08	7,07	6,59	0,57
Ease of purchase	82	2,00	7,00	6,29	0,78
Purchase confidence	82	3,68	6,93	6,19	0,56

From the information obtained the minimum value in the constructs does not vary significantly, except for the purchase confidence that has more points and the convenience of buying with two points more than the other constructs. At the maximum value there is no significant variation because all of these have values close to or equal to 7. The average ranges between 5.55 and 6.59 points, with a standard deviation close to 0, which means that the error is not significant.

Based on the research questions, 8 hypotheses were validated and 6 were not validated. Regarding the objective of verifying whether technology changes can offer potential online shoppers more confidence and convenience according to hypotheses H1, H1.1, H1.2 and H1.3, their validity is confirmed as respond clearly to the outcome of their correlations. Regarding the hypotheses H2, H2.1, H2.2 and H2.3, which relate the sociodemographic factors, with the intention to use the application under analysis, no significant differences were found for analysis. Significant age was found only for the age ranges between 18–24 years and 45–54 years, which is quite significant (,007) compared to the sample obtained, which is not enough to validate this hypothesis.

Hypothesis H3 demonstrates that based on the operating system used by users, there was no influence, because although Android has a smaller sample than iOS, both did not show a significant difference from the intention of using the application under analysis. The same may be true for hypothesis H4, because users made a purchase on a retail area, in this case IKEA was not considered valid, because there was no significant average difference from the intention to use the product. IKEA Place.

The opposite is true of hypothesis H5, which assesses whether a user, after trying an ecommerce application, is attracted or unwilling to buy with this new technology. Given the result of the correlation of the two constructs, it is clearly noted in relation to the sample obtained, a great attractiveness in using the application to make a next purchase. In hypothesis H6, the sample found that compared to the existing product sales channels, namely the traditional trade route, the interviewed individuals found it easier to buy a product with this type of technology. Regarding hypothesis H7, a significant existence has been found, which allows us to conclude that the user considers an AR application to be useful for future online product acquisition. Hypothesis H8 was valid, confirming that the perceived ease of use of an application using AR technology increases the attractiveness of the user to purchase.

6 Conclusions

The aim of this study was essentially to understand whether based on AR technology the user would feel greater buying confidence and greater purchasing convenience when purchasing a product online. From the results obtained in the study, users felt greater confidence and greater convenience of purchase when using the IKEA Place application.

Another conclusion of this study is that it is not clear whether the user after physically purchasing a product on a commercial surface could purchase a product online using an AR application from that organization. Thus, it is concluded that familiarity with IKEA is not at all necessary to purchase a product through the IKEA Place application.

With this study it was found that users find it easier to make purchases with the help of AR technology compared to the various existing channels. Another conclusion from this study is that respondents recognized the usefulness of the AR technology for purchasing products in the future, which validates that in the future they can use IKEA Place or another similar application to purchase IKEA products.

6.1 Limitations of the Study

One of limitations of study is the Educational Level, because the majority of respondents are BSc and MSc levels. Another limitation is about the studies on AR and MAR in e-commerce are scarce. Other important limitation is about the purchase are a simulation and not real and this factor may somewhat limit respondents' true intent.

6.2 Future Work

To study whether the assistance in assembling furniture using this application would be of value to both IKEA and the consumer. It would also be interesting to study the possibility of adding "Call to action" to see if the results of this study would be different.

References

1. Billinghurst, M., Clark, A., Lee, G.: A survey of augmented reality. Found. Trends® Hum.–Comput. Interact. **8**(2–3), 73–272 (2015). https://doi.org/10.1561/1100000049
2. Kotler, P., Kartajaya, H., Setiawan, I.: Marketing 3.0. Book (1a). Elsevier, Rio de Janeiro (2012). https://doi.org/10.1002/9781118257883
3. IKEA: IKEA apps (2018a). https://www.ikea.com/gb/en/customer-service/ikea-apps/. Accessed 22 Apr 2018
4. Dacko, S.G.: Enabling smart retail settings via mobile augmented reality shopping apps. Technol. Forecast. Soc. Change **124**, 243–256 (2017). https://doi.org/10.1016/j.techfore.2016.09.032
5. Rese, A., Baier, D., Geyer-Schulz, A., Schreiber, S.: How augmented reality apps are accepted by consumers: a comparative analysis using scales and opinions. Technol. Forecast. Soc. Change **124**, 306–319 (2017). https://doi.org/10.1016/j.techfore.2016.10.010
6. Raška, K., Richter, T.: Influence of augmented reality on purchase intention: The IKEA Case. Jönköping International Business School (2017)
7. IKEA Group: IKEA launches IKEA Place, a new app that allows people to virtually place furniture in their home (2017). http://newsroom.inter.ikea.com/news/ikea-launches-ikea-place–a-new-app-that-allows-people-to-virtually-place-furniture-in-their-home/s/f5f003d7-fcba-4155-ba17-5a89b4a2bd11. Accessed 12 Dec 2017
8. IKEA: IKEA place app launches on Android, allowing millions of people to reimagine home furnishings using AR. (2018b). https://www.ikea.com/us/en/about_ikea/newsitem/031918-IKEA-Place-app-launches-on-Android. Accessed 12 May 2018
9. Azuma, R.: A survey of augmented reality. Presence: Teleoper. Virtual Environ. **6**(4), 355–385 (1997). https://doi.org/10.1162/pres.1997.6.4.355
10. Reitmayr, G., Drummond, T.: Going out: robust model-based tracking for outdoor augmented reality. In: 2006 IEEE/ACM International Symposium on Mixed and Augmented Reality, pp. 109–118. IEEE (2006). https://doi.org/10.1109/ISMAR.2006.297801

11. Feng, Z., Duh, H.B.-L., Billinghurst, M.: Trends in augmented reality tracking, interaction and display: a review of ten years of ISMAR. In: 2008 7th IEEE/ACM International Symposium on Mixed and Augmented Reality, pp. 193–202. IEEE (2008). https://doi.org/10.1109/ISMAR.2008.4637362

12. Carmigniani, J., Furht, B., Anisetti, M., Ceravolo, P., Damiani, E., Ivkovic, M.: Augmented reality technologies, systems and applications. Multimed. Tools Appl. **51**(1), 341–377 (2011). https://doi.org/10.1007/s11042-010-0660-6

13. Porter, M., Heppelmann, J.: Why every organization needs an augmented reality strategy. Harvard Bus. Rev. **2017**(November-December), 1–13 (2017)

14. Davis, F.D.: Perceived usefulness, perceived ease of use, and user acceptance of information technology. MIS Q. **13**(3), 319 (1989). https://doi.org/10.2307/249008

15. Davis, F.D., Venkatesh, V.: A critical assessment of potential measurement biases in the technology acceptance model: three experiments. Int. J. Hum.-Comput. Stud. **45**(1), 19–45 (1996). https://doi.org/10.1006/ijhc.1996.0040

16. Gefen, D., Karahanna, E., Straub, D.: Trust and TAM in online shopping: an integrated model. MIS Q. **27**(1), 51 (2003). https://doi.org/10.2307/30036519

17. Huang, T.-L., Liao, S.: A model of acceptance of augmented-reality interactive technology: the moderating role of cognitive innovativeness. Electron. Commer. Res. **15**(2), 269–295 (2015). https://doi.org/10.1007/s10660-014-9163-2

18. Porter, C., Donthu, N.: Using the technology acceptance model to explain how attitudes determine Internet usage: the role of perceived access barriers and demographics. J. Bus. Res. **59**(9), 999–1007 (2006). https://doi.org/10.1016/j.jbusres.2006.06.003

19. Venkatesh, V.: Determinants of perceived ease of use: integrating control, intrinsic motivation, and emotion into the technology acceptance model. Inf. Syst. Res. **11**(4), 342–365 (2000). https://doi.org/10.1287/isre.11.4.342.11872

20. Venkatesh, V., Bala, H.: Technology acceptance model 3 and a research agenda on interventions. Decis. Sci. **39**(2), 273–315 (2008). https://doi.org/10.1111/j.1540-5915.2008.00192.x

21. Venkatesh, V., Davis, F.D.: A theoretical extension of the technology acceptance model: four longitudinal field studies. Manag. Sci. **46**(2), 186–204 (2000). https://doi.org/10.1287/mnsc.46.2.186.11926

22. Venkatesh, V., Morris, M.G., Davis, G.B., Davis, F.D.: User acceptance of information technology: toward a unified view. MIS Q. **27**(3), 425 (2003). https://doi.org/10.2307/30036540

23. Konzen, A.A., Pelegrini, P., Baggenstoss, S., Silva, R.T.P.: A realidade virtual aumentada como ferramenta de atratividade de compra no e-commerce. In: XXVIII Enangrad, Brasilia, p. 18 (2017)

24. Forsythe, S., Liu, C., Shannon, D., Gardner, L.C.: Development of a scale to measure the perceived benefits and risks of online shopping. J. Interact. Market. **20**(2), 55–75 (2006). https://doi.org/10.1002/dir.20061

25. Ghaffar Khan, A.: Electronic commerce: a study on benefits and challenges in an emerging economy. Global J. Manag. Bus. Res.: B Econ. Commerce **1**, 5 (2016)

26. Lim, W.M., Ting, D.H.: E-shopping: an analysis of the technology acceptance model. Modern Appl. Sci. **6**(4) (2012). https://doi.org/10.5539/mas.v6n4p49

27. Baptista, M.J., Sousa, C.S.: Como fazer investigação, dissertações, teses e relatórios segundo Bolonha (4a). Pactor, Lisboa (2011)

28. Malhotra, N.K.: Pesquisa de marketing: foco na decisão (3a). Pearson Education do Brasil, São Paulo (2011)

State-of-the-Art Applications of Spatial Data Infrastructure in the Provision of e-Government Services in Latin America

Mariuxi Bruzza[1](✉), Manuel Tupia[2], and Glenn Vancauwenberghe[3]

[1] Universidad Laica "Eloy Alfaro" de Manabí, Ciudadela Universitaria,
Manta, Ecuador
mariuxi.bruzza@uleam.edu.ec

[2] Department of Engineering, Pontificia Universidad Católica del Perú,
Av. Universitaria, 1801 San Miguel, Lima, Peru
tupia.mf@pucp.edu.pe

[3] Katholieke Universiteit Leuven, Spatial Applications Division
Leuven (SADL), Celestijnenlaan 200 E, 3001 Leuven, Belgium
glenn.vancauwenberghe@kuleuven.be

Abstract. Framed within the Open Government and Open Data initiatives and supported by data sharing and interoperability platforms at different administrative levels, the use of spatial data in the provision of e-government services strongly increased in recent years. Better understanding these practical implementations of spatial data contributes to the effective development of Spatial Data Infrastructures (SDI) that promote and facilitate the accessibility of spatial data. This paper introduces an in-depth review of the state-of-the-art applications of spatial data based on a literature review of scientific papers and articles on related topics. For the elaboration of this state of the art, the systematic review was used with the PICOC criteria commonly used for such purposes. Five databases of indexed papers of the first order were reviewed emphasizing the implementations of spatial data and open spatial data in both Ecuador and Peru.

Keywords: Spatial data · Open Government · Open data · eGovernment · Spatial Data Infrastructure

1 Introduction

Technical, scientific and even governmental communities deal with a great number of technical and non-technical problems related to data management. Particularly the access to a wide range of information brings higher complexity, needs of transparency, integrity and interpretation problems to explore said information [1]. The insufficient availability of spatial data has become an issue that affects many organizations of the public and private sphere when providing electronic government services that require this kind of information. Especially for public organizations data sharing between organization become a way to facilitate the access to this type of data [2].

The implementation of interoperability platforms for government data is required to promote the sharing of data, and of spatial data in particular. This paper uses the

© Springer Nature Switzerland AG 2020
Á. Rocha et al. (Eds.): ICITS 2020, AISC 1137, pp. 124–140, 2020.
https://doi.org/10.1007/978-3-030-40690-5_13

PICOC (Population-Intervention-Context-Outcomes-Comparison) method to review the state-of-the-art of the spatial data implementation literature in Latin America [6]. Better understanding the implementation of applications that make use of spatial data will contribute to the further development of Spatial Data Infrastructures required to ensure the availability and accessibility of spatial data. This paper is organized in the following manner: Sect. 1 introduces a brief description of open data providing a comprehensive overview on which spatial data are based; Sect. 2 details the systematic review process based on the PICOC method; Sect. 3 shows the outcomes of the state-of-the-art in a summarized manner; Sect. 4 introduces conclusions, recommendations and future work based on the review presented.

2 Systematic Review Based on the PICOC Method

2.1 A Systematic Review

A systematic review aims to bring evidence together to answer a pre-defined research question. This involves the identification of all primary research relevant to the defined review question, the critical appraisal of this research, and the synthesis of the findings. This process may combine data from different studies and repositories in order to produce new results o conclusions, or even new evidence about an existing (non-solved) problem [17]. Normally, systematic reviews are addressing by research questions and include reproduced methods for identification and refinement of primary research studies. There are many different frameworks for define systematic review questions as in Fig. 1, based primarily on the use of the PICOC criteria.

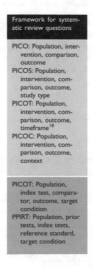

Fig. 1. Frameworks for systematic review [18]

A systematic review should be prompted by an interest in a topic, and a wish to answer a specific question. The question should clarify the problem to be addressed, specifying the population to which the question applies, as well as any intervention and outcomes of interest. Figure 2 refers to whole process:

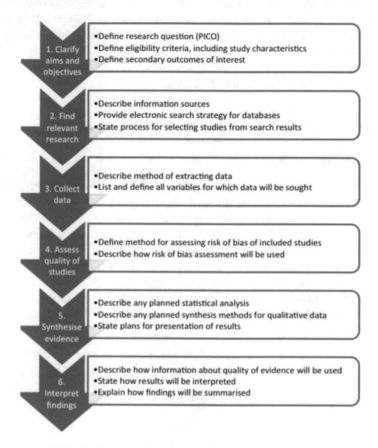

Fig. 2. Key stage for a protocol and systematic review

Table 1 lists PICOC criteria used in the systematic review:

Table 1. Attributes according to the PICOC criteria.

Attributes	Description
Population	Implementation of spatial data from 2010 to 2019
Intervention	Models used for the implementation of spatial data
Context	Implementation cases where spatial data has been used
Outcomes	Technological applications with spatial data
Comparison	Use of spatial data in Latin America

2.2 Research Questions

The purpose is to identify spatial data applications and the implementation of related technologies cited in the literature as well as determine its use through the application of models that work as guidelines for the implementation in Latin America. Table 2 details research questions for the systematic review:

Table 2. Research questions

No.	Research questions
1	What are the applications and/or technological solutions of spatial data that have been implemented?
2	What are the technological models or framework the spatial data implementation cases have been based on?
3	What collaboration models have been adopted in the provision of services in public administration?

2.3 Strategy and Search Chains

For the process of search through primary papers the following steps were taken:

1. PICOC criteria were used as search terms including both synonyms and abbreviations
2. For the search chains, Boolean operators have been used (OR, AND)

The search chain used for the systematic review of literature has been custom-made to adapt to the syntaxes of each database selected as shown in Table 3:

Table 3. Search string

Database	Search string	Filters	Total papers	Selected papers
Gale	Keyword (spatial data) Or Keyword ("Geospatial data") Or Keyword ("Spatial Data Technologies") Or Keyword ("spatial data infrastructure") Or Keyword ("spatial data analysis") Or Keyword ("Ecuador Peru")	2010–2019 Scientific journals Papers evaluated by experts	1197	30
IEEE	((((spatial data) OR spatial data infrastructures) OR spatial data applications) OR spatial data implementation models)	2010–2019 Scientific journals	7256	17
Proquest	("spatial data models") OR (spatial data infrastructure) OR (spatial data Ecuador) AND (spatial data Peru) OR (spatial data implementation)	2010–2019 Scientific journals	219.579	21
Springer	spatial AND data AND models AND "spatial data" AND (SDI)		56	27
Other sources	Spatial Data		8	5

2.4   Selection Criteria

Inclusion and exclusion criteria were specified according to the research questions as detailed in Table 2. For the final selection of papers reviewed, the following inclusion criteria were considered:

- Implementation cases of spatial data.
- Papers proposing technological applications of spatial data.
- Implementation cases that use the model and/or framework for (support/ management of) implementation of spatial data.
- Proposals of implementation models of spatial data.
- Critical factors of success or failure in the implementation of spatial data whether (or not) for specific cases.
- Papers presenting subjects related to the models used for technological applications of spatial data.
- Papers demonstrating improvements in open governments when adopting spatial data initiatives.

  Exclusion criteria were the following:

- Papers not indicating implementation cases of spatial data.
- Papers not containing any model or framework to implement spatial data.
- Papers not describing any technological application for the use of spatial data.

2.4.1   Selection Process Conducted

The survey identified a total of 228.096 primary papers. After verification, 227.996 papers were ruled out because they did not comply with inclusion criteria. The remaining 100 papers are the surveys approved for this review. Table 4 displays extracts of these papers:

Table 4. Selected Papers applying the inclusion and exclusion criteria.

Paper	Reference	Publication year
A framework for integrating multi-accuracy spatial data in geographical applications	[7]	2011
An interactive framework for spatial joins: a statistical approach to data analysis in GIS	[8]	2011
An online calculator for spatial data and its applications	[9]	2014
Co-clustering spatial data a generalized linear mixed model with application to the integrated pest management	[10]	2012
Design of two dimensional visualization system for embedded marine environment spatial data	[11]	2018
Others	[12–105]	Several years

2.4.2 Evaluation of Papers Selected

Primary studies were evaluated according to the proposal by Aznar, Mendes and Riddle [104], who put forward the so-called evaluation checklist of the quality of primary studies. This checklist provides a means to assess quantitatively the quality of the papers chosen for the systematic review where a three-tiered scale has been established:

- Yes, 1 point.
- No, 0 points
- Partially, 0.5 points.

Each paper may score between 0 to 10 points depending on the compliance of each of the research questions and inclusion criteria. The higher the global score of the paper, the higher the degree to which the paper covers the objectives of the research questions. In addition, the first quartile of papers will be selected (those scoring 2.5) to serve as a cutting point with the intention of establishing the final literature, excluding those with a lower score. For this study, out of the 100 papers selected, 13 were removed because they scored lower than 2.5, with 87 remaining papers after applying the evaluation process. Table 5 shows an extract of this evaluation process by score:

Table 5. Evaluation of papers according to quality criteria.

Paper	P1	P2	P3	P4	P5	P6	P7	P8	P9	P10	Total
A framework for integrating multi-accuracy spatial data in geographical applications	0	1	0	0	1	1	0	1	1	1	**6**
An interactive framework for spatial joins: a statistical approach to data analysis in GIS	0	1	1	0	1	0	0	1	1	1	**6**
An online calculator for spatial data and its applications	1	0	1	0	1	0	0	1	1	0	**5**
Co-clustering spatial data a generalized linear mixed model with application to the integrated pest management	1	1	1	0	1	0	0	0	1	1	**6**
Design of two-dimensional visualization system for embedded marine environment spatial data	1	1	1	0	1	0	0	1	1	0	**6**

2.5 Results of the Systematic Review

From the research questions that were proposed previously and following the PICOC methodology, the outcomes obtained from the papers selected are presented below according to the question.

2.5.1 Results of Question 1

For question 1, *What are the applications and/or technological solutions of spatial data that have been implemented?* 33 surveys were identified that propose solutions corresponding to common operations in GIS (Geospatial Information Systems) for the analysis and localization of data to support decision making; as well as technological application for environments of Spatial Data Infrastructures (SDI). All these cases of implementation aim at improving the recovery, availability, integration and maintenance of spatial data from different sources. Table 6 shows some results:

Table 6. Answers to question 1 (extract)

Paper	Applications and/or technological solutions
An interactive framework for spatial joins: a statistical approach to data analysis in GIS	Interactive framework that provides faster approximate answers/responses of spatial joins for data analysis and decision support
An online calculator for spatial data and its applications	Calculator to calculate spatial auto-correlations, to create maps of habitat aptitude and to generalize movement data in spatial-temporal clusters
Co-clustering spatial data a generalized linear mixed model with application to the integrated pest management	Relevant application to integrated pest management, combining the technique of spatial co-clustering with a statistical interference method to make the evaluation of pest densities more accurate
Design of two dimensional Visualization system for embedded marine environment spatial data	Design of two dimensional visualization system for the visualization of spatial data of the embedded marine environment
Improving geographic information retrieval in spatial data infrastructure	SESDI (Spatially Enabled Spatial Data Infrastructure), to reduce the important limitations that make it difficult for users to find the geospatial data they are looking for

2.5.2 Results of Question 2

For question 2, *What technological models or framework have the spatial data implementations been based on?* It seeks out to identify whether the applications and technological solutions implemented are based on any model or framework. Table 7 lists some of the papers answering this question:

Table 7. Results of Question 2

Paper	Framework or models used
A framework for integrating multi-accuracy spatial data in geographical applications	A model to represent a spatial database of multiple precision called MACS and a methodology to integrate two MACS databases containing metric and logic observations and discuss how these two types of information can be combined and be consistent in the resulting database
An interactive framework for spatial joins: a statistical approach to data analysis in GIS	Interactive framework providing faster approximate answers of spatial joins allowing users to interchange speed against a noted precision, thus providing a truly interactive data exploration
Co-clustering spatial data a generalized linear mixed model with application to the integrated pest management	Co-clustering method using a linear mixed model generalized for spatial data. To avoid high computational demands related to global optimization, we propose a heuristic optimization algorithm to seek an almost optimum co-clustering
Design of a two-dimensional visualization system for embedded marine environment spatial data	Design method of spatial data visualization system of integrated marine environment based on script language
Diffusion of spatial data infrastructure in Panamá	Diffusion theories of innovation may provide a highly useful framework for the survey and development of national and regional SDIs

2.5.3   Results of Question 3

Question 3 *What collaboration models have been adopted in the provision of services in public administration?* seeks out to whether the use of spatial data is enabled and/or supported by collaboration initiatives. Table 8 contains extracts of selected papers addressing this question:

Table 8. Results of question 3

Paper	Provision of services in public administration
Spatial distribution of a creativity index at municipal level in Mexico	Estimating an index at municipal level in Mexico, variables related to Talent, Tolerance and Technology are integrated so that regions can demonstrate how to become more competitive and how to improve their creativity index. To this end, getting to know their spatial distribution and showing best positioned municipalities through SDI solutions is a critical factor
Increasing access to and use of geospatial data by municipal government and citizens: The process of "Geomatization" in Rural Quebec	Access to sufficient high-quality geospatial data can be a major challenge at municipal level. In Canada and Quebec, this is particularly relevant given the uneven integration of municipal governments within provincial and federal spatial data infrastructures (SDIs). Recent initiatives have focused on light web-based methods to link municipal governments to these provincial and federal SDIs that increasingly collect more data for cloud-based delivery
Assessment of national policies of Geospatial information of Ecuador linked to the implementation of institutional IDE	The management of the country's territory is closely related to the geographical information (GI) available and its accessibility. These would be the fundamental principles of a Spatial Data Infrastructure (SDI) supported by a consolidated legal framework
Evaluating access to spatial data information in Rwanda	Access to spatial data is of increasing interest to professionals and the society, since the use of geospatial technology is present across all fields, and all economic sectors can use the same information in different applications. Means of access to data suitable for any given context must be found

3   Discussion

The procedure followed for the systematic review is shown in Fig. 3:

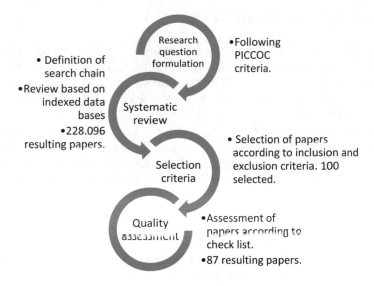

Fig. 3. Outline of the systematic review

As can be seen the search chain is perfectly adjustable to different databases indexed for scientific papers. PICOC criteria were applied depending on the used chain in order to fulfill the research objectives. A list of papers (228,096 papers) was subsequently refined in order to obtain a final list of papers that were included in the review. The first selection filter that uses an evaluation with exclusion and inclusion criteria yielded 100 papers, and the second filter that works according to a quality evaluation list yielded 87 papers.

Below are additional characteristics that have been identified on the literature:

- The results found show a difference between:
  - Use of open spatial data
  - Implementation of technological solutions based on open data
  - Implementation of a national infrastructure of open data, spatial data or interoperability.
  - In Latin America and in the two countries in which emphasis has been placed (Ecuador and Peru), the Open Government is underdeveloped, which directly impacts the application of spatial data.
- Papers referring to implementation cases have proposed applications and/or technological solutions with regards to spatial data infrastructures (SDI) as solutions to integrate and share spatial data coming from different open data repositories.
- For different cases of spatial data implementation, authors have built on or proposed frameworks and/or models enabling the incorporation of technology, standards and institutional plans from different government levels as well as facilitating availability and sharing of spatial data.
- The significance of spatial data infrastructures is remarkable, and many aspects could be considered as critical factors of success such as data, political communities, data suppliers, decision-making mechanisms, and so on.

4 Conclusions and Future Works

The main conclusion of the systematic review is that in Latin American Argentina, Colombia and Uruguay are the pioneers in adopting IDE initiatives, which will consolidate in future years, followed by Mexico and Cuba. In 2007, Peru's Spatial Data Infrastructure (IDEP), run by Secretary of Digital Government (https://www. gobiernodigital.gob.pe/) was created, and among 2012 and 2014 national policies regarding IDEP were enacted. Currently, in Ecuador new technologies and applications in the field of SIGs and SDIs are being developed, thus several institutions dedicated to geo-information are following the example of the Instituto Geográfico Militar (Military Geographic Institute or IGM) in the creation of an institutional SDI to publish, visualize and analyze spatial data generated in the different projects that are carried out [86, 88].

Significant restrictions exist related to spatial data infrastructure that make it harder for users to find the geospatial data they are interested in. Although there is literature on web services and spatial data infrastructure, research based on guideline data used for dynamic IDE is limited. For sharing and integration of heterogeneous spatial data from multiple sources, WebGIS is implemented.

New methodologies and/or models to acquire, integrate and update spatial datasets coming from different sources are required. It is not a common practice within the integration process to provide accurate information on spatial data stored within a spatial database.

Spatial data infrastructures require formal, legal and organizational structures as well as the presence of good personal contacts (informal) and people involved in the SDI concept and its benefits. To facilitate and coordinate spatial data exchange, different government levels everywhere in the world are developing spatial data infrastructure (SDI).

In future works we intend to develop a spatial data infrastructure model (SDI) focused on the provision of electronic government services in the G2C (government to citizen) and G2G (government to government) modalities in Peru and Ecuador. The first stage intends to develop a generic model and the second one will test two specific models for each country.

References

1. Schmidt, B., Gemeinholzer, B., Treloar, A.: Open data in global environmental research: the Belmont Forum's open data survey. PLoS ONE **11**(1), 1–29 (2016)
2. de Montalvo, U.W.: Mapping the Determinants of Spatial Data Sharing, 2nd edn. Routledge, London (2017)
3. Gough, D., Oliver, S., Thomas, J.: An Introduction to Systematic Reviews. SAGE Publications Ltd., London (2012)
4. Ubaldi, B.: Towards Empirical Analysis of Open Government Data Initiatives. OECD Working Papers on Public Governance, vol. 22, pp. 1–60 (2013)
5. EPPI Centre.: What is a systematic review? https://eppi.ioe.ac.uk/cms/. Accessed 26 June 2019

6. Petticrew, M., Roberts, H.: Systematic Reviews in the Social Sciences: A Practical Guide, pp. 1–352. Blackwell Publishing, Hoboken (2006)
7. Belussi, A., Migliorini, S.: A framework for integrating multi-accuracy spatial data in geographical applications. GeoInformatica 16(3), 523–561 (2011)
8. Alkobaisi, S., Bae, W.D., Vojtěchovský, P., Narayanappa, S.: An interactive framework for spatial joins: a statistical approach to data analysis in GIS. GeoInformatica 16(2), 329–355 (2011)
9. Remm, K., Kelviste, T.: An online calculator for spatial data and its applications. Comput. Ecol. Softw. 4(1), 22–34 (2014)
10. Zhang, Z., Jeske, D.R., Cui, X., Hoddle, M.: Co-clustering spatial data using a generalized linear mixed model with application to the integrated pest management. J. Agric. Biol. Environ. Stat. 17(2), 265–282 (2012)
11. Yan, L.: Design of two-dimensional visualization system for embedded marine environment spatial data. J. Coast. Res. 735–740 (2018)
12. Velazco, F., Yanet, S., Bal Calderon, E., Joyanes Aguilar, L., Abuchar Porras, A.: Difusión de la infraestructura de datos espaciales en Panamá. Revista Electronica Redes de Ingeniería 7(1), 94–103 (2016)
13. Villarreal González, A., Flores Segovia, M.A., Gasca Sánchez, F.M.: Distribución espacial de un índice de creatividad a nivel municipal en México. Estudios Demográficos y Urbanos 33(1), 149–153 (2018)
14. Narváez, R., León, F., Bernabé, M., Rubio, M.L.: Evaluación de las Políticas Nacionales de Información Geoespacial de Ecuador vinculadas con la implementación de IDE institucionales. Revista Cartográfica 1(92), 53–69 (2016)
15. Malinowski Gajda, E.: Evaluación de los sistemas de administración de bases de datos con extensiones espaciales. Revista de la Universidad de Costa Rica 24(2), 13–33 (2014)
16. Akinyemi, F.: Evaluating access to spatial data information in Rwanda. URISA J. 3(2), 39–47 (2011)
17. Shorbi, M., Wan, H.: Geospatial data infrastructure for natural disaster management. Adv. Environ. Biol. 257 (2014)
18. Stojanovic, N., Dragan, S.: High performance processing and analysis of geospatial data using CUDA on GPU. Adv. Electr. Comput. Eng. 14(4), 109 (2014)
19. De Andrade, F., De Souza, C., Davis, C.: Improving geographic information retrieval in spatial data infrastructures. GeoInformatica 18(4), 793–818 (2014)
20. Johnson, A., Sieber, R.: Increasing access to and use of geospatial data by municipal government and citizens: the process of "geomatization" in rural Québec. URISA J. 5(2), 53–60 (2012)
21. Memduhoglu, A., Basaraner, M.: Possible contributions of spatial semantic methods and technologies to multi-representation spatial database paradigm. Int. J. Eng. Geosci. 3(3), 108–118 (2018)
22. Coetzee, S.: Reference model for a data grid approach to address data in a dynamic SDI. GeoInformatica 16(1), 111–129 (2011)
23. Taktak, F., Demir, H.: Relations that show the network potential for spatial data sha-ring. Tehnicki vjesnik - Technical Gazette 26(2), 346–354 (2019)
24. Mahboubi, H., Bimonte, S., Deffuant, G., Chanet, J., Pinet, F.: Semi-automatic design of spatial data cubes from simulation model results. Int. J. Data Warehouse. Min. 9(1), 70–95 (2013)
25. Mukherjee, F.: Sociopolitical contexts and geospatial data-the case of Dane county. URISA J. 25(1), 39–46 (2013)
26. Wang, S., Yuan, H.: Spatial data mining: a perspective of big data. Int. J. Data Warehous. Min. 10, 50–70 (2014)

27. Tavra, M., Duplančić, T., Cetl, V.: Stakeholders needs requisite analysis: towards croatian marine spatial data infrastructure establishment. Tehnicki vjesnik - Technical Gazette **25**(1), 176–182 (2018)
28. Castelein, W., Bregt, A., Grus, L.: The role of collaboration in spatial data infrastructures. URISA J. **25**(2), 31–40 (2013)
29. Hong, T., Hart, K., Soh, L., Samal, A.: Using spatial data support for reducing uncertainty in geospatial applications. GeoInformatica **18**(1), 63–92 (2013)
30. Mir, U., Abbasi, U., Yang, Y., Bhatti, Z.A., Mir, T.: Spatial big data and moving objects: a comprehensive survey. IEEE Access **6**, 58835–58857 (2018)
31. Bao, L., Le, Y.: A spatial big data framework for maritime traffic data. In: 3rd International Conference on Computational Intelligence and Applications (ICCIA), pp. 244–248 (2018)
32. Zhang, Y., Kunqing, X., Xiujun, M., Dan, X., Cuo, C., Shiwei, T.: Spatial data cube: provides better support for spatial data mining. In: Major Project of National Natural Science Foundation, pp. 795–798 (2010)
33. Yu, X., Zhang, T.: The application of GML in spatial data conversion. In: 2013 International Conference on Computer Sciences and Applications, pp. 788–791 (2013)
34. Tang, J., Ren, Y., Yang, C., Shen, L., Jiang, J.: A WebGis for sharing and integration of multisource heterogeneous spatial data. In: IGARSS, pp. 2943–2946 (2011)
35. Yang, J., Lin, H., Xiao, Y., Fu, X., Xu, L.: Spatial data model for visualization system of GIS based urban pipe network. In: 2010 International Forum on Information Technology and Applications (2010)
36. Wu, X., Xia, L., Wu, L.: Service-based global spatial data directory in spatial information grid. In: Information Engineering, pp. 1–4 (2010)
37. Wang, X., Wu, X., Zhang, Y., Tian, Y.: A distributed spatial data integrated management prototype for geological applications. In: Development and Resource Exploration Application of Satellite-based Earth Observation Technologies, pp. 1–7 (2014)
38. Liu, F., Hao, F.: Study of storage and management system for mass spatial data. In: 2017 9th International Conference on Intelligent Human-Machine Systems and Cybernetics (IHMSC), pp. 89–92 (2017)
39. Maries, A., Mays, N., Olson, M., Wong, K.: GRACE: a visual comparison framework for integrated spatial and non-spatial geriatric data. IEEE Trans. Vis. Comput. Graph. **19**(12), 2916–2925 (2013)
40. Zeng, Y., Li, G., Guo, L., Huang, H.: An on-demand approach to build reusable, fast-responding spatial data services. IEEE J. Sel. Top. Appl. Earth Obs. Remote Sens. **5**(6), 1665–1677 (2012)
41. D'Amore, F., Cinnirella, S., Pirrone, N.: ICT methodologies and spatial data infrastructure for air quality information management. IEEE J. Sel. Top. Appl. Earth Obs. Remote Sens. **5**(6), 1761–1771 (2012)
42. Lee, K., Liu, L., Ganti, R.K., Srivatsa, M., Zhang, Q., Zhou, Y., Wang, Q.: Lightweight indexing and querying services for big spatial data. IEEE Trans. Serv. Comput. **12**(3), 343–355 (2019)
43. Silva, J.B., Giannotti, M.A., Larocca, A.P.C., Quintanilha, J.A.: Towards a spatial data infrastructure for technological disasters: an approach for the road transportation of hazardous materials. GeoJournal **82**(2), 293–310 (2015)
44. Boisvert, E., Brodaric, B.: GroundWater Markup Language (GWML) – enabling groundwater data interoperability in spatial data infrastructures. J. Hydroinf. **14**(1), 93–107 (2012)
45. Wan, Y., Shi, W., Gao, L., Chen, P., Hua, Y.: A general framework for spatial data inspection and assessment. Earth Sci. Inf. **8**(4), 919–935 (2015)

46. Bachurina, S.S., Belyaev, V.L., Karfidova, E.A.: Geoecological aspects of the development of a regional model of spatial planning: case study of Moscow. Water Res. **44**(7), 978–986 (2017)
47. Abramic, A., Kotsev, A., Cetl, V., Kephalopoulos, S., Paviotti, M.: A spatial data infrastructure for environmental noise data in Europe. Int. J. Environ. Res. Public Health **14**(7), 726 (2017)
48. Jaljolie, R.: Spatial data structure and functionalities for 3D land management system implementation: Israel case study. Ecol. Environ. Conserv. **233** (2018)
49. Bhattacharya, D., Painho, M.: Augmented smart cities integrating sensor web and spatial data infrastructure (SmaCiSENS). MDPI AG **1**(1), 1–15 (2017)
50. Anecka, K., Cerba, O., Jedlicka, K., Jezek, J.: Towards interoperability of spatial planning data: 5-steps harmonization framework. In: Sofia: Surveying Geology and Mining Ecology Management (SGEM) (2013). https://search-proquest-com.ezproxybib.pucp.edu.pe/docview/1464931378?accountid=28391
51. Paudyal, D.R., McDougall, K., Apan, A.: The impact of varying statutory arrangements on spatial data sharing and access in regional NRM bodies. ISPRS Ann. Photogram. Remote Sens. Spat. Inf. Sci. II-8, 193–197 (2014)
52. Meiner, A.: Spatial data management priorities for assessment of Europe's coasts and seas. J. Coast. Conserv. **17**(2), 271–277 (2011)
53. Elemen, C.A.G.: La infraestructura de datos espaciales (IDE) de México. modelo conceptual. Revista Geográfica (2016). https://search-proquest-com.ezproxybib.pucp.edu.pe/docview/2164476493?accountid=28391
54. Mathys, T., Kamel Boulos, M.N.: Geospatial resources for supporting data standards, guidance and best practice in health informatics. BMC Res. Notes **4**(1), 1–5 (2011)
55. Tavra, M., Cetl, V., Duplancic, T.L.: A framework for evaluation of marine spatial data geoportals using case studies. GeoSci. Eng. **60**(4), 9–18 (2014)
56. Corti, P., Lewis, B.G., Athanasios, T.K., Ntabathia, J.M.: Implementing an open source spatio-temporal search platform for spatial data infrastructures. PeerJ PrePrints (2016)
57. Corcoran, P.A.: An assessment of the accessibility of spatial data from the Internet to facilitate further participation with geographical information systems for novice indigenous users in South Australia. Webology **9**(2), 1–19 (2012)
58. Nakayama, Y., Nakamura, K., Saito, H., Fukumoto, R.: A web GIS framework for participatory sensing service: an open source-based implementation. Geosciences **7**(2), 22–28 (2017)
59. Jaroszewicz, J., Denis, M., Zwirowicz-Rutkowska, A.: Harmonization of spatial planning data model with inspire implementing rules in Poland. In: Sofia: Surveying Geology and Mining Ecology Management (SGEM) (2013). https://search-proquest-com.ezproxybib.pucp.edu.pe/docview/1464931742?accountid=28391
60. Chafiq, T., Groza, O., Oulid, H.J., Fekri, A., Rusu, A., Saadane, A.: Spatial data infrastructure. Benefits and strategy. Analele Stiintifice Ale Universitatii "Al.I.Cuza" Din Iasi.Serie Noua.Geografie **61**(1), 21–30 (2015)
61. Sterlacchini, S., Bordogna, G., Cappellini, G., Voltolina, D.: SIRENE: a spatial data infrastructure to enhance communities' resilience to disaster-related emergency. Int. J. Disaster Risk Sci. **9**(1), 129–142 (2018)
62. Abdolmajidi, E., Harrie, L., Mansourian, A.: The stock-flow model of spatial data infrastructure development refined by fuzzy logic. SpringerPlus **5**(1), 267 (2016)
63. Bhanumurthy, V., Rao, K.R.M., Sankar, G.J., Nagamani, P.V.: Spatial data integration for disaster/emergency management: an Indian experience. Spat. Inf. Res. **25**(2), 303–314 (2017)

64. Bychkov, I.V., Plyusnin, V.M., Ruzhnikov, G.M., Fedorov, R.K., Khmel'nov, A.E., Gachenko, A.S.: The creation of a spatial data infrastructure in management of regions (exemplified by Irkutsk oblast). Geograph. Nat. Res. **34**(2), 191–195 (2013)
65. Cooper, A.K., et al.: Exploring the impact of a spatial data infrastructure on value-added resellers and vice versa. In: Buchroithner, M., Prechtel, N., Burghardt, D. (eds.) Cartography from Pole to Pole. Lecture Notes in Geoinformation and Cartography. Springer, Heidelberg (2014)
66. Fugazza, C.: Toward semantics-aware annotation and retrieval of spatial data. Earth Sci. Inf. **4**(4), 225–239 (2011)
67. Fugini, M., Hadjichristofi, G. Teimourikia, M.: Adaptive security for risk management using spatial data. In: Decker, H., Lhotská, L., Link, S., Spies, M., Wagner, R.R. (eds.) Database and Expert Systems Applications. DEXA 2014. LNCS, vol. 8644 (2014)
68. Keenan, P., Miscione, G.: Spatial data: market and infrastructure. In: Encyclopedia of GIS, pp. 1–8 (2015)
69. Malvárez, G.C., Pintado, E.G., Navas, F., Giordano, A.: Spatial data and its importance for the implementation of UNEP MAP ICZM protocol for the mediterranean. J. Coast. Conserv. **19**(5), 633–641 (2015)
70. Niko, D.L., Hwang H., Lee Y. Kim C.: Integrating user-generated content and spatial data into web GIS for disaster history. In: Lee, R. (ed.) Computers, Networks, Systems, and Industrial Engineering 2011. Studies in Computational Intelligence, vol. 365. Springer, Heidelberg (2011)
71. Van Oort, P.A.J., Hazeu, G.W., Kramer, H., Bregt, A.K., Rip, F.I.: Social networks in spatial data infrastructures. GeoJournal **75**(1), 105–118 (2010)
72. Sinvula, K.M., Coetzee, S., Cooper, A.K., Nangolo, E., Owusu-Banahene, W., Rautenbach, V., Hipondoka, M.: A contextual ICA stakeholder model approach for the Namibian Spatial Data Infrastructure (NamSDI). In: Cartography from Pole to Pole, pp. 381–394 (2013)
73. Sridharan, N.: Can smart city be an inclusive city?—Spatial Targeting (ST) and Spatial Data Infrastructure (SDI). In: Advances in 21st Century Human Settlements, pp. 233–244 (2014)
74. Vancauwenberghe, G., Dessers, E., Crompvoets, J., Vandenbroucke, D.: Realizing data sharing: the role of spatial data infrastructures. In: Public Administration and Information Technology, pp. 155–169 (2014)
75. Würriehausen, F., Karmacharya, A. Müller, H.: Using ontologies to support land-use spatial data interoperability. In: Murgante, B., et al. (ed.) Computational Science and Its Applications – ICCSA 2014. ICCSA 2014. LNCS, vol. 8580. Springer, Cham (2014)
76. Zwirowicz-Rutkowska, A., Michalik, A.: The use of spatial data infrastructure in environmental management: an example from the spatial planning practice in Poland. Environ. Manag. **58**(4), 619–635 (2016)
77. Zwirowicz-Rutkowska, A.: A multi-criteria method for assessment of spatial data infrastructure effectiveness. Earth Sci. Inform. **10**(3), 369–382 (2017)
78. Crompvoets, J., Vancauwenberghe, G., Bouckaert, G., Vandenbroucke, D.: Practices to develop spatial data infrastructures: exploring the contribution to e-government. In: Assar, S., Boughzala, I., Boydens, I. (eds.) Practical Studies in E-Government. Springer, New York (2011)
79. Vancauwenberghe, G., Van Loenen, B.: Governing open spatial data infrastructures: the case of the United Kingdom. In: Rodríguez Bolívar, M., Bwalya, K., Reddick, C. (eds.) Governance Models for Creating Public Value in Open Data Initiatives. Public Administration and Information Technology, vol. 31. Springer, Cham (2019)

80. Coetzee, S., Steiniger, S., Köbben, B., Iwaniak, A., Kaczmarek, I., Rapant, P., Moellering, H.: The academic SDI—towards understanding spatial data infrastructures for research and education. In: Advances in Cartography and GIScience, pp. 99–113 (2017)

81. Vancauwenberghe, G., Van Loenen, B.: Exploring the emergence of open spatial data infrastructures: analysis of recent developments and trends in Europe. In: Integrated Series in Information Systems, vol. 23–45 (2017)

82. Schäffer, B., Baranski, B., Foerster, T.: Towards spatial data infrastructures in the clouds. Lecture Notes in Geoinformation and Cartography, pp. 399–418 (2010)

83. de Oliveira, W.M., Filho, J.L., de Paiva Oliveira, A.: A spatial data infrastructure situation-aware to the 2014 world cup. In: Computational Science and Its Applications – ICCSA, pp. 561–570 (2012)

84. Bareth, G., Doluschitz, R.: Spatial data handling and management. In: Precision Crop Protection - The Challenge and Use of Heterogeneity, pp. 205–222 (2010)

85. Dutta, D., Pandey, S.: Development of State Spatial Data Infrastructure (SSDI): Indian experience. In: Geospatial Infrastructure, Applications and Technologies: India Case Studies, pp. 31 44 (2018)

86. Espiritu, E.: Gestión para la creación de nodos institucionales en datos espaciales pata la temática forestal en el Perú. (Tesis de pregrado). Universidad Nacional Agraria La Molina. Perú (2018)

87. Oficina Nacional de Gobierno Electrónico e Informática-ONGEI: Gestión de la información geoespacial: guía de buenas prácticas para la implementación de infraestructuras de datos espaciales institucionales (no 2015-17681) (2015). Accessed 26 June 2019. www.giz.de/peru/pe

88. Ron, C., Chávez, F.: Infraestructura de datos espaciales de tipo biotico para los planes ecorregionales Pacífico Ecuatorial (PE) y Cordillera Real Oriental (CRO) de The Nature Conservancy TNC bajo políticas de geoinformación. Universidad de las Fuerzas Armadas del Ecuador, Ecuador (2012)

89. Avila, F.: Desarrollo e implementación de una infraestructura de datos espaciales para gobierno autónomo descentralizado municipal: aplicación particular cantón Guachapala Teses de maestria. Universidad de Cuenca, Ecuador (2014)

90. Ballari, D., Vilches, L., Randolf, D., Pacheco, D., Fernández, V.: Tendencias en infraestructuras de datos espaciales en el contexto latinoamericano. MASKANA, I+D +ingeniería 1(1), 177–184 (2014)

91. Sreedhar, M., Viswanath, R.: A schematic study and performance analysis of spatial data indexing. Int. J. Comput. Intell. Res. 6(3), 455–459 (2010)

92. Instituto Panamericano de Geografía e Historia.: Estudio de Factibilidad para la implementación de una infraestructura de Datos Espaciales (IDE) en la provincia de Mendoza. Revista Cartográfica 88(2), 5–147 (2012)

93. Haining, R.P., Kerry, R., Oliver, M.A.: Geography, spatial data analysis, and geostatistics: an overview. Geograph. Anal. 42(1), 7–31 (2010)

94. Hernández, I., Güiza, F.: Información Geográfica Voluntaria (IGV), estado del arte en Latinoamérica. Rev. Cartográfica 93(1), 35–55 (2016)

95. Comesaña, D.: Modelo conceptual de información geográfica para la IDE – Uruguay. Tesis de maestría. Universidad de la República (Uruguay). Facultad de Información y Comunicación (2015)

96. Oana, C., Staiculescu, S.: Nature conservation strategy through a spatial data infrastructure. Ann. DAAAM Proc. 1075–1079 (2011)

97. Liu, X., Chen, F., Lu, C.: On detecting spatial categorical outliers. GeoInformatica 18(3), 501–536 (2013)

98. Ayre, L.: Who's out there? The power of spatial data. Collab. Libr. 7(2), 96–101 (2015)

99. Eldawy, A., Mokbel, M.F.: The era of big spatial data: challenges and opportunities. In: 2015 16th IEEE International Conference on Mobile Data Management (2015)
100. Mooney, P., Corcoran, P., Sun, H., Yan, L.: Citizen generated spatial data and information: risks and opportunities. In: 2012 International Conference on Industrial Control and Electronics Engineering (2012)
101. Eldawy, A., Mokbel, M.F.: The era of big spatial data. In: 31st IEEE International Conference on Data Engineering Workshops (2015)
102. Parcher, W., Cuasapaz, M., Mora, R.: Comparing the advancement of the national spatial data infrastructure in the Americas (2000–2008). Rev. Cartográfica **87**(1), 21–39 (2011)
103. Barufi, A.M., Haddad, E., Paez, A.: Infant mortality in Brazil, 1980-2000: a spatial panel data analysis. BMC Public Health 12 (2012)
104. Choi, J., Hwang, M., Kim, H., Ahn, J.: What drives developing countries to select free open source software for national spatial data infrastructure? Spat. Inf. Res. **24**(5), 545–553 (2016)
105. Azhar, D., Mendes, E., Riddle, P.: A systematic review of web resource estimation. In: Proceedings of the 8th International Conference on Predictive Models in Software Engineering, PROMISE, vol. 12, pp. 49–58 (2012)

Detecting Representative Trajectories in Moving Objects Databases from Clusters

Diego Fernando Rodriguez[✉] and Alvaro Enrique Ortiz

Universidad Distrital Francisco José de Caldas, Bogotá D.C, Colombia
dfrodriguezl@correo.udistrital.edu.co

Abstract. The analysis of the information of moving objects is increasingly important and useful, thanks to the new technology that allows the acquisition, storage, and representation of movement over time. We can resort to techniques that allow discovering implicit information in the movement of objects through computational techniques, one of them is the discovery of trajectory patterns and a representative trajectory to summarize cluster in one path, and also the identification of anomalies, both allow to describe the movement of the trajectory. The representative trajectories found in this work, it could be useful to understand movement in moving objects like cars, airplanes, buses. In this work, we show results with real data of massive mobility transport like Transmilenio in Bogotá – Colombia, wherefrom raw data we can found a movement pattern using Spatio-temporal databases, these patterns found represent movement tendency of moving objects that can facilitate representation, visualization and decision making in transport systems.

Keywords: Trajectories · SDBMS · Moving object databases · Pattern discovery

1 Introduction

With the set of mobile devices and the integrated sensors, such as the reception of the signal from global positioning satellites – GNSS, which indicates the position in hundredths of a second, a range of possibilities has been opened to obtain behaviors of objects based on their movement. One of the options is the construction and analysis of the trajectories from the position data, etc., where the detection of anomalies plays a very important role since it allows us to discover the abnormal behaviors in a data set with respect to the rest of the set or the other one that is taken as a model.

One of the most important tasks in data mining is the detection of patterns that consists in the discovery of regular data behaviors. To find these patterns in trajectories, we start from the premise that movement is represented in 3 dimensions (location XY and time T), which adds some complexity to the design and execution of certain algorithms that must be adapted to the context of objects in motion and their trajectories.

Currently, there are several algorithms that propose ways to detect patterns, in this article we discuss a methodology of a previous work presented in [1], wherefrom the obtained results we want to characterize the movement of a cluster through

© Springer Nature Switzerland AG 2020
Á. Rocha et al. (Eds.): ICITS 2020, AISC 1137, pp. 141–151, 2020.
https://doi.org/10.1007/978-3-030-40690-5_14

representative trajectories on Spatio-temporal databases, such as Oracle Spatial, together with the Hermes extension for moving objects.

2 Methodology

Proposal methodology show in [1] is about pattern detection in spatiotemporal trajectories using DB-SCAN algorithm for trajectories over a platform of moving objects databases called HERMES. This work presents the next step that is detecting a representative trajectory from found groups in [1]. In detail, HERMES defines a trajectory data type and a collection of operations, which is further enhanced by a special trajectory preserving access method, namely TB-Tree for indexing [2]

3 Trajectories

Trajectories are those that represent the movement of an object from the beginning to the end of it. As a first definition to understand that it is a trajectory, it is necessary to understand that it is movement, which consists of the "position or orientation body change in time" [3], as can be deduced from definition, the movement of an object depends essentially on two components: position and time, which implies that in order to model it, it is necessary to take into account the aforementioned components [3, 4].

Construction derived from the previous definition assumes that movement is of continuous nature, which represents a challenge for its representation in computers, being necessary a discretization process from a series of ordered points composed of coordinates in a plane (p_i) and sampling time t, these points must be represented within trajectory to work with approximate data to reality, as it is observed:

$$T = \{\langle p_1, t_1 \rangle, \langle p_2, t_2 \rangle, \ldots, \langle p_n, t_n \rangle\} \tag{1}$$

Let $p_i \in \mathbb{R}^2, t_i \in \mathbb{R}, 1 \leq i \leq n y t_1 < t_2 < \ldots < t\_n.$

There is a question about a movement that makes us conclude that our knowledge about it is of a continuous nature, and the question is: Where was an object located in an instant time between two sampled points?. The answer includes certain uncertainty that depends on the rate of sampling and movement parameters. Interpolation techniques are used, the most popular being linear interpolation due to its easy implementation [5].

In this implementation technique, it is assumed that the speed and direction remain constant during a defined time interval. This model is known as "cut representation", which consists of the decomposition of the temporal development of movement into fragments called "slices" as shown in Fig. 1, where between each piece or point sampled there is a simple function that describes the movement between they can be linear or arc type [2].

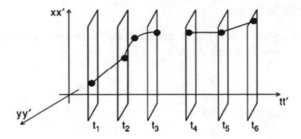

Fig. 1. Representation by slices for the definition of moving objects [2]

4 Discover Patterns in Trajectories

4.1 Pattern

A pattern that is defined as a regular behavior is a repetition in a systematic or predictable way. Specifically for the purpose of this paper, a temporal space pattern on a group of trajectories refers to the existence of objects that move and move in the same geographical space during the same time interval that a path must be along paths or representative trajectories of the movement of objects that are similar [6].

4.2 Cluster

Clustering is the process in which a data set is grouped into multiple groups or clusters, from the above it is deduced that the objects within the same group have high similarity, while they are very different from the objects in other groups. The measures of similarity and non-similarity are calculated from the values of the attributes that describe objects that often involve distance measurements [6, 7].

To define the clusters it is necessary to define what is the measure of similarity or non-similarity between the groups found, as well as the determination of the attributes involved in it [8].

In the case of the present work, the criteria that are taken into account to perform finding of the patterns are:

- The direction of the movement must be the same (spatial dimension)
- The time interval between two points must be similar (temporal dimension)

To find patterns of movement, it is required to perform the segmentation of the trajectories in sub-trajectories, where the concept of characteristic points is used to perform the partition in areas where the behavior of the movement changes rapidly [9].

4.3   Characteristic Points

As mentioned the characteristic points, are the points where the behavior of a trajectory changes rapidly. There is a trajectory:

$$TR_i = p_1 p_2 p_3 \ldots p_j \ldots p_{len} \tag{2}$$

Which have characteristic points set:

$$\left\{ p_{c_1}, p_{c_2}, p_{c_3}, \ldots, p_{c_{par_i}} \right\} \tag{3}$$

Let $c_1 < c_2 < c_3 < \ldots < c_{par_i}$

Then the trajectory TR_i is partitioned at each characteristic point, and a line segment between two successive characteristic points represents each segment of it. The trajectory is segmented into a set of $(par_i - 1)$ segments $\left\{ p_{c_1}, p_{c_2}, p_{c_3}, \ldots, p_{c_{par_i-1}} p_{c_{par_i}} \right\}$.

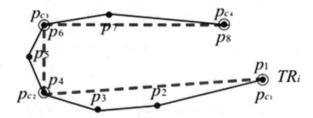

Fig. 2. Example of trajectory and its partitions.

In Fig. 2, the trajectory is observed in black line and the dotted blue line shows the partitions of the trajectory or sub-trajectories that are obtained between the characteristic points symbolized by a circumference around the point. The optimal partition must have two characteristics: precision and conciseness [9]

- Precision: It means that the difference between a trajectory and a set of the partitions of the same should be as small as possible.
- Conciseness: It means that the number of partitions in the path should be as small as possible.

To find the equilibrium point between conciseness and precision, the Minimum Description Length (MDL) method is adopted, as a general principle this means that any regularity in the data can be used to compress them, that is, describe them using the minimum number of symbols that the number of symbols necessary to describe the data literally, the more regularity there is higher the compression rate [10].

The definition of MDL includes two key concepts that are $d_\perp$ and d_θ, equivalent to the perpendicular distance and the angular distance, respectively. These distances form components of the distance function for trajectory segmentation using the characteristic point as shown in Fig. 3.

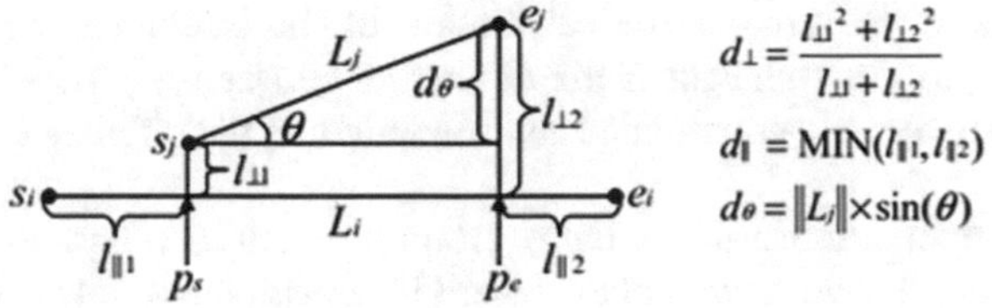

$$d_\perp = \frac{l_{\perp 1}^2 + l_{\perp 2}^2}{l_{\perp 1} + l_{\perp 2}}$$

$$d_\parallel = \mathrm{MIN}(l_{\parallel 1}, l_{\parallel 2})$$

$$d_\theta = \|L_j\| \times \sin(\theta)$$

Fig. 3. Components of the distance function to trajectory segmentation through characteristic points [9]

4.4 The Minimum Bounding Box (MBB)

The minimum bounding box refers to the minimum cube that is capable of enveloping the entire path segment found by characteristic points, that is, it is given by the coordinates of the two characteristic points that make up the segment as presented in Fig. 4.

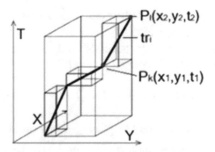

Fig. 4. Minimum bounding box [11]

It starts from the three-dimensional representation of movement from trajectories, where we have the two characteristic points P_i y P_k, composed by geographical coordinates (XY) and time, which defines a coordinate system in three dimensions as seen in Fig. 4. This way of representing the information allows summarizing the information in these cubes, which accelerate the processing speed that is required for subsequent procedures [11].

Once the concept of minimum envelope cube has been defined, it is possible to define the calculation of distances of similarity between two segments of trajectories, from volume and overlap:

$$Volume(MBBtr_i) = (x_2 - x_1) * (y_2 - y_1) * (t_2 - t_1) \qquad (4)$$

The formula is the same as the calculation of the traditional volume of a body where according to Fig. 4, height is the difference between t_2 y t_1, while the width is the difference between x's coordinates and the length is the difference between the y's coordinates.

The other important measure is the overlapping metric, which represents the volume of intersection between two cubes, where if the volume of intersection is equal to the volume of a cube, it follows that the movement has similar behavior, which means that they are neighbors in spatiotemporal dimensions. Regarding this measure, there can be two types of spatial relationships: an overlap and "Disjoint", where the first means that the cubes have some kind of spatial relationship and the second indicates that there is no type of relationship between two cubes.

Based on two previous measures, the measure of similarity is structured that can be calculated as follows [11]:

$$Sim = \frac{OV\left(MBB_i, MBB_j\right)}{Volume(MBB_i)} \tag{5}$$

Let $OV\left(MBB_i, MBB_j\right)$, is the overlapping area between two minimum bounding boxes and $Volume(MBB_i)$, cube volume.

Once the similarity measure has been defined, the measure of the distance of similarity between tr_i y tr_j can be defined as [11]:

$$SDist = (1 - Sim) * \beta * \left(len(tr_i) + len\left(tr_j\right)\right) \tag{6}$$

Let $len(tr_i)$ represents the segment length tr_i, β is a space-time weight to represent similarity distance. The weight is subject to range of $\left[0, \frac{\sqrt{2}}{2} * \left(len(tr_i) + len\left(tr_j\right)\right)\right]$.

4.5   DB-SCAN Algorithm for Trajectories

For the finding of the patterns, one of the most famous data mining tasks is used, clustering, i¿using the DB-SCAN (Density-Based Clustering based on connected regions with high density) algorithm oriented towards trajectories. In its original version, this algorithm finds core objects, where its main characteristic is the presence of dense neighborhoods, Two parameters are mainly used [6]:

- *E-neighborhood,* which is used to specify the radius of a neighborhood, the radius of an object or is the space within a radius E centered on o.
- *MinTrs,* which refers to the density limit of the regions, an object is a core object if and only if, the object's E-neighborhood contains at least MinTrs objects.

the current context of the trajectories, the proposed modifications to use this algorithm are the following [11]:

- A cluster of trajectories is made instead of points
- The distance measure previously found SDist is used to find neighboring sub trajectories.

The algorithm selected for the clustering of the trajectories works in the following way, initially, the sub trajectories are considered as unclassified. The algorithm consists of two phases; the first step is the calculation of the Eps-neighborhood parameter for each sub-trajectory not classified tr. If tr is a core sub-trajectory, $\left| N_{Eps}(tr_j) \right| \geq MinTrs$ is verified, if the proposed condition is met, for each tr_i that belongs to or is within the radius of the sub trajectory core tr, assign tr_i. to the specified cluster. In summary, the proposed methodology to find patterns in trajectories using Spatio-temporal databases is presented in the following diagram (Fig. 5):

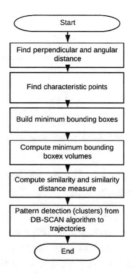

Fig. 5. Flowchart to find patterns in trajectories from DB-SCAN algorithm.

5   Representative Trajectories in Moving Object Databases

The representative trajectory of a cluster describes the general movement of the sub trajectories belonging to a cluster or a group. It can be considered a model for clusters. It is required to extract quantitative information of movement within a cluster so that experts in the domain area able to understand movement in trajectories [9]. Figure 6 shows the flow of the methodology to find representative trajectories in databases of moving objects:

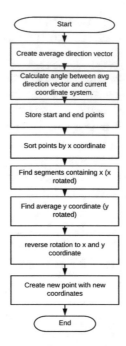

Fig. 6. Methodology to find representative trajectories from in moving object databases

Broadly speaking, and according to Fig. 6, to find representative trajectories from clusters, it shows the following procedure:

1. First is required to create an average direction vector in order to transform all the coordinates to a coordinate system where the x-axis is parallel with this vector in order not to repeat x coordinates that can alter the shape of the trajectory.
2. Then calculate the angle between the current coordinate system and the one that is required to arrive.
3. All the sub-traces found traversed to determine the cluster that stores all the points, both initial and final of each of the segments in an array.
4. The points sort from their x coordinates from least to greatest.
5. Go through the arrangement of ordered points to know which segments intersect with a certain coordinate x, where if the number of intersecting segments is less than MinTrs the point is discarded, on the contrary, if it is greater, the point will be taken to assemble the representative trajectory.
6. If the point is not discarded, the exact value of y is found with which the entered x coordinate is crossed.
7. The y-coordinate is found, which is the one associated with the x to construct a new point from the rotated coordinates, this new point is added to output array [9].

The procedure is summarized in the following pseudocode algorithm, contained in Fig. 7.

Algorithm 10: Find representative trajectories

Input: Cluster of line segments C_i
Input: $MinTrs$
Output: Representative trajectory - RTR_i
1: Calculate the average direction vector V;
2: Rotate the axes so that the X axis is parallel to V;
3: Let P be the set of the starting and ending points of the line segments in C_i;
4: Sort points in P by their $x's$ values;
5: **for** p *in* P **do**
   Let num_p be the number of the line segments that contain the $x's$ value of the point p;
   if $num_p \geq MinTrs$ **then**
      Calculate average coordinate avg'_p;
      Undo the roration and get point avg_p;
      Append avg_p to the end of RTR_i;
   end
end

Fig. 7. Pseudocode to find representative trajectories in moving objects databases.

Function to find representative trajectories done in PL/SQL programming language using Oracle 11 g together with HERMES extension to this database engine. The results shown in Fig. 8 correspond to trajectories described by a route (G43–K43) on 28th February 2018 in the bus system articulated called Transmilenio in Bogotá, Colombia. Where are the input parameters for the algorithm for the groups in general in [1], we have MinTrs = 50 and Eps = 500, which means that the group that you had at least neighborhood of the sub trajectory is taken into account core is 500 units (depends on the data coordinate system), the parameters that allow satisfactory results, as shown in Fig. 8.

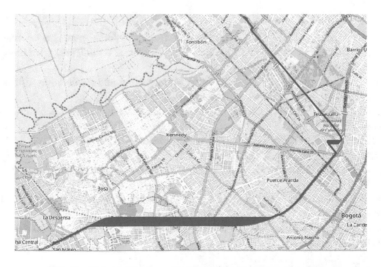

Fig. 8. The result of a representative trajectory algorithm.

In Fig. 8, the result obtained is shown where, in blue, the representative trajectory found for the set of trajectories that we entered into the algorithm with the parameters mentioned in the previous paragraph is observed. As an interpretation of results, we can say that it is the trajectory that most objects in motion follow, a there may be elements that are very distant from the pattern found that could be defined as anomalies.

From the information of the patterns and anomalies found, decisions can be made to normalize the behavior generated and make the necessary changes to make movement in the trajectories of the objects more efficient

6   Conclusions

The discovery of patterns in trajectories allows finding descriptions of collective behavior that allows describing different aspects associated with trajectories, remembering that different types of trajectories can be made that describe different aspects of elements that can change in space and time. Of course, anomalies are also found, which are also elements that help to better characterize the behavior of objects in motion. Both trends and anomalies will define the semantics and a future prospective behavior associated with the trajectories, in these two aspects of the use the patterns, future work is suggested to continue with the investigation of the use of the patterns on the trajectories of the objects in movement, and semantic association to enrich the trajectory information.

For future research, it could think in use data for predict movement behavior of moving objects, this work is only to characterize movement through find representative trajectories form movement clusters but no to predict future movements. Other future research is use online "mode", it's trying get data in real time and predict future movements for which it must modify proposal algorithms in this works.

References

1. Rodriguez Lamus, D.F., Ortiz Davila, A.E.: «Metodología para hallar patrones de movimiento en trayectorias sobre DBMS,» Revista ibérica de sistemas y tecnologias de la información, pp. 636–647 (2019)
2. Pelekis, N., Theodoridis, Y.: Mobility Data Management and Exploration. Springer, New York (2014)
3. Ortiz Dávila, A.E., Medina Daza, R.J.: «Representación de la dinámica espacio-temporal de objetos geográficos mediante trayectorias en base de datos,» In: 2017 12th Iberian Conference on Information Systems and Technologies (CISTI), Lisboa, 2017. LNCS. http://www.springer.com/lncs. Accessed 21 Nov 2016
4. Open Geospatial Consortium: «OGC Moving Features Access,» Open Geospatial Consortium (2017)
5. Güting, R.H., Schneider, M.: Moving Objects Databases. Morgan Kaufmann Publishers, San Francisco (2005)
6. Han, J., Kamber, M., Pei, J.: Data Mining Concepts and Techniques. Morgan Kaufmann Publishers, Waltham (2012)

7. Panagiotakis, C., Pelekis, N., Kopanakis, I., Ramasso, E., Theodoridis, Y.: Segmentation and sampling of moving object trajectories based on representativeness. IEEE Trans. Knowl. Data Eng. **24**(7), 1328–1343 (2012)
8. Li, Z., Ji, M., Lee, J.-G., Tang, L.-A., Yu, Y., Han, Y., Kays, R.: MoveMine: mining moving object databases. In: SIGMOD, Indianapolis, Indiana (2010)
9. Jae-Gil, L., Han, J., Whang, K.-Y.: Trajectory clustering: a partition-and-group framework. In: SIGMOD 2007, Beijing, China (2007)
10. Grünwald, P.D., Myung, I.J., Pitt, M.A.: Advances in Minimum Description Length: Theory and Applications. Massachusetts Institute of Technology, Cambridge (2005)
11. Ying, X., Xu, Z., Yin, W.G.: Cluster-based congestion outlier detection method on trajectory data. In: Sixth International Conference on Fuzzy Systems and Knowledge Discovery (2009)

Walk-Based Diversification for Data Summarization

Samuel Zanferdini Oliva[1]([⊠]) [ID] and Joaquim Cezar Felipe[2] [ID]

[1] Bioengineering Interunit Graduation Program, University of São Paulo,
São Carlos, Brazil
samuel.oliva@usp.br
[2] Department of Computing and Mathematics, University of São Paulo,
Ribeirão Preto, Brazil
jfelipe@ffclrp.usp.br

Abstract. Due to the large amount of data stored in current information systems, new strategies are required in order to extract useful information from databases. Hereupon, data summarization is an interesting process that allows reducing a large database maintaining just the relevant parts of the whole collection. In this study, we propose a new approach for data summarization based on a recently proposed tourist walk diversification method. This approach allows setting two ways of selecting elements considering density and hyper volume of each class. In order to evaluate the proposed approach, we compared it with two known methods of the literature considering one real world dataset and one artificial dataset. The artificial dataset was created considering different data distribution aspects. The conducted experiments outcomes demonstrate that our proposed data summarization approach is a promising alternative for addressing the problem of selecting elements from large databases considering different aspects of distribution.

Keywords: Data summarization · Tourist walk · Query diversification

1 Introduction

The continuous increase of data caused by the popularization of data storage technologies requires strategies to deal with the problem of extracting and interpreting useful information from large databases [1]. Hence, one of the main challenges regarding the large amount of data is to analyze it in a simple and faster way in order to obtain knowledge as prompt as possible [2].

For the purpose of handle this problem summarization has been explored in the literature considering different data domains such as networks data streams [3], plain texts [4], Intrusion Detection Systems (IDS) [5], and so on. Data summarization allows obtaining the relevant part of the content of a large collection of data. Thus, providing the ability to make useful inferences from the whole collection [6]. An example of summarization explored in the literature is the monitoring of the activity of the network. In this scenario, a network manager needs to recognize different types of anomalous events that may occur in the network in a period of time (e.g., month). For a

© Springer Nature Switzerland AG 2020
Á. Rocha et al. (Eds.): ICITS 2020, AISC 1137, pp. 152–161, 2020.
https://doi.org/10.1007/978-3-030-40690-5_15

monthly time spectrum, the size of network traffic data is unmanageable for human analysis. Hence, data summarization allows reducing the volume of data while holding the main aspects of the original data, then enabling human analysis and also decreasing the computational time for anomaly detection techniques [7].

Data summarization approaches can be classified into different taxonomies and main categories are: summarization for structured data and summarization for unstructured data. Summarization for structured data is related to the information that is organized in a pre-defined manner such as rows and columns. And summarization for unstructured data concerns the information that doesn't have a pre-defined model, for example, a plain text [8].

Thereby, data summarization can be employed for downsizing a very bulky database, by creating a representative subset of it. This process can provide an interesting way to relieve queries and analysis that could be complex and time-consuming. However, the summarization of a data collection is not an easy task, because extract a subset from an original dataset requires the preservation of the main aspects of it, such as class distribution, density, and space occupation. Accordingly, the study of deterministic walk has also called attention of the scientific community [9–11]. Deterministic walks can be defined as a process that describes a trajectory based on a sequence of steps performed on a mathematical metric space. These steps are defined by a specific movement rule. Hereupon, tourist walk (TW) is a deterministic partially self-repulsive walk.

Lima et al. [11] defines TW as a tourist that whishes to visit cities (data points) distributed on a d-dimensional map. To perform its trajectory, the tourist starts from a given city of that map and moves to the nearest city that still has not been visited in the last μ steps. The parameter μ is called memory and represents the number of steps required to repeat cities already visited [12]. The resulting trajectory of a TW is composed of a non-periodic initial part of t steps, called transient, and ends in a stable cycle period of p steps, called attractor, where the same data points are visited repeatedly in the same order [13].

Different TW methods were already proposed in the literature and are already applied to several computational contexts such as image analysis, pattern recognition; and classification. However, in the summarization context as far as we know this technique is still unexplored. Therefore, in this work, we use the TW concept as a bottom line to create an approach to handle the problem of data summarization motivated by the capability of it to capture structural and semantical information. The proposed approach aims to generate a database sampling that reproduces key characteristics of the original database keeping, for example, the distribution of data points, the local density, and the volume of the feature space for each class.

2 Related Work

Regarding solutions and approaches for data summarization, Kleindessner et al. [14] proposed a method based on a centroid-based clustering algorithm that aims to output a representative subsets from large datasets.

In another study [15], a data summarization algorithm based on sampling is proposed that is also able to create representatives subsets of large datasets. In this case, the algorithm is applied to dataset input of anomaly detection systems in order to provide similar or the same performance as an anomaly detection used on the original data.

Shou and Li [16] proposed a new method for summarizing large datasets in order to assist local outlier detection. Moreover, the study also proposed a new automatic parameter optimization approach and a method for parallel processing to improve the performance of the summary process.

Celis et al. [17] proposed a novel method based on fairness constraints, which includes fairness concerning sensitive attributes of data in sampling based on determinantal point process (DPP) for data summarization.

Besides that, other study [18] proposes a method that uses tensor for real time structural pattern summarization of dynamic graphs. This method is called tenClustS and it uses tensor decomposition to acquire the dynamics of networks and to reduce the dimensionality of the networks representation. Also in the domain of dynamic graph usage, another study [19] proposes two algorithms for the summarization of dynamic large-scale graphs, where one is based on cluster and the other uses micro-clusters concept in order to handle the limitation of memory requirements.

Different methods have been proposed considering graph summarization - in Shah et al. [20] propose a new technique to deal with graph summarization considering the compression problem; Fan et al. [21] propose a summarization method that treats reachability queries; and Hernández and Navarro [22] take into account neighbor and community queries.

3   Proposed Approach

In this section, we describe the proposed approach for data summarization that is based on clustering and tourist walk-based diversification proposed on a former study. The aim of the proposed approach is to select a compact but meaningful representation of a large database.

Hence, the approach has as input a given dataset that is further transformed into smaller set of summaries, focusing on retaining the maximum relevant information and features of the original dataset. he first approach can be divided in three parts: (i) cluster detection; (2) calculation of the amount of instances in which each class must keep in the summaries; (3) selection of instances that are dispersed on each class.

The main part of the algorithm is the function SummarizationWalk (Algorithm 1), which calculates the amount of elements of each class to be sampled. This function takes as input the ratio to be sampled and the lists of elements of each cluster. The algorithm requires that the dataset be already classified by some clustering method (part 1). In the function, there are also two parameter of control, one that allows to calculate the density of each cluster and the other that allows to adjust the samples in case the sample value is too high and the class doesn't have enough elements to be selected, then it can adjust the algorithm to get elements from another class.

Algorithm 1

1:	**procedure** SummarizationWalk(Clusters, Ratio, ByHVm = V, AdjustSamples = F)
2:	clustersSize ← *list with the size of each cluster*
3:	nClusters ← *number of clusters*
4:	ndim ← *dimension of the dataset*
5:	N ← *number of elements (rows) of the dataset*
6:	**for each** cluter **in** Clusters **do**
7:	distMat ← *distance matrix for each dataset's cluster*
8:	meanRadius ← **sum**(distMat) / **prod**(**dim**(distMat))
9:	radiusClusters ← {radiusClusters, meanRadius}
10:	hVm ← {hVm, calculateHypervolume(meanRadius, ndim)}
11:	**if** ByHVm == F **then**
12:	hVm ← clusterSize
13:	minRadius ← **min**(hVm)
14.	weight ← hVm/minRadius
15:	Nam ← N * Ratio
16:	Nmr ← Nam / **sum**(weight)
17:	Ncam ← weight * Nmr
18:	**if** AdjustSamples == *F* **then**
19:	X = (Nam − sum(Ncam)/**sum**(weight))
20:	**if** X > 1 **then**
21:	**for** i **in** 1: nClusters **do**
22:	**if** Ncam[i] <= $Nc[i]$ **then**
23:	Ncam[i]=Ncam[i]+X*weight[i]
24:	**for** i **in** 1: nClusters **do**
25:	**if** Ncam[i] >= $Nc[i]$ **then**
26:	Ncam[i] = clustersSize[i]
27:	**else**
28:	Ncam[i] = Ncam[i]
29:	**else**
30:	**for** i **in** 1: nClusters **do**
31:	**if** Ncam[i] >= $Nc[i]$ **then**
32:	Ratio ← Nc[i] * **sum**(weight)/weight[i] * N
33:	Nam ← N * Ratio
34:	Nmr ← Nam / **sum**(weight)
35:	Ncam ← weight * Nmr
36:	resultSet ← {}
37:	**for** i **in** 1: nClusters **do**
38:	k = **round**(Ncam[i])
39:	cluster ← Clusters[i]
40:	q ← centerOfMass(Clusters[i]) /* *It gets the class center element* */
41:	resultSet ← {resultSet, **walk**(clusters, k, q)}
42:	**return**(resultSet)

The approach has also two configuration options that allow calculating the amount of elements from each cluster, one proportional to the hyper volume of each cluster (maintaining the space ratio occupied by each cluster) and the other proportional to the amount of elements of each cluster (maintaining the cluster density).

The SummarizationWalk function pseudocodes presented in Algorithm 1 have functions that are native of the R platform, which allows controlling the clusters through list and nested lists. In Algorithm 1, we can see that line 41 calls the "walk" procedure (Algorithm 2), which is grounded on a walk-based diversification technique proposed in [23]. However, there are some slight differences regarding how the tradeoff parameter is defined and incremented.

Algorithm 2

1:     **procedure** walk(S, k, q)
2:         $Iti \leftarrow \{q\}$
3:         $\lambda \leftarrow 0$
4:         $s_n \leftarrow findNext(S, Iti, \lambda)$
5:         $Iti \leftarrow enqueue(Iti, s_n, k)$
6:         $Iti \leftarrow \{Iti \backslash q\}$
7:         **while** $|Iti| < k$
8:                 $\lambda \leftarrow \lambda + 1/k$
9:                 $s_n \leftarrow findNext(S, Iti, \lambda)$
10:                 $Iti \leftarrow enqueue(Iti, s_n, k)$ /* It queues elements in the itinerary */
11:         **return**(Iti)

This parameter is used for balancing how much dispersed the selected elements must be on the dataset. In this case, the tradeoff parameter is auto adjustable thus the user cannot change it because it is initiated as 0 and it is automatically incremented in each step. This increment is defined by the max value allowed for λ (1) divided by the size of the desired result set (k), which in the "walk" procedure is calculated as $\lambda \leftarrow \lambda + 1/k$.

Thereby, in each step the tradeoff parameter is incremented when selecting the next element to be visited. This element is selected by the procedure "findNext" (Algorithm 3), which was proposed in [23]. This procedure allows selecting elements from the dataset considering different degrees of diversity hence this is capable of getting elements that are well dispersed on the dataset.

Algorithm 3

1:     **procedure** findNext(S, Iti, λ)
2:         $Sum \leftarrow \{\}$
3:         **for each** $s_i \in S$ **do**
4:                 $agg \leftarrow null$
5:                 **for each** $s_j \in Iti$ **do**
6:                         **if** $s_i \notin Iti$ **then**
7:                                 $agg \leftarrow agg + dist(s_j, s_i)$
8:                 $Sum \leftarrow \{Sum, agg\}$
9:         $\lambda \leftarrow 0.8 * \lambda^2 + 0.2 * \lambda$
10:         $M \leftarrow \lambda * (max(Sum) - min(Sum)) + min(Sum)$
11:         $near \leftarrow which.min(abs(Sum - M))$
12:         **return**($near$)

---

4   Experiments

4.1   Datasets

The proposed summarization approach was implemented using the R platform and we have used two datasets in order to evaluate it. Among them there are one artificial dataset that we created and the other is a real-world dataset.

Iris dataset is a known real-world dataset widely used in the pattern recognition literature. However, for the purpose of a better visualization of the data distribution, we have used a version of this dataset that contains just two dimensions of the four original ones. Iris dataset contains 150 instances and 3 classes, one class is linearly separable from the two; and the others are not linearly separable from each other.

Artificial dataset contains 375 instances and 4 clusters with the same density. In order to maintain the density, the number of instances of each cluster was configured proportionally to its area. One of the clusters has 25 instances, the other 50, the other has 100 and the last one has 200. From this dataset, we've planned to evaluate the approach considering the situation when the size of the cluster increases, but it keeps the same density.

4.2   Evaluation Metrics

In order to evaluate the results returned by the methods, we've calculated two metrics. The first metric (Eq. 1) aims to demonstrate that the amount returned objects of the class is proportional to the amount of objects of the class.

$$distr\_by\_qty = \frac{quantity\_returned\_class}{total\_amount\_class} \qquad (1)$$

The second metric (Eq. 2) aims to demonstrate that the amount of returned object of the class is proportional to the volume occupied by the class.

$$distr\_by\_vol = \frac{quantity\_returned\_class}{hypervolume\_class} \qquad (2)$$

4.3   Experimental Results

In this section, we compare the proposed walk-based summarization approach that has two possible variations, one for selecting elements considering the hyper volume and the other that considers the amount of objects per class. And we compare this approach with two summarization methods of the literature.

One of the literature methods is the RandomSampling, which selects samples from the dataset at random. The other method is the SystematicSampling that selects samples from the dataset at regular intervals considering the size of the dataset.

Figures 1 and 2 present the summarization approaches applied to the Iris and artificial datasets. In the figures, classes are distinguished by different colors or can be visually recognized by the space between them and the selected elements are highlighted by red star-shape (*) points. These classes were separated through a hierarchical cluster technique based on centroids. In each class, there are two metrics that we've described in Eqs. 1 and 2. For all methods we considered a sample of 20% of the original dataset.

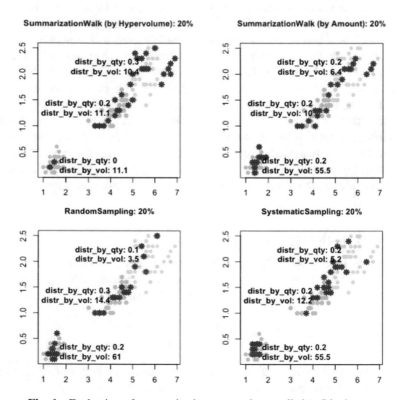

Fig. 1. Evaluation of summarization approaches applied to Iris dataset.

In SummarizationWalk (by Hypervolume) plot (Fig. 1), we can notice that the distr_by_vol values of the 3 classes are very close, i.e., the results show a distribution of selected elements according to the space occupied by each class. This is because the approach aims to sample elements considering the volume occupied by the class in the space.

In SummarizationWalk (by Amount) plot (Fig. 1), we can notice that the values of ditr_by_qty of the 3 classes are very close, i.e., sampled elements are distributed in the classes and they maintain the density of each class.

In the other hand, the plots (Fig. 1) and the metrics have shown that RandomSampling and SystematicSampling methods do not maintain the density between classes when select the elements, and we can see it through the two metrics that present quite different values for each class. This occurs because both methods do not take into account the construction of the result set neither by maintaining the distribution by class volume, nor distribution by class density.

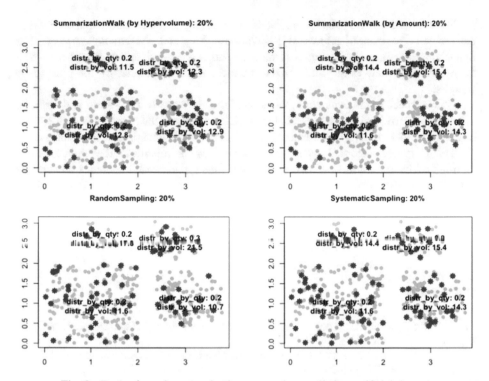

Fig. 2. Evaluation of summarization approaches applied to artificial dataset.

In SummarizationWalk (by Hypervolume) plot (Fig. 2), we can notice that distr_by_vol values for each class is very close to each other thus we can say that the algorithm selects the amount of elements by class according to the space occupied by the class, even when the number of elements is doubled.

In SummarizationWalk (by Amount) plot, as the classes maintain the same density the method results also keep the same amount of elements per class when selects the sample. In this case, RandomSampling and SystematicSampling methods also select the same amount of elements per class.

5 Conclusion

In this study, we've proposed a new approach for the creation of summaries from databases. This approach is motivated by the tourist walk heuristic, which is capable of selecting elements in different positions of a dataset distribution. Hence, we developed an algorithm with two possible configurations for summarizing considering different aspects of data points distribution in databases. One of the configurations aims to select data points that are distributed according to their space occupied in each class. While, the other setting focuses on selecting data points that are distributed according to the density of each class.

For comparison, we used two methods of sampling of the literature that are used to perform summarization, known as RandomSampling and SystematicSampling. The first chooses samples of a dataset at random and the second selects samples starting from a random data point until a fixed, periodic interval defined by the ratio to be sampled.

With respect to the evaluation, we applied the methods to real-world and artificial datasets. The artificial one was created with aspects that test the algorithms with respect to the capability of selecting elements considering the distribution of the data points.

The proposed approach presented positive results regarding the characteristics highlighted by the datasets. Thus, we can say that the approach provides advantages over the other methods.

Acknowledgments. This study was financed in part by the Coordination for the Improvement of Higher Education Personnel (CAPES) - Finance Code 001.

And Grant #2016/17078-0, São Paulo Research Foundation (FAPESP).

References

1. Zins, C.: Conceptual approaches for defining data, information, and knowledge. J. Am. Soc. Inf. Sci. Technol. **58**(4), 479–493 (2007)
2. Hoplaros, D., Tari, Z., Khalil, I.: Data summarization for network traffic monitoring. J. Netw. Comput. Appl. **37**, 194–205 (2014)
3. Hu, Y., Chiu, D.M., Lui, J.: Entropy based adaptive flow aggregation. IEEE/ACM Trans. Netw. (TON) **17**(3), 698–711 (2009)
4. Yu, L., Ren, F.: A study on cross-language text summarization using supervised methods. In: 2009 International Conference on Natural Language Processing and Knowledge Engineering, pp. 1–7. IEEE (2009)
5. Zhu, R.: Intelligent rate control for supporting real-time traffic in WLAN mesh networks. J. Netw. Comput. Appl. **34**(5), 1449–1458 (2011)
6. Yager, R.R.: A new approach to the summarization of data. Inf. Sci. **28**(1), 69–86 (1982)
7. Ahmed, M., Mahmood, A.N., Hu, J.: A survey of network anomaly detection techniques. J. Netw. Comput. Appl. **60**, 19–31 (2016)
8. Ahmed, M.: Data summarization: a survey. Knowl. Inf. Syst. **58**(2), 249–273 (2019)
9. Kinouchi, O., Martinez, A.S., Lima, G.F., Lourenço, G.M., Risau-Gusman, S.: Deterministic walks in random networks: an application to thesaurus graphs. Phys. A Stat. Mech. Appl. **315**(3–4), 665–676 (2002)
10. Bunimovich, L.A.: Deterministic walks in random environments. Phys. D Nonlinear Phenom. **187**(1–4), 20–29 (2004)
11. Lima, G.F., Martinez, A.S., Kinouchi, O.: Deterministic walks in random media. Phys. Rev. Lett. **87**(1), 010603 (2001)
12. Terçariol, C.A.S., Martinez, A.S.: Analytical results for the statistical distribution related to a memoryless deterministic walk: dimensionality effect and mean-field models. Phys. Rev. E **72**(2), 021103 (2005)
13. Stanley, H.E., Buldyrev, S.V.: Statistical physics: the salesman and the tourist. Nature **413**(6854), 373 (2001)
14. Kleindessner, M., Awasthi, P., Morgenstern, J.: Fair k-center clustering for data summarization. arXiv preprint arXiv:1901.08628 (2019)

15. Ahmed, M.: Intelligent big data summarization for rare anomaly detection. IEEE Access **7**, 68669–68677 (2019)
16. Shou, Z., Li, S.: Large dataset summarization with automatic parameter optimization and parallel processing for local outlier detection. Concurr. Comput. Pract. Exp. **30**(23), e4466 (2018)
17. Celis, L.E., Keswani, V., Straszak, D., Deshpande, A., Kathuria, T., Vishnoi, N.K.: Fair and diverse DPP-based data summarization. arXiv preprint arXiv:1802.04023 (2018)
18. Fernandes, S., Fanaee-T, H., Gama, J.: Dynamic graph summarization: a tensor decomposition approach. Data Min. Knowl. Discov. **32**(5), 1397–1420 (2018)
19. Tsalouchidou, I., Bonchi, F., Morales, G.D.F., Baeza-Yates, R.: Scalable dynamic graph summarization. IEEE Trans. Knowl. Data Eng. **32**(2), 360–373 (2018)
20. Shah, N., Koutra, D., Zou, T., Gallagher, B. and Faloutsos, C.: Timecrunch: interpretable dynamic graph summarization. In: Proceedings of the 21th ACM SIGKDD International Conference on Knowledge Discovery and Data Mining, pp. 1055–1064. ACM (2015)
21. Fan, W., Li, J., Wang, X., Wu, Y.: Query preserving graph compression. In: Proceedings of the 2012 ACM SIGMOD International Conference on Management of Data, pp. 157–168. ACM (2012)
22. Hernández, C., Navarro, G.: Compression of web and social graphs supporting neighbor and community queries. In: Proceedings 5th ACM Workshop on Social Network Mining and Analysis (SNA-KDD). ACM (2011)
23. Oliva, S.Z., Felipe, J.C., Ribeiro, M.X.: Deterministic tourist walk for effective diversification of query results. In: 16th International Conference on Applied Computing (AC 2019), Cagliari. Proceedings of the 16th International Conference on Applied Computing, pp. 163–170 (2019)

The Analysis of Competency Model for a Performance Appraisal System in the Management of Food Service Industry

João Paulo Pereira[1,2(✉)], Efanova Natalya[3], and Ivan Slesarenko[3]

[1] Instituto Politécnico de Bragança, Campus de Santa Apolónia,
5300-302 Bragança, Portugal
jprp@ipb.pt
[2] UNIAG (Applied Management Research Unit), Bragança, Portugal
[3] Kuban State Agrarian University, Krasnodar, Russian Federation
efanova.nv@gmail.com, one.concealed.light@gmail.com

Abstract. This paper presents a competency model as an instrument for restaurant management. The research showed that existing appraisal methods, such as KPI and 360-degree feedback method, are difficult to apply during evaluating the performance of non-managerial employees (waiters). Test method comes with several shortcomings too but using competency model can make this method more effective. Competency model allows for combining various assessment results and forming an employee profile, which is constructing and storing in a computer system. Previous research demonstrated effectiveness of model and some problems of its using like source data purity. In this paper test questions' balance and the reasons of collective mistakes are researched. Experiments were conducted using employee profiles data. Results from these experiments confirmed that the model can improve restaurant management but there are some problems and features of its using. It is important to provide test questions balance and source data purity.

Keywords: Competency model · Performance appraisal · Performance level · Food service industry · Employee profile · Computer program · Experiment

1 Introduction

Employee appraisal became an important duty of the HR department of any major organization.

With the development of the food service industry and spread of restaurant chains, choosing the right performance appraisal method [10, 11] for a large number of employees has become a pressing issue. So much attention now needs to be devoted not only to managerial staff but also to non-managerial employees, because there are many dependencies between financial efficiency of restaurant and non-managerial employees' skills.

Various evaluation methods are used depending on the structure and capabilities of that particular company. The review is centred on a competency-based approach.

Á. Rocha et al. (Eds.): ICITS 2020, AISC 1137, pp. 162–171, 2020.
https://doi.org/10.1007/978-3-030-40690-5_16

Presently, this approach becomes more popular – it evaluates not only the professional but also the personal qualities of an employee [1, 3, 16].

It is effective to use popular methods like 360-degree feedback and KPI to evaluate managerial staff. But these methods are not effective to evaluate non-managerial employees.

Although 360-degree feedback method has several advantages like comprehensive opinion [15] or some successful experience of using [7], it has major shortcomings [5, 6]. It is near impossible to evaluate the specific performance of an employee. This drawback makes the effectiveness of the 360-degree feedback method in the food service industry disputable. For example, the ability of a waiter to serve a customer or to apply sales methods may not always be noticeable to his colleagues and superiors. The same applies when that waiter lacks some skills in this area. One more problem is negative reaction of employees on negative feedback that may cause prejudice of facilitator [1]. Because of this, the 360-degree feedback method can only be an addition to any already existing performance appraisal technique.

KPI can be the interface between scheduling and control which will be used as a strategy for maximizing the plant performance [2]. Also KPIs monitoring can be used for decision support [14]. But organizations often have their own system of performance indicators and the formulas for calculating them. That is why there is presently no uniform standard for applying the KPI method [9]. Companies are not willing to share their achievements with business rivals.

There are little or no publicly available information on the best practices of using the KPI method. To apply this technique in the food service business, one would need to take a number of actions aimed at determining the performance indicators of departments and employees, and the formulas for calculating them. Also testing of this method is needed after initializing. This could lead to costs.

At the moment, tests represent such a simple method. Implementing a test does not require major expenses. Where resources are limited, one would only need to create the test, upload it to a free testing platform and provide access to it for employees. The total score gained by the staff will be their performance assessment.

With all the simplicity, testing still has a number of significant shortcomings [8, 12]:

1. It checks only theoretical knowledge. A person's ability to put his theoretical knowledge into practice is hard to assess.
2. Test answers can be memorized. Moreover, an employee can decide to cheat during the test. These undermine the reliability of result obtained.
3. Since only the final grade is most often taken into account, the manager may not always get a complete picture of the abilities of an employee.

These disadvantages are critical in the catering industry – a waiter's ability to provide quality service to a customer determines whether that customer will visit again or not in the future. This ability cannot be evaluated through a test. Therefore, in order to evaluate such ability, mystery shoppers are recruited. This requires additional resources. In addition, if attention is paid only to the total score earned by the employee, testing will not be able to determine areas where the employee is stronger or weaker.

This complicates the process of distributing employees in shifts and restaurants. Here, the growth/decline dynamics of certain skills is not taken into account.

Thus, choosing a suitable appraisal method for non-managerial employees becomes a challenge. Testing alone cannot solve this problem due to its number of shortcomings. Also, there is no standard employee competency card because internal standards vary from restaurant to restaurant.

Competency model was tested during previous research [13]. Staff's testing results were used as a data source. As experiments demonstrated, speaking of waiters, there is correlation between profile data, based on competency model, and some performance indicators data. Relations between waiter's knowledge represented with profile and skills represented with performance indicators were confirmed. Also some problems with usage of competency model were detected, such as "purity" of profile source data and test's questions balance. First problem is the massive cheating during test. Second problem affects models' informativeness.

The influence of test's questions balance on models' components' levels' distribution will be researched in this paper. Using competency model not only as the method of single employee's dynamic appraisal but as the method of group of employees' (restaurant) appraisal altogether will be researched too. It will let us understand efficiency of restaurant staff management on macro level.

2 Materials and Methods

In this section, the proposed competency model for evaluation of non-managerial employees in catering enterprises is considered in more detail.

This model represents a weighted graph [13]:

$$G = (V, R), \tag{1}$$

where $V = \langle P, Q, C \rangle$ is a set of vertices of the graph:

$P = \{p_i\}$ – positions, $i = 1..N_P$;
$Q = \{q_j\}$ – competencies, $j = 1..N_Q$;
$C = \{c_k\}$ – competence components; $k = 1..N_K.$;
$R = \{\{r_{ij}\} \cup \{r_{jk}\}\}$ – a set of edges describing the connections between the vertices.

Due to the "here and now" model assembly principle and storing data only of leaf vertices, lower layer of graph is the core of model: higher levels' construction and its adequacy depends on it. So, it's necessary to choose right methods of low layer components' levels calculation: these components' levels should be equally informative. In another words, if two components have numerically equal levels, semantic interpretation of this levels should be equal too. For example, with levels permissible range [0..1] we should avoid situations when cause of level "0.25" of component A is weaker than cause of same level of component B.

Lower layer of model, for its part, is a component of employee profile. Profile is a special structure for storing information about the employee himself, his professional and personal qualities, and the dynamics of changes in these qualities. In total, there are three elements of the profile:

1. Personal data (full name, place of work, position, etc.).
2. Information on level of competence (leaf vertices).
3. History of changes in the level of competence.

The profile can be formed using various sources.

So, an employee profile contains both static (1) and dynamic (2 and 3) elements. Of interest in this study are the dynamic elements. This paper will focus on them.

In order to test the competency model developed, a software system was designed to track the professional growth of employees at a restaurant chain. Employee profile is a key element of the software system. Test results serve as data sources in the study. Having a clear structure of results and being able to relate them to competence allows for automatic processing.

The system can fill in the database, which includes importing data from sources, editing employee profiles, and obtaining results of analysis of competency change dynamics. Analysis of the dynamics of changes in competencies involves evaluating the nature of changes in the level of each competence over a selected period of time. It enables one to automate not only the process of identifying problem-plagued employees, but also the process of pinpointing their specific shortcomings. This can help in optimization when planning on how to eliminate the shortcomings.

Test results are used to form employee profile as follows. In the process of compiling a test, each of the questions is assigned a competence or its component, thereby creating a correspondence table for questions and competencies. In the course of processing the test results, the automated system counts the number of correct answers A for an employee and the total number of questions T for each competency or component. The level of competence or component L_{leaf} is the ratio of the number of correct answers A to the total number of questions T:

$$L_{leaf} = \frac{A}{T} \tag{2}$$

Formula (2) can only be used to calculate the levels of leave vertices.

As mentioned above, the aspect of lower layer components' level calculation is important. Since all levels are normalized and theirs permissible range is [0..1], cost of mistake becomes a main indicator: how much the component's level reduction will be if employee makes mistake in one question of a certain category.

After the levels are defined, they are saved in the database, while the previous profile state is saved in the profile history.

The software system is implemented in the training and development department of a pizza restaurant chain, whose employees are trainers. Their responsibility includes training and monitoring employees, as well as solving tasks aimed at developing the company. Trainers are tasked with compiling tests, scheduling visits by mystery shoppers, and processing the results obtained.

To test possibility of using model during staff management process, we will conduct two experiments in 30 restaurants of the southern region in Russia. Average amount of appraising employees per restaurant is 8.

First experiment is valuation of test's questions balance (or misbalance) efficiency. During experiment we will choose two competences with high cost of mistake. Because of this cost there are two opinions: "first competence is hard and has high priority in monitoring" and "second competence is easy because many employees have a high level in it".

We will rebalance count of questions and this action will reduce cost of mistake value from 0.3–1 to 0.2–0.25 units. Then we will compare levels' distribution before rebalancing and after. If opinions mentioned above are wrong, balance of questions is an important aspect in using tests as data source for employee profile.

Phenomenon of "collective mind" was discovered during system test. It means that more than a half of restaurant employees made the same mistakes in one question. Search of this phenomenon cause is important for effective decisions making in restaurant management.

Second experiment is analysis of dynamic of profiles' group. Profiles are merged by territorial attribute (employee's place of work). Source data will contain all cases of restaurants negative dynamic for three months of observation. Experiment's objective is determination of main causes of "collective mind" phenomenon and what decisions should be made when it appears.

3   Results

The following charts show results obtained from the experiments.

3.1   Comparing the Distribution of Competence Levels

It is worth noting that the principle of compiling the tests was changed in May. Prior to this, competences "Regulations knowledge" and "Main menu knowledge: Asian cuisine" were allocated an average of 1–2 question and had cost of mistake 0.5–1 units.

A bubble chart was used to present the results of the experiment. In this chart, bubbles mean the number of employees having a certain level of competence. The bubbles were categorised into groups, each group representing the distribution of levels of competence in a particular month. Competence levels were placed on the Y-axis, while the X-axis featured the months when the levels were determined.

Figure 1 shows the distribution of levels of competence "Regulations knowledge" for 6 months. According to the trainers, the waiters had no problems with this competence.

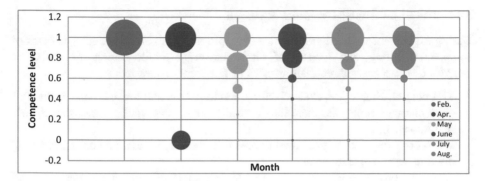

Fig. 1. Distribution of levels of competence "Regulations knowledge"

After balancing the questions, distribution of levels changed as follows: from May to July, at least half of the employees had a fairly high competence level, and then the number of employees per level decreased with this level in a geometric progression. In August, the distribution centre shifted to 0.8, indicating that a majority of those tested gave some wrong answers. Under the old balancing of questions, the distribution could be similar to the April one.

Figure 2 shows the distribution of levels of competence "Main menu knowledge: Asian cuisine". According to the trainers, this competence was the most problematic in the restaurant chain.

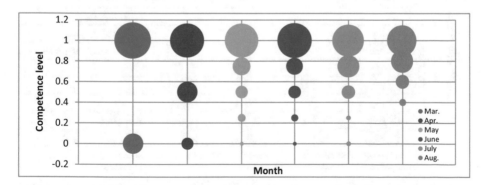

Fig. 2. Distribution of levels of competence "Main menu knowledge: Asian cuisine"

After balancing, the distribution dynamics had the same tendency as in the previous comparison. There were some employees who were unable to acquire this competence. However, they were not as many as previously assumed.

These charts approved that opinions about competencies were wrong. Many waiters don't know regulations well enough and knowledge of Asian cuisine is not so problematic.

3.2 The Restaurants' Dynamics Research

During experiment we used profiles data of employees from 30 restaurants, 40 cases of competence falling, and 3 cases of whole restaurant profile falling. According to analysis of these cases three main reasons of fall were discovered:

1. Test control. Employees cheated during previous tests and had the best results. But once manager controlled them they couldn't cheat. Bad results demonstrated their real knowledge.
2. Wrong cheating. Employees tried to pass test collectively and had common wrong opinion about some questions.
3. Operation problems at restaurant. Negative dynamic of competence demonstrated restaurant's problems related to competence.

In general, all three reasons boil down to restaurant's problems. Main question is when these problems were discovered and how fatal they are.

In first case we see employees' real knowledge level, but discover it later than could. In second case we see only part of real knowledge level. Some problems are concealed by true answers which were made with cheating. So this case is the most dangerous because we know about cheating but don't know what problems were concealed by this. In third case we find out in time restaurant's problems.

Figure 3 shows experiment results. Falling competences are placed on the X-axis. Count of fall cases are placed on the Y-axis. Reasons are presented as the columns.

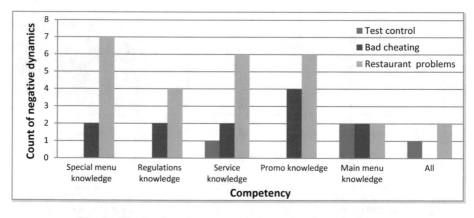

Fig. 3. Distribution of competencies' negative dynamics' reasons

Experiment showed, that if the source data is clean, monitoring the dynamics of restaurant allows you to discover in time restaurant's problems. Usually it happens due to low quality of the manager's work:

1. Weak control of employees' knowledge. It is represented in the dynamics of competences related with memorizing (menu knowledge, promo knowledge).

2. Bad employees' discipline or poor restaurant service. These problems are represented in the dynamics of competences related with regulations knowledge and service knowledge. Sometimes they appear together.

The cheating frequency is still big enough and it is the result of poor source data purity. Discovering of this problem demonstrates the low level of restaurant management and the lack of possibility to find out in time problems and solve them.

4 Discussion

Experiments allowed us to make conclusions about features of using competency model for restaurant management.

Firstly, it is important to provide balance of test questions if test results are a source data for employee profile. Otherwise following problems may appear:

1. Informativeness of model: very high or very low cost of mistake can misrepresent employee problems.
2. Poor productivity of the dynamics analysis: cardinal change of questions count will lead to different interpretation of the same mistake.

Since all competences may have different amount of knowledge we should do following actions during model initialization:

1. The division of large competences into components and the assignment a group of questions for each of them.
2. Adaptation of dynamics analysis methods to high cost of mistake.

Perspective evolution of testing as a source of employee profile is the development of individual testing system. This system will monitor employee's progress and create favorable conditions for work on mistakes. Competency model, for its part, should be divided to two parts: operative part (represents results of last test) and permanent part. Permanent part will be constructing during whole model's cycle of life and will represent full information about employee knowledge.

Secondly, competency model can be used as an instrument for effective monitoring of the dynamics not only singles employee but the whole restaurant. During experiment we discovered not only restaurant management problems but employees training problems. This proves necessity of the automatic dynamics monitoring and operative alerts instruments development.

Perspective evolution of this aspect is the developing of system which will detect potential reasons of negative dynamics according to statistics. Also this system should predict restaurant growth. The restaurant profile, for its part, should be developed too. The employee profile will be a component of the restaurant profile.

But we should pay attention to problems of competency model usage:

1. Reconstruction of testing system. Balance of question can break up some foundations of organization so it can become a barrier to using the model.

2. Data purity. As experiment demonstrated, cheating may be territorial. This case can signal about low management quality in restaurant. So it is important to provide enough management staff competence.

5 Conclusion

Using of competency can enhance management of restaurant or restaurants chain. It allows operatively getting information about weaknesses and strengths of employees and making decisions based on this information. Also results of employees testing demonstrate quality of their training and managing, so top management has more opportunities in analysing of problems' reasons.

However, competency model is vulnerable to quality of processes making influence on it like testing organization. It is important to adapt these processes, for example balance count of tests' questions and provide high quality of these questions, otherwise informativeness'es quality of model will be low and it may to not enhance management process at all.

Acknowledgment. UNIAG, R & D unit funded by the FCT - English Foundation for the Development of Science and Technology, Ministry of Science, Technology and Higher Education. Project n. UID/ GES/ 4752/2019.

References

1. Ampion, M.A., Fink, A.A., Ruggeberg, B.J., Carr, L., Phillips, G.M., Odman, R.B.: Doing competencies well: best practices in competency modeling. Pers. Psychol. **64**, 225–262 (2011). https://doi.org/10.1111/j.1744-6570.2010.01207.x
2. Bauer, M., Lucke, M., Johnsson, C., Harjunkoski, I., Schlake, J.C.: KPIs as the interface between scheduling and control. IFAC-PapersOnLine **49**(7), 687–692 (2016)
3. Bernikova, O.: Competency-based education: from theory to practice. In: Proceeding of the 8th International Multi-conference on Complexity, Informatics and Cybernetics (IMCIC), pp. 316–319 (2017)
4. Brett, J.F., Atwater, L.E.: 360 degree feedback: accuracy, reactions, and perceptions of usefulness. J. Appl. Psychol. **86**(5), 930–942 (2001)
5. Carson, M.: Saying it like it isn't: the pros and cons of 360-degree feedback. Bus. Horiz. **49**(5), 395–402 (2006)
6. Jackson, E.: The 7 Reasons Why 360 Degree Feedback Programs Fail. https://www.forbes.com/sites/ericjackson/2012/08/17/the-7-reasons-why-360-degree-feedback-programs-fail. Accessed 06 Nov 2019
7. Karkoulian, S., Assaker, G., Hallak, R.: An empirical study of 360-degree feedback, organizational justice, and firm sustainability. J. Bus. Res. **69**(5), 1862–1867 (2016)
8. Loban, N.E., Dongak, H.G.: The problem of testing personnel in the enterprise. Ekonomika I socium **5–1**(36), 806–809 (2017). [In Russian]
9. Malysheva, M.A.: KPI: advantages and disadvantages of implementing. Informaciya kak dvigatel nauchnogo progressa: sbornik statey Mezhdunarodnoy nauchno - prakticheskoy konferencii, pp. 134–137 (2017). [In Russian]

10. Mayhew, R.: How to Develop a Performance Evaluation System for Fast Food Restaurants. Small Business - Chron.com. http://smallbusiness.chron.com/develop-performance-evalua tion-system-fast-food-restaurants-20675.html. Accessed 10 Nov 2019
11. Mayhew, R.: Performance Appraisal in the Food and Drink Industry. Small Business - Chron.com. http://smallbusiness.chron.com/performance-appraisal-food-drink-industry-124 52.html. Accessed 06 Nov 2019
12. Nagaeva, I.A.: The organization of electronic testing: advantages and disadvantages. Online J. "Naukovedenie" 5(18), 119 (2013). [In Russian]
13. Pereira J.P., Efanova N., Slesarenko I.: A new model for evaluation of human resources: case study of catering industry. In: Rocha, Á., Adeli, H., Reis, L., Costanzo S. (eds.) New Knowledge in Information Systems and Technologies. WorldCIST 2019. Advances in Intelligent Systems and Computing, vol. 930. Springer, Cham (2019)
14. Pérez-Álvarez, J.M., Maté, A., Gómez-López, M.T., Trujillo, J.: Tactical business-process-decision support based on KPIs monitoring and validation. Comput. Ind. 102, 23–39 (2018)
15. Popova, N.V.: Features of the personnel assessment method "360 degrees". Sbornik nauchnyh trudov, pp. 383–387 (2018). [In Russian]
16. Sliter, K.: Assessing 21st century skills: competency modeling to the rescue. Ind. Organ. Psychol. 8(2), 284–289 (2015). https://doi.org/10.1017/iop.2015.35

Smart University: An Architecture Proposal for Information Management Using Open Data for Research Projects

Marlon Santiago Viñán-Ludeña[1,3](✉) ,
Luis Roberto Jacome-Galarza[2] , Luis Rodríguez Montoya[1] ,
Andy Vega Leon[1] , and Christian Campoverde Ramírez[1]

[1] Universidad Nacional de Loja, Loja 110103, Ecuador
marlon.vinan@unl.edu.ec
[2] Escuela Superior Politécnica del Litoral, Guayaquil, Ecuador
[3] Universidad de Granada, 18071 Granada, Spain

Abstract. The ubiquity of information is traditionally framed with Smart Cities in which, information is based on the integration of information, efficiency, sustainability and participatory governance. Ubiquity is present in all areas and domains, that is; why the "Smart University architecture" is proposed. The main goal of this work is to present the proposal for the information integration related to research projects of the "Universidad Nacional de Loja" (Ecuador). This proposal refers to the implementation of a three-layer architecture: (i) raw data collected manually or automatically from research projects; (ii) data structuration and data enrichment using external sources following the Open Data philosophy and (iii) exploitation and visualization of the information. The use of web services allows this architecture to automatically manage and share all the necessary information that serves as a basis to have the main characteristics of a smart university: adaptation, sensing, inferring, self-learning, anticipation and self-organization; and in addition, generate new business and research ideas.

Keywords: Smart university · Integration of information · Open Data · Architecture · Web services

1 Introduction

The IoT (Internet of Things) revolution has allowed that any device can be connected through the web; i.e. cameras, sensors, motion detectors, wearables, etc. This generates large volumes of information that must be captured, structured and stored. Research projects in the academy, generates large volumes of data that after being processed, generate useful information for the benefit of the university and the society in general.

Each university requires a general-purpose system architecture that can incorporate repositories with heterogeneous data, as it is the case of this research projects. The system must be of general purpose in which the information or research results can be accessed, allowing the exploitation of information by entrepreneurships and new

© Springer Nature Switzerland AG 2020
Á. Rocha et al. (Eds.): ICITS 2020, AISC 1137, pp. 172–178, 2020.
https://doi.org/10.1007/978-3-030-40690-5_17

professionals. It is important to highlight that the information will not only be available to the university's community (people) but, when it is processed it will follow the Open Data philosophy and it can be accessed automatically (machines).

1.1 Related Works

In [1], they present Octopus, an Open Source, dynamically extensible system that supports information management for internet of things applications in which its architecture is based on web services.

Another interesting work [2], uses an ubiquitous computing platform based on NFC, where these devices and applications interact with each other to provide an intelligent environment.

In [3], they use an architecture with three layers; sensing, data storage and analytics, they focus on the classroom attendance, student study space usage, parking lot occupancy and bus-stop waiting-times.

A framework is proposed in [4], that involves (a) smart people-support staffs, academic staffs and students; (b) big data analytics; (c) smart classroom-smart band, smart board and blended learning- and smart faculty-e-learning, automated building management and smart attendance. They mention that there are a lot of benefits in becoming a smart university, such as increasing the education quality, the research output by implementing the right analysis and understanding lots of information through the use of big data, improved ranking and performance.

In this study [5], a daily analysis of internet traffic is carried out, in addition, four learning approaches were used to analyze and compare the collected information, however, they don't mention any architecture for the treatment of that data.

In [6], they consider the self-learning property as a basic constitutive feature of a smart university and validate mathematically this criteria, in addition, they consider the smart university as a self-learning organization functioning on the basis of a team of like-minded people that has an established mechanism, such as: learning ability, relation convergence, the transfer and increase of knowledge, meeting the society requirements, adapting promptly to contemporary economic conditions.

In [7], they developed and tested models of defining a smart university infrastructure development level, taking into account some indices of assessment like: the use of smart platforms, smart technologies, smart knowledge management systems, the teaching staff using smart technologies, the use of mobile devices during the learning process and the use of the e-learning tools.

Another study [8], shows the technologies that are implemented in smart campuses and smart universities which are: big data, cloud computing, internet of things and artificial intelligence. They indicate by "Smart campus" four thematic axes: infrastructure, governance and management, services and education; and for "smart university" highlights the development of an architecture that contributes to the active application of smart technologies and devices in the educational process.

In [9], they propose a system architecture for smart universities that provides smart building monitoring and management. The proposed a solution that integrates heterogeneous geographically disparate sensor networks and devices and enables optimal operations of the building while reducing its energy footprint.

In [10], they propose the concept of Smart University describing needs and advantages and ending with a possible architecture based on smart objects.

In [11], they present a framework architecture for integrating various types of wireless networks into a smart university campus to enhance communication among students, instructors, and administration.

In [12], they focuses on how to leverage IoT technologies to build a modular approach to smart campuses. The work identifies the key benefits and motivation behind the development of IoT-enabled campus. Then, it provides a view of general types of smart campus applications.

In the present work, we proposed an architecture that permits establishing an infrastructure that allows collecting data and visualize the processed information. This architecture establishes the basis for a university to become smart. We believe that the first step is the collection, processing and visualization of information that comes from the results of research projects, so that, in the future, inferences can be made, optimization process can be carried out, new research-ideas can be proposed and provide new tools for the generation of new business ideas.

This work is divided into four sections. Section 1 presents an introduction and literature review about "Smart University". Section 2 presents a theoretical background. Section 3 presents the architecture proposal and the final section summarizes the conclusions, recommendations and future work.

2 Theoretical Background

The concept of smart university requires universities to provide students with suitable software/hardware systems and assistive technologies that will help them to succeed in technological learning environments such as smart classrooms and laboratories, smart libraries, and smart campuses [13].

The Smart University involves a comprehensive modernization of all educational process [14]. The architecture involves several layers, where the number of layers decides the architecture complexity; five-layer architecture is the ideal from the perspective of security and complexity and should integrate the main technologies that actually exist: (a) big data, (b) cloud computing, (c) internet of things and (d) data analysis.

In [15], they present the outcomes of an ongoing research project at the InterLabs Research Institute, Bradley University (Peoria, IL, U.S.A.) aimed to validate "smartness-features" for a smart university; in this study, they include some components:

- Smart software and hardware systems: Smart learning analytics systems, web-lecturing systems, collaborative web-based audio/video one-to-one and many-to-many communication systems, interactive whiteboards, panoramic video cameras, robotic controllers and actuators, etc.
- Smart technologies: Internet-of-things technologies, cloud computing technology, web-lecturing technology, smart agents technology, augmented and virtual reality technology, sensors, and so on.

- Smart pedagogy: Collaborative learning, crowdsourcing-based learning, serious games and gamification-based learning, smart robots-based learning, etc.
- Stakeholders: Students (local, online), lifelong learners, students with disabilities, professional staff, etc.

Finding that, the main features of a smart university are:

1. Adaptation that consists in the ability to automatically modify its teaching/learning strategies, administrative workflows, safety, technological and other characteristics.
2. Sensing, that deals with the ability to automatically use various sensors/control devices to understand the university's operation, infrastructure or well-being of its components.
3. Inferring, that means the ability to automatically make logical conclusions from the basis of raw data.
4. Self-learning, that consists in the ability to automatically obtain acquire or formulate new knowledge or modifying it.
5. Anticipation, that deals with the intelligence and predictive analytics in software systems that have the ability to automatically collect raw data.
6. Self-organization, with the ability to automatically change its internal structure without the intervention of an external agent.

3   Architecture Proposal

The main objective of this work is to create and implement a "Smart University" architecture that allows storing data from research projects-data before processing and information generated after that data has been processed to generate visualizations, reports and so on. The Fig. 1 shows the proposed architecture.

3.1   Capturing Layer

In this layer are identified sensors, social networks, data generated from the web that could be extracted and stored, and data that have been manually captured-surveys, interviews, physical records, etc.

The raw data measured from sensors will first pass through an interface that allows that information to be transformed into readable data and then stored in the general database. We decided to use JSON-JavaScript Object Notation- because it is a light-weight format for data exchange. Each observation or measurement must be transformed to this format that must have: "sensor identifier", "research project identifier" and "content". Then, this data is transferred through the web service so that it is stored on "Raw data" server.

In this work, it is essential that each research project that incorporates sensors, must take into account and eliminate those commercial sensors that send data to the provider's cloud, since it is indispensable that the data is kept in the databases and don't leave the campus or our proposed infrastructure, in addition, this allows us to have independence and don't depend on a provider to access our data.

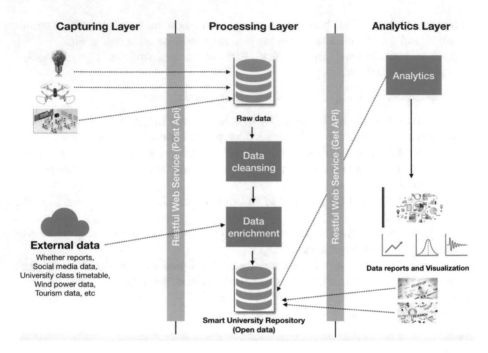

Fig. 1. Smart University architecture

3.2 Processing Layer

It stores all data that have been captured in the previous layer, specifically on a raw data server. After being stored they go through a cleaning process and then through an enrichment process-if necessary. Then, the data is stored in another centralized and global server where all the processed information will be kept. The information in this server will be aligned with the Open Data philosophy.

Linked data describes a method of publishing structured data that can be interconnected and more useful. Despite the efforts made in this line of research, it is still not possible for the web to be interconnected, in such way that necessary metadata can be used to obtain significant conclusions from this data. There are efforts made by CKAN, Dbpedia, GeoNames, etc.; which create semantic data that can be linked and interconnected but there is still no massification of this process [16]. One of the first steps is to propose architectures that follow this philosophy.

3.3 Analytics Layer

This layer allows getting processed data from the central repository to generate analysis, reports and visualizations. This new source of information will allow the university community (especially students and graduates) to generate business ideas.

This architecture is composed by a Restful Web Service that interconnects all layers and will allow the entire university community, to obtain manually the information of the research projects, but it can also work so that machines can automatically access the data-machine to machine communication.

With the proposed architecture, it is established the basis for the management of information from research projects, since all of them are based on data capture either manually or automatically-sensing-; while adaptation, inferring, self-learning antici-pation and self-organization are tasks that will be developed later.

This work tries to establish an architecture for information management for research projects. This is aligned with the open science initiative that has the aim of improving science, with some benefits such as: improving the interpretation, under-standing and reproducibility of research results; greater transparency in the life cycle of science avoiding fraud; reduction of the cost of research projects; promotion of the reuse of results and better identification and evaluation results [17]. This will allow to obtain the raw data, the processing and the results of the researches through platforms such as Rpubs o github; with that, students will be able to propose new uses of data through degree works, innovate through startups or continue researching based on these results.

4 Conclusions

In this work, we propose an architecture that allows to manage all the collected information from research projects, and to efficiently access the results of that research. This open data repository will have a web service to access the information, so that anyone can access and use it to generate new business ideas for the benefit of the society. This will allow transparent processes, the reuse of information and universal participation.

In the other hand, it is remarkable that Smart Universities will be persuaded to automate common processes for being more efficient and optimize the consumption of resources in order to be environmentally friendly. Moreover, in the coming years, universities will have to adapt their infrastructure to the needs of remote students.

One of the biggest challenges is the privacy of the data, especially when we are using cameras to make facial or vehicle license recognition; that information cannot be shared between different entities. In addition, each sensor provider has a format in which the collected data is presented for processing. Another major challenge is to standardize heterogeneous information in JSON format and then it can be processed in a homogeneous way.

As a future work, we plan to share the experience of having an open data platform for research projects and to contribute with the ideas that have been generated from processed information. Initially, the proposed architecture will follow the Open Data philosophy without linking and interconnecting data; subsequently, it is intended to identify all the ontologies for each of the research projects to get interconnecting the data that has been stored.

References

1. Firner, B., Moore, R.S., Howard, R., Martin, R.P., Zhang, Y.: Smart buildings, sensor networks, and the Internet of Things. In: Proceedings of the 9th ACM Conference on Embedded Networked Sensor Systems - SenSys 2011, p. 337 (2011)
2. Bueno-Delgado, M.V., Pavon-Marino, P., De-Gea-Garcia, A., Dolon-Garcia, A.: The smart university experience: an NFC-based ubiquitous environment. In: 2012 Sixth International Conference on Innovative Mobile and Internet Services in Ubiquitous Computing, pp. 799–804 (2012)
3. Sutjarittham, T., Gharakheili, H.H., Kanhere, S.S., Sivaraman, V.: Realizing a smart university campus: vision, architecture, and implementation. In: 2018 IEEE International Conference on Advanced Networks and Telecommunications Systems (ANTS), pp. 1–6 (2018)
4. Shamsuddin, N.T., Aziz, N.I.A., Cob, Z.C., Ghani, N.L.A., Drus, S.M.: Big Data Analytics Framework for Smart Universities Implementations, pp. 53–62. Springer, Cham (2019)
5. Adekitan, A.I., Abolade, J., Shobayo, O.: Data mining approach for predicting the daily Internet data traffic of a smart university. J. Big Data 6(1), 11 (2019)
6. Glukhova, L.V., Syrotyuk, S.D., Sherstobitova, A.A., Pavlova, S.V.: Smart University Development Evaluation Models, pp. 539–549. Springer, Singapore (2019)
7. Mitrofanova, Y.S., Sherstobitova, A.A., Filippova, O.A.: Modeling the Assessment of Definition of a Smart University Infrastructure Development Level, pp. 573–582. Springer, Singapore (2019)
8. Rico-Bautista, D., Medina-Cárdenas, Y., Guerrero, C.D.: Smart University: A Review from the Educational and Technological View of Internet of Things, pp. 427–440. Springer, Cham (2019)
9. Stavropoulos, T.G., Tsioliaridou, A., Koutitas, G., Vrakas, D., Vlahavas, I.: System Architecture for a Smart University Building, pp. 477–482. Springer, Heidelberg (2010)
10. Cata, M.: Smart university, a new concept in the Internet of Things. In: 2015 14th RoEduNet International Conference - Networking in Education and Research (RoEduNet NER), pp. 195–197 (2015)
11. Khamayseh, Y., Mardini, W., Aljawarneh, S., Yassein, M.B.: Integration of wireless technologies in smart university campus environment. Int. J. Inf. Commun. Technol. Educ. 11(1), 60–74 (2015)
12. Abuarqoub, A., et al.: A Survey on Internet of Things enabled smart campus applications. In: Proceedings of the International Conference on Future Networks and Distributed Systems - ICFNDS 2017, pp. 1–7 (2017)
13. Bakken, J.P., Varidireddy, N., Uskov, V.L.: Smart University: Software/Hardware Systems for College Students with Severe Motion/Mobility Issues, pp. 471–487. Springer, Singapore (2019)
14. Uskov, V.L., Bakken, J.P., Pandey, A., Singh, U., Yalamanchili, M., Penumatsa, A.: Smart University Taxonomy: Features, Components, Systems, pp. 3–14. Springer, Cham (2016)
15. Uskov, V.L., Bakken, J.P., Gayke, K., Jose, D., Uskova, M.F., Devaguptapu, S.S.: Smart University: A Validation of 'Smartness Features—Main Components' Matrix by Real-World Examples and Best Practices from Universities Worldwide, pp. 3–17. Springer, Singapore (2019)
16. Larkou, G., Metochi, J., Chatzimilioudis, G., Zeinalipour-Yazti, D.: CLODA: a crowd-sourced linked open data architecture. In: 2013 IEEE 14th International Conference on Mobile Data Management, pp. 104–109 (2013)
17. Assante, M., et al.: Enacting open science by D4Science. Futur. Gener. Comput. Syst. 101, 555–563 (2019)

GRAY WATCH: An Extended Design Process

Jorge-Luis Pérez-Medina$^{(\boxtimes)}$ (iD)

Intelligent and Interactive Systems Lab (SI2 Lab),
Universidad de Las Américas (UDLA), Quito, Ecuador
jorge.perez.medina@udla.edu.ec

Abstract. The process of developing a software product must have
mechanisms to verify the quality of the product and the process under
development. Talking about the quality of the resulting software requires
taking into account a series of parameters that allow defining the minimum levels that a software product must meet to be categorized of
quality. Thus, quality concerns corresponds to properties that characterize an architectural solution. The GRAY WATCH (GW) method is
framed in the development of business software under the component
reuse paradigm. This method lacks mechanisms that allow obtaining
a quality architecture and needs to be improved. In order to resolve
this deficiency, this article proposes to define and incorporate the activities and products that allow to expand the design process of GW. The
aim of the proposals is to consider the ISO/IEC 25010 product quality
standard, which defines criteria to stipulate the quality requirements for
a software product, their measurement criteria and evaluation. It take
into consideration a quality model composed of characteristics and sub-
characteristics. The extension is supported by the Domain Engineering
process, based in Software Quality "InDoCaS" and the "InDoCaSE" process as the methodologies for the definition of all products and activities
in the Design process.

Keywords: GRAY WATCH · Design process · Software quality ·
ISO/IEC 25010 · InDoCaS · InDoCaSE

1 Introduction

Method engineering (IM) aims to assist in the creation of new methods and techniques of information systems engineering. Among its fundamental principles are
optimization, reuse and adaptation [25]. [4] defines IM as "a discipline of conceptualization, construction and adaptation of methods, techniques and tools for
the development of information systems". Basically, it deals with the definition
of new engineering methods for the domain of Information Systems (SI). Other
definitions highlight the notion of IM to the construction of new methods, for
example: [22] defines IM as "an approach to building methods combining different (parts of) methods to develop an optimal solution of the given problem". On

© Springer Nature Switzerland AG 2020
Á. Rocha et al. (Eds.): ICITS 2020, AISC 1137, pp. 179–188, 2020.
https://doi.org/10.1007/978-3-030-40690-5_18

the contrary, Kumar [11] proposes a more general definition that does not impose the use of existing methods as a starting point for IM when defining the latter as "a mechanism for the design and development of a meta-methodology for the design of information systems development methods". We define the IM as the discipline that points its efforts in providing effective solutions, focused on the design, the improvement, and also, the evolution of the development methods of the IS. In the domain of the SI, the references, norms and standards generally represent the state of the art and the know-how of a given context. Beside the product standards, management standards appear more present; These introduce an organizational level to the technical aspects naturally taken into account by computer services. ISO 9001 specifies the requirements required by a quality management system applicable to all activities and businesses [24]. Talking about the quality of the resulting software requires having the parameters that allow us to establish certain minimum levels that a software product must have in order to be categorized as a quality product [21]. In practice, the quality parameters of software products are often described in general terms [28], it being difficult for the project group to validate the quality of the software product against the requirements. Several authors have created several models to validate its quality, among which are: Boehm [1] McCall [13], and ISO/IEC [16]. The ISO/IEC 25000 standard incorporates a series of international standards called "Requirements and Quality Evaluation of Software Products" (SQuaRE) [16]. SQuaRE establishes a servie of criteria for the correct definition of quality requirements for software products, their goals and their evaluation SQuaRE is formed by the division of the quality model through Standard 25010 that recommends the use of a quality model through eight quality characteristics such as: Functionality, Performance, Safety, Compatibility, Usability, portability, reliability, and maintainability.

In this context, we refer to the WATCH [15] method in its latest version GRAY WATCH (GW) [14]. The latter is framed in the development of business software under the component reuse paradigm. GW considers in the Requirements Engineering phase, the identification of functional and non-functional requirements, without specifying the specific activities to obtain the requirements and the quality architecture. It method has been only extended in the analysis process [18]. However, this method does not have a mechanism to guarantee software quality for the other phases. Consequently, in order to advance the component-based and requirements-driven design of (software) systems and to ensure quality in the final artifact or software product, a process that considers the specification of quality requirements, the design of quality models and architectures based on quality attributes, which allow the software to be implemented using quality artifacts must be considered. This article proposes to define and incorporate the activities and products that allow to expand the design process of the GW method. The extension applies the standard of quality "ISO/IEC 25010" and couples the artifacts generated in the Design process with the implementation process. We used the processes for Domain Engineering based on Software Quality called InDoCaS [5] and the InDoCaSE [23] process as a methodology for the specification of the activities and products. Both processes are based on the standard of quality "ISO/IEC 25010". The article is organized as follows. Section 2, presents a conceptual framework of the fundamental principles and concepts related to the research context. Next, the proposal to extend the GW method in its Design process is presented. Finally, the conclusions and perspectives are discussed.

2   Related Works

The term method has its origin from Greek "methodos" which refers "means of investigation". Harsem [9] defines the means of investigation as a set of procedures, techniques, product descriptions and tools for the efficient, effective and consistent assistance of the engineering process of an information system. Other definitions establish a clear separation between the product that the method produces and the process that produces the product [3, 8, 10, 12, 19, 26, 27, 29] and [4]. This acceptance is synthesized by Booch [2] that defines a method as a rigorous process allowing to generate a collection of models describing different aspects of a software product under construction. In other words, a method addresses the two aspects of engineering, the product and the process, and is based on two elements: one or several models of products and one or several models of processes [20, 26]. This approach is used by the GW method that is methodologically based on a process model, a product model and an actor model.

2.1   The GRAY WATCH Method

GRAY WATCH [14] is a methodological framework that describes the technical, managerial and support aspects that must be used by the work teams that will be responsible for the development of business software applications. A methodological framework is a pattern that must be instantiated, that is, adapted every time [14] is used. Each work team must use the method as a methodological template, from which said team must develop the specific process of developing the application that is produced.

GRAY WATCH covers the entire application life cycle; from modeling the domain of the application, through the definition of user requirements, to the implementation of the application. It method is composed of three fundamental models: the product model, the actor model and the process model [14]. The latter establishes the processes necessary to manage business application development projects and carry out the technical and support activities required by these projects. The technical processes of the method are divided into three groups: Analysis, Design and Implementation Processes. The application analysis processes cover the Business Modeling and Requirements Engineering processes. The design phase consists of the Architectural Design and Component Design processes, while the Implementation processes group the Programming and Integration, Testing and Application Delivery processes.

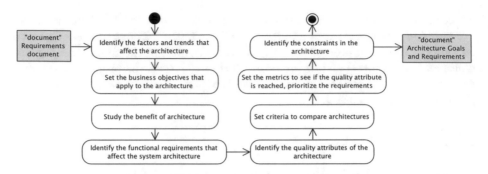

Fig. 1. The sub-process "Definition of design goals and Architectural Requirements". Adapted from: [14].

The GRAY WATCH Design Processes have as a general objective to specify the structure and the set of components that must conform the business application so that it satisfies the established requirements [14]. For this, appropriate methods, techniques and tools are used to define both the general design of the application (its architecture) and to describe in detail each of its components; that is, the user/application interface, databases, programs, documentation and procedures. This group of processes is composed of the Architectural Design (DA) and Detailed Application Design (DD) processes.

The DA produces the structure of the application that shows the components of the application, its connectors and architectural constraints, in addition to identifying the subsystems that the application has, the relations and restrictions of interaction between them and with other applications and physical distribution of each of the components. The DD describes how each of these architectural components should be implemented. This consists in allowing to specify precisely each of the components of the architecture; including the programming interfaces (APIs) of each of its components, the user/system interface, the data model and the connections provided in the architecture. The DA process is divided into four complementary sub-processes: Definition of architectural goals and requirements, the identification of subsystems, the development of architectural views and the evaluation of architecture.

The sub-process "Definition of Architectural requirements and goals" allows to identify the functional and non-functional requirements that affect the selection and design of the application software architecture, such as the attributes to evaluate the quality of the architecture and the restrictions that will serve as a guide to specify and design the software architecture. The final product of this process is the document of architectural design goals and requirements, being its workflow illustrated in Fig. 1.

The sub-process, called "Identification of Subsystems" guides the definition of the different subsystems that make up the application, the way of relating to each other and with other applications (or systems) and their objectives. The sub-process is based on some decomposition criteria according to the rules of the

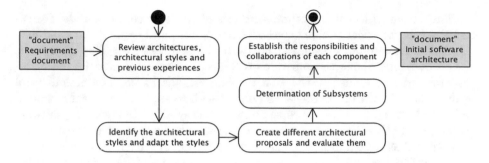

Fig. 2. Workflow of the "Subsystems Identification" sub-process Adapted from: [14].

business (by functionalities, by users, etc.) to divide the system into subsystem and relate these subsystems (components) using one or more architectural styles. The definition of the subsystems should consider the definition of criteria of the characteristics of the application and the hardware and software technology platform The final product of this process is the Initial Software Architecture document (see Fig. 2).

2.2 Software Quality

The term quality is frequently used in various areas of society. Pressman [21] indicates that the quality of the system is the agreement that must exist between the requirements, the specifications and the design process of the system. The ISO/IEC 25010 [16] standard describes a bipartite model for software product quality. This consists of international standards that present detailed quality models for software products and computer systems, including features of internal and external functionality, and quality in use. It also provides guidance on the use of quality models. Requirements Classification Model (RECLAMO) [6] is a model for requirements classification considering a quality perspective. It supports the process for specifying the requirements. Its objective is to facilitate the identification of several kinds of requirements involved in the definition of a software system, especially, non-functional requirements that related to quality requirements.

2.3 The Domain Engineering Process Based on Software Quality

InDoCaS [5] is a process for Domain Engineering based on Software Quality. It can be used in software product line development approaches, which can be instantiated for a specific domain and the produced software assets can be reused to generate a product from a particular family domain. The domain is considered as a family of products that have common characteristics. The design phase of the InDoCaS domain focuses on the synthesis and architectural evaluation of the domain. The entries to this phase are the minimal set of functional and non-functional requirements, architectural styles and catalogs of architectural patterns obtained in the domain analysis discipline.

The output of the architectural synthesis identified a set of candidate architectural solutions that respond to the set of architectural requirements and may be alternative solutions, even partial (parts of architecture). They reflect the design decisions about the software structure, initially reflected in the architectural styles, obtained in the Domain Analysis discipline. The architectural evaluation activities of the InDoCaS domain seek to analyze and identify potential risks in its structure and properties, which may affect the resulting software system, verify that the non-functional requirements are present in the architecture, as well as determining the degree to which quality attributes are satisfied.

The InDoCaSE [23] process is a combination of InDoCaS [5] and the WATCH-COMPONENT [7] method. Its objective is to support the development of software product lines and is oriented to the domain Implementation discipline. The implementation of the domain of this process results in the release of the component in an internal or external repository that can be accessed by developers. This release includes the publication of the component specification, together with the results of the tests performed.

3   The GRAY WATCH Extension in Its Design Process

Our proposal was to apply quality in the GW method design processes, for which the activities of the InDoCaS [5] process were incorporated, which allowed extending and coupling the resulting threads. For the representation of the activity diagrams that show the extension of the activities and artifacts in the GRAY WATCH method threads, the diagramming was used using SPEM [17].

To extend the GW method, the sub-processes "Definition of architectural goals and requirements" and "Identification of subsystems" contained in the Architectural Design process were selected (see Figs. 1 and 2). Both sub-processes bring together the activities related to the selection and design of the architectural components. For the extension of the process, the activities defined in the InDoCaS process were taken into account, which allow obtaining quality requirements, and also an architecture based on software requirements considering the ISO/IEC 25010 product quality standard. It was necessary to carry out a re-engineering of the processes to couple the activities and the artifacts according to the determined objective, which means that, is to apply product quality by means of the ISO/IEC 25010 standard.

Regarding the sub-process "Definition of architectural goals and requirements", three activities were incorporated that were taken from the InDoCaS process. Likewise, the rest of the activities and artifacts were modified and coupled, in order to obtain a quality model, based on the architectural requirements and compliant to the quality standard ISO/IEC 25010. For its part, the generated quality model allows the choice of architectural styles, to later choose the architectural patterns that best suit, defining the design accordingly initial of the architecture, based on the considerations, that specify the initial architectural design. This process has been called "Architecture design definition". Figure 3 illustrates the sub-process resulting from the definition of the architecture design, note that activities and artifacts have been incorporated and coupled.

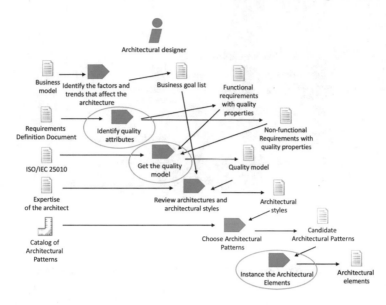

Fig. 3. Proposal for the "Architecture design definition" sub-process of the GRAY WATCH method.

The identification of quality attributes allows obtaining the functional and non-functional quality requirements by identifying the characteristics and sub-characteristics of the ISO/IEC 25010 model. **The activity of obtaining the quality model** takes into consideration the requirements of the application along with its quality properties. This activity allows to generate a quality model as an instance of the ISO/IEC 25010 standard. **The review of architectures and architectural styles** is an activity carry out by a software architect. This task consists in using the instantiated quality model and, according to the list of requirements, identify and provide the architectural solutions. **The choice of Architectural Patterns** refers to evaluating the application of architectural patterns to each element of quality requirements. Its purpose is to obtain the most appropriate patterns for architecture. **The activity of instantiating the architectural elements** consists in selecting the patterns and types of components that meet the requirements. This activity is necessary since it is necessary to adapt the patterns to both functional and non-functional requirements.

The sub-process "Identification of subsystems", presented in the Fig. 2, is responsible for reviewing architectural styles and architectural patterns, creating architectural proposals and based on this, dividing the system into subsystems and describing each of them, its components and interactions with other subsystems. To extend this sub-process, it is necessary to add the activities "validate solutions with quality model" and "Decide the architectural solution" of the InDoCaS process which will allow validating the generated architectural proposals and then select the best proposal that satisfies the set of Requirements and quality model. The extended sub-process is called the "Architecture selection"

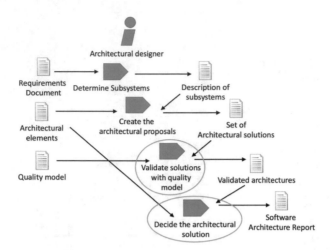

Fig. 4. Proposal for the "Architecture selection" thread for the GRAY WATCH method.

and is represented in the activity diagram illustrated by Fig. 4. The diagram of the proposed sub-process "Architecture selection" obtains the report that contains the system software architecture. Its importance derives from the fact that this device allows observing the system from different points of view. The execution of this sub-process allows the products generated in the technical design processes of the application of the GW method that we propose to be coupled.

The activity of validating the solutions with the quality model consists in ensuring compliance with the quality model in the architectural solutions. The output of this activity is a validated architecture. **The activity of deciding the selection of architecture as an architectural solution** aims to choose the architecture for the application. The document resulting from this activity is a detailed report of the software architecture. This report must contain the evaluation resulting from compliance with the quality model.

In summary, the architectural design process guides the selection of the system architecture, by identifying the characteristics and sub-characteristics of the ISO/IEC 25010 quality model with respect to the functional and non-functional requirements of the system. After carrying out this work, the different architectural styles and architectural patterns are observed in order to create the various architectural proposals. Subsequently the proposed architectures are validated with the quality model in order to choose the one that best suits.

4   Conclusions and Perspectives

Software development is a complex process, it requires the application of principles, methods, models and techniques of Software Engineering and Project Management. New research [7, 18] consider the GW [14] method as a software development method, whose methodological framework describes the technical, managerial and support processes that work teams should use, to develop the architectural components of an application and integrate it into the business system for which it is developed. The proposal presented in this study incorporates to the GW method the application of the ISO/IEC 25010 standard, thus allowing to deepen in obtaining the most appropriate product model that describes the artifacts generated in the quality process when applying this standard. Throughout the extension process, we were able to know the steps to follow to obtain quality models and architectural design. For this, sub-processes of the GW Method Design process were modified and extended. The processes and products generated were adapted to the rest of the processes in the GW method chain.

Future work will focus efforts on presenting an experimental case study that shows how to apply the proposed extension, and also, reviewing other technical processes of the GW method in which quality aspects could be incorporated by applying another series of the ISO/IEC 25000 product quality division standard, for example: we consider it possible to expand the processes Test of the Application of the GW method by incorporating product evaluation processes that can be performed by applying the ISO/IEC 25040 quality evaluation standard as it provides a guide for software quality evaluation. Likewise, we consider examining the possibility of incorporating quality metrics that allow us to evaluate the quality in use of the System developed, we will also consider applying the ISO/IEC 25020 quality measurement standard, which presents metric applications for software quality from the internal perspective, external and in use. Finally, the Software Engineering community is invited to consider other aspects or elements that could evolve the extension of the GW method and create new experiences in terms of software development with quality standards.

References

1. Boehm, B.W., Brown, J.R., Kaspar, H., Lipow, M., MacLeod, G., Merritt, M.: Characteristics of software quality (1978)
2. Booch, G.: Object Oriented Analysis and Design with Applications, the Benjamin. Cummings, Redwood City (1994)
3. Brinkkemper, J.N.: Formalisation of information systems modelling. Ph.D. thesis, Amsterdam, Thesis (1990)
4. Brinkkemper, S.: Method engineering: engineering of information systems development methods and tools. Inf. Soft. Technology **38**(4), 275–280 (1996)
5. Canelón, R.: Un Proceso para la Ingeniería del Dominio Basado en Calidad de Software. Una aplicación al dominio del aprendizaje móvil sensible al contexto. Ph.D. thesis, Tesis Doctoral. Universidad Central de Venezuela. Caracas (2010)
6. Chirinos, L., Losavio, F., Matteo, A.: Identifying quality-based requirements. Inf. Syst. Manag. **21**(1), 15–26 (2004)
7. Hamar, V., Montilva, J.: Aspectos metodológicos del desarrollo y reutilización de componentes de software. Ph.D. thesis, Tesis doctoral, Universidad de los Andes, Mérida (2003)
8. Harmsen, A.F., Brinkkemper, J.N., Oei, J.H.: Situational Method Engineering for Information System Project Approaches. Citeseer (1994)

9. Harmsen, A.F., Ernst, M., Twente, U.: Situational Method Engineering. Moret Ernst & Young Management Consultants Utrecht (1997)
10. Kronlöf, K.: Method Integration: Concepts and Case Studies. Wiley, Hoboken (1993)
11. Kumar, K.: Methodology engineering: a proposal for situation-specific methodology construction. Challenges and strategies for research in systems development (1992)
12. Lyytinen, K., Smolander, K., Tahvainen, V.: Modelling case environments in systems work. In: CASE 1989 Conference Papers, Kista, Sweden (1989)
13. McCall, J.A., Richards, P.K., Walters, G.F.: Factors in software quality. Volume I concepts and definitions of software quality. Technical report, General Electric Co., Sunnyvale, CA (1977)
14. Montilva, J., Barrios, J., Rivero, M.: Gray watch método de desarrollo de aplicaciones empresariales versión preliminar proyecto methodius fonacit 2005000165 mérida venezuela: Fonacit (2008)
15. Montilva, J.A., Hazam, K., Gharawi, M.: The watch process model for developing business software in small and mid-size organizations. In: Proceedings of the IV World Multiconference on Systemics, Cybernetics, and Informatics, vol. 12, pp. 263–268 (2000)
16. de Normalización, O.I.: ISO-IEC 25010: 2011 Systems and Software Engineering-Systems and Software Quality Requirements and Evaluation (SQuaRE)-System and Software Quality Models. ISO (2011)
17. OMG Standard, Notation OMG: Software & systems process engineering meta-model specification. OMG Std. Rev. **2**, 18–71 (2008)
18. Pérez-Medina, J.L., Sánchez, I.C.: Hacia la extensión del método gray watch basado en el estándar de calidad iso/iec 25010, vol. 6, pp. 5–19 (2012)
19. Prakash, N.: A process view of methodologies. In: 6th International Conference on Advanced Information Systems Engineering, CAISE 1994 (1994)
20. Prakash, N.: On method statics and dynamics. Inf. Syst. **24**(8), 613–637 (1999)
21. Pressman, R.S.: Software Engineering: A Practitioner's Approach. Palgrave Macmillan, London (2005)
22. Punter, T., Lemmen, K.: The mema-model: towards a new approach for method engineering. Inf. Softw. Technol. **38**(4), 295–305 (1996)
23. Rivero, L.: Una línea de producción de software para sistemas transaccionales. una aplicación al proceso de desarrollo de software en la coordinación nacional de tecnología de información de la u.n.e.x.p.o. Trabajo de Grado, Maestría en Ciencias de la Computación. Universidad Centroccidental Lisandro Alvarado (UCLA), Barquisimeto, Venezuela (2011)
24. Rivet, A.: Normes de qualité et systèmes d'information. les journées réseaux, pp. 20–23 (2007)
25. Rolland, C.: A guided tour of method engineering. Electron. J. Inf. Technol. (1) (2005)
26. Seligmann, P., Wijers, G., Sol, H.: Analyzing the structure of is methodologies, an alternative approach. In: Proceedings of the First Dutch Conference on Information Systems, pp. 1–28. Amersfoort, The Netherlands (1989)
27. Smolander, K., Lyytinen, K., Tahvanainen, V.P., Marttiin, P.: Metaedit—a flexible graphical environment for methodology modelling. In: International Conference on Advanced Information Systems Engineering, pp. 168–193. Springer (1991)
28. Van Veenendaal, E., Hendriks, R., Van Vonderen, R.: Measuring software product quality. Softw. Qual. Prof. **5**(1), 6 (2002)
29. Wynekoop, J.L., Russo, N.L.: System development methodologies: unanswered questions and the research-practice gap. In: ICIS, pp. 181–190 (1993)

Suitable Electric Generator for Pico-Hydro Power Plant in Ambato–Huachi–Pelileo Water Irrigation Channel

Myriam Cumbajín[1] , Patricio Sánchez[1] , Andrés Hidalgo[2] ,
and Carlos Gordón[2(✉)]

[1] Facultad de Ingeniería, Tecnologías de la Información y la Comunicación,
Universidad Tecnológica Indoamérica, UTI, 180103 Ambato, Ecuador
{myriamcumbajin, patriciosanchez}@uti.edu.ec
[2] Facultad de Ingeniería Civil y Mecánica,
Facultad de Ingeniería en Sistemas, Electrónica e Industrial,
Universidad Técnica de Ambato, UTA, 180150 Ambato, Ecuador
{andresshidalgo, cd.gordon}@uta.edu.ec

Abstract. We present the analysis of the selection of electric generator suitable for Pico-Hydro Power Plant in the Ambato-Huachi-Pelileo irrigation channel in Ecuador. For which a meticulous literary review has been carried out about the main characteristics of the Pico-Hydro Power Plant as it is the height, the hydric resource, the turbine and the generator. Considering the water resource, which in our case is provided by the Ambato-Huachi-Pelileo irrigation channel, the parameters of low height and slow speed have been taken around 1.6 m^3/s, which has allowed us to analyze the generator suitable for the required Pico-Hydro Power Plant. Therefore, the study conducted has allowed us to conclude that the electric generator suitable for the Pico-Hydro Power Plant in the Ambato-Huachi-Pelileo irrigation channel is a radial flux permanent magnet generator that is capable of providing sufficient electrical power with a simple structure, low revolutions, and high efficiency. Also, a Matlab simulation was developed in order to predict the power which is capable of provide the radial flux permanent magnet generator which in this case is 3.8 KW. This work aims to provide several benefits to the society bordering the irrigation channel, such as generating energy for self-consumption as a contribution to the lighting of the channel at no cost to the public in order that it is not dangerous for the passers at night and providing the ability for maintenance works at any time.

Keywords: Electric generator · Pico-Hydro Power Plant · Water Irrigation Channel · Matlab

1 Introduction

Electric power is one of the most important and indispensable resources worldwide, most services and components require energy for efficient operation and performance [1]. Among services that require electricity, we can mention: Hospitals, Clinics, Banks, Educational centers, all the activities of the general public, among others. While

© Springer Nature Switzerland AG 2020
Á. Rocha et al. (Eds.): ICITS 2020, AISC 1137, pp. 189–199, 2020.
https://doi.org/10.1007/978-3-030-40690-5_19

between the electrical and electronic components, they necessarily need electrical power for their operation and performance. So electric power is essential for several applications. In addition, considering the need exposed, it is important to emphasize that the demand for electricity is at all levels and that many rural, remote and poor sectors lack this necessary resource which reduces their chances of improving their lifestyle [2]. Fuels are a way to generate electricity in remote and poor places, but they would be limited by the economic need to dispose of them. Currently, the generation of electrical energy is framed in the use of alternative energies to replace conventional energy sources, which allows a marked reduction in the use of fossil fuels [3]. There are several ways to generate clean energy such as: Photovoltaics, Wind turbines, Hydro-electric and Biomass [4]. All of them demand a cost per year, but the small-scale generation is the most attractive due to its cost in development, implementation and maintenance, among which we can mention the Pico-Hydro Power Plant [5].

A Pico-Hydro Power Plant (PHP) is a small-scale electric power generation system with a capacity of less than 5 KW of power [6]. Pico-Hydro Power Plant technology is not new, rather it corresponds to a technology that has been maturing in parallel in the last 30 years to hydroelectric plants of other scales [7]. Pico-Hydro Power Plant have been widely used, as can be mentioned in rural electrification in developing countries, particularly in China, and other countries in the Middle East such as the Philippines, India, Laos, etc. As well as its use in Latin America has been considered: Bolivia, Colombia, Ecuador and Peru [8]. PHPs are commonly called family hydroelectric, because they are widely used in individual houses or in a group of houses. Considering a recent study, it is estimated that there are around four million Pico-Hydroelectric systems installed in the world which confirms the degree of applicability and use that can be given to the Pico-Hydro Power Plant [9]. One of the most attractive aspects of the PHPs is the low cost of the system, the continuous availability, and the reduced maintenance requirements, being a technology much desired by the most vulnerable sectors that would have access and availability without the need for so much money. Considering the Pico-Hydro Power Plant systems from the most economical, you can see the Chinese systems that range between $ 20 and $ 50. Also, you can take into account the more robust systems like the Vietnamese that range between $ 50 and $ 100. The low costs of PH systems are those that result in a reduced cost of life cycle per year that are in a range of $ 74 to $ 150/year and become more attractive than solar, wind and hybrid systems with the lowest life cycle cost per year that is at least $ 140/year [10].

The most important and determining characteristics in the development and implementation of the PHPs are the height, the water resource, the turbine and the generator. About height, high-rise sites and low-rise sites are considered. In high altitude sites, much flow is not required and an efficient turbine for this case is the Pelton turbine [11]. On the other hand, in low-lying sites, a large flow is needed, in which the Kaplan turbines are the most efficient [12]. Considering the water resource, PHPs are generally used in rivers or in water channels. In the case of rivers, there may be a large flow, but the behavior is very dependent on the seasons of the year, with a lot of variation in electricity generation. On the other hand, when the PHPs are housed in a water channel there is a relatively constant flow, but it is generally very low, requiring in this case the analysis of another component such as the generator [13]. In the case of generators, it is necessary to consider the torque and the number of revolutions that the turbines are capable of providing. It is important to emphasize that besides carrying out turbine selection studies suitable for the height and flow available, also, the selection of a suitable generator for the Pico-Hydro Power Plant must be made.

The objective of this work is to present a detailed study of the most appropriate generators to be used in a pico-hydroelectric plant that is intended to be implemented in the Ambato-Huachi-Pelileo irrigation channel and the power which is capable of provide by the Matlab simulation. The determining factors of the system have been considered, such as the flow of the irrigation channel, the height and the torque of the turbine that will be implemented in the system, so that the study of the selection of the required generator will be carried out in accordance with the parameters available in the area. This work is intended to contribute additionally to the studies of turbines suitable for Pico-Hydroelectric so that also suitable generators are available so that in the future you will have the knowledge and experience for the selection of the most appropriate and efficient components for the Pico-Hydro Power Plant development.

2   Electric Generators

Generators are electrical machines that have become essential in recent times and are widely used from the large units of generators located in power plants producing electricity to machines used in passenger transport and industry. The generator is a rotating instrument that is responsible for transforming mechanical energy into electrical energy, therefore, it is an electrical machine. The fundamental principle of its operation is based on Faraday's law, which mentions that when a coil is rotated in the internal zone of a magnetic field, a variation of the flow of said field is generated, producing an electric current. It therefore maintains a potential difference between two points called poles. The figure shows the rectangular loop that rotates within a magnetic field, so that the flow of the field through it varies. Then, a current is created that circulates through the loop, so that between the terminals represented in green, a potential difference ΔV appears, which indicates that the generator, with the movement of its rotor, generates electrical energy [14]. The generator is mainly constituted by a rotor and a stator. The stator is the fixed external part of the generator in which the coils are located which, when induced, produce the electric current. The stator is placed on a metal casing that also serves as a support. The rotor, on the other hand, is the mobile component that turns inside the stator and causes an inductor magnetic field that allows to induce the stator coils and generate electrical energy [15]. There are several types of generators that can be broadly divided into two groups: DC machines and AC machines.

2.1   DC Machines

The DC machines, better known as dynamos, are characterized in that the inductor is in the stator, which is of salient poles, and the armature is in the rotor. Both windings are connected to continuous voltages, but the induced winding receives its voltage through a delta collector, so the current flowing through it is alternating [16]. When acting as a generator, alternating currents are generated in the armature, which are rectified by the Delgas collector, so that continuous voltage is supplied to the outside. One of the most outstanding DC machines can be mentioned the brush machine that has great advantages such as being powerful and flexible. But it also has its disadvantages because, being of sliding brushes, tend to wear out causing voltage drops and several problems at high speeds and load.

The DC machines acting as a generator shown in Fig. 1 can be classified according to the arrangement of the inductor and induced windings. Of which, we can mention the DC generators of: Independent excitation since the windings are separated. Shunt excitation by its winding of said configuration. Series excitation for its windings arranged in series. Compound excitation Long derivation by its inductor winding and shunt winding in series to the armature forming a mesh, the voltage output being that of the shunt winding. Compound excitation Short shunt in which the shunt winding is parallel to the induced winding and this configuration arranged in series with the inductor winding generates the total output voltage [17].

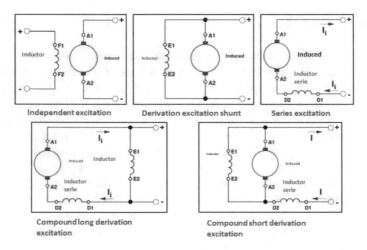

Fig. 1. DC machines classification.

2.2 AC Machines

Considering the AC machines better known as alternators, two types can be distinguished: AC Asynchronous or Induction Machines and AC Synchronous Machines.

AC Asynchronous machines have a magnetic circuit without salient poles, with both the stator and the rotor being slotted, which will be subjected to the action of rotating magnetic fields [18]. The rotor can be of two types: squirrel cage or short circuit and rotor coil or with rings. While the stator generally has a three-phase winding but at low power can be a single-phase or two-phase winding.

When the asynchronous AC machine acts as a generator, the condition must be fulfilled that the speed of the motor must be greater than that of synchronism. The sense of these powers is the opposite of that of functioning as an engine. In this case the pair is negative, and the speed is positive. Therefore, the torque of the induction machine is opposed to the speed and it is a braking torque, which requires another torque that moves the group and is the one that is forcing it to turn at a speed higher than the speed of synchronism, for our case the component responsible for generating the torque is the turbine. In addition, it is important to emphasize the greatest advantage of asynchronous AC or induction machines which is the simplicity. They only have one moving part, the rotor, which makes them low cost, silent, long lasting and relatively robust free of problems. But, it is also necessary to evoke its disadvantages, as it is that the speed of an induction motor depends on the frequency requiring a variable frequency drive to be able to control the speed, being also less efficient than its similar synchronous AC machine.

The synchronous AC machines are characterized because the armature is located in the stator and the inductor is located in the rotor. This architecture is built, because it is important that the voltages and currents in the rotor and, above all, in the ring collector be as small as possible [19]. Other, low power synchronous AC machines lack winding inductor and collector because they are replaced by permanent magnets as shown in Fig. 2. The air gap of these machines is usually greater than in the induction.

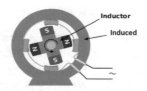

Fig. 2. Synchronous AC machine with permanent magnets

When the synchronous AC machine acts as a generator, when rotating the rotor the stator phases see a moving magnetic field. This results in these phases being subjected to a variable flux in time and inducting in them alternating electromotive forces generating electrical output energy [20]. Considering the advantages of synchronous AC machines it can be mentioned that their dominant advantage is that they can function as an isolated generator since they do not require a network that supplies reactive power. The frequency is stable fixed by the axis. Likewise, the synchronous AC machine has the advantage of efficient coupling, flexibility and simplicity by having a rotor with some remaining magnetization. But as a disadvantage, it can be mentioned that they are not 100% reversible requiring complex circuitry to achieve it.

3   Suitable Generator for Pico-Hydro Power Plant

For the PHP, considering its low power generation, it is required generators to produce power in that range and proportions advantages in this operating regime that in our case is intended to reach around 2 KW. Considering the advantages and disadvantages of the generators, it can be determined that a candidate for our purpose is the synchronous

AC machine with permanent magnets acting as a generator. For that, it is necessary to make a detailed study of the configurations of this type of generators.

Within synchronous AC machines with permanent magnets acting as a generator, two configurations can be highlighted, as follows: Axial Flow Permanent Magnet Generator (AFPMG) [21] shown in Fig. 3(a) and Radial Flow Permanent Magnet Generator (RFPMG) which is shown in Fig. 3(b) [22].

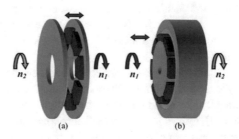

Fig. 3. Permanent magnet generators. (a) Axial Flux. (b) Radial Flux.

3.1   Axial Flux Permanent Magnet Generator

The AFPMG is characterized in that the flow passes from the rotor to the stator axially, in which the windings are oriented in the radial direction. The magnets can be located on the surface or be embedded, which allows reaching high speeds. The AFPMG are the generators of great application for electricity generation and those that are most available in the market. The main characteristic of the AFPMG is its capacity to deliver electrical energy at low revolutions, with an acceptable efficiency [23].

The structure of an AFPMG consists of a pair of thin iron disks that revolve around an axis perpendicular to it, and in whose contour are placed a set of magnets with their respective coils, which create a magnetic field parallel to the axis of rotation [24]. The axial arrangement for cutting the magnetic field is parallel to the winding, this kind of arrangement allows an excitation of the electrons without friction and magnetic opposition, the cut that is made is 180 generating lower losses than the radial generation.

In the architecture of an AFPMG the coils fulfill a primary function. For its elaboration, in the case of coil with air core, the conductor is wound on a hollow support and later it is removed with an aspect like that of a spring. A variant of the previous coil is called a solenoid and differs in the insulation of the coils and the presence of a support that does not necessarily have to be cylindrical. The output voltage of this generator depends on the way the coils are connected in series or parallel. The series connection consists of joining the end of the first coil with the beginning of the next so that the voltages are added. On the other hand, for parallel connection the initial and final terminals are connected to each other and the resulting voltage is the same as that of a coil getting a high current. In parallel, it is extremely difficult for the coil voltage to be equal, which leads to the generation of parasitic currents that waste energy.

Another fundamental component in the AFPMG are the permanent magnets that are developed on the basis of neodymium-iron-boron (NdFeB) that has low resistance to corrosion due to its coating with thin layers of nickel and chromium to isolate the base material from the environment, the disadvantage of demagnetizing at lower temperatures like other compounds. Magnets with a high degree of magnetization are required, which make it possible to locate several magnets in each rotor with the direction of their polarization pertinent to the application, being the most suitable neodymium magnets grade 40 [25].

3.2   Radial Flux Permanent Magnet Generator

The RFPMG is characterized in that the flow passes from the rotor to the stator radially in which the windings are oriented in the axial direction [26]. Among the main advantages of the RFPMG are the compact structure, the high torque capacity, high efficiency due to the rotor windings and excitation losses and above all the high power density compared to the AC or induction asynchronous machines. Considering the structure of the RFPMG, they usually have the rotor inside and the stator on the outside, although in some cases they are positioned upside down, which allows a better evacuation of the heat. One of the main components are the permanent magnets made in neodymium that are in the rotor. These magnets can be located on the surface or be embedded which allows reaching high speeds [27]. The coils in the RFPMG also fulfill a primary function similar to the AFPMG. Depending on its serial or parallel connectivity we will have the output voltage, even with the appropriate configuration you get generators of one, two or three phases. In addition, it is important to emphasize that serial or parallel connectivity have the same advantages and disadvantages previously mentioned in the AFPMG. Next, Table 1 is presented, which shows a comparison between the AFPMG and RFPMG, which allows to have a very detailed and specific vision of these two types of permanent magnet AC synchronous generators.

Table 1. Comparison permanent magnet generators.

Parameters	RFPMG	AFPMG
Low Speed Operation	Possible	Possible
Gear Box	Not Required	Not Required
Power Converters	Full scale converters	Full scale converters
Excitation	Not Required	Not Required
Starting Torque	Low	Medium
Cogging Torque	High	High
Grid Interface	Easy to control	Easy to control
Construction	Simple	Complex
Power Factor	Good	Good
Efficiency	Higher	Moderate

4   Generator for Ambato – Huachi – Pelileo Water Irrigation Channel

At this point it is necessary to analyze the characteristics of the Ambato-Huachi-Pelileo irrigation channel in Ecuador as shown in Fig. 4. The Ambato – Huachi – Pelileo Water Irrigation Channel is located at 2952 meters above sea level in Ambato town in the province of Tungurahua in Ecuador. A detailed study and hydraulic characterization of the AHPWIC has been carried out in 2.7 km length and 0.0017 degrees slope [28]. The results reveal that the Ambato – Huachi – Pelileo Water Irrigation Channel mainly has trapezoidal sections of geometry. Also, the Ambato – Huachi – Pelileo Water Irrigation Channel has an average stable flow Q = 1.6 m^3/s, speed V = 1,66 m/s and a head H = 2.5 m. obtained from the measurements. It is important to highlight that the head and flow are the primary features considered in the turbine selection, and the development of the electric generator by using a multi-criteria analysis. Therefore, considering the flow rate and generator analysis, it is determined that our most preferred candidate is the RFPMG radial flow permanent magnet generator because of its low starting torque, simple construction and high efficiency, which is perfectly suited to the characteristics of the channel, as it is the low speed. Also, it is important to mention from the characterization of the AHPWIC that the required torque and the speed for the development of the RFPMG is 5 N/m and 1.66 m/s, respectively.

Fig. 4. Characterization of the Ambato – Huachi – Pelileo Water Irrigation Channel.

Also, the Matlab simulation was performed by using Simulink, in order to predict the power which can provide the RFPMG. In this case, we have used the required torque which is 5 N/m. The Matlab simulation depicted in Fig. 5, reveals that the RFPMG can provide 3.8 KW with 60 V that arrive to 120 V by using a transformer.

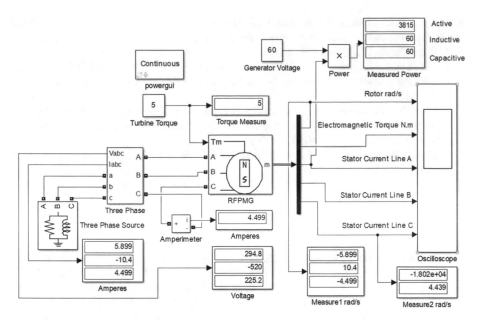

Fig. 5. Matlab simulation of RFPMG.

5   Conclusions

After having made a detailed analysis of the characteristics provided by the channel, together with the characteristics of the generators, it is considered to develop a radial flux permanent magnet generator with the following features: speed of 1.66 m/s, torque at 5 N/m, flow rate of Q = 1.6 m³/s and capable of providing 3.8 KW of power. The main characteristic that has led us to select, it is the low starting torque since the channel speed is low, simple construction and high efficiency. Therefore, it is determined that the use of a suitable generator in the Ambato - Huachi - Pelileo Water Irrigation Channel is feasible.

Acknowledgements. The authors thank the Technological Indoamerica University in Ambato – Ecuador and the "Departamento de Investigación", for their support in carrying out this research, in the execution of the project "Estudio de Energía Eléctrica de Baja Potencia, en los Canales de Riego como Fuentes Hídricas.", project code: 151.100.2018.

References

1. Gomez, A., Conejo, A.J., Canizares, C.: Electric Energy Systems: Analysis and Operation. CRC Press, Boca Raton (2018)
2. Hartvigsson, E., Stadler, M., Cardoso, G.: Rural electrification and capacity expansion with an integrated modeling approach. Renew. Energy **115**, 509–520 (2018)
3. Iniyan, S., Jebaraj, S., Suganthi, L., Samuel, A.A.: Energy models for renewable energy utilization and to replace fossil fuels. Methodology (2019)

4. Lozano, F.J., Lozano, R.: Assessing the potential sustainability benefits of agricultural residues: biomass conversion to syngas for energy generation or to chemicals production. J. Clean. Prod. **172**, 4162–4169 (2018)
5. Lahimer, A.A., Alghoul, M.A., Sopian, K., Amin, N., Asim, N., Fadhel, M.I.: Research and development aspects of pico-hydro power. Renew. Sustain. Energy Rev. **16**(8), 5861–5878 (2012)
6. Kapoor, R.: Pico power: a boon for rural electrification. Int. J. Sci. Res. **2**(9), 159–161 (2013)
7. Williams, A.A., Simpson, R.: Pico hydro – reducing technical risks for rural electrification. Renew. Energy **34**(8), 1986–1991 (2009)
8. Taylor, S.D., Fuentes, M., Green, J., Rai, K.: Stimulating the market for pico-hydro in Ecuador. IT Power, UK (2003)
9. Desai, A., Mukhopadhyay, I., Ray, A.: Theoretical analysis of a pico-hydro power system for energy generation in rural or isolated area. In: 2014 IEEE PES Asia-Pacific Power and Energy Engineering Conference (APPEEC), pp. 1–4. IEEE (2014)
10. Howey, D.A.: Axial flux permanent magnet generators for pico-hydropower. In: EWB-UK Research Conference, pp. 1–8 (2009)
11. Cobb, B.R., Sharp, K.V.: Impulse (Turgo and Pelton) turbine performance characteristics and their impact on pico-hydro installations. Renew. Energy **50**, 959–964 (2013)
12. Janjua, A.B., Khalil, M.S., Saeed, M.: Blade profile optimization of Kaplan turbine using CFD analysis. Mehran Univ. Res. J. Eng. Technol. **32**(4), 559–574 (2013)
13. Zainuddin, H., Yahaya, M.S., Lazi, J.M., Basar, M.F.M., Ibrahim, Z.: Design and development of pico-hydro generation system for energy storage using consuming water distributed to houses. World Acad. Sci. Eng. Technol. **59**(1), 154–159 (2009)
14. Bezerra, A.Z.L.N., Cabreira, F.M., Freitas, W.P.S., Cena, C.R., Alves, D.C.B., Reis, D.D., Goncalves, A.M.B.: Using an Arduino to demonstrate Faraday's law. Phys. Educ. **54**(4), 043011 (2019)
15. Kenfack, P., Matt, D., Enrici, P.: Mover guide in a linear electric generator with double-sided stationary stators. J. Eng. **2019**(17), 3986–3990 (2019)
16. Sun, L., Zhang, Z., Gu, X., Yu, L., Li, J.: Analysis of reactive power compensation effect of a new hybrid excitation brushless DC generator. IEEE Trans. Ind. Electron. (2019)
17. Zhang, Z., Yan, Y., Tao, Y.: A new topology of low speed doubly salient brushless DC generator for wind power generation. IEEE Trans. Magn. **48**(3), 1227–1233 (2011)
18. Enache, M.A., Campeanu, A., Enache, S., Vlad, I.: Dynamic state of starting for a high-power asynchronous motor used for driving a surface mining excavator. In: 2019 11th International Symposium on Advanced Topics in Electrical Engineering (ATEE), pp. 1–6. IEEE, March 2019
19. Lu, C., Abshari, M., Pellegrino, G.: Design of two PM synchronous machines for EV Traction Using Open-Source Design Instruments (2019)
20. Demir, Y., Yolacan, E., El Refaie, A., Aydin, M.: Investigation of different winding configurations and displacements of a 9-phase permanent magnet synchronous motor with unbalanced AC winding structure. IEEE Trans. Ind. Appl. (2019)
21. Rostami, N.: High efficiency axial flux permanent magnet machine design for electric vehicles. Gazi Univ. J. Sci. **32**(2), 544–556 (2019)
22. Qu, R., Lipo, T.A.: Dual-rotor, radial-flux, toroidally wound, permanent-magnet machines. IEEE Trans. Ind. Appl. **39**(6), 1665–1673 (2003)
23. Wirtayasa, K., Irasari, P., Kasim, M., Widiyanto, P., Hikmawan, M.: Design of an axial-flux permanent-magnet generator (AFPMG) 1 kW, 220 volt, 300 rpm, 1 phase for pico hydro power plants. In: 2017 International Conference on Sustainable Energy Engineering and Application (ICSEEA), pp. 172–179. IEEE (2017)

24. Claudio-Medina, C., Mayorga-Pardo, A.: Characterization of an axial flow generator for applications in wind energy. Ingenius. Revista de Ciencia y Tecnología **19**, 19–28 (2018)
25. Brad, C., Vadan, I., Berinde, I.: Design and analysis of an axial magnetic flux wind generator. In: 2017 International Conference on Modern Power Systems (MPS), pp. 1–7 (2017)
26. Lee, G.C., Jung, T.U.: Cogging torque reduction design of dual stator radial flux permanent magnet generator for small wind turbine. In: IEEE 2013 Tencon-Spring, pp. 85–89. IEEE (2013)
27. Ashraf, M.M., Malik, T.N., Zafar, S., Raja, U.N.: Design and fabrication of radial flux permanent magnet generator for wind turbine applications. Nucleus (Islamabad) **50**(2), 173–181 (2013)
28. Buñay, J.: Estudio Y Caracterización Hidráulica Del Óvalo 10 Al 13 Del Canal De Riego Ambato - Huachi - Pelileo, Cantón Ambato, Provincia De Tungurahua, Engineering dissertation, Technical University of Ambato, Ecuador (2018)

Multi-criteria Analysis of Turbines for Pico-Hydro Power Plant in Water Irrigation Channel

Myriam Cumbajín[1] (ID), Patricio Sánchez[1] (ID), Andrés Hidalgo[2] (ID),
and Carlos Gordón[2(✉)] (ID)

[1] Facultad de Ingeniería y Tecnologías de la Información y la Comunicación,
Universidad Tecnológica Indoamérica, UTI, 180103 Ambato, Ecuador
{myriamcumbajin,patriciosanchez}@uti.edu.ec
[2] Facultad de Ingeniería Civil y Mecánica, Facultad de Ingeniería en Sistemas,
Electrónica e Industrial, Universidad Técnica de Ambato, UTA,
180150 Ambato, Ecuador
{andresshidalgo,cd.gordon}@uta.edu.ec

Abstract. We present the Multi-criteria Analysis of turbines for Pico-Hydro Power plant in Ambato–Huachi–Pelileo water irrigation channel. Not only, we have been able to perform a turbine selection using the common requirements like head and flow rate. But also, we have taken into account a multi-criteria analysis by considering: Quantitative criteria, such as Efficiency and Size, combined with Qualitative criteria, such as Maintainability & Serviceability, Portability, Modularity, and Installation Environment requirements. The criteria were analyzed by using the analytical hierarchy process (AHP). We have reviewed different turbine options for being used in a pico-hydropower plants, from the analytical hierarchy process. We have concluded that Michell Banki turbine is the recommendable option in terms of easy build and highly efficient. The study of the requirements for the development of Pico-Hydro Power plant in Ambato–Huachi–Pelileo water irrigation channel in Ecuador will be a promising method for meeting the energy needs using renewable resources and avoiding the conventional fossil fuels or any other approach that affects the environment.

Keywords: Turbines · Pico-Hydro Power Plant · Water Irrigation Channel · Analytical hierarchy process

1 Introduction

Energy access is limited in different areas around the world. One of these situations is in south America, in which there are around 2.4 million total of non-electrified rural households in many countries like: Ecuador, Bolivia, Perú, Colombia, and Venezuela [1]. Facing this reality, it is important to consider the most suitable approach for energy generation. One of the most required and productive methodology to generate Electricity is by using green and renewable resources in order to reduce the contamination with conventional fossil fuels or any other approach that affects the environment [2]. There are different renewable resources like solar photovoltaic [3], wind power

© Springer Nature Switzerland AG 2020
Á. Rocha et al. (Eds.): ICITS 2020, AISC 1137, pp. 200–209, 2020.
https://doi.org/10.1007/978-3-030-40690-5_20

generation [4], Biomass [5] and hydropower [6]. All of them, are excellent approaches in order to provide enough energy. But, considering actual and significant problems faced by different rural communities in order to obtain energy, like: lack of economic resources, remote, hilly and inaccessible locations [7]. It is necessary to take advantage of the natural resources available in the areas with the energy needs. The natural resource available in the study areas is the water of rivers or irrigation channels which is a valuable resource for energy generation [8].

Hydropower is an eco-friendly clean power generation method. They are plants which generate energy from the water flow [9]. Hydropower captures the energy of moving water for energy generation purpose, as an example we consider the Ambato–Huachi–Pelileo water irrigation channel which is the resource available in the study area in Ambato, in the province of Tungurahua - Ecuador [10]. The hydropower plant is employed for converting around 90% of the energy of moving water into electricity/power around the world. The hydropower takes advantage of the available renewable resources comparable to the fossil fuels-based power plants, which provide around 60% efficient [11]. Also, an actual report mentions that the hydropower is strong important for the energy generation around the world, since it provides about 20% of world total power consumption. As an example, China has become the leader in the hydropower sector with a capacity of 146 GW [12]. Besides, the capacity of the hydropower plays an important role when considering the size of electrical power produced by them. So, the hydropower capable of providing more than 10 MW is named large hydropower [13]. Then, hydropowers are named Small [14], Mini [15], Micro [16] and Pico [17] when they reach less than 10 MW, 1 MW, 100 KW and 5 KW of power, respectively. An efficient, reliable and cost effective alternative power sources is a Pico-Hydro Power Plant (PHP). Many researches have payed attention in the development of PHPs in order to avoid the high expenditure and adverse environmental concerns associated with the use of large hydro power plants [18]. Also, the demand for Pico-Hydro Power in the market shows that PHP is the best and low-priced choice for rural or remote area electrification in developing countries [19, 20]. Considering the available resource in Ecuador: the water flow in the irrigation channels, the energy needs, capability of energy generation and the requirement of reducing contamination and not being aggressive with civil works in the irrigation channel, the energy generation should focus in the Pico-Hydro Power approach [21].

The aim of this work is to present a study of Turbines which are suitable for Pico-Hydro Power Plant in Ambato–Huachi–Pelileo Water Irrigation Channel. Not only, we carried out a selection using a multi–criteria analysis by seeing the common requirements like head and flow rate. But also, we considered Quantitative criteria, such as Efficiency and Size, combined with Qualitative criteria, such as Maintainability & Serviceability, Portability, Modularity, and Installation Environment requirements [22]. We performed a study of different turbine options for being used in a pico-hydropower plants which revealed the recommendable option based on Qualitative and Quantitative criteria. The study of the requirements for the development of Pico-Hydro Power plant in Ambato–Huachi–Pelileo water irrigation channel Ecuador will be a promising method for meeting the energy needs using renewable resources and avoiding the conventional fossil fuels or any other approach that affects the environment.

2 Pico-Hydro Power Plant (PHP) - Fundamentals

The PHP is a very small hydropower suitable for a single household or a small group of households. It is a plant capable of generating energy up to 5 KW by using the water flow from a river or an irrigation channel. These features make PHPs the suitable method for rural electrification. PHP power stations are usually installed in remote areas as off-grid or stand-alone power stations. The arrangement of a Pico-Hydro Power station depends on the hydraulic characteristics of the study area. Thus, we mainly consider two requirements the head and water flow that we observe in the structure of a PHP, which is depicted in Fig. 1. Also, we identify two main components of the PHP, which are the turbine and the generator [23]. In this work, we have mainly focused in the study of the turbine. We will review the development of the generator and the matching between the turbine and generator in future studies.

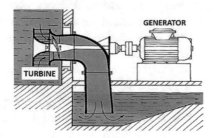

Fig. 1. Structure of PHP.

A Turbine is a hydraulic machine which converts hydraulic energy into mechanical energy. It is an important component which allows providing mechanical energy to an electrical generator in order to produce electricity [24]. The turbine is simplified to make cheap energy and uses inexpensive components for widespread utilization. Then, turbine is an assembly which comprises a nozzle or stator, runner, and shaft, among others. Each component plays an important role. The nozzle or stator directs the flow to the runner it may be an orifice that creates a high-speed jet, or it may be a set of vanes. Then, runner converts the hydraulic energy into mechanical power by redirecting fluid flow. Mainly, the runner is typically equipped with cups or blades that interact with the moving water and cause the runner to rotate. So, the mechanical work is transferred by the shaft to a generator. Thus, all the components collectively convert momentum and pressure in a water flow into rotational mechanical work. There are different types of turbines which are selected for hydro Power implementation, according to a wide range of requirements. Then, we consider the multi–criteria analysis carried out in this work with the intention of choosing the most suitable turbine for Pico-Hydro Power Plant in Ambato–Huachi–Pelileo Water Irrigation Channel. According to the Quantitative analysis, we take account the efficiency which is the ratio of the power developed by the runner of a turbine to the power supplied at the inlet of a turbine [25], and size which provides the idea of the dimensions of the turbine and the capability of power generation [26]. Also, we have applied the Qualitative approach, which refers the

Maintainability & Serviceability, Portability, Modularity, and Installation Environment requirements.

Maintainability & Serviceability is defined as the probability of performing a successful repair action within a given time and the ease of maintaining and servicing the unit, especially with unskilled labour [27]. While, Portability refers to the minimized volume for easy transportation [28]. And Modularity is related to the line replaceable units and to disassemble the unit for ease of portability [29]. Then, the installation environment refers to environmental damages when civil works are carried out when implementing the hydro power [30]. As a result, the present work faces the selection of a turbine suitable for a PHP with efficiency, size, maintainability, serviceability and installation environment that provide enough energy by reducing at the minimum the environmental impact.

3 Review of Micro-Pico-Hydro Power Plant in Ecuador

Actually, Ecuador is a country which mainly have an energy production based on Hydro – Power generation, so the green and clean energy generation is one of the most important directives for the actual government which is in charge of the national plan of development named "Plan Nacional para el Buen Vivir", due to the fact that they want to take advantage of the renewable resource available in the country. But, unfortunately most of the hydro – power plants operating in Ecuador are large scale, which produce high expenditure and adverse environmental concerns. For these reasons, Ecuador is paying more attention in the development of mini – hydro power plants. Within this category of mini plants are currently 10 projects on the way, in accordance with the National Electrification Plan 2012–2021 approved by CONELEC "Consejo Nacional de Electrificación". The new mini plants are: "1 - Topo, 2 - San José de Minas, 3 - Victoria, 4 - Sigchos, 5 - Pilaló 3, 6 - Apaqui, 7 - Río Luis, 8 - La Merced de Jondachi, 9 - Sabanilla y 10 - Huapamala". The new plants will involve a total power of 170 megawatts (MW) with an investment that will reach close to USD 300 million. This new energy will represent around 4% of the total installed capacity in Ecuador. Also, as good news, recent reports are related to the Stimulation of the Market for Pico-hydro in Ecuador. Where, the technical capacity to install and maintain a PHP in the Andean Region in Ecuador was developed. Also, it was presented the commercial opportunities arising from the sale of PHPs within the country.

4 Ambato–Huachi–Pelileo Water Irrigation Channel

The Ambato–Huachi–Pelileo Water Irrigation Channel (AHPWIC) is located at 2952 m above sea level in Ambato town in the province of Tungurahua in Ecuador [10]. A detailed study and hydraulic characterization of the AHPWIC has been carried out in 2.7 km length and 0.0017 degrees slope [31]. The results reveal that the AHPWIC mainly has trapezoidal sections of geometry as shown in Fig. 2, which depicts the dimensions and the high speed with yellow color. Also, the AHPWIC has an average stable flow $Q = 1.6 \text{ m}^3/s$, speed $V = 1,66$ m/s and a head $H = 2.5$ m. obtained from the measurements. It is important to highlight that the head and flow are the primary features considered in the turbine selection, to then continue by using a multi-criteria analysis.

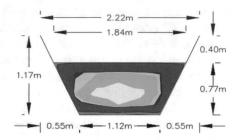

Fig. 2. Speed distribution curves of the AHPWIC [34].

5   Turbines for Hydro - Power Plants

There are different turbines for being used in a hydropower. They are classified as impulsive and reaction. Among impulsive turbines we have Pelton, Turgo, and Crossflow-Michell Banki. While, as reaction turbines we have Francis, Kalpan-Propeller and Archimedes. Also, turbines are mainly differentiated by head. From high head to low head, they are as follows: Pelton, Turgo, Crossflow-Michell Banki, Francis, Kalpan-Propeller and Archimedes. The analysis of each turbine was carried out by considering the low power requirement and the multi-criteria analysis as described in previous reports [22].

A Pelton turbine is suitable for large head and low flow sites. The Pelton turbines have one or multi-jets, in the case of small and micro-hydropower configuration it is used a single jet. Generally, a Pelton turbine has high efficiency rate of 70–90%. The Pelton turbine is sketched in Fig. 3, which demands on complex and well elaborated design. Turgo turbines have been optimized for being used in Micro- and Pico-Hydro Power Plants, which can be used for heads between 3 m and 150 m [22]. Usually, the efficiency of the Turgo turbine is from 87–91% and depends on many factors, such as nozzle or jet inclination, cup design and speed ratio. Also, Turgo turbines require complex and well elaborated design, which is sketched in Fig. 4.

Fig. 3. Pelton turbine.   **Fig. 4.** Turgo turbine.

Crossflow-Michell Banki turbine depicted in Fig. 5 is typically used at higher flow rate and lower head than the Pelton and Turgo turbines. The average efficiency of CF-MB turbines is usually 80% for Small, Micro and Pico-Hydro Power plants. The efficiency mainly depends on geometrical parameters like the number of blades, runner diameter, nozzle entry arc and angle of attack. As an advantage of CF- MB turbine, we can mention the easy build and highly efficient. Francis turbine can be used for micro, medium or large hydropower. The operating range of Francis turbine is between 1 m and 900 m. The efficiency is 70–74%, which depends on the development of the helical vortex, or the so-called as vortex rope downstream the runner, in the draft tube cone. The Francis turbine is depicted in Fig. 6, in which is evident the complex and well elaborated requirement design.

Kaplan-Propeller turbine is more efficient for low water heads sites. The efficiency

Fig. 5. Michell Banki turbine.

Fig. 6. Francis turbine.

of Kaplan-Propeller turbine is around 70% and can be improved with adjustment of the turbine blades and guide vanes angles. Figure 7 shows the Kaplan-Propeller turbine which demands on complex and well elaborated requirement design. The Archimedes turbine depicted in Fig. 8, is more attractive for lower head sites, as its heads can be set as low as 1 m. Also, they are especially suited to sites with large flows. The efficiency of Archimedes turbine is around 86% and depends on the geometry. So, the development of this turbine requires complex and well elaborated design.

Fig. 7. Kaplan-Propeller turbine.

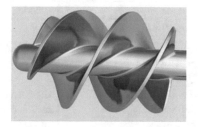

Fig. 8. Archimedes turbine.

6   Turbine Selection for Pico-Hydro Power Plants

Considering the main features of the AHPWIC we can select the turbines which are suitable for those features. Figure 9 depicts different turbines in terms of head and flow. Also, we observe the red point which depicts the area of interest in function of the head and flow of the AHPWIC. From Fig. 9, we estimate the three main turbines suitable for the PHP in the AHPWIC which are: Crossflow or Michelle Banki, Propeller or Kaplan and Archimedes.

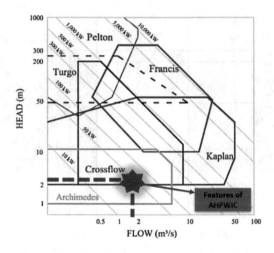

Fig. 9. Turbines in terms of head and flow [22].

The analytical hierarchy process (AHP) in Excel was carried out thanks the Quantitative and Qualitative analysis of the three turbines mentioned before. Table 1 depicts the main consideration of the selection criteria for the turbines.

Table 1. Selection criteria [22].

Criteria	Meaning
Quantitative	
Efficiency	Efficiency of the unit at rated flow/head
Size - Power	Power of the unit required 2 kW
Qualitative	
Maintenance & Serviceability	The ease of maintaining and servicing the unit, especially with unskilled labour
Portability	Minimized volume for easy transportation
Modularity	Scope to incorporate modularity into the design for line replaceable units and to disassemble the unit for ease of portability
Civil works – Installation Environment	Minimized civil works - concrete sparsely available in site locations

Table 2. Scored regime for the turbines.

Criteria	Michelle Banki	Kaplan	Archimedes
	Score	Score	Score
Quantitative			
Efficiency	4	4	4
Size - Power	4	4	3
Qualitative			
Maintenance & Serviceability	5	4	3
Portability	5	4	3
Modularity	5	4	3
Civil works	4	4	3
Total	**27**	**24**	**19**

Next, we provided a scored regime for each turbine in Table 2. The scale is considered 5 for the best and 1 for the worst performance [22]. Also, the engineering process in the design and the construction was considered in order to assign the score to each criterion. Then, Table 2 summarizes the multi-criteria of the Crossflow - Michelle Banki turbine, Propeller – Kaplan turbine, and Archimedes turbine, respectively.

Considering the score obtained and depicted in Fig. 10, from the turbines by using the analytical hierarchy process in the AHP Excel Template, which consists of 20 input worksheets for pair-wise comparisons of the multi-criteria. As a result, the Crossflow - Michelle Banki turbine is the most suitable turbine for being used in the Pico-Hydro Power Plant in Ambato–Huachi–Pelileo Water Irrigation Channel, due to the fact that this turbine presents the better performance achieving the higher score.

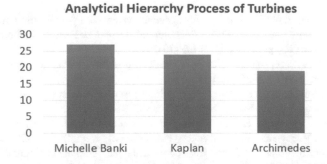

Fig. 10. Multi-criteria analysis of turbines.

7 Conclusions

We have presented the study of turbines for pico-hydropower plant in Ambato–Huachi–Pelileo water irrigation channel. We have reviewed different turbine options for being used in a pico-hydropower plants, from them we have concluded that Michell

Banki turbine is the recommendable option in terms of easy build, highly efficient and the performed multi-criteria analysis. The study of the requirements for the development of Pico-Hydro Power Plant in Ambato–Huachi–Pelileo water irrigation channel Ecuador will be a promising method for meeting the energy needs using renewable resources and avoiding the conventional fossil fuels or any other approach that affects the environment due to the fact that we look for the suitable components for PHPs.

Acknowledgements. The authors thank the Technical University of Ambato - Ecuador and the "Dirección de Investigación y Desarrollo (DIDE)", for their support in carrying out this research, in the execution of the project "Obtención de electricidad a través de canales de riego como fuentes energéticas" code of the project: PFICM15.

References

1. Terrapon-Pfaff, J., Gröne, M., Dienst, C., Ortiz, W.: Productive use of energy–Pathway to development? Reviewing the outcomes and impacts of small-scale energy projects in the global south. Renew. Sustain. Energy Rev. **96**, 198–209 (2018)
2. Benedek, J., Sebestyén, T., Bartók, B.: Evaluation of renewable energy sources in peripheral areas and renewable energy-based rural development. Renew. Sustain. Energy Rev. **90**, 516–535 (2018)
3. Al Shafeey, M., Harb, A.M.: Photovoltaic as a promising solution for peak demands and energy cost reduction in Jordan. In: 2018 IEEE 9th International Renewable Energy Congress (IREC), pp. 1–4 (2018)
4. El-Arroudi, K., Joós, G.: Performance of interconnection protection based on distance relaying for wind power distributed generation. IEEE Trans. Power Delivery **33**(2), 620–629 (2018)
5. Jones, M., Kansiime, F., Saunders, M.J.: The potential use of papyrus (Cyperus papyrus L.) wetlands as a source of biomass energy for sub-Saharan Africa. GCB Bioenergy **10**(1), 4–11 (2018)
6. Choi, K., Yeom, J.W.: Modeling of management system for hydroelectric power generation from water flow. In: 2018 Tenth International Conference on Ubiquitous and Future Networks (ICUFN), pp. 229–233. IEEE (2018)
7. Williams, A.A., Simpson, R.: Pico hydro–reducing technical risks for rural electrification. Renew. Energy **34**(8), 1986–1991 (2009)
8. Mishra, S., Singal, S.K., Khatod, D.K.: Optimal installation of small hydropower plant—a review. Renew. Sustain. Energy Rev. **15**(8), 3862–3869 (2011)
9. Gielen, D.: Renewable Energy Technologies: Cost Analysis Series, Hydropower. In: International Renewable Energy Agency (IRENA), vol. 1, no. 3, pp. 1–32 (2012)
10. Ambato–Huachi–Pelileo Water Irrigation Channel. http://rrnn.tungurahua.gob.ec/red/estaciones/estacion/53a2fdf8bd92ea542c000001. Accessed 30 June 2019
11. Kadier, A., Kalil, M.S., Pudukudy, M., Hasan, H.A., Mohamed, A., Hamid, A.A.: Pico hydropower (PHP) development in Malaysia: potential, present status, barriers and future perspectives. Renew. Sustain. Energy Rev. **81**, 2796–2805 (2017)
12. Sopian, K., Razak, J.A.: Pico hydro: clean power from small streams. In: Proceedings of the 3rd World Scientific and Engineering Academy and Society International Conference on Renewable Energy Sources, Tenerife, Spain, vol. 13 (2009)

13. Mo, W., Chen, Y., Chen, H., Liu, Y., Zhang, Y., Hou, J., Gao, Q., Li, C.: Analysis and measures of ultralow-frequency oscillations in a large-scale hydropower transmission system. IEEE J. Emerg. Sel. Top. Power Electron. **6**(3), 1077–1085 (2018)
14. Ur Rehman, U., Riaz, M.: Design and implementation of electronic load controller for small hydro power plants. In: IEEE 2018 International Conference on Computing, Mathematics and Engineering Technologies (iCoMET), pp. 1–7 (2018)
15. Alnaimi, F.B.I., Ziet, F.W.: Design and development of mini hydropower system integrated for commercial building. In: AIP Conference Proceedings AIP Publishing, vol. 2035, no. 1, p. 070006 (2018)
16. Bilgili, M., Bilirgen, H., Ozbek, A., Ekinci, F., Demirdelen, T.: The role of hydropower installations for sustainable energy development in Turkey and the world. Renew. Energy **126**, 755–764 (2018)
17. Mhlambi, B.A., Kusakana, K., Raath, J.: Voltage and frequency control of isolated pico-hydro system. In: IEEE 2018 Open Innovations Conference (OI), pp. 246–250 (2018)
18. Nimje, A.A., Dhanjode, G.: Pico-Hydro-Plant for small scale power generation in remote villages. IOSR J. Environ. Sci. Toxicol. Food Technol. (IOSR-JESTFT) **9**, 59–67 (2015)
19. Maher, P., Smith, N.P.A., Williams, A.A.: Pico hydro power for rural electrification in developing countries. Int. J. Ambient Energy **19**(3), 143–148 (1998)
20. Kapoor, R.: Pico power: a boon for rural electrification. Adv. Electron. Electr. Eng. **3**(7), 865–872 (2013)
21. Fortaleza, B.N., Juan, R.O.S., Tolentino, L.K.S.: IoT-based Pico-hydro power generation system using Pelton turbine. J. Telecommun. Electron. Comput. Eng. (JTEC) **10**(1–4), 189–192 (2018)
22. Williamson, S.J., Stark, B.H., Booker, J.D.: Low head pico hydro turbine selection using a multi-criteria analysis. Renew. Energy **61**, 43–50 (2014)
23. Carravetta, A., Houreh, S.D., Ramos, H.M.: Pumps as Turbines: Fundamentals and Applications. Springer, Heidelberg (2017)
24. Kramer, M., Terheiden, K., Wieprecht, S.: Pumps as turbines for efficient energy recovery in water supply networks. Renew. Energy **122**, 17–25 (2018)
25. Louie, H.: Off-grid wind and hydro power systems. In: Off-Grid Electrical Systems in Developing Countries. Springer, Cham (2018)
26. Powell, D., Ebrahimi, A., Nourbakhsh, S., Meshkahaldini, M., Bilton, A.M.: Design of pico-hydro turbine generator systems for self-powered electrochemical water disinfection devices. Renew. Energy **123**, 590–602 (2018)
27. Okafor, C., Atikpakpa, A., Irikefe, E.: Maintainability evaluation of steam and gas turbine components in a thermal power station. Am. J. Mech. Ind. Eng. **2**(2), 72–80 (2017)
28. Azhar, A.A., Shahran, M.A.M., Arifin, M.N., Yahaya, I., Rahim, R.A., Pusppanathan, J.: Pocket-hydro turbine into capsule hydro turbine. J. Tomogr. Syst. Sens. Appl. **1**(1) (2018)
29. Uhunmwangho, R., Odje, M., Okedu, K.E.: Comparative analysis of mini hydro turbines for Bumaji Stream, Boki, Cross River State, Nigeria. Sustain. Energy Technol. Assess. **27**, 102–108 (2018)
30. Ikeda, T., Lio, S., Tatsuno, K.: Performance of nano-hydraulic turbine utilizing waterfalls. Renew. Energy **35**(1), 293–300 (2010)
31. Buñay, J.: Estudio Y Caracterización Hidráulica Del Óvalo 10 Al 13 Del Canal De Riego Ambato-Huachi-Pelileo, Cantón Ambato, Provincia De Tungurahua, Engineering dissertation, Technical University of Ambato, Ecuador (2018)

Big Data Analytics and Applications

Analysis of Node.js Application Performance Using MongoDB Drivers

Leandro Ungari Cayres[(✉)], Bruno Santos de Lima, Rogério Eduardo Garcia, and Ronaldo Celso Messias Correia

Faculty of Science and Technology, São Paulo State University (UNESP), Presidente Prudente, SP, Brazil
{leandro.ungari,bruno.s.lima,rogerio.garcia,ronaldo.correia}@unesp.br

Abstract. At the last few years, the usage of NoSQL databases has increased, and consequently, the need for integrating with different programming languages. In that way, database drivers provide an API to perform database operations, which may impact on the performance of applications. In this article, we present a comparative study between two main drivers solutions to MongoDB in Node.js, through the evaluation of CRUD tests based on quantitative metrics (time execution, memory consumption, and CPU usage). Our results show which, under quantitative analysis, the MongoClient driver has presented a better performance than Mongoose driver in the considered scenarios, which may imply as the best alternative in the development of Node.js applications.

Keywords: Performance · Node.js application · MongoDB · Drivers · NoSQL databases

1 Introduction

At the last few years, the growth of data volume has changed the perspective of how organizations behavior, from simple data recording to potential advantage in competitive markets.

This event, known as Big Data, not only implies in large storage but also perspectives related to variety, velocity, and value [1]. The traditional architecture of relational databases based on ACID (atomicity, consistency, isolation, and durability) properties, which affect the aspects related to availability and efficiency directly in Big Data environments [2]. The non-relational databases (NoSQL) have been proposed oo solve the side-effects, and allowing more structural flexibility, scalability, and support to replication and eventual consistency [3].

In this context of development environments and programming languages, a variety of database drivers aim to support the execution of the internal database operations. However, in many cases, the development of these drivers are very recent and may present defects or limitations, which results in side-effects to the access and manipulation of data [4]. Thus, the usage decision of which NoSQL

© Springer Nature Switzerland AG 2020
A. Rocha et al. (Eds.): ICITS 2020, AISC 1137, pp. 213–222, 2020.
https://doi.org/10.1007/978-3-030-40690-5_21

database and driver may impact on the performance, due to unknown factors previously.

In this work, we conduct a comparative study of performance between MongoClient[1] and Mongoose[2], both solutions of database drivers to MongoDB[3] in Node.js applications. The main difference between them is the predefinition of schema, a factor which is not mandatory in the majority of NoSQL databases, and MongoDB too; but in one of these drivers is required. In that way, this experimental study analyses the impact of each driver at CRUD (create, read, update, and delete) operations.

The choose of MongoDB based on a crescent number of studies in the research community, and also it is the main option of the document-oriented database. In about Node.js, despite recent development, it presents technical viability to implement robust applications. Also, the database system and application lead to uniformization, because both are implemented in JavaScript.

This study is unique in the investigation of effects of performance in different database drivers in Node.js application since the other works have analyzed performance between database [5–7] or the modeling impact on the performance in databases [8].

The remaining of this article as follows: Sect. 2 presents relevant topics related to NoSQL databases and MongoDB. Section 3 presents the conception of an experimental project. Section 4 describes the obtained results, which analysis is in Sect. 5. Finally, Sect. 6 presents the final remarks of the presented study.

2   Background

2.1   NoSQL Databases

NoSQL databases were developed to fulfill storage requirements in big data environments. In that way, their schemaless data structure provides more flexibility to many applications, such as e-mails, documents, and social media content [9,10].

The NoSQL term refers to wide variety of storage systems, which are non-strict ruled by ACID properties, to allow better data structure and horizontal performance [4], join operations, high scalability, and data modeling by simplified queries [10]. The relational databases are divided into four categories: document-oriented, column-oriented, key/value-oriented, graph-oriented, and multimodal.

This work focus on oriented-documents databases, which modeling is similar to object-oriented data definition using registers with fields and complex operations [6]. Each database contains collections, each collection defines similar content groups, and each item corresponds to a document structured as JSON (JavaScript Object Notation) or XML (Extensible Markup Language).

---

[1] https://mongodb.github.io/node-mongodb-native/.

[2] https://mongoosejs.com/.

[3] https://www.mongodb.com/.

2.2   MongoDB

MongoDB is an open-source document-oriented database, which provides additional storage functionalities such as data sorting, secondary indexing, and interval queries [11]. It does not require a defined schema, despite the similarity of elements into a collection [8,12]. There are two main approaches to document modeling:

- **Embedded data modeling:** the data are defined in a unique data structure or document, which results in a high concentration of data.
- **Normalized data modeling:** the data have references among documents to represents relationships.

The adopted format is JSON, which passes through a codification to binary format BSON[4]. About the integration to programming languages, several drivers are available to Java, C++, C#, PHP and Python [12], and Node.js too.

Concerning to Node.js drivers to MongoDB, the first is the MongoClient[5], the official distributed solution, which provides an API to manipulation of data. The main feature is the implicit document-object modeling, which discards any need for data description.

The second option is the Mongoose[6], a database driver that provides data modeling based on the object-relational model. It implies that all data elements must describe the definition of attributes, which allows verification of types and validation, nullity checking by a defined schema.

3   Experimental Setup

In this section, we present the definitions of the experimental project which intents to compare the performance of each driver with MongoDB. Figure 1 presents the structure of experiment.

The analysis aimed to identify which couple (driver-database) presents better performance in a Node.js application based on quantitative parameters.

We developed a tool to run the test with each driver. In that way, to perform the comparison, the execution flow of application receives a set of parameters, such as driver and number of elements, in follow the database connection is performed and the respective operations. During the tests, we extracted some metrics related to CPU (Central Processing Unit) and memory usage. The tool is available in the following open-source repository: https://github.com/leandroungari/database-driver.

We also defined the performance metrics to evaluate the conducted tests:

- **Average Execution Time:** it defines the average time (in milliseconds) in each operation.

---

[4] http://bsonspec.org/.
[5] https://mongodb.github.io/node-mongodb-native/index.html.
[6] https://mongoosejs.com/.

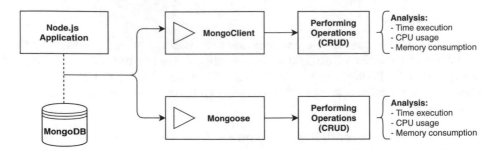

Fig. 1. Comparison between MongoDB drivers.

- **Average CPU Usage:** it defines the average time (in milliseconds) usage of CPU during each operation.
- **Average Memory Usage:** it defines the average variation of RAM memory usage (in kilobytes) in each operation.

Finally, we formulate a set of research questions to lead the analysis of the results:

- **RQ1** – *Does the driver selection impact on application performance under time execution?*
- **RQ2** – *Does the driver selection affect application performance under CPU usage?*
- **RQ3** – *Does the driver selection impact application performance under RAM memory consumption?*

3.1 Dataset

We used a dataset of 18 thousands of instances in the conducting of the experiment. Initially, all registers have 89 attributes (mainly textual) with an average size of 1.37 KB. From the original dataset, we also built a second dataset with reduced instances (only six attributes) using the same instances, which resulted in an average size of 0.13 KB.

We applied both datasets in the experimental process, in which reduce dataset intent to compare the drivers about the relation between the number of attributes and data modeling impact. In Table 1, we summarize the main characteristics of datasets:

3.2 Execution Environment

The execution environment consisted of a machine running Ubuntu 18.04.2 Operating System, Intel i3 3217U processor, and 4 GB DDR3 RAM. During the execution of the tests, the Node.js application execution environment was set to use the maximum size 3 GB heap, thereby restricting the maximum operations of each test.

Table 1. Description of experimental datasets.

	Dataset I	Dataset II
Number of instances	18,000	18,000
Number of attributes	89	6
Average size of instance	1.37 KB	0.13 KB

In each execution scenario, data regarding the runtime, CPU usage time, and RAM usage were extracted. Scenarios with different quantities of CRUD operations were analyzed, ranging from 1,000, 10,000, 100,000, and 200,000; each repeated 10 times and recorded the average of the executions. Performance metrics were obtained through the JSMeter library.[7].

4   Experiment Results

The results are presented from the perspective of each of the CRUD operations, where 100% of the records are reached in each operation. Each result refers to a specific operation of combining a driver with MongoDB in an application manipulating the large data set (with all attributes) or small data set (with a reduced number of attributes).

Figure 2 graphically illustrates the execution time when performing CRUD operations contrasting the use of both drivers. Considering the execution time for insert operations - Fig. 2a, it was identified that the execution time with the use of driver Mongoose was longer for both sets, compared to the use of MongoClient, which showed no significant differences between the sets. It can also be noted that when manipulating the small dataset using Mongoose, there was a drop in execution time from 100,000 insert operations. A possible factor that justifies this behavior is the occurrence of set splitting in the insert operation when the quantity exceeds 100,000 items, according to MongoDB documentation, however, this does not occur for the large data set.

Considering the execution time for Search operations - Fig. 2b, using MongoClient resulted in lower execution time for both sets compared to using Mongoose. Using Mongoose in this case, the executions for both sets exhibit increasing and proportional behavior to the detriment of the record size difference. Finally, from a processing time perspective, when analyzing test results for both update operations - Fig. 2c and deletion operations - Fig. 2d, similar behaviors were observed with the use of both drivers.

Figure 3 graphically illustrates CPU consumption when performing CRUD operations by contrasting the use of both drivers. Similarly to the previous analysis, for insert operations - Fig. 3a and fetch - Fig. 3b, MongoClient usage provides significantly lower average CPU consumption time in both sets, to the detriment of the high consumption time presented by Mongoose. For insertion,

---

[7] https://github.com/wahengchang/js-meter.

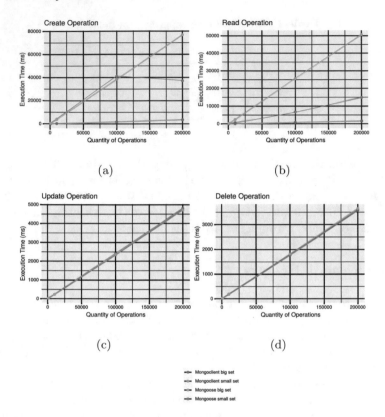

Fig. 2. Comparison of the use of drivers with the application of CRUD operations in relation to the runtime.

it also presents the exception case for 200,000 operations, whose possible justification is similar to the previous analysis.

Figure 3c illustrates the CPU consumption when performing update operations, where it can be observed that using the Mongoose driver with a larger record set had a longer CPU consumption time, unlike the others, which were similar and with a shorter time, even with oscillations. Importantly, the processing time of all runs was less than 250 ms. In Fig. 3d, CPU consumption is shown when performing deletion operation, each execution presented relatively unstable behavior, in which the use of driver MongoClient had a higher CPU consumption time, however, there is no significant difference because all executions obtained processing time of less than 10 ms.

Figure 4 graphically illustrates RAM consumption when performing CRUD operations contrasting the use of both drivers. Figures 4a and b respectively show the memory consumptions for insert and search operations, in which a pattern of memory usage cannot be identified, however it can be identified that predominantly driver Mongoose has a higher memory consumption in operations

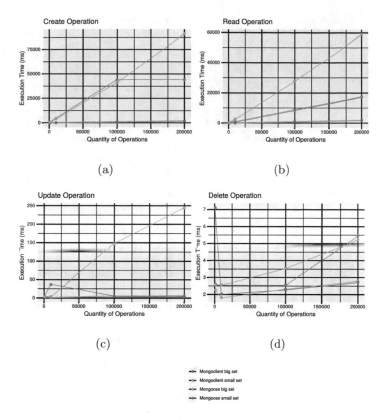

Fig. 3. Comparison of the use of drivers with the application of CRUD operations in relation to the CPU consumption.

performed on both sets to the driver MongoClient cases. For both operations, there is additional memory usage of around 30 to 40 MB.

Finally, Figs. 4c and d respectively show the memory consumption when performing an update and delete operations, do not show standardization for both drivers and sets. Despite having some points of instability, it is possible to notice that such operations consume little additional memory, approximately 1MB or less, even considering the oscillation peaks.

5   Analysis

This section presents the analysis of the results obtained, which are described from the perspective of the descriptive research questions in the Sect. 3.

5.1   *RQ1 – Does the Driver Selection Impact on Application Performance Under Time Execution?*

Regarding RQ1, in general, the tests performed using the Mongoose driver had a higher execution time compared to using MongoClient in two operations,

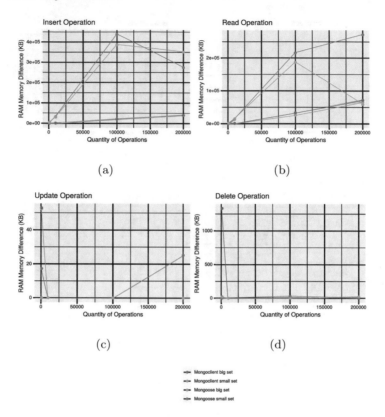

Fig. 4. Comparison of the use of drivers with the application of CRUD operations in relation to the Memory consumption.

while in the other operations the performance was similar. Thus, from the average execution time perspective of each operation performed on the MongoDB database, driver choice can impact performance, with MongoClient being the best choice for a Node.js application that makes use of MongoDB if time execution is a critical factor for this application.

5.2   *RQ2 – Does the Driver Selection Affect Application Performance Under CPU Usage?*

As in RQ1, tests performed using MongoClient had lower CPU consumption time compared to Mongoose in both sets, mainly for insert and fetch operations, while in the other operations (update and delete), it was also recorded better performance, but to a lesser extent. In short, in terms of processing time, driver choice can influence performance, also presenting MongoClient as the best option for a Node.js application that makes use of MongoDB.

5.3   RQ3 – Does the Driver Selection Impact Application Performance Under RAM Memory Consumption?

Considering RQ3, we have that for insert and search operations, in most execution cases, driver Mongoose has higher memory consumption, while for update and delete operations there are no significant differences. However, it is noteworthy that in none of the comparisons there was stable behavior Thus, in terms of memory consumption, the choice of driver does not significantly impact all operations, despite the lower consumption by driver MongoClient.

5.4   Overview

In general, the data obtained indicate that insertion and search operations are the most costly in execution time, in the worst cases approximately 70,000 to 80,000 ms, while the others are less than 5,000 ms. From the perspective of CPU consumption time the same analysis is also used, the operations indicate approximate proportionality compared to the total execution time. As for memory usage, both operations are also costly, even if nonlinear, at levels close to 30 to 40 MB, to the detriment of other operations with usage close to or less than 1 MB.

Considering the average difference in record size of the adopted dataset, driver MongoClient was indifferent, showing no significant oscillations, while Mongoose presented performance directly proportional to the record size.

In terms of driver comparison, MongoClient performed more stable from the perspective of all adopted metrics than Mongoose. Therefore, it is concluded that, under the exclusive performance criterion, MongoClient presents itself as the best option for the Mongoose. It should be noted that if any additional resources provided by one of the options, such as data verification or ease of implementation, are relevant factors, the choice should be reevaluated, however, this study is quantitative and is not intended to measure the use of additional resources. which may vary from context and application used.

6   Final Remarks

This paper presents a study that analyzes the influence of using drivers, contrasting the use of two distinct drivers, MongoClient and Mongoose, on the performance of a Node.js application that makes use of the NoSQL MongoDB database. A performance analysis study was conducted considering aspects such as runtime, CPU time, and RAM usage in performing CRUD operations. Also, the impacts resulting from the variation in the average size of the records in the dataset were compared.

In general, from the quantitative results, it was found that the performance of a Node.js application integrated with MongoDB when using driver Mongo-Client is generally better compared to using driver Mongoose, especially under the runtime and CPU consumption and, less significantly, RAM consumption, considering an application developed in a Node.js environment.

As an additional contribution, there is the implementation and availability of the testing tool, in order to enable the execution of future analyzes in Node.js environments.

References

1. Ward, J.S., Barker, A.: Undefined by data: a survey of big data definitions. arXiv preprint arXiv:1309.5821 (2013)
2. González-Aparicio, M.T., Younas, M., Tuya, J., Casado, R.: A new model for testing CRUD operations in a NoSQl database. In: 2016 IEEE 30th International Conference on Advanced Information Networking and Applications (AINA), pp. 79–86 (2016)
3. Han, J., Haihong, E., Le, G., Du, J.: Survey on NoSQL database. In: 2011 6th International Conference on Pervasive Computing and Applications, pp. 363–366. IEEE (2011)
4. Rafique, A., Van Landuyt, D., Lagaisse, B., Joosen, W.: On the performance impact of data access middleware for NoSQL data stores a study of the trade-off between performance and migration cost. IEEE Trans. Cloud Comput. **6**(3), 843–856 (2018)
5. Jung, M., Youn, S., Bae, J., Choi, Y.: A study on data input and output performance comparison of MongoDB and PostgreSQL in the big data environment. In: 2015 8th International Conference on Database Theory and Application (DTA), pp. 14–17 (2015)
6. Patil, M.M., Hanni, A., Tejeshwar, C.H., Patil, P.: A qualitative analysis of the performance of MongoDB vs MySQL database based on insertion and retrieval operations using a web/android application to explore load balancing—sharding in MongoDB and its advantages. In: 2017 International Conference on I-SMAC (IoT in Social, Mobile, Analytics and Cloud) (I-SMAC), pp. 325–330 (2017)
7. Ongo, G., Kusuma, G.P.: Hybrid database system of MySQL and MongoDB in web application development. In: 2018 International Conference on Information Management and Technology (ICIMTech), pp. 256–260 (2018)
8. Kanade, A., Gopal, A., Kanade, S.: A study of normalization and embedding in MongoDB. In: 2014 IEEE International Advance Computing Conference (IACC), pp. 416–421. IEEE (2014)
9. Mohamed, M., Altrafi, O.G., Ismail, M.O.: Relational vs. NoSQL databases: a survey. Int. J. Comput. Inf. Technol. (IJCIT) **3**, 598 (2014)
10. Ramesh, D., Khosla, E., Bhukya, S.N.: Inclusion of e-commerce workflow with NoSQL DBMS: Mongodb document store. In: 2016 IEEE International Conference on Computational Intelligence and Computing Research (ICCIC), pp. 1–5 (2016)
11. Membrey, P., Plugge, E., Hawkins, D.: The Definitive Guide to MongoDB: The NoSQL Database for Cloud and Desktop Computing. A press, New York (2011)
12. Lutu, P.: Big data and NoSQL databases: new opportunities for database systems curricula. In: Proceedings of the 44th Annual Southern African Computer Lecturers' Association (SACLA), pp. 204–209 (2015)

Classify Ecuadorian Receipes
with Convolutional Neural Networks

Luis Soria[1], Gabriela Alejandra Jimenez Cadena[2],
Carlos Eduardo Martinez[3], and David R. Castillo Salazar[4](✉)

[1] Facultad de Arquitectura e Ingenierías, Universidad Internacional SEK,
Quito, Ecuador
Luis.soria@uisek.edu.ec
[2] Facultad de Arquitectura, Diseño y Artes, Carrera de Arquitectura,
Pontificia Universidad Católica del Ecuador, Quito, Ecuador
gajimenez@puce.edu.ec
[3] Carrera de Sistemas, Universidad Regional Autónoma de los Andes,
Km 5 1/2 vía a Baños, Ambato, Ecuador
ua.carlosmartinez@uniandes.edu.ec
[4] Universidad Indoamérica, Ambato, Ecuador
davidcastillo@uti.edu.ec

Abstract. This work is a proposal to resolve the problem of identification plates of food through photographs. It involves using a large set of pictures which are processed by convolutional neural networks and parallel processing TensorFlow. The results show a 90% greater accuracy in training and between 63% and 80% in the test. The reason is that Ecuadorian dishes are very similar in the images of some recipes.

Keywords: Food dishes · Photos · Parallel

1 Introduction

The classification of images, allows to see the process of fermentation of a cocoa with the purpose is to develop a multiplatform application that recommends cooking recipes, based on the ingredients that the user accepts or discards as available or not. This application has access to the user to discover new and interesting recipes, but always because they can be made at that time, without the need to buy additional ingredients. To use this application, the user must first select a main ingredient for their recipe, after which the application specifies several additional ingredients for the user. For each of these ingredients, the user must indicate whether the user has this ingredient or not. While doing this, the application will select prescriptions that are specific to residents' availability and discard those that are not what they are. As soon as all the necessary ingredients for the preparation of a recipe are available, the user can consult the desired recipe and obtain all the necessary information for its preparation [1]. The process of identifying food items from an image is quite an interesting field with various applications [2]. Automatic food image recognition systems are alleviating the process of

© Springer Nature Switzerland AG 2020
Á. Rocha et al. (Eds.): ICITS 2020, AISC 1137, pp. 223–229, 2020.
https://doi.org/10.1007/978-3-030-40690-5_22

food-intake estimation and dietary assessment. However, due to the nature of food images, their recognition is a particularly challenging task, which is why traditional approaches in the field have achieved a low classification accuracy.

A real-world test on a dataset of self-acquired images, combined with images from Parkinson's disease patients, all taken using a smartphone camera, achieving a top-five accuracy of 55. An online training component was implemented to continually fine-tune the food and drink recognition model on new images. The model is being used in practice as part of a mobile app for the dietary assessment of Parkinson's disease patients. [3]. Currently, food image recognition tasks are evaluated against fixed datasets. To conduct realistic experiments, we made use of a new dataset of daily food images collected by a food-logging application. In this study, a small-scale dataset consisting of 5822 images of ten categories and a five-layer CNN was constructed to recognize these images. Further improvements can be expected by collecting more images and optimizing the network architecture and hyper-parameters [4]. Food image recognition and classification is a challenging task in image recognition. Due to a variety of food dishes available, it becomes a very complicated task to correctly classify the food image as belonging to one of the predefined classes. Significant work has been carried out on food recognition and classification using Computer Vision. Food recognition is important to assess diet of people with diabetics and people suffering from various food allergies. Food recognition also helps in finding the calorie value of foods, its nutrition value, food preferences. This survey paper covers some of the work done in food image recognition and classification using Deep Convolutional Neural Networks (DCNN) using various parameters and models, and other machine learning techniques [5]. The classification accuracy of various models is also mentioned. The on-going work on Indian food image classification is also mentioned, which aims to improve the classification accuracy by choosing the suitable hyperparameters [6]. The process of identifying food items from an image is quite an interesting field with various applications. In this paper, an approach has been presented to classify images of food using convolutional neural networks. An accuracy of 86.97% for the classes of the FOOD-101 dataset is recognized using the proposed implementation [7]. Automatic food understanding from images is an interesting challenge with applications in different domains. Since retrieval and classification engines able to work on food images are required to build automatic systems for diet monitoring (e.g., to be embedded in wearable cameras), we focus our attention on the aspect of the representation of the food images because it plays a fundamental role in the understanding engines. The food retrieval and classification is a challenging task since the food presents high vari-ableness and an intrinsic deformability. It was composed of 4754 food images of 1200distinct dishes acquired during real meals. Finally, we propose a new representation based on the perceptual concept of Anti-Textons which is able to encode spatial information between Textons outperforming other representations in the context of food retrieval and Classification [8]. The features and their combinations for food image analysis and a classification approach based on k-nearest neighbors and vocabulary trees was evaluated on a food image dataset consisting of 1453 images of eating occasions in 42 food categories which were acquired by 45 participants in natural eating conditions. Experimental results indicate that using our combination of features and vocabulary trees for classification improves the food classification performance about 22% for the Top 1 classification accuracy and 10% for the Top 4 classification accuracy [9]. Automatic food understanding from images is an interesting challenge with applications in different domains. In this paper, we address the study of food image processing from the perspective of Computer Vision. As first contribution we present a survey of the studies in the context of food image processing from the early attempts to the current state-of-the-art methods. Since retrieval and classification engines able to work on food images are required to build automatic systems for diet monitoring (e.g., to be embedded in wearable cameras), we focus our attention on the aspect of the representation of the food images because it plays a fundamental role in the understanding engines. The food retrieval and classification is a challenging task since the food presents high variableness and an intrinsic deformability. It was composed of 4754 food images of 1200 distinct dishes acquired during real meals. Finally, we propose a new representation based on the perceptual concept of Anti-Textons which is able to encode spatial information between Textons outperforming other representations in the context of food retrieval and Classification [8]. "Food" is an emerging topic of interest for multimedia and computer vision community. In the last couple of years, advancements in the deep learning and convolutional neural networks proved to be a boon for the image classification and recognition tasks, specifically for food recognition because of the wide variety of food items. The experimental results show a high accuracy of 99.2% on the food/non-food classification and 83.6% on the food category recognition [10]. Experimental results show a high accuracy of 99.2% on the food/non-food classification and 83.6% on the food category recognition [11]. In this paper convolutional neural network is used, which work with parallel mechanisms, which can work with many images, in this case of food dishes. The following section describes the methodology used for the experiment and then describe the findings of the case is explained.

2 Methodology

1. The first step is to collect photos of recipes, an example of this collection is displayed in Fig. 1.
2. The second step is to describe the recipes and search features Fig. 2.
3. The next step is to experiment with the ingredients in order to check whether one or the other prevails. The chi-square is obtained, in order to determine that influences you the recipe if number of ingredients, which gives way to a dish can be identified according to the ingredients found in it. Finish apply convolutional neural network.

Fig. 1. Receipts images

Table 1. Receipts features

RECETAS	INGR1	INGR2	INGR3	INGR4
CEVICHE PERUANO	pescado	cebolla	limon	aji
TIRAMISU	huevo	azucar	queso	bizcochos
PAELLA	arroz	carne	conejo	tomate
LASAGNA	cerdo	queso rallado	aceite	tomate
TORTILLA ESPAÑOLA	papas	huevos	cebollas	jamon
KIMCHI COREANO	col	sal	harina de arroz	chile coreano
ARROZ CON GRIS CUBANO	frijoles	arroz	ajo	pimiento rojo
ALMEJA CON CREMA DE HUEVO	almejas	huevos	caldo de gallina	aceite
SHAWARMA	pierna de cordero	cebollo	perejil	nuez
AREPAS VENEZOLANAS	harina	agua	sal	carne de pollo
TACOS MEXICANOS	tortillas de trigo	carne	ajo	tomates
camarones al ajillo	camarones	acite de oliva	ajo	cebolleta
PAPAS CON CUY	papas	cuy	aceite	aji molido
GUATITA	mantequilla	leche	mantequilla de mani	cebolla
CHUGCHUCARA	carne de res	platanos	maiz	papas

3 Experimentation

The following classes are used:

1ArrozCaldoso
2Arrozconcamarón
5Bolonesa
6Chuletóndeávila
Arrozmarinero
Bunuelosdebacalao
Caldode31
Caldodegallina
Churrasco
Dulcedenaranja
Guatita
Hronado
JULIIO
Llapingacho
Moussedezarzamora
Naranjasconfitadas
pie
Piedelimon
Sopadebolasdeverde
Tartadenaranja

Fig. 2. Class 1

Process and outcome are shown in the table of chi-square, with a total of 1700 photos were obtained (Table 1).

```
INFO:tensorflow:2019-05-21 17:02:28.313121: Step 3980: Cross entropy = 0.359621
INFO:tensorflow:2019-05-21 17:02:28.516834: Step 3980: Validation accuracy = 46.0% (N=100)
INFO:tensorflow:2019-05-21 17:02:29.994682: Step 3990: Train accuracy = 97.0%
INFO:tensorflow:2019-05-21 17:02:29.994682: Step 3990: Cross entropy = 0.353355
INFO:tensorflow:2019-05-21 17:02:30.172685: Step 3990: Validation accuracy = 38.0% (N=100)
INFO:tensorflow:2019-05-21 17:02:31.424478: Step 3999: Train accuracy = 98.0%
INFO:tensorflow:2019-05-21 17:02:31.424478: Step 3999: Cross entropy = 0.343117
INFO:tensorflow:2019-05-21 17:02:31.608866: Step 3999: Validation accuracy = 50.0% (N=100)
INFO:tensorflow:Final test accuracy = 68.9% (N=334)
```

Fig. 3. Results to classification

The outturn shows high accuracy in training and median in the test. This is necessary to classify individual images (Figs. 3 and 4):

```
naranjasconfitadas (score=0.85182)
tartadenaranja (score=0.08427)
dulcedenaranja (score=0.02516)
sopadebolasdeverde (score=0.02113)
piedelimon (score=0.01103)
```

```
arrozmarinero (score=0.62635)
churrasco (score=0.25211)
hronado (score=0.10364)
naranjasconfitadas (score=0.00539)
sopadebolasdeverde (score=0.00520)
```

```
guatita (score=0.63830)
caldodegallina (score=0.08025)
dulcedenaranja (score=0.07816)
tartadenaranja (score=0.07027)
naranjasconfitadas (score=0.05452)
```

```
piedelimon (score=0.74088)
tartadenaranja (score=0.18802)
guatita (score=0.03721)
naranjasconfitadas (score=0.00877)
dulcedenaranja (score=0.00649)
```

Fig. 4. Which mostly match the image was tested.

4 Conclusions and Future Work

In this investigation, we have inspected the efficiency of a CNN-based method for receipts Ecuadorian classification with 20 datasets. The datasets were composed from openly available images and social media. The model for this job can be applied to pre-processing of food item recognition or screen the search result of questions related to food, meals or dishes. In the future, food/non-food classification could be applied to more complex processing of food images.

References

1. Parra, P., Negrete, T., Llaguno, J., Vega, N.: Computer vision techniques applied in the estimation of the cocoa beans fermentation grade. In: Andescon 2018, Santiago de Cali, Colombia (2018)
2. Attokaren, D.J., Fernandez, I., Sriram, A., Murthy, S., Koolaguidi, S.: Food classification from images using convolutional neural networks. In: TENCON, Malaysia (2017)

3. Mezgec, S., Korousic, B.: NutriNet: a deep learning food and drink image recognition system for dietary assessment. Nutrients **9**, 657 (2017)
4. Horiguchi, S., Amano, S., Ogawa, M., Aizawa, K.: Personalized classifier for food image recognition. Latex Class Files **20**, 2836–2848 (2015)
5. Lu, Y.: Food image recognition by using convolutional neural networks (2016)
6. Shamay, J., Rekha, S., Quadri, A.S.: Bird's eye review on food image classification. In: IJLTEMAS (2018)
7. Attookaren, D., Fernandes, I., Murthy, Y., Koolagudi, S.: Food classification from images using convolutional neural networks. In: TENCON, Penang (2017)
8. Farinella, G., Allegra, D., Moltisanti, M., Stanco, F., Battiato, S.: Retrieval and classification of food images. Comput. Biol. Med. **77**, 23–39 (2016)
9. He, Y., Xu, C., Khanna, N., Boushey, C., Delp, E.: Analysis of food images: features and classification. In: ICIP, Paris (2014)
10. Kagaya, H., Aizawa, K.: Highly accurate food/non-food image classification based on a deep convolutional neural network. In: ICIAP, Cham (2015)
11. Singla, A., Yuan, L., Ebrahimi, T.: Food/non-food image classification and food categorization using pre-trained GoogleNet model. In: MADiMa, Amsterdam (2016)

Software and Systems Modeling

Inverse Kinematics of a Redundant Manipulator Robot Using Constrained Optimization

José Varela-Aldás[1]([✉]) [iD], Manuel Ayala[1] [iD], Víctor H. Andaluz[2] [iD],
and Marlon Santamaría[3] [iD]

[1] SISAu Research Group, Universidad Indoamérica, 180103 Ambato, Ecuador
{josevarela,mayala}@uti.edu.ec
[2] Departamento de Eléctrica y Electrónica,
Universidad de las Fuerzas Armadas – ESPE, 171103 Sangolquí, Ecuador
vhandaluz1@espe.edu.ec
[3] Facultad de Ingeniería en Sistemas, Electrónica e Industrial,
Universidad Técnica de Ambato, 180104 Ambato, Ecuador
ma.santamaria@uta.edu.ec

Abstract. Redundant manipulative robots are characterized by greater manipulability improving performance but complicating inverse kinematics, on the other hand, optimization techniques allow solving complex problems in robotics applications with greater efficiency. This paper presents the inverse kinematics of a redundant manipulative robot with four degrees of freedom to track a desired trajectory, and considering constraint in manipulability. The optimization problem is proposed using the quadratic position errors of the operative end and the constraint is established by a manipulability index, for this the kinematic model of the robot is determined. The results show the points of singularity of the robot and the performance of the proposal implemented, observing the positional errors and the manipulability for each point of the trajectory. In addition, the optimization is evaluated for two desired manipulability values. Finally, it is concluded that the implemented method optimizes the inverse kinematics to track the desired path while constraining the manipulability.

Keywords: Inverse kinematics · Constrained optimization · Manipulability · Trajectory tracking

1 Introduction

1.1 Background

Manipulators robots in recent decades have been implemented in industries for various human-robot collaborative applications [1], these mainly cover operations in hazardous environments, repetitive actions, intelligent machining and recently are used in medical applications, so controls have been developed for optimal and accurate performance [2, 3], this has implied a detailed study of the kinematics of robots.

© Springer Nature Switzerland AG 2020
Á. Rocha et al. (Eds.): ICITS 2020, AISC 1137, pp. 233–242, 2020.
https://doi.org/10.1007/978-3-030-40690-5_23

Inverse kinematics is a technique used for robot control that is based on the conversion of the position and orientation of an end effector of the manipulator of the Cartesian space to the joint space [4]. This method stands out from current approaches for its accuracy and efficiency [5, 6]. Basically there are three types of techniques to model the inverse kinematic problem: complete analytics (closed form solution), numerical and semi-analytical, with numerical optimization being one of the most applied in redundant robotic systems [7].

Redundant robotic systems offer several advantages over conventional ones, these have more degrees of freedom than are necessary to perform a task, increase the workspace and eliminate singular configurations [8, 9]. The singular configurations limit the movement of the manipulator robot, so, an efficient way to solve the control problem is to maximize its manipulability [10], generating an optimal movement that corresponds to the shortest path that satisfies the smoothing constraints [11, 12].

In this sense, the development of methods to solve the inverse kinematics in redundant robots is a wide area of study in which techniques are sought that optimize a parameter during the positioning of the manipulator such as the distance of movement, the evasion of obstacles, which allows to have additional factors to solve a problem of infinite possibilities [13, 14]. It is important to highlight that the methods for the control of manipulators have optimized the configurations, minimizing efforts, torques and the kinematic and dynamic performance, and generating greater adaptability and flexibility.

1.2   Related Works

At the kinematic level, studies have been carried out by selecting a redundancy resolution for inverse kinematics, which performs a human-like movement [15]. On the other hand, Wan et al. they present a method for the resolution of the inverse kinematics of redundant manipulators with articulation limits that is able to maintain the restrictions [16]. In addition, the optimization methods propose techniques to choose the correct parameterization, depending on the kinematic characteristics and the desired trajectory [17].

In optimization, the resolution of the singularity problem has been solved using the least squares method, which includes the closed loop inverse kinematics to ensure tracking accuracy [18]. Kelemen et al. present a model of inverse kinematic optimization based on a Jacobian method that considers weight matrices in order to prioritize particular tasks, for example, avoiding cinematic singularities [19].

This paper presents the inverse kinematics of a redundant manipulator robot using constrained optimization. The rest of the contribution is organized as follows: in Sect. 2 the formulation of the problem; in Sect. 3 the kinematic modeling is performed; Sect. 4 describes the optimization; and finally the results and conclusions are presented in Sects. 5 and 6, respectively.

2 Problem Formulation

The objective is to track a trajectory using the inverse kinematics of each point, Fig. 1 describes the required elements. The direct kinematics allows to find the position of the operative end from the angles of the robot, the objective function is constructed by means of the current position and the desired position of the operative end, in addition, the constraints are generated from the angles of the robot to control manipulability. The objective function and constraint is part of the optimization problem that determines the correct angles for trajectory tracking.

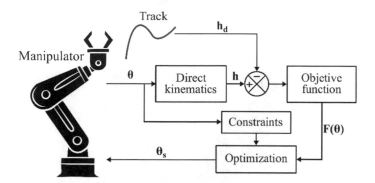

Fig. 1. Diagram of optimization of inverse kinematics.

The optimization function $F(\theta)$ is expressed by the robot angles, using the equations of the kinematics of the robot, according to the degrees of freedom of the robot, and form presented in the expression (1). The resolution of the minimization problem must find a solution that satisfies (2), that is, find the minimum value of all possible input values.

$$min\ F(\theta)\ :\ \theta \in R^n \tag{1}$$

$$F(\theta_s) \leq F(\theta) \tag{2}$$

The optimization problem with restrictions is solved using sequential quadratic programming generating a succession of points according to (3), where s_k is the search direction and α_k is the step length, these parameters are determined using a linear search method (Quasi-Newton or BFGS) [20], which solves a minimization subproblem by approximating the Hessian of the Lagrangian, when the value of θ reaches a minimum the iterative search stops.

$$\theta_{k+1} = \theta_k + \alpha_k\ s_k \tag{3}$$

3   Modelling

3.1   Kinematics

The kinematics of the robot describe the position of the operating end from the input parameters of the system, Fig. 2 shows the manipulator robot to be modeled mathematically., Where l_1, l_2, l_3 and l_4 are lengths and θ_1, θ_2, θ_3 and θ_4 are the angles of the kinematic chain joints.

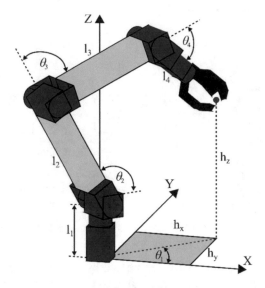

Fig. 2. Parameters of the redundant manipulator robot.

Equations (4–6) describe the position of the final effector through geometric projections, the manipulator unfolds in space $\mathbf{R}^3$ using four degrees of freedom, so it is considered a redundant robot.

$$h_x = cos(\theta_1)\,(l_2\,cos(\theta_2) + l_3\,cos(\theta_2 + \theta_3) + l_4\,cos(\theta_2 + \theta_3 + \theta_4)) \tag{4}$$

$$h_y = sin(\theta_1)\,(l_2\,cos(\theta_2) + l_3\,cos(\theta_2 + \theta_3) + l_4\,cos(\theta_2 + \theta_3 + \theta_4)) \tag{5}$$

$$h_z = l_1 + l_2\,sin(\theta_2) + l_3\,sin(\theta_2 + \theta_3) + l_4\,sin(\theta_2 + \theta_3 + \theta_4) \tag{6}$$

In matrix format the kinematic model of the manipulator is summarized in (7) and the gradient is presented in (8), where $\mathbf{J(\theta)}$ is the Jacobian and $\mathbf{\Delta\theta}$ are the angular velocities of the robot.

$$\mathbf{h(\theta)} = [\,h_x \quad h_y \quad h_z\,]^{\mathrm{T}} \tag{7}$$

$$\mathbf{\Delta h(\theta)} = \mathbf{J(\theta)}\,\Delta\theta \tag{8}$$

3.2 Manipulability

The manipulability in a manipulative robot is a component to control in the trajectory tracking by inverse kinematics, the Eq. (9) allows to find the determinant of the rectangular matrix of the Jacobian, this is a famous index of manipulability. Due to the redundancy, the system has several configurations so that the operating end reaches the same position, this allows to solve the reverse kinematics using the greater manipulability or avoid singular positions in the robot.

$$M = \left| \mathbf{J}(\boldsymbol{\theta})\,\mathbf{J}(\boldsymbol{\theta})^{\mathrm{T}} \right|^{1/2} \tag{9}$$

4 Optimization

4.1 Objective Function

The problem of inverse kinematics is solved for a sequence of points that make up a desired trajectory. Each desired position is solved independently of the others, so that the optimization problem is solved as many times as the number of points contained in the trajectory.

The desired position for the operating end is a three element vector as expressed in (10), and the error function (11) should tend to zero, as shown in Fig. 3.

$$\mathbf{h_d} = \begin{bmatrix} h_{dx} & h_{dy} & h_{dz} \end{bmatrix}^{\mathrm{T}} \tag{10}$$

$$f(\boldsymbol{\theta}) = \mathbf{h}(\boldsymbol{\theta}) - \mathbf{h_d} \tag{11}$$

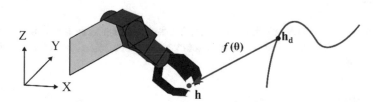

Fig. 3. Objective of the inverse kinematics of the manipulator robot

It is not possible to use the expression (11) as a minimization objective due to the negative values, which is why the quadratic error function (12) that minimizes is required through the manipulator angles. In addition, the gradient of the objective function (13) is determined because it is a parameter for optimization.

$$min\,\mathbf{F}(\boldsymbol{\theta}) = 1/2 f(\boldsymbol{\theta})^{\mathrm{T}} f(\boldsymbol{\theta}) \;:\; \boldsymbol{\theta} \in \mathbf{R}^4 \tag{12}$$

$$\varDelta \mathbf{F}(\boldsymbol{\theta}) = f(\boldsymbol{\theta})^{\mathrm{T}} \mathbf{J}(\boldsymbol{\theta}) \tag{13}$$

4.2   Constraints

The first restriction is to limit the values of the angles between $-\pi$ and π as described in (14), additionally, it is proposed to restrict the manipulability using Eq. (9) to pose (15), so that the manipulability does not be less than a desired value M_d.

$$- \le \theta_i \le \pi, i = 1, 2, 3, 4 \tag{14}$$

$$\left| \mathbf{J}(\boldsymbol{\theta})\, \mathbf{J}(\boldsymbol{\theta})^{\mathbf{T}} \right|^{1/2} \ge M_d \tag{15}$$

Table 1. Simulation parameters.

Parameter	Value	Parameter	Value
l_1	12.5 [cm]	n	100
l_2	27.5 [cm]	k	[0 1 2, ..., n]
l_3	27.5 [cm]	t	$-2\pi + 4\pi k/n$
l_4	17.5 [cm]	h_{xd}	r cos(t) + 25
θ_1 [0]	0.5 [rad]	h_{yd}	r sin(t) + 25
θ_2 [0]	2 [rad]	h_{zd}	t + 25
θ_3 [0]	−1 [rad]	M_{d1}	15000
θ_4 [0]	−1 [rad]	M_{d2}	30000
r	15 [cm]		

5   Results

To generate all the results of this proposal, the data presented in the Table 1 are used, which contains all simulation parameters such as: lengths, initial search conditions and desired values.

5.1   Manipulability Analysis

An analysis of the manipulability of the robot is performed using Eq. (9) and the data in Table 1, Fig. 4 shows the results of the manipulability sweep according to the degree of freedom.

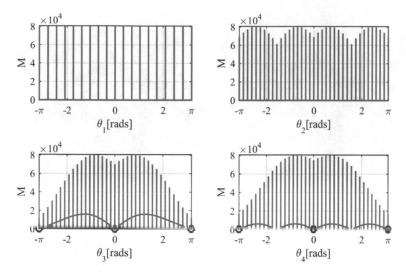

Fig. 4. Manipulability of manipulator robot for each degree of freedom.

The following information is obtained from the manipulability analysis: θ_1 does not affect the manipulability, θ_2 does not have manipulability points equal to zero, θ_2 and θ_3 do not have independent values that generate manipulability equal to zero, the manipulability is zero when θ_2 and θ_3 is equal to 0 or multiple of, and the maximum manipulability is 80000.

5.2 Inverse Kinematics

The optimization problem for inverse kinematics is implemented in Python using the sequential quadratic programming method and using the parameters of Table 1, which proposes the follow-up of a spiral-shaped path in space, in addition, the animation of the simulation using Matlab, as seen in Fig. 5. The simulation is configured for two cases: (a) with desired manipulation constraint greater than 15000, where the robot does not have problems in tracking the trajectory maintaining this manipulability, and (b) with desired manipulation constraint greater than 30000, where the operating end has positions outside the trajectory.

Figure 6 shows the trajectory tracking errors generated by the values of the objective function in the two cases mentioned. In Fig. 6a the error is kept zero in all cases of optimization of the inverse kinematics and in Fig. 6b the objective function has two divergence events, which is the position closest to the trajectory for maintain the desired manipulability, the maximum error produced is 23 [cm].

Figure 7 shows the manipulability for each point of the trajectory performed in the two cases of desired manipulability, in Fig. 7a the values are always above the desired manipulability, and in Fig. 7b the manipulability reaches the limit allowed in two segments of the desired trajectory.

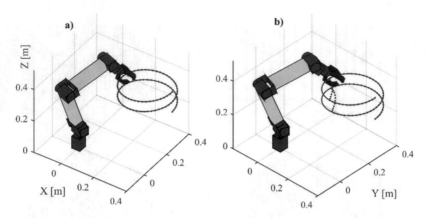

Fig. 5. Simulation of trajectory tracking with inverse kinematics for: (a) M_d = 15000, y (b) M_d = 30000

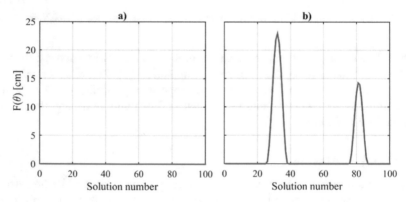

Fig. 6. Values of the objective function in trajectory tracking for: (a) M_d = 15000, y (b) M_d = 30000

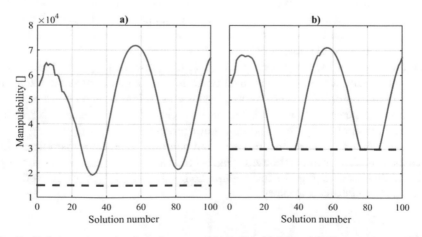

Fig. 7. Robot manipulation in trajectory tracking for: (a) M_d = 15000, y (b) M_d = 30000

6 Conclusions and Discussions

This paper proposes a reverse kinematics technique for a redundant manipulator robot through a constrained optimization problem. The proposed objective function is based on the quadratic errors of the position of the operative end where the direct kinematics of the robot is considered, on the other hand, the manipulability is restricted in the solution of the inverse kinematics by an inequality with the determinant of the Jacobian, and the optimization is solved by sequential quadratic programming using Python.

The optimization technique with constraints allows designing robotics problems using mathematical models and performance indices to establish constraints. In addition, the manipulability analysis determined singularity or manipulability points equal to zero when θ_2 and θ_3 equals $m\pi$ (for any integer value of m) and the redundant robot can reach a maximum manipulability of 80000. The results of trajectory tracking they show an ideal performance of the proposal for $M_d = 15000$, but in the case of $M_d = 30000$ it does not complete all the points of the trajectory because it maintains the manipulation constraints, finding a minimum distance to the trajectory that meets the optimization problem.

Comparing this work with the bibliography found, this method stands out for the flexibility of adding restrictions while solving inverse kinematics, because the optimization methods have been developed using formats that favor scalability. On the other hand, the objective function guarantees a local or global minimum of all possible redundant system solutions. Finally, the use of Phyton allows solving the optimization problem for each position, in short execution times (4 ms).

References

1. Glogowski, P., Lemmerz, K., Schulte, L., Barthelmey, A., Hypki, A., Kuhlenkötter, B., Deuse, J.: Task-based simulation tool for human-robot collaboration within assembly systems. In: Tagungsband des 2. Kongresses Montage Handhabung Industrieroboter (2017)
2. Staicu, S.: Dynamics of Parallel Robots. Springer Nature Customer Service Center Gmbh (2019)
3. Kyrarini, M., Haseeb, M.A., Ristić-Durrant, D., Gräser, A.: Robot learning of industrial assembly task via human demonstrations. Auton. Robots 43, 239–257 (2019). https://doi.org/10.1007/s10514-018-9725-6
4. Zaplana, I., Basanez, L.: A novel closed-form solution for the inverse kinematics of redundant manipulators through workspace analysis. Mech. Mach. Theory 121, 829–843 (2018). https://doi.org/10.1016/j.mechmachtheory.2017.12.005
5. Kofinas, N., Orfanoudakis, E., Lagoudakis, M.G.: Complete analytical forward and inverse kinematics for the NAO humanoid robot. J. Intell. Robot. Syst. 77, 251–264 (2015). https://doi.org/10.1007/s10846-013-0015-4
6. Iliukhin, V.N., Mitkovskii, K.B., Bizyanova, D.A., Akopyan, A.A.: The modeling of inverse kinematics for 5 DOF manipulator. Procedia Eng. 176, 498–505 (2017). https://doi.org/10.1016/j.proeng.2017.02.349
7. Kim, S., Xu, W., Ren, H.: Inverse kinematics with a geometrical approximation for multi-segment flexible curvilinear robots (2019). https://doi.org/10.3390/robotics8020048

8. Chiaverini, S., Oriolo, G., Walker, I.D.: Kinematically redundant manipulators. In: Springer Handbook of Robotics (2008). https://doi.org/10.1007/978-3-540-30301-5_12

9. Varela-Aldas, J., Andaluz, V.H., Chicaiza, F.A.: Modelling and control of a mobile manipulator for trajectory tracking. In: Proceedings of the - 3rd International Conference on Information, Systems and Computer Science, INCISCOS 2018, pp. 69–74 (2018). https://doi.org/10.1109/INCISCOS.2018.00018

10. Jin, L., Li, S., La, H.M., Luo, X.: Manipulability optimization of redundant manipulators using dynamic neural networks. IEEE Trans. Ind. Electron. **64**, 4710–4720 (2017). https://doi.org/10.1109/TIE.2017.2674624

11. Li, K.-L., Yang, W.-T., Chan, K.-Y., Lin, P.-C.: An optimization technique for identifying robot manipulator parameters under uncertainty. Springerplus **5**, 1771 (2016). https://doi.org/10.1186/s40064-016-3417-5

12. Azad, M., Babič, J., Mistry, M.: Effects of the weighting matrix on dynamic manipulability of robots. Auton. Robots **43**, 1867–1879 (2019). https://doi.org/10.1007/s10514-018-09819-y

13. Hwang, S., Kim, H., Choi, Y., Shin, K., Han, C.: Design optimization method for 7 DOF robot manipulator using performance indices. Int. J. Precis. Eng. Manuf. **18**, 293–299 (2017). https://doi.org/10.1007/s12541-017-0037-0

14. Husak, E., Karabegović, I.: Heuristic optimization methods in industrial robotics. In: Advanced Technologies, Systems, and Applications II (2018)

15. Liu, W., Chen, D., Steil, J.: Analytical inverse kinematics solver for anthropomorphic 7-DOF redundant manipulators with human-like configuration constraints. J. Intell. Robot. Syst. **86**, 63–79 (2017). https://doi.org/10.1007/s10846-016-0449-6

16. Wan, J., Wu, H., Ma, R., Zhang, L.: A study on avoiding joint limits for inverse kinematics of redundant manipulators using improved clamping weighted least-norm method. J. Mech. Sci. Technol. **32**, 1367–1378 (2018). https://doi.org/10.1007/s12206-018-0240-7

17. Ferrentino, E., Chiacchio, P.: Redundancy parametrization in globally-optimal inverse kinematics. In: Advances in Robot Kinematics 2018 (2019)

18. Wan, Y., Kou, Y., Liang, X.: Closed-loop inverse kinematic analysis of redundant manipulators with joint limits. In: Advances in Mechanical Design (2018)

19. Kelemen, M., Virgala, I., Lipták, T., Miková, Ľ., Filakovský, F., Bulej, V.: A novel approach for a inverse kinematics solution of a redundant manipulator (2018). https://doi.org/10.3390/app8112229

20. Nocedal, J., Wright, S.J.: Numerical Optimization. Springer, Heidelberg (2006)

User Profile Modelling Based on Mobile Phone Sensing and Call Logs

Alexander Garcia-Davalos[1,2]([✉]) [iD] and Jorge Garcia-Duque[2] [iD]

[1] Universidad Autonoma de Occidente, Cali, Colombia
agdavalos@uao.edu.co
[2] Universidad de Vigo, Vigo, Spain
jgd@det.uvigo.es

Abstract. There are remaining questions concerning user profile modelling in the mobile advertising domain. The research question addressed in this paper is how to design a specific user profile model, that is a simplified model in terms of the amount of user data to be collected, that considers relevant aspects of mobile advertising such as social and personal context, and user privacy preservation. To address this question, a new user profile model consisting of three phases was proposed: (1) data collection, (2) integration and normalization of collected data, and (3) inference of knowledge about the mobile user's profile. The most significant contributions of the proposed model are a simplified user profile model approach which tackles the dependency on other data sources like OSN platforms and local data gathering and storage that contributes to the user privacy-preserving since the user can exert more control over his/her personal data.

Keywords: User profile · User model · Social context · Personal context · Mobile advertising · Mobile phone sensing

1 Introduction

The purpose of the user profile depends on its application domain. E-commerce uses it to recommend products/services or offers that interest the user/client. Various application domains, including advertising, traditionally use user profile as input for personalisation of services/applications.

The user profile is defined by Mohamad and Kouroupetroglou [10] as "an instantiation of a user model representing either a specific real user or a representative of a group of real users". In this definition, the user profile is associated with the concept of the user model, which these authors define as "explicit representations of the properties of an individual user, including the needs, preferences and physical, cognitive and behavioural characteristics". The terms "user profile" and "user model" are used in the literature as synonyms for referring to the concept of user representation. Based on the former definition, in this paper, we assume the user profile as an instance of the model. The user profile can be obtained by two basic ways [14] (1) explicitly, where the user registers the data through forms or graphics interfaces designed for this purpose; (2) implicitly, that is, the actions/activities of the user are observed/monitored through applications/services and the data is recorded. The second way, gives rise to a

© Springer Nature Switzerland AG 2020
Á. Rocha et al. (Eds.): ICITS 2020, AISC 1137, pp. 243–254, 2020.
https://doi.org/10.1007/978-3-030-40690-5_24

permanent profile updating, making it a "dynamic profile" [5], instead of generating a "static profile" as in an explicit way.

Schiafino and Amina [14] state that the most common contents of the user profile are: user interests, user's knowledge, user's goals, user behaviour, user's interaction preferences, user's individual characteristics, and user's context (e.g. personal and social context). Technological advances in software and hardware have allowed the emergence of new techniques for obtaining the user profile. This has benefited advertising applications/services which can generate more relevant advertising strategies and better user experience for its target audience by having a more dynamic user profile.

However, with the increase of personal data collected by applications/services, users' concern over privacy preservation has become a very sensitive issue and users are demanding more control over their user profile. Therefore, the users' privacy concern and their claim to reduce the amount of personal data collected are interdependent requirements for addressing the modelling of the user profile, and designing new models, particularly for mobile advertising domain. The research question addressed in this paper is how to design a specific user profile model, that is a simplified model in terms of the amount of user data to be collected, and which considers relevant aspects of mobile advertising and contributes to user privacy preservation.

The remainder of the paper is organised as follows: in Sect. 2 we present the review of previous work related to user models: in Sect. 3, we propose a new user profile model and describe its design: in Sect. 4 we present the software testing of the new model and discuss its results; finally, in Sect. 5, we summarize the contributions of this paper and state the future work.

2 Related Work

Initially, the user models designed were generic and applicable to different application domains. Subsequently, specific models whose design focuses on particular needs and its domain of application is previously defined were proposed. Table 1 shows a summary of related works that propose generic and specific user models.

Table 1 Summary of user models

Model	Authors	Description
Generic	Cena et al. [4]	Real World User Model (RWUM): characterised for greater coverage in terms of data collection and the ability to model more features of the user, greater precision about users' behaviour
	Smailovic [15]	Advanced User Profile Model: contemplates four stages for user profiling: data acquisition, aggregation or combination of data from three sources, consolidation of data, and analysis and reasoning on the user's profile

(continued)

Table 1 (*continued*)

Model	Authors	Description
	Musto et al. [11]	Holistic User Model: can be described through different facets of people, such as their interests, activities, habits, mood, social connections, etc, and these facets can be modelled by collecting and merging data from OSN platforms and user devices
Specific	Qiang Ma [9]	User Modeling for Online Advertising: a reference framework for advertisement tracking based on users profile and the relation between users profiles and advertisements to which they are exposed is empirically addressed
	Qaffas and Cristea [13]	Adaptive E-Advertising Delivery System; includes a new user model which uses two fundamental data sources to generate the user profile: explicit data supplied by users and data gathered from OSN
	Asif and Krogstie [3]	Interactive User Model: outsource the user model to alleviate the problems of invisibility and inconsistency of the user profile

The growth and evolution of mobile advertising have been strengthened with technological advances and have allowed for the deployment of mobile advertising that seeks to be increasingly relevant and less "disturbing" for the mobile user. Indeed, the basic premise of advertising is the "right person, right place, right message, at the right time".

The user profile has been exploited intensely in advertising; moreover, significant progress has been made in generating user profiles based on new data sources, such as mobile devices and OSN platforms. For the specific case of mobile advertising, data of the mobile phone (e.g. messages), sensors (e.g. GPS) and in some cases wearable devices have been used. However, there are possibilities to continue exploring questions related to user profile modelling in mobile contexts and particularly in the search for mobile advertising solutions, which take advantage of data on user social aspects (e.g. social context) that are implicitly and dynamically collected by mobile phones.

3   New User Profile Model for Advertising Domain

The design of the proposed user profile model is inspired by two previous research works: (a) the Advanced User Profile Model (AUP Model) proposed by Smailovic [15], which uses various data sources to generate the user profile and the inference of knowledge for social aspects of the user; (b) the personalisation of the mobile services proposed in the PhD thesis of Asif [2], in which user profile is considered as a fundamental aspect since it improves the user's understanding of user profile and user experience with mobile services/applications [3].

The main characteristics that identify the new user profile model are:

- Specific model oriented to mobile advertising: the proposed model has been designed considering four relevant aspects of mobile advertising - social and personal context, privacy-preserving, and updating profile. These aspects seek to contribute to the improvement of the user experience in advertising solutions/services.
- Implicit data collection using mobile phones: the data gathered is: (1) data of the user's social interaction (calls and personal contacts), and (2) readings from specific mobile phone sensors (e.g. accelerometer, gyroscope, GPS) that allow identifying the user's activity t (e.g. walking, sitting, stopped, etc.) location.
- Contribute to user privacy-preservation: the user profile data is saved locally in the mobile phone users are allowed to view their profile and are enabled to validate what personal data they wish to share with their personal social network.
- Generating a dynamic user profile: mobile advertising requires a permanent update of data about the user in order to offer ads and advertising strategies that are more relevant for people. In the proposed model, data is collected permanently, and the user profile is continuously updated using this data.
- Generate an input to determine the personal social network (PSN) of the mobile user based on the content of the user profile. The user's social context data is used to identify who makes up his/her PSN.

3.1   Phases of User Modeling

The proposed new model was designed with a simplified and three-phase approach, wherein each phase, the data about the user is processed and serves as input for the next until obtaining the user profile. The three phases that comprise the modelling of the user profile are:

1. Phase 1 - Data collection: it comprises the process of implicit data collection from two main sources in the mobile phone:

   - Device sensors: smartphones are equipped with a set of sensors and their operating systems offer some APIs to access/use these sensors. Considering that Android is the dominant operating system in the mobile phone market it was taken as the referring operating system for the present work. The selected API to obtain the data about user personal context (location, activity) was the so-called "Awareness API" of Android.
   - Data of user's social interaction: this is obtained by accessing the list of contacts from the mobile phone and the call records. Data from other mobile apps, like chat, is not gathered because it is not easy to access data since the apps owned companies have been forced to enhance the user's privacy and the data is encrypted.

   The data collected is classified according to its purpose (user activity, location, interaction through calls) and stored in its respective temporary file in the mobile phone's memory to be processed during the next phase of the user profile modelling.

2. Phase 2 - Integration and normalization of data: because the data is in a "raw" form, data is normalised, integrated, and stored for later use in the next modelling phase of the user profile model. The actions carried out in this phase are (a) the data of the user's location is converted from raw form (coordinates) into addresses which facilitate the processing of the data. Likewise, the records are classified according to place and time; (b) the phone calls data is classified by contact or person, frequency, and call duration. As a result of this process, a set of new files is generated. The data obtained from the integration and normalization process is separated into two categories according to the context type: (a) social context data: this is related to social aspects, such as mobile user relationships (friends, colleagues, acquaintances, and family); (b) Personal context data: this is related to the user's state, such as the situation in which the mobile user is in (running walking, driving, etc.) and his/her current location.

3. Phase 3 - Inference of knowledge: in this phase, two different approaches are used to obtain knowledge about the user. These approaches allow determining the two main attributes of the mobile user according to what is stated in the present model: the social relationship and the typical situations in what the user is. The two approaches to be used are:

(a) Partitioning clustering: cluster partitioning algorithms [16] are applied to determine how the mobile user's social relationships are composed, namely, which are the people close to the user based on social interaction through the mobile phone, and basic concepts of personal social networks. The goal is to obtain the concentric circles of the user's PSN based on the frequency of communication with other people through the mobile phone (calls). These circles were identified from the contact frequency of ego interactions with other people [1] and they are called: core network or support clique, sympathy group, affinity group, and active network.

Regarding the frequency of communication, La Gala et al. [7] classify communication between the user and other people in two types:

(i) Incoming frequency: represents the frequency with which a person interacts with the user (ego). This frequency can be calculated as the number of direct communications received by the user from other people, divided by the total duration of incoming communications originated by other people.

$$Fic = \frac{Nir}{Ti} \tag{1}$$

Where,

Fic, is the frequency of the incoming communication between the user and a given person.

Nir, is the number of communications received from a given person.

Ti, is the total duration of communications received from the user with a given person.

(ii) Outgoing frequency: represents the frequency with which the user (ego) contacts each of the people and is socially related. This frequency can be calculated as the total number of interactions (calls) between the user and a given person, divided by the duration of these communications.

$$Foc = \frac{Ni}{Ti} \qquad (2)$$

Where,

Foc, is the frequency of outgoing communication between the user and a given person (*i*).
Ni, is the number of user interactions with a given person.
Ti, is the total duration of the user's social interactions with a given person.
The authors also state that by modelling each user's social relationship as bidirectional links, it is possible to demonstrate reciprocity in the communication and characterize the concentric circles around the user (ego).

To demonstrate the reciprocity of communications between users, it means when the two communication frequencies are greater than zero. La Gala et al. [7] propose an index called the Adjacency Frequency (AdjFreq) that integrates the two frequencies. This index of the frequency of communications between users was the base for grouping user's contacts according to what is specified in ego social networks by Dunbar [6].

According to the new user model, the user's personal social network of a specific mobile user is determined as follows:

$$PSNi = \sum Ci \qquad (3)$$

Where,

Ci, is the user's social circles and is composed of *x* users. The user contacts are grouped in social circles based on the adjacency frequency and this is expressed in terms of limits as follows:

$$Ci = LimAdjia < AdjFreqij < LimAdjib \qquad (4)$$

Where,

LimAdjia, is the minimum frequency of adjacency for the social circle *i*.
AdjFreqij, is the frequency of adjacency of the mobile user with a given person (j).
LimAdjib, is the maximum adjacency frequency for the social circle *i*.

The clustering algorithm selected to identify the concentric circles was the so-called Jenks Natural Breaks algorithm [12]. This is an algorithm for dividing a dataset into several homogeneous classes.

(b) Bayesian inference: it is used to infer knowledge about the most common situations in which the user finds himself since this technique can be used to make inferences in contexts where a large volume of data is not available. Bayesian inference is a technique of statistical inference, which is based on Bayes' theorem to update the probability of a given hypothesis based on obtaining more data or evidence about it [8].

In this phase of the new model, the personal context and mobile user location are the input data to carry out the Bayesian inference process. Two basic hypotheses were posed to infer knowledge about the mobile user, and they are focused on identifying which are the most common user situations. These hypotheses are based on the data obtained from the mobile sensing process: (a) H1 - a user is a person who tends to be an itinerant; (b) H2 - the user often drives a vehicle (bicycle or car).

Figure 1 shows a diagram of the two approaches used in this phase to infer knowledge about the mobile user profile.

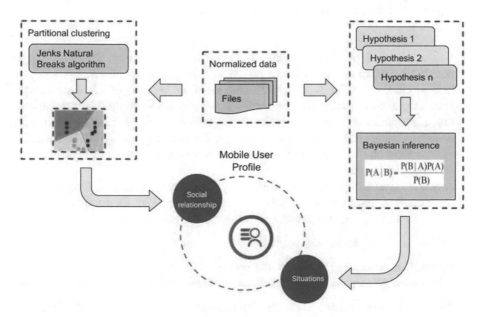

Fig. 1. Inference knowledge phase

4　Testing of the New User Profile Model

To test the new user profile model, a mobile application was developed to run on Android mobile phones. This app implemented the first phase of the proposed model and was tested with a group of 10 users for two weeks. The general data about the users that participated in the test is: four men, six women, and the average age is 36.8 years.
　The implementation of the phases of the user model is described below.

　　Phase 1 - Data Collection: the user's social interaction data (contacts and calls) and personal context was collected through the mobile application. The application executes an algorithm for collecting data implicitly through two Android APIs: Awareness (location) and Contacts (contacts). The data collected by the mobile app was stored in its respective temporary file on the mobile phone's memory.

　　Phase 2 - Integration and Normalization: data stored in the temporary files was taken as input by a Python script that generates a set of normalised data which were grouped into two categories according to the type of context: personal (location and situation) and social (social relationship). GPS coordinates were converted to addresses using Google Maps API and the incoming/outgoing call records were sorted by the call duration and its frequency.

　　Phase 3 - Inference of Knowledge: data normalised in Phase 2 was used to infer knowledge about two fundamental mobile user's attributes: social relationship and typical situations in which the user finds himself.

– Social relationships: Jenks Natural Breaks algorithm was used for clustering of people who make up the circles of the PSN. People (alters) were grouped in concentric circles (Dunbar's circles) around the user or ego, which were defined according to contact frequency between the ego and other people. The data input of this algorithm was: (1) the Adjacency frequency which reflects the interaction between the user and his/her contacts through the mobile phone, (2) and the number of classes or groups, which in this case is four, that is the number of social circles according to Dunbar [6].
– Personal situations (Personal context): Bayesian inference was used to determine the situations that identify the user in terms of the profile. Two hypotheses were posed for testing personal context detection in the proposed model:

　　H1: A user is a person who tends to be an itinerant.

　　Proposition A: the user is outside the house.
　　Proposition A': number of user activities per week that are different from "sitting".

　　H2: The type of mobility most used by the user is the vehicle (bicycle, car).

　　Proposition A: the user drives a vehicle.
　　Proposition A': The user drives the vehicle frequently.

4.1 Results and Discussion

Data from the test of the new model was processed as explained in the previous section and the two main attributes of the mobile user profile, user social relationship and personal situations in what the user is, were obtained for each test user. The test results for each attribute are presented below.

Social relationships: Jenks Natural Breaks algorithm was applied to find the four circles of the user's personal social network. Figure 2 shows the social circles obtained for the ten users where $c3$ stands for the core network of the user, $c2$ for sympathy group, $c1$ for the affinity group, and $c0$ for the active network. The contacts of each user were normalised based on the frequency of communications (calls), meaning that only the Adjacency frequencies greater than zero were taken for grouping the users in these four circles.

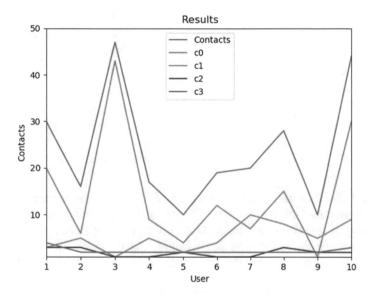

Fig. 2. Social circles of the users

The obtained data evidence that the number of people (average 19) in users' PSN is smaller than the number defined by Dunbar (150). The whole user's personal network could be seen as a network composed of face-to-face contacts and virtual contacts (OSN and by phone). The results show that there is a direct correlation between the number of user contacts and the social circle called "active network" (c0 - contacts at least once per year). On the other hand, for the smaller social circle which is the so-called "core network" (c3), the number of people grouped in this circle was between two and four, that is in the range defined by Dunbar.

These results were compared with the test users' social circles of face-to-face social interactions in two dimensions: the global coincidence (GC - contacts that interact face-to-face and by mobile phone), and the coincidence inside each social circle (CSC -

contacts inside virtual and face-to-face circles). In terms of percentage, the median of global contacts coincidence was 79%. Likewise, the median of coincidence inside each social circle was 68%. Figure 3 illustrates the number of coincidences for each user in the two dimensions.

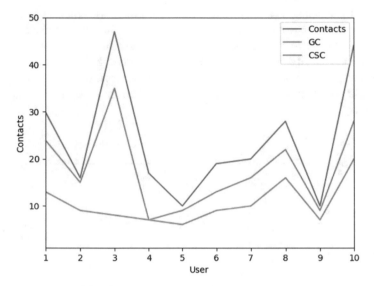

Fig. 3. The coincidence of users' contacts

Personal situation: the hypotheses proposed focus on two basic user situations: the user's itinerancy and driving vehicles. The probability for each hypothesis was calculated using Bayes theorem. The results show that the probability that users were itinerant was on average 67.2% and the probability that the users drive a vehicle was on average 5.7%, which is relatively low compared to the percentage of the user's itinerancy.

The collection of data on users personal situation by the mobile app was consistent in terms of data recorded for 50% of the users (more than 100 records), but in the rest of the users was intermittent (less than 100 records per user). Probably this issue is related to some technical aspects like the mobile phone CPU load and the number of apps that were running at the same time. This is one of the items to improve in the mobile app for gathering user location and personal situations.

One limitation of this research is that the test group was small. Clearly, 10 users are not enough to make a generalization about user profile modelling. However, it was enough to prove the concept of the proposed model: moreover, the test outcomes are evidence of the feasibility of implementing this model. Although, it is necessary to scale the test with a greater number of people that enhance the inference of social circles of use's personal network, and to obtain results that will help strengthen the validation of hypotheses about the user's personal context.

Finally, attributes of the mobile user profile obtained for each user of the test were saved in a file that represents the mobile user profile: the social relationship (social circles), personal situation (hypothesis H1 and H2), and privacy aspects (permission).

5 Conclusions

The new proposed model is focused on two main aspects: user's claims to reduce the amount of collected personal data, and user's privacy concern. These are key elements to address the user profile modelling for domains like mobile advertising.

The proposed user profile model was tested by implementing a mobile app and 10 people were enrolled as volunteering users for the test. Inference of user profile main features was achieved through Bayesian inference (personal context) and Jenks Natural Breaks algorithm (social context). Despite the small sample of users, the test of the proposed model was suitable for validating the proof-of-concept of a simplified user profile model. However, to make a generalization about user profile modelling, it is necessary to scale the test.

The test results regarding the user's social context made evident that there are differences between inferred social circles and circles of face-to-face social interactions. Concerning the user's personal context, two hypotheses focused on two basic user situations were proposed: the user itinerancy and driving vehicles. The probability for each hypothesis was calculated using Bayes theorem and the results showed that the probability that users drive a vehicle is low (5.7%) compared to the percentage of the user's itinerancy (67.2%). However, the mobile app must be improved to increase the user's location records and his/her personal situations data.

The main contributions of this research are a simplified user profile model approach that use mobile phones as the only source of personal data and tackles the dependency of user modeling on other data sources like OSN platforms; also, data is collected and stored only on mobile phones which gives the user more control over his/her personal data and contributes to the user's privacy preservation. These contributions are closely related since users demand more control over their personal data.

The future research work will focus on a new mobile advertising model, which is mobile user-centered and based on personal and social context from the proposed user profile model that will be integrated as one of the components in the design of this new model.

References

1. Arnaboldi, V., Guazzini, A., Passarella, A.: Egocentric online social networks: analysis of key features and prediction of tie strength in Facebook. Comput. Commun. **36**(10–11), 1130–1144 (2013)
2. Asif, M.: Personalization of mobile services. Ph. D. thesis, Norwegian University of Science and Technology, Trondheim (2014). https://tinyrl.com/y8utlhqg. Accessed 29 Aug 2019
3. Asif, M., Krogstie, J.: Externalization of user model in mobile services. Int. J. Interact. Mobile Technol. (IJIM) **8**(1), 4–9 (2014)

4. Cena, F., Likavec, S., Rapp, A.: Real world user model: evolution of user modeling triggered by advances in wearable and ubiquitous computing. Inf. Syst. Front. **20**, 1–26 (2018). Springer, US
5. Cufoglu, A.: User profiling - a short review. Int. J. Comput. Appl. **108**(3), 1–9 (2014)
6. Hill, R.A., Dunbar, R.I.M.: Social network size in humans. Hum. Nat. **14**(1), 53–72 (2003)
7. La Gala, M., Arnaboldi, V., Passarella, A., Conti, M.: Ego-net digger: a new way to study ego networks in online social networks. In: Proceedings of the First ACM International Workshop on Hot Topics in Interdisciplinary Social Networks Research, HotSocial 2012, Beijing, China, 12–16 August 2012, New York, USA, pp. 9–16 (2012)
8. López, J., Krzywinski, M., Altman, N.: Points of significance: Bayes' theorem. Nat. Methods **12**(4), 277–278 (2015)
9. Ma, Q.: Modeling user for online advertising. Ph. D. Thesis, The State University of New Jersey (2016). https://rucore.libraries.rutgers.edu/rutgers-lib/51361/PDF/1/play. Accessed 30 Aug 2019
10. Mohamad, Y., Kouroupetroglou, C.: W3C: user modeling. Research and development working group (2013). http://www.w3.org/WAI/RD/wiki/User_modeling. Accessed 28 Aug 2019
11. Musto, C., Semeraro, G., Lovascio, C., de Gemmis, M., Lops, P.: A framework for holistic user modeling merging heterogeneous digital footprints. In: 26th Conference on User Modeling, Adaptation and Personalization, Singapore, 8–11 July 2018, New York, USA, pp. 97–101 (2018). https://doi.org/10.1145/3213586.3226218
12. North, M.A.: A method for implementing a statistically significant number of data classes in the Jenks algorithm. In: Sixth International Conference on Fuzzy Systems and Knowledge Discovery (FSKD 2009), Tianjin, China, 14–16 August 2009, Los Alamitos, CA, USA, pp. 35–38 (2009)
13. Qaffas, A.A., Cristea, A.I.: An adaptive E-advertising user model: the AEADS approach. In: 13th International Joint Conference on e-Business and Telecommunications (ICETE 2016), Lisbon, Portugal, 26–28 July 2016, pp. 124–131 (2015)
14. Schiaffino, S., Amandi, A.: Intelligent User Profiling. Artificial Intelligence and International Perspective, Lecture Notes in Computer Science Book Series, vol. 5640, pp. 193–216 (2009)
15. Smailovic, V.: User provisioning profile for information and communication services based on user influence. Ph. D. thesis, University of Zagreb, Zagreb (2016). https://bib.irb.hr/datoteka/856944.Vanja_Smailovic_doktorska_disertacija_-_FER_-_2016.pdf. Accessed 2 Aug 2019
16. Xu, D., Tian, Y.: A comprehensive survey of clustering algorithms. Ann. Data Sci. **2**(2), 165–193 (2015)

Analysis of the Performance of Wireless Sensor Networks with Mobile Nodes Under the AODV Protocol

Néstor Zamora Cedeño[✉], Orlando Philco Asqui, and Emily Estupiñan Chaw

Faculty of Technical Education for Development,
Catholic University Santiago de Guayaquil, Guayaquil 090101, Ecuador
{nestor.zamora, luis.philco}@cu.ucsg.edu.ec,
emily.estupinan.chaw@gmail.com

Abstract. Due to recent technological advances, the manufacture of tiny and low-cost sensors has become technically and economically feasible. The censor electronics measure the environmental conditions related to the environment surrounding the sensor and transforms them into an electrical signal. The processing of such a signal reveals some properties on localized objects and/or events that occur in the vicinity of the sensor. Wireless sensor networks (WSN) contain hundreds or thousands of these network-connected sensor nodes that can communicate with each other or directly to an external base station. Currently there are countless areas where WSNs find application, with telemedicine, industrial process control and precision agriculture being some of the most relevant. In this work the fundamental theoretical aspects of this technology are studied, paying special attention to routing protocols. In addition, modeling and simulation of two network scenarios is performed using Y comparing two softwares, in a network simulator (NS-3) and Omnet++ together with the Castalia module to analyze their respective performances with and without mobility in the nodes, taking into account it counts the energy efficiency, the error rate of reception of the packets and the network load due to the AODV protocol.

Keywords: WSN · AODV · NS-3 · OMNeT++ · Castalia · Protocol

1 Introduction

Wireless technology has had a dizzying growth in recent decades. In 1993, the "Infrared Data Association" was created, which defined a physical standard for the transmission and reception of point-to-point data using infrared rays [1]. A few years after surgery, short-range and multipoint wireless personal area networks (WPAN) such as the well-known "Bluetooth" standard or multi-hop mid-range networks such as "ZigBee". Among other wireless technologies that we can name are the WIFI standard for wireless local area networks (WLAN), "WIMAX" for wireless metropolitan area networks (WMAN). We can also highlight the progress of cellular telephony and the development of M2M (machine to machine) communications with wireless technology

© Springer Nature Switzerland AG 2020
Á. Rocha et al. (Eds.): ICITS 2020, AISC 1137, pp. 255–264, 2020.
https://doi.org/10.1007/978-3-030-40690-5_25

[2]. One of the technologies with an interesting development and that is being used more and more are the wireless sensor networks (WSN) due to its multiple applications. The sensors link the physical with the digital world, capturing and revealing real-world phenomena and then converting them so that they can be processed and stored [3]. Wireless sensor technology is present in different sectors, for example: security, environmental monitoring, industry, agriculture, telemedicine, among others. Among the manufacturers that have launched research lines in this technology are Motorola, Microsoft, IBM, Intel, Texas Instruments, among others [4].

The WSN are based on low cost and consumption devices, generally known as "nodes", which obtain information from their environment, which can then be processed locally and transmitted via wireless links. The nodes act as elements of the communications infrastructure by forwarding the messages transmitted by farther nodes to the coordination center [5]. The wireless sensor network is made up of numerous small spatially distributed devices that use sensors to control various conditions at different points, including temperature, sound, vibration, pressure and movement or contaminants. The sensors can be fixed or mobile. The coordinating node or sink is an essential link in the network, because it is responsible for collecting all the information from the rest of the sensor nodes to then process it and give it a certain treatment. These sensor devices are autonomous units consisting of a microcontroller, a power source (almost always a battery), a radio transceiver and a sensor element. According to the authors [6], the costs of implementing these types of networks are decreasing and the reliability of software applications is increasing.

A sensor node has two components. The first, called mote, is responsible for storage, calculation and communication. The second component, called a sensor, is responsible for detecting physical phenomena such as temperature, light, sound and vibration, to name a few. A sensor is always connected to a nickname. The sensor nodes collect data and can perform network processing on the data collected at the intermediate nodes before sending them to a central collection point, called a sink (or base station), for further analysis and processing [7]. The topologies of wireless sensor networks determine the way in which the nodes are going to communicate, therefore, they define the physical or logical map of the network to exchange the data. The authors [8] affirm that mobility is a prominent research area in the field of wireless communications due to the variations that are introduced in the radio channel and the need for management carried out by a central node or coordinator, which requires that every node, whether mobile or landline, have continuous communication with him.

Works such as [9] experimentally evaluate the impact of propagation conditions in various environments with mobile wireless sensor nodes. And [10] address the design of new communication protocols to improve connectivity in wireless sensor networks with mobile nodes. In the work of [11], they indicate that the evaluation of the different metrics: performance, network load, bit error rate (BER), received power, signal-to-noise ratio (SNR) and end-to-end delay for 40 nodes in topologies in star and mesh using the AODV protocol it obtained 0.002 J/s in the received power.

In this work, a network of 49 mobile nodes is evaluated and the energy consumption of each of them is obtained. The objective of this research paper compares the behavior of two models of wireless sensor networks with and without mobility in the nodes in terms of energy efficiency and quality in the transmission of information.

2   Wireless Sensor Networks

The WSN is composed of wireless modules called sensor the archive node in Fig. 1. Sensor nodes in unattended networks can have a profound effect on the effectiveness of many military and civil applications such as image capture of a target, intrusion detection, time monitoring, security and tactical surveillance, distributed computing, detection of ambient conditions such as temperature, movement, sound, light or the presence of certain objects and disaster management. The deployment of a sensor network in these applications can process randomly, for example, it can drop from a plane in a disaster management operation. Manuals can also be used, for example, by locating fire alarm sensors in an installation. These red sensors can help rescue operations by locating survivors, identifying risk areas and making rescue teams better informed of the overall situation in the disaster area [11].

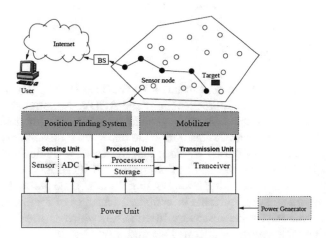

Fig. 1.   Sensor node architecture [12].

Each of these sensor nodes can collect and route data to other sensors or back to an external base station (BS). A base station can be a fixed node or a mobile node capable of connecting the sensor network to an existing network, to a communications infrastructure or to the Internet, where a user can access the reported data [13]. The nodes in a WSN can vary widely, that is, simple nodes can monitor a single physical phenomenon, while more complex devices can combine many different detection techniques, for example, acoustic, optical, magnetic. They may also differ in their communication capabilities, for example, using ultrasound, infrared or radio frequency technologies with variable data rates and latencies [14]. At the time of design and deployment, the cost of the nodes should be kept low to minimize the total cost of the network, and a good compromise must be found between the power of the sensor, that is, the amount of features provided and its cost [15].

The main features of wireless sensor networks are the following [16–18]:

- They are scalable, being able to have a high number of nodes.
- Does not use network infrastructure. A sensor network does not need any infrastructure to operate, since its nodes can act as transmitters, receivers or routers (routers).
- The cost and implementation time are lower compared to wired networks, and they can also be implemented in hostile sites, for example: outer space or battlefields.
- WSNs are independent of external power sources since they are normally powered using batteries.
- Nodes with reduced energy and memory consumption. These networks are battery operated and have a long operating autonomy. We can operate without maintenance for several months or years.
- Short-range wireless technologies are used, routing between two nodes without direct vision is done through multi-hop communications.

3   AODV Protocol

Among the reactive routing protocols used in the WSN are DYMO (Dynamic MANET On-demand), DSR (Dynamic Source Routing) and AODV. Although the performance of the protocol depends on the scenario, in general, the scientific community agrees that AODV has a better performance compared to the other two protocols mentioned under high mobility scenarios, increased network density and low traffic loads. Therefore, AODV is one of the best-known demand protocols [19]. In AODV, unlike the OLSR proactive protocol (Optimized Link State Routing Protocol), the nodes do not maintain any routing information or participate in periodic exchanges of routing tables. AODV is based on a broadcast route discovery mechanism, which is used to dynamically establish routing table entries at intermediate nodes [20].

The route discovery process in the AODV begins now when a source node needs to transmit data to another node and does not have the routing information in its table. To this end, the source node disseminates a route request packet (RREQ) to its neighbors, which contains the source and destination addresses, a hop count value, a broadcast ID and two sequence numbers. The broadcast ID is incremented each time the source issues a new RREQ packet and is combined with the source address to uniquely identify in RREQ. Upon receiving an RREQ packet, a node that has a current route to the specified destination responds by sending a unicast route (RREP, route response) directly to the neighbor from which the RREQ was received. Otherwise, the RREQ is retransmitted to the neighbors of the intermediate node. Duplicate RREQs are discarded as they are identified by their source address and broadcast ID [21, 22].

Each node in the network maintains its own sequence number. A source that issues an RREQ packet also includes its own sequence number and the most recent sequence number it has for the destination. Therefore, intermediate nodes respond to an RREQ only if the sequence number of their route to the destination is greater than or equal to the destination sequence number specified in the RREQ packet. When a RREQ is retransmitted, the intermediate node registers the address of the neighbor from which the RREQ was received, thus establishing a reverse path from the destination to the source. As the RREP packet returns to the source, each intermediate node establishes a forwarding pointer to the node from which the RREP originates and registers the last destination sequence number for the requested destination [23].

RREP packets contain the addresses of the source and destination nodes, the destination sequence number and a hop count. An intermediate node that receives a RREP propagates this packet to the source only if: (1) this is the first copy of this RREP, (2) the RREP contains a destination sequence number greater than the previous RREP, or (3) the destination sequence number is the same as in the previous RREP, but the jump count is lower. This decreases the number of RREPs that travel to the source and ensures that the routing information for the shortest route (in terms of number of hops) reaches the source. An example of this route discovery process is shown in Fig. 2 [24].

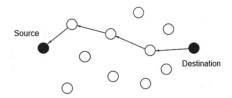

Fig. 2. Route discovery process in AODV [24].

A timer for each entry in the routing table of each node limits the useful life of unused routes. Neighboring nodes also exchange periodic HELLO messages to monitor the status of their links. When a link is broken along the route (for example, due to a moving node), the intermediate node closest to the source that realizes the broken link issues a route error packet (RERR) that travels in direction to the source. Upon receiving a RERR packet, the source node can restart the route discovery process [25, 26].

4   Analysis in the Simulator

In this part, modeling and simulation of two scenarios of wireless sensor networks is carried out to determine how the mobility of the nodes affects the behavior of parameters such as energy efficiency, the error rate of reception of the packets and the load of the network due to the AODV protocol. This research work will use two programs such as OMNET+ Castalia and the NS-3 simulator to simulate the performance results of the wireless sensor network using the AODV protocol.

OMNeT++ is a discrete event simulation tool designed to simulate computer networks, multiprocessors and other distributed systems. Your applications can be expanded to model other systems as well. OMNeT++ attempts to fill the gap between open source simulation software NS-2 and commercial payment alternatives such as OPNET [29].

The simulations in OMNeT++ are made from two main files. One written in NED language, which describes the structure of the network and its links. The other file has a

".ini" extension and contains the simulation configuration to be performed on the network described in the ".ned" file. The configuration files can contain several scenarios from which you can select the ones you want to simulate at any given time. This work uses version 4.6 of OMNeT++ due to compatibility reasons with Castalia.

Castalia is a realistic, fast and flexible simulator. The speed is achieved because all its modules are written in C++. Flexibility is at the cost of not having a graphical user interface (GUI-Graphic User Interface).

The results are also given in text files but can be accessed with convenient parser scripts. Instead, the NS 3 is a modular network simulator that connects to a real network. It has the ease of registration to debug and track to get the output. Most users focus on wireless simulation, including models for Wi-Fi, LTE or WiMAX and routing protocols such as AODV and OLSR.

This research work is used in the Linux platform (Ubuntu) using the NS-3.25 simulator. The objective of our work is to know how the mobility of the nodes affects: (1) the error rate of reception of the packets, (2) the energy efficiency and (3) the load of the network, using the routing protocol on demand AODV. For this, the simulation of two scenarios has been conceived, both have the same parameters which will be described later in this chapter. The difference is that in the first simulation the nodes are fixed in a mesh-shaped field and in the second the nodes move in groups (Table 1).

Table 1. Simulation characteristics.

Parameters	
Number of nodes	49
Simulation time	100 s
Routing protocol	AODV
Logic connection topology	Scenario 1: Mesh, Scenario 2: Star
Field size	70 × 70 m
Initial energy of each node	18 720 J, which is the typical energy of 2 AA batteries
Transmission power of radio modules	−5 dBm

5   Results

This part describes the results obtained from the simulations with the OMNET + Castalia programs and the NS-3 simulator.

5.1   Packet Loss Rate Per Node

Figures 3a and b show the graphs corresponding to the packet loss per node of each simulator. The packet loss rate is shown on the Y axis. The X axis represents the 49 nodes (from 0 to 48). The orange line represents the network of mobile nodes, while the blue line represents the network of fixed nodes. These values are calculated by the successful sending and receiving of all nodes with node 0. The loss rate of node 0 is 0 because communication with it is always successful.

As expected, the loss of packets in the mobile node network is high due to its own nature. In the fixed network with distribution in the form of a mesh, the packet loss rate is very low since the mesh structure allows the packets to jump from node to node until they reach their destination. As can be seen in the figures, there are similarities between the results obtained between the two simulators. There is a margin of error of 12.765% between the two simulators.

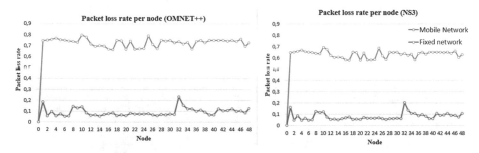

Fig. 3. Packet loss rate per node (a) the OMNET + Castalia simulator. (b) The NS·3 simulator.

5.2   Energy Consumed Per Node

Figures 4a and b show the graphic results of the energy consumed per node. The value of the energy consumed in Joules is shown on the Y axis. The X axis represents the 49 nodes (from 0 to 48). The orange line represents the network of mobile nodes while the blue line represents the network of fixed nodes.

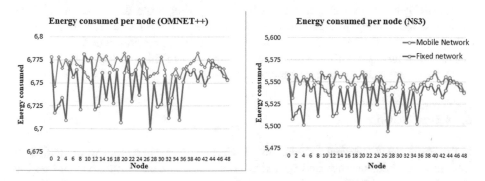

Fig. 4. Energy consumed per node (a) the OMNET + Castalia simulator. (b) The NS·3 simulator.

As can be appreciated, the consumption of both networks is very similar, but the mobile node network has a slightly higher consumption. This can be explained because there is a higher packet loss rate, the nodes remain active for a longer time trying to retransmit the packet that was not sent successfully.

5.3   Number of Routing Packets Per Node

The graphical results of the number of routing packets generated per node are shown in Figs. 5a and b. The value of the number of routing packets is shown on the Y axis. The X axis represents the 49 nodes (from 0 to 48). The orange line represents the network of mobile nodes while the blue line represents the network of fixed nodes.

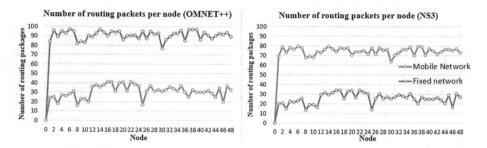

Fig. 5. Number of routing packets per node (a) the OMNET + Castalia simulator. (b) The NS·3 simulator.

The number of routing packets per node gives the measure that both the network is loading a specific routing protocol. This research paper presents this graph to get an idea of how the AODV routing protocol behaves in fixed node networks and in mobile node networks. Keep in mind that a high number of routing packets can overload the network making it work unstably. As can be seen, the fixed node network has a smaller number of generated routing packets, which is because, since the nodes are in fixed positions, the routing information does not change over time. In the mobile node network, on the other hand, since the nodes are always in motion, a node may not be accessible at any given time when using a route and in this case, it is necessary to calculate a new route generating routing packets.

6   Conclusion

In this work the theoretical foundations of wireless sensor networks were studied, as well as their characteristics and operation. The research carried out showed that this technology is constantly growing and that it has a great impact on cutting-edge applications such as precision agriculture and telemedicine. The OMNeT++ network simulation software was explored, especially the Castalia module and the NS-3 simulator specialized in the simulation of wireless sensor networks with AODV as routing protocol. Two simulations were carried out, one in an environment of fixed nodes and another in an environment of mobile nodes in each simulation program where the results of the two simulation programs were compared and it was concluded that the results obtained in each scenario were similar, having a margin of error of 12,765%. In the simulations it was found that the fact that the nodes are mobile generates an energy consumption 3% higher than in the case in which the nodes are fixed. In addition, the packet reception rate behaves considerably better in fixed sensor networks. The generation of packets by the AODV routing protocol was 68% higher in the case of mobile sensors, that is, in this scenario increased traffic was generated.

All the previous analysis allows to verify what is established in the theory of wireless sensor networks. The fact that a single node or several nodes of a wireless network are mobile, poses new challenges such as having dynamic network topologies, since mobility can completely modify the topology of the network and therefore its performance. In addition, with mobile nodes, the devices of these networks must be able to discover new routes and create new topologies, in order to ensure that all transmitted packets are received by the destination reliably and efficiently. In the simulations that have been carried out, it can be verified that mobility is a key aspect to be considered at the time of design as it directly influences the performance of wireless sensor networks. The study presented in this paper serves as the basis for the selection of a certain model depending on where it will be applied, in other words, in certain cases mobility can be favorable or not, making it viable depending on its application.

References

1. Kintzig, C.: Communicating with smart objects: developing technology for usable persuasive computing systems (innovative technology series. information systems and networks). epdf.pub (2003). https://epdf.pub/communicating-with-smart-objects-developing-technology-for-usable-persuasive-com95164.html. Accessed 15 Aug 2019
2. Baviskar, A., Baviskar, J., Mulla, A., Jain, N., Radke, A.: Comparative study of various wireless technologies for smart grid communication: a review. Int. J. Recent Innov. Trends Comput. Commun. **4**(4), 8 (2016)
3. Muriira, L.M., Zhao, Z., Min, G.: Exploiting linear support vector machine for correlation-based high dimensional data classification in wireless sensor networks. Sensors **18**(9), 2840 (2018)
4. Güngör, Ç., Hancke, G.P.: Industrial Wireless Sensor Networks: Applications, Protocols, and Standards. CRC Press (2013). https://www.crcpress.com/Industrial-Wireless-Sensor-Networks-Applications-Protocols-and-Standards/Gungor-Hancke/p/book/9781138076204. Accessed 15 Aug 2019
5. Ullah, I., Safi, A., Arif, M., Azim, N., Ahmad, S.: Wireless Sensor Network Applications for Healthcare. ResearchGate (2017). https://www.researchgate.net/publication/317338965_Wireless_Sensor_Network_Applications_for_Healthcare. Accessed 16 Feb 2019
6. Campoverde, A., Hernandez, D., Mazon-Olivo, B.: Cloud computing for the Internet of Things. Case of study for precision agriculture (2015)
7. Khan, S., Pathan, A.-S.K., Alrajeh, N.A.: Wireless Sensor Networks: Current Status and Future Trends. CRC Press (2016). https://www.crcpress.com/Wireless-Sensor-Networks-Current-Status-and-Future-Trends/Khan-Pathan-Alrajeh/p/book/9781138199873. Accessed 15 Aug 2019
8. Montero, S., Gozalvez, J., Sepulcre, M., Prieto, G.: Efecto de la Movilidad en Redes Inalámbricas de Comunicaciones Industriales, p. 4 (2012)

9. Willig, A.: Recent and emerging topics in wireless industrial communications: a selection. IEEE Trans. Ind. Inf. **4**(2), 102–124 (2008)
10. Zinonos, Z., Vassiliou, V.: S-GinMob: soft-handoff solution for mobile users in industrial environments. In: 2011 International Conference on Distributed Computing in Sensor Systems and Workshops (DCOSS), Barcelona, Spain, pp. 1–6 (2011)
11. Singh, S.P., Sharma, S.C.: A survey on cluster based routing protocols in wireless sensor networks. Procedia Comput. Sci. **45**, 687–695 (2015)
12. Al-Karaki, J., Kamal, A.: Routing techniques in wireless sensor networks: a survey. IEEE Wirel. Commun. **11**, 6–28 (2005)
13. Kumar, M.S., Gopinath, M.: Routing issues in wireless sensor networks: a survey (2013)
14. Dahane, A., Berrached, N.-E.: Mobile, Wireless and Sensor Networks: A Clustering Algorithm for Energy Efficiency and Safety. CRC Press, Boca Raton (2019)
15. Rault, T.: Energy-Efficiency in Wireless Sensor Networks. Universidad de Tecnología de Compiègne, Francia (2017)
16. Bildea, A.: Link Quality in Wireless Sensor Networks, p. 146 (2015)
17. Mancilla, M., López-Mellado, E., Siller, M.: Wireless sensor networks formation: approaches and techniques. J. Sens. (2016). https://www.hindawi.com/journals/js/2016/2081902/. Accessed 16 Feb 2019
18. Sarkar, A., Murugan, T.S.: Routing protocols for wireless sensor networks: what the literature says? Alexandria Eng. J. **55**(4), 3173–3183 (2016)
19. Perkins, C.E., Royer, E.M.: Ad-hoc on-demand distance vector routing. In: Proceedings WMCSA 1999. Second IEEE Workshop on Mobile Computing Systems and Applications, pp. 90–100 (1999)
20. Gómez Puerta, P.A., Posada Pérez, G.A., Vallejo Velásquez, M.A.: Performance evaluation of AODV routing protocol for different scenario in wireless sensor networks. Ing. Desarrollo **32**(1), 80–101 (2014)
21. Kumar, K.: A New Comparative Study of AODV & DSR Routing Protocols in Mobile Ad-Hoc Networks, p. 6 (2015)
22. Kulla, E., Ikeda, M., Barolli, L., Xhafa, F., Younas, M., Takizawa, M.: Investigation of AODV throughput considering RREQ, RREP and RERR packets. In: 2013 IEEE 27th International Conference on Advanced Information Networking and Applications (AINA), Barcelona, pp. 169–174 (2013)
23. Saini, A.: Analysis of Security Attacks and Solution on Routing Protocols in MANETs, p. 8 (2016)
24. Dargie, W., Poellabauer, C.: Fundamentals of Wireless Sensor Networks: Theory and Practice. Wiley, Hoboken (2011)
25. Das, S.R., Belding-Royer, E.M., Perkins, C.E.: Ad hoc on-demand distance vector (AODV) routing. https://tools.ietf.org/html/rfc3561. Accessed 16 Feb 2019
26. Taha, A., Alsaqour, R., Uddin, M., Abdelhaq, M., Saba, T.: Energy efficient multipath routing protocol for mobile ad-hoc network using the fitness function. IEEE Access **5**, 10369–10381 (2017)
27. Varga, A., Hornig, R.: An overview of the OMNeT ++ simulation environment. In: Proceedings of the 1st International Conference on Simulation Tools and Techniques for Communications, Networks and Systems & Workshops, p. 60. ICST (2008)

An Empirical Analysis of IFPUG FPA and COSMIC FFP Measurement Methods

Christian Quesada-López[(✉)], Denisse Madrigal-Sánchez, and Marcelo Jenkins

University of Costa Rica, San José, Costa Rica
{cristian.quesadalopez,denisse.madrigal,marcelo.jenkins}@ucr.ac.cr

Abstract. The accuracy of functional size measuring is critical in software project management, because it is one of the key inputs for effort and cost estimation models. The functional size measurement (FSM) process is performed based on standardized methods; however, the accuracy of the FSM results is still based mostly on the knowledge of the measurers. In this paper, an empirical study was conducted to analyze the accuracy, reproducibility, and acceptance properties of the IFPUG FPA and COSMIC FFP functional size measurement methods. Results show that the performance of participants in measuring the requirement specifications using IFPUG FPA and COSMIC FFP did not differ significantly in terms of accuracy and reproducibility. Likewise, acceptance properties such as perceived ease of use, perceived usefulness, and intention to use did not present significant differences. Our results suggest that novice measurers could apply both methods with similar results.

Keywords: Function points · COSMIC · IFPUG · Controlled experiment

1 Introduction

Functional size measurement (FSM) has become widely used in industry. Size and complexity are important attributes to be measured to evaluate a developed software product [15]. The accuracy of FSM is critical in software project management, because it is one of the key inputs for effort and cost estimation models [18]. The FSM process is performed based on standardized methods; however, the accuracy of the FSM results is still based mostly on the knowledge of the measurers. Differences in measuring the same product may occur, because FSM involves subjectivity in the application of some of the counting rules [8]. Likewise, when FSM is conducted by different measurers it can provide different results [8].

Albrecht presented the first Functional Size Measurement (FSM) method to size software quantifying the functional requirements [2]. FSM concepts and principles are defined in the ISO/IEC 14143-1 standard. There are five standardized FSM methods, where two of the most predominant are COSMIC-FFP (ISO/IEC 19761) and IFPUG FPA (ISO/IEC 20926).

© Springer Nature Switzerland AG 2020
Á. Rocha et al. (Eds.): ICITS 2020, AISC 1137, pp. 265–274, 2020.
https://doi.org/10.1007/978-3-030-40690-5_26

This paper presents the results of a controlled experiment conducted to identify and analyze the accuracy, reproducibility, and acceptance properties in the measurement process of IFPUG FPA and COSMIC FFP. The study was conducted with 13 participants. Each participant measured two software applications, one with IFPUG FPA and the other one with COSMIC FFP. Our motivation was to contribute to increment of the reliability of FSM measurement results by reporting common errors made during size measurement, and identifying assignable causes. Furthermore, choosing a counting method has significant implications for organizations that wish to implement software size measurement programs, hence our motivation to study and compare two of them in terms of accuracy, reproducibility and ease of use. The study was performed to answer the following research questions:

RQ1. What is the accuracy and reproducibility of the IFPUG FPA and COSMIC FFP measurement processes?

RQ2. What is the perceived ease of use, perceived usefulness, and intention to use of the IFPUG FPA and COSMIC FFP measurement processes?

RQ3. What are the errors made by participants using IFPUG FPA and COSMIC FFP measurement processes?

The remaining of the paper is organized as follows. Section 2 presents an overview of the FSM methods. Section 3 presents the related work. Section 4 describes the empirical study, and Sect. 5 presents and discusses the results of the experiment. Finally, Sect. 6 outlines the conclusions.

2   Background

Albrecht presented the first Functional Size Measurement (FSM) method to size software from the user point of view [2], quantifying the functional requirements. FSM is the process of measuring functional size, its concepts and principles for application are defined in the ISO/IEC 14143-1 standard [7]. There are five standardized methods for FSM based on functional user requirements: COSMIC-FFP (ISO/IEC 19761), IFPUG FPA (ISO/IEC 20926), MkII (ISO/IEC 20968), NESMA (ISO/ IEC 24570) y FiSMA (ISO/IEC 29881).

The International Function Point Users Group (IFPUG) refined Albrecht's proposal and presented the FPA counting practice manual. IFPUG FPA measurement method classifies user requirements using a set of basic functional size components (BFC) called transactional (TF) and data functions (DF). Data functions (DF) are classified into internal logic files (ILF) and external interface files (EIF). Transactional functions (TF) are classified into external inputs (EI), external outputs (EO), and external inquires (EQ). Functions are then counted according to a defined complexity criterion based on the data element types (DET), the file types referenced by a transaction (FTR), and the record element types (RET) within a data group. The measurement process in FPA involves the identification and count of these BFC types: EI, EO, EQ, ILF, and EIF, in order to obtain the unadjusted function points (UFP).

COSMIC FFP measurement method was presented in 1997 and accepted as standard in 2002. It is based on the identification of data movements and it uses the generic software model which has three main principles: (1) User functional requirements are analyzed as unique functional processes that have sub processes, (2) each functional process starts with an Entry data movement, and (3) a data movement moves an attribute data group that represents a unique object of interest. COSMIC FFP establishes four data movement sub processes. Entry (E) is a data movement from the functional user into a functional process. Exit (X) is a data movement from a functional process to the functional user (out of the boundaries of the application). Write (W) is a data movement from a functional process to a persistent storage. Finally, Read (R) is a data movement from the persistent storage to a functional process. The functional size of a piece of software consists on the total number of data movements representing the COSMIC function points (CFP).

3   Related Work

The reliability of the functional size measurement methods has been studied since the FSM concept was introduced [18,21]. For example, Low and Jeffrey [10] assessed the consistency of Function Points (FP) method among measurers. Kemerer [8,9] identified the major sources of variation that impact the reliability of FP. Abrahao et al. [1] evaluated the accuracy, reproducibility, and acceptance properties of FSM methods. Turetken et al. [19,20] evaluated the effect of different interpretations of the measurers on the measurement results, and the difficulties faced in the COSMIC FFP, IFPUG FPA, and MkII FPA. Top et al. [18] and Ungan et al. [21] analyzed the variances in COSMIC FFP among individuals, and identified the frequency of errors during the measurement process. Robiolo [15] presents a comparison that evaluates the simplicity of FSM measurement processes.

In recent years, protocols have been proposed to evaluate the consistency and verify the accuracy of the FSM results. For example, Yilmaz et al. [24] developed a tool for automatic defect detection in COSMIC FFP measurements, and identified error categories associated to measurer, process, and requirements specification. Soubra et al. [17] developed a protocol that allows the verification of accuracy in COSMIC FFP. In a previous work, we adapted Soubra's protocol to verify the accuracy of IFPUG FPA [11,13] and evaluate the accuracy, reproducibility, and acceptance properties of IFPUG FPA [12,14]. In this work, we elaborate in our previous studies by providing a more in-depth statistical analysis of the results.

4   Empirical Study Design

Our study was conducted as part of a graduate Software Estimation course at the University of Costa Rica in the first half of 2018. Participants were randomly assigned to two groups (Group 1 and Group 2). They attended four training sessions for each FSM method, where the measurement task materials and the experimental objects were presented and experimental tasks were explained. The experiment was performed over a 6-week period, during which four training sessions (3 h each one) on functional size measurement were conducted. The participants were trained in the concepts of software size and complexity, and software functional size measurement in order to measure a particular software application. Participants carried out two measurement tasks, counting the functional size of two software applications, one with each functional size measurement method. The first measurement task was conducted on April 13, 2018 (4 h, in class task). The second measurement task on May 4, 2018 (4 h, in class task). After the two measurement tasks, they filled out the acceptance properties survey and conducted an analysis of their measurement results using an accuracy verification protocol [13,17]. The accuracy verification protocol process was conducted on May 18, 2018 (take home task). In total, 26 measurements were done by the 13 participants (all of them were professionals working on the Costa Rican software industry). To report our study, we followed the guidelines in [23].

Goal. The *aim* of our study was to "*Analyze* the measurement process of IFPUG FPA and COSMIC FFP methods *for the purpose of* evaluating them with respect to the accuracy, reproducibility, and acceptance properties (perceived ease of use, perceived usefulness, and intention to use) *from the point of view of* the researchers and practitioners *in the context of* graduate students in a Software Estimation master degree course". This objective was achieved comparing the measurement results and collecting the perceptions on the usage of the methods.

Participants. The participants were graduate students in Computer Science. Before the experiment took place, the participants filled out a pre questionnaire to gather their demographic information. Five of them were junior professionals (less than 2 years experience), and the rest had 2 or more years of experience in industry. They were not experts in functional size measurement, only one of the 13 participants had previously applied either IFPUG FPA or COSMIC FFP in an academic course. The participants were not aware either of the aspects under study or the research questions. From their perspective, they were solving a course exercise. All participants were guaranteed anonymity.

Experimental Material. The instruments include the requirement specification document of the target software applications (experimental objects), the training materials (a set of instructional slides that describe the FSM methods, a measurement example used in the training sessions, the measurement guideline for each FSM method, and the accuracy verification protocol), and the survey questionnaire with closed questions to collect data related to the perceived acceptance properties. All experimental instruments were prepared in advance, and experimental materials had been used in previous studies [1,12–14].

Target Software Applications. The experimental objects were two software applications (Domain A: Registration System [3,4], Domain B: Warehouse Management System [6]). The applications are transactional systems that mostly perform create, read, update, delete and list (CRUDL), and assignment, search and filter operations. Besides, these applications implement more complex functionalities that involve reports and business processes associated with each problem domain. The applications were counted independently by participants based on the requirements specified using the IEEE 830 Recommended Practice for Software Requirements Specifications standard.

Metrics and Hypotheses. The dependent variables were the accuracy, reproducibility, and acceptance properties (perceived ease of use, perceived usefulness, intention to use) [1,5]: (1) *Accuracy:* the closeness of the agreement between the functional size result of a measurement by one participant (p) and the true value calculated by an expert (e). We collected the magnitude of relative error $MRE_i = |FP_e - FP_{p_i}|/FP_e$. (2) *Reproducibility:* the closeness of the agreement between the results of successive measurements of the same product carried out under the same conditions. It refers to the use of the method on the same product and environment by different participants. The reproducibility is defined as $Rep_i = |AverageOthers - FP_{p_i}|/AverageOthers$. (3) *Perceived ease of use:* the degree to which a person believes that using a particular method would be ease of use. This construct measures the perceptual judgment about the effort required to learn, use and apply the FSM method. (4) *Perceived usefulness:* the degree to which a person believes that a particular method will be effective in achieving its intended objectives. This construct measures the perceptual judgment about the effectiveness of the FSM method. (5) *Intention to use:* the degree to which a person intends to use a particular method. This construct measures the perceptual judgment about the performance of the FSM method. The acceptance properties items were formulated as a five-point Likert scale using opposing statements question format. Each of these were measured using several questions, as proposed in [1,5].

For each dependent variable (metric), the hypotheses tests whether there is no difference between the accuracy and reproducibility of the measurement results of IFPUG FPA and COSMIC FFP processes. We also test whether there is no difference between the perceived ease of use, perceived usefulness, and intention to use of IFPUG FPA and COSMIC FFP processes.

Experiment Design. Our experiment used a crossover design [22]. Participants were split into two experimental groups and administered with every treatment only once. The number of participants in Group 1 and Group 2 was 6 and 7, respectively. The method is the experiment factor, with two treatments (or levels): IFPUG FPA and COSMIC FFP. The subjects applied both methods. Therefore, there are two periods in the experiment (one to apply the first method, and another to apply the second method). Each period takes place in a different session, held tree weeks apart from one another (this was the wash-out period to leave time for the treatment, and thus possibly minimize carryover effects [16]). We consider two sequences of method application, one for each experimental group. Based on the differences of the methods, we do not think that there is a chance of either of the sequences improving the experimental results of the other. Subjects are given the specifications in each treatment.

Analysis Procedure. We analyze our study according to [16, 22]. First, the variation between the accuracy (MRE) and reproducibility (Rep) of the IFPUG FPA and COSMIC methods was tested. We used the analysis procedure detailed in [16]. Besides, we followed the recommendations detailed in [22]. First, we performed a pre-test to check the presence of a carryover effect. The within-participant sums of the dependent variable values in both periods were computed. Next, we ran an unpaired two-sided t-test/Mann-Whitney U test to verify the hypothesis (the expected mean values of within-participant sums are the same). If the t-test/Mann-Whitney U test does not reject the null-hypothesis, the carryover effect is not statistically significant [16]. When carryover effect is not statistically significant, we performed the following steps [16]: (1) We computed descriptive statistics for each dependent variable. (2) We tested whether the effect of the measurement method was statistically significant on the dependent variable. For each participant per group, we computed the within-participant differences of the dependent variable in both periods. Then, an unpaired two-sided t-test/two-sided Mann-Whitney U test was run to test if expected mean values are the same ($\alpha = 0.05$) [16]. Second, we tested the acceptance properties: perceived ease of use (PEOU), perceived usefulness (PU), and intention to use (ITU). Participants filled out the survey instrument about perceived acceptance properties after both measurement tasks. We tested the hypothesis with the two-sided Mann-Whitney U test. Finally, we analyzed measurement results and reported common errors made during size measurement.

Threats to Validity. Internal Validity. We selected students in a course to take part in the experiment, this could influence the results since they could be motivated by the graded activity. To prevent that participants exchanged information, we monitored them during the execution of the measurement tasks. Group1 and Group2 performed the tasks at the same time. The use of a crossover design might affect the validity. We dealt with this threat with a wash-out period. Also, we studied the carryover in the data analysis. This study identified the factors that cause the variation in the FSM methods, noise introduced by the diversity of the participants, requirements, and projects should be considered. To avoid a possible learning effect the applications came from different domains.

Construct Validity. We used methods and metrics well known and adopted in the literature and industry. Although, we did not inform the participants about our research goals, there could be the risk of hypotheses guessing. We communicated to the participants that the collected data would be shared anonymously and used for research purposes only. To avoid social threats due to evaluation apprehension, the participants were graded only based on completeness.

Conclusion Validity. We used standard measurements methods for the treatments. We validated the experimental materials in a pilot study. Two researchers collected and validated the IFPUG FPA and COSMIC FFP functional sizes used as "true values", but they are not certified as function point specialists. They have more than 10 years of experience in measurement, and both applications had been previously measured and reported in [3, 4, 6].

External Validity. The participants were sampled by convenience. Generalizing the results to a different population poses a threat of interaction of selection and treatment. The sample offered a homogeneous prior knowledge. The participants were not counting experts; however, they are familiar with software engineering practices. Although the lack of experience could cause variances in counting, we expected similar performance among participants. Both software applications are small but their requirement specifications were documented based on standards. The effect of experiment artifacts can influence the study outcomes. The applications were selected by convenience and were not a random selection. The size of the applications ranged from 118–146 UFP and 111–118 FFP and results could not be generalized to other domains or larger applications.

5   Analysis of Results

5.1   Carryover Analysis

The Shapiro test suggested that the data was normally distributed for accuracy (MRE: $p = 0.263$). Reproducibility was not normally distributed (Rep: $p = 0.045$). We applied the t-test/Mann-Whitney U test to check the presence of a carryover effect. The obtained p-values are: accuracy (MRE: $p = 0.079$), reproducibility (Rep: $p = 0.153$). The results indicate that the carryover effect is not statistically significant for the dependent variables.

5.2   Accuracy and Reproducibility (RQ1)

Table 1 summarizes descriptive statistics of the dependent variables grouped by treatment (IFPUG FPA, COSMIC FFP). The accuracy (MRE) of the participants measuring the requirement specifications using IFPUG FPA present a mean of 5.1%. The reproducibility (Rep) mean is 5.8%. In the case of COSMIC FFP, the accuracy (MRE) mean is 7.4% and the reproducibility (Rep) of 5.2%. The accuracy (mean of MRE) and reproducibility (mean of Rep) of the total function points count for IFPUG FPA and COSMIC FFP could be considered acceptable in industry ($\pm 10\%$) [10]. The variation between the accuracy (MRE) and reproducibility (Rep) of the IFPUG FPA and COSMIC measurement methods was tested. We used the t-test for accuracy and reproducibility because the data was normally distributed (Shapiro test, MRE: $p = 0.431$, Rep: $p = 0.326$). The obtained p-values are: accuracy (MRE: $p = 0.059$), reproducibility (Rep: $p = 0.109$). The results indicate that for both dependent variables, we could not reject the null hypothesis. There is no statistically significant difference between the accuracy, and reproducibility of the two FSM methods. These results support the claim that IFPUG FPA and COSMIC FFP could produce similar performance in measurement results.

Table 1. Accuracy and reproducibility results.

Metric		IFPUG	COSMIC	Metric		IFPUG	COSMIC
MRE	Mean	5.1%	7.4%	Rep	Mean	5.8%	5.2%
	Median	4.1%	8.1%		Median	4.1%	4.1%
	Std	4.9%	3.5%		Std	6.1%	4.5%
	Min	0.0%	0.0%		Min	0.0%	0.3%
	Max	17.1%	11.7%		Max	20.0%	18.0%

5.3   Acceptance Properties (RQ2)

We analyzed the acceptance properties of COSMIC FFP and IFPUG FPA methods: perceived ease of use (PEOU), perceived usefulness (PU), and intention to use (ITU) [1, 14]. The participants' perceptions show similar results of PEOU for COSMIC and IFPUG. Similarly, PU and ITU show similar results. All properties indicate slightly higher values than the neutral value (Neutral = 3), indicating moderate acceptance of both methods. We tested the acceptance properties: PEOU, PU, and ITU. The results from the Mann-Whitney U test indicate that the IFPUG FPA and COSMIC FFP measurement methods' acceptance properties were not significantly different. There is not enough evidence to reject the null hypothesis for PEOU ($p = 0.181$), PU ($p = 0.599$), and ITU ($p = 0.637$).

5.4   Errors in Measurements (RQ3)

After analyzing measurement results and reporting common errors made, we grouped errors in two levels from finest granularity level of error categorization (L2) to higher granularity categories (L1). We based our error categorization on [13, 17, 18, 21]. Error categories at level 1 (L1) are: (1) Measurement and (2) Inputs. For COSMIC FFP, level 2 errors were: (1a) OOI Identification: incorrect identification of external data groups and data groups considered attributes, (1b) FP Identification: data validation disregarding, action selection disregarding, action confirmation disregarding, and multiple FPs condensed into single FP, and (1c) Errors in applying the measurement rules: cascading operations omission and functional users misunderstanding. (2a) Interpretation: requirements misunderstanding, and (2b) Assumptions: calculated fields assumed stored.

For IFPUG FPA, level 2 errors were: (1a) DF Identification: incorrect identification of EIFs, RETs considered attributes, DETs definition misunderstanding, (1b) TF Identification: omission of confirmation controls, incorrect BFC classification EO vs EQ, multiple TF condensed into single TF, and (1c) Errors in applying the measurement rules: incorrect requirement subdivision, missing DETs and RETs when multiple entities are involved. (2a) Interpretation: incorrect measurement for requirements report, and (2b) Assumptions: assuming extra functions. These results have two direct consequences: (1) More in-depth subject training could be needed for them to achieve better accuracy in counting the BFCs. (2) Accuracy at the BFC level must be improved in order to achieve industry strength results.

6 Conclusion

This study reported a controlled experiment to compare the IFPUG FPA and COSMIC FFP functional size measurement methods. Our experiment shows that the performance of participants in measuring the size of two small software applications from their requirement specifications using the IFPUG FPA and COSMIC FFP methods did not differ significantly in terms of accuracy, and reproducibility. Acceptance properties (perceived ease of use, perceived usefulness, and intention to use) did not present significant differences. Our results suggest that novice measurers could apply both methods with similar results. This has a significant implication for software organizations that wish to implement software size measurement programs since their developers might deem both methods to be equal in terms of these three characteristics. Further research is required for projects in different application domains and different sizes. In the near future we plan to conduct further experimentation with more professionals to reaffirm or refute these initial results. In addition, we would like to analyze more in detail the assignable causes of errors.

Acknowledgments. This work was partially supported by the University of Costa Rica No. 834-B8-A27. We thank all participants of the study and the Empirical Software Engineering Group at UCR for the valuable feedback.

References

1. Abrahao, S.: On the functional size measurement of object-oriented conceptual schemas: design and evaluation issues. Universidad Politecnica de Valencia (2004)
2. Albrecht, A.: Measuring application development productivity. In: Proceedings of the Joint Share, Guide, and IBM Application Development Symposium (1979)
3. Bundschuh, M., Dekkers, C.: The IT Measurement Compendium: Estimating and Benchmarking Success with Functional Size Measurement. Springer, Heidelberg (2008)
4. COSMIC: The COSMIC Functional Size Measurement Method Version 4.0.1 Course Registration (C-REG) System Case Study. Version 2.0 (2015)
5. Davis, F.: User acceptance of information technology: system characteristics, user perceptions and behavioral impacts. J. Man Mach. **38**(3), 475–487 (1993)
6. Fetcke, T.: The warehouse software portfolio: a case study in functional size measurement (1999)
7. ISO: Information Technology, Software Measurement, Functional Size Measurement: Definition of Concepts. ISO/IEC (2007)
8. Kemerer, C.: Reliability of function points measurement: a field experiment (1990)
9. Kemerer, C., Porter, B.: Improving the reliability of function point measurement: an empirical study. IEEE Trans. Software Eng. **18**(11), 1011–1024 (1992)
10. Low, G., Jeffery, R.: Function points in the estimation and evaluation of the software process. IEEE Trans. Software Eng. **16**(1), 64–71 (1990)

11. Madrigal-Sánchez, D., Quesada-López, C., Jenkins, M.: Towards the automation of a defect detection protocol for functional size measurements. In: CibSE, pp. 354–367 (2018)
12. Quesada-López, C., Jenkins, M.: An evaluation of functional size measurement methods. In: CibSE, pp. 151–165 (2015)
13. Quesada-López, C., Jenkins, M.: Applying a verification protocol to evaluate the accuracy of functional size measurement procedures: an empirical approach. In: PROFES, pp. 243–250. Springer (2015)
14. Quesada-López, C., Madrigal-Sánchez, D., Jenkins, M.: An empirical evaluation of automated function points. In: CibSE, pp. 151–165 (2016)
15. Robiolo, G.: How simple is it to measure software size and complexity for an it practitioner? In: ESEM, pp. 40–48. IEEE (2011)
16. Romano, S., et al.: The effect of noise on software engineers' performance. In: ESEM, p. 9. ACM (2018)
17. Soubra, H., Abran, A., Ramdane-Cherif, A.: Verifying the accuracy of automation tools for the measurement of software with COSMIC–ISO 19761. In: IWSM-MENSURA, pp. 23–31. IEEE (2014)
18. Top, O.O., Demirors, O., Ozkan, B.: Reliability of COSMIC functional size measurement results: a multiple case study on industry cases. In: Euromicro, pp. 327–334. IEEE (2009)
19. Turetken, O., et al.: The effect of entity generalization on software functional sizing: a case study. In: PROFES, pp. 105–116. Springer (2008)
20. Turetken, O., et al.: The impact of individual assumptions on functional size measurement. In: IWSM-MENSURA, pp. 155–169. Springer (2008)
21. Ungan, E., et al.: An experimental study on the reliability of COSMIC measurement results. In: IWSM-MENSURA, pp. 321–336. Springer (2009)
22. Vegas, S., Apa, C., Juristo, N.: Crossover designs in software engineering experiments: benefits and perils. IEEE Trans. Software Eng. 42(2), 120–135 (2015)
23. Wohlin, C., et al.: Experimentation in Software Engineering. Springer, Heidelberg (2012)
24. Yilmaz, G., Tunalilar, S., Demirors, O.: Towards the development of a defect detection tool for COSMIC functional size measurement. In: IWSM-MENSURA, pp. 9–16. IEEE (2013)

Support Vector Machine as Tool for Classifying Coffee Beverages

José Varela-Aldás[1](✉) (iD), Esteban M. Fuentes[1,2](✉) (iD),
Jorge Buele[1] (iD), Raúl Grau Meló[2] (iD), José Manuel Barat[2] (iD),
and Miguel Alcañiz[3] (iD)

[1] SISAu Research Group, Universidad Indoamérica, 180103 Ambato, Ecuador
{josevarela, estebanfuentes, jorgebuele}@uti.edu.ec
[2] Departamento de Tecnología de Alimentos, Grupo CUINA,
Universidad Politècnica de Valencia, Valencia, Spain
{rgraume, jmbarat}@tal.upv.es
[3] Instituto de Reconocimiento Molecular y Desarrollo Tecnológico (IDM),
Centro Mixto Universitat Politècnica de València, Universidad de Valencia,
Camino de Vera s/n, 46022 Valencia, Spain
mialcan@upvnet.upv.es

Abstract. Classifiers are tools widely used nowadays to process data and obtain prediction models that are trained through supervised learning techniques; there is a wide variety of sensors that acquire the data to be processed, such as the voltammetric electronic tongue, as a device employed to analyze food compounds. This paper presents a normal and decaffeinated coffee beverage classifier using a Support Vector Machine with a linear separation function, detailing the classification function and the model optimization method; to train the model, the data measured by 4 electrodes of a voltammetric tongue that is excited by a predetermined sequence of positive pulses is used. In addition, the results graphically show the measurements obtained, the support vectors and the evaluation data, the values of the classifier parameters are also presented. Finally, the conclusions establish an acceptable error in the classification of coffee drinks according to caffeine presence at the sample analyzed.

Keywords: Classifier · Voltammetric tongue · Supervised learning · Support Vector Machine

1 Introduction

Technological evolution has motivated the development of new tools and concepts that human beings has linked to their daily lives. One of them is artificial intelligence (AI), which is a combination of several algorithms that aim to provide machines with characteristics similar to those of humans [1]. Although science fiction relates it directly to robots that have total autonomy (without any limitation), the reality is that all currently experience a "weak AI", designed to perform a single specific task [2]. Its main applications include data mining, virtual environments, facial recognition, robotic, decision-making systems, machine learning, among others [3]. Based on computational,

© Springer Nature Switzerland AG 2020
Á. Rocha et al. (Eds.): ICITS 2020, AISC 1137, pp. 275–284, 2020.
https://doi.org/10.1007/978-3-030-40690-5_27

mathematical and logical algorithms, machine learning aims to present methods and processes that provide a computer or computational algorithm with the ability to predict or take decisions [4, 5]. The demand for computer systems that filter and search for specific information in huge databases motivated the creation of unsupervised learning, through which this task was made easier [6]. Characteristics are given to the algorithm in order to group the information based on their similarities for later analysis. On the other hand, there is supervised learning, through which in addition to providing the characteristics (questions) their respective labels (answers) are established, so that the something-rhythm combines them and predictions can be made [7].

This has led to the creation of a completely new industry that demands the development of intelligent, personalized interfaces, capable of predicting and solving any problems that may arise. For this purpose, the Support Vector Machine (SVM), a term introduced in 1995, is based on the principle of minimization of structural risks to avoid problems of excessive adjustment [8]. SVM is a supervised learning algorithm, of the family of linear classifiers, which achieves a maximum separation between classes and a minimum limit on the expected generalization error [9]. Resuming the algorithm focuses on the general problem of learning to discriminate between positive and negative members of a given class of n-dimensional neighbors. In general, SVMs allow to find an optimal hyperplane that separates the classes, obtaining the maximum performance of generalization in classification. Its most relevant application is related to classification (examining two classes without loss of generality); It is therefore important that it works properly in future data, i.e. generalize the properly classification [10, 11].

It can be guaranteed that the technology provides the opportunity to improve processes, including the adequate intake of food and beverages. To promote health care, it is necessary to control in a better way the quality of the products that consumers acquire [12]. Currently, systems for food recognition can be built and their main properties determined automatically, which can be tested in domestic or commercial environments (restaurants) [13]. That is why electronic sensors have been developed, such as the electronic nose, a system that detects volatile organic compounds (VOCs) by discriminating them from a set of odorous substances [14]. Another clear example is the voltammetric tongue, a device based on pulse voltammetry that works with an electrode system made from noble metals (platinum, gold, iridium, rhodium and palladium) [15]. Considering the five basic flavors of the tongue: sour, salty, bitter, sweet and "umami" a tool similar to the real one can be precisely designed. This demonstrates that the development of these electronic sensors will allow obtaining complex systems that encompass all the senses as a whole, facilitating the authentication of food in a more reliable way.

2 State of the Art

Using machine learning in the food industry allows better performance by merging technology with conventional methods [16], Supervised Machine Learning represents an effective alternative for agriculture. This document establishes an algorithm that allows predicting the performance of a cocoa crop using SVM and a Generalized Linear

Model (GLM) to determine its main characteristics. Similarly, research on food classifiers has taken importance because of its contribution to the choice of a better diet for human beings. A system is presented, that by means of an SVM classifier of multiple labels allows the identification of food thanks to a visual recognition. To evaluate this proposal, 50 food categories are used and each category contains 100 photographs from different sources (own or internet) with a general accuracy of 68.3%. Similarly, a system for food recognition is briefly described through algorithms that can be used for dietary monitoring [17]. With convolutional neural networks, three different classification strategies using several visual descriptors. Experimental tests demonstrate approximately 79% accuracy in the recognition of 3616 foods, grouped into 73 classes [13].

For the development of this proposal, several investigations related to sensors applied to the food industry have been taken, as a basis, describes the simple design of a voltammetric tongue using a single voltammetric sensor. Its functionality has been evaluated by drinking tea for immediate intake (a simulation of homemade tea). Using the algorithms of successive projections algorithm (SPA), genetic algorithm (GA) and stepwise (SW) the respective experimental tests are performed; where SPA provides a prediction close to 100%. The utility of voltammetric languages for the detection of characteristics and anomalies in wine and coffee [18, 19]. The discrimination algorithms developed in these proposals demonstrate high percentages of efficacy in the detection of adulterated foods. For more bibliographic support, the most representative works on electronic sensors (nose and electronic tongue) used to assess the quality of food in the last 13 years [20]. Between the most important and relevant is a study which focuses on applications that monitor the quality of food such as meat, dairy and beverages, for which different data processing algorithms are developed [14].

This paper presents the implementation of a classifier of two coffee drinks, with caffeine and without caffeine, using a 3 electrode configuration and with 4 noble metal working electrodes. The method used operates by a Support Vector Machine with a linear function. The proposal is divided as follows: Sect. 3 describes the collection of voltammetric signals, Sect. 4 details the elements of the classifier, Sect. 5 presents the results obtained and Sect. 6 contains the main conclusions of this work.

3 Data Collection

To obtain the data for the present study, a voltammetric electronic tongue was employed; the Inter-University Institute of Molecular Recognition and Technological Development of the Polytechnic University of Valencia and the University of Valencia (UPV-UV IDM) built the equipment.

The voltammetric electronic tongue (VET) equipment consists of a software employed to set up the pulse sequences as to generate data files as a result of the analysis, and also to configure the VET equipment to work as 2 (WE and RE) or 3 (WE, RE, CE) electrodes. The VET machine consists of three electrodes, a calomelanos (Ag/AgCl) due to its stability and works as a reference electrode (RE) because can maintain a constant current throughout the test, four noble metals (Ir, Rh, Pt and Au) and four non-noble metals (Ag, Cu, Co and Ni), which function as working electrodes (WE). The WE are contained in two cylinders of grade 316L stainless steel, separated into noble and non-noble metals, the cylinder which houses the working electrodes functions also as a counter electrode (CE), the configuration for noble metals is shown in Fig. 1.

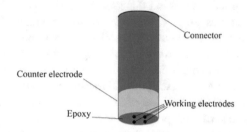

Fig. 1. Noble metal electrode configuration [12]

For the development of this experience, only noble metals were used as working electrodes, for this a generic configuration of the pulse sequence consisting of 10 pulses ranging from −1000 mV to 1000 mV and with a duration of 20 ms each pulse was applied. So the total for each essay was 1800 data, due to the large amount of data generated from the trials and their repetitions, statistical treatment is used through multivariate analysis [21]. Two sort of samples were analyzed 5 times each one, de basic difference between them was the presence or not of caffeine.

4   Classifier

4.1   Classification Function

The Support Vector Machine classification technique requires a separation function, the linear function being more common and facilitating the problem. Two groups of data are assumed linearly separated by a function, so that a hyperplane separates two opposite classes.

By having two classification options (normal coffee and decaffeinated coffee) is it can label the classes by the sign $(1, -1)$, obtaining the function of classification of the expression (1), where $\mathbf{X}$ is the data obtained by the four electrodes $(X_1, X_2, X_3$ and $X_4)$ and the reference pulse applied to all electrodes (X_5). Classification by Support Vector Machine can reduce the number of training data if the support vectors containing all the necessary characteristics for the classification of these two groups are known.

$$f(X) = sign\left(\mathbf{w}^\mathrm{T}\mathbf{X} + b\right) \tag{1}$$

4.2   Optimization Problem

The classifier using Support Vector Machine is reduced to an optimization problem with restrictions that fits all training values according to Eq. 2, where the parameters to be determined are $\mathbf{w}$ and b. The method to be used is based on the dual Wolfe problem that determines the Lagrange multipliers (**u**) which best fit the classification model, according to the minimization proposal of (2), where **Q** is found by (3) and **A** are the labels of supervised learning, this method is described in [22].

$$Min : 1/2\,u^{T}Qu - u^{T}1 \tag{2}$$

$$Q = A\,X^{T}\,X\,A \tag{3}$$

4.3 Programming

The software used in the training and evaluation is Python, it has the optimization tool of the SciPy library (*scipy.optimize*) that allows to solve the minimization problem.

The function implemented is *minimize(objective function, initial search values, optimization method, gradient, constraints, bounds)*; where the objective function is (2), the initial values are set to zero, the optimization method used is Sequential Least Squares Programming 'SLSQP', the constraints specified in [22], and bounds for the response are not established.

5 Results

5.1 Data Acquisition

To generate the voltammetric data using the device described in Sect. 3, a succession of 10 positive pulses of different magnitudes is used, Fig. 2 shows the reference pulses applied to the four electrodes.

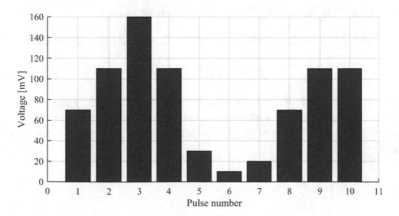

Fig. 2. Pulses taken for the four electrodes

Figure 3 shows the voltammetric response of the 4 electrodes (X_1, X_2, X_3 and X_4) when the 10 reference pulses are applied.

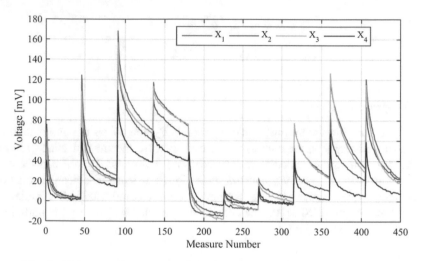

Fig. 3. Voltammetric response of the four electrodes to the reference pulses.

5.2   Training

For the supervised learning training detailed in Sect. 4, the data of 6 experiments, 3 experiments with the normal coffee drink and the remaining 3 with decaffeinated coffee are used, each experiment uses the 4 electrodes with the pulse sequence of the Fig. 2, in total 60 pulses are generated for training. Next, the proposed classifier is implemented obtaining the parameters of the linear model of (4) and (5).

$$\mathbf{w} = [0.793533 \ -0.141245 \ -0.286532 \ -0.183178 \ -0.183425] \tag{4}$$

$$b = -0.89818815717 \tag{5}$$

In addition, the hyperplane found has 20 support vectors (V_s) shown in Fig. 4, the position of said vectors is marked with red circles, all peak responses of the 60 pulses are also plotted, observing that the most support vectors are found in pulses of lesser magnitude.

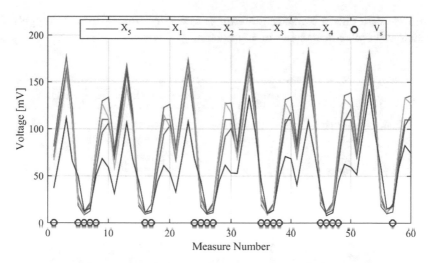

Fig. 4. Support vectors of the voltammetric data provided in training.

5.3 Evaluation

The classifier is tested with new measured data, starting with normal coffee 2 new sequences of 10 pulses are generated, Fig. 5 shows the data obtained to evaluate the classification function with normal coffee, from these data only a classification error is generated, the vector not recognized as normal coffee is enclosed in a red circle.

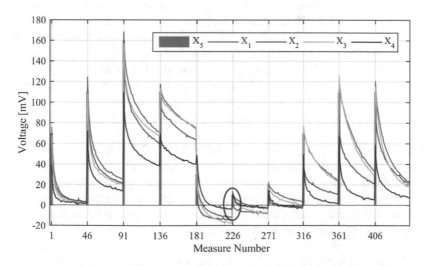

Fig. 5. Voltammetric measurements in normal coffee.

Additionally, 2 new 10-pulse sequences are generated in decaffeinated coffee, Fig. 6 shows the data obtained to evaluate the classification function with decaf coffee, from these data only a classification error is generated, the vector not recognized as decaffeinated coffee is enclosed in a red circle.

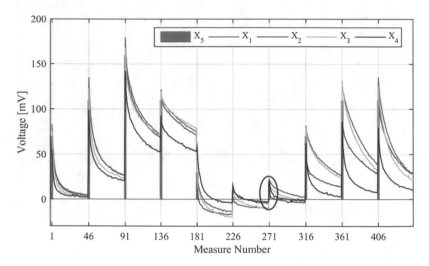

Fig. 6. Voltammetric measurements in decaffeinated coffee.

6   Conclusions

This paper presents a two-label classifier to distinguish the type of coffee drink (regular coffee and decaffeinated coffee), using a Support Vector Machine using a linear sorting function. The data is acquired using voltammetric tongues with three electrode configuration, besides of the 4 working electrodes (noble metals) and a sequence of 10 positive pulses per electrode is applied. The supervised learning problem is designed using a 5-variable hyperplane and Wolfe's dual problem is used to find the parameters of the optimal linear function that separates the two classes. Measurements are used for training and evaluation of the proposed classifier using Python as programming software.

Voltammetric tongues are devices that offer the possibility to study and identify the characteristics of food compounds, generating the possibility of recognizing food through data processing. On the other hand, the Support Vector Machine used in this work to classify two coffee drinks is flexible to adapt new data that provide new label separation characteristics. In addition, including the sequence of input pulses generates a reference that provides more classification characteristics.

In conclusion, the results show a classification model with a 5% error in the evaluation carried out, and 20 support vectors of the 60 data used in the training were found, in general the classifier meets the objective of the proposal, but it could analyze the performance using a more complex classification function.

Relating this work with the literature found, this proposal stands out in the use of 4 electrodes voltammetric tongues as a data acquisition tool, as well as the transparency of the classification method; the linear function is very flexible, because it allows adding new variables to the system, how for example more electrodes be in the voltammetric tongues. Finally, in future works it is intended to incorporate the eight electrodes and classify more features in a coffee drink.

References

1. Salah, K., Rehman, M.H.U., Nizamuddin, N., Al-Fuqaha, A.: Blockchain for AI: review and open research challenges. IEEE Access (2019). https://doi.org/10.1109/ACCESS.2018. 2890507
2. Lu, H., Li, Y., Chen, M., Kim, H., Serikawa, S.: Brain intelligence: go beyond artificial intelligence. Mob. Networks Appl. (2018). https://doi.org/10.1007/s11036-017-0932-8
3. Çaliş, B., Bulkan, S.: A research survey: review of AI solution strategies of job shop scheduling problem. J. Intell. Manuf. (2015). https://doi.org/10.1007/s10845-013-0837-8
4. Nickel, M., Murphy, K., Tresp, V., Gabrilovich, E.: A review of relational machine learning for knowledge graphs (2016). https://doi.org/10.1109/JPROC.2015.2483592
5. Jara Estupiñan, J., Giral, D., Martínez Santa, F.: Implementación de algoritmos basados en máquinas de soporte vectorial (SVM) para sistemas eléctricos: revisión de tema. Rev. Tecnura. **20**, 149–170 (2016). https://doi.org/10.14483/udistrital.jour.tecnura.2016.2.a11
6. Ghahramani, Z.: Probabilistic machine learning and artificial intelligence (2015). https://doi. org/10.1038/nature14541
7. Conneau, A., Kiela, D., Schwenk, H., Barrault, L., Bordes, A.: Supervised learning of universal sentence representations from natural language inference data (2018). https://doi. org/10.18653/v1/d17-1070
8. Zendehboudi, A., Baseer, M.A., Saidur, R.: Application of support vector machine models for forecasting solar and wind energy resources: a review (2018). https://doi.org/10.1016/j. jclepro.2018.07.164
9. Yang, D., Liu, Y., Li, S., Li, X., Ma, L.: Gear fault diagnosis based on support vector machine optimized by artificial bee colony algorithm. Mech. Mach. Theory (2015). https:// doi.org/10.1016/j.mechmachtheory.2015.03.013
10. He, H., Kong, F., Tan, J.: DietCam: multiview food recognition using a multikernel SVM. IEEE J. Biomed. Health Inform. (2016). https://doi.org/10.1109/JBHI.2015.2419251
11. Guevara, C., Sanchez-Gordon, S., Arias-Flores, H., Varela-Aldás, J., Castillo-Salazar, D., Borja, M., Fierro-Saltos, W., Rivera, R., Hidalgo-Guijarro, J., Yandún-Velasteguí, M.: Detection of student behavior profiles applying neural networks and decision trees. In: Advances in Intelligent Systems and Computing, pp. 591–597 (2020). https://doi.org/10. 1007/978-3-030-27928-8_90
12. Fuentes, E., Alcañiz, M., Contat, L., Baldeón, E.O., Barat, J.M., Grau, R.: Influence of potential pulses amplitude sequence in a voltammetric electronic tongue (VET) applied to assess antioxidant capacity in aliso. Food Chem. (2017). https://doi.org/10.1016/j.foodchem. 2016.12.076
13. Ciocca, G., Napoletano, P., Schettini, R.: Food recognition: a new dataset, experiments, and results. IEEE J. Biomed. Health Inform. (2017). https://doi.org/10.1109/JBHI.2016.2636441
14. Loutfi, A., Coradeschi, S., Mani, G.K., Shankar, P., Rayappan, J.B.B.: Electronic noses for food quality: a review (2015). https://doi.org/10.1016/j.jfoodeng.2014.07.019
15. Arrieta, Á.A., Rodríguez-Méndez, M.L., De Saja, J.A.: Aplicación de una lengua electrónica voltamétrica para la clasificación de vinos y estudio de correlación con la caracterización química y sensorial. Quim. Nova **33**(4), 787–793 (2010)

16. Gamboa, A.A., Cáceres, P.A., Lamos, H., Zárate, D.A., Puentes, D.E.: Predictive model for cocoa yield in Santander using Supervised Machine Learning. In: 2019 22nd Symposium on Image, Signal Processing and Artificial Vision, STSIVA 2019 - Conference Proceedings (2019). https://doi.org/10.1109/STSIVA.2019.8730258
17. Chen, M.Y., Yang, Y.H., Ho, C.J., Wang, S.H., Liu, S.M., Chang, E., Yeh, C.H., Ouhyoung, M.: Automatic Chinese food identification and quantity estimation. In: SIGGRAPH Asia 2012 Technical Briefs, SA 2012 (2012). https://doi.org/10.1145/2407746.2407775
18. Rodrigues, D.R., de Oliveira, D.S.M., Pontes, M.J.C., Lemos, S.G.: Voltammetric e-tongue based on a single sensor and variable selection for the classification of teas. Food Anal. Methods (2018). https://doi.org/10.1007/s12161-018-1162-9
19. de Morais, T.C.B., Rodrigues, D.R., de Carvalho Polari Souto, U.T., Lemos, S.G.: A simple voltammetric electronic tongue for the analysis of coffee adulterations. Food Chem. (2019). https://doi.org/10.1016/j.foodchem.2018.04.136
20. Peris, M., Escuder-Gilabert, L.: Electronic noses and tongues to assess food authenticity and adulteration (2016). https://doi.org/10.1016/j.tifs.2016.10.014
21. Alcañiz Fillol, M.: Diseño de un sistema de lengua electrónica basado en técnicas electroquímicas voltamétricas y su aplicación en el ámbito agroalimentario, p. 295 (2011)
22. Cristianini, N., Shawe-Taylor, J.: An Introduction to Support Vector Machines and Other Kernel-based Learning Methods. Cambridge University Press (2013). https://doi.org/10.1017/CBO9780511801389

People with Disabilities' Needs in Urban Spaces as Challenges Towards a More Inclusive Smart City

João Soares de Oliveira Neto[1](✉), Sergio Takeo Kofuji[2], and Yolaine Bourda[3]

[1] Universidade Federal do Recôncavo da Bahia (UFRB), Cruz das Almas, BA 40150-080, Brazil
jneto@ufrb.edu.br
[2] Universidade de São Paulo (USP), São Paulo, SP 05508-010, Brazil
kofuji@lsi.usp.br
[3] LRI, CentraleSupélec, Université Paris-Saclay, 91190 Gif-sur-Yvette, France
yolaine.bourda@lri.fr

Abstract. Cities are the new frontier for computational platforms. As urban spaces are equipped with hardware and software solutions, inhabitants benefit from several Smart Cities applications that promise to improve their quality of life, while dealing with problems of our contemporary society, such as transportation, health, education, environmental issues, among others. People with disabilities (PwD) are a very active part of the urban population and, hence, they should be benefited by the advancements brought by Smart City initiatives. We designed an online questionnaire that led us to investigate what for and how people with different types of disabilities use technology in urban spaces, as well as the problems these users have in this context. With the results found, it is possible to observe that the specific needs and requirements that PwD have in urban spaces can help professionals to develop innovative projects that engage Smart City enabling technologies, and will induct a positive social impact.

Keywords: Requirements · People with disabilities · Inclusive Smart Cities · Urban spaces · Assistive technologies

1 Introduction

The increasing potential of Smart Cities (SC) initiatives motivated researches and industry forecasters to point out the use of Information and Communications Technology (ICT) in the urban context as the basis technology for digital economy in the coming years. In addition, technology is becoming the catalyzer of deep changes on the way we interact with each other and with the environment around us (Hernández-Muñoz et al. 2011). Enabling technologies, such as Ubiquitous Computing, Internet of Things (IoT), Big Data, and Cloud Computing are some of the key concepts responsible for enhancing the delivery of services and information on safety and security, transportation, education, citizens' health, tourism, and other public services (Khatoun and Zeadally 2016). IoT technology, for instance, plays a key role: (a) identifying

© Springer Nature Switzerland AG 2020
Á. Rocha et al. (Eds.): ICITS 2020, AISC 1137, pp. 285–293, 2020.
https://doi.org/10.1007/978-3-030-40690-5_28

devices – also known as things; (b) connecting them to each other and to other SC nodes through a heterogeneous network infrastructure; and (c) adding new features and capabilities to these things, such as performing tasks/services and informing their states to users, or even to other things (Want et al. 2015). However, the lack of social aspects is one of the pitfalls of SC initiatives and it can lead them to failure (Hollands 2008).

On the other hand, assistive technology has been developed aiming to balance the limitation or special needs that people with disabilities (PwD) may have. From computer softwares to magnifying glasses, assistive technologies bring to PwD the ability of doing activities that otherwise they could not do – and usually are ordinary to people with no disabilities. In fact, advocating for inclusiveness and for increasing the development of assistive technology is not a cause of a specific group of citizens or even a being corporative. Developing assistive technology benefits everybody, since sooner or later everyone will get older; also, at any time in life one may suffer an accident or even be temporarily impaired (Field et al. 2007). To consider PwD as potential users of the technology in general is to admit the diversity as part of the target audience and to reflect it as a component of the solution development process. Therefore, it would be feasible to broaden the range of customers and deliver technological solutions successfully in a competitive market.

The goal of this paper is to understand how PwD interact with digital solutions, specifically when they are in urban spaces and with GPS-based applications, aiming to provide a set of problems and issues that this interaction causes in these users. We believe that capturing the users' perspectives can provide SC professionals with directions and insights to develop applications that are able to address the real needs of this active range of citizens.

The remainder of this paper is organized as follows: Sect. 2 presents a topic on accessibility in urban spaces; in Sect. 3 we discuss the methodology and the experiment we have conducted; Sect. 4 presents the results based on the data collected from real users; in Sect. 5 we discuss these results; and finally, the last section presents the conclusions and future works.

2 Background and Related Work

Around 15% of the world's population lives with some form of disability. This global estimate for disability is on the rise due to the population aging and the rapid spread of chronic diseases (WHO 2011). Contemporary societies have approved laws and regulations recognizing the rights of these citizens to have education, work, leisure, and health, among all other civil rights that any other ordinary citizen has (Soegaard 2017). However, PwD must move around the city, namely urban spaces, in order to benefit and make use of these rights.

Moving around the city remains a problem for PwD (de Oliveira Neto and Kofuji 2016): most urban information is visual or auditory, which prevents blind and deaf people from receiving crucial information about the environment around them; obstacles, such as potholes and gaps, are not indicated; complex environments, such as a subway station, are not fully accessible, and PwD do not have a simple and easy way to contact human assistance, even though there are employees to help these individuals;

if a person with disability wants to go to a specific place, such as a bus stop or a hospital, he/she must memorize the route from where he/she is until the desired place or he/she needs to rely on someone else's help. In summary, autonomy and independence, two key values for PwD, are constantly threatening them when it comes to urban spaces, especially because access and mobility are important dimensions of quality of life, as (Matthews et al. 2003) states. For most PwD, daily trips are often fraught with problems, with many barriers imperceptible even to people without disabilities, hindering or totally restricting them to circulate in the cities.

Some academic works explore different approaches to capturing PwDs experience in urban spaces. In order to maintain and improve the accessibility level to wheel-chaired people in urban areas, (Matthews et al. 2003) conducted a study with 400 wheelchair users and a focus group aiming to develop a system that could quantify existent access levels and to model the changes that will occur as the result of any intervention. This proposal relates to the work of urban planners and suggests how urban physical barriers should be considered in urban intervention plans.

The authors developed and tested a GPS-based system for wheelchair users in urban environments. Notably, this work provides urban planners with key information to assist in the evaluation, building and development of mobility options and urban facilities adapted to the access and mobility of wheelchair users. However, the authors remark that the system is not accessible to other categories of PwD, such as blind people, or even people with severe physical disabilities.

These previously cited works have some limitations, once they focused only on field observations, automatic detection of urban problems, development of technologies, or concentrating only on a specific type of disability. Our work, however, focuses on getting in contact with potential real users of assistive technologies for urban spaces, keeping in mind the Smart City context. Understanding user needs and investigating the activities and contexts relevant to the application domain can help designers (Benyon 2013): (a) understand the system requirements they are developing, (b) have insights and perceive opportunities, and also (c) anticipate constraints provided by technologies.

3 Methodology

In order to reach a wider number of PwD, which would reflect in considering different contexts, cities and backgrounds, we first developed an online survey in the form of a questionnaire. This research instrument mainly aimed to anonymously identify problems and barriers that PwD face when they are in public spaces, the strategies they use to solve these problems, what are their feelings concerning urban assistive technology, and how favorably disposed they are to adopt digital technologies.

The questionnaire development was carried out in two steps. In the first step, a preliminary version was created as a pilot study. In the second step, an improved version was developed, considering the issues pointed out by the subjects in the pilot study. A total of five individuals took part in the pilot study. After the adjustments signalized in the pilot study, the questionnaire with 14 questions was published in a web service. We sent messages with the questionnaire hyperlink to mailing lists, specific discussion groups, and social media networks that had PwD as members or that was related to accessibility issues inviting participants. The average time to fill out the questionnaire was about 15 min.

Respondents were invited to anonymously answer different kinds of questions: open, scenario-based, multi-options, and Likert Scale style. Moreover, the web-based questionnaire was provided in English, Portuguese, and French. The questionnaire was available online to respondents from January to July 2017. Respondents were invited, without financial incentives, through social networks, Special Interest Groups (SIG) in accessibility or in related subjects, and online advertisement. The participants' inclusion criteria were: (1) be a PwD, specifically, motor/mobility, visual, auditory or any combination of the previous disabilities, i.e., multiple disabilities; and (2) older than 18 years.

4 Results

We obtained 186 participant responses – 91 in Portuguese, 84 in French, and 11 in English. In order to guarantee the validity of the responses, we examined and cleansed the data verifying the conformance with the inclusion criteria, resulting in the exclusion of 115 participants who did not meet one or more of these criteria. As Table 1 shows, after dropping out the excluded responses, the final dataset amounted to a sample size of 71 usable responses.

Table 1. Responses to the online survey.

Language	Total responses	Excluded responses	Valid responses
Portuguese	91	36	55
French	84	77	7
English	11	2	9
Total	186	115	71

All data collected in this study were first translated into English before being submitted to the Qualitative Content Analysis (QCA) method, a careful, detailed, specific for systematic examination and interpretation of a material corpus to identify patterns, themes, biases, and meanings (Neuendorf 2016). The QCA comprises the following stages for the data analysis procedure (Bengtsson 2016): decontextualization (the researcher must familiarize him or herself with the data), categorization (extended meaning units must be condensed), and compilation (the analysis and writing up process begin).

The majority of the respondents were between 18 and 44 years of age – therefore, a sample characterized by young adult and adult individuals (33.8% each). Most of them were male (56.3%), while 38.0% were female. Still concerning the variable Gender, the alternative "other" was assigned by 5.7% of the sample, referring to non-binary, genderfluid, gendervague, genderqueer, and demigirl conditions. In despite of our efforts to distribute the participants among the predefined set of disabilities in the inclusion criteria, the most predominant disabilities were Visual (45.1%), followed by Motor/Mobility (42.2%). The countries with more responses were Brazil (76.0%), followed by France and United States (9.9% each), and finally Canada, Switzerland, and Germany (1.4% each).

A multiple-choice question was asked about how they most often use the Internet on mobile devices – such as smartphones and tablets. Most of the respondents said that the main activity performed is sending and/or receiving e-mails, SMS and WhatsApp messages (89%), followed by using social networks (79%), playing music, videos or games (70%), making or receiving phone calls (68%), and reading daily news (61%), as may be observed in Fig. 1.

When asked about the mobile Internet usage frequency, 22.5% of the sample pointed out they use from one to five times a day, whereas the majority of 69% informed they use more than five times a day. The lower number of indication, 8.5% of the sample, signalized not having any mobile device but that they connected to the Internet using other kind of equipment. Specifically about the use of Internet when they are outside or in public spaces, 74.6% of the respondents confirmed that they do connect to the Internet in such situations; while 25.4% answered they do not connect to the Internet in public spaces.

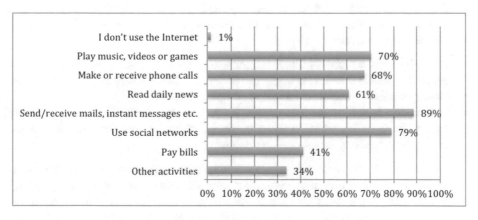

Fig. 1. Internet usage on mobile devices.

Next, in a multiple choice question, we presented to respondents a set of activates and asked them which ones they would perform outside of his/her private places – such as, home and office – based on the GPS current position. Accessing online maps and contacting a cab were extensively chosen by 68% of the sample, followed by checking the weather forecast and finding information about public transportation (61% each). The "other activities" option was chosen by 6% of respondents, who described the following extra activities: to access search engines, connect to location-based educational software, obtain indoor information when he/she is in a shopping mall, and use Ariadna[1].

---

[1] Ariadna is a mobile navigation system for visually impaired and blind people.

Aiming at exploring which problems PwD face in urban spaces, as well as the strategies they adopt to solve these problems, we idealized a situation described as a scenario. Jacko (2012) argues that using scenarios creates a situation in which the user can experience what he/she would need and would feel like in a realistic context. The scenario we generated was: "You have an extremely important appointment in a neighborhood where you have never been before and you aren't familiar with the area".

Firstly, in an open question, respondents were asked "how they would get information about the neighborhood and how would they arrive at the meeting on time?". 53.5% stated they would use GPS-based mapping and routing applications – namely, Google Maps, CityMapper, and Waze. Even considering only people with visual disability, 40% of the respondents answered they would use GPS-based solutions in order get in the meeting on time. Other alternatives cited were: to call a cab; ask help to parents and friends; search for wheatear forecast; access applications that provides panoramic views along the streets, such as Google Street View; call to public agencies, city hall, or public transportation companies in order to discover if the route is accessible; and use instant messaging apps.

Secondly, still working with an open question, we asked "what kind of information about the neighborhood would be useful before leaving his/her home to go to the meeting?". The majority informed that transportation-related information would be valuable: public transportation availability and schedule; distances; routing and description of the whole route; and traffic conditions. Respondents also pointed out other types of information:

- The neighborhood accessibility level: the existence of accessible facilities – such as adapted sidewalks and crossings, tactile floors, lifts, escalators, bus stops and subway stations – and their condition of operation; and
- Landmarks: ideally those that remain for a long period of time and are widely known, such as hospitals, drugstores, public agencies, gas stations, bus stops, subway stations, city monuments or statues.

According to the responses of this study, landmarks stand out as a support instrument in which people with disabilities rely on. One of the respondents, a young adult man with visual disability, stated that since "people usually give wrong information", he has established a strategy of combining the use of a GPS-based app and landmarks to locate precisely where he wants to go.

Thirdly, we questioned them about the types of problems respondents could face on the way to the appointment's place. 54.5% indicated that transportation problems would have more chance to happen: public transportation delays; traffic jams; traffic collisions; missing a bus/train or a connection; public transportation no-show; demonstrations; and public transportation emergency or mechanical failures. Moreover, respondents informed that getting lost would also be a problem. This could happen in the case of losing the mobile network connection and automatically the access to GPS-based apps, for instance, or even the malfunctioning of these apps occurring in routing the user to a complete undesirable destination or to dead-end streets. Respondents also expressed that lack of accessibility would be a significant problem in the proposed situation, once they could find: complete lack of accessibility, such as not having adapted public transport or ramps, or having lots of stairs, sidewalks under constructions or streets in maintenance; not accessible buildings; narrow and heavy doors; inaccessible traffic lights; large open areas to cross, such as parks and squares; and finding the exact entrance, for example, crossing large company areas to get to the main entrance.

The respondents were then asked to provide more details about what would help them solve the problems they might have in public spaces. Therefore, we grouped respondents' suggestions into five main categories of plausible solutions ordered here according to the number of suggestions of each category:

- Technology: improvements on GPS-based apps and the mobile network, especially providing alternative routes;
- Information sources: need for information describing what is happening on the route, such as obstacles and barriers, traffic, and if there is availability of accessible facilities;
- Governmental actions: some actions to solve their problems should be ruled by government authorities, for example: to improve the quality of public transportation services and adapt them according to accessibility requirements;
- Human assistance: human aid has also been reminded as a crucial factor in helping PwD to find a solution if they have any problems in the suggested scenario;
- Personal attitude: respondents admitted that part of the solution to possible problems that may occur in the suggested scenario depends on their own behavior.

5 Discussion

This paper provides SC professionals and public authorities with real testimonials from users to encourage the emergence of SC initiatives that are more socially oriented. Our effort can make a considerable difference, especially since one of the most frequent criticisms of SC initiatives is to have a more technology-and-entrepreneurial-driven perspective (Hollands 2008). In addition, SC initiatives are still concerned with future scenarios, rather than focusing on real problems of real citizens.

Our study points outs that PwD are ready to interact with mainstream technology: instant messaging applications, GPS-based systems, peer-to-peer ridesharing and transportation network company apps, social networks, as well as the most widespread solutions that citizens have in their smartphones. Such information indicates that the majority of SC initiatives that would benefit PwD would be easily adopted by these users.

SC covers an urban area with technology that can efficiently distribute information (Gutiérrez et al. 2013). The capillarity of SC actions can help to provide PwD with the most cited source of problems they have in cities: the adaptation of technology according to their individual needs, and the lack of relevant information in the right place and in the appropriate specificity of each citizen. In this context, IoT, one of the SC key technologies, can play a strategic role. Smart things can identify objects, which can communicate with devices of users with disabilities, sending appropriate and adapted information. These smart things can act as reference points by providing navigation and guidance information. In this way, PwD can take advantage not only of the network of human beings, but also of the smart objects that some contemporary cities are implementing. In this context, humans and objects communicate with each other using the network infrastructure to perform a variety of tasks (Latour 2005).

Our study also points out that PwD are already consumers of the most common ICTs and mobile applications. Consequently, we believe that these users would also adopt SC solutions if they took into account the real needs of the PwD. Hindering PwD access to SC solutions motivates the digital divide, at least when it comes to digital solutions that have already positively changed the lives of common citizens. The idea of mainstreaming disability needs to be continuously strengthened, especially since, despite not addressing the specific needs of the PwD, most mainstream activities and technological solutions exclude the participation and use of PwD.

6   Conclusions and Future Work

Smart Cities and their enabling technologies can be used as tools for the inclusion of minority groups suffering from problems common to all city inhabitants. For People with Disabilities, for example, navigating the urban space and finding information about objects and places are activities that usually require the help of other people.

We believe that the PwD's needs can be seen as potential opportunities that would confer SC initiatives a more social-driven approach. The first step in this direction is to understand the users and their needs. We enumerated in this paper a set of activities that the analyzed sample of PwD performed connected to the Internet and with GPS-based systems. Next, the data collected indicated that PwD felt the absence of adapted technologies in urban spaces that could: (i) provide personalized and real-time information; (ii) allow them to connect with other people; and (iii) help PwD manage their own strategies to deal with the barriers in these urban spaces.

As future work, we propose to take a step forward by meeting with other stakeholders in the accessibility field to obtain their impressions in order to develop a set of requirements from multiple perspectives. Furthermore, we propose to design an architecture system that would support the implementation system and, consequently, would address the needs of PwD in urban spaces, once they are the reason why this work was developed.

References

Bengtsson, M.: How to plan and perform a qualitative study using content analysis. NursingPlus Open **2**, 8–14 (2016). https://doi.org/10.1016/j.npls.2016.01.001

Benyon, D.: Designing Interactive Systems: A Comprehensive Guide to HCI, UX and Interaction Design. Pearson Higher Education, London (2013)

de Oliveira Neto, J.S., Kofuji, S.T.: Inclusive smart city: an exploratory study. In: Antona, M., Stephanidis, C. (eds.) Universal Access in Human-Computer Interaction. Interaction Techniques and Environments: 10th International Conference, UAHCI 2016, Held as Part of HCI International 2016, Proceedings, Part II, Toronto, ON, Canada, 17–22 July 2016, pp. 456–465 (2016). http://dx.doi.org/10.1007/978-3-319-40244-4_44

Field, M.J., Jette, A.M. (eds.): Assistive and Mainstream Technologies for People with Disabilities (2007). https://www.ncbi.nlm.nih.gov/books/NBK11418/

Gutiérrez, V., Galache, J.A., Sánchez, L., Muñoz, L., Hernández-Muñoz, J.M., Fernandes, J., Presser, M.: SmartSantander: internet of things research and innovation through citizen participation. In: The Future Internet, pp. 173–186 (2013). https://doi.org/10.1007/978-3-642-38082-2_15

Hernández-Muñoz, J.M., Vercher, J.B., Muñoz, L., Galache, J.A., Presser, M., Gómez, L.A.H., Pettersson, J.: Smart cities at the forefront of the future internet. In: Domingue, J., Galis, A., Gavras, A., Zahariadis, T., Lambert, D. (eds.) The Future Internet, pp. 447–462 (2011). http://dl.acm.org/citation.cfm?id=1983741.1983773

Hollands, R.G.: Will the real smart city please stand up? City 12(3), 303–320 (2008). https://doi.org/10.1080/13604810802479126

Jacko, J.A.: Human Computer Interaction Handbook: Fundamentals, Evolving Technologies, and Emerging Applications, Third Edition. CRC Press (2012)

Khatoun, R., Zeadally, S.: Smart cities: concepts, architectures, research opportunities. Commun. ACM 59(8), 46–57 (2016). https://doi.org/10.1145/2858789

Latour, B.: Reassembling the Social an Introduction to Actor-Network-Theory. Oxford University Press, Oxford, New York (2005)

Matthews, H., Beale, L., Picton, P., Briggs, D.: Modelling access with GIS in urban systems (MAGUS): capturing the experiences of wheelchair users. Area 35(1), 34–45 (2003). https://doi.org/10.1111/1475-4762.00108

Neuendorf, K.A.: The Content Analysis Guidebook. SAGE, Thousand Oaks (2016)

Soegaard, M.: Accessibility: usability for all (2017). The Interaction Design Foundation website: https://www.interaction-design.org/literature/article/accessibility-usability-for-all. Accessed 9 Dec 2017

Want, R., Schilit, B.N., Jenson, S.: Enabling the internet of things. Computer 48(1), 28–35 (2015). https://doi.org/10.1109/MC.2015.12

WHO: World report on disability (2011). http://apps.who.int/iris/bitstream/10665/70670/1/WHO_NMH_VIP_11.01_eng.pdf

Human-Computer Interaction

Websites and Social Networks. A Study of Healthcare SMEs in Andalusia

Irene Rivera-Trigueros[1]([✉]) [iD], Juncal Gutiérrez-Artacho[1] [iD],
and María-Dolores Olvera-Lobo[2,3] [iD]

[1] Department of Translation and Interpreting,
Faculty of Translation and Interpreting,
University of Granada, C/Buensuceso, 11, 18003 Granada, Spain
{irenerivera, juncalgutierrez}@ugr.es
[2] Department of Information and Communication, University of Granada,
Colegio Máximo de Cartuja, Campus Cartuja s/n, 18071 Granada, Spain
molvera@ugr.es
[3] CSIC, Unidad Asociada Grupo SCImago, Madrid, Spain

Abstract. Small and medium sized enterprises face numerous challenges, including opening up to globalized markets, and internationalization. The modern information society provides great opportunities for SMEs, allowing them to communicate their message globally at low cost through Web 2.0 tools like websites and social networks. The objective of this study is to analyze how SMEs in Andalusia from the dentistry sector use the Web 2.0 tools within their reach, and to determine whether relationships exist between the different variables associated with their usage, focusing on the existence or not of websites and/or social networks, the visibility of profiles on websites, the number of followers and the frequency of social network profile updates, among others.

Keywords: Digital marketing · SMEs · Social networks · Websites · Web 2.0

1 Introduction

Information and Communication Technologies (ICTs) have meant great change at all levels of society, including the business sphere. Among the opportunities provided to companies by the modern information society, it is worth mentioning the possibility of the global transmission of a message at considerably reduced cost, thus allowing entry into new markets and the exploration of different business models. ICTs have therefore transformed traditional marketing and are a good indicator of the capacity of businesses to modernize and compete in globalized environments [1]. Nevertheless, the economic sphere is in constant flux, which demands that companies, especially small and medium sized ones (SMEs), have a great capacity for adaptation in order to face the new challenges of an ever more globalized and difficult market. To this end, it is essential for them to take advantage of the elements provided by ICTs when taking their brands to their target audience via the creation of a solid internet presence thanks to the opportunities offered by Web 2.0 tools, including social networks [2, 3].

© Springer Nature Switzerland AG 2020
Á. Rocha et al. (Eds.): ICITS 2020, AISC 1137, pp. 297–306, 2020.
https://doi.org/10.1007/978-3-030-40690-5_29

Digital marketing is today one of the cornerstones for building relationships between consumers and businesses, as it promotes the creation of communities and interaction, favouring loyalty [4, 5]. It can be defined as the efforts on the part of a company to promote and sell products and services over the internet [6]. The application of digital technologies in order to achieve profitability and retain clients is essential, to which it is necessary to recognise their strategic importance, and develop a planned approach to improve knowledge about customers, gain effective communication and offer online services adapted to individual needs [7]. A website is a public relations vehicle for a business and allows it to inform, promote and commercialize its products and services [6]. For this to take place, one of the first steps for companies when implementing digital marketing strategies should be the creation of an attractive website in line that shows their objectives and interests, creates engagement and gains the attention of its users although it should, above all, be useful [1, 2, 6].

We are currently seeing a transfer of power from companies to consumers, to which the former should pay special attention to the opinions and demands of the latter [8]. In this manner, digital marketing also includes other tools such as social networks that, via the multiple platforms they offer, are a great medium for the interaction and expansion of social circles, and for sharing information, organizing events and communities, and even improving competitive positions [1, 8]. In addition, user interaction may be a source of innovation for SMEs [9, 10]. Thus, the appropriate employment of social networks can mean a big advantage in terms of business expansion without translating into excessive costs, to which they are especially advantageous for SMEs. Social networks have been proven to have a positive impact as regards aspects such as: reputation and brand image [11]; business organisation and reduction of marketing and advertising costs [12–14]; reaching a greater number of potential customers [15], and opening new commercialization channels [16].

However, despite all of the advantages offered by Web 2.0 tools like websites and social networks, there is still a long way to go. A number of studies point to the fact that, despite the quality of the corporate websites of Spanish SMEs being acceptable [17–19], their social network presence and activity is, in many cases, merely testimonial; in addition, in general terms, the objective of this presence is not communication and interaction with users [20, 21].

In Europe, practically the totality (99.8%) of companies operating in the economy —save for the financial sector—are SMEs [22]. 77% of European businesses have a website, but only 43% of companies between 10 and 49 employees make use of social networks [23, 24]. As regards the situation in Spain, the information offered in the latest official report [25] coincides with the European situation. Thus, 99.8% of the Spanish business fabric is comprised of SMEs. In addition, as is the case for Europe, 77% of Spanish SMEs have a website and 43% use social networks [26]. Catalonia, Madrid, Valencia and Andalusia are the autonomous regions that contain the most SMEs, over 60% of the total in Spain. Specifically, 501,097 small and medium enterprises in Andalusia make up 99.93% of the business fabric of this region [25]. In terms of Web 2.0 tool usage, Andalusia follows the national and European trend, with 70% of SMEs having a website and 48% using social networks in order to develop their company image or for advertising purposes [25].

It is beyond question that SMEs are a key element in both the Spanish and Andalusian economy; therefore, different initiatives have been set up. At national level, amongst the objectives set by the Agenda Digital para España driven by the Ministry of Industry, Energy and Tourism [27] are the use of ICTs to improve SME productivity and competitiveness. This has translated over to autonomous regional level, in the case of Andalusia, in the development of the Horizon 2020 Strategic Plan for the Internationalization of the Andalusian Economy [28]. For its part, the main objective of the 2020 Stimulus Strategy for the Andalusia ICT Sector is to favor the development and consolidation of the ICT sector in Andalusia [29]. Finally, one of the challenges of the Plan de Acción Empresa Digital (PAED) is to improve the establishment of ICTs in Andalusian SMEs, [30].

If we focus on business activity by sector, according to data from the latest report from the General Secretariat for Industry and Small and Medium Enterprises [31], 81.4% of Spanish undertakings carry out their activity in the services sector. In the case of Andalusia, the figure is somewhat higher, with over 83% belong to this sector [32]. Within it, healthcare is the second biggest group as regards business creation in net terms [31]. These companies belong to what is referred to as Group Q: Healthcare and social services activities according to the CNAE (National Classification of Economic Activities)-2009 classification, [33], which is divided into various subgroups: hospital activities; medical and dentistry activities; and other healthcare activities. Medical and dentistry activities represent around 53% of the total of group Q companies at national level, with approximately 55% companies in Andalusia belonging to this sector [34]. In this paper we will focus on companies who carry out their activity in the dentistry sector, as this has undergone considerable expansion in the last two decades and is currently in continuous growth [35, 36].

The objective of the study is to analyze how Andalusian companies in the dentistry sector employ Web 2.0 tools and to determine if there are relationships between the variables associated with their use, such as the existence of websites and/or social networks, the number of followers and social profile update frequency, among others.

2   Methodology

2.1   Description of the Object of Study

The object of study was selected using information from the Sectoral Ranking of Companies by Turnover offered by the Spanish source elEconomista.es. The data from this Company Ranking comes from the INFORMA D&B S.A.U. (S.M.E.) database (which boasts the AENOR quality certificate) and is fed from a number of public and private sources, including the Official Companies Register Gazette, the Official Accounts Records, the Official State, Autonomous Regional and Provincial Gazettes, national and regional press, ad hoc studies and other publications.

The selected companies belong to the CNAE sector: (8623) Dentistry activities. This sector contains 4,344 businesses listed in the national ranking, of which 4,307 are SMEs—3,402 small and 905 medium companies. For this study we selected the Andalusian SMEs belonging to this sector. 516 companies were initially examined, after filtering those companies forming part of franchises or large business groups, and a number whose description of the company purpose did not coincide with the activities specific to the sector, with the final sample comprising 498 SMEs in Andalusia, which in this case corresponds to the population of this sector in the region, and which represent 11.5% of the sector at national level according to the Sectoral Ranking of Companies by Turnover. Given that for the ordering of the companies in the ranking their financial situations filed in the Companies Register are used, with the close of registration date being July of the year for the object of study and June of the following year, the data obtained and employed in this paper correspond to 2017. We compiled the company data between the months of May and June 2019.

2.2 Analysis Criteria and Instruments

Firstly, we compiled the data from the companies obtained from the information provided by the Sectoral Ranking of Companies by Turnover. These data were as follows: Company name; Commercial or brand name (where available); Size according to turnover; Location; Website address (where available). Secondly, we adopted the criteria established in prior studies [37, 38] for compiling data relating to corporate websites and the use of social networks. The social networks analyzed were Facebook, Twitter, LinkedIn, Instagram and YouTube, with the use of other social networks on the part of SMEs being anecdotal and therefore insignificant for analysis. The indicators employed were:

- Existence or not of website. If yes: Contact information and social network links visible on website
- Existence or not of social network profiles. If yes: Existence or not of profile on Facebook, Twitter, LinkedIn, Instagram and YouTube; number of followers for each network and updating of information thereon (daily, very frequently, frequently, not very frequently, sporadically, rarely or social network abandoned).

In all cases we located the company websites and social networks—both using their company name and commercial name, where this appeared in the information from elEconomista.es—via Internet searches. In the cases where it was not possible to locate the link to the website or social networks of the company, or it was not possible to verify the effective belonging of a specific website to a company (consulting information on the legal notice, address, etc.), it was determined that the company did not have, or it was impossible to locate, a website or social network.

In the case of the visibility or lack thereof of links to social networks on the company website, we omitted broken links (17 cases), those that failed to include all profiles of the company present on social networks (9 cases), or those that pointed to links to non-existent social network profiles (3 cases). Thus, we only considered as valid those links that correctly redirected to all company social network profiles.

To analyze follower numbers, as this involved large data groups, we calculated the class interval corresponding to each social network profile, which allowed us to group the data into intervals in order to facilitate their subsequent analysis. To do so, once the total amount of data was determined (number of followers on each social network), we applied the Sturges rule to calculate the number of necessary intervals and then calculated their amplitude.

In the case of the last indicator—social network updating—we took into account posts made in the last month to determine whether the updating was daily (over 20 publications/month); very frequent (16–20 publications/month); frequent (11–15 publications/month); not very frequent (6–10 publications/month) or sporadic (1–5 publications/month). We determined the frequency as rarely if the companies had posted at least once in the last year. Lastly, it was decided that the social network profile had been abandoned if no type of post had appeared in the last year.

The data were treated with the SPSS (v. 22) statistical package via the analysis of frequencies and correlations. As the data did not follow a normal distribution, the Spearman coefficient was calculated—a non-parametric version of Pearson's correlation coefficient—to evaluate the possible associations between the different variables. The degree of correlation for the variables was determined according to the scale proposed by [39].

3   Results

Firstly, of the 498 companies analyzed, 252 had a website, 50.6% of the total. In the case of social network profiles, 244 companies, 49% of the total, had at least one profile (see Table 1).

Table 1. Existence or not of website and social networks.

	Website		Social networks	
	Frequency	Percentage	Frequency	Percentage
Yes	252	50.6	244	49
No	246	49.4	254	51

After calculating the Spearman coefficient, the existence of a mean positive linear directly proportionate correlation between the website and social networks variables can be confirmed (Spearman Rho = 0.679, p = 0.000). This indicates that the fact a company has a social network or not is associated with it having a website. Furthermore, no linear association was found between the company size and the website (Spearman Rho = −0.023, p = 0.615) and social networks (Spearman Rho = −0.18, p = 0.695) variables. Thus, it cannot be concluded that a relationship exists between the size of the company and whether or not it has a website or social networks.

If we look separately at the profiles of each of them with Spearman's rho we find a weak, positive and directly proportionate linear correlation between the website and existence of a Facebook profile (Spearman Rho = 0.183, p = 0.004) and existence of an Instagram profile (Spearman Rho = 0.260, p = 0.000) variables. Despite there being an association between the variables, as it is weak it cannot be concluded that fact that a company has a website determines whether it has a Facebook and Instagram profile, and it must be presumed that there are further reasons for explaining this dependence. For the rest of the social networks analyzed—Twitter, LinkedIn and YouTube—there is no correlation between the web and social network variables.

In addition, we analyzed whether there were correlations between the different social networks, that is, if the fact that a company had a specific social network account was associated with also having other social networks. In this regard, we found a weak, directly proportional positive linear correlation between YouTube and Instagram (Spearman Rho = 0.295, p = 0.000), Twitter and Instagram (Spearman Rho 0.242, p = 0.000) and LikedIn and YouTube (Spearman Rho = 0.224, p = 0.000). This tells us that, to a certain extent, a company having a YouTube channel is associated with it also having profiles on LinkedIn and Instagram. The same occurs in the case of Twitter and Instagram. Nevertheless, given that the correlation is weak, it is presumed that there are more reasons for explaining this dependency (Table 2).

Table 2. Correlation between the different social networks

Social network	Facebook		Twitter		LinkedIn		YouTube		Instagram	
	Rho	Sig.	Rho	Sig.	Rho	Sig.	Rho	Sig.	Rho	Sig.
Facebook			−.0080	0.218	−0.01	0.838	0.029	0.657	0.093	0.160
Twitter	−.0080	0.218			0.105	0.104	0.117	0.104	**0.242**	**0.000**
LinkedIn	−0.01	0.838	0.105	0.104			**0.224**	**0.000**	0.147	0.026
YouTube	0.029	0.657	0.117	0.071	**0.224**	**0.000**			**0.295**	**0.000**
Instagram	0.093	0.160	**0.242**	**0.000**	0.147	0.026	**0.295**	**0.000**		

Table 3 shows social network visibility and website contact information. That is, if companies had visible links to their social networks on their websites and a section where contact information was clearly specified or not. Over half the websites analyzed (51.2%) had links to company social networks and almost all of the companies analyzed (99.2%) made their contact information available on their websites.

Table 3. Social network visibility and website contact information.

	Social network links		Contact information	
	Frequency	Percentage	Frequency	Percentage
Yes	129	51.2	250	99.2
No	123	48.8	2	0.8

The Spearman rank correlation coefficient indicates the existence of a mean positive linear correlation, directly proportionate between the website and social network visibility variables (Spearman Rho = 0.584, p = 0.000). For their part, there is a very strong positive linear and directly proportional correlation between the website and contact information variables (Spearman Rho = 0.992, p = 0.000). This shows us that there is an almost perfect association between the fact that a company has a website and the publication of its contact information.

Additionally, to analyze the fact that a company had links on its website to its social networks we calculated the Spearman's rho and verified the existence of a weak negative and inversely proportional linear correlation between the social network visibility and Facebook followers variables (Spearman Rho = −0.289, p = 0.000). This is noteworthy, as it would indicate that the fact that a company's social networks are visible and accessible from its website is not necessarily associated with an increase in followers. Regarding the rest of the social networks, no linear association is found between social network visibility and followers on each.

Moreover, we analyzed whether the updating of social network profiles was associated with follower numbers. The Spearman's rho demonstrated that there is a negative and inversely proportional linear correlation between the Update frequency and Facebook followers variables (Spearman Rho = −0.535, p = 0.000), and also occurs in the case of Twitter, which has a weak correlation (Spearman Rho = −0.297, p = 0.003). These results create dissonance as they do not lead to the conclusion that more frequent updating on social network profiles is linked to a higher number of followers. That is, a very high update frequency could lead to a loss of followers. In the case of the other social networks analyzed, no significant correlations were found between update frequency and follower numbers.

Finally, we analyzed whether the number of followers on a specific social network was associated with that of other social networks. Hence, Table 4 shows a mean positive directly proportionate linear correlation between Facebook followers and YouTube followers (Spearman Rho = 0.614, p = 0.000) and a mean positive linear correlation between Twitter followers and Facebook followers (Spearman Rho = 0.298, p = 0.004). On the other hand, no significant correlations were found between followers on the other social networks analyzed.

Table 4. Correlation between followers on the different social networks.

Followers	Facebook		Twitter		LinkedIn		YouTube		Instagram	
	Rho	Sig.	Rho	Sig.	Rho	Sig.	Rho	Sig.	Rho	Sig.
Facebook			**0.298**	**0.004**	0.117	0.596	**0.614**	**0.000**	0.303	0.15
Twitter	**0.298**	**0.004**			−0.358	0.230	0.250	0.228	0.185	0.281
LinkedIn	0.117	0.596	−0.358	0.230			0.694	0.018	0.300	0.370
YouTube	**0.614**	**0.000**	0.250	0.228	0.694	0.018			0.110	0.608
Instagram	0.303	0.15	0.185	0.281	0.300	0.370	0.110	0.608		

4   Conclusions

This study analyzes the use of two of the most significant digital marketing tools—websites and social networks—on the part of SMEs in Andalusia dedicated to dentistry activities. Firstly, it should be pointed out that the SMEs analyzed are almost twenty points behind the European, Spanish and Andalusian average in terms of website availability. The use of social networks by the SMEs analyzed is, however, one point above the average for Andalusia and six above the Spanish and European average. These data show that they are still not making use of all of the advantages offered by websites and social networks. Notwithstanding, the positive correlation between these variables—having a website and social networks—tells us that those companies that do use the tools seem to be aware of their combined potential, given that the fact that a company has a website is positively associated with it also having social profiles. This association does not however determine on which specific social network the SMEs have an account or profile. Likewise, having a profile on a specific social network is not meaningful in terms of influencing if there are other social network accounts.

Companies are aware of the importance of providing clear and visible contact information on their websites, which is indicated by the near perfect correlation between the website and contact information variables. There is, in addition, a positive correlation between the website and social network visibility variables, which indicates that a significant proportion of SMEs appear to be aware of the need to have links to their social networks on their websites, given that 51.2% of those analyzed had links that effectively redirected to their profiles. However, the fact that a company has links to its social networks on its website is not associated with follower numbers, telling us that a mere digital presence could not be enough to generate engagement among users. It is also insufficient to frequently update social networks; moreover, a high frequency of posts can have negative repercussions on followers in the case of Twitter and Facebook. There are, though, associations indicating that the number of Facebook followers has an influence on Twitter and YouTube followers, and vice-versa.

This study serves as a starting point for further exploration of Web 2.0 tool usage on the part of SMEs, and the different relationships established between the variables involved. Thus, future lines of research should consider an expansion of the object of study including small and medium enterprises in Andalusia, Spain and Europe in the healthcare and social services sector, as well as other economic sectors. It would also be convenient to elaborate on both the perception of companies and users regarding the use of these resources through mixed approaches that include interviews or focus groups, in order to determine the reason for the occurrence of phenomena such as the negative association between frequency of posting and number of followers, and what other reasons would explain the weak correlations found. Furthermore, given that they face a globalized market, it would be necessary to research the degree of the internationalization of companies, paying particular attention to aspects such as website and social network translation, localization and transcreation.

Acknowledgements. Study supported by the Spanish Ministry of Science, Innovation and Universities (MCIU), the State Research Agency (AEI) and the European Regional Development Fund (ERDF) via the RTI2018.093348.B.I00 project and by the MCIU via the University Staff Training Program (FPU17/00667).

References

1. Peris-Ortiz, M., Benito-Osorio, D., Rueda-Armengot, C.: Positioning in online social networks through QDQ media: an opportunity for Spanish SMEs? In: Gil-Pechuán, I., Palacios-Marqués, D., Peris-Ortiz, M., Vendrell, E., Ferri-Ramirez, C. (eds.) Strategies in E-Business. Springer, Boston (2014)
2. Alcaide, J.C., Bernués, S., Díaz-Aroca, E., Espinosa, R., Muñiz, R., Smith, C.: Marketing y Pymes. Las principales claves de marketing en la pequeña y mediana empresa (2013)
3. Ferreira, T., Gutiérrez-Artacho, J., Bernardino, J.: Freemium Project management tools: asana, freedcamp and ace project. In: Rocha, Á., Adeli, H., Reis, L.P., Costanzo, S. (eds.) Trends and Advances in Information Systems and Technologies. WorldCIST 2018. Advances in Intelligent Systems and Computing, pp. 1026–1037. Springer, Cham (2018)
4. Lahuerta Otero, E., Cordero Gutiérrez, R.: Using social media advertising to increase the awareness, promotion and diffusion of public and private entities. In: Omatu, S., et al. (eds.) Distributed Computing and Artificial Intelligence, 12th International Conference. Advances in Intelligent Systems and Computing, vol. 373, pp. 377–384. Springer, Cham (2015)
5. Piñeiro-Otero, T., Martínez-Rolán, X.: Understanding digital marketing—basics and actions. In: Machado, C., Davim, J.P. (eds.) MBA. Management and Industrial Engineering, pp. 37–74. Springer, Cham (2016)
6. Kotler, P., Armstrong, G.: Marketing. Versión para Latinoamérica. Pearson Educación, Atlacomulco (2007)
7. Chaffey, D., Smith, P.R.: EMarketing EXcellence : Planning and Optimizing Your Digital Marketing. Butterworth-Heinemann, Oxford (2013)
8. Shaltoni, A.M., West, D., Alnawas, I., Shatnawi, T.: Electronic marketing orientation in the Small and Medium-sized Enterprises context. Eur. Bus. Rev. **30**, 272–284 (2018)
9. Cheng, C.C., Shiu, E.C.: How to enhance SMEs customer involvement using social media: the role of social CRM. Int. Small Bus. J. Res. Entrep. **37**, 22–42 (2019)
10. Ioanid, A., Deselnicu, D.C., Militaru, G.: The impact of social networks on SMEs' innovation potential. Proc. Manuf. **22**, 936–941 (2018)
11. Ahmad, S.Z., Ahmad, N., Abu Bakar, A.R.: Reflections of entrepreneurs of small and medium-sized enterprises concerning the adoption of social media and its impact on performance outcomes: Evidence from the UAE. Telemat. Informatics. **35**, 6–17 (2018)
12. Ainin, S., Parveen, F., Moghavvemi, S., Jaafar, N.I., Mohd Shuib, N.L.: Factors influencing the use of social media by SMEs and its performance outcomes. Ind. Manag. Data Syst. **115**, 570–588 (2015)
13. Parveen, F., Jaafar, N.I., Ainin, S.: Social media's impact on organizational performance and entrepreneurial orientation in organizations. Manag. Decis. **54**, 2208–2234 (2016)
14. Tajudeen, F.P., Jaafar, N.I., Ainin, S.: Understanding the impact of social media usage among organizations. Inf. Manag. **55**, 308–321 (2018)
15. Franco, M., Haase, H., Pereira, A.: Empirical study about the role of social networks in SME performance article information. J. Syst. Inf. Technol. **18**, 383–403 (2016)
16. Elghannam, A., Mesías, F.J.: Las redes sociales como nuevo canal de comercialización de alimentos de origen animal: un estudio cualitativo en España. Arch. Zootec. **67**, 260–268 (2018)
17. Gutiérrez-Artacho, J., Olvera-Lobo, M.: Web localization of Spanish SMEs: the case of study in chemical sector. J. Inf. Syst. Eng. Manag. **2**, 15 (2017)

18. Zumba-Zuniga, M.-F., Torres-Pereira, G., Aguilar-Campoverde, B., Martinez-Fernandez, V.-A.: Social media: a new tool for managing innovation in SMEs: analysis in the service sector in the southern region of Ecuador. In: 11th Iberian Conference on Information Systems and Technologies (CISTI), pp. 1–6. IEEE (2016)
19. Gutiérrez-Artacho, J., Olvera-Lobo, M.D.: Web localization as an essential factor in the internationalisation of companies: an approximation of Spanish SMEs. In: Rocha, Á., Correia, A., Adeli, H., Reis, L., Costanzo, S. (eds.) Recent Advances in Information Systems and Technologies. WorldCIST 2017. Advances in Intelligent Systems and Computing, vol. 569. Springer, Cham (2017)
20. Marín Dueñas, P.P., Lasso de la Vega González, M.C.: La efectividad de las páginas web en la comunicación empresarial de las pequeñas y medianas empresas. Un estudio en PYMES de la provincia de Cádiz. ZER. Rev. Estud. Comun. (2017)
21. Sixto García, J., Aguado Domínguez, N., Riveiro Castro, R.: Presencia 2.0 de las pymes gallegas: niveles de participación y engagement con los usuarios. RLCS Rev. Lat. Comun. Soc. **72**, 47–68 (2017)
22. Muller, P., Julius, J., Herr, D., Koch, L., Peucheva, V., McKiernan, S.: Annual report on European SMEs 2016/2017: Focus on self employment. EU, Brussels (2017)
23. Eurostat: Social media - statistics on the use by enterprises - Statistics Explained. https://bit.ly/2MuQGVJ
24. Eurostat: Internet advertising of businesses-statistics on usage of ads Statistics Explained. https://bit.ly/2Z6kuyB
25. Ministerio de Economía, Industria. y Competitividad.: Estadísticas Pyme. Evolución e indicadores (2018)
26. Urueña, A., Ballestero, M.P., Prieto Morais, E.: Informe e-Pyme 2016. Análisis sectorial de la implantación de las TIC en las empresas españolas (2016)
27. Ministerio de Industria, Energía. y Turismo.: Agenda Digital para España (2013)
28. Junta de Andalucía: PAIDI Plan Andaluz de Investigación Desarrollo e Innovación 2020 (2016)
29. Junta de Andalucía: Estrategia de Impulso del Sector Tic. Andalucía 2020 (2017)
30. Junta de Andalucía: Plan de Acción Empresa Digital 2020. https://bit.ly/2Z8VhmZ
31. Dirección General de Industria y de la Pequeña y Mediana Empresa: Retrato de la PYME. DIRCE a 1 de enero de 2018 (2019)
32. Instituto de Estadística y Cartografía de Andalucía: Nota Divulgativa. Directorio de Empresas y Establecimientos con Actividad Económica en Andalucía. 1 de enero de 2017. https://bit.ly/2XHOu21
33. Instituto Nacional de Estadística: Clasificación Nacional de Actividades Económicas (CNAE-2009). https://www.cnae.com.es/
34. Instituto Nacional de Estadística: Empresas por CCAA, actividad principal (grupos CNAE 2009) y estrato de asalariados (298). http://www.ine.es/jaxiT3/Tabla.htm?t=298
35. Llodra-Calvo, J.C.: La Demografía de los Dentistas en España (2010)
36. Sevilla, B.: Odontología: número de dentistas colegiados España 2005–2018 | Estadística. https://es.statista.com/estadisticas/627543/dentistas-colegiados-en-espana/
37. Olvera-Lobo, M.D., Castillo-Rodríguez, C., Gutiérrez-Artacho, J.: Spanish SME use of web 2.0 tools and web localisation processes. In: International Conferences WWW/Internet 2018 and Applied Computing (2018)
38. Olvera-Lobo, M.D., Castillo-Rodríguez, C.: Dissemination of Spanish SME information through web 2.0 tools. J. Transnatl. Manag. **23**(4), 178–197 (2018). https://doi.org/10.1080/15475778.2018.1509422
39. Hernández-Sampieri, R., Fernández Collado, C., Baptista Lucio, P.: Metodología de la Investigación. McGraw-Hill, New York (1991)

Users Preferences Regarding Types of Help: Different Contexts Comparison

Eduardo G. Q. Palmeira[1] , Abiel Roche-Lima[2] , and André B. de Sales[3(✉)]

[1] Federal University of Uberlândia, Uberlândia, Minas Gerais 38400-902, Brazil
eduardo.palmeira@ufu.br
[2] University of Puerto Rico, San Juan, PR 00936, USA
abiel.roche@upr.edu
[3] University of Brasília, Gama, Brasilia, Distrito Federal 72444-240, Brazil
andrebdes@unb.br

Abstract. Despite the significant advance in usability and intuitiveness of software interfaces, users still use help systems to seek solutions concerning doubts, whether they are beginners or advanced in the use of the software. Therefore, entirely intuitive, self-communicating interfaces that, regardless of task and user, do not require help systems, still do not exist. This study, as an exploratory type, aimed to verify if users refer to types of help significantly at different levels of intensity and preference when compared an initial user contact with software to a specific moment of doubt concerning any tool or task after being familiar. Thus, a user preference study was conducted by applying a questionnaire to 104 students of Software Engineering undergraduate course at the University of Brasília, Brazil. After that, the results from the two contexts were compared. Analysis of the data suggested that users refer more to types of help at an initial contact with software, and usually, they do not search directly for help documentation, referring more to search engines.

Keywords: User preferences · Help systems · Help documentation

1 Introduction

The use of software is a frequent activity in modern daily life associated with various contexts, and as such provides support across various aspects. McKita [1] believed that in the future software would be so intuitive, with user friendly interfaces, that users would be capable of performing any task without needing training or help documentation. Nevertheless, Nielsen [2], in one of his ten usability heuristics, states that help documentation needs to be provided, although it is better when a software can be used intuitively without its help.

Despite the significant advance in usability and intuitiveness of software interfaces, users still use help systems to seek solutions concerning doubts, whether they are beginners or advanced in the use of the software. Therefore, entirely intuitive, self-communicating interfaces that, regardless of task and user, do not

© Springer Nature Switzerland AG 2020
Á. Rocha et al. (Eds.): ICITS 2020, AISC 1137, pp. 307–314, 2020.
https://doi.org/10.1007/978-3-030-40690-5_30

require help systems, still do not exist. Hence, help systems are designed to predict and address communication failures caused by insufficiencies in interfaces, acting as a metacommunication to the user concerning the software interface itself (existing communication) [3].

In the book "Guide to the Software Engineering Body of Knowledge" (SWE-BOK) version 3.0 [4], when approaching software testing, it is stated that documentation that helps users is part of the process for evaluating how easy it is to learn to use the software, and along with other evaluations (such as testing the software functions that help the user in terms of tasks, as well as the systems ability to recover from user errors) are the main tests and usability tasks of Human-Computer Interaction (HCI). Furthermore, the help documentation provided by the software developer is no longer sufficient and unique; thus, third party help systems provide additional help to fill that gap [5].

McKita [1] also believed that interactive online tutorials would eliminate the need for printed tutorials, as users could learn how to use the software directly from the system itself. However, it is well known that each help system has its advantages and disadvantages depending on the user and the context, so there is effectively no substitution of some help systems with others, but an addition that encourages coexistence between such systems [3]. Thus, different types of help do not replace any other but serve different needs and audiences. This is noted, for example, in Price [6], where a situation was described in which users used their offline documentation for an overview of software ability when it came to initial contact, and online for quick access to information in order to perform a given task. Furthermore, in Earle et al. [5], two types of articles and videos are shown, descriptive when the user refers to these as being introduced to the software in general terms, and "how-to" when seeking a description or explanation of how to operate a specific task.

This study aimed to verify if users refer to types of help in significantly different ways, from the point of view of intensity and preference when compared an initial user contact with software to a specific moment of doubt concerning any tool or task after being familiar with the software, and then compared the results from these two contexts. Hence, the article is organized as follows: Sect. 2 presents the related work; Sect. 3, the methodology; Sect. 4, the research questions; Sect. 5 describes the results, Sect. 6, the discussion; Sect. 7, conclusions, limitations, and future work; and finally, in the last section, the references are listed.

2   Related Work

In Sales et al. [7], a study concerning user preferences for most types of help documentation and help alternatives found in the literature is presented, and these are listed on Table 1.

From the theoretical analysis of the study, it was understood that the help documentation can be provided by the software developer company or provided by third parties. Some documentation can be found both ways, such as articles, tutorial videos, online forums, and training. The study showed that while

Table 1. The eight types of help documentation and the three types of help alternatives identified on literature.

Source of help	Type of help
Help documentation	Manuals
	Online help
	Articles
	Frequently asked questions (FAQ)
	Interactive online tutorials
	Tutorial videos
	Online forums
	Training
Help alternatives	Autonomous research
	Search engines
	Ask experienced users for help

attaching a high level of importance to the help documentation provided by the software developer, which a few decades ago was essential and unique, and now is no longer used to the same degree, users prefer to seek help from communities and online documents provided by third parties who share information and offer help. In addition, more than just the use of help documentation, to which users give priority, they prefer help alternatives, and as such show that there is no culture of referring to help systems.

Among the help alternatives, those which are most commonly used are search engines. This form of help in many situations leads the user to some help documentation provided by the software developer company or third parties. Thus, it is evident that in most cases the user does not search directly for help documentation, but usually refers to search engines, trusting that the most relevant help documentation will be recommended. Autonomous research, popularly known as "trial and error", is how the user learns to use the software by interacting with the interface, exploring, experimenting, making mistakes, until he can perform the desired task. This is because some users prefer to avoid searching for help documentation that could aid them. Also, many users prefer to ask more experienced users for help, with whom they share the same work environment, for example.

In the reported study, users preferences were not presented by differentiating contexts, such as an initial user contact with software or a specific moment of doubt regarding any tool or task after being familiar with the software.

3   Methodology

As an exploratory type, this study was developed "with the objective of providing an overview, of approximate type, concerning a certain fact" [8]. As for the approach, it is of the qualitative type, because "qualitative research is not concerned with numerical representativeness, but with the deepening of the understanding of a social group" [9]. The social group analyzed here consisted of undergraduate students of Software Engineering at the Gama College of Engineering (FGA) of the University of Brasília (UnB) in which is the first undergraduate level course of this nature in Brazil [10,11]. Also, FGA supports around 560 new students each year and offers 5 undergraduate level courses in engineering: Aerospace Engineering, Automotive Engineering, Electronic Engineering, Energy Engineering and Software Engineering.

In order to test the aim of the present study, a user preference study was conducted by applying a questionnaire to 104 respondents. All participants were informed and agreed to participate in the study after reading and signing the free consent form. At the beginning of the questionnaire, the types of help identified in the literature of the related work were introduced. Respondents profile (n = 104) was as follows: 92 men and 12 women, with a majority aged between 20 to 22 years (67.3%). The respondents were students who had been studying until at least the 4th semester of Software Engineering undergraduate course, and the questionnaire was applied at the beginning of some classes of the course. The questionnaire consisted of listing Table 1 types of help in two different contexts: at an initial user contact with software and at a specific moment of difficulty using the software. Respondents were free to check none, one or more of an option (types of help in which they refer) in each of the two contexts indicated. Articles, tutorial videos, online forums, and training had two available markings: provided by the software developer and provided by third parties.

4   Research Questions

The basic research questions that led the questionnaire were:

1. What type of help do you usually refer to when learning how to use a software for the first time?
2. What types of help do you usually refer to when learning how to operate a specific tool or task when you need a quick response?

5   Results

Based on an initial question that asked the respondents for their five most commonly used software, most cited were programming software as their main use tools, and communication and organization software for team projects as support tools.

From Fig. 1, for the purpose of organizing the analysis of the obtained data, the results section was divided into two contexts: at an initial user contact with the software, and at a specific moment of difficulty using the software.

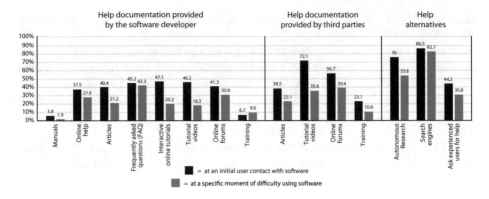

Fig. 1. Frequency of respondents preferences for software types of help in two different contexts: at an initial contact with the software (black chart bars), and at a specific moment of difficulty using the software (gray chart bars). The types of help are divided into three groups: help documentation provided by the software developer, help documentation provided by third parties, and help alternatives.

5.1   At an Initial User Contact with the Software

More than half of respondents, at an initial contact with the software, said they refer to search engines (86.5%), autonomous investigation (76%), third-party tutorial videos (72.1%), and online forums also provided by third parties (56.7%). None of the help documentation provided by the software developer company obtained more than 50% of respondents stating the referring.

Among the help documentation provided by the software developer, the most obsolete, with referring rates below 10%, are training (6.7%) and manuals (5.8%). Excluding these, comparing the referring rate of other help documentation, there is a variation of less than 10%, with interactive online tutorials (47.1%) being the most frequently referred to, and online help (37.5%), the least.

In the help documentation provided by third parties, there was a more significant variation of referring to help documentation. Tutorial videos (72.1%) were referred to the most, followed by online forums (56.7%), articles (38.5%), and finally, training (23.1%).

Among the help alternatives, two were marked as the most frequently used at the initial contact with the software: search engines (86.5%) and autonomous research (76%). Asking a more experienced user for help, even though it was the least marked alternative among help alternatives, represented a significant referring rate of 44.2%.

5.2   At a Specific Moment of Difficulty Using the Software

Only search engines (82.7%) and autonomous research (53.8%) were marked by more than half of the respondents. No help documentation, whether provided by the software developer or by third parties, was claimed to be referred to by more than half of respondents, only frequently asked questions (FAQs) came close to that with 42.3%.

Among the help documentation provided by the software developer, the training (9.6%) and the manuals (1.9%) were marked as the least referred (less than 10%), and the FAQs (42.3%) and the online forums (30.8%) as the most. The rest of the help documentation has a variety of less than 10% difference, with online help (27.9%) being the most referred, and tutorial videos (18.3%) the least.

Among the help documentation provided by third parties, the online forums (39.4%) and the tutorial videos (35.6%) were marked as the most referred, and the training (10.6%) as the least.

Finally, among the help alternatives, two were marked with referring rate higher than 50%: search engines (82.7%) and autonomous research (53.8%). Asking a more experienced user for help (30.8 %) was the least marked alternative among help alternatives.

6 Discussion

Analysis of the data suggested that users refer more to types of help at an initial contact with software than at a specific moment of doubt, only in training provided by the software developer was there noted different, but not very significant behavior. In addition, the types of help that have a significantly higher reference at an initial contact with the software are help alternatives and help documentation provided by third parties, while at a specific moment of doubt using the software, only help alternatives are significantly more referred to.

Consistent with results from related work [7], it may be suggested that most users do not search directly for help documentation, referring more to search engines, confident that these will be directed to the most relevant help documentation of their search. In addition, it is clear that the use of autonomous research is a feature that is prioritized over the help documentation in general, suggesting that users prefer to learn how to operate the software with "trial and error".

In addition, users are significantly more likely to refer to tutorial videos provided by third parties than by the software developer. There is also a subtle preference for referring to help documentation provided by third parties regarding online forums and training. However, the data are insufficient to support statistically significant conclusions regarding the difference in referring to articles, thus suggesting that users use both sources in similar fashion.

Noteworthy also was that when analyzing the two different contexts, there is not much difference in referring to search engines (3.8% difference), while there is a lot in autonomous research (22.2%) and a considerable difference when asking a more experienced user for help (13.4%). In addition, it is clear that some help documentation had significant referencing differences that depended on the context, all of these possessing a higher referral rate in terms of initial contact with the software. Examples of such initial contact can be tutorial videos provided by third parties with 36.5% difference, tutorial videos provided by the software developer with 27.9%; interactive online tutorials with 26.9%; the articles provided by the software developer company with 19.2%, online forums provided by third parties with 17.3% and articles provided by third parties with 15.4%. The remainder of help documentation has a referral rate with differences of less than 15% when compared to context, so no significant differences were noted. Thus, these results reinforce differences in intensity concerning the referral to most types of help when the context changes.

7 Conclusion

In this study, 104 respondents answered a questionnaire requesting their prefer-ences for the types of help sought after concerning the software they frequently use, differentiating two context, the moment of initial user contact with software and at a specific moment of doubt regarding any tool or task after becoming familiar with the software. The questionnaire covered help documentation pro-vided by the software developer, provided by third parties, and help alternatives. Thus, in this paper, through a user study, evidence showed that in different con-texts, users refer significantly at different levels of intensity and preference to the types of help available. In addition, this study reported consistent and com-plementary results from related work [7].

Thus, this is a brief exploratory study that intended to isolate the contexts and present the differences found in user preferences for the types of software help. By obtaining greater knowledge of such behavior, software developers are assisted in user-centered decision making, when dealing with the creation of improved and diverse help documentation, where greater effort can be given to thinking about the context and problem-solving in which it is most likely to be used.

7.1 Limitations

Emphasis is here given to the fact that the results found are specific to the group of users under analysis in this paper - undergraduate students of Soft-ware Engineering at the University of Brasília, Brazil. If these research are to be applied to another context and with other users, the results could present differences. Also, the results showed users preferences concerning the software types of help often used on a daily basis by the selected student group, so there was no controlled user group with similar levels of experience in common for specific software. Maybe is possibly hard to make strong conclusions in this very specific area based only on presented data, therefore, the present exploratory study presented brief results that suggest subjects for further investigation in future works.

7.2 Future Work

In future studies, we will select a sample of users who deal with specific software, use a significantly bigger database of the questionnaire recipients, analyze their preferences when referring to types of help, and compare the results with the present study. In addition, these studies intend to investigate why in initial contact with the software users refer more intensively to help systems.

References

1. McKita, M.: Online documentation and hypermedia: designing learnability into the product. In: IPCC 1988 Conference Record 'On the Edge: A Pacific Rim Conference on Professional Technical Communication', pp. 301–305 (1988). https://doi.org/10.1109/IPCC.1988.24055
2. Nielsen, J.: Enhancing the explanatory power of usability heuristics. In: Proceedings of the SIGCHI Conference on Human Factors in Computing Systems (CHI 1994), pp. 152–158. ACM, New York (1994). https://doi.org/10.1145/191666.191729
3. Silveira, M.: Metacomunicação Designer-Usuário na Interação Humano-Computador: design e construção do sistema de ajuda. Doctoral thesis. Informatics Department, Pontifícia Universidade Católica do Rio de Janeiro, Rio de Janeiro, Brazil (2002)
4. IEEE Computer Society, Bourque, P., Fairley R.: Guide to the Software Engineering Body of Knowledge (SWEBOK(R)): Version 3.0, 3rd edn. IEEE Computer Society Press, Los Alamitos (2014)
5. Earle, R., Rosso, M., Alexander, K.: User preferences of software documentation genres. In: Proceedings of the 33rd Annual International Conference on the Design of Communication (SIGDOC 2015). ACM, New York (2015). https://doi.org/10.1145/2775441.2775457. Article 46, 10 pages
6. Price, L.: Using offline documentation online. In: Proceedings of the Joint Conference on Easier and More Productive Use of Computer Systems (Part - II): Human Interface and the User Interface - Volume 1981 (CHI 1981), pp. 15–20. ACM, New York (1981). https://doi.org/10.1145/800276.810955
7. Sales, A., Palmeira, E.: Use preferences in help documentation. In: 2019 14th Iberian Conference on Information Systems and Technologies (CISTI), Coimbra, Portugal, pp. 1–6. IEEE (2019). https://doi.org/10.23919/CISTI.2019.8760947
8. Gil, A.: Métodos e Técnicas de Pesquisa Social, 5th edn. Atlas, São Paulo (1999)
9. Métodos de Pesquisa, 1st edn. In: Gerhardt, T., Silveira, D. (eds.) Coordenado pela Universidade Aberta do Brasil - UAB/UFRGS e pelo Curso de Graduação Tecnológica - Planejamento e Gestão para o Desenvolvimento Rural da SEAD/UFRGS, Porto Alegre, Brazil (2009)
10. Figueiredo, R., Sales, A., Ribeiro, L., Laranjeira, L., Rocha, A.: Teaching software quality in an interdisciplinary course of engineering. In: 7th International Conference on the Quality of Information and Communications Technology (QUATIC), pp. 144–149 (2010). https://doi.org/10.1109/QUATIC.2010.28
11. Figueiredo, R., Ribeiro, L., Sales, A., Canedo, E., Chaim, R., Rocha, A., Santos, G., Ramos, C.: Graduação em Engenharia de Software: uma proposta de flexibilização e interdisciplinaridade. In: Congresso Brasileiro de Software: Teoria e Prática, Salvador. III Fórum de Educação em Engenharia de Software. XXIV SBES - Simpósio Brasileiro de Engenharia de Software (SBES), pp. 43–50 (2010)

Electronic System for Memory Process in Children with Difficulties

Diana Lancheros-Cuesta$^{(\boxtimes)}$, Yulian Humberto Triana Garca,
Stefan Nicolas Aponte, Maick Peter Marin Rektemvald,
Jose Luis Ramirez, Mario Fernando Castro Fernandez,
Martha Patricia Fernandez-Daza, and Cristian Camilo Arias-Castro

Universidad Cooperativa de Colombia, Bogota, Santa Marta,
Monteria, Colombia
{diana.lancheros,yulian.triana,
stefan.apontem}@campusucc.edu.co

Abstract. Education teachers need to improve the training in the field of Learning Difficulties. The present paper show a system electronic that allow improve the process of memory in the children with dyslexia. The system allow at the students play in different levels and improve the attention process.

Keywords: Electronic device · Short term memory · Mobile application · Attention deficit · Human computer interface

1 Introduction

With respect to the use of technology in educational applications, Gonzales et al. [2] mentions that technology revolutionizes all fields of life, providing new techniques to solve problems. The research use a kinect [1] for tests and stimulate memory in children from 5 to 8 years. The results show the generation of high levels of motivation generated by the children when interacting with the electronic device. [4] develops an expert system in order to perform tests on the development of memory in cognitive processes. The system allow to know the impact of psychological tests for memory processes, when use technology. Taking into account the above, this paper show the development of a electronic system that integrates an electronic board and mobile application to stimulate memory process in children with attention difficulties. This paper show the design of the system in session 2, followed by a description of the validation. Finally, the conclusions and future work are mentioned.

2 System Design

The electronic system is composed of two systems, the system mechanical and electronic design, and the second system is the development of the mobile application.

Mechanical and Electronic Design. The mechanical device consists of 16 luminous buttons of different colors in a white board-like structure, supported by a solid tripod-

© Springer Nature Switzerland AG 2020
Á. Rocha et al. (Eds.): ICITS 2020, AISC 1137, pp. 315–319, 2020.
https://doi.org/10.1007/978-3-030-40690-5_31

like base that was designed to be easily transported. The system mechanics the option to adjust the height so that it can be adapted according to the height of children between 5 and 8 years old. The average height of these ages is 110 cm to 134 cm. The structure is composed for two essential parts which will be shown below.

Luminous Buttons. This pushbutton the button has the advantage of having a led bulb that can be controlled individually (Fig. 1).

Fig. 1. Electronic device.

Electronic Circuit. For the dynamic visualization was necessary the used the operation of a LED array, the design which will control the electronic part of the system, this is show in Fig. 2. The operation of this system have 4 columns, 4 rows and 16 buttons, when lighting a column it will behave an enabler and will give way connecting each of the LED's or buttons to ground, in one cycle of time. Then in a short time lapse the system activate the rows and receive a data, then turn off and continue with the same process with each of the columns. The system is executed at a high speed so that the human eye don't perceive the blinking. When the participant presses the buttons, the system always ready to receive a piece of information. The control device used is Arduino Mega [3].

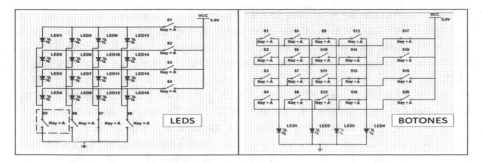

Fig. 2. Circuit dynamic visualization

Development of the Mobile Application. The mobile application has a screen for registration for the users (students) that interact with the game. It have a record button where the connection to the database is enabled and the user's data is stored. The child's stored data are: name, surname and age. After registering the child who will participate in the game, it is necessary to select the bluetooth device of the micro-controller that controls the game board (an arduino card). This list of devices is

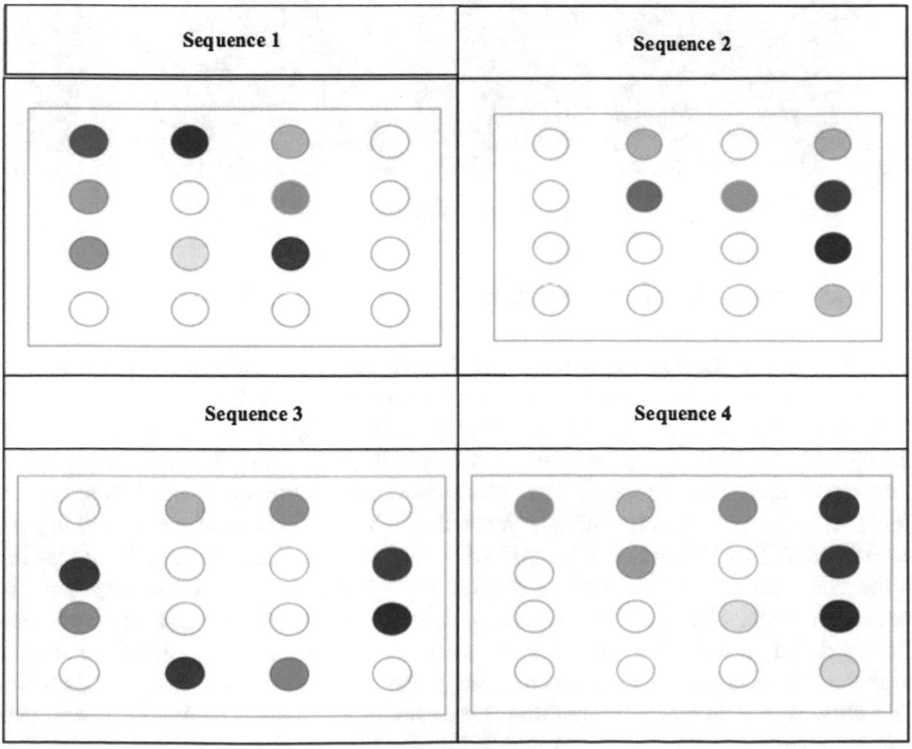

Sequence 1	Sequence 2
Sequence 3	Sequence 4

Fig. 3. Sequences

generated so that another bluetooth device can be easily selected in case of any damage. The application contains two options, interactive sequences for children of 5 years, and sequences for children of 6.7 and 8 years. The first option shows in the mobile device the matrix of circles that different sequences of colors are randomly displayed (Fig. 3). The child must first observe the sequence in the system electronic, then he press the buttons of the same sequence and then the compare button. The next button stores a level indicator and attempts in case of failure. Another button allow return to the previous view to start a new level. Screen for game of levels 6, 7 and 8 years: Fig. 3 shows the interface that present randomly numbers which must be replicated in the electronic system. In the same way the activity described above, the compare button verifies the successes or failures of the child (Fig. 4).

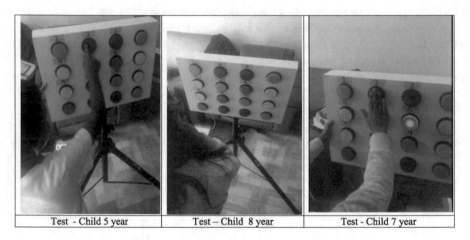

| Test - Child 5 year | Test – Child 8 year | Test - Child 7 year |

Fig. 4. Electronic device.

3   Validation Electronic System

The method for the validation the electronic system is a study case, because it is necessary test if the electronic board allow the attention child. For the validation three children with 3 cases of the ages established in the design of the system. After executing the game correctly, a interview with the children is made, this instrument shows their experience with the game. The questions of the instrument were: From 1 to 10 being 1 very easy 10 very difficult What difficulty did you have when making the levels? From 1 to 10 being 1 little fun and 10 very fun. Indicate if the activity was fun What was your feeling when doing this activity? Figure 4 shows photographs of the experimentation. The results of validation showed: in the first case of study, a 5-year-old child finds it difficult to follow a series of instructions, among which the handling of the application. In this case was necessary the support of a psychologist specialist. The child was a difficulty in attention processes in the school. In this case was necessary to generate short sequences with a longer response time in the application. The child complete the sequences and he has fun doing it. In the second case, an 8-year-old girl do the sequences quickly, first a pilot test is carried out to familiarize her with the activity. The sequences was assigned taking into accounting the age 6-year-old, she manages to pass it easily, then she proceeds to perform the tests with the level for age 8-year-old but she can not do the sequences. Next in the second level, she feels that this level requires a greater effort. She felt that the way to use the game is very didactic and fun. The attention process was achieved. The third case a 7-year-old girl manages to take the attention quickly, starts with a test level of a 6-year-old child who manages to pass if no problem, then assigned the correct level which takes practically double the time to do it completely and to finish it is assigned the test with the level of a child of 5 years which indicates that it is the easiest of all. The girl from the beginning shows an attitude of concentration and passes each of the proposed levels. It is important to mention that the girl did not have attention problems detected.

4 Conclusions

The MemoryBoard system allows the random generation of sequences in order to stimulate memory in children with or without attentional problems. This will allow them to perform better in teaching and learning processes. The interaction between a mechatronic system and a mobile application generates greater attention and motivation in children, which was evidenced in the case studies analyzed. As future work we have contemplated the use of MemoryBoard in experiments with greater population and determining the attentional levels whit the brain signals.

References

1. qu es el dispositivo kinect? | kinect for developers. http://www.kinectfordevelopers.com/es/2012/11/06/que-es-el-dispositivo-kinect/
2. Gonzlez, J.L.C.: Test wisc IV: una mirada desde la herramienta kinect
3. Manhes de Oliveira, C., Soares, P.J., Morales, G., Arica, J., Matias, I.: State of the art on arduino and RFID, pp. 213–220 (2019)
4. Poma Mamani, C.: Aplicacin del test WISC-III y sistema experto para la estimulacin cognitiva. http://localhost:8080/xmlui/handle/123456789/8745

Feeling Younger? An Investigation of Cognitive Age on IT Use

Maximilian Haug$^{(\boxtimes)}$ and Heiko Gewald

Neu-Ulm University of Applied Sciences, Wileystraße 1,
89231 Neu-Ulm, Germany
{maximilian.haug, heiko.gewald}@hs-neu-ulm.de

Abstract. Seniors show remarkable differences in IT use among their own peers. Even though IT use among seniors is increasing, viewing them as a homogenous group is insufficient to predict use behavior. This research investigates how influences on finding new ways to use IT is moderated by seniors' perceived age. We build on the post-adoption perspective to evaluate whether cognitive age moderates the key traits of computer self-efficacy, computer anxiety, and personal innovativeness in IT in the context of innovation attempts. The results show that based on the cognitive age, influences on innovation behavior differ significantly amongst seniors.

Keywords: Cognitive age · Chronological age · Older adults

1 Introduction

With increasing access to digital devices such as smartphones or tablets, as well as to the Internet, the general population has started to develop increasingly sophisticated capabilities towards using IT. However, in terms of their usage behavior, there are inherent differences in distinctive cluster of the population, when it comes to using digital offerings. When we look at online health information, elderly people (in the following defined as people 60+) do not seek this information as frequently as their younger counterparts [1]. Seniors apparently refrain from using the Internet as a source of health information and are more prone to using traditional media [2]. However, as previous research suggests, there are also clusters of elderly people who frequently use the Internet to acquire health information and therefore explore IT in new ways.

Research comes to the point to acknowledge, that chronological age does not serve as a good predictor towards the behavioral characteristics of an individual. The perceived, subjective or cognitive age much better reflects the individual's perception of his or her own age [3] and thus the individual's behavior. This implies that if seniors show behavior that is normally associated with a younger segment, they might regard themselves as younger than they physically are. To date, cognitive age remains mostly unexplored in IS research [4], but studies utilizing the technology acceptance model indicate that seniors who perceive themselves as younger show a higher use of the Internet than their respective counterparts [5]. Therefore, we want to investigate the following research question: *How does cognitive age influence IT-related behavior, as opposed to physical age?*

© Springer Nature Switzerland AG 2020
Á. Rocha et al. (Eds.): ICITS 2020, AISC 1137, pp. 320–329, 2020.
https://doi.org/10.1007/978-3-030-40690-5_32

Online health information may be of great benefit to seniors, so it is of great importance to understand and identify triggers that make elderly people reach out for such information. We conducted a quantitative study to investigate the role of cognitive age in finding new ways of using IT among elderly people in the US. 120 complete datasets could be collected, which were then separated into two groups—one comprising people who feel younger than they actually are and the other in which the perceived age equals the chronological age.

The paper is structured as follows: First, we present the background of IT-related traits and the concept of cognitive age. The research model, including the development of hypotheses, follows. The research methodology and the results are then presented. The paper closes with the discussion section, further research propositions, and the conclusion.

2 Background

2.1 Cognitive Age vs. Chronological Age

Age usually refers to the chronological age of an individual—i.e. the number of years that have passed since birth [4]. However, in the field of gerontology, the concept of chronological age is seen as problematic, since age as a number does not reflect the specific situation of heterogeneous individuals [4]. This means that individuals may have the same chronological age but show very different patterns of behavior or character traits. Lindberg, Näsänen [6] suggest that when observing user behavior – specifically in the context of digital resources researchers should not solely rely on the chronological age. Therefore, in gerontology and psychology, a self-perceived age measure has been constructed [7, 8], which is referred to as perceived or cognitive age. According to Barak and Schiffman [8], the cognitive age consists of four dimensions— how the individual feels, how old the individual looks, how the individual behaves in comparison to other groups (e.g. younger segments), and the interests of the individual compared to other groups. Several studies show that there is a significant difference between chronological age and the self-perception of seniors.

The concept of cognitive age has been gaining more and more attention in the fields of psychology and marketing, but studies in IS research remain scarce (e.g. Hong, Lui [4]). In the context of Internet use, Eastman and Iyer [5] show that people of younger cognitive age use the Internet more often than seniors who do not perceive themselves as younger than their chronological age. Hong, Lui [4] elaborate upon cognitive age and suggest that since cognitively younger and older individuals show strong differences, variances may be better explained by applying the concept of cognitive age instead of the classic chronological age. In the literature on technology acceptance and use, an overwhelming number of articles only rely on the chronological age [4].

2.2 Innovative Use of IT

The construct "trying to innovate in IT," defined as finding new ways of using the existing IT, has gained attention in the context of working environment and is grounded in the theory of trying [9]. To gain a better understanding of post-adoptive behavior, studies focus on how individuals find novel uses of IT. Most of the existing research is in an organizational context [9, 10]; the private sector has not been considered much.

To understand how new ways of using IT emerge, post-adoptive IT-behavior has been studied in various contexts. IS research proposes different kinds of post-adoptive use to understand differences in the usage and ultimately in the outcome of IT use [11–13]. The exploratory behavior of individuals has been shown to be a promoter for enhanced system use, through which people find new ways of using IT [10]. IT-related traits such as computer self-efficacy (CSE), computer anxiety (CA), and personal innovativeness in IT (PIIT) have been shown to explain technology-related behavior [14, 15]. CSE, which refers to whether individuals think they are capable of operating the existing IT, is a key component of behavioral control [16]. CA refers to the stress or fear faced by people when confronted with computers or IT and explains phenomena such as "computer phobia" or "techno stress," which can negatively affect performance and outcome [17]. PIIT describes a personal trait—namely, the willingness to try out new IT, which is rooted in innovation diffusion theory [18].

3   Research Model

3.1   Hypotheses Development

Based on the literature review, we use the post-adoption perspective to evaluate whether cognitive age moderates the key traits of CSE, CA, and PIIT in the context of innovation attempts (Fig. 1).

Trying to innovate, the dependent variable, refers to the individual's willingness to try out new ways of using the existing IT [9]. In this context, we want elderly people to use the Internet for health information, since studies show that younger segments use the Internet more often in this regard. Here, we focus on seniors who have capabilities with and access to the Internet. All seniors included in the study operate computers or smartphones/tablets.

PIIT describes a personal trait—namely, the willingness to try out new IT [19]. Unlike attempt to innovate, PIIT involves exploring new IT or innovations and therefore can also be associated with early adopters [20]. Due to the exploratory characteristic of this trade, we assume that people who like to risk using new IT are also likely to risk using the existing IT in different ways and therefore innovate in their usage behavior. On the other hand, individuals who report lower levels of PIIT have less risk tolerance and show less confidence regarding technology [21]. Thus, we hypothesize:

H1: *PIIT positively influences trying to innovate.*

Furthermore, Davis and Mun [22] propose that PIIT influences CSE, since early adopters show confidence in their capabilities when using new IT. Therefore:

H2: *PIIT positively influences CSE.*

CSE refers to an individual's own judgement about the extent to which he or she can use computers or IT, as already described in the background. Self-efficacy is a key regulating mechanism in human behavior [16]. Individuals with high CSE develop a positive attitude toward IT. Positive attitude, in turn, can encourage deeper exploration of IT [22]. Therefore, we conclude a positive effect on attempt to innovate.

H3: *CSE positively influences trying to innovate.*

CA is the feeling of fear or uneasiness when using a computer or IT, due to the possibility of making mistakes [17]. Individuals with high levels of CA tend to have negative attitudes toward IT. Furthermore, the relationship with CSE is reported to be negative in several studies [23–25]. Due to these negative relationships, we hypothesize:

H4: *CA negatively influences to innovate.*
H5: *CA negatively influences CSE.*

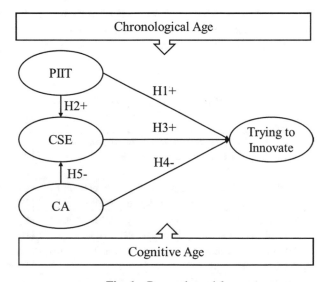

Fig. 1. Research model

3.2 Moderating Effects

To answer the stated research question regarding the different effects of physical and chronological age, the calculations need to be carried out with two different age constructs. Thus, following the research by Hong, Lui [4], we separate the analysis into two groups; one comprising seniors who view themselves as younger than they are, and another comprising seniors who perceive their age to match their chronological age. The items for the cognitive age are averaged and a cognitive age score is determined. The model is calculated using the two data sets in order to evaluate differences in the behavior of the respondents.

4 Data Collection

To test our research model, we conducted an interview-based quantitative study in the US, targeting elderly people of 60+ years. Data was gathered at public places such as senior citizen centers, adult schools, and pedestrian zones. The questionnaire was delivered in person so that participants' questions could be addressed directly and assistance be given if needed. After data cleaning, 120 records from the initial 249 records could be used for data analysis. Records were dropped if they had too much missing data.

Of the participants, 35.83% were male and 64.17% female. The chronological age distribution across the sample is relatively even; thus, no specific age group among the seniors has a stronger impact.

5 Methodology

The questionnaire was constructed using established measurements which are shown in Table 1. All items were measured reflectively on a seven-point Likert scale, except for cognitive age, where 10 year intervals were used, as suggested by Barak and Schiffman [8].

Table 1. Measurement

Construct	Items	Source
Trying to Innovate	2 items; e.g., 'I try to find new uses of computer-technology for tasks important to me'	Ahuja and Thatcher [9] (adapted)
Computer Self-Efficacy	3 items; e.g., 'I could use new computer- technology if there was no one around to tell me what to do'	Compeau and Higgins [23]
Computer Anxiety	4 items; e.g. 'I hesitate to use computer technology for fear of making mistakes I cannot correct'	Compeau, Higgins [26]
Personal Innovativeness	3 items; e.g. 'Among my peers, I am usually the first to try out new computer-technologies'	Thatcher and Perrewe [14]
Cognitive Age	4 items; e.g. 'I feel as though I am in my … 20s; 30s; 40s; 50s; 60s; 70s; 80s	Barak and Schiffman [8]

The items for the cognitive age refer to age intervals (10 years). For the classification regarding whether seniors see themselves as younger or not, their real age is also coded into the range groups of cognitive age. A 67-year-old senior therefore belongs to the "60 s" group. After that, the mean value of the four items is calculated and compared to the chronological age group. All participants who perceived themselves in a younger age group are classified as "younger," whereas everyone else is classified as "same." This way, the sample is split in the subgroups of 74 "younger" seniors (cognitive age < chronological age) and 56 "same" seniors (cognitive age = chronological age).

Based on the data, we cannot identify a single participant who viewed herself/himself as older than his/her actual age.

6 Results

6.1 Measurement Model

To evaluate the data, we perform structural equation modelling (SEM) using the software SmartPLS 3.2.7 [27].

Cronbach's Alpha and composite reliability (CR) are used to assess internal consistency. Cronbach's Alpha is a conservative criterion that underestimates the internal consistency, while composite reliability takes into account the different outer loadings and generally overestimates the internal consistency. Therefore, both can be taken as boundaries for the consistency, since the true consistency should lie within these boundaries. Values above 0.70 are desirable [28].

Table 2. Cronbach's alpha, composite reliability, average variance extracted, Fornell–Larcker criterion

	Cronbach's alpha	CR	AVE	1. TTI	2. PIIT	3. CSE	4. CA
1. Trying to Innovate	0.928	0.965	0.933	0.966			
2. PIIT	0.954	0.970	0.916	0.647	0.957		
3. CSE	0.807	0.886	0.723	0.512	0.593	0.850	
4. CA	0.936	0.955	0.840	−0.435	−0.485	−0.382	0.917

Convergent validity is assessed using the average variance extracted (AVE) and by observing the factor loadings. For the AVE, values of above 0.5 indicate convergent validity [28], which is the case seen in Table 2. In addition, the factor loadings are all above the threshold of 0.708 [28] and significant at the 0.001 level. Based on these criteria, we conclude convergent validity for this research model.

Discriminant validity is assessed by observing two criteria. Fornell–Larcker criterion supports discriminant validity. In addition, cross-loadings are observed to verify that each item correlates the most with its respective construct. Based on these criteria, we conclude discriminant validity. Furthermore, all quality criteria were applied to the subgroups of the study and met all thresholds.

6.2 Structural Model

The following Fig. 2 depicts the results of the structural equation modelling for the two groups of seniors—those who perceive themselves as younger (bold) and those for whom cognitive age fits chronological age (not bold).

Results show that influences differ, depending on the group. For the group of seniors who perceive themselves as younger, PIIT shows a significant influence on attempt to innovate ($\beta = 0.514$, $p < 0.001$) and on CSE ($\beta = 0.519$, $p < 0.001$).

However, surprisingly, CSE and CA have no significant influence on attempt to innovate, which means that H1 and H2 can be supported, but H3–H5 are not supported in this subgroup. The model explains 46.3% of the variance in attempt to innovate and 34.4% in CSE.

For the group of seniors who perceive themselves to be at the same age as their physical age, PIIT shows a significant influence on attempt to innovate ($\beta = 0.367$, $p < 0.05$) and on CSE ($\beta = 0.519$, $p < 0.001$). Also, CSE shows a significant influence on attempt to innovate ($\beta = 0.375$, $p < 0.01$). CA significantly influences CSE ($\beta = -0.233$, $p < 0.05$) but does not influence attempt to innovate, according to our data. Therefore, H1, H2, H3 and H5 are supported, but H4 cannot be supported. The model explains 48.1% of the variance in attempt to innovate and 42.7% in CSE.

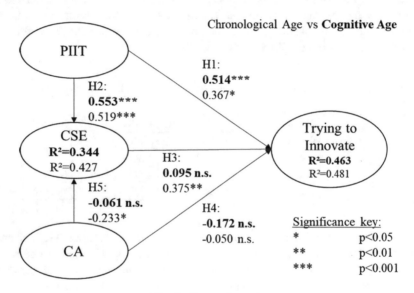

Fig. 2. Research results

7   Limitations and Further Research

Due to splitting the sample, the subgroups became comparatively small, though statistically valid. Also, only older adults in the US were part of the study, so results may differ for other countries. Moreover, the study took place in public places, so seniors who are less outgoing or are bound to their homes could not be reached.

In further research, the sample size should be increased to evaluate whether influences stay insignificant and to elaborate more on the effect of size differences of the two groups.

8 Discussion

As a rather counterintuitive result, CA is not found to be as influential on attempt to innovate in either of the groups or in the complete sample. However, the path coefficient direction (negative) is in line with the results of previous research [23–25]. The insignificance can be explained by the relatively low sample size and also by the rather strong impact of PIIT. Therefore, a definite statement about the influence of CA cannot be made. For both groups, the influence of PIIT on attempt to innovate and on CSE is as expected, as also suggested by the literature [21, 22].

However, the role of CSE shows quite strong differences between the two groups. The seniors of younger cognitive age do not seem to be influenced by their own judgement when they try to find new ways in IT. Only the character trait of personal innovativeness is seen to be a determinant for attempt to innovate. For the group of seniors who do not see themselves to be younger, CSE has a rather strong impact. When comparing the effect sizes, CSE has a medium effect [28] in the "same" group, while it remains almost negligible in the "younger" group. It can be argued that seniors who see themselves as younger may be more open to taking risks compared to the other subgroup. Greater risk-taking may be one of the attributes that seniors take into account when they report their cognitive age.

8.1 Theoretical Implications

This research has several implications for research on post-adoption behavior. Research on older adults is relatively scarce [25] and needs to receive more attention due to the inevitable demographic change in literally all developed societies.

First of all, we contribute to the body of literature on research on older adults. We explicitly do not treat older adults as a homogenous group but want to take into account difference among elderly people. We show that there are subsegments among seniors, which show inherent different characteristics and therefore different usage behavior and outcomes. As the introduction suggests, there are seniors who already make use of IT in the context of online health. We go on to propose that cognitive age is a characteristic that reveals information among seniors, which was hidden before. Due to the differences of influences, further research should take into account the self-perception, especially in the context of seniors, since the difference between cognitive age and chronological age tends to increase with advancing age. The segmentation shows that not all seniors perceive themselves as younger than they are, although the majority does.

Second, the results give evidence that seniors of younger cognitive age are not influenced by their own judgement to use computers. When trying to innovate, the only determinant shows to be PIIT. Therefore, the only influence to try out new ways in using IT is reflected by a character trait to explore new things, such as the adoption of new technology. For this case we showed that attempt to innovate is not a matter of the perception of own capabilities or confidence, but more rooted in the general motivation to explore new things and the willingness to take risk.

8.2 Practical Implications

This research has various practical implications. With the further penetration of the Internet, also seniors have more and more access to all kinds of information. Especially health-related information may benefit elderly people the most. The problem lies within the senior group which tends to see themselves as not younger than they are. Due to doubts in their capabilities, they do not try out new ways of using IT. Therefore, seniors should be introduced to many more innovations, so that they can adapt to them and use them more often. This fosters capabilities, which in return may be the origin of their feeling younger due to the use of new technology. Elderly people may refrain from using new technology at first, but the task is to show them the benefits of IT. Due to this, seniors may be more open to new IT and try to innovate their usage behavior.

References

1. Wei, K.-K., et al.: Conceptualizing and testing a social cognitive model of the digital divide. Inf. Syst. Res. **22**(1), 170–187 (2011)
2. Nimrod, G.: Older audiences in the digital media environment. Inf. Commun. Soc. **20**(2), 1–17 (2017)
3. Wilkes, R.E.: A structural modeling approach to the measurement and meaning of cognitive age. J. Consum. Res. **19**, 292–301 (1992)
4. Hong, S.-J., et al.: How old are you really? Cognitive age in technology acceptance. Decis. Support Syst. **56**, 122–130 (2013)
5. Eastman, J.K., Iyer, R.: The impact of cognitive age on Internet use of the elderly: an introduction to the public policy implications. Int. J. Consum. Stud. **29**(2), 125–136 (2005)
6. Lindberg, T., Näsänen, R., Müller, K.: How age affects the speed of perception of computer icons. Displays **27**(4–5), 170–177 (2006)
7. Schiffman, L.G., Sherman, E.: Value orientations of new-age elderly: the coming of an ageless market. J. Bus. Res. **22**(2), 187–194 (1991)
8. Barak, B., Schiffman, L.G.: Cognitive age: a nonchronological age variable. Adv. Consum. Res. **8**(1), 602–606 (1981)
9. Ahuja, M.K., Thatcher, J.B.: Moving beyond intentions and toward the theory of trying: effects of work environment and gender on post-adoption information technology use. MIS Q. **29**(3), 427–459 (2005)
10. Liang, H., et al.: Employees' exploration of complex systems: an integrative view. J. Manag. Inf. Syst. **32**(1), 322–357 (2015)
11. Jasperson, J.S., Carter, P.E., Zmud, R.W.: A comprehensive conceputalization of the post-adoptive behaviors associated with IT-enabled work systems. MIS Q. **29**(3), 525–557 (2005)
12. Burton-Jones, A., Straub, D.W.J.: Reconceptualizing system usage: an approach and empirical test. Inf. Syst. Res. **17**(3), 228–246 (2006)
13. Bagayogo, F.F., Lapointe, L., Bassellier, G.: Enhanced use of IT: a new perspective on post-adoption. J. Assoc. Inf. Syst. **15**(7), 361–387 (2014)
14. Thatcher, J.B., Perrewe, P.L.: An empirical examination of individual traits as antecedents to computer anxiety and computer self-efficacy. MIS Q. **26**(4), 381–396 (2002)
15. Maier, C., et al.: Using user personality to explain the intention-behavior gap changes in beliefs: a longitudinal analysis. In: International Conference on Information Systems (2012)
16. Bandura, A.: Self-Efficacy: The Exercise of Control. Freeman, New York (1997)

17. Heinssen, R.K., Glass, C.R., Knight, L.A.: Assessing computer anxiety: development and validation of the computer anxiety rating scale. Comput. Hum. Behav. **3**, 49–59 (1987)
18. Rogers, E.M.: The Diffusion of Innovation. Free Press, New York (2003)
19. Agarwal, R., Prasad, J.: A conceptual and operational definition of personal innovativeness in the domain of information technology. Inf. Syst. Res. **9**(2), 204–215 (1998)
20. Agarwal, R., Prasad, J.: Are individual differences germane to the acceptance of new information technologies? Decis. Sci. **30**(2), 361–391 (1999)
21. Chou, S.-W., Chen, P.-Y.: The influence of individual differences on continuance intentions of enterprise resource planning (ERP). Int. J. Hum. Comput. Stud. **67**(6), 484–496 (2009)
22. Davis, J.M., Mun, Y.Y.: User disposition and extent of web utilization: a trait hierarchy approach. Int. J. Hum. Comput. Stud. **70**(5), 346–363 (2012)
23. Compeau, D.R., Higgins, C.A.: Computer self-efficacy: development of a measure and initial test. MIS Q. **19**(2), 189–211 (1995)
24. Marakas, G.M., Yi, M.Y., Johnson, R.D.: The multilevel and multifaceted character of computer self-efficacy: toward clarification of the construct and an integrative framework for research. Inf. Syst. Res. **9**(2), 126–163 (1998)
25. Wagner, N., Hassanein, K., Head, M.: Computer use by older adults: a multi-disciplinary review. Comput. Hum. Behav. **26**(5), 870–882 (2010)
26. Compeau, D., Higgins, C.A., Huff, S.: Social cognitive theory and individual reactions to computing technology: a longitudinal study. MIS Q. **23**(2), 145–158 (1999)
27. Ringle, C.M., Wende, S., Becker, J.-M.: SmartPLS, vol. 3. SmartPLS GmbH, Boenningstedt (2015)
28. Hair, J.F.J., et al.: A Primer on Partial Least Squares Structural Equation Modeling (PLS-SEM). Sage Publications, Thousand Oaks (2013)

Virtual Environment Application that Complements the Treatment of Dyslexia (VEATD) in Children

Jorge Buele[1,2(✉)] , Victoria M. López[2] , José Varela-Aldás[1] ,
Angel Soria[3] , and Guillermo Palacios-Navarro[4]

[1] SISAu Research Group, Universidad Tecnológica Indoamérica,
180212 Ambato, Ecuador
{jorgebuele,josevarela}@uti.edu.ec
[2] Universidad de las Fuerzas Armadas ESPE, 050104 Latacunga, Ecuador
vmlopez2@espe.edu.ec
[3] Purdue University, Lafayette 47907, USA
asoriach@purdue.edu
[4] University of Zaragoza, 50009 Zaragoza, Spain
guillermo.palacios@unizar.es

Abstract. The educational disorders that children present at an early age can cause them to not fully develop throughout their lives. In this research work a 3D virtual system that allows the child who has been diagnosed with dyslexia to complement the exercises performed in a conventional therapy is described. To achieve this an application was developed, the app consists of two games (each with three levels of difficulty), and that are part of the rehabilitation program. In each of these games virtual objects are combined with auditory messages to provide the user with an immersive experience, and to train more than one sense at a time. In the first game task, the activity asks the children to correctly locate the syllables that compose a word and for the second activity the children will listen to a word, after the games asks the children to select the correct word. This tool has been tested by a group of children (eight), with ages ranging from 8 to 12 years old, whose development can be supervised at home by their parents, since it is an intuitive and easy to use interface. The results obtained are stored in a database and in this way the medical specialist can monitor the progress of the child throughout his treatment. For the validation of this proposal the SUS usability test was used.

Keywords: Dyslexia · Educational rehabilitation · Training · Virtual reality

1 Introduction

The constant advance of science allows the human being to look for new ways to improve their quality of life, through the development of technological tools [1–3]. These tools are used in various fields such as industry, military combat, entertainment, medicine, and education among others [4, 5]. Allowing people to adapt to technology as an everyday part of their lives, regardless of their sex or age. The development of applications showed an accelerated growth. In order to, meet the needs and demands of

© Springer Nature Switzerland AG 2020
Á. Rocha et al. (Eds.): ICITS 2020, AISC 1137, pp. 330–339, 2020.
https://doi.org/10.1007/978-3-030-40690-5_33

the users. A clear example is the medical field where devices can preserve the lives of patients [6]. Acting as elements of assistive technology, and providing greater autonomy for people with disabilities [7, 8].

When talking about disability, the physical disabilities should not be the only ones taken into account [9]. There are other types of disabilities that affect a person emotionally and psychologically [10]. The population of the early age requires greater understanding and emotional support during this stage when they are building their identity and self-esteem. One of the most common disorders in children when starting their academic training is dyslexia [11]. Dyslexia is a specific learning disability that has a neurobiological origin. It is characterized by difficulties with the precise and/or fluid recognition of words and by their poor spelling and decoding ability [12, 13]. These difficulties are usually the result of a deficit in the phonological component of language, in relation to other cognitive skills and the provision of effective classroom instruction. This language learning disability affects one in five people in the world and can produce anxiety, when imagining situations of mockery or social rejection. It is also associated with the difficulty to distinguish words that have similar sounds and in memorization activities [14].

Cases of dyslexia occur in all known languages, although it differs according to the spelling, which triggers labor problems as well. That is why identifying it at an early age, and carrying out an adequate therapeutic process is a high priority. In [15] we describe a cognitive behavioral therapy based on mindfulness called Mindfulness Based Rehabilitation of Reading, Attention & Memory (MBR-RAM ©). It uses techniques of visual meditation to improve attention focusing on formats assisted by therapists, and practice at home to rehabilitate reading deficits, lack of visual attention, lack of visual motor coordination and visual memory. The proposal was tested in 3 children between 8 and 10 years old, obtaining significant clinical and statistical improvements after a trial period of 6 months. In [16] several technological proposals that seek to remedy the reading difficulties of children who have this disability are described. "Friendly" features are incorporated for the user, such as altering the size and format of the text and converting text to speech. The main tools currently available are virtual reality (VR) and interactive systems [17, 18]. Therefore, in [19] a virtual psychometric tool for the rehabilitation of patients with dyslexia is presented. Such tool based on the Nintendo Wii video game system. The results show that it is a tool that improves attention, but it has no immediate effect on reading performance, suggesting a longer protocol as future work.

This work is composed of 6 sections; Sect. 1 describes the introduction and the works related to the subject. Section 2 shows the treatment observations and Sect. 3 the multilayer design. The development of the applications is presented in Sect. 4; the tests and results are presented in Sect. 5 and finally, the conclusions shown in Sect. 6.

2 Treatment Observations

As a previous stage, the conventional treatment received by children suffering from dyslexia has been supervised. The observation was carried out in a foundation, a center that provides comprehensive care to people with low economic resources and who possess some type of physical, psychological or mental syndrome. As a basis to develop the applications, two children of different genders that have this educational disorder and are within the age range have been observed. The girl had a higher degree of the disease that was associated with the presence of dysgraphia. For the case of the older child, he had already been part of a rehabilitation program for a period of almost a year and presented an improvement on his condition. This observation stage was carried out for 4 weeks, for 2 days each. The medical center has a great demand due to its reduced costs. It does not have advanced technology and the treatment sessions last only 30 min, which directly affects the patients.

In addition, the repeatability of the tasks and activities carried out and the limited resources that they possess generate another barrier that hinders the full development of the children. This should be complemented at home, where parents motivate their children to perform additional exercises suggested by the specialists. Based on the literature review it has been noted that the use of technological tools such as tablets increases the interest and concentration of children.

Therefore, with the help of speech therapists, objectives that must be met in the applications have been determined, such as:

- Develop an interactive application that establishes levels of difficulty and progress, given that this syndrome is not found in all people in the same percentage. In addition, it must be intuitive, so that a relative that does not possess specialized knowledge can perform it.
- Provide a complementary tool to conventional sessions that allow parents to encourage and promote better results.
- Generate a report of the results obtained by the patient, in order to perform an effective control of the progress achieved.

3   Multilayer Design

The design of this virtual system was made based on the multilayer diagram shown in Fig. 1 and is described below:

(i)   This layer defines the resources and materials used in the design process, which starts from the sociocultural theory of Lev Vygotsky. This theory suggests that children develop their learning capabilities through social interaction. They acquire new and better cognitive abilities as a logical process of immersing themselves in a way of life. For this reason, designs of objects and colors that the child experiences in his usual environment were used.

(ii)   The design of some 3D objects that are in the virtual environment have been developed in the 3DS MAX software, which presents a wide compatibility with the VR programs.

(iii)   The Unity 3D software is used for the development of the virtual environment, which incorporates several tools. For the implementation of some objects assets of this software framework were used to give a greater realism to the application. Each virtual object was assigned the respective properties according to the function that must be fulfilled within the virtual environment. In addition, textures, colors and depth are added according to the requirements of each of the games, finally the rendering process is performed. All games incorporate sound effects to complement the immersive experience of the patient.

(iv) The control of video games in the virtual environment is done through the programming scripts that each interface has. In this way the application is executed in a hierarchical and organized manner, achieving the objectives set out above. The scripts contain the add-ons that enable control of the virtual environment and sounds as system output. Based on the input data provided by the user, there are buttons that when pressed execute a script, and each game starts with its respective elements. Allowing the functions to be managed in an organized way to link the different environments and access each one. When the game starts, these scripts activate the corresponding audio source to inform the purpose of the activity to be performed to the player.

(v) This layer defines the resources and materials used in the design process, which starts from the sociocultural theory of Lev Vygotsky. This theory suggests that children develop their learning capabilities through social interaction. They acquire new and better cognitive abilities as a logical process of immersing themselves in a way of life. For this reason, designs of objects and colors that the child experiences in his usual environment were used.

Fig. 1. General diagram of the implemented system.

4 Development of the Applications

The virtual environment was designed using the Unity 3D graphic engine, the two video games were developed under the supervision of a specialist doctor. It should be noted that dyslexia affect people in different degrees, and therefore the symptoms may differ from one child to another. The objective of this application is to interact with patients, generate rehabilitation, and improve their language skills, writing, spelling and reading comprehension. As a first step for the use of the application the patient must register their data: name, surname and age, once the system has the data the video game is chosen, and after this the patient or the person in charge of the rehabilitation must choose the degree of difficulty for the video game.

As shown in Fig. 2, the first video game presents an image and the syllables that are part of the word that describes the image. To complement the patient's experience and facilitate their learning, the patient can hear the word associated with the image. The person who is performing the exercise has to order the syllables in a correct way to obtain a virtual score, and if the patient fails, the patient can try again with a decrease in the score points. In video game 2 an image is presented, and the patient must listen to the sound, and choose the word that corresponds to the indicated image, with this the user will develop auditory and visual memory. Since the patient must choose between two words that have a similar sound, but different meaning.

Fig. 2. Design of the video games.

When a new patient is registered in the application a database is generated, which records the user's data, the degree of difficulty chosen and the number of correct answers and errors during the entire session. If the patient has frequent use of the system, them it is not recorded as the entry of a new patient, but the new data is added to the user's record.

In the database the specialist or the person in charge of the rehabilitation can enter with his password to observe the progress that the children achieved during the treatment. The database indicates the number of correct answers and errors that the patient has during each session enabling the evaluation of the progress.

5 Tests and Results

5.1 Tests

The participants were subjected to sessions of 60 min, 3 times a week, coinciding with the days that they did not carry out their conventional therapy, with the intention of covering in full the weekdays of a week. During the first stages of the prototype development, user tests began as part of a participatory development. The first week the tests were carried out in a foundation with the presence of the therapists, so that the parents learned how to guide their children during the execution of the games. The initial tests were intended to determine the effectiveness of the exercises in each game. Observing the way in which the children reacted, and to verify if the mechanics of the game were adjusted to the desired age. Regarding the application as a whole, it is important to observe the adequacy of navigability, the narrative and the aesthetic choice. When the need for a change was identified, the prototype was modified and tested again as explained above.

During the tests, the diagram of the use mode presented in Fig. 3 was used as a basis; the games were fairly simple until the child became familiar with the developed interface. Therefore, when the difficult level was low, and the execution of shorter duration, the feedback on failure and success was faster. Thus, the child had greater autonomy and learned the game with ease. However, by increasing the complexity of the games, the tasks were more extensive and the child had difficulties, for which the speech therapist gave advice to overcome these barriers. In the Fig. 4 the execution of both video games with a low complexity are presented.

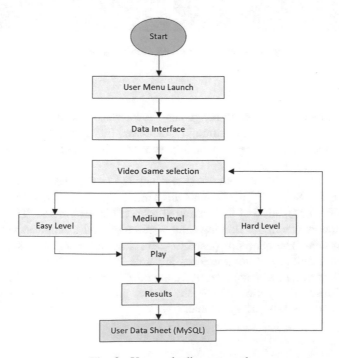

Fig. 3. Use mode diagram used.

It could be denoted that, the motivation to win virtual rewards had results. Because the children were motivated to play until they got the reward they wanted. Games with greater complexity represented a challenging activity. Even when there were difficulties the child repeated the activity several times to overcome it. Regarding the system that includes the set of screens, its operation and ease of use for the user had an important acceptance.

Fig. 4. Tests performed by children in real time.

5.2   Results

After completing the execution of the games, the interface has an option that allows the visualization of data in real time through a database, Fig. 5 displays its contents. This report details the name of the patient and the score obtained, so the specialist can verify if there was an improvement.

In addition, the SUS evaluation test was applied to the users, also known as the System Usability Scale. By performing this simple test, whose questions are predefined, the user's acceptance of the submitted proposal is determined. This information is of great importance in determining safety, sensation and discomfort when using an interactive system. The questions posed to users about the interactive system, and the results of the usability test are presented in Table 1. The result of the SUS test performed by eight users (patients) after using the interactive system is: (82.5 ± 0.61). As the obtained result is superior to 68%, the system presents an acceptable usability for the treatment of this learning syndrome.

Table 1. SUS test.

Question	Result (N = 8)	
	Mean	SD
1. I think I would like to use this Virtual Reality system frequently	4	0.63
2. I found the Virtual Reality system to be unnecessarily complex	2	0.63
3. I thought that the Virtual Reality system was easy to use	4.2	0.75
4. I think I would need the support of a technical person to be able to use this Virtual Reality	2.6	0.48
5. I found that the various functions in this Virtual Reality System were well integrated	4.4	0.8
6. I thought there was too much inconsistency in this Virtual Reality system	1.8	0.4
7. I would imagine that most people would learn to use this Virtual Reality system very quickly	4.6	0.49
8. I found the Virtual Reality system to be very cumbersome to use	1.8	0.75
9. I felt very confident using the Virtual Reality system	4.6	0.49
10. I needed to learn a lot of things before I could get going with this Virtual Reality system	2.2	0.75
Global score (total)	82.5	0.61

Nombre	Edad	Fecha	No. aciertos	No. errores	Tiempo	Videojuego
Danna Paredes	8	10-07-2019	1	3	120	Videojuego1
Danna Paredes	8	10-07-2019	2	2	120	Videojuego2
Isaac López	10	12-07-2019	2	3	180	Videojuego1
Isaac López	10	12-07-2019	2	3	180	Videojuego2
Juan León	12	12-07-2019	1	3	180	Videojuego1
Juan León	12	12-07-2019	2	2	180	Videojuego2
Danna Paredes	8	20-07-2019	2	3	120	Videojuego1
Danna Paredes	8	20-07-2019	2	2	120	Videojuego2
Juan León	12	22-07-2019	1	2	180	Videojuego1
Juan León	12	22-07-2019	2	4	180	Videojuego2
Juan León	12	20-07-2019	3	1	180	Videojuego1
Juan León	12	20-07-2019	3	2	180	Videojuego2

Fig. 5. Database created in MySQL.

6 Conclusions

The design of appropriate interfaces for infants requires a prior process, in which constant feedback and experimental tests can deliver a product of quality and satisfaction for the user. Based on the design approach, certain needs and particular characteristics can be identified. Those must be taken into account in order to provide more personalized solutions. Providing the parent with the opportunity to guide the rehabilitation sessions, and complement the conventional therapy received in a specialized center demonstrates the validity of this proposal. In addition, this work shows that the introduction of virtual rewards motivates children more by challenging their abilities, and to meet the goals proposed throughout the games, which increase in difficulty as the level advances.

For a better feedback of the satisfaction of the users with the implemented proposal, the SUS test was applied, which have the following result 82.5 ± 0.61. Which denotes that it has represented a valid tool that does not seek to eclipse the work of an educational therapist, but rather to complement its work, in the search for better results. In this way, it is a future work to evaluate from the medical point of view the contribution of this research work in the long-term improvement of patients.

References

1. Bennett, S., Agostinho, S., Lockyer, L.: Technology tools to support learning design: implications derived from an investigation of university teachers' design practices. Comput. Educ. (2015). https://doi.org/10.1016/j.compedu.2014.10.016
2. Sharma, A., Khosla, A., Khosla, M., Yogeswara Rao, M.: Technological tools and interventions to enhance learning in children with autism. In: Supporting the Education of Children with Autism Spectrum Disorders (2016). https://doi.org/10.4018/978-1-5225-0816-8.ch011
3. Waight, N., Abd-El-Khalick, F.: Technology, culture, and values: implications for enactment of technological tools in precollege science classrooms (2018). https://doi.org/10.1007/978-3-319-66659-4_7
4. Kuoppamäki, S.M., Taipale, S., Wilska, T.A.: The use of mobile technology for online shopping and entertainment among older adults in Finland. Telematics Inform. (2017). https://doi.org/10.1016/j.tele.2017.01.005
5. Cheung, T.M.: Innovation in China's defense technology base: foreign technology and military capabilities. J. Strateg. Stud. (2016). https://doi.org/10.1080/01402390.2016.1208612
6. Varela-Aldás, J., Palacios-Navarro, G., García-Magariño, I.: Immersive virtual reality app for mild cognitive impairment. Rev. Ibérica Sist. e Tecnol. Informação. **E19**, 278–290 (2019)
7. Salazar, F.W., Núñez, F., Buele, J., Jordán, E.P., Barberán, J.: Design of an ergonomic prototype for physical rehabilitation of people with paraplegia, 28 October 2020. https://doi.org/10.1007/978-3-030-33614-1_23
8. Chicaiza, F.A., Lema-Cerda, L., Marcelo Álvarez, V., Andaluz, V.H., Varela-Aldás, J., Palacios-Navarro, G., García-Magariño, I.: Virtual reality-based memory assistant for the elderly. In: Lecture Notes in Computer Science (including subseries Lecture Notes in Artificial Intelligence and Lecture Notes in Bioinformatics) (2018). https://doi.org/10.1007/978-3-319-95270-3_23
9. Andrea Sánchez, Z., Santiago Alvarez, T., Roberto Segura, F., TomásÂ Núñez, C., Urrutia-Urrutia, P., Franklin Salazar, L., Altamirano, S., Buele, J.: Virtual rehabilitation system using electromyographic sensors for strengthening upper extremities. In: Smart Innovation, Systems and Technologies (2020). https://doi.org/10.1007/978-981-13-9155-2_19
10. Tophoven, S., Reims, N., Tisch, A.: Vocational rehabilitation of young adults with psychological disabilities. J. Occup. Rehabil. (2019). https://doi.org/10.1007/s10926-018-9773-y

11. Frith, U.: Beneath the surface of developmental dyslexia. In: Surface Dyslexia: Neuropsy-chological and Cognitive Studies of Phonological Reading (2017). https://doi.org/10.4324/9781315108346

12. D'mello, A.M., Gabrieli, J.D.E.: Cognitive neuroscience of dyslexia. Lang. Speech Hear. Serv. Sch. (2018). https://doi.org/10.1044/2018_LSHSS-DYSLC-18-0020

13. Snowling, M.J., Melby-Lervåg, M.: Oral language deficits in familial dyslexia: a meta-analysis and review. Psychol. Bull. (2016). https://doi.org/10.1037/bul0000037

14. Buele, J., López, V.M., Franklin Salazar, L., Edisson, J.H., Reinoso, C., Carrillo, S., Soria, A., Andrango, R., Urrutia-Urrutia, P.: Interactive system to improve the skills of children with dyslexia: a preliminary study. In: Smart Innovation, Systems and Technologies (2020). https://doi.org/10.1007/978-981-13-9155-2_35

15. Pradhan, B., Parikh, T., Sahoo, M., Selznick, R., Goodman, M.: Current understanding of dyslexia and pilot data on efficacy of a mindfulness based psychotherapy (MBR-RAM) model. Adolesc. Psychiatry (Hilversum) (2017). https://doi.org/10.2174/2210676607666170607160400

16. Caute, A., Cruice, M., Marshall, J., Monnelly, K., Wilson, S., Woolf, C.: Assistive technology approaches to reading therapy for people with acquired dyslexia. Aphasiology (2018). https://doi.org/10.1080/02687038.2018.1489119

17. Ali, M., Han, S.C., Bilal, H.S.M., Lee, S., Kang, M.J.Y., Kang, B.H., Razzaq, M.A., Amin, M.B.: iCBLS: an interactive case-based learning system for medical education. Int. J. Med. Inform. (2018). https://doi.org/10.1016/j.ijmedinf.2017.11.004

18. Renganathan, S.M., Stewart, C., Perez, A., Rao, R., Braaten, B.: Preliminary results on an interactive learning tool for early algebra education. In: Proceedings of the Frontiers in Education Conference, FIE (2017). https://doi.org/10.1109/FIE.2017.8190594

19. Pedroli, E., Padula, P., Guala, A., Meardi, M.T., Riva, G., Albani, G.: A psychometric tool for a virtual reality rehabilitation approach for dyslexia. Comput. Math. Methods Med. (2017). https://doi.org/10.1155/2017/7048676

Assistive Technological Tools to Strengthen Interaction, Communication and Learning in Children with Different Abilities

Jácome-Amores Ligia[1(⊠)],
Amaluisa Rendón Paulina Magaly[1(⊠)],
Sánchez Sánchez Richard Patricio[2(⊠)],
and Sánchez Sánchez Paulina Elizabeth[1(⊠)]

[1] Universidad Tecnológica Indoamérica, 180103 Ambato, Ecuador
{ligiajacome,paulinasanchez}@uti.edu.ec,
pauli_ar4@hotmail.com
[2] 180103 Ambato, Ecuador

Abstract. This document first presents an overview of the current situation of children with special educational needs in the province of Tungurahua, Ecuador and the development of assistive technology tools (ATT) in response to the particular needs in the interaction, communication, and learning of these children. It also explains about the implemented tools: TEVI (Virtual Keyboard) with natural language processing; to strengthen interaction and communication in children with language problems because of their motor disability. HETI@DOWN, a playful software to help develop cognitive skills in down children. GAMSEMAT, a serious adaptive game developed to strengthen mathematical learning in children with intellectual disabilities. ALBITIC, a repository of learning objects for preschoolers, based on a children's story. Finally, it shows the results of the investigations where these were applied; evidencing a significant contribution in interaction, communication and learning in children with different abilities.

Keywords: Assistive technology · Playful learning · Interaction and communication

1 Introduction

A child with an intellectual disability brings frustration for not being able to normally develop social and educational activities on a daily basis. Assistive technology is vital to help children with disabilities in order for them to develop the skills necessary to solve the problems themselves. According to UNICEF, assistive technology is one of the key elements to promote the inclusion of children with disabilities and can be essential for their development and health, as well as for participation in various stages of their lives. These include communication, mobility, personal care, housework, family relationships, education and participation in play and recreation. Assistive technology can improve the quality of life of children but also of their families [1].

Information and Communication Technologies (ICTs) in general and Assistive Technological Tools in particular, to promote educational inclusion, accessibility,

© Springer Nature Switzerland AG 2020
Á. Rocha et al. (Eds.): ICITS 2020, AISC 1137, pp. 340–350, 2020.
https://doi.org/10.1007/978-3-030-40690-5_34

communication and learning; making the participation of boys and girls with disabilities a reality; fulfilling their rights and changing their lives; Obtaining personal assistance, natural language interpreters, augmentative and alternative communication (AAC) or serious games for helping to develop cognitive skills not only at school but also at home as an inclusive environment for children with disabilities should start at home [2, 3].

This article is prearranged in the following manner: First, in the background section, the problem in Ecuador is analyzed. Then, some research and work related to the integration of assistive technology in special education inclusion are quoted. Afterwards, the methodology that was used in the investigation is explained. Next, the characterization and architectures of the four help tools developed are detailed. Then, the developed assistance technologies evaluation results are reported in the contexts and realities addressed in the Province of Tungurahua and finally, the conclusions and future works are presented.

2 Background

According to data from the World Health Organization (WHO) and the World Bank (WB), of October 2018 about 1000 million dwellers who represent 15% of the world population, live with some type of disability. In Ecuador 460,586 habitants have registered some type of disabled; the most frequent being physical disability with 46.65%, followed by intellectual disability with 22.35%. The male population has a higher percentage of disability with 56.17% [4].

On the other hand, although the Organic Law of Intercultural Education in art. 47 states: "The Ecuadorian State will guarantee the inclusion and integration of persons with disabilities by removing the barriers of their learning", so that regular education institutions are obliged to receive students with disabilities by creating physical and curricular adaptations [5, 16].

Nevertheless, the teachers of these regular institutions are not prepared to work with students with different abilities at the same time, since these students require adaptations in the curriculum, methodologies, of didactic and technological material according to their type of disability. In addition, few students have been included in regular education centers and the vast majority continues to attend special education centers, still feeling discriminated or marginalized.

This fact has provided the opportunity to develop four alternative technologies that help children, teachers and parents to strengthen the communication, interaction and learning of young children because of the different types of disabilities such as physical, intellectual and language for that can develop fully and completely as their right.

3 Method

This article compiles the results of four investigations that took place in the province of Tungurahua. In these tasks, helping software tools were developed regular children, with physical, language and intellectual disabilities; where four educational institutions participated and more than 100 children with these three types of disabilities.

Each of these investigations was completed in three phases: In the first phase, exploratory research was done to know the methods applied in the classroom, the technological needs, the requirements and characterization of the tools to be created. In the second phase, the software tools were developed; and in the third phase, these tools are evaluated in a real context and with representative samples.

3.1   Designing Assistive Technologies

The design of technologies for children with disabilities is highly complicated due to the diversity of needs, information of interest, the scale of the user population concerned, the research philosophy and also the experience of the researchers involved [6]. However, several researchers who have competencies in the area of software development want to contribute with the creation of assistive technology to assist children with disabilities improve their quality of life and go on in their daily and educational activities in the best possible way.

The following describes four assistive technology tools (ATT) developed by a group of collaborators creating interdisciplinary groups involving systems engineers, psychologists, psych pedagogues, teachers, authorities of foundation centers and special education centers; and, parents. These tools were developed and evaluated between 2013 and 2018 (see Fig. 1).

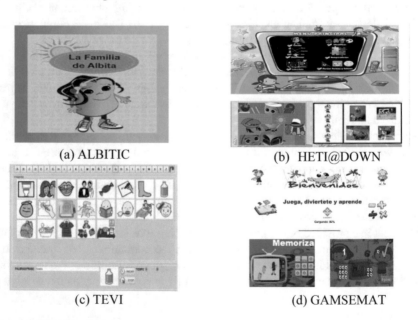

(a) ALBITIC   (b) HETI@DOWN

(c) TEVI   (d) GAMSEMAT

Fig. 1. (a) ALBITIC Cover of the didactic repository based on a youngster's story that works as a repository of recreational activities which served as the basis for the development of (b) HETI@DOWN is an educational software that strengthens the development of cognitive skills of children with Down Syndrome who began their school stage, (c) TEVI is a virtual keyboard with natural language processing to help children with language problems due to their motor disability. (d) GAMSEMAT is a serious game with an adaptive mechanism that teaches these children math with intellectual disability.

ALBITIC. A repository of tangible objects that teach children who are beginning their school stage to develop cognitive skills making possible the development of hearing, retentive, concentration, uptake, reasoning, oral expression, creativity, creativity, affectivity, elements necessary to function in different contexts [7]. The story tells about the family of Albita who is the main character and who recounts the daily activities of his family: Charito his mother, Pepe his father, Panchito his second brother and Sara his younger sister [8]. The story tells about the family of Albita. The 2014 initial education curriculum of the Ministry of Education of Ecuador, defines the emotional and social linkage as the first characterization of the areas of development and learning, trying to improve the emotional aspect of the child in the first instance and the metacognitive development later; from the interpersonal relationship with his family and with his surroundings [5]. Figure 2 shows the architecture of the technological help tool developed.

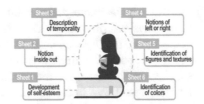

Fig. 2. Architecture of the ATT developed: ALBITIC

HETI@DOWN. By its acronym, Interactive Technological Tool to help develop cognitive skills in children with Down's Syndrome, an interactive educational software developed from the characterization and characters of ALBITIC, to strengthen the development of cognitive skills in children with Down Syndrome (DS) who are starting their school stage. The input module that is composed of the web user interface that shows a navigation menu between the different learning objects. It is important to mention that each of the learning activities consisted of a group of grouped recreational activities; among those that included: a story in two languages (Spanish and Quechua), puzzles, identification of objects, relating, matching by shapes and colors and discrimination through mazes.

The story allows the child to create, know, sensitize, communicate, share, play and solve problems. So a teaching-learning process through the story makes it possible to integrate the development of different areas, in addition to capturing the attention of children [9]. In this module the software provides some registered statistical data of each activity carried out by the child such as the number of attempts, the number of hits and the time it took for the child to assemble the activity, this to help the teacher put together adaptive strategies of learning in the child [10]. Finally, the output module composed of Digital Signal Processing (PSD) and the TTS (Text-To-Speech) voice synthesizer [11]. Figure 3 shows the architecture of the technological help tool developed.

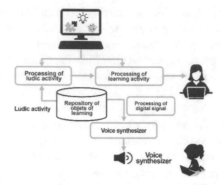

Fig. 3. Architecture of the ATT developed: HETI@DOWN

TEVI. By its acronym Virtual Keyboard, an Augmentative and Alternative Communication System (SAAC), designed to facilitate the communication of children with language problems due to a physical disability. This assistive technology tool (ATT) consists of three modules: first, the input module that is composed of a virtual keyboard, a pictogram panel, and a webcam. Finally, the output module composed of the Natural Language Processing (PLN) component, the Natural Language Generator (GLN) and the voice synthesizer. The TEVI interface has two components, one that adapts to children who do not yet know how to read or write and the information is entered through a pictogram panel, the other is a virtual keyboard with predictive text functions.

The panel interface shows simple and easy to interpret emocards, that is, images or pictograms familiar to the child. Each of them describes words or phrases about mood, desire or an action to be performed, which are useful for more agile communication; for example: 'I am hungry', 'I want to take a shower', 'I want to rest', etc. The predictive text of the interface makes the child's communication faster and more efficient [12], as it anticipates the word or phrase minimizing the number of keys pressed to form it. Figure 4 shows the architecture of the developed technological aid tool.

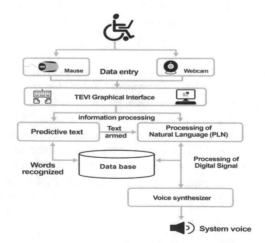

Fig. 4. Architecture of the ATT developed: TEVI

GAMSEMAT. A serious game developed presenting a web environment that consists of a small train, in which a child must help it to reach its destination; loaded with eight wagons, each of which carries a Learning Result (RAP) to develop, which is the new challenge or challenge that the child must face; and that has as learning units the sense of number, counting, addition, subtraction, addition, subtraction, greater-less than; and build basic mathematical operations for himself (where he combines the previous ones) [17, 18].

In its structure, the serious game consists of an adaptive mechanism that ensures it is always entertaining, challenging and sustainable for the child [13, 19]. Figure 5 shows the architecture of the technological help tool developed.

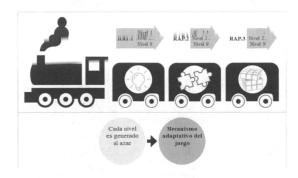

Fig. 5. Architecture of the ATT developed: GAMSEMAT

This mechanism uses Dynamic Difficulty Adaptation (DDA), which is based on the performance or mastery of the game, and also includes adaptive workflow [14]. The system changes the difficulty of the game by feedback without the child noticing [14]. This feedback can be positive or negative and consists of reducing or increasing the difficulty between two related levels according to the domain that has reached the current level [15].

The following Table 1 gives a summary of the usefulness of the ATT developed:

Table 1. Summary of usefulness

Tool	Target	Advantage
ALBITIC	Regular education children	Children who are beginning their school stage can develop cognitive skills making possible the development of hearing, retentive, concentration, uptake, reasoning, oral expression, creativity, affectivity
HETI@DOWN	Children with Down syndrome	Children can practice cognitive and rehabilitation activities while learning to read
TEVI	Children with motor disabilities and language disabilities	Children can communicate with others using a SAC while strengthening their reading and writing skills
GAMESEMAT	Children with intellectual disabilities	Children can learn math through serious adaptive play

3.2   Social Impact

The participants, instrument, procedure and analysis of the experimental phase are described below.

Benefited. The study concerned ordinary children or those with some type of physical, language or intellectual disability with 80% of them with problems of school learning problems; from seven education centers in the province of Tungurahua. The data of the participating children are summarized in Table 2.

Table 2. Data of children benefited from AHT

Tool	Institution	Benefited	Education
TEVI	Ambato special school	75 children with physical and language disabilities	Special
	Maximiliano Spiller special school		
HETI@DOWN	San José de Huambaló Fundation	13 Children with Down syndrome	Special
GAMESEMAT	San José de Huambaló Fundation	25 Children with Intellectual Disability	Special
	Unión Nacional School	25 first graders children	Special
ALBITIC	Early Education Centers: Plaza Mayorista, Los Pitufos y Amigos del alma	50 kindergarden children	Special
Total of Benfited Children		**188**	

Experimentation Instrument. As experimentation instruments, the HTAs developed in distinct developed IDEs were used, the identical ones that had been set up on computing, educational institutions and portable computers of teachers and researchers. In the cases of children with greater difficulty to handle the mouse, laptops with touch screen were used to facilitate the work. Desktop computers were used for the rest of the children. Children are shown in Fig. 6 during the experiments, using the HTA.

Process. For the experimentation stages, carried out from 2014 to 2019; First, the training of the teachers of the educational institutions (regular/special) was carried out, then several ATT performance tests were carried out. With the authorization of the authorities and together with the teachers, all the consents of the parents were collected, and an outreach activity was carried out with the children.

With the trained teachers and the children familiar with the use of the HTA, the experiments and evaluations were carried out. Each group used interchangeably the traditional method that teachers used to use in the classroom, on the one hand; and the developed HTA.

During the experiments, the research teams recorded events and observations, in addition to all these, the teachers together with the researchers recorded the achievements, response times, attempts and failures in the realization of the learning activities of the HTA.

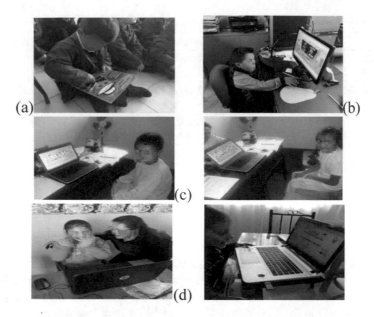

Fig. 6. Photos of the children during the experiments, using the ATT: (a) Regular child using ALBITIC activity, (b) Child Down using HETI@DOWN, (c) Children with language and motor disabilities using panel pictograms of TEVI, (d) Girls with intellectual/regular disability using GAMESMAT.

4   Statistic Analysis

For the quantitative analysis of the data, it was checked in the first instance if the data behaved normally; for this and by the sample size, the Shapiro Wilks test was applied ($p > 0.05$). Subsequently, statistical tests and models were applied to determine if the ATT had a statistically significant contribution to learning in children. The statistical tests used with a 95% confidence level were: To evaluate ALBITIC and HETI@-DOWN, the Student's parametric T was used ($p < 0.05$); to evaluate TEVI it was necessary to use the general univariate linear model; and finally to evaluate GAMESMAT the parametric test of Student's T was used.

5   Results

During the experiments were important events, observations, learning outcomes, interaction, communication, and concentration levels were recorded for further analysis. The following Table 3 shows the results of the statistically significant differences in the communication in case of TEVI, and the learning for the rest of the HTA; with the statistical models applied:

Table 3. Results of the statistical method applied with the data

HTA	Test normality	Method of data analysis applied	GL.	SIG.
ALBITIC (learnings)	ShapiroWilks (p > 0.05) (0.101)	Tstudent (p < 0.05) (−4,588)	49	0,000
HETI@DOWN (learnings)	ShapiroWilks (p > 0.05) (0.165)	Tstudent (p < 0.05) (6,580)	77	0,000
TEVI (communication)	ShapiroWilks (p > 0.05) (0.547)	Modelo Lineal General Univariante (p < 0.05) (31,414)	1	0,000
GAMESMAT (Aprendizajes)	ShapiroWilks (p > 0.05) (0,1385)	Tstudent(p < 0.05) (−8,76)	49	0,000

Furthermore, during the ATT evaluation in real contexts, groundbreaking qualitative results were reported:

In all the studies, having worked directly with the teachers as part of the development teams helped to produce technological tools custom-made to their educational needs in all centers. They also mentioned the tools developed will support their teaching strategies in the classroom and that could be used interchangeably in regular children or children with some type of language, motor and intellectual disability in any type of context in the Ecuadorian school system. Throughout the experiments, it was observed that the children paid attention to the teacher's indications and showed interest and motivation when learning "playing". The interaction, the awards to reach each of the challenges and the positive messages of the tools to their small achievements encouraged them; this was evident in their expression of joy and state of concentration.

6   Conclusions

To carry out these investigations, the collaborative work of the study team in coordination with the authorities, teachers, parents and children was crucial. The accompaniment of teachers and specialists as members of the development team allowed the implementation of technological solutions adapted to the particular needs of these children. The principles of agile methodologies and psychopedagogical criteria relevant to the production of quality recreational and educational tools were applied, which were designed with a friendly and accessible interface, with the support of visual and auditory elements to strengthen interaction and perceptual sensory stimulation. The evaluation of the help tools implemented in a real context, allowed us to identify their potential in aspects of interaction and motivation, their usefulness in aspects of learning. The results showed a remarkable contribution of these tools in children with respect to interaction, motivation and degree of concentration. On average, the levels of achievement for the two groups (Regular/Special) improved significantly with a 29.5% increase. Being the group of children of special education that could show a greater increase in their levels of achievement with 49.95%, compared to the increase of the group of children of regular education in only 9.03%. These results and the statistical model applied showed that in these studies there were significant statistical differences in learning with the use of developed tools versus traditional methods; with a significance level of (p = 0.000).

Future Work

Finally, as continuous future work we will work and improve the assisted tools developed for compartments with more families and schools where it has not yet been implemented. We want to follow assisted technology to work with other types of disabilities such as visual and auditory, because in Ecuador there are many assistance needs for blind and deaf people who have assistive technology to improve their quality of life.

Gratitude

A recognition to the Technological University Indoamérica, for the financial support granted to research project numbers: 078.025.2012 and 38.084.2017. Special thanks to all the children, authorities, specialists and teachers of the educational institutions who collaborated in these studies.

References

1. UNICEF: Assistive Technology for Children with Disabilities: Creating Opportunities for Education, Inclusion and Participation A discussion paper. World Health Organiztaion (2017)
2. McConnell, D., Breitkreuz, R., Uditsky, B., Sobsey, R., Rempel, G., Savage, A., Parakkal, M.: Children with disabilities and the fabric of everyday family life (2013)
3. Avila-pesantez, D.F., Padilla, N.P., Rivera, L.A.: Design of an augmented reality serious game for children with dyscalculia: a case study. In: Technology Trends, vol. 895. Springer (2019). https://doi.org/10.1007/978-3-030-05532-5
4. CONADIS: Agenda Nacional para la Igualdad en Discapacidades 2013–2017. LNCS (2018). http://www.planificacion.gob.ec/wp-content/uploads/downloads/2014/09/Agenda-Nacional-para-Discapacidades.pdf. Accessed 23 de abril del 2019
5. Ecuador: Ley Orgánica de educación intercultural y vilingue (2011)
6. Druin, A.: The role of children in the design of new technology. Behav. Inf. Technol. **21**(1), 1–25 (2002)
7. García, E.R.: El cuento infantil como herramienta socializadora de género. Cuestiones Pedagógicas. Revista de Ciencias de la Educación (22), 329–350 (2013)
8. Amaluisa, P., Jácome, L., Jadán, J., Amaluisa, D., Nuñez, M., Valarezo, A.: Material Didáctico Basado en un Cuento Infantil para el Desarrollo Cognitivo de los Niños Preescolares, Revista Investigar, Séptima Edición. LNCS (2018). http://investigar.cimogsys.com/Articulos_edicion/7tarticulo%205%20.pdf
9. Bruder, M.B.: Family-centered early intervention: clarifying our values for the new millennium. Top. Early Child. Spec. Educ. **20**(2), 105–115 (2000)

10. Jácome, L., Amaluisa, P., Suarez, N., Chango, G.: HETI@DOWN: Herramienta Tecnológica Interactiva de Ayuda para desarrollar destrezas cognitivas en niños DOWN. Investigar 25264(8), 7–19 (2018). http://investigar.cimogsys.com/Ediciones/OctavaEdicion.pdf
11. Asterisk: Text-To-Speech (TTS) and Automatic Speech Recognition (ASR). LNCS (2009). http://www.wikiasterisk.com/index.php?title=TTS_y_ASR. Accessed 19 de junio del 2017
12. Amores, L.J., Jadán-Guerrero, J.: TEVI: Teclado virtual como herramienta de asistencia en la comunicación y el aprendizaje de personas con problemas del lenguaje vinculados a la discapacidad motriz. Káñina 40(4), 105–122 (2016)
13. Stanitsas, M., Kirytopoulos, K., Vareilles, E.: Facilitating sustainability transition through serious games: a systematic literature review. J. Clean. Prod. 208, 924–936 (2018)
14. Wilson, A.J., Dehaene, S., Dubois, O., Fayol, M.: Effects of an adaptive game intervention on accessing number sense in low-socio-economic-status Kindergarten children. Mind Brain Educ. 3(4), 224–234 (2009)
15. Jácome-Amores, L.: Adaptación dinámica en los Juegos Serios para el desarrollo de destrezas cognitivas de la matemática en niños con problemas de aprendizaje. RIST Revista Ibérica de Sistemas y Tecnologías de Información (20), 217–228 (2019)
16. Jadán-Guerrero, J., Guerrero, L.A.: Tic@ula: Diseño de una Herramienta Tecnológica para Fortalecer la Alfabetización de Niños con Capacidades Intelectuales Diferentes. Revista: VAEP-RITA, Versión Abierta Español–Portugués, 1, 2, 3 (2014)
17. Ramani, G.B., Siegler, R.S.: Promoting broad and stable improvements in low-income children's numerical knowledge through playing number board games. Child Dev. 79(2), 375–394 (2008)
18. Whyte, J.C., Bull, R.: Number games, magnitude representation, and basic number skills in preschoolers. Dev. Psychol. 44(2), 588 (2008)
19. Tremblay, J., Bouchard, B., Bouzouane, A.: Adaptive game mechanics for learning purposes-making serious games playable and fun. In: CSEDU, vol. 2, pp. 465–470, April 2010

Intelligent and Decision Support Systems

Multi-robots Trajectory Planning Using a Novel GA

Killdary A. Santana[1]([✉]), Vandilberto P. Pinto[1,2]([✉]),
and Darielson A. Souza[3]([✉])

[1] Electrical and Computer Engineering, Federal University of Ceará - UFC, Sobral,
CE 62010-560, Brazil
killdary@alu.ufc.br
[2] Institute for Engineering and Sustainable Development - IEDS,
University for the International Integration of the Afro-Brazilian
Lusophony - UNILAB, Redenção, CE 62700 000, Brazil
vandilberto@unilab.edu.br
[3] Department of Electrical Engineering, Federal University of Ceará - UFC,
Fortaleza, CE 60455-760, Brazil
darielson@dee.ufc.br
http://www.ufc.br/

Abstract. One of the biggest challenges encountered in mobile robots is
in the planning of their trajectory, as their development is directly related
to the greater autonomy of robots. In this work a solution for multi-robot
path planning problem is presented, the problem modeling is performed
using the combination of the team orienteering problem and the problem
of the multiple backpack, this combination allows each robot to have an
individual limitation, the proposed solution was developed using genetic
algorithms.

Keywords: Multi-robot · Path planing · Metaheuristic algorithm ·
Team orienteering problem

1 Introduction

One of the biggest challenges encountered in mobile robots is in the planning of
their trajectory, as their development is directly related to the greater autonomy
of robots. The complexity of the route planning problem for robots has motivated
the development of several algorithms. This comes from the need to integrate
robot navigation with sensing, efficiency and route planning, as well as the need
to save significant resources and the participation of multiple robots.

The routing problem of multiple collaborative robots can be modeled through
Team Orienteering Problem (TOP). TOP is defined as a set of vertices with
awards that must be visited by a x number of identical agents that have a
movement cost limitation in which prize collection must be maximized. Several approaches have been proposed in the literature to solving the TOP among

© Springer Nature Switzerland AG 2020
Á. Rocha et al. (Eds.): ICITS 2020, AISC 1137, pp. 353–363, 2020.
https://doi.org/10.1007/978-3-030-40690-5_35

them can be ant colony [9], guided local search [2], linear programming [3], evolutionary algorithms [4,5], among others. But most of the solutions are mono-objective, which aim to maximize premium collection and cost minimization is a consequence of the higher premium collected, except for [3,4] and [6] being the last two applied to single agent issues.

The multiple knapsack problem (MKP) consists of an item set each with a value and weight and a set of knapsack each with capacity, the goal is to collect the items and place them in the knapsack so that they are collected. the highest possible value without exceeding the capacity of each knapsack [7].

This research aims to provide a solution for the calculation of routes for multiple energy-restricted robots using a genetic algorithm based on the [4] solution. Our method differs from [4] by combining MKP to perform different cost constraint insertion for each robot, making the solution closer to reality. The objective function will be modified by the function used in [3] which calculates fitness points by assigning weights to the cost and award objective, making the solution more adaptable to various applications.

The validation of the proposed solution was performed in a test environment with symmetrical maps of forty to eighty points to be visited, each point has an award. The tests were performed with different robot numbers and different costs applied. In the tests is performed the execution of the routes for multiple robots using a algorithm particle swarm optimization (PSO), and with the proposed solution of this work, in which a comparison between the obtained results is performed.

2   Mathematical Formulation

Consider a graph $G(N, A)$, where N represents the set ($|N| = N_c$) and A the set of edges. Consider also a symmetric matrix with minimal costs or distances between network nodes while considering that $c_{ij} = +\infty \forall i \in N_c$. The solution can use N_k robots, each robot having a maximum restriction of energy C_{kmax}. The x_i waypoints that make up the N set have a p_i prize. The TOP uses binary variables x_{ik} to indicate if a waypoint x_i was visited by a k robot, where 1 indicates that the waypoint was visited and 0 indicates not. The T_k variable is a set of waypoints belonging to a k robot route. The matrix $X[x_{ijk}]$ is composed of binary variables that indicate whether a x_{ij} edge is part of the route taken by the robot k.

Thus, the mathematical formulation can be defined as [3,4,7]:

$$min \left(\alpha \sum_{i=1}^{N_c} \sum_{j=1, i \neq j}^{N_c} \sum_{k=1}^{N_k} c_{ij} x_{ijk} - \beta \sum_{i=1}^{N_c} \sum_{k=1}^{N_k} p_i x_{ik} \right) \tag{1}$$

$$\sum_{i=1}^{N_c-1} x_{ni} = 0 \tag{2}$$

$$\sum_{i=1}^{N_c} x_{ij} \leq 1 \qquad j = 2, 3, \ldots, N_c \tag{3}$$

$$\sum_{j=1}^{N_c} x_{ij} \leq 1 \qquad i = 2, 3, \ldots, N_c \tag{4}$$

$$\sum_{i=1}^{N_c} \sum_{j=1}^{N_c} \sum_{k=1}^{N_k} c_{ij} x_{ijk} \leq C_{k_{max}} \tag{5}$$

$$1 \leq u_{ik} \leq N_c \qquad \forall i \in T_k \tag{6}$$

$$i_i - u_j + 1 \leq (1 + -x_{ijk}) N_c \qquad 2 \leq i \neq j < N_c \tag{7}$$

In this problem the objective function (1) consists of two objectives, maximizing premium collection and minimizing the cost of all agents. This proposal uses a modified version of the objective function used in [3], in which the change made was multi-agent adaptation, the variables α and β continue to have their functions unchanged as weights for minimal route and maximization of awards. The constraint (2) requires no waypoints to be visited after reaching the endpoint, the Eqs. (3) and (4) restrict each vertex from being visited. Once the Eq. (5) limits the maximum cost of each agent, sub-routes are prevented from being generated by the Eqs. (6) and (7). The result of this optimization will be a set of ordered vertices to be visited by agents.

3   Genetic Algorithm

Genetic Algorithms (GA) are optimization techniques inspired by the principle of survival and reproduction of the fittest individuals, proposed by Charles Darwin [8].

The GA proposed in this work uses the chromosome representation presented in [4], which is formed by a set of vectors, or genes, that represent a route to be traveled by a single robot. Each gene stores the index of the waypoint to be visited by its respective robot. Each waypoint is unique, being used in only one gene. Source and destination deposits can be shared. The chromosome is combined with a maximum cost list C, where the number of elements is the same number of robots N_k, for each k robot there is a maximum cost C_{kmax}. Thus the chromosome represents a valid solution to the problem. The cost of each robot represents its battery charge in minutes until its complete discharge.

The operation of the proposed GA can be visualized in the Algorithm 1a.

The process of chromosome generation is performed in two moments, at the initialization of the population and its crossing and mutation. Population initialization uses a random nearest neighbor algorithm, which consists of selecting a random waypoint for the gene and initiating the choice of the nearest neighbor to the last waypoint inserted, the process ends when it exceeds the maximum value

Algorithm 1 Genetic Algorithms
1: start population;
2: **while** Stop Test **do**
3:    **while** population is not complete **do**
4:        crossover
5:        mutation
6:        evaluation
7:        selection
8:        update
9: Finishing

Genetic Algorithms

Algorithm 2 Random nearest neighbor
IN: CostList, ListDepositBegin, ListDepositEnd, NumberRobots, AvailableWaypoints
OUT: Chromosome
1: **for each** R ∈ NumberRobots **do**
2:    Gene ← New
3:    Begin ← ListDepositBegin(R)
4:    End ← ListDepositEnd(R)
5:    W ← SelectRandom(AvailableWaypoints)
6:    TmpG ← Join(Begin, Gene, End)
7:    GeneCost ← MeasureCost(TmpGene)
8:    N ← AvailableWaypoints
9:    C ← CostList(R)
10:    **while** GeneCost < C e Size(N) > 0 **do**
11:        Remove(W, AvailableWaypoints)
12:        Insert(W,Gene)
13:        TmpG ← Join(Begin, Gene, End)
14:        W ← NearestNeighbor(W)
15:        GeneCost ← MeasureCost(TmpG)
16:        Gene ← TmpG
17:        Insert(Gene, Chromosome)
18: Finishing

Random nearest neighbor

Algorithm 3 Fitness
IN: Chromosome, CostsMap, Prizes, WeightCost, WeightPrizes
OUT: Fitness
1: Fitness ← 0
2: **for each** G ∈ Chromosome **do**
3:    A ← Prizes(G)*WeightPrizes
4:    C ← CostsMap(G)*WeightCost
5:    FitnessGene ← A - C
6:    Fitness ← Fitness + FitnessGene
7: Finishing

Fitness

Fig. 1. GA Algorithm 1, 2 and 3

of the respective robot, C_{kmax} or when there are no more waypoints available. The process can be seen in the Algorithm 1b.

The population is finalized when the desired amount of chromosomes are created, after generating the population its fitness is calculated, the evaluation method can be seen in Algorithm 1c (Fig. 1).

The fitness calculation is the same as used by [3] in which a weight for the prize and a weight for the cost is determined, for each gene the fitness calculation is performed, and the sum of the fitness of all genes is the fitness of the complete chromosome.

After the initial population is generated, the generation process is started. The generation process will only end when the maximum number of generations, G_{size}, is reached or when the best fitness chromosome remains unchanged for $BestG_{size}$ generations. The first step when beginning the generations is to select the individuals to cross, this selection is performed on a percentage of the population through the Tournament method. In this method n individuals are randomly selected and whichever has the best fitness will be selected.

For the crossing operation 2 chromosomes are selected, these chromosomes referred to as parents will generate 2 new chromosomes referred to as offsprings. Because the chromosome can have several genes, one for each valid robot route, the crossover process will select only one gene for the crossover. At the beginning of the crossover will be randomly selected which gene will be crossed, with the selected gene is removed its deposits and then a clipping operation will be performed on the selected genes, the clipping operation is called of cut. At the end of the clipping operation a slice of the gene will be obtained. The clipping operation differs for each gene, as the genes may have different sizes so their slices may also. Then the resulting slice is removed from its respective gene. Finally, the Gene1 slice will be inserted into Gene2 and vice versa, the insertion position will be based on the resulting Gene size after clipping, and the deposits will be reinserted. Children are generated by receiving the unmodified genes from their father and the new gene generated by the operation.

Algorithm 4 Crossover
IN: Parent1, Parent2, ListDepositBegin, ListDepositEnd
OUT: Offspring1, Offspring2

1: idxGene ← SelectRandom(Pai1)
2: Start ← ListDepositBegin(indexGene)
3: End ← ListDepositBegin(indexGene)
4: P1 ← Parent1
5: P2 ← Parent2
6: Gene1, Gene2 ← P1(idxGene), P2(idxGene)
7: Gene1, Gene2 ← RemoveDeposits(Gene1, Gene2)
8: Slice1, Slice2 ← Cut(Gene1), Cut(Gene2)
9: pos1, pos2 ← SelectRandom(Gene1,Gene2)
10: Gene1 ← Remove(Gene1, Slice1)
11: Gene2 ← Remove(Gene2, Slice2)
12: Gene1 ← Join(Gene1, Slice2, pos1)
13: Gene2 ← Join(Gene2, Slice1, pos2)
14: Gene1, Gene2 ← InsertDeposits(Gene1, Gene2, Start, End, indexGene)
15: Offspring1, Offspring2 ← New
16: **for each** g ∈ Indexes(P1) **do**
17:     **if** g = idxGene **then**
18:         Insert(Gege1, Offspring1)
19:         Insert(Gege2, Offspring2)
20:     **else**
21:         Insert(SelectGene(P1, g), Offspring1)
22:         Insert(SelectGene(P2, g), Offspring2)
23: RemoveRepeatedElements(Offspring1)
24: RemoveRepeatedElements(Offspring2)
25: Finishing

Crossover

Algorithm 5 Insertion and Deletion Mutation
IN: Chromosome, CostList, ProbMut, AvailableWaypoints
OUT: Chromosome

1: Prob ← RandomValue(0,1)
2: Ways ← AvailableWaypoints
3: **if** Prob >= ProbMut **then**
4:     **for each** idx ∈ Index(Chromosome) **do**
5:         MaxCost ← CostList[idx]
6:         Gene ← Chromosome[idx]
7:         Cost ← MeasureCost(Gene)
8:         **if** Cost < MaxCost **then**
9:             W ← SelectRandom(Ways)
10:            P ← PositionBestFitness(Gene, W)
11:            NewGene ← Insert(Gene, W, P)
12:            NewCost ← MeasureCost(NewGen)
13:            **if** Cost < NewCost **then**
14:                Gene ← NewGene
15:                Ways ← Remove(W,Ways)
16:        **else**
17:            WP ← WorstWaypoint(Gene)
18:            Gene ← Remove(WP, Gene)
19:            Ways ← Insert(WP, Ways)
20:        Chromosome[idx] ← Gene
21: Finishing

Insertion and Deletion Mutation

Fig. 2. GA Algorithm 4 and 5

After performing the crossover, in a part of the population will be performed a mutation operation to ensure its evolution. Individuals that will mutate are selected based on mutation probability *PropMut*. The mutation process is performed on a randomly selected gene. In chromosome are performed 3 mutation operations an insertion operation, a removal operation, and the Swap worst gene locally mutation (SWGLM) [1] (Fig. 2).

Insertion and deletion mutations are performed to ensure better population diversity by inserting new waypoints into the population. In the insertion mutation a random waypoint is selected from the waypoints not present on the chromosome, then a permutation is performed at each position of the gene to define which position generates the best fitness for the gene. As for the deletion mutation the gene is resorted to and a waypoint removed and its fitness measured, the waypoint that achieves the least loss of fitness will be removed from the gene. As for the deletion mutation the gene is traversed and a waypoint is removed and its fitness is measured, the waypoint that achieves the least loss of fitness will be removed from the gene. The SWGLM mutation was selected based on a study by [1] that demonstrates a good fitness gain in PCV, this mutation consists of finding the lowest fitness waypoint *wp* and performing two exchanges an exchange between *wp* and your right-hand neighbor and measure your fitness and an exchange between the 2 left-most neighbors, without moving the *wp*, the best fitness found determines the new positions. The insertion and deletion mutation process can be viewed in the Algorithm 3b.

4 Particle Swarm Optimization

Particle Swarm Optimization (PSO) was first mentioned by James Kennedy and Russell Eberhart in 1995 [10]. PSO has received a lot of attention for its efficiency in solving complex optimization problems.

PSO is inspired by the flock behavior observed in birds, where they move in a N dimensional space in search of the best solution. Each particle in a population has two properties, its current position and its velocity. Using these parameters, the particles move through the search space according to a mathematical formula in order to converge on a desired location. In order for particles to change their positions, each particle has its best position in memory, called *pbest*. The particle swarm also keeps track of the best values and its location obtained from any particle in the population, the best particle of all generations being called *gbest*. During movement within the search space the particles adjust their positions based on the best position obtained and the best population particle. The particles are displaced from their current position by applying the velocity vector. The size and direction of your vectors is based on a function of your position and the best position found.

The PSO developed in this work assumes that a particle has the same format used in the proposed GA of this work and explained in the previous section, using position is a valid route for all robots within the imposed energy restrictions. The particle velocity will be proposed by inserting, removing and reorganizing the waypoints present in the route of each particle robot. The fitness calculation is the same as [3].

Population initialization is performed using a random nearest neighbor algorithm, the same method used in GA and can be viewed in the Algorithm 1b. Soon after the best particle is selected in the population to be *gBest*.

With the initial population generated, the iterations process begins. The first step is to select the best particle present in the current iteration population, this particle will be *pBest*. After selecting *pBest*, each particle will be scanned where the current particle will be crossed with *pBest*. The crossover used in the proposed PSO is Partially Matched Crossover (PMX) [11].

After the combination of the particle and *pBest* will be performed the particle movement operation in the sample space, normally this activity is performed with the particle velocity change, but in the developed PSO the particle position change will be performed using three different operations insertion and deletion used in the proposed GA and were displayed in the Algorithm 3b.

Immediately after moving the particle a comparison is made between the fitness of the new position and the old position, if the new position has better fitness the particle takes the new position. The last step is to compare if the new particle has a better fitness than *gBest*, if it has the new particle it takes the place of *gBest* and the operation and process starts again until the number of interactions has been completed.

5 Test Environment

The following section presents the environment in which the solution tests were performed. Planning is required to perform a sampling simulation for a vehicle team. In the tests were used 20 symmetric auto-generated scenarios with different number of robots and different battery time for each robot, the test instances can be seen in Table 1 and the battery time for instance can be seen in Table 2. The cost of locomotion between each point was calculated using Euclidean distance. The overall objective of the experiment is to demonstrate the effectiveness of robot calculation using the modified version of TOP compared to its traditional version. For the experiment it is assumed that the available battery time for each robot will be given in minutes, the speed of movement of the robots is considered constant.

For a better understanding of the instances used in this work, that are presented in Table 1, will be made a detail of instance *I*1. The instance *I*1 is made up of 40 waypoints and 5 depots, where robots can have their origin and return defined in the depots. Deposits have no punctuation and are available to the robot only if they are set as source or destination. The instance *I*1 has a total prize pool of 218 points, where the score ranges from 2 to 9 points.

The tests will be performed in 3 experiments:

- **Experiment 1:** Scenario with one base and all robots maintains their battery level.
- **Experiment 2:** The scenario with one base and all robots maintain the battery level, at which the lowest battery level robot is removed from the route calculation.
- **Experiment 3:** Scenario with one source base and multiple target bases in which all robots maintain battery level;

The tests were performed on a computer with an Intel i5-7600 processor, 8 GB RAM and Ubuntu 16.04 operating system. The proposed GA was developed in Python language version 3.6 using the Numpy 1.14.5 package.

6 Results

The tests were performed based on the instances and experiments proposed in the Sect. 5. The tests were performed using two algorithms for route calculation, the first PSO algorithm that calculates routes for multiple robots with different energy constraints and different bases, and a genetic algorithm proposed by this work that combines the traditional TOP with the PMM. Each instance was combined with all experiments and performed a total of 30 executions.

Table 1. Instances

Instance	Robots	Waypoints	Total prizes	Battery time robots (min)
I_1	4	40	218	20, 23, 25, 30
I_2	4	45	222	23, 25, 27, 30
I_3	4	50	241	23, 25, 27, 30
I_4	4	60	313	20, 23, 25, 30
I_5	4	70	387	20, 23, 25, 30
I_6	4	80	436	20, 23, 25, 30
I_7	5	40	206	18, 22, 24, 25, 28
I_8	5	50	236	19, 23, 25, 26, 29
I_9	5	60	309	20, 22, 25, 27, 30
I_{10}	5	70	383	21, 23, 26, 27, 30
I_{11}	5	80	441	22, 24, 26, 28, 32
I_{12}	6	50	239	18, 20, 22, 24, 25, 27
I_{13}	6	60	310	19, 22, 23, 24, 25, 28
I_{14}	6	70	379	20, 22, 24, 25, 27, 30
I_{15}	6	80	440	21, 23, 24, 26, 29, 30
I_{16}	7	60	310	15, 20, 23, 25. 25, 27, 30
I_{17}	7	70	377	19, 20, 21, 23, 24, 25, 29
I_{18}	7	80	443	19, 21, 23, 24, 25, 26, 30
I_{19}	8	70	388	15, 20, 23, 25, 25, 27, 30, 30
I_{20}	8	80	428	15, 20, 23, 25, 25, 27, 30, 30

The Table 2 presents the results obtained, the calculated route performance is given by the total points collected by all robots involved, where the displayed data shows the average prize collection from the execution of each experiment. Experiment 1 and 2 tests were run with a single source and destination base, the robots should leave one base and return to the same base, experiment 3 was run with one source base and different target bases, in which each robot goes to a different base.

The performance of the solution presented in this paper is compared in 3 different ways with PSO. First we compare the generated route with the same number of robots for each model, this comparison can be seen in the Table 2, in which correspond the columns of experiments 1. The model proposed in this article obtained a better score, it is possible to note that the score collected in the model proposed in some scenarios was able to exceed 40.3% to the method using PSO.

The second comparison is performed by reducing the number of robots of each instance in a unit, the results of these tests can be seen in the Table 2, in which correspond the columns of experiments 2. The model using the PSO achieved an improvement in premium collection in some instances, $I13$ and $I16$, where in some cases point collection is close to its first execution, experiment 1 compared to experiment 2, due to increased locomotion of the remaining robots, but the solution proposed in this work still got better performance compared to PSO, where in some scenarios managed to surpass the results obtained by the PSO by up to 33.3% in the collection of prizes.

The third comparison is made by changing target deposits for each scenario, where each robot goes to a different base, the results of these tests can be seen in the Table 2, in which correspond the columns of experiments 3. PSO achieved a performance drop in experiment 3 compared to its performance in experiment 1, this method did not show good performance with different target deposits. Comparing PSO Experiments 3 and GA, it is possible to see that GA was able to achieve great results by exceeding prize collection by up to 49%. In addition to the performance improvement between the solutions, the result obtained in GA experiment 3 surpassed the results obtained by himself in the execution of experiment 1, where up to 13% improvement in prize collection, this demonstrates that the proposed method has a good generalization in solving problems with different deposits.

Table 2. Results

I.	E. 1	E. 2	E. 3	E. 1	E. 2	E. 3	I.	E. 1	E. 2	E. 3	E. 1	E. 2	E. 3
	PSO			GA				PSO			GA		
I_1	156,23	135,66	166,16	180,23	153,13	207,5	I_{11}	311,06	276,4	315,96	340,8	312,4	379,2
I_2	166,33	148,4	161,13	205,36	175	208	I_{12}	189,5	172,3	181,8	210,8	197,1	228,1
I_3	171,13	150,5	161,03	207,5	177	223,4	I_{13}	244,93	231,33	239,33	270,9	255,6	294,2
I_4	197,36	175,36	194,2	254,8	210	265,36	I_{14}	296,83	276,6	288,8	337,4	313,5	360
I_5	223,26	197,93	219,06	299,3	262,53	329,5	I_{15}	338	309,23	330,73	365,8	344	396,6
I_6	231,43	205,06	233,03	324,7	273,4,	347,7	I_{16}	248,13	336,6	244,6	274,7	267	294,6
I_7	173,2	164,76	178,13	193,6	186,1	199,3	I_{17}	299,8	279,23	287,53	336,4	318	361,1
I_8	183,4	171,4	191,6	202,4	191,9	224,4	I_{18}	348,46	327,63	331,96	388,4	359	401,1
I_9	235	216,7	236	265,3	246,4	283,7	I_{19}	291,26	281,33	276,16	329,2	317	363,7
I_{10}	275,2	249,1	275,16	306	284,6	334,7	I_{20}	324,46	313,06	305,6	370,5	358	402,1

7 Conclusion

This work presented a method of calculation of multiple collaborative robots. Such a problem can be solved by using a team orienteering problem, however, the proposed solution was modeled based on the team orienteering problem combined with the multiple backpack problem, in order to make full use of the restrictions of each robot. This paper presents a heuristic for the modified TOP solution using genetic algorithms. To validate the solution, a comparison was made between a model developed using PSO and the solution proposed in this research. In terms of performance in the largest prize collection, the purpose of this research proved superior in all scenarios tested. Multiple base modification has been shown to be effective by demonstrating good generalization.

The tests achieved good perfomances in the comparisons between the TOP experiments performed and the proposed solution. In all experiments the proposal of this paper was superior, calculating the average of the results achieved. The proposal of this research achieved 40% higher results compared to the traditional model. The results are still better for the research proposal, where the deposit change achieved 49% higher results compared to the PSO, where the PSO showed a decrease in its performance compared to its execution with a single deposit.

Although the results are satisfactory the symmetrical nature of the tests makes their use limited to environments without cost volatility between the waypoints to be visited, where its use for route calculation for marine robots and unmanned autonomous vehicles (UAVs) is not optimal, as they have a more dynamic environment due to tidal force and direction and wind speed and direction, however its use can be applied in industrial environments in the calculation of Automated Guided Vehicle (AGV) routes, extending its use to industry 4.0.

For future work would involve testing the performance of the solution on asymmetric maps to extend its use. The multiple filing nature of the proposal makes it ideal for testing together with battery fault diagnosis systems, as the cost of getting around can be reduced quickly, requiring a recalculation of the route to get the most out of the mission. Finally it would be worth extending its future use to maps with task execution times and time window for their completion.

Acknowledgment. The authors acknowledge the support of FUNCAP (BP3-0139-00241.01.00/18) BPI 03/2018.

References

1. Hassanat, A., Alkafaween, E., Alnawaiseh, N., Abbadi, M., Alkasassbeh, M., Alhasanat, M.: Enhancing genetic algorithms using multi mutations. PeerJ Comput. Sci. **14** (2016). https://doi.org/10.7287/PEERJ.PREPRINTS.2187V1
2. Vansteenwegen, P., Souffriau, W., Berghe, G.V., Oudheusden, D.V.: A guided local search metaheuristic for the team orienteering problem. Eur. J. Oper. Res. **196**, 118–127 (2009). https://doi.org/10.1016/j.ejor.2008.02.037
3. Candido, A.S.: Sistema de gerenciamento do voo de quadrirotores tolerante a falhas. Instituto Tecnológico de Aeronáutica (2015)
4. Bederina, H., Hifi, M.: A hybrid multi-objective evolutionary algorithm for the team orienteering problem. In: 4th International Conference on Control, Decision and Information Technologies (CoDIT), pp. 0898–0903. IEEE Press (2017). https://doi.org/10.1109/CoDIT.2017.8102710
5. Ferreira, J., Quintas, A., Oliveira, J.A., Pereira, G.A.B., Dias, L.: Solving the team orienteering problem: developing a solution tool using a genetic algorithm approach. Soft Comput. Ind. Appl., 365–375 (2014). https://doi.org/10.1007/978-3-319-00930-8_32
6. Schilde, M., Doerner, K.F., Hartl, R.F., Kiechle, G.: Metaheuristics for the biobjective orienteering problem. Swarm Intell. **3**, 179–201 (2009). https://doi.org/10.1007/s11721-009-0029-5

7. Kellerer, H., Pferschy, U., Pisinger, D.: Multiple knapsack problems, pp. 285–316. Springer, Heidelberg (2004). https://doi.org/10.1007/978-3-540-24777-7_10
8. Goldberg, D.E.: Genetic Algorithms in Search, Optimization, and Machine Learning, vol. 1. Addison-Wesley Professional, Reading (1989)
9. Ke, L., Archetti, C., Feng, Z.: Ants can solve the team orienteering problem. Comput. Ind. Eng. **54**, 648–665 (2008). https://doi.org/10.1016/j.cie.2007.10.001
10. Eberhart, R., Kennedy, J.: A new optimizer using particle swarm theory. In: Proceedings of the Sixth International Symposium on Micro Machine and Human Science, pp. 571–579 (2017). https://doi.org/10.1016/B978-0-12-811318-9.00030-2
11. Crossover of Enumcrated Chromosomes. http://www.wardsystems.com/manuals/genehunter/crossover_of_enumerated_chromosomes.htm

Intelligent Support System
for the Provision of Inpatient Care

Sónia Faria[✉], Daniela Oliveira, António Abelha, and José Machado

Algoritmi Center, University of Minho, Campus Gualtar, 4710 Braga, Portugal
{sonia.faria,daniela.oliveira}@algoritmi.uminho.pt
{abelha,jmac}@di.uminho.pt

Abstract. Inpatient care is seen as a rigorous healthcare environment, as several daily tasks are performed to provide adequate treatment to inpatients and a minor flaw in these tasks may result in irreversible damage to patients. It is therefore required that the information related to the patient is always updated and available to all health professionals. Thus, comes up the motivation of the project described in this paper, which presents an intelligent system to support the practice of inpatient healthcare through a Web platform that allows the monitoring of patients admitted to a health facility. Thus, the developed system culminates in an application where all relevant information is gathered to monitor the different hospitalization episodes, presenting this information in a simplistic and intuitive way and alerting the professionals to the occurrence of events related to medical exams and analysis, surgical procedures, among others. This paper presents the architecture, the requirements and a SWOT analysis of the solution proposed, the main conclusions and a proposed future work.

Keywords: Inpatient support system · Multi-Agent System · Clinical Decision Support System · Healthcare

1 Introduction and Contextualization

Nowadays, the major concern of health institutions is to improve their quality, focusing on the patient security and his health-being and thus providing more efficient and effective health care.

During hospitalization episodes there are several and highly heterogeneous teams of care providers that monitor the patients' evolution. And it's necessary to transmit information about patient treatment at various times, making a correct and assertive communication among the professionals essential. An incorrect or weak communication between professionals can lead to clinical errors and compromise patient safety. What makes hospitalization considered a place conducive to the occurrence of Adverse Events (AE), that is, the occurrence of unwanted, unforeseen and unintentional complications resulting from the care provided to the patient [1].

© Springer Nature Switzerland AG 2020
Á. Rocha et al. (Eds.): ICITS 2020, AISC 1137, pp. 364–374, 2020.
https://doi.org/10.1007/978-3-030-40690-5_36

Worldwide, it is estimated that 1 in 10 patients is affected by one or more AE and 50.3% of the AE occurred were considered avoidable [2]. In Portugal, one or more AE occur in public hospitalizations, which results in a consequent increase of hospital death risk from 5% to 7% [3]. Communication failures are pointed as one of the main contributing factors for the occurrence of AE, jeopardizing the continuity of the patient's treatment [4].

In order to overcome these events, Medical Informatics focus on the development and implementation of Hospital Information Systems (HIS). HIS aim to facilitate and improve the performance of all functions conducted by the hospital in patient care, taking into account human resources, technological resources, economic resources and legal requirements such as data security, among others [9,10]. With the increasing improvement of HIS, Clinical Decision Support Systems (CDSS) have emerged, which are computer systems that allow quick access to a set of pertinent information and automatically extract knowledge from a high volume of data and also allow analysis and comparison of the same data [11,12].

To overcome failures in communication, mismanagement of hospitalization episodes and to prevent failures in shift handovers, emerged the motivation to develop an Intelligent Clinical Decision Support System (ICDSS), presented in this paper, intending to assist the professionals during their tasks through an automated process.

This paper is divided in six sections. The first section introduces and contextualizes the work. In the second section are presented the background and related work. Section three presents the methodologies used to conduct the research, namely the Design Science Research methodology. The solution proposed is explained in the fourth section. A proof of concept is accomplished in the fifth section. Finally, in the last section the conclusions are drawn and future work is suggested.

2   Background and Related Work

In hospitalisation episodes, failures occur that are often caused by the difficulty in accessing the patient's current clinical record and by the lack of recorded information. Thus, communication failures or failures in the transmission of information at the shift handover are constantly reported and demonstrated to be factors that influence the delivery of health care, which can endanger the life of a patient and thus reduce the quality of care delivery.

At the most health institutions the relevant information related to the hospitalization episodes is decentralized, dispersed by several HIS and several databases, making its access and consequent interpretation difficult. It is therefore important to consider the concept of interoperability, which can be broadly defined as the ability of two or more systems to exchange and use data efficiently [5,6].

In order to focus data on a single source, a manual mechanism for inserting information into a static prototype was introduced into Centro Hospitalar

Universitario do Porto (CHUP). In some health units of CHUP the prototype is developed on a white board, while in other health units it is developed on Excel sheets, however, in both cases, entering the required data is based on a manual process. Therefore, the information held in this prototype is daily recorded manually by the professionals of each health unit and should be updated whenever there is any change in patient' data, whether administrative or clinical.

The mechanism mentioned above has serious limitations and failures, one of the main being that it requires time for healthcare professionals, which would be fundamental for the effective delivery of patient care. In addition, this system presents failures due to the dynamism of hospitalization, since, in this context, there is a great flow of tasks and a constant alteration of patients' condition. It is, therefore, necessary to update the information contained in the prototype several times a day, what doesn't always happen and may mislead other professionals. Also, subjectivity in completing the data may lead to ambiguities and misinterpretations of the information included in the prototype. At last, this mechanism still presents another flaw that results from the deletion of information or the filling of erroneous information, although by mistake, that when omitted may compromise the safety and well-being of patients. For these reasons, this system was considered an obsolete system and does not correspond to the needs of the professionals.

Thus arises the need to automate the prototype mentioned before and thereby the need to create a computerized board for monitoring inpatients. The automation of services within this institution has already proved to be an asset, and is achieved through increased interoperability between systems [7]. For this purpose, Agency for Integration, Diffusion and Archive of Medical Information (AIDA), was created, which is a Multi-Agent System (MAS) developed by a group of researchers from the University of Minho, which is implemented in several portuguese health institutions, such as CHUP [10]. AIDA is an agency that provides software agents that exhibit proactive behavior and are responsible for certain tasks, namely, communication between different systems [5,6].

Therefore, the objective of the project presented in this paper is to automate the services present in the inpatient environment, joining all the pertinent information to the patient, such as their location (bed number), scheduled medical exams, medical analysis and surgical procedures and associated alerts, in one platform. For this purpose a web platform was created to allow the consultation and facilitate the visualization of this information and thus reducing the need to access multiple data sources, which would be more time consuming.

3   Methodology Design Science Research

The methodology used to conduct the research was problem oriented, which means that it was intended to develop a solution to a given problem. The research was developed according to the needs and emerging problems of CHUP and taking into account the information systems present in it, as well as the results obtained by the static prototype mentioned above. The resulting system was inserted in a real context, so the research was carried out cyclically and continuously, and not through a linear method, since this system has to meet the needs of professionals and therefore has to overcome challenges or consequent problems that may arise. That said, the methodology adopted was Design Science Research, whose main objective is to support the development and evaluation of information technologies artifacts [8].

A schematic of the Design Science Research (DSR) methodology is presented in Fig. 1, where the essential steps for building scientific IT artifacts are specified.

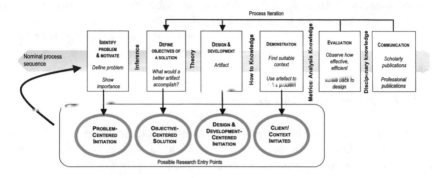

Fig. 1. DSR methodology and its essential steps 1.

That said, the first step in this project development was the identification of the problem, which translates into the lack of a single platform for monitoring hospitalization. Thus, in CHUP, health professionals who monitor hospitalization episodes, encounter difficulties when accessing information. This information is scattered across several IS and in multiple formats. The main objective of this research proposal is to find a solution that allows health professionals to access relevant and unambiguous clinical information quickly and intuitively. Thus, the artifact developed is an ICDSS that gathers all the information considered crucial for the episodes of hospitalization and presents it to health professionals, using a web platform. Throughout the development of the solution, its practical demonstration in CHUP and its performance is evaluated through an analysis and an usability study with the elaboration of interviews and questionnaires to the target users. Depending on the assessment, adjustments are made and new objectives are defined. In the last phase the artifact obtained is communicated and produced in the different health units of CHUP.

4 Intelligent Support System for the Provision of Hospitalization Care

For the development of the resultant ICDSS it was first necessary to evaluate and define its clinical and technical requirements, using the help of physicians, nurses and health technicians.

Therefore, at a technical level it is intended that the system has the following characteristics:

– **Interoperable** Interoperable with all HIS present in the health organization that contain the relevant information for this platform.
– **Automated** Works automatically, with the minimum input of data by users. Thus, information regarding patient hospitalization episodes is collected through an automated mechanism and through cycles programmed with predefined time periods.
– **User-friendly** Intuitive, user-friendly, unambiguous user interface that enables simplified access to key patient information.
– **Distributed and Concurrent** Simultaneous access for multiple users and on multiple devices.
– **Integrable** Its implementation and production needs to integrate into the CHUP information systems without compromising their performance.
– **Data Confidentiality** Ensuring access to data by authorized professionals only.
– **Adaptable** Adaptable at different data sources and at various health units.
– **Ubiquitous Access** Access can be done anytime and anywhere. System needs to operate in any operating system or any type of device.

As for the clinical requirements, it was necessary to evaluate the main needs, once multiple and different data is generated during hospitalization episodes, such as therapeutic attitudes, examinations, medical devices, among many others. Therefore, the following clinical requirements were considered:

– **Bed, Unit and Patient Identification** "Bed Number", "Unit Description", "Patient Name", "Episode Number" and "Responsible Relative"
– **Physician and Nurse Responsible**
– **Medical Analysis and Exams**
– **Medical Devices** Device type identifier and change need alert.
– **Medical/Clinical Alerts** "Fasting", "Risk of Fall", "Physical Containment", among others.
– **Allergies and Isolation**
– **Medical/Clinical Notes**

Throughout the development of the system, the authors of this article were granted CHUP approval to extract the information needed in order to test the system and thus meeting the clinical requirements.

4.1   System Architecture

The resultant Intelligent System to Support Inpatient Healthcare, presented in Fig. 2, displays an architecture based on three essential components: Web Platform, Database and MAS.

The web platform is based on client–server model. The Client refers to the User Interface (UI), that is, the web application where user interacts with the platform. Therefore, the Client communicates with the Server through *HTTP Requests*, which sends requests to a secondary Web Service (used only to manage access to multiple databases present in the institution), which in turn, queries the databases, receives the data in *JSON* format and returns it. The Server receives the data and sends the *HTTP Responses* to the Client.

The database created to support the new system, stores the data needed for the application in related tables through a specific key, in this case the episode number. In order to have the tables always filled with updated information, a cyclical mechanism based on MAS was implemented. The MAS is fundamental to the synchronization between the system Database and the different data sources regarding different HIS. Thus, agents were developed to query the different hospital databases, collect the data, process and transmit it and fill the new system database, repeating this process in certain periods of time. This process is essentially the Extract, Transform and Load (ETL) process, a type of data integration used to combine data from multiple sources [13].

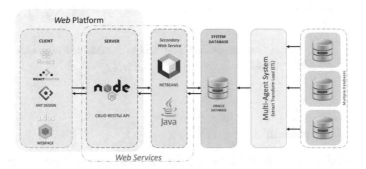

Fig. 2. Intelligent System to Support Inpatient Healthcare architecture and the technologies used.

4.2 MAS and Interoperability

The MAS plays a key role in updating the data and increasing interoperability between different information systems. That's because its agents work accordingly to collect and transform data from different data sources, loading the resultant data into the system database tables, allowing it to be permanently updated [10]. For this purpose, some agents were created, which sequentially execute their cycles.

The MAS process is initialized by Agent number 1, as shown in Fig. 3, that performs the ETL process to the general episode information, such as bed, unit and patient identification. Also, this agent provides variables, such as patient episode and sequential numbers, used as parameters in the next agents.

Each next agent will sequentially perform its ETL process, inserting at last the information in the intended table.

The process, outlined in the Fig. 3, is sequential, each agent runs sequentially and its actions are performed in a chain, eliminating the possibility to occur a lock in the process and eliminating the possibility of its actions jeopardize the actions of the following.

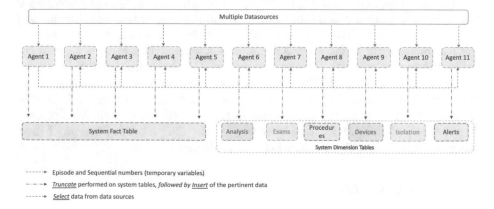

Fig. 3. Multi-Agent System schema developed and representation of tables created to support the intelligent system.

4.3   Web Application

The developed web application consists of a computerized internment board. This platform displays all the relevant information referent to patients admitted at a given moment in one of the CHUP health units. Figure 4 shows the resultant hospitalization chart of the Surgery health unit and it's possible to verify the different information represented by graphic elements, namely icons, which allows to gather a large number of information in a small space, still allowing its interpretation.

5   Proof of Concept

A project developed must be evaluated before making it available to its target users. So the application was subjected to a proof of concept, through a SWOT analysis, in order to prove the viability and usability of the solution proposed in this project. A technology acceptance study using questionnaires was also developed.

The Strengths, Weaknesses, Opportunities and Treats (SWOT) analysis, as the name implies, is an analysis that evaluates the strengths (advantages and qualities), weaknesses (disadvantages that prevent a particular artifact from

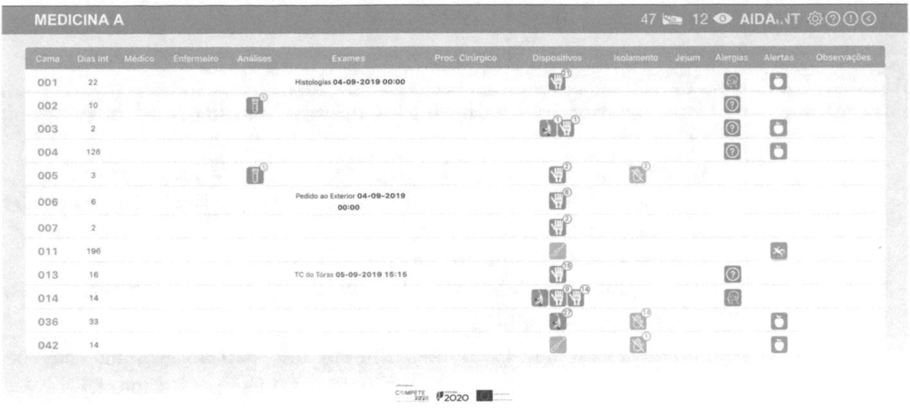

Fig. 4. Web application presenting the inpatient board of the Surgery health unit.

achieving its objectives), opportunities (conditions of the implementation environment that positively influence the success of the artifact), and threats (conditions of the implementation environment that may compromise or negatively influence the success of a given product). Strengths and Weaknesses correspond to factors internal to the developed artifact, while Opportunities and Threats relate to external factors [7].

The Table 1 presents the SWOT analysis performed on the product under study in this article.

In order to understand the level of acceptance of health professionals regarding the use of the tool presented in this dissertation, a questionnaire was developed and interviews were conducted, following the methodology and constructs of TAM, a model that evaluates a number of factors that influence their decision to use, how to use and when to use this technology [14].

The questions evaluated were the follow:

1 Do you consider important to use new technologies to support health care?
2 How much do you attribute to the need for an interactive platform that replaces the manual completion tables in each service?
3 Do you think this table optimizes your search for patient information?
4 The application is user friendly and intuitive.
5 The clinical terms used are correct and are used appropriately.
6 The icons illustrating the various clinical conditions are appropriate and easy to interpret.
7 The information contained in the table is that necessary to the context of hospitalization.
8 Inpatient information is now easier to consult and is always up to date.
9 The application improves communication and assists daily actions in the health unit.
10 Overall rate the developed application.

Table 1. SWOT analysis

Parameter	Analysis
Strengths	– Centralization and gathering of information dispersed in different HIS; – Appropriate and timely sharing of vital clinical information; – Provision of better information and evidence, thus increasing the quality of decision making; – Increased security and protection of data; – High usability, scalability and adaptability to various health units, multiple users and multiple devices
Weaknesses	– The data obtained from the different information systems may be incorrect, due to its incorrect fulfillment by the professionals; – Synchronization of data is not immediate
Opportunities	– Improvement of the quality and effectiveness of the services provided in the context of hospitalization; – Decrease in associated clinical errors; – Continuous increase of the interoperability of the HIS present in the health organization.
Treats	– Changes and failures in the structure of the databases that constitute the data source system may jeopardize the operation and purpose of the proposed system; – Low level of acceptance of professionals regarding a new technology, since the implementation of new technologies requires them to change their habits and to adapt to a new tool

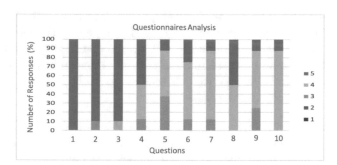

Fig. 5. Analysis of results obtained from questionnaires based on TAM.

The results obtained for the questionnaires are presented in the graph in Fig. 5. These results are presented as a percentage and the answers are in accordance with the *Likert* scale, where number 5 represents a very positive degree in relation to the statement or question and number 1 represents a negative degree, while the rest are intermediate degrees. To date, it has only been possible to collect the responses from a small sample of the target audience, since the application, although already functional in the institution, is not yet inserted in the daily lives of professionals. Thus, the answers obtained were given by professionals who already had contact with the application during training.

6   Conclusions and Future Work

This project had as main initial motivations the existence of communication failures, among professionals and between the different shifts, the occurrence of clinical errors due to incorrect information or the complexity of access to information. Also was motivated by the lack of interoperability among several HIS that contain inpatient relevant information and the lack of computer tools to support the practice of health care, specific to the context of hospitalization.

The main contribution of this application is the effective support to health professionals when providing health care, providing them with better real-time evidence for the accomplishment of their tasks and decisions, which results in the reduction of errors, better clinical outcomes, increased patient safety, reduced costs associated with clinical errors, and increased quality of care.

Also, the development of this project also improved the interoperability of the CHUP HIS, as it opened the door for the transmission and grouping of information crucial to the treatment of patients, which until now were stored in isolated IS, which makes access to this information harder.

As future work can be pointed out the need to insert a new functionality to normalize the exchange of information from shift to shift. Thus, it is intended to develop a module in the application where professionals can fill in the specific information regarding the episode of hospitalization, noting throughout the day the tasks performed, the tasks to be done and filling in more detailed observations. Thus obtaining a complete daily report of the patient's condition, presenting in detail examinations, analyzes, diagnoses and care, thus facilitating shift handovers. It is also intended to develop different profiles depending on the different types of health professionals. And finally, the implementation of sockets is also intended, which allows immediate synchronization.

Acknowledgements. This work has been supported by FCT – Fundação para a Ciência e Tecnologia within the Project Scope: UID/CEC/00319/2019.

References

1. Abdellatif, A., Bagian, J.P., Barajas, E.R., Cohen, M., Cousins, D., Denham, C.R., Horvath, D.: Communication during patient hand-overs: patient safety solutions, volume 1, solution 3, May 2007. Jt. Comm. J. Qual. Patient Saf. **33**(7), 439–442 (2007)
2. Schwendimann, R., Blatter, C., Dhaini, S., Simon, M., Ausserhofer, D.: The occurrence, types, consequences and preventability of in-hospital adverse events-a scoping review. BMC Health Serv. Res. **18**(1), 521 (2018)
3. Sousa-Pinto, B., Marques, B., Lopes, F., Freitas, A.: Frequency and impact of adverse events in inpatients: a nationwide analysis of episodes between 2000 and 2015. J. Med. Syst. **42**(3), 48 (2018)

4. Roh, H., Park, S.J., Kim, T.: Patient safety education to change medical students' attitudes and sense of responsibility. Med. Teach. **37**(10), 908–914 (2015)
5. Peixoto, H., Santos, M., Abelha, A., Machado, J.: Intelligence in Interoperability with AIDA. In: International Symposium on Methodologies for Intelligent Systems, pp. 264–273. Springer, Heidelberg, December 2012
6. Cardoso, L., Marins, F., Portela, F., Abelha, A., Machado, J.: Healthcare interoperability through intelligent agent technology. Procedia Technol. **16**, 1334–1341 (2014)
7. Guimarães, T., Coimbra, C., Portela, F., Santos, M.F., Machado, J., Abelha, A.: Step towards Multiplatform Framework for supporting Pediatric and Neonatology Care unit decision process. Procedia Comput. Sci. **63**, 561–568 (2015)
8. Peffers, K., Tuunanen, T., Rothenberger, M.A., Chatterjee, S.: A design science research methodology for information systems research. J. Manag. Inf. Syst. **24**(3), 45–77 (2007)
9. Haux, R., Ammenwerth, E., Winter, A., Brigl, B.: Strategic Information Management in Hospitals: An Introduction to Hospital Information Systems. Springer, New York (2004)
10. Machado, J.M., Alves, V., Abelha, A., Neves, J.: Ambient intelligence via multiagent systems in the medical arena. Eng. Intell. Syst. Electr. Eng. Commun. Spec. Issue Decis. Support. Syst. **15**(3), 151–157 (2007)
11. Berner, E.S.: Clinical Decision Support Systems, vol. 233. Springer Science, New York (2007)
12. Castaneda, C., Nalley, K., Mannion, C., Bhattacharyya, P., Blake, P., Pecora, A., Suh, K.S.: Clinical decision support systems for improving diagnostic accuracy and achieving precision medicine. J. Clin. Bioinform. **5**(1), 4 (2015)
13. Machado, J., Abelha, A. (eds.): Applying Business Intelligence to Clinical and Healthcare Organizations. IGI Global, Hershey (2016)
14. Portela, F., Aguiar, J., Santos, M.F., Abelha, A., Machado, J., Silva, Á., Rua, F.: Assessment of technology acceptance in intensive care units. Int. J. Syst. Serv. Oriented Eng. (IJSSOE) **4**(3), 26–45 (2014)

Automatic Processing of Histological Imaging to Aid Diagnosis of Cardiac Remodeling

Rogério Adriano de Sousa[1]([⊠]) [iD], Ana Carolina Mieko Omoto[2] [iD],
Rubens Fazan Junior[2] [iD], and Joaquim Cezar Felipe[1] [iD]

[1] Department of Computing and Mathematics, FFCLRP - USP,
Ribeirão Preto, Brazil
rogerio-sousa@usp.br
[2] Department of Physiology, FMRP - USP, Ribeirão Preto, Brazil

Abstract. Despite the growing level of research in the medical field, heart diseases are still predominant and recurrent, regardless of the degree of economic and social development of different populations. The absence of oxygen and nutrient intake causes cardiac cells to die, which are replaced by nonfunctional fibrotic tissue, leading to an accumulation of proteins in the extracellular matrix, usually replaced by collagen. The amount of interstitial collagen in myocardial fibers plays an important role in identifying changes in the heart after cardiac remodeling. This paper aims to model, implement and evaluate a method for automatic analysis of microscopic images of cardiac tissue, aiming to quantify the presence of interstitial collagen in myocardial fibers, using image processing and feature extraction techniques, in order to train a classifier that provides a reference to assist in the identification of reactive fibrosis. The method was evaluated by statistical measures, taking as reference the quantification previously performed by specialists. In the experimental tests our method achieved accuracy rate of 92.4%, showing that it is reliable and promising.

Keywords: Computer aided Diagnosis · Medical image processing · Heart remodeling

1 Introduction

According to the World Health Organization[1], acute myocardial infarction is responsible for more than 15 million deaths a year worldwide and most of these occurrences are due to the increase of life expectancy of the population and the prevalence of morbidities such as obesity and metabolic syndrome.

The process that leads to myocardial infarction stems from the lack of oxygen and nutrient input, which causes death of the cardiac cells. The death cells are replaced by nonfunctional fibrotic tissue, leading to ventricular dysfunction and marked changes in cardiac autonomic balance [1]. Fibrosis is characterized by an accumulation of extracellular matrix proteins, in particular collagen. The remodeling process occurs in death

---

[1] https://www.paho.org/bra/index.php?option=com_content&view=article&id=5253:doencas-cardiovasculares&Itemid=839.

© Springer Nature Switzerland AG 2020
Á. Rocha et al. (Eds.): ICITS 2020, AISC 1137, pp. 375–382, 2020.
https://doi.org/10.1007/978-3-030-40690-5_37

cardiac cell to ensure heart integrity without cell loss, and in response to stretching (reactive fibrosis) [14]. There are other pathophysiological conditions associated to cardiac fibrotic remodeling, including hypertension [2], diabetes type 1 and type 2 [3] or obesity [4], which, as well as myocardial infarction, leads to the formation of dense fibrous scabs [5].

Acute myocardial infarction usually causes myocardial ischemia/reperfusion injury (RI), which represents a major damage to the heart. The main therapeutic interventions used to minimize the damage caused by acute myocardial infarction are related to the restoration of myocardial perfusion [6–8].

The search for therapeutic strategies and the study of the mechanisms involved in RI represent important research issues. Histological analysis of cardiac tissue is a common method that provides not only the opportunity for quantification but also for the differentiation between morphological patterns and fibrotic accumulation in perivascular spaces (perivascular fibrosis) or interstitial fibrosis (fibrotic accumulations in the interstitium between cardiomyocytes) [17]. Figure 1 shows a typical image of cardiac tissue, with a resolution of 2048×1536 pixels, stained with red picrosirius and acquired through a microscope at 40x magnification.

Bioimaging Informatics consists of the retrieval, extraction, comparison and management of biomedical knowledge [15]. Bioimaging Informatics focuses on the application of image processing, database, mining and data visualization methods based on medical image processing and analysis [9, 16].

The Systems for Computer-aided Diagnosis (CAD) are used by specialists to identify abnormal patterns (abnormalities, structural changes of tissues, detection of lesions, among others) and further diagnosis of the medical professional, increasing their accuracy in the observation process. CADs make use of automatic or semi-automatic procedures. Semi-automatic procedures require human intervention to indicate some parameters or assist in image preprocessing, so this intervention often generates significant process restrictions in addition to the inherent subjectivity. Targeted at image analysis, CADs can provide features such as automatic morphometric evaluation, pattern classification from image scanning, image retrieval based on their content (texture characteristics, color, shape, size, etc.), or algorithms based on statistical and heuristic methods [10]. This has been made possible by developments in medical imaging technologies, such as radiological, nuclear medicine, MRI, endoscopic and microscopic imaging [11, 19].

Stereology is considered very important in the evaluation of quantitative information from microscopic sections, providing methods to evaluate different morphometric forms, including collagen volume, and can thus be used for quantification of fibrosis [18].

Experts have made use of generic computational image processing tools that provide visual score-based collagen quantification algorithms, where pre-processing of the image from strategic visual marking is possible for possible semi-automated analysis or stereology. These procedures vary considerably and are strongly dependent on

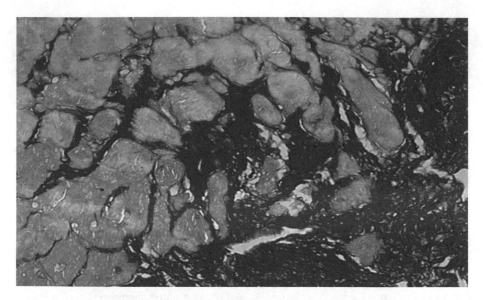

Fig. 1. Cardiac tissue. The red área represents cells with high concentration of colagen.

subjective parameters, thus limiting comparability between studies [12, 20]. Analysis of only a few sections is done quickly, but entails the risk of obliquity. Quantification by visual score depends on the analyst and is therefore very subjective and difficult to reproduce.

Supporting medical decision-making through CADs specifically designed for particular case studies and analysis has been of great benefit to the medical industry because CAD adds a rich layer of standardization and concentration of diagnostics in a single point of consultation. This procedure contrasts with the use of generic tools.

Thus, this work aims to develop a method for automatic analysis of myocardial tissue images that can support the medical diagnosis, providing support to the histological analysis procedure of the cardiac tissue in verifying the damage caused by acute myocardial infarction, more specifically the fibrosis process [13]. The system should offer specialists the identification and quantification of collagen and fibrotic accumulation in perivascular spaces [21].

The general objective of this study is to analyze myocardial tissue biopsy images automatically to characterize structural changes resulting from morbidities, which lead to the appearance of nonfunctional fibrotic tissue. This analysis consists of quantifying collagen present in cardiac tissue in images stained with red picrosirius by classifying fibrosis levels. Through image processing and machine learning techniques it is possible to separate the color channels, isolating areas where there is a higher concentration of collagen and thus suggesting the level of fibrosis present in the tissue. To consolidate the analysis, a classifier was designed based on the characteristics extracted from the images, following the classification standards used regularly by the specialists to analyze the cardiac tissue.

2 Material and Methods

For the implementation of the project we used the library OPENCV[2], a computer vision library and Machine Learning, which is written in C++ programming language, which was also the language used for the development of this work.

Another tool that we used for the development of this work is WEKA[3], which contains tools for preparation, classification, regression, clustering, association rule mining and data visualization.

To perform the tests and to validate the method, 10 images of picrosirius red stained animal heart biopsies were used to quantify interstitial collagen (fibers in red), provided by the Department of Physiology of a Medical School. These images have a resolution of 2048 × 1536 pixels, and were acquired through a microscope at 40x magnification.

The images are subdivided into: RI (animals that underwent surgery for induction of infarction followed by reperfusion) and Sham (control/healthy animals). Each animal has images of two regions of the heart, the left ventricular free wall (PL) and the septum.

After the implementation of the algorithm, the validation was performed by statistical analysis comparing the values obtained using manual segmentation of 4 images and the automatic segmentation performed by our algorithm. For the classification evaluation we used a set of 30 images.

3 Proposed Method

Fibrotic remodeling of the heart is a frequent condition caused by various diseases related to cardiac dysfunction. An important finding in cardiac fibrosis studies is the quantification of collagen in tissue, and the accumulation of this element is fundamental for a more accurate diagnosis in several patient's conditions and contexts.

One of the widely applied techniques is the semiautomatic threshold analysis, which has a significantly lower degree of accuracy, mainly due to the inclusion of non-collagen structures or the lack of detection of certain collagen deposits, increasing the error curve of a diagnosis due to the amount of noisy information.

This work aims to develop a method for automatic analysis through image processing and machine learning algorithms applied to cardiac tissue to quantify interstitial collagen in myocardial fibers.

First we extract the color channels from the RGB image and check the channel in which the collagen fibers stand out best. Then segmentation techniques are applied to isolate regions of the image where there is a higher concentration of collagen.

For better segmentation accuracy the threshold parameters were determined through a decision tree generated and classified by the WEKA software. Using 30 different images, each pixel pre-classified as collagen and non-collagen were used to train and test the classifier.

---

[2] https://opencv.org.

[3] https://www.cs.waikato.ac.nz/ml/weka/.

Finally, applying a new image to the classifier, each pixel can be classified after segmentation and the pixels representing the collagen in the total area of the image are computed.

4 Results and Discussion

To carry out the initial proposal related to the quantification of collagen present among cardiac tissue fibers, slides containing tissue samples stained with red picrosirius were used to highlight the amount of collagen in the tissue, considering that the regions containing collagen will have a chemical reddish color. From an image processing point of view, it is clear that it is necessary to extract from the image the pattern that makes the collagen present in the fiber as visible as possible.

For the implementation of the techniques, a C++ algorithm was developed using the OpenCV library and parameters extracted from an analysis using WEKA to classify tissues with positive or negative characteristics for collagen.

Extractions of the R, G and B (Red, Green, Blue) color channels were performed for testing purposes and to verify which delivers the best possible result. Examples of the original image and images resulting from channel extractions are shown in Fig. 2.

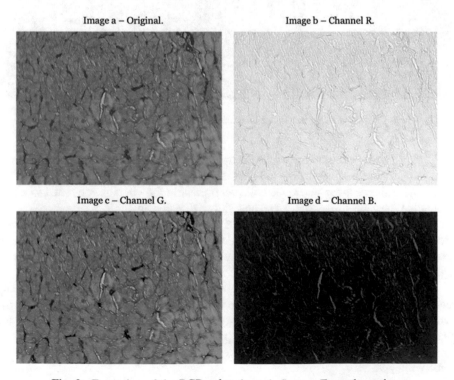

Fig. 2. Extraction of the RGB color channels Source: From the author.

Considering that collagen should be the most highlighted point in the images after the extraction of the channels, we identified the G channel image as allowing the greatest differentiation of the regions of collagen occurrence. Therefore, the further techniques are applied on the G channel image. Figure 3 represents the image after G channel extraction and subsequent application of the thresholding technique.

Fig. 3. G channel image thresholding Source: From the author.

The threshold parameters used for the application of the technique were determined from the decision tree generated and classified by WEKA.

Figure 3 shows areas where the collagen present in the fibers is converted to white, facilitating the next processing steps and aiding decision making.

After thresholding, the algorithm counts the ratio of white pixels to the total pixels in the image, which is the percentage of the image area that is occupied by the collagen concentration. With this result it is already possible to determine the amount of collagen present in the tissue in relation to the total proportion of the image.

To evaluate the proposed method, we used 10 images of different animals manually segmented by a specialist and obtained the percent quantification of collagen over the total area of tissue analyzed. Subsequently, these same images were processed by the algorithm and compared with the manual process.

The results achieved were: Precision = 92.4%, Accuracy = 86.8% and Sensitivity = 92.9%.

5 Conclusion

During the development of this work we extracted features of the images aiming at isolating and quantifying the collagen present in the cardiac tissue fibers. After testing and bibliographic studies, an algorithm with operations based on machine learning and digital image processing was assembled to classify and isolate the regions of interest.

The results achieved were satisfactory considering the original parameters. So, we can assert that the process automation proposal is reliable and accurate. We conclude that the developed algorithm is a tool that can quickly and accurately assist the decision making support for collagen quantification and, consequently, the degree of cardiac remodeling.

Acknowledgments. Grant #2016/17078-0, São Paulo Research Foundation (FAPESP).

References

1. Webb, S.W., Adgey, A.A.J., Pantridge, J.F.: Autonomic disturbance at onset of acute myocardial infarction. Br. Med. J. **3**, 89–92 (1972)
2. Diez, J.: Mechanisms of cardiac fibrosis in hypertension. J. Clin. Hypertens. (Greenwich) **9**, 546–550 (2007)
3. Russo, I., Frangogiannis, N.G.: Diabetes-associated cardiac fibrosis: cellular effectors, molecular mechanisms and therapeutic opportunities. J. Mol. Cell. Cardiol. **90**, 84–93 (2016)
4. Cavalera, M., Wang, J., Frangogiannis, N.G.: Obesity, metabolic dysfunction, and cardiac fibrosis: pathophysiological pathways, molecular mechanisms, and therapeutic opportunities. Transl. Res. **164**, 323–335 (2014)
5. Talman, V., Ruskoaho, H.: Cardiac fibrosis in myocardial infarction-from repair and remodeling to regeneration. Cell Tissue Res. **365**, 563–581 (2016)
6. Braunwald, E., Kloner, R.A.: Myocardial reperfusion: a double-edged sword? J. Clin. Invest. **76**(5), 1713–1719 (1985)
7. Piper, H.M., Garcıa-Dorado, D., Ovize, M.: A fresh look at reperfusion injury. Cardiovasc. Res. **38**(2), 291–300 (1998)
8. Yellon, D.M., Hausenloy, D.J.: Myocardial reperfusion injury. N. Engl. J. Med. **357**, 1121–1135 (2007)
9. Giger, M.L.: Computer-aided diagnosis of breast lesions in medical images. Comput. Sci. Eng. **2**(5), 39–45 (2000)
10. Müller, H., et al.: A review of content-based image retrieval systems in medical applications—clinical benefits and future directions. Int. J. Med. Inform. **73**(1), 1–23 (2004)
11. Miranda, G.H.B., Felipe, J.C.: Computer-aided diagnosis system based on fuzzy logic for breast cancer categorization. Comput. Biol. Med. **64**, 334–346 (2015)
12. Rao, R.B., et al.: Mining medical images. In: Proceedings of the Third Workshop on Data Mining Case Studies and Practice Prize, Fifteenth Annual SIGKDD International Conference on Knowledge Discovery and Data Mining (KDD 2009), Stanton, vol. 1895 (2009)
13. Eirin, A., Zhu, X.Y., Ferguson, C.M., Riester, S.M., van Wijnen, A.J., Lerman, A., Lerman, L.O.: Intra-renal delivery of mesenchymal stem cells attenuates myocardial injury after reversal of hypertension in porcine renovascular disease. Stem Cell Res. Ther. **6**, 7 (2015)

14. Akoum, N., Marrouche, N.: Assessment and impact of cardiac fibrosis on atrial fibrillation. Curr. Cardiol. Rep. **16**(8), 518 (2014). https://doi.org/10.1007/s11886-014-0518-z
15. Felipe, J.C.: Desenvolvimento de métodos para extração, comparação e análise de características intrísecas de imagens médicas, visando à recuperação perceptual por conteúdo (2005). https://doi.org/10.11606/T.55.2005.tde-12082006-000758. 2008, 176
16. Gonzalez, R.C., Woods, R.C.: Processamento digital de imagens. Pearson Educación, São Paulo (2009)
17. Jensen, E.C.: Quantitative analysis of histological staining and fluorescence using ImageJ. Anat. Rec. (2013). https://doi.org/10.1002/ar.22641
18. Júnior, J.S.: Morfologia dos cardiomiócitos e quantificação do colágeno no miocárdio de ratas tratadas com isoflavonas ou estrogênios. Cep, 0–5 (2012). http://www.scielo.br/pdf/rbgo/v34n10/a03v34n10.pdf
19. Pasqualin, C., Gannier, F., Malécot, C.O., Bredeloux, P., Maupoil, V.: Automatic quantitative analysis of t-tubule organization in cardiac myocytes using ImageJ. Am. J. Physiol. Cell Physiol. (2015). https://doi.org/10.1152/ajpcell.00259.2014
20. de Queiroz, J.E.R., Gomes, H.M.: Introdução ao Processamento Digital de Imagens. Rita **8**(1), 1–31 (2001). https://doi.org/10.1590/S0102-261X1998000100035
21. Schipke, J., Brandenberger, C., Rajces, A., Manninger, M., Alogna, A., Post, H., Mühlfeld, C.: Assessment of cardiac fibrosis: a morphometric method comparison for collagen quantification. J. Appl. Physiol. (2017). https://doi.org/10.1152/japplphysiol.00987.2016

Expectation Differences Between Students and Staff of Using Learning Analytics in Finnish Universities

Jussi Okkonen[✉], Tanja Helle, and Hanna Lindsten

Tampere University, 33014 Tampere, Finland
jussi.okkonen@tuni.fi

Abstract. The aim of this paper is to examine and discuss the dissonance between expectations and hopes towards utilising learning analytics in Finnish universities. The analysis is based on data collected among Finnish university students and staff in spring 2019. As a key result we present, that the university staff found it important that all data and information should and could be used for various planning, management and counselling purposes. At the same time student found it unnecessary or even harmful to allow university staff examine their personal data. We therefore propose that universities should develop and implement specific policy for using of analytical data.

Keywords: Learning analytics · Higher education · User expectations · Survey

1 Introduction

Using learning analytics in higher education is a game of contractionary hopes, motives and goals. In general, the literature draws an idealistic picture of well-functioning apparatus that supports studying process, enhances learning outcomes and eases the managerial burden. The praxis is different as there are several different conceptions of learning analytics and motivations of using it. Also, the personal role in the organisation affects to attitudes.

Extensive use of educational technology, especially digital learning environments, digital curricula, and digital managerial systems have brought about the need for analytics to monitor the use of the learning environments and the learning itself. Technology that is more sophisticated is provided by educational technology industry to better serve teachers, end users and primary customers. There are four main trends in enhancing teaching and learning practices. On micro level (learning event), analytics is for assessment of achieving certain goals. On meso level (subject level), i.e. implementing the curricula analytics are for achievements, adaptivity and general assessment. On macro level analytics if for promoting management by knowledge, risk assessment and measuring key performance indicators on different levels. The fourth trend is compliance, privacy, and security issues on the level mentioned and can be considered as the primary prerequisite for utilising learning analytics.

© Springer Nature Switzerland AG 2020
Á. Rocha et al. (Eds.): ICITS 2020, AISC 1137, pp. 383–393, 2020.
https://doi.org/10.1007/978-3-030-40690-5_38

The levels of learning analytics could be elaborated as follows:

- Micro level concentrates on student and individual learning processes. This is enabled by gathering extensive, high quality data on student transactions and subject under study.
- Meso level concentrates on tuition provider and development of teaching practices. This level includes the levels of subject, curricula and teaching events as aggregate.
- Macro level learning analytics aims to managing learning by knowledge, recognition of risks and acknowledging hindrances of optimal flow in learning processes.

On all three levels learning analytics are somewhat overlapping and complementary.

Learning analytics aims to visualise, analyse, and interpret the actions of users in the context of learning and teaching (e.g. [1–3]). By achieving these aims new ways of teaching, learning, organisational functioning, and decision making could be achieved. To fully utilise learning analytics also ethical and privacy issues must be concerned [4, 5]. The gathering and integration of the data, providing consent, anonymisation of data, transparency, information security, interpretation of the data, data management are issues of national and international laws and statutes that should be taken into account when learning analytics is integrated into educational technology. In order to reach beyond the state of the art learning analytics needs to be developed in cooperation with key stakeholders, e.g. students, institutions and teachers [6]. Moreover, practices should be developed, tested, and validated in wild among practitioners. The development should be conducted in dialogue on the levels of technology, studying and teaching, and administrative practices.

In recent studies ethical issues have been considered. In particular, there have been concerns about misusing the data by individuals and that decisions are made based on incorrect information or that learning analytics itself is based on false techniques. However, in a study implemented in 2017, it was found that students are rarely asked what they think about using learning analytics [4]. This deficiency itself can be seen as an ethical issue which was strongly raised in the result of our study. It is important to note, that approximately 40% on responded students answered, that there are some ethical issues to consider.

Ethical issues are studied in the study of Arnold and Sclater [5]. They explored student perceptions of their privacy in learning analytics applications. The study showed that the majority of the students accepted the use of their data but also approval varies depending on suggested aim of the analytics. The variations between students enrolled at British and American educational institutions were recognized as well. In that study more than half of American respondents and only a quarter of the respondents from United Kingdom would like to have a possibility to compare with other students in the learning analytics application. The researchers explained that students were unclear of exactly how the comparison would happen, for instance how anonymously it would have implemented. Addition to ethical issues, students' expectations were also taken account into Whitelock-Wainwright et al. [7] who present a descriptive tool for measuring student expectations of learning analytics services. Expectations can be distinguished between ideal and predicted expectations. Through literature the researchers produced four identified expectation themes: ethical and privacy, agency, intervention, and meaningfulness expectations [7, 8].

Using learning analytics in higher education is a game of contractionary hopes, motives and goals. In general, the literature draws an idealistic picture of well-functioning apparatus that supports studying process, enhances learning outcomes and eases the managerial burden. The praxis is different as there are several different conceptions of learning analytics and motivations of using it. Also, the personal role in the organisation affects to attitudes. The aim of this paper is to examine and discuss the dissonance between expectations and hopes towards utilising learning analytics in Finnish universities. The analysis is based on data collected among Finnish university students and staff in spring 2019.

2   Data on Finnish Students and University Staff

The data of this study has been collected from six Finnish universities during the spring of 2019. The universities are Tampere University, Aalto University, LUT University, University of Turku, University of Eastern Finland and University of Oulu. The inquiry was disseminated through news bulletins of university intranet sites and targeted mailing lists. The purpose was to reach the widest possible range of respondents. The respondents were divided according to the user groups of the analytical data, into groups of students, teachers, teacher tutors, i.e. those who are responsible for the academic counselling of individuals, study coordinators and those responsible for education. Each of these groups had their own form to answer, questions targeted to take in to account their possible, special needs on utilizing learning analytics. Heads of study affairs, heads of degree programmes, deans and vice deans responsible for education were instructed to respond to the survey of those responsible for education. In one of the six universities (University of Oulu), the inquiry was only distributed to a teacher tutor. The way in which a poll was distributed in each university varied, so that it is very difficult to estimate the number of respondents to the inquiry and thus the response rate. It is known that there are a total of 77 430 Bachelor's or Master's degree students in the six universities. We assume that the different ways in which universities distributed the questionnaire has also had a strong impact on the number of respondents. In the analysis and interpretation of the material, it should be noted that the respondents have had experience with the registry systems used at their own university and in their responses they reflect the experience of that particular system.

Table 1 displays the amount of all students of universities as well as the amounts and percentages of different groups participated in this survey.

Table 1. Respondents of the inquiry

University	Amount of students	Student responded	Teachers responded	Teacher tutors responded	Study coordinators responded	Responsible for education responded
Tampere University	18 402	150 66%	41 44%	11 13%	21 44%	12 40%
Aalto University	14 712	50 22%	14 15%	7 8%	2 4%	1 3%
LUT University	4 767	6 3%	6 7%	–	3 6%	10 33%
University of Turku	14 178	8 4%	17 18%	7 8%	15 31%	5 17%
University of Eastern Finland	13 884	13 6%	15 16%	7 8%	7 15%	2 7%
University of Oulu	11 487	–	–	45 52%	–	–
Total number of respondents of each group		227	93	77	48	30

The purpose of the five user surveys was primarily to find out what the different user groups consider to be important goals in utilizing analytical information. The survey was not collected primarily for research purpose, it was mainly to be used in the development of applications and information systems utilizing analytical information. This places limitations on the analysis of the material, for example the questions asked to different user groups are not always comparable.

The questionnaires consisted of both multiple choice and open-ended answer questions. Multiple choice answers were directly coded into SPSS which was used for statistical analysis. Open-ended answers were processed by outlining the key topics and creating a variety of categories which would describe responses as closely as possible. It is also important to notice that a certain response could include in severe codes.

In the questionnaires there were also open-ended questions in order to help us collect a lot of information about the reasons for different answers. Open-ended answers collected rich information especially in ethical considerations and utilization of the analytical data. Those two questions were presented in all the forms in order to make the comparison possible. In addition, there were also some questions which appeared on some forms. There were 86 responses from the students for question which explored the ethical issues and 81 responses for a question related to how the university should utilize student data. That indicates that more than a third of student respondents answered to these two pure open-ended questions.

The surveys were based on three group interviews for teachers, students, and for study managers and study coordinators together. Based on these background interviews we found that coordinators and study managers have a very different perspective on the use of register and the information based on that. Therefore, separate forms were prepared for them. It was also found that teachers who work as teacher tutors should also have a form their own.

At the same time, it was found that the work of coordinators and teacher tutors, the guidance of studies, often lies at the interface between micro and meso levels, the coordination of teaching and study requirements and the student's personal curriculum. It can be said that the levels of utilization of analytical data are not so much divisible by the actors, but rather the processes that can be supported by analytical data. The levels and processes concerning the analytical information are presented in Table 2.

Table 2. Different agents' processes concerning the analytical information and levels

Agent	Process, concerning the analytical information	Level
Student	Studying and planning studies	Micro
Coordinator/Tutor teacher	Counselling, guidance	Micro
Teacher	Planning education	Meso

These findings, based on three group interviews as mentioned above, were taken into account in the preparation of the questionnaires. The student's form focused on particular on ways of utilizing analytical knowledge that could support the student's planning, guidance and follow-up. The study coordinator and teacher tutor surveys focused in particular on the needs of study guidance and in the teachers' questionnaire on the information to be used in planning teaching of individual courses. In particular, the survey for those who are responsible for the studies focused on information management issues. To collect unpredictable information the question how the university should use the information collected in the systems was asked in all forms.

Ethical issues, especially with regard to the private information about the individual student, were strongly raised in the students' interview. As a result, surveys attempted to identify students' readiness to allow different user groups to utilize student personal information as well as the need for other user groups to utilize such an information. In addition, an open-ended question asked the respondents about the possible ethical problems associated with the use of analytical information.

In this study we found ethical concerns which are partly complementary with those presented, and may be divided into the following three categories: (1) in individual behavior, (2) in policy of institution using learning analytics or (3) in validity and reliability of the data. The ethical considerations are reflected by the results of the study as those the key for justifying the use of learning analytics, successfully implement learning analytic, and executing learning analytic policy of the respective institution.

3 Findings Related to Expectation Differences

The respondents were either staff members or students that are high achievers in sense of how their studies have progressed. At this point, we were interested in whether the staff and students shared their expectations on learning information system. The expectations about access to students' data and combining that data were studied in Likert type questions. The perceptions of utilising the learning information system was found out by open-ended questions. The results on these issues were unanimous, based on the view of the majority.

There were several different insights on how data and information from different systems should be used for analytical purposes or even for counselling purposes. The general expectations for information systems were dichotomic. The staff found it almost self-evident that all data and information should and could be used for various planning, management and counselling purposes, yet the students were more critical. Especially use of personal data without permission on consent was the divider. The students had clear idea that the use of any personal data should be authorized by informed consent and the use should also be justified and bound to certain use context. According to the staff, combining the data from different sources would be useful. However, the students didn't appreciate that feature of data usage. Although there were different insights, both staff and students had common sense that analytical data should be used for management and development of studies. Staff's and students' expectations are described in Table 3.

Table 3. Staff and students' expectations of learning information system

Staff's expectation	Students' expectation
Access to students' data	Critical attitude, permission on consent required for personal data
Combine the data from different sources	Not combining the data from difference sources without per-mission
Data utilized for management and development of studies	Data utilized for management and development of studies

Teachers having role as a tutor and studying coordinators had explicit wish to have detailed and up to date information on students' personal studying plans, current degree programs and progress of the studies. Staff members also had need for easier access to data and information from different sources. Those in the role of administration had distinctly versatile information need from the systems. Evidently their role requires statistics of attendance, success, and other course or degree related issues. However, coordinators are somewhere in between as they had need for information on individuals as well as on different groups. The staff had needs to drill into student groups, and sometimes even to individuals, in order to maintain view on operational status. There was unanimity that information from different sources could and should be utilised to management and development of studies. However, the extent of using the data and scope varied distinctly among the respondents. Staff's specified expectations are presented in Table 4.

Table 4. Staff's specified expectation of learning information system

Role of staff	Expectation
Tutor teacher	Students' personal studying plan
Studying coordinator	Current degree program details Progress of studies
Responsible for education	Statistics of attendance and success Course or degree related issues
Teacher	List of students registered

There are differences among the students between the disciplines and some trend on those discrepancies could be drawn on Table 5. There was somewhat uniform opinion that the system should send notifications to student if progress does not meet the studying plan or if one should enroll to an exam or a course. There was also strong agreement on that system should provide schedule according to planned studies. Those studying in Medical school students were more willing to have ready, general schedule as they have less options in their studies.

Students did not want to be compared with other students, yet there was variation according to the field of studies. The students of humanities were most strictly against comparing individuals yet those studying medical had less objection. Moreover, the students of technology had wish for difference performance indicators. The students of medicine and education were most critical toward extensive use of performance indicators. However, there was unanimity on using feedback on good performance to motivate even beyond. The science students seemed to need less this kind of feedback. The students of social sciences were the most unsure on additional features of information systems as they had insight that those should contain curriculum, studying plans and registry for passed courses.

Table 5. Shared and different expectations of the students

Students' shared expectations (%)	Students' different expectations (%)
Notification to enroll to an exam or a course (app. 87%)	Compared with other students Humanities (85%) Medical (57%)
Ready structured schedule related to studying plan (74%)	Use of performance indicators Technology (72%) Medical (50%) Education (47%)
Notification when progress doesn't meet the studying plan (app. 40%)	

There were some distinctions when comparing according to the year of the studies. Students starting their studies and fourth year students had wish for ready structured schedule. The student between had wish for a ready, yet personalised schedule. This reflects the notion of progress and general performance. Less than half of all students evenly in all programs had wish for notification of underperformance and lagging, but on the other hand also half of them were against such notifications. This seems to divide students, as there were only a few indifferent. Most important system feature was notifications on enrolling certain course or enrolling to exam.

Using information to compare individual performance to peers was mostly welcome by fifth year students, i.e., those about to graduate. Those already behind the designated schedule were mostly against such comparison, obviously for being unwilling to be stigmatized or notified for underperformance. However, there was ambivalence towards performance indicators as first year and fifth year student were in favor of them, but fourth year students were most against. The same groups also wished for motivational feedback for good performance and the analogically the same groups found it unnecessary.

In general, the information system should be beyond of being just repository for passed courses and studying plans. Especially the first, fourth, and sixth year students recognized the need for additional features. However, the respondents were divided as almost one third had no opinion on most additional features.

The ethical considerations varied between different respondent groups. The most distinct difference was between the way the domain and scope of learning analytics activities are justified and accepted. The managerial perspective, i.e. all other respondents than students, see learning analytics as a meso-level and macro-level activity that mostly deals with anonymized data and information. The students have totally different notion as they see themselves as subjects of learning analytics activities and they take it quite personally. From the ethics perspective this kind on distinction was established by the authorization about the use of individual data. Students are on micro-level and for them learning is personal and related to achieving their degree. Staff is somewhat liberal on granting or gaining access to student data, even non-aggregated data. The students on the other hand are only willing to grant access only based on justified reasons, e.g. certain staff member is their tutor, or they need some kind of special service provided by the administration. Taking the evident differences between rationale of using learning analytics as well as different scopes and domains of using it, the differences on expectations raise from the presented levels of learning analytics and the sources of ethical issues.

Table 6 indicates the major ethical considerations of different groups. The results show that the policy of institution is seen here as the major ethical consideration. Yet the more accurate reasons behind the policy of institution reveal the different views of the users. The students would like to have strict instructions and transparency of the data while studying coordinators see the data combining as an ethical risk. Those who are responsible for education suggest that only statistical data should be used. In this context the policy of institution means the process related to learning analytics. In the open-ended responses, both the students and staff underlined the concern about how the data is used and they don't necessarily even know how it is used. The majority of teachers pointed out that they don't see any ethical considerations using data. Tutor teachers considered the validity and reliability of the data as the major ethical problem. They were especially worried about the use of sensitive and personal data of students.

Table 6. Ethical considerations

The role of respondent	The major ethical consideration (%)
Student	Policy: Strict instructions and transparency (31%)
Teacher	Individual: No ethical considerations (24%)
Tutor teacher	Validity and reliability: Use of sensitive and personal data (39%)
Studying coordinator	Policy: Combining the data (55%)
Responsible for education	Policy: Only statistical data should be used (27%)

4 Discussion

The results presented above implicate that the ethical issues raise from using learning analytics as vehicle and support for tutoring as students and staff members have distinct expectations of how data is gathered, analysed, used and distributed within course, degree program or institution (cf. [9–11]). The privacy issues are considered as students are reluctant on indifferent to grant access to their personal information for random staff members. On the contrary the managerial perspective for learning analytics calls for accessible data for certain processes or even to ad hoc purposes. This is analogical to all other data privacy discussion as vendors and service providers justify their expanding data consumption by better and well targeted services and products. People, especially those with some degree of privacy awareness, tend to think just the opposite. Especially when analytics is implemented in full extension, the privacy issue, or how my data is used, takes new form. As presented above only half of the students were willing to receive benchmarking or other algorithm powered data on their performance. This reflects to finding that university students considered themselves as part of academic community, not raw material, or livestock, refined for the needs of the society.

Institutions have established policies for using learning analytics and staff mostly comply with those. It is not that divergent expectations exist, but mostly it is about what to staff, and especially teaching staff, is willing to allocate their scarce resources. The survey brought about the issue that teaching staff has low expectations towards learning analytics as they need mostly up to date contacts to students and related information. The realistic expectations of teaching staff seem to dilute the high hopes of data driven tutoring and counselling of students.

Since most of the studying related processes are digitalised and large amount of activities can be carried out in platforms universities provide, the window of opportunities is open for utilising learning analytic, say data driven management, more throughly. The results point out the dichotomy in expectations as students and teachers seek to manage daily activities on micro-level, but those in managerial position seek extensive leverage effect on meso- and macro-levels. These are not contradictory, yet they might not be served by the same development schemes and policies. The students and teachers are willing to maintain personal relationships and utilise digital tools for process management. The managerial perspective [over] emphasises openness of the data, analytical approach, and seeking gains on mass as well as putting effort on what brings reward to the institution.

It should be noted that the Finnish university system has so far used very little analytical information in students guidance. As previous research has shown, students tend to be more positive about a system that they already use and whose logic they are familiar with, cf. [7]. Some of the critical attitude of the respondents to the information produced by the hypothetical system may be explained by its unfamiliarity.

5   Conclusions

The study was based on simple sample data and therefore the results are explorative by the nature. When assessing the results one should take into account that the original aim was not to conduct an exhaustive survey, but verify the findings of a qualitative user needs study. The results also serve as verification for certain assumptions regarding to attitudes and insight in form of expectations. The results are context sensitive as Finnish higher education has its unique features, yet applicable to those contexts where university education is publicly produced private commodity, i.e. not driven by supply and demand but public steering and financing. Also, important note in contextualising the results is to note that such a system operates in a very different logic than one based on large tuition fees and offering tuition to all with sufficient means to acquire university education.

As stated earlier, the universities have had different ways of inviting to participate in this research. This has affected not only the number of respondents but also the quality of them. Based on the number of credits reported by students, it is noted that 70% of respondents complete their studies within the target time, while we know that only 20% of Finnish university students complete their degree in target time. Thus, the sample does not represent the population in this respect.

The future research on the theme will provide additional evidence on user centric approach to learning analytics. The results of this survey are operationalised to design targets and functionalities of a student service as well as how the data could be utilised on micro-, meso- and macro-level in the respective institutions. The main result is, that different groups of users have different aims and needs in using analytical data. The needs of the different groups vary and might even be conflicting. Therefore, it is important to evaluate how the objective of different users of analytical information can be combined. University policy should pay a special attention to the ethical aspects of data use and gaps that may allow individual and sensitive information to fall into the wrong hands. In each situation a student should have an opportunity to decide where his/her personal information will be used and who it is to be seen.

References

1. Ferguson, R., Brasher, A., Clow, D., Cooper, A., Hillaire, G., Mittelmeier, J., et al.: Research evidence on the use of learning analytics: implications for education policy. European Union, Centre for Research in Education and Educational Technology (2016)
2. Greller, W., Drachsler, H.: Translating learning into numbers: a generic framework for learning analytics. J. Educ. Technol. Soc. **15**(3), 42–57 (2012). Learning and Knowledge Analytics

3. Viberg, O., Hatakka, M., Bälter, O., Mavroudi, A.: The current landscape of learning analytics in higher education. Comput. Hum. Behav. **89**, 98–110 (2018)
4. Tsai, Y.-S., Gaševic, D., Whitelock-Wainwright, A., Munoz-Merino, P.J., Moreno-Marcos, P.M., Fernández, A.R., Kollom, K.: SHEILA: supporting higher education to integrate learning analytics (2018)
5. Arnold, K.E., Sclater, N.: Student perceptions of their privacy in learning analytics applications. In: LAK 2017 (2017)
6. Ferguson, R.: Learning analytics: drivers, developments and challenges. Int. J. Technol. Enhanc. Learn. **4**(5/6), 304–317 (2012)
7. Whitelock-Wainwright, A., Gašević, D., Tejeiro, R., Tsai, Y.-S., Bennett, K.: The student expectations of learning analytics questionnaire. J. Comput. Assist. Learn. **35**, 633–666 (2019)
8. Elouazizi, N.: Critical factors in data governance for learning analytics. J. Learn. Anal. **1**(3), 211–222 (2014)
9. Howell, J.A., Roberts, L.D., Seaman, K., Gibson, D.C.: Are we on our way to becoming a "helicopter university"? Academics' views on learning analytics. Technol. Knowl. Learn. **23**, 1 (2018)
10. Roberts, L.D., Howell, J.A., Seaman, K., Gibson, D.C.: Student attitudes toward learning analytics in higher education: "the fitbit version of the learning world". Front. Psychol. **7**, 1959 (2016)
11. Roberts, L., Chang, V., Gibson, D.: Ethical considerations in adopting a university- and system-wide approach to data and learning analytics. In: Kei Daniel, B. (ed.) Big Data and Learning Analytics in Higher Education, pp. 89–108. Springer, Cham (2016)

Software Systems, Architectures, Applications and Tools

Multiagent System for Controlling a Digital Home Connected Based on Internet of Things

Pablo Pico-Valencia[1(✉)], Belkix Requejo-Micolta[1],
and Juan A. Holgado-Terriza[2]

[1] Pontifical Catholic University of Ecuador, Esmeraldas, Ecuador
{pablo.pico,belkix.requejo}@pucese.edu.ec
[2] University of Granada, Granada, Spain
jholgado@ugr.es

Abstract. This paper presents a Multiagent System (MAS) oriented to proactively control a connected digital home scenario based on Internet of Things (IoT). The objects of IoT have been deployed using the openHAB tool and the agents that control these objects have been programmed using Java Agent DEvelopment Framework (JADE). The agents incorporate a behavior that enables them to be able to execute automatic request/response processes that allow them sharing data of interest and executing rules that define complex behaviors. As a result, we obtained a system developed in Java that proactively controls the objects of IoT according to the changes raised in a digital home scenario.

Keywords: Multiagent System · Internet of Things · JADE · openHAB

1 Introduction

The Internet of Things (IoT) is a paradigm aimed at creating a network of intelligent objects that have the capacity to self-organize, share information, data and resources, reacting and acting on the basis of changes in the environment [15]. This opens up a wide range of applications that can solve problems in which human beings develop daily [21].

Within the context of the IoT, scenarios such as home, city, health, university and industry have begun a transformation process to become smart environments [3]. In these new scenarios, technologies based on Internet are capable of measuring data transforming them into useful information for decision making that improves people's quality of life and optimizes the management of services. In this particular case, Artificial Intelligence (AI) techniques such as Machine Learning [18], Data Mining [13], Multiagent Systems (MASs) [20] and Genetic Algorithms [16] have been applied. All these techniques (individually or hybrids) support modeling more intelligence processes [1].

© Springer Nature Switzerland AG 2020
Á. Rocha et al. (Eds.): ICITS 2020, AISC 1137, pp. 397–406, 2020.
https://doi.org/10.1007/978-3-030-40690-5_39

In the particular case of MASs, software agents are one of the AI techniques that integrated with IoT allow this type of ecosystems to scale in order they acquire characteristics that IoT objects do not own. These characteristics are autonomy, intelligence, proactivity and collaboration [7]. Then, software agents, integrated in IoT ecosystems, enable connected objects to be able to operate in a more active way; that is, by proactively capturing data and controlling actuators. Moreover, through software agents, IoT objects can modify their behavior sharing data with external objects, through the agent that manages them. Thus, actions based on collective intelligence are enabled. Consequently, smarter and more complex actions on IoT can be achieved by groups of agents that share data with their counterparts.

Today, many of the commercialized IoT objects are marketed as smart objects. However, their degree of intelligence is limited to the parameterization of the data used by a set of rules upon which they direct their behavior. Also, many of these objects are limited to executing actions based on self-captured data. In general, few IoT objects are capable to support a proactive behavior and act based on external data. A particular case that adopts these good practices is Netatmo, an IoT weather station equipped with indoor and outdoor modules that measure weather factors such as temperature, humidity or air quality, and analyze past readings, observing the present and making predictions to adapt to users' lifestyles [17]. Against this background, the definition of specific agent models for IoT from a practical perspective is relevant. This motivates us to use the Java Agent DEvelopment Framework (JADE) [2,8] to develop software agents oriented to model a behavior focused on the automatic control of IoT objects as well as their associated resources [12].

The purpose of this study is to present an application that integrates technologies aimed at managing IoT platforms with agents. The JADE framework has been used to create agents that proactively control objects in an IoT ecosystem. Agents base their behavior on a request/response mechanism so that they can establish automatic collaborative processes to execute more complex processes than those carried out by isolated objects. In order to validate our proposal, a case study has been conducted in a digital home accessed using openHAB, a tool specialized in the control of connected digital home systems [11].

This paper is structured in five sections. Section 2 presents the general concepts oriented to the management of IoT objects from the MAS perspective. Section 3 describes the main design and development aspects of the proposed MAS for controlling IoT devices using a REST interface. Section 4 details the results of the scope of a MAS aimed at providing comfort in a connected digital home scenario. Finally, Sect. 5 summarizes the conclusions and future works.

2   Related Works

The agent-based modeling of the IoT involves two technologies: agent-oriented and IoT technologies. In both, several middlewares/frameworks have been proposed for enabling the creation of agents and accessing IoT objects, respectively.

Regarding tools already proposed for the management of IoT objects and their resources, the study developed by Razzaque et al. [23] defines a taxonomy of the middlewares to do it. An IoT middleware plays a important role since it facilitates the interaction between a multitude of devices and the sensed data [4]. Thus, the taxonomy proposed in several studies includes middleware based on events, services, virtual machine, agent, tuple space, database oriented and based on specific applications [23].

Additionally, studies such as those proposed by Razzaque et al. [23], Chelloug ct al. [4] and Savaglio et al. [24] evidence that the integration of IoT and agent-oriented technologies constitutes an alternative for controlling IoT. Thus, agent-based modeling is defined as a paradigm of modeling, programming and simulation of IoT ecosystems. Some of the agent-based IoT middlewares are the following: IMPALA, Smart messages, ActorNet, Agila, Ubiware, UbiROAD, AFME, MAPS, MASPOT, TinyMAPS, ACOSO [23]. These middlewares are mainly oriented to improve the interoperability in IoT ecosystems establishing communication with agents linked to IoT running in distributed scenarios.

The agent-based modeling of the IoT has gained importance in recent years [20]. In this sense, several models of agents running on IoT's own objects have been proposed (e.g., smart objects, IoT-a, agents of things, iota, REST and HTML5 agents) [20]. However, as this study focuses mainly on the integration of agents in IoT ecosystems using the MAS approach, relevant works that implement MASs running over personal computer or servers have also been analyzed.

Table 1 lists some approaches of MASs aimed at IoT management proposed to date. In general, MASs include heavyweight and lightweight agents developed in JADE. However, in particular cases, some research projects propose a specific agent model/tool for the development of the MAS. The use of any model/tool depends on the scope of the agents, the execution environment, the application domain and the problem to be solved. Likewise, the available technological resources is another determining aspect when developing a MAS for the IoT.

The integration of MASs with IoT technologies has been done using middleware and specific tools that depend on the domain of the problem to be solved. Practical cases that support this assertion are the models of agents employed to create MASs described in column six of Table 1. Six of the ten approaches analyzed in this table developed their own agent model (items 4–9). However, some of them used JADE as the basis for implementation (items 6, 8) as did the approaches listed in items 1–3. Less common ones are approaches for general purpose (item 10) that can be implemented in the agent tool that developer team considers appropriate.

On the other hand, with respect to the reach acquired in IoT using MASs, most focus on introducing an intelligence component to IoT technologies. They use rules, knowledge base and in some cases AI algorithms. Finally, as summarized in Table 1, agents also model behaviors that establish collaborative processes among MASs.

Table 1. Main approaches of MASs for the management of IoT. P = Proactivity, I = Intelligence and C = Collaboration.

#	MAS approach	P	I	C	Agent tool	IoT tool
1	Context-based [25]		*	*	JADE	General
2	Knowledge-based [27]		*	*	JADE	RFID
3	Machine Learning-based [18]		*		JADE	PC/smartphones
4	Recommendation-based [9]		*		Cyber agent	General
5	Distributed mobility-based [5]			*	DMA	General
6	Cloud Computing-based [10]			*	ACOSO	General
7	Edge Computing-based [26]	*		*	COSAP	PC/smartphones
8	Reasoning-based [19]		*		UmU-Act	Raspberry PI
9	Semantic-based [22]	*	*	*	LOA	PC/SBC
10	Service-based [14]		*	*	General	General

3   Our Approach

3.1   Architecture Overview

Our MAS links an agent to each of the IoT objects of an ecosystem (agents of things). From agents of things and agent coordinators that organize them a MAS for IoT is formed. Thus, a MAS in the context of IoT consists of a set of agent coordinators that manage communications between agents of things that handle objects and their data captured in order to achieve specific targets.

The main elements making up the proposed architecture are illustrated in Fig. 1. We have used JADE agents in the proposed architecture because they can interoperate with external MASs compatible with FIPA (Foundation for Intelligent Physical Agents) standard communication. However, our model can be extended to be implemented using another agent framework.

The proposed architecture is a 4-layer architecture: physical IoT, virtual IoT, service and agent layer. The lower layer is the physical IoT layer. This layer manages low-level communications with the IoT devices connected to the network. However, in order to abstract the complexity when accessing IoT devices in this layer are performed, a second layer, virtual IoT layer, has been introduced. This particular layer can be implemented by a middleware oriented to the management of the IoT. Next layer, service layer, can be implemented with a existing middlewares similarly the previous layer. This layer can be incorporated by a service-based IoT middleware such as those described in [23]. However, in order to facilitate accessing to the IoT layers, it is recommended to use an IoT middleware compatible with the Resource Oriented Architecture (ROA) [6], which enables accessing data and objects via uniform resource identifiers (URIs).

Finally, a layer of perception modeled by agents has been introduced at a higher level. These agents communicate with each other to create MASs. From these agents data gathered from the IoT layer is managed in order to provide services or operations demanded.

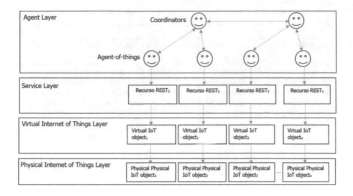

Fig. 1. Proposed 4-layer architecture.

3.2 Selected Middlewares

Two software tools have been selected to implement the architecture shown in Fig. 1. For agent technologies, the JADE tool has been used due to the fact that, it is one of the most widespread frameworks used by developers of MASs. Additionally, JADE has been used because agents can run on non sophisticated machines interoperating with external MASs that run in a distributed way.

On the other hand, regarding the management of the IoT, openHAB has been used because the domain of the application of this study is the digital home. OpenHAB has compatibility to connect to heterogeneous physical devices (e.g., TVs, light bulbs, HVAC, temperature sensors) of multiple brands (e.g., Samsung, LG, Xiaomy, Nest, Pioneer, Hue). Additionally, openHAB is an open source tool that keeps compatibility with ROA and consequently, accessing and controlling devices is possible to be achieved by using URIs. This aspect abstracts the complexity of accessing to physical devices and therefore, agents can compose business processes for controlling the IoT by invoking RESTful services [6].

3.3 Agent Model

Figure 1 shows the use of agents of things and coordinator agents to create the MAS for IoT. These agents save particularities such as using behaviors to model their tasks for achieving their goals. However, they have also some differences because they have specialized actions in the IoT ecosystem.

Figure 2 illustrates the structural elements that make up both the agents of things (a) and the coordinator agents (b). Both agents have three common elements: FIPA communication interface (enabling communication with external agents), IoT management interface (accessing to resources and data captured by the object) and a knowledge base (input data for the agent to operate). However, some of these elements have been specialized according each type of agent.

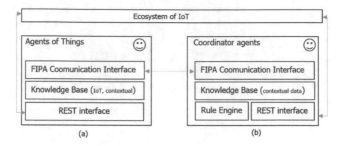

Fig. 2. Architecture of agents of things and coordinator agents.

As far as agents of things entities are concerned, by having linked an IoT object, in their arguments, they store information related to the linked IoT object (network, object, resource) as well as the data of the context where the object is operating (room). Likewise, in the case of the coordinator agents, they store information about the context in which the agents of things with whom they must communicate to carry out control actions must be found. In addition, these types of agents have an additional element called rule engine. This element enables them to analyze data delivered by IoT objects, through the agents of things, and evaluate a set of IF-THEN-THAT rules to control actuators of the IoT ecosystem to meet the functional requirements.

4   Results

4.1   Scenario

In order to carry out the evaluation of the developed MAS, a set of twenty-five virtual objects of IoT were created in openHAB, eleven sensors and fourteen actuators. These objects belong to three categories, among which are: lighting comfort (light bulbs, lighting sensors, shutters), thermal comfort (temperature sensors, HVAC) and security (windows, alarms). All them were deployed within a digital home scenario. Each one of the rooms of the scenario (room1, kitchen, living room and outdoor) are illustrated in Fig. 3.

4.2   MAS Communication Model

Figure 4 shows how agents of things and coordinator agents have been employed to create a MAS for controlling the IoT ecosystem described in Fig. 3.

At the lowest level it is observed that agents of things have been created and linked to each of the IoT objects that make up the IoT ecosystem under study. To abstract this process, a virtual object has been created for each physical object of IoT. This was done using openHAB. Thus, for the case of the object light-bulb

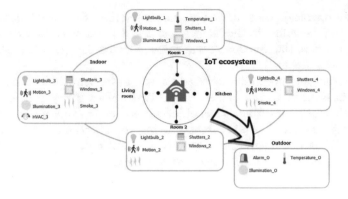

Fig. 3. Digital home scenario based on IoT.

(LB) of the $ROOM\_1$ the agent AOT_LB was created, for the object motion (MT) of the same room, the agent AOT_MT was created, and so on with the remaining objects. Linking an agent of thing to an IoT object was applied for each object of the rooms in the digital home shown in Fig. 3.

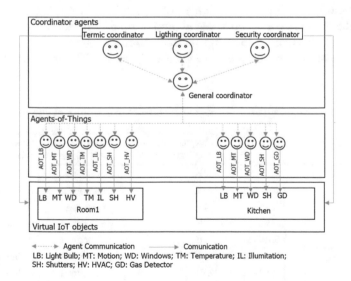

Fig. 4. Model of communication of agents.

At a higher level coordinator agents were included as part of the MAS. In our MAS three specialized coordinator agents have been introduced: thermal, lighting and security. Moreover, a *general_coordinator* agent was integrated. The main function of this agent is making a request/response to all agents-of-things for sending data that each specialized coordinator agent needs (context where they operate); that is, this agent distributes the useful data to carry out the thermal comfort to the *termic_coordinator*, the necessary data to apply lighting comfort to the *lighting_coordinator* and thus for the case of the security control (*security_agent*).

Finally, the general coordinator agent executes its behavior cyclically in order to maintain comfort in the IoT scenario. This makes it possible for agents to apply continuous monitoring to the objects installed in each of the rooms of the modeled digital home (room1, kitchen, living room and outdoor) according to changes in the environment. Consequently, it is not required that the user must be controlling each of the IoT objects at any time.

4.3   MAS Evaluation

The proposed MAS was implemented on a Lenovo branded personal computer with a 2.50 GHz i7 processor and 16 GB of RAM. The computer had installed the 64-bit Windows 8.1 operating system and the Java 1.8 virtual machine and the openHAB 2.0 server ran on it.

The evaluation of the MAS developed has consisted mainly in the evaluation of the communications between the agents that constitute the system; that is, the validation of the request/response actions carried out by the general coordinator agent with the agents of things. Likewise, the data transmission process between the general coordinator agent and the specialized coordinator agents was validated: thermal comfort agent, light comfort agent and security control agent. The results obtained were satisfactory in each of the executions as shown in Table 2.

Table 2. Evaluation of agents of things and coordinator agents of the proposed MAS.

Coordinator agents	Context	Request/response counterparts	Rule execution	Goal achieved
General	All	OK	Executed	Achieved
Thermal	Room 1	OK	Executed	Achieved
	Room 2	OK	Executed	Achieved
	Living room	OK	Executed	Achieved
	Kitchen	OK	Executed	Achieved
Lighting	Room 1	OK	Executed	Achieved
	Room 2	OK	Executed	Achieved
	Living room	OK	Executed	Achieved
	Kitchen	OK	Executed	Achieved
Security	Outdoor	OK	Executed	Achieved

Table 2 also shows data that each of the coordinator agents received from their counterpart agents to execute control actions on the actuators installed in the rooms of the modeled digital home. The shared data were made based on the contextual information of each object, and the control actions executed were based on rules that evaluated data provided by multiple IoT objects. These allowed the achievement of the targets of each coordinator agent in a collaborative way. In summary, the coordinator agent for thermal comfort, lighting and safety successfully met their goals. None of the agents had loss of data that would affect the execution of their rules.

5   Conclusions

The control of IoT ecosystems using MASs enables objects to acquire proactive behaviors. These objects do not depend on the user to adapt to the environment as current objects do today. Additionally, the fact that IoT objects are controlled by agents allows these objects, through their linked agent, to establish collaboration with other agents to model more complex behaviors. This means that IoT ecosystems can distribute the intelligence of objects using agents and take advantage of the resources that their objects do not have, using data that can be shared with their counterparts of heterogeneous MASs.

References

1. Arsénio, A., Serra, H., Francisco, R., Nabais, F., Andrade, J., Serrano, E.: Internet of intelligent things: bringing artificial intelligence into things and communication networks. In: Inter-cooperative collective intelligence: Techniques and applications, pp. 1–37. Springer (2014)
2. Bellifemine, F., Caire, G., Poggi, A., Rimassa, G.: JADE: a software framework for developing multi-agent applications. Lessons learned. Inf. Softw. Technol. **50**(1–2), 10–21 (2008)
3. Borgia, E.: The internet of things vision: key features, applications and open issues. Comput. Commun. **54**, 1–31 (2014)
4. Chelloug, S.A., El-Zawawy, M.A.: Middleware for internet of things: survey and challenges. Intelligent Automation & Soft Computing pp. 1–9 (2017)
5. Choi, S.I., Koh, S.J.: Distributed coap handover using distributed mobility agents in internet-of-things networks. J. Inf. Commun. Converg. Eng. **15**(1), 37–42 (2017)
6. Dar, K., Taherkordi, A., Baraki, H., Eliassen, F., Geihs, K.: A resource oriented integration architecture for the internet of things: a business process perspective. Pervasive Mob. Comput. **20**, 145–159 (2015)
7. Dorri, A., Kanhere, S.S., Jurdak, R.: Multi-agent systems: a survey. IEEE Access **6**, 28573–28593 (2018)
8. Filgueiras, T.P., Rodrigues, L.M., Rech, L.d.O., de Souza, L.M.S., Netto, H.V.: RT-JADE: a preemptive real-time scheduling middleware for mobile agents. Concurr. Comput.: Pract. Exp. **31**(13), e5061 (2019)
9. Forestiero, A.: Multi-agent recommendation system in internet of things. In: 2017 17th IEEE/ACM International Symposium on Cluster, Cloud and Grid Computing (CCGRID). pp. 772–775. IEEE (2017)
10. Fortino, G., Guerrieri, A., Russo, W., Savaglio, C.: Integration of agent-based and cloud computing for the smart objects-oriented iot. In: Proceedings of the 2014 IEEE 18th international conference on computer supported cooperative work in design (CSCWD). pp. 493–498. IEEE (2014)

11. Heimgaertner, F., Hettich, S., Kohlbacher, O., Menth, M.: Scaling home automation to public buildings: A distributed multiuser setup for openhab 2. In: 2017 Global Internet of Things Summit (GIoTS). pp. 1–6. IEEE (2017)
12. Iounes, D., López-Torralba, J.M., Pico-Valencia, P., Holgado-Terriza, J.A.: Construyendo un dispositivo de internet de las cosas para el hogar conectado. In: Jornadas SARTECO 2019 (2019)
13. Li, H., Ota, K., Dong, M.: Learning IoT in edge: deep learning for the internet of things with edge computing. IEEE Netw. **32**(1), 96–101 (2018)
14. Lin, D., Murakami, Y., Ishida, T.: Integrating internet of services and internet of things from a multiagent perspective. In: International Workshop on Massively Multiagent Systems. pp. 36–49 (2018)
15. Madakam, S., Ramaswamy, R., Tripathi, S.: Internet of Things (IoT): a literature review. J. Comput. Commun. **3**(05), 164 (2015)
16. Mardini, W., Khamayseh, Y., Yassein, M.B., Khatatbeh, M.H.: Mining internet of things for intelligent objects using genetic algorithm. Comput. Electr. Eng. **66**, 423–434 (2018)
17. Meier, F., Fenner, D., Grassmann, T., Otto, M., Scherer, D.: Crowdsourcing air temperature from citizen weather stations for urban climate research. Urban Clim. **19**, 170–191 (2017)
18. do Nascimento, N.M., de Lucena, C.J.P.: Fiot: An agent-based framework for self-adaptive and self-organizing applications based on the internet of things. Information Sciences 378, 161–176 (2017)
19. Nieves, J.C., Andrade, D., Guerrero, E.: Maiot-an iot architecture with reasoning and dialogue capability. In: Applications for Future Internet, pp. 109–113. Springer (2017)
20. Pico-Valencia, P., Holgado-Terriza, J.A.: Agentification of the internet of things: A systematic literature review. International Journal of Distributed Sensor Networks 14(10) (2018)
21. Pico-Valencia, P., Holgado-Terriza, J.A., Herrera-Sánchez, D., Sampietro, J.: Towards the internet of agents: an analysis of the internet of things from the intelligence and autonomy perspective. Ingeniería e Investigación **38**(1), 121–129 (2018)
22. Pico-Valencia, P., Holgado-Terriza, J.A., Senso, J.A.: Towards an internet of agents model based on linked open data approach. Auton. Agents Multi-Agent Syst. **33**(1–2), 84–131 (2019)
23. Razzaque, M.A., Milojevic-Jevric, M., Palade, A., Clarke, S.: Middleware for internet of things: a survey. IEEE Internet Things J. **3**(1), 70–95 (2015)
24. Savaglio, C., Ganzha, M., Paprzycki, M., Bădică, C., Ivanović, M., Fortino, G.: Agent-based internet of things: State-of-the-art and research challenges. Future Generation Computer Systems 102, 1038–1053 (2020)
25. Sorici, A., Picard, G., Florea, A.M.: Multi-agent based context management in ami applications. In: 2015 20th International Conference on Control Systems and Computer Science. pp. 727–734 (2015)
26. Suganuma, T., Oide, T., Kitagami, S., Sugawara, K., Shiratori, N.: Multiagent-based flexible edge computing architecture for IoT. IEEE Netw. **32**(1), 16–23 (2018)
27. Zhou, L., Lou, C.X.: Intelligent cargo tracking system based on the internet of things. In: 2012 15th international Conference on Network-Based information systems. pp. 489–493 (2012)

Evaluation of Edge Technologies Over 5G Networks

Kourtis Michail-Alexandros[1]([✉]), Christinakis Dimitris[2],
Xilouris George[1], Thanos Sarlas[1], Soenen Thomas[3],
and Kourtis Anastasios[1]

[1] National Center of Scientific Research "Demokritos", Athens, Greece
`akis.kourtis@iit.demokritos.gr`
[2] Orion Innovations P.C., Athens, Greece
[3] Ghent University – imec, Ghent, Belgium

Abstract. The emerging technologies in the field of telecommunications have evolved into the paradigm of 5G. The 5G technology suggests that communication networks can become sufficiently flexible to handle a wide variety of network services from various domains. One of the aspects envisaged by 5G is the virtualization of small cells, which allows enhanced mobile edge computing capabilities, thus enabling network service deployment and management near the end user. This paper presents a cloud-enabled small cell architecture for 5G networks developed within the 5G-ESSENCE project. The paper demonstrates a set of experiments for different platform architectures in 5G, based on a prototype Deep Packet Inspection (DPI) Virtualized Network Function (VNF).

Keywords: Network Function Virtualization · Software defined networking · Edge cloud computing · 5G

1 Introduction

The rapid growth of mobile data networks and various multimedia content provider services has introduced the need for novel paradigms in support of efficient content delivery from an operator's perspective. The trade-off between increased capacity and cost reduction for the provider (i.e. CAPEX, OPEX) is one of the key challenges for mobile network operators. An initial approach to addressing this challenge was investigated by the EU H2020 funded SESAME project [1]. A solution based on a flexible Radio Access Network (RAN) with enhanced with advanced management and deployment capabilities, delivered Network Function Virtualization (NFV) was developed by the project. The joint radio-cloud architecture realized the concept of placing computing resources at the network edge, using NFV as an enabler to build a cost effective and energy-efficient RAN. The ongoing 5G ESSENCE project continues to evolve the small cell concept by integrating compute capabilities (i.e., a low-cost micro server) and as a result is able to execute of applications and network services, in accordance to the Mobile Edge Computing (MEC) paradigm [2]. A key enabled in these efforts is the Network Function Virtualization (NFV) [3], as it can provide flexible methods on network service management [4].

© Springer Nature Switzerland AG 2020
Á. Rocha et al. (Eds.): ICITS 2020, AISC 1137, pp. 407–417, 2020.
https://doi.org/10.1007/978-3-030-40690-5_40

This paper centers on an evolved Cloud Enabled Small Cell (CESC) architecture proposed by the H2020 5GPPP project 5G ESSENCE. The project is focused on enhancing the processing capabilities for data that have immediate value beyond locality. The challenges of managing processing-intensive management functions, such as Radio Resource Management (RRM)/Self Organizing Network (SON) on small cells is also be addressed by the project Finally the project is delivering real-world demonstrations of the solution based on a number of relevant use cases. In order to realize these challenging ambitions research in wireless access, network virtualization, and end-to-end (E2E) service delivery is necessary.

Under-utilized virtualized resources in small cells will be exploited to their full potential and in a dynamic way, in order to support ultralow-latency, high-performance services. The approach being investigated is also expected to improve network resiliency, and to provide substantial capacity gains at the access network for 5G related applications. To achieve these goals, a distributed edge cloud environment (designated t the 'Edge Data Centre' -Edge DC-) is developed [5, 6], based on a two-tier architecture.

The first tier, i.e., the Light DC, will remain distributed inside the CESCs in order to provide latency-sensitive services to users directly from the network's edge. The second tier will be a more centralized, 'high-scale' cloud, namely the Main Data Centre (Main DC), which will provide higher processing capabilities for computing intensive network applications. It will also have a more centralized view so as to host efficient Quality of Service (QoS) enabled scheduling algorithms.

In the domain of hardware technologies, adding new computational capabilities to small cells brings a variety of new opportunities to the network, as well as new challenges. In addition, the placement of low power/low cost processors within small cells, coupled with hardware acceleration, will be a key focus within the proposed architecture.

Firstly, infrastructure programmability is achieved by leveraging the virtualized computation resources available at the Edge DC. These resources will be used for hosting VNFs tailored according to the needs of each tenant, on a per-slice basis. Second, the Main DC allows centralizing and software-based control plane small cell functions to enable more efficient utilization of radio resources coordinated among multiple CESCs.

At the network's edge, each CESC is able to host a number of VNFs comprising a service, which is available to users of a specific operator. The Light DC can be used to implement different functional splits within the Small Cells as well as supporting end user mobile edge applications. At the same time, the paper proposes the development of small cell management functions implemented as VNFs, which run in the Main DC and coordinate a fixed pool of shared radio resources, instead of considering each small cell station as a set of unique and dedicated resources.

It should be noted that this paper does not only propose the development and adaptation of a multitenant CESC platform. It also addresses the performance evaluation of different NFV platforms for a vDPI network service.

The remainder of the paper is organized as follows. We initially describe the overall 5G-ESSENCE system architecture and focus on the management modules in Sect. 2. Subsequently, Sect. 3 presents an experimental performance evaluation for different virtualization platforms. Finally, Sect. 4 concludes the paper and draws future research lines.

2   5G ESSENCE Overview

This section, firstly, details the overall 5G ESSENCE architecture, and then, reviews the definition of a network service in the context of the architecture.

2.1   5G ESSENCE Overall Architecture

In the 5G ESSENCE approach, the Small Cell concept is evolved not only to provide multi-operator radio access, but also, to achieve an increase in the capacity and the performance of current RAN infrastructures, and to extend the range of the services provided while maintaining its agility. To achieve these goals, the paradigms of RAN scheduling need to be leveraged and additionally provides an enhanced, edge-based, virtualized execution environment attached to the small cell, taking advantage and reinforcing the concepts of MEC and network slicing.

The current instantiation the of 5G ESSENCE architecture Fig. 1 combines the current 3GPP framework for network management in RAN sharing scenarios and the ETSI NFV framework for managing virtualized network functions. The CESC offers virtualized computing, storage and radio resources and the CESC cluster is considered as a cloud by the upper layers. This cloud can also be 'sliced' to enable multi-tenancy. VNFs which implement the different Small Cells features as well as supporting end user mobile edge applications are supported by the execution platform.

As shown in Fig. 1, the architecture allows multiple network operators (tenants) to provide services to their users through a set of CESCs deployed, owned and managed by a third party (i.e., the CESC provider). In this way, operators can extend the capacity of their 5G RAN in areas where the deployment of their own infrastructure could be expensive and/or inefficient, as would be the case (e.g., highly dense metropolitan areas) where massive numbers of Small Cells would be required to provide expected high quality services within a given geographical area.

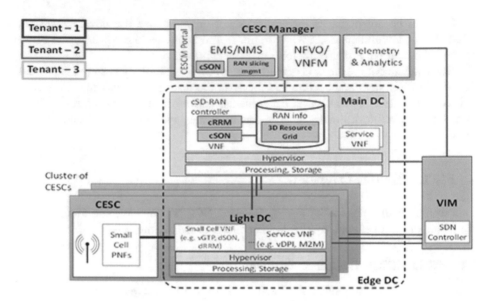

Fig. 1. 5G ESSENCE high level architecture

In addition to capacity extension, the described platform is equipped with a two-tier virtualized execution environment, materialized in the form of an Edge DC, which supports the provisioning of MEC capabilities for mobile operators in order to provide an enhanced user experience and agile service delivery. The first tier, i.e., the Light DC hosted inside the CESCs, is used to support the execution of VNFs for carrying out the virtualization of the Small Cell access. In this regard, network functions supporting traffic interception, GTP encapsulation/decapsulation and some distributed RRM/SON functionalities are expected to be executed therein. VNFs that require low processing power, e.g., a Deep Packet Inspection (DPI), could also be hosted here. The connection between the Small Cell Physical Network Functions (PNFs) and the Small Cell VNFs can be realized through, e.g., the network Functional Application Platform Interface (nFAPI). Finally, backhaul and fronthaul transmission resources are part of the CESC, allowing for the required connectivity.

The second cloud tier, i.e., the Main DC, will host more computation intensive tasks and processes that require centralization in order to have a global view of the underlying infrastructure. This includes the cloud-enabled Software Defined – Radio Access Network (cSD-RAN) controller which provides the control plane decisions for all the radio elements within the geographical area of a CESC cluster, including the centralized Radio Resource Management (cRRM) over the entire CESC cluster. Other potential VNFs that could be hosted by the Main DC include security applications, traffic engineering, mobility management, and in general, any additional network E2E services that can be deployed and managed on virtual networks, effectively and on demand.

The management modules for the operation of the CESC platform and service provisioning within the CESCM framework are shown in Fig. 1. The following subsections provide a more detailed description of each architecture component in Fig. 1.

2.2 Main Architectural Components of 5G ESSENCE

In our scope, a CESC consists of a Multi-RAT 5G small cell with its standard backhaul interface, standard management connection (TR069 interface for remote management) and with the necessary modifications to the data model (TR196 data model) to allow Multi-Operator Core Network (MOCN) radio resource sharing. The CESC is composed by a physical small cell unit attached to an execution platform based on x86, or ARMv8 architectures. Each specific CPU architecture is evaluated for its performance in Sect. 3 based on a prototype vDPI VNF. Edge cloud computing and networking are realized through sharing the computation, storage and network resources of thee micro servers present in each CESC and from the Light DC. Therefore, the CESCs become a neutral host for network operators or virtual network operators that want to share IT and network resources at the edge of the mobile network.

The CESC is envisioned to accommodate multiple operators (tenants) by design, offering Platform as a Service (PaaS), capable of providing the deployed physical infrastructure among multiple network operators. Different VNFs can be hosted in the CESC environment for different tenants. This also provides support for mobile edge computing applications deployed for each tenant that, operating in proximity to the end users, may significantly reduce the service delivery time and deliver composite services in an automated manner.

The CESC exposes different views of the network resources: per-tenant small cell view, and physical small cell substrate, which is managed by the network operator, decoupling the management of the virtual small cells from the platform itself. In the CESC, rather than providing multiple S1 connections from the physical small cell to different operators' EPC network elements such as Mobility Management Entity (MME) and SGW, such fan-out is done at the Light DC. The CESC provides termination of multiple S1 interfaces connecting the CESC to multiple MME/SGW entities as in S1-Flex. The interconnection of multiple CESCs forms a 'cluster' which can facilitate access to a broader geographical area with one or more operators (even virtual ones), extending the range of their service footprint, while maintaining the required agility for on demand extensions.

The Edge DC encompassing Main DC and Light DC
The proposed architecture is combining the concepts of MEC and NFV with Small Cell virtualization in 5G networks and enhancing them for multi-tenancy support. The purpose of the Edge DC is to provide Cloud services within the network infrastructure and to facilitate by promoting and assisting the exploitation of network resource information. To this end, all the hardware modules of the Light DC and the Main DC will be delivered as abstracted resources using novel virtualization techniques. The combination of the proposed Edge DC architecture coupled with the concepts of NFV and SDN facilitates greater levels of flexibility and scalability.

As seen in the architecture presented in Fig. 1, the Main DC executes different Small Cell and Service VNFs under the control of the CESCM. In particular, the Main DC hosts the cSD-RAN controller which performs cRRM decisions for handling efficiently the heterogeneous access network environment composed of different access technologies such as 5G RAN, LTE, and Wi-Fi. These radio access networks can be programmable and under the supervision of the centralized controller.

The CESCM

The CESC Manager (CESCM) is responsible for coordinating and supervising the consumption, performance, and delivery of radio resources and services. It controls the interactions between the infrastructure elements (CESCs, Edge DC) and network operators. It also has responsibility for Service Level Agreements (SLAs) compliance. From an architectural perspective, the CESCM encompasses telemetry and analytics as fundamental capabilities to enable efficient management of the overall infrastructure landscape and network. The Virtualized Infrastructure Manager (VIM) is responsible for controlling the NFV Infrastructure (NFVI), which includes the computing, storage and network resources of the Edge DC.

Management and orchestration of the proposed uniform virtualized environment, which can support both radio connectivity and edge services, is a challenging task by itself. Management of the diverse lightweight virtual resources is of primary importance, in order to enable a converged cloud-radio environment and efficient placement of services and extend current solutions in the NFV field [7, 8]. For this purpose, the CESCM is the central service management and orchestration component within the architecture. It integrates collectively the traditional network management elements, and the new functional blocks necessary to realize NFV operations. A single instance of CESCM is able to operate over several CESC clusters at different Points of Presence (PoP), each constituting an Edge DC through the use of a dedicated VIM per cluster.

An essential component of the CESCM is the Network Functions Virtualization Orchestrator (NFVO). It is responsible for realizing network services on the virtualized infrastructure and includes the interfaces to interact with the CESC provider for service management (e.g., exchange of network service descriptors and SLAs for each tenant). The NFVO composes service chains and manages the deployment of VNFs at the Edge DC. The NFVO uses the services exposed by the VNF Manager, which are in charge of the instantiation, updating, querying, scaling and termination of VNFs. Moreover, NFVO may include features to enhance the overall system performance, e.g., to improve energy efficiency.

The CESCM hosts also the Element Management System (EMS), which provides a package of end-user functions for the management of both physical network functions (PNFs) and VNFs at the CESCs. In particular, the EMS carries out key management functionalities such as Fault, Configuration, Accounting, Performance, Security (FCAPS) operations. The EMS has responsibility for partitioning the single whole-cell management view into multiple virtual-cell management views, one per tenant. In this way, a virtualized Small Cell with a set of (limited) management functionalities can be made visible to, e.g., the Network Management System (NMS) of each tenant in order to, e.g., collect performance counters, configure neighbor lists for a proper mobility management, etc. In addition to the NMSs of each tenant the CESCM can also incorporate an NMS for managing the whole set of CESCs deployed by an operator. This may be appropriate for example in scenarios where existing CESCs belonging to different vendors in the same deployment, each one with its own EMS. The EMS/NMS will also host the cSON functionalities (e.g. self-planning, Coverage and Capacity Optimization (CCO), etc.) and the functionalities for the lifecycle management of RAN slicing (i.e. for the creation, modification or termination of RAN slices).

As shown in Fig. 1, the CESCM encompasses a telemetry and analytics module that capture and analyzes relevant indicators of infrastructure and network operations. This capability provides the CESCM with accurate operational knowledge that characterizes the behavior of the network and its users in relation to the utilization of both cloud and radio resources. This will facilitate the realization of effective optimization approaches based on e.g. machine learning techniques for service placement, which can dynamically adapt to the context of the provided services and their execution environment and to enable automated enforcement of SLAs. Finally, the CESCM also incorporates the CESCM portal. It is a control panel with web Graphical User Interface (GUI) that serves as the entry point for all users, including the CESC provider and tenants, to CESCM functionalities and constitutes the main graphical frontend to access the 5G ESSENCE platform. The CESCM Portal in general provides visual monitoring information of the platform, the agreed SLAs, and the available network services/VNFs, allowing parameters' configuration.

The VIM
The CESCM functions will be built upon the services provided by the VIM which has responsibility for appropriately managing, monitoring and optimizing the overall operation of NFVI resources (i.e. computing, storage and network resources) at the Edge DC. The role of VIM is essential for the deployment of NFV services and to form and provide a layer of NFV resources for CESCM functions. NFVI resources will be ultimately offered as a set of APIs that will allow the execution of network services over the decentralized CESCs, located at the edge of the network. As seen in Fig. 1, the VIM relies on an SDN controller for interconnecting the VNFs and for offering SFC on the data-plane by establishing the path for the physical connections.

3 Edge Technology Performance Evaluation Results

In this last section, we consider as an indicative performance evaluation of the deployment of an NFV service comprising a virtual Deep Packet Inspection (vDPI) which is designed to analyze in real-time network traffic and to recognize specific applications.

The network stack, on which the vDPI has been developed is based upon the open source nDPI [9] and is commonly used as a basis for networking monitoring solutions. Its primary goal is the provisioning of a general-purpose network stack for a fully functional operating system rather than a stack that is specifically designed for high packet throughput performance. Therefore, a standard Linux network stack cannot scale to the performance level often required for a software network appliance. Nevertheless, the implementation used in this experimental process has been extensively tested in virtualized environments for high packet throughput [10, 11], and showcased high performance results.

The proposed Traffic Classification solution is based upon a DPI approach, which is used to analyze a small number of initial packets from a flow in order to identify the flow type. After the flow identification step no further packets are inspected. The Traffic Classifier follows the Packet Based per Flow State (PBFS) in order to track the respective flows. This method uses a table to track each session based on the 5-tuples (source address, destination address, source port, destination port, and the transport protocol) that is maintained for each flow.

The vDPI VNF utilized in the proposed experimental testbed is based upon a variety of technologies to achieve advanced packet capturing as well as traffic identification. The use case under test for vDPI was to test it under low-power platforms to assess its performance, when computing resources are not available in ampleness. The platforms selected where Intel's DPDK-in-a-box mini-PC [12], NXP's LS2085 ARM board, and the Raspberry Pi 3, also an ARM-based system. As the hosting platforms do not offer significant computing power, but rather provide a power efficient environment, Docker containerization was selected for virtualization implementation. Containers have a significantly lower resource fingerprint in comparison the standard virtual machine of cloud computing environments such as OpenStack.

Docker is a form of Linux container which provides a self-contained execution environment, that provides isolated CPU, memory, block I/O, and network resources based on sharing the kernel of a host operating system. In order to investigate the pros and cons of a container based approach, the vDPI was implemented as a container application. F providing forwarding and inspection of network traffic., The results for the 3 different platforms are shown in Fig. 2.

The results provided in Fig. 2 provide an shown the comparative performance between NXP's ARM board and Intel's x86 based DPDK-2220. The Raspberry board demonstrated approximately 50% of the performance level that was achieved by the other 2 platforms, but still maintains a throughput rate which can support r the requirements of a telecom backhaul an Edge Light DC serve. It is important to mention NXP's board performance, which covers the needs of a Light DC, and can even approach x86 numbers.

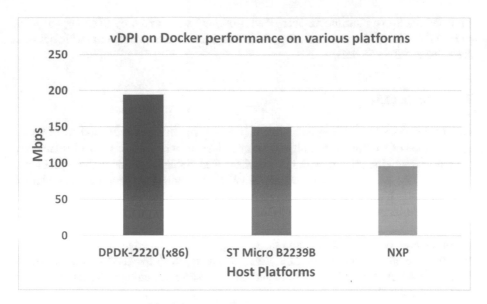

Fig. 2. vDPI on Docker performance results for packet processing in Mbps, based on 3 different platforms.

The Light DC is capable of instantiating network services and performing also the appropriate network traffic steering, in order to support service chaining (i.e. the forwarding of the traffic seamlessly to each VNF) and finally to the UE.

For the needs of the paper, a GTP decapsulation and re-encapsulation software has been implemented as an Edge enabler [13], running on top of the widely used packet processing library PF_RING [14]. The implemented module enables the forwarding of packets to VNF at the edge, as packets transmitted back and forth the EPC and the eNB are GTP encapsulated and cannot be recognized or processed by other endpoints. The Edge enabler runs as a VNF, in the Edge DC forwarding the traffic both directions. It also passes through all the signaling and data traffic between EPC and the small cell, thus preserving the connectivity between them. The data traffic is filtered from the rest of the control traffic, the GTP header is removed from the filtered packets, which are then forwarded to the OVS.

As the decapsulation and re-encapsulation operations can introduce a penalty to the performance of the system, they are performed in a parallel manner to the rest of the process. The signaling and the rest of the data traffic is forwarded using the zero-copy PF_RING library, and only the video service packets are copied to memory, as they need to be further processed by a GTP agnostic mechanism. The GTP header storage for the re-encapsulation is considered insignificant as the GTP header, merely allocates 8 bytes to the memory.

The functionality provided by the Edge enabler is vital to the proposed framework, as it handles GTP traffic and delivers it in a valid IP format to the vDPI VNFs to process it. The Edge enabler enables the integration of the small cell architecture and environment into a multimedia-over-IP environment seamlessly, by handling the small cell GTP traffic. GTP traffic is used by telco providers to secure network traffic between the EPC and the eNodeB.

4 Conclusions

This paper presented a 5G system architecture, which aims to evolve and improve the performance of existing RAN infrastructures. The manuscript not only presented the current issues in the related field, but also discussed viable solutions to these problems.

Firstly, the presented NFV enabled Small Cell architecture proposes a converged ecosystem for 5G, which is flexible, programmable and efficient. Regarding, small cell management functions, this manuscript described a mapping paradigm for RRM in 5G New Radio.

In the case of platform selection for hosting VNFs at the edge, a set of results was presented based on a prototype DPI VNF, comparing x86 to ARM platforms. Additionally, in order to be able to forward and process packets at the edge a prototype Edge enabler was developed.

As future line a wider set of use cases will be exploited in the frame of 5G, along with their corresponding challenges and novelties. Furthermore, the concept of cSON small cells [15] will be investigated in relevance to RRM and how it can enhance 5G in terms of performance and efficiency.

Acknowledgements. The research leading to these results has been supported by the 5G-ESSENCE H2020 5G-PPP project (no. 761592) and 5GENESIS H2020 5G-PPP (no. 815178).

References

1. Khodashenas, P.S., Blanco, B., Kourtis, M.-A., et al.: Service mapping and orchestration over multi-tenant cloud-enabled RAN. IEEE Trans. Netw. Serv. Manag. **14**, 904–919 (2017)
2. ETSI: Mobile-edge computing: Introductory technical white paper (2014)
3. Network Function Virtualization (NFV): Use Cases. ETSI GS NFV 001 v1.1.1 (2013-10)
4. Twamley, P., Muller, M., Bok, P.-B., et al.: 5GTANGO: an approach for testing NFV deployments. In: 2018 European Conference on Networks and Communications (EuCNC). IEEE (2018)
5. Small cell virtualization functional splits and use cases, Small Cell Forum (2016)
6. Integrated HetNet architecture framework, Small Cell Forum (2016)
7. Riccobene, V., et al.: Automated generation of VNF deployment rules using infrastructure affinity characterization. In: 2016 IEEE NetSoft Conference and Workshops (NetSoft). IEEE (2016)
8. Soenen, T., et al.: Empowering network service developers: enhanced NFV DevOps and programmable MANO. IEEE Commun. Mag. **57**, 89–95 (2019)
9. Deri, L., Martinelli, M., Bujlow, T., Cardigliano, A.: nDPI: open-source high-speed deep packet inspection. In: 2014 International Wireless Communications and Mobile Computing Conference (IWCMC). IEEE (2014)

10. Kourtis, M.-A., McGrath, M.J., et al.: T-NOVA: an open-source MANO stack for NFV infrastructures. IEEE Trans. Netw. Serv. Manag. **14**, 586–602 (2017)
11. Kourtis, M.-A., Xilouris, G., Riccobene, V., McGrath, M.J., Petralia, G., Koumaras, H., Gardikis, G., Liberal, F.: Enhancing VNF performance by exploiting SR-IOV and DPDK packet processing acceleration. In: 2015 IEEE Conference on Network Function Virtualization and Software Defined Network (NFV-SDN). IEEE (2015)
12. DPDK development team: Data Plane Development Kit. http://www.dpdk.org
13. Kourtis, M.-A., Koumaras, H., Xilouris, G., et al.: An NFV-based video quality assessment method over 5G small cell networks. IEEE MultiMedia **24**, 68–78 (2017)
14. https://www.ntop.org/products/packet-capture/pf_ring/
15. Kourtis, M.-A., Blanco, B., Perez-Romero, et al.: A cloud-enabled small cell architecture in 5G Networks for broadcast/multicast services. IEEE Trans. Broadcast. 1–11 (2019)

Design and Development of an Interactive Community-Driven Information System for Rural Artisans: CISRA Framework

Jayanta Basak[1]([✉]), Parama Bhaumik[2], and Siuli Roy[3]

[1] Social Informatics Research Group, Indian Institute of Management Calcutta, Kolkata, India
lettertojayanta@gmail.com
[2] Department of Information Technology, Jadavpur University, Kolkata, India
[3] Department of Information Technology, Heritage Institute of Technology, Kolkata, India

Abstract. Most e-commerce platforms have been developed as sales and marketing channels, offering artisans of all skill levels equal opportunities for growth and provide very low-cost worldwide visibility. These strategies never allow any artisans to give opinions about willingness to accept/reject orders due to personal inconvenience, wages, item shipping deadlines, etc. Extant exogenous approaches undertaken to provide rural producers with direct access to market have mostly done so sporadically, by incorporating measures that are not designed in accordance with rural context and needs. In this context, this paper attempts to securing active participation of rural producers in the process of production by considering the opinion and purposive collaboration between relevant rural-urban agents by allocating an order to the rural artisans. This paper will propose a Community-driven Information System for Rural Artisans (CISRA) which identifies the selection of artisans in fairness manner as a strategy to boost market opportunities of rural producers. We used a simulator called AnyLogic to ensure fairness and scalability in the artisan selection method and to measure the efficiency of our allocation model. The simulator will optimize the allocation policies of a job to the artisans based on the opinion value.

Keywords: Information system · Order distribution · Rural artisan · AnyLogic

1 Introduction

The handloom and handicraft sector is the second-largest unorganized employment sector in India where most of rural crafts are produced in small household workspaces and sales are generally limited to the inconsistent and unpredictable local markets [1]. Artisans are perennially trapped in the vicious cycle of low investment capacity, low productivity, weak market linkages, low-value proposition leading to inconsistent revenues and low risk-taking ability [2].

To address these deep-rooted problems of artisan industry, we investigate the current opportunity structure and workflow of the crafts industry in West Bengal, India by interviewing the artisans directly at different local handicraft fairs and were able to

© Springer Nature Switzerland AG 2020
Á. Rocha et al. (Eds.): ICITS 2020, AISC 1137, pp. 418–428, 2020.
https://doi.org/10.1007/978-3-030-40690-5_41

identify some of the major stumbling blocks that hinder the growth of handicrafts sectors in India. These are poor accessibility of rural crafts villages from urban markets, lack of awareness about market-oriented product design, quality management, packaging, branding, certification etc., inappropriate and unorganized markets, exploitation by middleman, lack of timely and qualitative input supplies, high input prices, inappropriate pricing, stiff competition and lack of innovations.

In this context, we began our fieldwork to understand the baseline practices of the artisans. It consisted of in-depth semi-structured interviews and structured surveys. An analysis of the field data explicitly reveals how various reasons (like global market connects; less profit margin, training needs etc.) sustain marginalization of rural communities. A total of 66% responses show that rural artisans are not able to connect with the global market due to the lack of proper ICT based infrastructural support.

Many e-commerce platforms, mobile commerce, online social media sites like Facebook, Twitter and Pinterest etc., has become an important driver of e-commerce [3] which offer equal growth opportunities to artisans of all skill levels and provide global visibility at a very low cost. However, these passive e-commerce sites adopt a "push" approach that only allows publishing and selling of catalogue based products where buyer-seller interaction for customization of products and price negotiations never happens. Furthermore, online social media-based businesses are mostly owned by urban sellers/entrepreneurs, not by the rural artisans possibly because the rural artisans are not aware of its benefits and not digitally equipped to handle online platforms. It is now important to digitally empower the artisan community so that they can use these online platforms and expand their business beyond local markets.

In this context, our interaction with urban entrepreneurs reveals that a potential market opportunity still remains underexplored in crafts industry where the artisans could offer their skills to create innovative products with contemporary designs for the urban entrepreneurs using traditional art forms.

Our objective, in this paper, is to create an online platform which will provide the artisans and urban buyers/entrepreneurs a collaborative co-creation/co-production environment where artisans and customers may work together to ideate, design and develop customize their products. Such platforms adopt a combination of "push" and "pull" mechanisms where the buyers/entrepreneurs publish ("push") job prospects (customization/co-creation) for artisans mentioning desired skill level and other criterions. The system allows the artisans with matching criterion to explore ("pull") the business prospects of the posted jobs and bid for the suitable jobs. Based on bidding price, customer feedback, production capacity and delivery commitment and entrepreneur's choice, each job is dynamically assigned to one or more artisans. Besides dynamic job assignment, the system should also allow the artisans to regularly share their work-in-progress and sample work and get customer feedback at every stage of their work through online interactions. It ensures the quality and timely delivery of products. Fairness in the job allocation process is ensured by considering only those artisans in a job allocation process who are located in the close geographical proximity of the entrepreneur, matching the desired criteria and not getting order for a long time.

The rest of this paper is organized as follows. Section 2 summarizes related work in this field and Sect. 3 describes our system architecture. In Sect. 4, we describe the simulation model of selecting artisans and comparative analysis between two well-known numerical methods for the same purpose. We conclude this paper in Sect. 5.

2   Related Work

This literature review proposes and validates a design theory for digital platforms which effectively support and maintains social interactions and user activities among various actors. Drawing upon literature review on information system design theory and platforms, we have derived the mechanisms for designing effective Community-driven Information System in the rural context.

Among the various kinds of theories in information systems [4], the previous study largely led to four kinds of theories: theories of analysis [5], theories of explanation [6], theories of prediction [7], and theories of explanation and prediction [8]. Few previous studies fall into the tradition of 'design and action' theories that concentrate on knowledge building through information systems design and assessment [9]. Design and action theories vary from descriptive or predictive theories by providing specific prescriptions on how to design and develop an information system [4, 10].

Our literature review shows that few design theories have been developed for platforms that support Community-driven Information Systems in rural context. Given the omnipresence and growing significance of information systems, such design theories have convincing requirements. In order to fill this gap, building upon [10] and the literature on information systems and platforms, we propose a framework guiding the development of design theory of our platform (Fig. 1).

On the basis of the above study, we have observed that three main characteristics of an information system which will be depicted as follows,

First, it is necessary to support the sharing of digital material in various formats and various devices. The platform should promote the dissemination of codified data (information that can be compressed into codes) and abstract data (information that can be allocated to a particular phenomenon) among participants from various societies [11]. There must be external control (system administrator) to supervise the overall activities of the actors. By pursuing their own objectives, each actor is free to engage.

Second, there is a need to support cooperation among members of the society who share common interests, values and trust. It is necessary to favour limited reciprocity among anonymous actors contributing to the platform. Registration in the platform is based on self-selection, involvement is free, and surveillance is small. Coordination is self-regulating and together with some type of negotiation, formal elements must be present [11].

Third, to be successful, a platform should provide mechanisms for negotiating goals and loyalty among members. It is important that any registered members can freely access all types of active modules in the digital platform have to support all three social interaction structures to their full extent.

3   System Architecture

Please From the lens of a literature survey on the design theories of information systems, we have proposed a Community-driven Information System for Rural Artisans (CISRA) to bridge the rural-urban market separation.

3.1 System Design

CISRA facilitates communication, collaboration and trade between rural artisan, urban entrepreneurs and other actors in the system. Specifically, it promotes transactions between producers, consumers and other associated stakeholders by providing a standardised, flexible and open platform that not only improves productivity but also ensures fairness and financial benefits to all. The collaboration allowed by this platform ensures transparency and helps to optimise the positions of all stakeholders in the business. This platform provides a virtual space through which multiple actors in this system can exchange information between them. In our work, we are considering this virtual space as 'community'. Thus, our platform can be defined as a temporary association between all associated stakeholders who will establish dynamic peer-to-peer connections to collaborate with each other through a coordinated sharing of skills, resources, information, risks, costs and benefits in order to satiate a given business opportunity.

3.2 Proposed Architecture of CISRA

Our framework is consisting of different functional entities that enable the micro-manufacturing unit to establish peer to peer coordination in a decentralized manner. In this section we will explain the architectural description of our CISRA system.

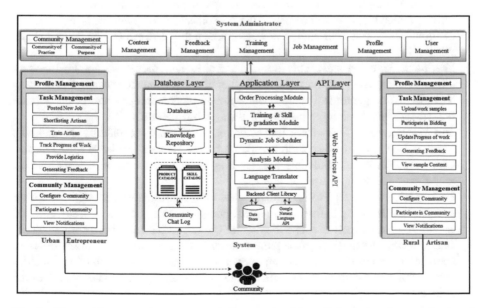

Fig. 1. The architecture of the proposed Community-driven Information System for Rural Artisans (CISRA)

3.2.1 Users

There are mainly three types of users in the system: (i) Rural Artisans; (ii) Urban Entrepreneurs; (iii) System Administrator. They are connected to this platform via a mobile app or through a web-based interface.

Rural Artisan: The rural artisans can advertise their product details, contact details, skill sets, production capacity and product feedback, which will be maintained in a digital catalogue.

Urban Entrepreneur: The urban entrepreneur can view profiles of rural artisans in digital catalogues and 'post an order' for prospective rural artisans, where the order details, budget and timeline are mentioned.

System Administrator: The system administrator is responsible for the upkeep, configuration, and reliable operation of the platform. The functional description of each activity of the system administrator is given below:

- **Community Management:** Community management is one of the most important activities in our platform. Here we deal with two types of virtual communities: (i) Community of Practice: a group of actors who share a concern or a passion for something they do, and learn how to do it better as they interact regularly, and (ii) Community of Purpose: is a group of actors in our platform people who are going through the same process of job execution.
- **Content Management:** This module is a set of processes that supports the collection, managing, and publishing of digital content in our platform in any form or medium. Modification or adding new contents in the platform can be done through this module.
- **Feedback Management:** This module captures and processes the feedback of different actors and performs analytical operations to produce meaningful results.
- **Training Management:** Training Management module keeps track of all training related activities, which are conducted through our platform.
- **Job Management:** There are a few activities which are performed by this module. They are (a) Organize and search by open jobs. (b) Manage the list of rural artisans, who are available to handle any job. (c) Flag unassigned work orders (d) Manage jobs through an easy to use job console.
- **Profile Management:** Profile management auto-consolidates and optimizes user profiles in order to minimize management and storage demands and needs minimal administration, support and facilities while offering enhanced options for users.
- **User Management:** User management describes the ability for administrators to manage user access to various components of the systems. It maintains a directory service that has the capacity to authenticate, authorize, and audit user access in the platform.

3.3   Salient Features of the System

Order Processing Module: Rural artisan can advertise their product details, contact details, skill sets, production capacity and product feedback, which will be maintained in a digital catalogue. This module also tracks the prices in every step in the production process which ensures fairness in the platform. On the other hand, urban entrepreneurs can order products online and after receiving the products they can provide online feedback and rating for the product; if a single artisan fails to deliver the ordered quantity, there is a provision in our system which allows buyers to buy from multiple artisans who are producing similar types of products.

Training and Skill Up-Gradation Module: It will be linked to a video-conferencing platform through which urban entrepreneurs can train the selected rural artisans online regarding their skill up-gradation and other assistance.

Analysis Module: The order distribution, product demands, customer comments, feedback, product ratings etc. for individual rural artisans will be served into this module which will be analyzed to suggest the scope of future improvement in the business of rural artisans.

Dynamic Job Scheduler (DJS) Module: This module is responsible for carrying out the scheduling activity between urban entrepreneurs and rural artisans. The DJS module examines the availability of resources, coordinates and selects relevant actors to form an instance of a supply chain to process any particular activity...
DJS may aim at one or more of the following goals, for example:-

- *Efficient Load Balancing*: keeping all available resources uniformly busy
- *Ensuring quality of service*: Selecting resources to ensure QoS
- *Minimizing Response Time*: Time from work becoming enabled until it is finished
- *Maximizing fairness*: Granting equal opportunity to all users according to the priority and workload

Bidding Activity: It is an important event in the platform that allows rural artisans to quote their own prices and bargain effectively for maximum profit and healthy competition. Every registered rural artisan in the platform is allowed to bid on each and every order posted by an urban entrepreneur.

Language Translator Module: This module uses a backend client library and integrates google API to translate the content of this platform from one language to another language. Since we have designed this platform keeping in mind rural context, the platform provides provisions for rural artisans to navigate using the native language.

3.4   Workflow

The chain of events and responses in the platform start with registering of individuals and terminates at allocation of the job. In order to make the final allocation of a job to a particular artisan or a group of artisan it is necessary to follow some specific procedure.

The whole process of job allocation is more clearly illustrated in the following diagram (Fig. 2).

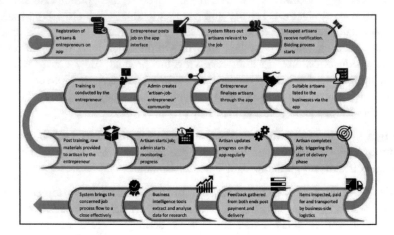

Fig. 2. Illustration of workflow

4 Simulation Environment – AnyLogic

AnyLogic is an innovative modeling tool which mainly aiming at modeling of hybrid systems based on JAVA [12–14]. It allows the user to combine different techniques and approaches such as differential equations, discrete events and agent-based systems.

4.1 Description and Hypothesis

Usually, any e-commerce systems only consider the skill set of an artisan to distribute an order. Apart from the skill set, we have tried to consider *Category of Order*, *Cost Quote* and *Artisan Review Score* for artisan selection. The top-layer model of 'artisan selection' is shown in Fig. 3. In the model, there are three modules: Urban Entrepreneur, the Platform itself and Rural Artisan. The platform accepts requests from urban entrepreneur and searches for a list of prospective artisans for execution of the order. The list of potential artisan is prepared by the platform based on a number of specifications associated with the artisans as mentioned in the next section (Sect. 4.3). Based on the parameters for a particular order, the platform searches in the large pool of artisan records and prepare a list of potential artisans to execute that order within the deadline and the information returns back to the urban entrepreneurs.

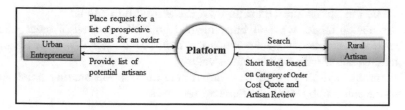

Fig. 3. Top-layer of artisan selection model

In this paper, we have used Euler method and Runge–Kutta fourth-order method to find out the potential list of artisans with considering three broad factors like *Category of Order, Cost Quote* and *Artisan Review*. Customers' order is random and has a time constraint. This model assumes that the skill set of an artisan varies from 1 to 4 and the maximum number of the artisan is set to 50000 (this is the maximum no of permissible agents in AnyLogic Personal Learning Edition 8.2.4 version).

4.2   Simulation Model Setup

In our simulation model, we actively collected the opinions of artisan and entrepreneurs in different regards as mentioned in this section and used it in two separate differential equations to compare the performance of our proposed model.

We used two different differential equations (Euler method and Runge-Kutta Method of Order 4) to measure the performance of '*artisan selection*' mechanism for a particular job. Figure 4 shows the 'artisan_selection' mechanism in AnyLogic.

4.3   Simulation Result and Analysis

A comparison is made between two separate differential equations (RK4 and Euler method) in job allocation strategies among a large number of artisans with different skill levels. Simulation experiment verifies the exactness of this model and concludes that Euler method is better than RK4 method of order 4 with respect to artisan selection as more number of artisans can get a job.

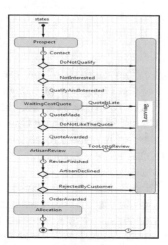

Fig. 4.

At last, optimization experiments are made to testify the optimize artisan selection strategy under certain conditions as mentioned in Sect. 4.3.2.

4.3.1   The Functions and Operations of the Simulation Model

In the simulation model we used Euler method and RK4 method of order 4 for our artisan selection problem which are numerical methods to solve first order first-degree differential equation with a given initial value.

We first used Euler method with initial condition as our starting point and we generate the solution by using the iterative formulas:

$$x_{n+1} = x_n + h$$
$$y_{n+1} = y_n + hf(x_n, y_n)$$

To find the solution to our problem we terminate this process when we have reached the right end of the desired interval.

Next we used RK4 of order 4 as another strategy in our simulation model. The RK4 method takes to extremes for correcting the predicted value of the next solution point and the summary of this method is:

$$x_{n+1} = x_n + h$$
$$y_{n+1} = y_n + (1/6)(k_1 + 2k_2 + 2k_3 + k_4) \, where$$
$$k_1 = hf(x_n, y_n)$$
$$k_2 = hf(x_n + h/2, y_n + k_1/2)$$
$$k_3 = hf(x_n + h/2, y_n + k_2/2)$$
$$k_4 = hf(x_n + h, y_n + k_3)$$

4.3.2   The Analysis of the Simulation Results

One of the objectives of our simulation is to discover the optimal combination of conditions so that one can find the best possible solution. Optimization in AnyLogic is built based on the OptQuest Optimization Engine which can find the best parameters of a model with respect to certain constraints. In this model Euler method and RK4 method is adopted in order to find prospective list of artisans where the total number of artisans is maximized.

Table 1 records the results of different selection strategies of artisan based on different skillset and other associated factors as mentioned in Sect. 4.2. The results of two optimization experiments are shown in Table 2. As the skill set of every individual artisan vary from 1 to 4, so the optimization in our case indicates that the total number of selected artisans is not more than 15% of total artisans.

Table 1 shows that the Euler method with a minimum skill set for an artisan will return slightly better result than RK4 method of order 4.

Table 1. Comparison of two different artisan selection strategies

	Euler method				RK4 method of order 4			
	Skill = 1	Skill = 2	Skill = 3	Skill = 4	Skill = 1	Skill = 2	Skill = 3	Skill = 4
Do not qualify	9900	10100	10100	10100	9987	10000	9900	10000
Not interested	12050	11950	11950	12000	12150	12050	12050	11900
Quote too late	22050	4150	24	280	21970	11850	25	18
Do not like quote	1750	7100	8350	8300	1680	4800	8500	8400
Too long review	1800	7150	8350	8400	1874	4650	8400	8500
Order declined	650	2800	3350	3350	587	1950	3350	3300
Customer rejects	200	700	750	800	182	450	800	800
Selected	**1600**	**6050**	**7126**	**6770**	**1570**	**4250**	**6975**	**7082**

Table 2. Results showing the number of artisans selected with varying skill set of each and every individual artisan

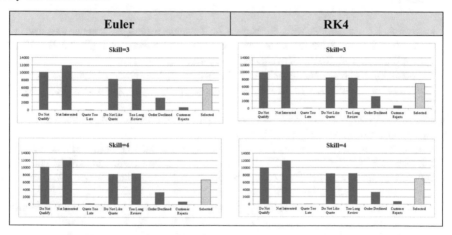

5   Conclusion

Most of the e-commerce platforms have been created as selling and marketing channels and offer equal growth opportunities to artisans of all skill levels and provide global visibility at a very low cost. These e-commerce sites adopt a "push" approach that only allows publishing and selling of catalogue based products where buyer-seller interaction regarding customization of products and price negotiations never happens. The main disadvantage of this kind of approach is that this approach will never allow any artisan to provide opinions regarding willingness to accept/reject orders due to unavoidable circumstances, wages, product delivery deadline etc.

From this context first time our CISRA platform will give an opportunity to the artisans to express their willingness regarding the acceptance of an order with respect to wages, deadline etc. This mechanism has been introduced in the system in the form of opinion dynamics where the system allows the artisans to bid in their native language for jobs posted by urban entrepreneurs. The entire allocation processes will be based on the bidding price, production capacity and delivery commitment offered by the artisans and the jobs will be dynamically assigned to one or more suitable artisans accordingly which ensures the quality and timely delivery of products. Fairness in the job allocation process is one of the important features of this platform.

To ensure fairness and scalability in the job allocation process and to test the performance of our allocation model, we have used a simulator named AnyLogic. Based on the opinion value the simulator will optimizes the allocation strategies of the artisan. We found that prospective list of artisans gets optimized with respect to the given constraints and we observe that rejection of artisans under each constraint in Euler method is less compared to that of RK4 which in another way can be considered as more humanitarian.

References

1. Parthiban, R., Bandyopadhyay, S., Basak, J.: Towards a nex-gen cottage industry in the digital age: insights from an action research with rural artisans in India. In: PACIS 2018 Proceedings, p. 196 (2018). https://aisel.aisnet.org/pacis2018/196
2. Abirami, P., Velavan, M., Arunkumar, S., Anand, V.V., Sivasumbramanian, J.: Indian handicrafts and its challenges faced by artisan community. Int. J. Econ. Res. **14**(7), 431–438 (2017)
3. Shah, A., Vidyapith, G., Patel, R., Vidyapith, G.: Problems and challenges faced by handicraft artisans. Voice Res. **6**(1), 57–61 (2017)
4. Gregor, S.: The nature of theory in information systems. MIS Q. **30**(3), 611–642 (2006)
5. Stanoevska-Slabeva, K., Schmid, B.: A typology of online communities and community supporting platforms. In: Proceedings of the 34th Hawaii International Conference on System Sciences, Maui, HI (2001)
6. Jones, Q., Rafaeli, S.: Time to split, virtually: 'discourse architecture' and 'community building' create vibrant virtual publics. Electron. Mark. **10**(4), 214–223 (2000)
7. Rothaermel, F.T., Sugiyama, S.: Virtual internet communities and commercial success: individual and community-level theory grounded in the atypical case of TimeZone.com. J. Manag. **27**(3), 297–312 (2001)
8. Balasubramanian, S., Mahajan, V.: The economic leverage of the virtual community. Int. J. Electron. Commer. **5**(3), 103–138 (2001)
9. Hevner, A.R., March, S.T., Park, J., Ram, S.: Design science in information systems research. MIS Q. **28**(1), 75–105 (2004)
10. Schreieck, M., Wiesche, M., Krcmar, H.: Design and governance of platform ecosystems-key concepts and issues for future research. In: ECIS, ResearchPaper76 (2016)
11. Saarikko, T., Westergren, U.H., Blomquist, T.: The inter-organizational dynamics of a platform ecosystem: exploring stakeholder boundaries. In: 49th Hawaii International Conference on System Sciences, pp. 5167–5176 (2016)
12. https://en.wikipedia.org/wiki/AnyLogic
13. https://www.anylogic.com/resources/educational-videos/introduction-to-anylogic-software/
14. https://mosimtec.com/anylogic/

Comparison Between Fuzzy Control and MPC Algorithms Implemented in Low-Cost Embedded Devices

Jorge Buele[1,2(✉)] ⓘ, José Varela-Aldás[1] ⓘ, Marlon Santamaría[2] ⓘ,
Angel Soria[3] ⓘ, and John Espinoza[4] ⓘ

[1] SISAu Research Group, Universidad Tecnológica Indoamérica,
Ambato, Ecuador
{jorgebuele, josevarela}@uti.edu.ec
[2] Universidad Técnica de Ambato, Ambato, Ecuador
ma.santamaria@uta.edu.ec
[3] Purdue University, Lafayette, USA
asoriach@purdue.edu
[4] Universidad de las Fuerzas Armadas ESPE, Sangolquí, Ecuador
jjespinoza6@espe.edu.ec

Abstract. This paper presents the design of two advanced control algorithms applying low-cost embedded boards. To obtain experimental results a flow plant design was used. For the first part of the work detailed in this paper, a fuzzy controller was designed and implemented in different embedded boards such as Arduino Uno, Raspberry Pi 3, BeagleBone Black and Udoo Neo Full. Next, an MPC was developed in the Raspberry Pi 3 board, being the only device that provides the necessary technical requirements for the design. Further, the signal conditioning before and after the controller integration with industrial equipment is described. To demonstrate the performance of each board used, and to acquire the data in real time two interfaces were created, for the first controller the interface was developed in LabVIEW software, and for the second one a GUIDE interface was developed in MATLAB software. The tests carried out validate these proposals, which despite the computational cost they require, they present high performance, robustness and good response, when tested in the flow system design.

Keywords: Fuzzy control · MPC · Embedded devices · Process control

1 Introduction

At the end of the decade of the 30s, the derivative action was included to the pneumatic controllers that existed at the time, which gave rise to the Proportional, Integral and Derivative (PID) controller. The empirical tuning of the PID controller was derived from the formulas that the engineers of Taylor Instruments: Ziegler and Nichols introduced in 1942 [1]. These rules for the adjustment of this type of controllers are still used, and have great acceptance as they provide effective results with a minimum investment of time and calculations. With technological development, the term

© Springer Nature Switzerland AG 2020
Á. Rocha et al. (Eds.): ICITS 2020, AISC 1137, pp. 429–438, 2020.
https://doi.org/10.1007/978-3-030-40690-5_42

"Advanced Control" is formally born, an engineering system that puts together a set of control strategies that are applied to a process, and encompasses a variety of disciplines for its structuring [2]. For the design of advanced control algorithms, it is required that both hardware and software comply with the minimum requirements of the system, since they involve a robust computational platform. In addition, they are based on the complete understanding of the process, its phenomenology, dynamics, possible disturbances and other technical characteristics. Based on the mathematical model that allows an approximation of the system, simulations of its behavior can be performed. As a result, it is not very common in the market because the design generally must be performed by an expert in control systems. Among the commonly used are: neural networks; robust control; expert control, where the fuzzy control stands out, and optimal control with the predictive control model [3–6].

Together with the emergence of the digital revolution at the end of the 1960s, new devices that facilitate the execution of complex control algorithms were developed. One of these devices is the PLC (Programmable Logic Controller), which is commercially popular for its remarkable presence in the industrial field. Although, the coupling of additional modules increases the cost [7]. In response to the high cost offers, currently several manufacturers introduce the term "low-cost technologies", to identify the embedded development boards, as an alternative engineering solution depending on the budget, and user needs. Among the main options in the market: Arduino (the best known); Raspberry Pi, Orange, BeagleBone, Udoo, etc. [8–12]. This research paper aims to demonstrate that these embedded boards can be used in an industrial environment, performing the adequate conditioning of the signals.

2   Related Works

Innovating automation processes is a task that requires a study that demonstrates its feasibility and links to the industrial field. Therefore, investigations that have dealt with this subject have been developed and are presented below. In [13] and [14] fuzzy control algorithms are implemented in the Raspberry Pi board to perform process control and monitoring. In addition, tools were added to visualize the data obtained in real time and save it to analyze it afterwards. Similarly, in [15] the Raspberry Pi board is used as a control unit to implement an MPC (Model Based Predictive Control) algorithm. By determining the mathematical model of the plant, it is possible to perform control tasks in a plant of the industrial type. Despite the high computational cost generated, this proposal provides a high-performance design that generate a high-speed response in the final control element. Moreover in [16], a system that obtains data through Kinect to monitor the gesture of the user's hand and face, while eating to recognize and count the bites made using the Udoo board is described.

Ensuring the best performance of the process depends on the control strategy used, and the proper selection of the tuning parameters. Following this context, this paper proposes the development of two advanced control algorithms implemented in low-cost embedded boards, which control the water flow of a flow plant that contains industrial elements and devices. For the first case a fuzzy controller was developed, consisting of 7 membership functions, operating on a SISO system (simple input/simple output), whose input is the value of the accumulated error product of the difference between the desired Set-Point value (SP), and the process variable value in which case the output from this variable acts on the frequency inverter of the process. As for the second controller an MPC was implemented, and for its development the experimental mathematical model of the process was required, either discrete or continuous. In this case the model was obtained through the MATLAB software.

Therefore, this paper is organized as follows: Sect. 1 presents the introduction and Sect. 2 the related works. In Sect. 3 the design of the controllers and Sect. 4 describes the case study and in Sect. 5 the implementation of proposal. Sections 6 and 7 present the results of the experimental tests carried out and the conclusions.

3 Study Case

The proposed case study describes a flow plant system, and the objective is to simulate an industrial process on a smaller scale. This station was designed and built for academic and training purposes in the field of automation and process control. The closed loop control is composed of a cylindrical metal tank with a capacity of 25 gallons, ¾ galvanized steel pipe (requirement for the operation of the transmitter), and a manual passage valve that handles the passage of fluid from the tank to a centrifugal THEBE pump of ½ Hp. It also has a graduated rotameter; whose plunger shows the flow rate between 0 to 40 liters per minute (LPM) and Delta frequency inverter [17].

For the fuzzy controller, the embedded boards Arduino UNO R3, Raspberry Pi 3, BeagleBone Black and Udoo Neo Full are used as control devices. Despite their relative low costs, they allow the development of engineering solutions that are applied in various fields of science, as previously mentioned. Similarly, in the area of process control, these embedded systems showed good performance in the implementation of both classical (PID) and advanced control algorithms. For a detailed explanation of their functions, a comparison of them is made in Table 1.

Table 1. Characteristics of the boards used.

Component	Arduino UNO R3	Raspberry Pi 3	BeagleBone Black	Udoo Neo Full
Processor	Microcontroller ATmega328P	Broadcom BCM2387 (1.2 GHz) ARMv8-A53	SITARA AM335x (1 GHz) ARM Cortex-A8	NXP i.MX 6SoloX
Operating voltage	3.3–5 V	3.3–5 V	3.3–5 V	3.3–5 V
Memory	Flash: 32 KB SRAM: 2 KB	1 GB LPDDR2	512 MB DDR3L 800 MHz	1 GB
Serial ports	1 UART port	2 UART ports	1 UART port	3 UART port
Other interfaces	1 I^2C interfaces 1 SPI interface	1 I^2C interface 2 SPI interfaces	1 I^2C interface 1 SPI interface	3 I^2C interface 1 SPI interface
Internal ADC	5 V to 10 bits	Does not have	1.8 V to 12 bits	3.3 V to 12 bits

4   Controllers Design

4.1   Fuzzy Controller

The elaboration of the set of fuzzy logic rules is based on the "general" knowledge of variation of fluid flow in the station. Based on this prior knowledge, the probable combinations of the input sets are associated with an output value. These input and output rules are presented in Fig. 1. For each rule, the inference engine finds the membership value where the vertical line intersects a membership function.

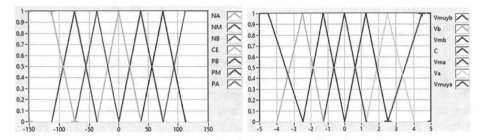

Fig. 1. Fuzzy sets: fuzzification (left) and defuzzification (right).

The resulting fuzzy set must become a unique number that represents a control signal that will be introduced in the plant to be controlled. To achieve this, the degree of belonging (between 0 and 1) is defined, which has an input value in the corresponding fuzzy input sets, and is projected in the output sets, building a flat area. To calculate the corresponding numerical value, the calculation of the area center or gravity center was carried out with the defuzzification method. The implemented controller is of Mamdani inference type, using the eFLL (Embedded Fuzzy Logic Library), compatible with the Arduino IDE and C language compilers. Table 2 shows the basis of rules with which the fuzzy controller will act. The structure to create the objects of the antecedent and consequent of the algorithm simulates the structure of the Mamdani fuzzy rules in the form IF …… THEN …… conditions, in this way the antecedent is represented by the FuzzyRuleAntecedent object, followed by the joinSingle (Linguistic tag) method to point to the case according to the rule to be formed.

Table 2. Fuzzy rules base.

	EN	O	EP
ENA	T	T	T
ENM	VA	VA	VMA
ENB	VMA	VMA	VMA
Z	VMA	C	VMB
EPB	VMB	VMB	VB
EPM	VB	VB	VB
EPA	VB	VmoreB	VmoreB

4.2   MPC Controller

To design and implement the MPC controller it is necessary to obtain an optimal mathematical model of the plant represented in the state space, being possible after the appropriate definition of the work ranges at the entrance and at the exit. After that, using the MATLAB system identification toolbox this model is generated, and imported into the tool that the software itself provides for the simulation and implementation of an MPC. Through the suitable selection of the parameters, the operation of the process can be simulated in a way that is closest as possible to a real case scenario, attributed to the characteristics of the tool, which works with models represented in state matrices, work restrictions and even delays caused by external factors. For this reason, it is necessary to select a prediction horizon, and a suitable control horizon for the tuning of the control model. In Fig. 2 displays the block diagram of the controller implemented in Simulink.

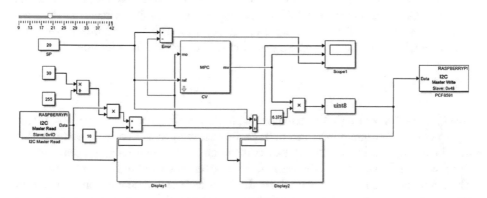

Fig. 2. Block diagram MPC.

5   Implementation

5.1   Hardware

Signal Conditioning Before Entering the Controller. The standard signal that the majority of industrial transmitters work with is 4 to 20 mA, but the boards that will be used do so with voltage signals. Therefore, it is necessary to implement a current to voltage converter (I/E) by placing a resistor connected in series with the process. Thus, the signal to be operated now is a voltage signal. Two ranges have been used: the first goes from 1 to 5 V and the second goes from 0.66 to 3.3 V (only for Udoo case), obtained by the means of voltage dividers. Additionally, to eliminate unwanted high-frequency noise signals, an RC low-pass filter with a 100 Hz cut-off frequency is performed.

When working with industrial instruments, it has been demonstrated that the boards exert charge effect and the control signal does not produce the desired effect. To improve the signal transmission from this stage a voltage follower circuit has been developed with the high impedance amplifier I.C. LF353 after the filter. According to the technical characteristics of the boards only Udoo Neo Full has an internal analog-digital converter (ADC) with a resolution of 16 bits at 3.3 V. On the other hand, BeagleBone Black supports 1.8 V and Arduino 5 V with 10-bit resolution, while Raspberry Pi in any of its versions does not offer an internal ADC. Thus, in order to standardize this parameter, in the case of Udoo Neo Full its own converter will be used for its high resolution (using only 12 bits), and for the other boards the integrated circuit (IC) MCP3202 has been selected, MCP3202 offers 12-bit resolution converter working on SPI communication.

Signal Conditioning After the Controller. None of the boards used has an internal digital-to-analog converter (DAC). To solve for this, the external converter I.C. PCF8591 with an 8-bit resolution and I2C communication was used. The signal that will reach the actuator (frequency inverter) operates in a voltage range of 0 to 10 V. Next, the signal passes through a voltage follower circuit to avoid loading effects in the process. The maximum voltage of the control signal is given by the V_REF of the external DAC used that is equal to 5 V. A non-inverting amplifier circuit of gain 2 was implemented to meet the requirements of the inverter. The I.C LF353 has two operational amplifiers in the same integrated circuit, thus optimizing resources.

5.2   Software

HMI for the Fuzzy Logic Controller. The HMI was designed in LabVIEW software, where the block diagram has three parts on its structure, as can be seen in Fig. 3. This has three progress bar indicators that correspond to set-point (SP), process variable (PV) and control variable (CV); symbolized in green, blue and red colors respectively. The communication is serial and allows the transmission and acquisition of data from and to the boards, in which the programming algorithm is implemented and directed to the indicators (progress bars and trends) of the front panel. Finally, the storage of the trend data is established, also it exports the data to Excel.

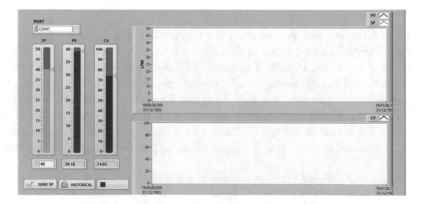

Fig. 3. Front panel of the HMI for the fuzzy controller.

HMI for the MPC Controller. To implement the controller a graphical user interface was developed, using MATLAB's GUIDE tool. The main screen consists of 5 buttons, as shown in the Fig. 4. The button Connecting to Raspberry Pi 3 establishes the connection of MATLAB with the Raspberry Pi3 board. The second button allows importing the data to the workspace, this step is essential since the model developed in Simulink needs these values to execute the control algorithm. Among these variables is the plant that was imported in the design of the model, process restrictions, etc. Pressing Open MPC model- Simulink opens the block diagram developed in Simulink, where the mathematical model of the previously obtained plant is shown. With the option Save Excel Data the generated data is exported to Excel, to perform the data analysis later and Close cleans the variables in the workspace and closes the main window.

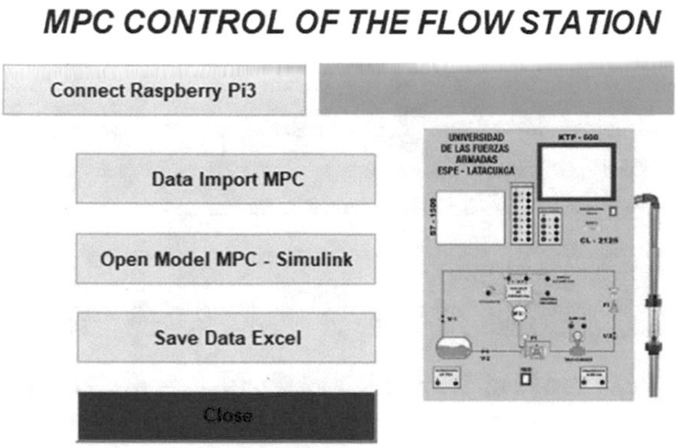

MPC CONTROL OF THE FLOW STATION

Connect Raspberry Pi3

Data Import MPC

Open Model MPC - Simulink

Save Data Excel

Close

Fig. 4. Main window of the MPC developed in MATLAB's GUIDE.

6 Results

The response curves produced by the process variables, after implementing the control algorithms in the embedded boards are shown in Fig. 5. This starts with the process in the lower limit (10 LPM) and in 10 s it increases to 25 LPM; subsequently it is modified to 35, 15, 35 and 20 LPM every 20 s respectively.

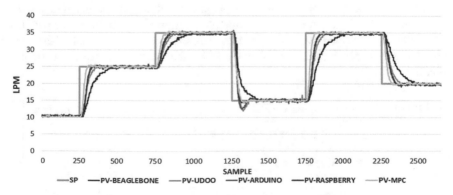

Fig. 5. Response curves of SP vs. PV variable with different SP values.

Although both controllers present high efficiency results, it can be observed that the response of the MPC controller is faster, and with less overshoot than the one achieved with the fuzzy controller, the results vary depending on the board in which the controller was implemented. The percentage of the control action is less than 62%, and without oscillations, thus lengthening the useful life of the final control element. Next, the temporal and over-oscillation characteristic values of the system are presented before the excitation of a unit step that occurs when changing the value of the Set Point from 10 to 30 [LPM], the results are shown in Fig. 6.

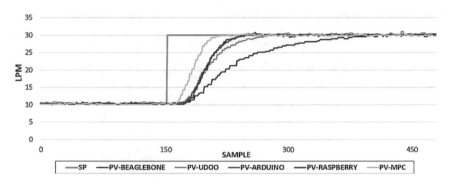

Fig. 6. Response curves of SP vs. PV variable when is excited with a unit step.

The delay time (Dt) represents the interval that the controller takes to perform the necessary corrective action, and is calculated from the moment the SP variable changes. The variation range of the SP is equal to 20 [LPM]. Therefore, the settlement time (St) is determined when the response curve reaches a range equal to or less than 5% around the desired value. In this case, said range will be given by 20 ± 5%, that is between 29 and 31 [LPM]. The rise time (Rt) is determined at the moment the response curve hits the 10% until it reaches 90% for this case it goes from 12 to 28 [LPM]. Table 3 shows these parameters and when performing a global analysis of the recorded data, it is evident that within the implementation of the fuzzy driver, Raspberry Pi 3 presents the best performance among the other embedded board options, with a settling time equal to 3.2 s before a unit step excitation. Complying with the expectations, the MPC controller provided even better results, demonstrated in the signal of the control variable that does not present oscillations, a factor that seeks to increase the life of the actuator.

Table 3. Temporal and over-oscillation characteristic values that each of the board.

	Arduino	Raspberry	BeagleBone	Udoo	MPC
D_t	0.8 s	0.76 s	0.88 s	0.8 s	0.44 s
R_t	1.92 s	1.8 s	4.12 s	2.49 s	1.36 s
S_t	3.16 s	3.2 s	7.11 s	4.24 s	2.16 s
%O	3%	0.25%	0.25%	2%	0.75%

7   Conclusions

The variety of embedded cards that is on the market grows exponentially, not only in brands, but also in new versions of existing ones, depending on the needs and budget of the user. In this work it has been corroborated that the so-called low-cost technologies despite their technical limitations allow the implementation of advanced control. Through the use of high-level programming, libraries and toolboxes of mathematical software developed specifically for this type of device.

The design of the fuzzy control algorithm with Mamdani inference type is relatively easy, since it does not require the mathematical model of the process to be controlled. Despite this, it provided efficient results and lower computational requirements. Making it possible to implement in all the proposed boards. In the case of the MPC controller design, Simulink was used due to their adequate support packages for Raspberry Pi 3. Specifically, from the 2017 version it includes blocks to read and write data through I2C communication and simplifies the SPI communication block.

Through the experimental tests carried out, the embedded board that presented the best results on the implementation of a fuzzy controller, and the only one that allowed the development of an MPC driver was the Raspberry Pi 3. Impressive since it only has digital signal inputs and outputs available, demonstrating that using external elements, and signal conditioning can develop control applications that offer competitive results to those that provide industrial control devices. This research proposal present two main issues. The first one focuses on the loading effect the that embedded boards cause in the process, and for the second the lack of robust communication protocols.

References

1. Ziegler, J.G., Nichols, N.B.: Optimum settings for automatic controllers. J. Dyn. Syst. Meas. Control **115**(2B), 220–222 (1993). https://doi.org/10.1115/1.2899060
2. Zanoli, S.M., Barboni, L., Cocchioni, F., Pepe, C.: Advanced process control aimed at energy efficiency improvement in process industries. In: Proceedings of the IEEE International Conference on Industrial Technology (2018). https://doi.org/10.1109/ICIT.2018.8352152
3. Kouro, S., Perez, M.A., Rodriguez, J., Llor, A.M., Young, H.A.: Model predictive control: MPC's role in the evolution of power electronics. IEEE Ind. Electron. Mag. **9**, 8–21 (2015). https://doi.org/10.1109/MIE.2015.2478920
4. Wibisono, R.P., Suwastika, N.A., Prabowo, S., Santoso, T.D.: Automation canal intake control system using fuzzy logic and Internet of Things (IoT). In: 2018 6th International Conference on Information and Communication Technology, ICoICT 2018 (2018). https://doi.org/10.1109/ICoICT.2018.8528756
5. Bagyaveereswaran, V., Mathur, T.D., Gupta, S., Arulmozhivarman, P.: Performance comparison of next generation controller and MPC in real time for a SISO process with low cost DAQ unit. Alexandria Eng. J. (2016). https://doi.org/10.1016/j.aej.2016.07.028
6. Pinsker, J.E., Lee, J.B., Dassau, E., Seborg, D.E., Bradley, P.K., Gondhalekar, R., Bevier, W.C., Huyett, L., Zisser, H.C., Doyle, F.J.: Randomized crossover comparison of personalized MPC and PID control algorithms for the artificial pancreas. Diabetes Care **39**, 1135–1142 (2016). https://doi.org/10.2337/dc15-2344

7. Amir, S., Kamal, M.S., Khan, S.S., Salam, K.M.A.: PLC based traffic control system with emergency vehicle detection and management. In: 2017 International Conference on Intelligent Computing, Instrumentation and Control Technologies, ICICICT 2017 (2018). https://doi.org/10.1109/ICICICT1.2017.8342786

8. Aftab, M., Chen, C., Chau, C.K., Rahwan, T.: Automatic HVAC control with real-time occupancy recognition and simulation-guided model predictive control in low-cost embedded system. Energy Build. **154**, 141–156 (2017). https://doi.org/10.1016/j.enbuild.2017.07.077

9. Govindan, K., Diabat, A., Madan Shankar, K.: Analyzing the drivers of green manufacturing with fuzzy approach. J. Clean. Prod. **96**, 182–193 (2015). https://doi.org/10.1016/j.jclepro.2014.02.054

10. Bermúdez-Ortega, J., Besada-Portas, E., López-Orozco, J.A., Bonache-Seco, J.A., La Cruz, J.M.D.: Remote web-based control laboratory for mobile devices based on EJsS, Raspberry Pi and Node.js. IFAC-PapersOnLine **48**, 158–163 (2015). https://doi.org/10.1016/j.ifacol.2015.11.230

11. Bin Kassim, M.F., Haji Mohd, M.N.: Tracking and counting motion for monitoring food intake based-on depth sensor and UDOO board: a comprehensive review. IOP Conf. Ser. Mater. Sci. Eng. **226**, 012089 (2017). https://doi.org/10.1088/1757-899X/226/1/012089

12. Saá, F., Varela-Aldás, J., Latorre, F., Ruales, B.: Automation of the feeding system for washing vehicles using low cost devices (2020). https://doi.org/10.1007/978-3-030-32033-1_13

13. Buele, J., Espinoza, J., Pilatásig, M., Silva, F., Chuquitarco, A., Tigse, J., Espinosa, J., Guerrero, L.: Interactive system for monitoring and control of a flow station using labVIEW. In: Advances in Intelligent Systems and Computing (2018). https://doi.org/10.1007/978-3-319-73450-7_55

14. Sittakul, V., Chunwiphat, S., Tiawongsombat, P.: Fuzzy logic-based control in wireless sensor network for cultivation. In: Advances in Intelligent Systems and Computing (2017). https://doi.org/10.1007/978-981-10-1645-5_23

15. Espinoza, J., Buele, J., Castellanos, E.X., Pilatásig, M., Ayala, P., García, M.V.: Real-time implementation of model predictive control in a low-cost embedded device. In: IMCIC 2018 - 9th International Multi-conference on Complexity, Informatics and Cybernetics, Proceedings (2018)

16. Kozák, Š., Pytel, A.: MPC controller as a service in IoT architecture. In: Proceedings of the 29th International Conference on Cybernetics and Informatics, K and I 2018 (2018). https://doi.org/10.1109/CYBERI.2018.8337550

17. Pruna, E., Andaluz, V.H., Proano, L.E., Carvajal, C.P., Escobar, I., Pilatasig, M.: Construction and analysis of PID, fuzzy and predictive controllers in flow system. In: 2016 IEEE International Conference on Automatica, ICA-ACCA 2016 (2016). https://doi.org/10.1109/ICA-ACCA.2016.7778493

Virtual Goniometer Using 3 Space Mocap Sensors for Lower Limbs Evaluation

Jorge Buele[1,2(✉)] , Marco Pilatásig[2] , Hamilton Angueta[2] ,
Belén Ruales[1] , and José Varela-Aldás[1]

[1] SISAu Research Group, Universidad Tecnológica Indoamérica,
Ambato 180212, Ecuador
{jorgebuele,belenruales,josevarela}@uti.edu.ec
[2] Universidad de las Fuerzas Armadas ESPE, Latacunga 050104, Ecuador
{mapilatagsig,haangueta}@espe.edu.ec

Abstract. The deterioration of motor skills in lower limbs represents a reduction in people's quality of life. This paper presents a virtual goniometer developed by the 3 Space Mocap sensors and Unity 3D software. The system shows lower limbs angles in real-time. This system helps physiatrists determine the movements of hip and knee and was compared with a goniometer to corroborate the system performance. To validate this prototype, comparative tests are carried out present a standard deviation of $1.86°$ and $2.12°$ in the left hip and knee respectively. Therefore, it confirms to be a reliable proposal for physical evaluations of lower limbs.

Keywords: 3 Space Mocap · Goniometer · Lower limbs · Rehabilitation · Unity 3D

1 Introduction

Continuous and unavoidable process of scientific evolution experienced by modern societies is reflected in the development of information and communication technology (ICT), internet of things (IoT), telecommunications equipment, storage and data management. This has provided innovative environments of perception, stimulation and interaction in the process of acquisition, transmission and dissemination of information [1–3]. Among the management and information environments include: internet, machine learning, big data, artificial intelligence, virtual and augmented reality, among others [4]. Development and implementation meet the human necessity of evolve and transform the conventional way of sharing knowledge [5–7]. The situation described shows the development of new scenes; where new technologies have emerged that allow the management and exchange of knowledge [8].

Using these technologies is possible to interconnect most devices and thus provide more and better services. Thus, virtual reality appears as a necessary complement to all areas of knowledge [9]. Acceptance lies in the presentation in immersive 3D graphics environments that use I/O devices (gloves, goggles, helmets, mobile devices). This is in pursuit of greater interaction with the virtual environments. [10, 11]. "Telepresence" or illusion of "being there" is achieved by using sensors that capture movement and

© Springer Nature Switzerland AG 2020
Á. Rocha et al. (Eds.): ICITS 2020, AISC 1137, pp. 439–448, 2020.
https://doi.org/10.1007/978-3-030-40690-5_43

sensation of the user, in order to create a list of what is displayed in real time on the screen. Thus, it manages to live 3D virtual experiences in simulated environments, perceive sensations of touch, hold, and maneuver everything that is being observed [12–14]. In health, applications appear from creating 3D animations for the study and treatment of people with different physical and mental disorders, to complex devices that allow rehabilitation processes [15].

In the field of rehabilitation after accidents, many patients develop pathologies that reduce their ability to perform movements and displacements and losing their autonomy [16]. This is the reason that health professionals suggest beginning the respective rehabilitation after the process of wound healing. In order to evaluate the body movement, we have developed several medical instruments, such as the goniometer. A device for determining the angle at which a limb is moved to perform an action [17]. Regularly in the area of physiotherapy, this process using conventional invasive instruments, which have a certain degree of imprecision. In addition, this tool does not provide the possibility of keeping a statistical record to review the percentage of recovery. Trying to provide a better service, they have been implemented variants of these devices, including: In [18] the development of an application for smartphone a goniometer used in the measurement of user foot dorsiflexion is presented; to validate its performance is compared to conventional one, getting over 80% efficiency in tests.

In [19] an application for smart phone that seeks to evaluate the movement of the elbow joint, presenting unstable results anatomical variations and different load angles elbow. A digital goniometer to measure the position of the knee joint is presented in [20], where experimental results of a prototype made with a microcontroller and compared with a conventional goniometer, acceptable results at a low-cost system is presented. In this context, this document presents the implementation of a virtual goniometer through 3 Space Mocap sensors. This system fuses a routine and monotonous session of exercises with the immersion that virtual reality environments allow. This prototype demonstrates a high performance when compared to a conventional instrument, as it allows a strong interaction between the patient and his therapist, during a rehabilitation session of her lower extremities.

The paper is organized in six sections, including the introduction, in Sect. 2 Methodology used for system development, in Sect. 3 mode of use is explained, Sect. 4 presents tests, Sect. 5 shows the goniometer virtual validation and finally the conclusions and future work are presented in Sect. 6.

2 Methodology

This section presents the virtual goniometer stages, the same as shown in Fig. 1.

2.1 Input Peripherals

The 3 Space Mocap sensors has nine degrees of freedom. They were used to detect the angles of the hip and knee by the data generated by the gyroscope, accelerometer and magnetometer. The communication between the sensors (three on the right leg and three on the left leg) and the computer is wireless. Data entered into the computer were processed using an algorithm that allows viewing angles of the lower limbs in the virtual environment developed in Unity 3D.

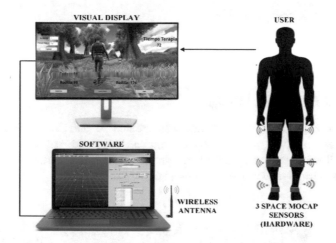

Fig. 1. General diagram of the proposal.

2.2 Script Development

The sensors provide data quaternion format (q), and are given in (1) and (2).

$$q = [q_0 \quad q_1 \quad q_2 \quad q_3]^T = [q_w \quad q_x \quad q_y \quad q_z]^T \tag{1}$$

$$|q|^2 = q_0^2 + q_1^2 + q_2^2 + q_3^2 = q_w^2 + q_x^2 + q_y^2 + q_z^2 = 1 \tag{2}$$

A quaternion can be associated with a rotation around an axis by the expressions presented in (3), (4), (5) and (6).

$$q_0 = q_w = \cos(\alpha/2) \tag{3}$$

$$q_1 = q_x = \sin(\alpha/2)\cos(\beta_x) \tag{4}$$

$$q_2 = q_y = \sin(\alpha/2)\cos(\beta_y) \tag{5}$$

$$q_3 = q_z = \sin(\alpha/2)\cos(\beta_z) \tag{6}$$

Where: α is the rotation angle in radians, and $\cos(\beta_x)\cos(\beta_y)\cos(\beta_z)$ can locate the axis of rotation. To present different angles in the Unity software, we must convert the data format quaternion to Euler by the following matrix describes in (7).

$$\begin{bmatrix} \phi \\ \theta \\ \psi \end{bmatrix} = \begin{bmatrix} \arctan\frac{2(q_0q_1+q_2q_3)}{1-2(q_1^2+q_2^2)} \\ \arcsin(2(q_0q_2+q_3q_1)) \\ \arctan\frac{2(q_0q_3+q_1q_2)}{1-2(q_2^2+q_3^2)} \end{bmatrix} \tag{7}$$

Where:

ϕ It represents the rotation in x axis.
θ It represents the rotation in y axis.
ψ It represents the rotation in z axis

2.3 Virtual Environment Design

To capture user movements an avatar was created in the Unity 3D platform, the respective configurations to produce an animated humanoid without a preset pattern using the following steps are performed: model, rigging and animation. In Fig. 2 the design of the virtual environment is presented.

Fig. 2. Virtual display environment for hip and knee angles.

- **Model.** This is the process to create a humanoid own or mesh, the 3DSMax and Blender software was used.
- **To rig.** A skeleton is created with joints to control avatar movements in this process. In this case, the movements will be controlled according to the coordinates of the sensors 3 Space Mocap.
- **Animations.** Animations of the movements of the hip, right and left knee. The corresponding angles of the limb measurement is also displayed.

3   How to Use

  i The user must wear sports clothes, if possible, shorts and shirt, to be placed correctly sensors.

  ii The user must be at a distance of 3 to 4 m from Gateway to sensors placed in the lower limbs are detected.

  iii In the main menu interface, first left clicking on the button "calibrated" so that the sensor calibration is performed.

  iv Place physical goniometer in the limb to be measured.

  v Specialist indicate the posture that the user must perform to detect the angle of the limb to be measured.

  vi In the virtual environment angles corresponding to the hip and knee are displayed in real time.

  vii When the test is generated a database with information of each user, the same can be subsequently analyzed by the Specialist.

4   Tests

Experimental tests were conducted to evaluate the response of the 3 Space Mocap sensors measurement of hip and knee angles with a specialist supervision. A physical goniometer was placed on the limb to be measured to compare angles acquired by the virtual goniometer as shown in Figs. 3 and 4 respectively.

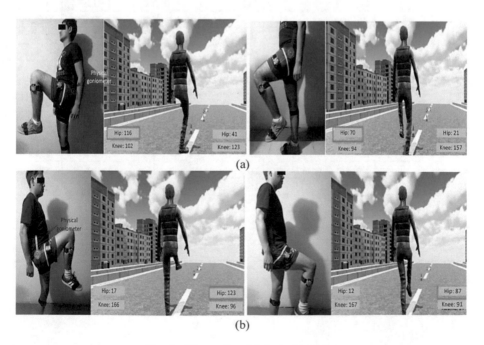

(a)

(b)

Fig. 3. Hip angles: (a) Left. (b) Right.

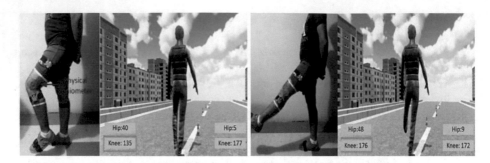

Fig. 4. Left knee angles.

5 Validation of Virtual Goniometer

To confirm the operating of the virtual goniometer implemented several tests was developed at different angles of flexion and extension. These measurements were made with the physical and virtual goniometer considering the same position; error, variance and standard deviation, are presented in Table 1.

Table 1. Left hip angles.

Number of samples	Physical goniometer (°)	Virtual goniometer (°)	Error	u-Xi	(u-Xi) ^ 2
1	118	111	7	−1.87	3.50
2	92	87	5	0.13	0.02
3	73	69	4	1.13	1.28
4	51	48	3	2.13	4.54
5	121	112	9	−3.87	14.98
6	98	95	3	2.13	4.54
7	57	52	5	0.13	0.02
8	22	18	4	1.13	1.28
9	124	117	7	−1.87	3.50
10	127	119	8	−2.87	8.24
11	109	102	7	−1.87	3.50
12	105	100	5	0.13	0.02
13	84	79	5	0.13	0.02
14	81	75	6	−0.87	0.76
15	36	33	3	2.13	4.54
16	31	28	3	2.13	4.54

To calculate the variance used the formula in (8) and standard deviation in (9).

$$s^2 = \frac{\sum (u - x_i)^2}{(n - 1)} = 3.45 \tag{8}$$

$$s = \sqrt{3.45} = 1.86 \tag{9}$$

Table 2 presents the angles of flexion and extension of the left knee, considering 16 samples.

Table 2. Left knee angle

Number of samples	Physical goniometer (°)	Virtual goniometer (°)	Error	or-Xi	(U-Xi) ^ 2
1	26	21	5	−0.47	0.22
2	75	67	8	−3.47	12.04
3	136	130	6	−1.47	2.16
4	178	175	3	1.53	2.34
5	27	20	7	−2.47	6.10
6	75	70	5	−0.47	0.22
7	148	142	6	−1.47	2.16
8	179	177	2	2.53	6.40
9	179	178	1	3.53	12.46
10	180	179	1	3.53	12.46
11	147	144	3	1.53	2.34
12	145	141	4	0.53	0.28
13	81	75	4	0.53	0.28
14	82	77	5	−0.47	0.22
15	36	30	6	−1.47	2.16
16	32	25	7	−2.47	6.10

Data variance and standard deviation are presented below:

$$s^2 = 4.52$$

$$s = \sqrt{4.52} = 2.12$$

In Figs. 5 and 6 graphs showing the error, the average error, the upper control limit, the lower control limit, considering 16 samples in the left hip and knee are presented respectively. The data obtained with the left knee have an average error of 5.13 and 4.53 with the left hip.

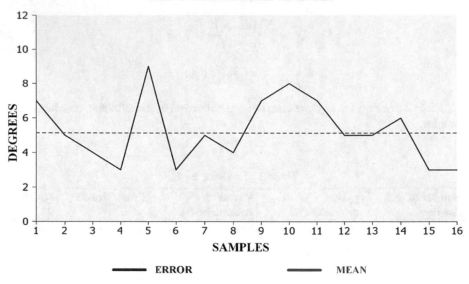

Fig. 5. Error of left hip angle.

All samples have a positive error with different values. By analyzing this data, it is determined that the error hip increases at big angles and knee error increased by small angles.

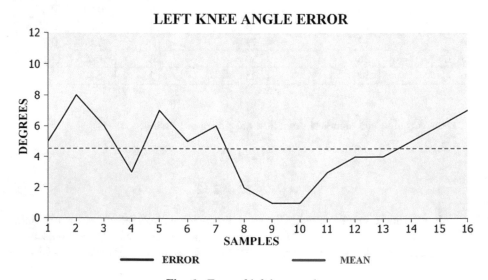

Fig. 6. Error of left knee angle.

All samples have a positive error with different values. By analyzing this data, it is determined that the error hip increases at big angles and knee error increased by small angles.

6 Conclusions and Future Work

3 Space Mocap sensors are inertial tracking units with high precision and high reliability, being useful in human body assessment projects. Algorithm generates the acquired data quaternion format to Euler to show the angles in the virtual environment developed on the Unity 3D software.

By tests it was determined that the data obtained with the virtual goniometer has a standard deviation of 1.86° and 2.12° for the hip and knee left respectively with respect to physical goniometer. From the medical point of view, these values obtained are considered acceptable. This validates the presented prototype and can be used in therapy sessions. In this way, the user is given a technological tool that allows him to have a more interactive recovery process thanks to the implementation of virtual environments.

Future work is planned to make a virtual goniometer for upper and lower extremities using 3 Space mocap sensors.

References

1. Li, S., Xu, L.D., Zhao, S.: The internet of things: a survey. Inf. Syst. Front. **17**(2), 243–259 (2015)
2. Gretzel, U., Sigala, M., Xiang, Z., Koo, Ch.: Smart tourism: foundations and developments. Electron. Mark. **25**(3), 179–188 (2015)
3. Gretzel, U., Werthner, H., Koo, Ch., Lamsfusd, C.: Conceptual foundations for understanding smart tourism ecosystems. Comput. Hum. Behav. **50**, 558–563 (2015)
4. Pourzolfaghar, Z., Helfert, M., Melo, V.A.B., Khalilijafarabad, A.: Proposing an access gate to facilitate knowledge exchange for smart city services. In: 2017 IEEE International Conference on Big Data (Big Data), Boston, pp. 4117–4122. IEEE (2017)
5. Salazar, F.W., Núñez, F., Buele, J., Jordán, E.P., Barberán, J.: Design of an ergonomic prototype for physical rehabilitation of people with paraplegia. In: Advances in Intelligent Systems, vol. 1078. Springer (2020)
6. Bozcuoğlu, A.K., et al.: The exchange of knowledge using cloud robotics. IEEE Robot. Autom. Lett. **3**(2), 1072–1079 (2018)
7. Gang, K., Ravichandran, T.: Exploring the determinants of knowledge exchange in virtual communities. IEEE Trans. Eng. Manage. **62**(1), 89–99 (2015)
8. Takenouchi, H., Tokumaru, M.: Wrist watch design system with interactive evolutionary computation. In: Stephanidis, C. (ed.) HCI International 2017 – Posters' Extended Abstracts, HCI 2017. Communications in Computer and Information Science, vol. 714. Springer, Cham (2017)
9. Tong, X., Kitson, A., Salimi, M., Fracchia, D., Gromala, D., Riecke, B.: Exploring embodied experience of flying in a virtual reality game with Kinect. In: 2016 IEEE International Workshop on Mixed Reality Art (MRA), Greenville, pp. 5–6. IEEE (2016)

10. Nguyen-Vo, T., Riecke, B.E., Stuerzlinger, W.: Moving in a box: improving spatial orientation in virtual reality using simulated reference frames. In: 2017 IEEE Symposium on 3D User Interfaces (3DUI), Los Angeles, pp. 207–208. IEEE (2017)
11. Wu, C.M., Hsu, C.W., Lee, T.K., Smith, S.: A virtual reality keyboard with realistic haptic feedback in a fully immersive virtual environment. Virtual Reality **21**(1), 19–29 (2017)
12. Fairchild, A.J., Campion, S.P., García, A.S., Wolff, R., Fernando, T., Roberts, D.J.: A mixed reality telepresence system for collaborative space operation. IEEE Trans. Circuits Syst. Video Technol. **27**(4), 814–827 (2017)
13. Pilatásig, M., et al.: Interactive system for hands and wrist rehabilitation. In: Advances in Intelligent Systems and Computing, vol. 721. Springer (2018)
14. Varela-Aldás, J., Palacios-Navarro, G., García-Magariño, I.: Immersive virtual reality app for mild cognitive impairment. Rev. Ibérica Sist. e Tecnol. Informação. **E19**, 278–290 (2019)
15. Galarza, E.E., Pilatasig, M., Galarza, E.D., López, V.M., Zambrano, P.A., Buele, J., Espinoza, J.: Virtual reality system for children lower limb strengthening with the use of electromyographic sensors. In: Lecture Notes in Computer Science, vol. 11241. Springer (2018)
16. Andrea Sánchez, Z., Santiago Alvarez, T., Roberto Segura, F., Tomás Núñez, C., Urrutia-Urrutia, P., Franklin Salazar, L., Altamirano, S., Buele, J.: Virtual rehabilitation system using electromyographic sensors for strengthening upper extremities. In: Rocha, Á., Pereira, R. (eds.) Developments and Advances in Defense and Security. Smart Innovation, Systems and Technologies, vol. 152. Springer, Singapore (2020)
17. Pruna, E., et al.: Implementation of a multipoint virtual goniometer (MVG) trough Kinect-2 for evaluation of the upper limbs. In: Rocha, Á., Correia, A., Adeli, H., Reis, L., Costanzo, S. (eds.) Recent Advances in Information Systems and Technologies, WorldCIST 2017. Advances in Intelligent Systems and Computing, vol. 570. Springer, Cham (2017)
18. Otter, S.J., et al.: The reliability of a smartphone goniometer application compared with a traditional goniometer for measuring first metatarsophalangeal joint dorsiflexion. J. Foot Ankle Res. **8**(30), 1–7 (2015)
19. Resende, T., et al.: Universal goniometer and smartphone app for evaluation of elbow joint motion: reproducibility analysis. In: 2017 International Conference on Virtual Rehabilitation (ICVR), Montreal, pp. 1–2. IEEE (2017)
20. Domínguez, G., Cardiel, E., Arias, S., Rogeli, P.: A digital goniometer based on encoders for measuring knee-joint position in an orthosis. In: 2013 World Congress on Nature and Biologically Inspired Computing (NaBIC), Fargo, pp. 1–4. IEEE (2013)

Building a Web Tracking Browser Information System: The Online Panel as a Research Method in Internet Studies

Filipe Montargil[1]([✉]), Branco Di Fátima[2], and Cristian Ruiz[1]

[1] ESCS (School of Communication and Media Studies), Lisbon, Portugal
fmontargil@escs.ipl.pt
[2] CIES/IUL (Centro de Investigação e Estudos de Sociologia), Lisbon, Portugal

Abstract. Internet research requires internet-based research methods and tools. Since the commercial adoption and the rapid expansion of internet use in civil society, social scientists have tried to understand the impact and contribution of this technology in the shaping of contemporary human existence. However, academic research has been based mainly in mature and traditional data collection and analysis methods, previously established, like survey research or social network analysis. Some authors suggest that this trend represents a challenge for the social sciences' ability to champion innovative methodological resources, underlining the need to create also native digital methods. The authors developed, considering this challenge, an online panel of internet users – a method already used in the market research industry to monitor and characterize audiences and consumer behaviour, but not usually explored in social science academic research. Following a case study approach, the current paper presents the information system and technological infrastructure developed to support this online panel.

Keywords: Information system · Online panel · Internet Studies · Internet use · Native digital methods

1 Introduction

The internet is today an essential part of contemporary life, mostly in developed and middle-income countries. We use it now, in our everyday life, for almost everything, from looking up for information, entertainment, communicating with family, friends or people with common interests, to buying products and services.

With an amazing amount of information produced every second it is easy to assume that, one way or the other, this technology is changing human habits and social practices. This called the attention of social scientists who, since the very beginning of commercial use of the internet, began to study its different manifestations in society, shaping an area that many refer now to as *Internet Studies*.

With the growing interest of several scientific areas, quickly emerged the need to find adequate data collection and analysis methods for this field. Two data collection methods have been widely used and are probably becoming central, in the field: on one side, we can find data collected through self-report methods, relying on the individual's

Á. Rocha et al. (Eds.): ICITS 2020, AISC 1137, pp. 449–455, 2020.
https://doi.org/10.1007/978-3-030-40690-5_44

own report of their opinion or behaviour (through surveys, individual qualitative interviews, focus groups or diary studies, for instance) and, on the other side, data obtained online and treated through social network analysis. Both have their roots in more traditional and conventional research areas, in the social sciences.

Some authors suggest, however, that the proliferation of other sources of data ('social' transactional data) and resources for its analysis is challenging the social sciences' ability to champion innovative methodological resources [1, 2].

The LLMCP (Living Lab for Media Content and Platforms), a research project from ESCS (School of Communication and Media Studies), based in Lisbon (Portugal), intends to contribute to this debate and to develop a native digital method using an online panel of internet users, which is already used in the industry, to monitor and characterize audiences and consumer behaviour [3], but not usually explored in academic research.

Following a case study approach, the current paper aims to present the information system and technological infrastructure developed to support this online panel, in order to describe and revisit a year of development in our exploratory research (September 2018–September 2019).

2   Internet Studies and Online Panels

Many have been the initiatives to develop methods for the Internet Studies in Social Sciences, with different areas working in new concepts and perspectives. In media studies, for example, several authors have carried out efforts to develop methods adapted to the analysis of Computer-Mediated Communication (CMC) [4–6].

Rogers [5, 6] identifies the web, websites, hyperlinks, search engines, wikis and social network sites as entirely digital objects, with its specific ontology, and proposes the expression *native digital methods* to identify methodological resources that are originally conceived and developed in the digital context, as opposed to existing methods that are translated to online spaces.

Another reality can be found in the industry, with profit-oriented research, not aiming to understand, discuss and explore social phenomena, but instead to describe and predict consumer behaviour, in order to make more efficient and profitable business decisions.

The online panel methodology was developed in the internet business landscape, as a consequence of the convergence of the audience measurement industry with the new digital era. According to Kent, "In market research, a panel is a representative sample of individuals, households or organizations that have agreed to record, or permit the recording of, their activities or opinions in respect of an agreed range of products, services or media-use behaviours on a continuous or regular basis" [7] (p. 10). While it is true that this modality already existed before the study of media audiences, like television and radio, and that there are digital panels that follow the same classical model, it is also true that there is an entirely digital approach in which a group of users accept to be monitored, over a period of time[1].

Our goal, with the LLMCP – Living Lab on Media Content and Platforms project, is to bring this method and the corresponding resources to academic research.

3   The LLMCP Information System

By the end of 2018, the team developed an information system as an organizational structure for an online panel of Internet users.

The project consortium was led by ESCS - School of Communication and Media Studies, from the Lisbon Polytechnic Institute, and included also the Aveiro University, the Leiria Polytechnic Institute, the Santarém Polytechnic Institute and Innovation Makers, a private company operating in the IT sector. The project received public funding through FCT - Fundação para a Ciência e a Tecnologia, the Portuguese national science foundation.

According to the project's cost-benefit assessment, a Google Chrome extension was the most efficient solution to collect systematically data from the web browsing history, in this exploratory stage [8].

Despite of its limitations, that must be considered natural, mostly in a period of development of native digital methods, Chrome is the most used browser by the Portuguese and has the largest market share in the global browser industry [9].

In the first stage of the project, the team looked for similar experiences that could support the panel's methodological decisions. The development of the system was planned in three major stages: (i) project design; (ii) development of user form and; (iii) computer programming of the extension. The development of the system took about two months, which also featured information collection tests [10]. After several tests, the system was launched in October 2018, initiating recruitment of members in November and regular and ongoing data collection in January 2019.

The recruitment process works in the following sequence: the candidate to participate in the panel applies through a form in a HTML page, with JavaScript and CSS components, divided into four main sections, (i) personal information, (ii) labour situation, (iii) issues related with internet access and (iv) account authentication data (See Fig. 1). The registration creates a REST (Representational State Transfer) to the server developed in Java, that at the same time processes information and communicate with a database in MySQL and Linux.

---

[1] See, for example:
Nielsen: https://computermobilepanel.nielsen.com/cmp/landingeng.jsp.
ComScore: https://www.comscore.com/comScore-Panel-Services.
Marktest: http://netpanel.marktest.pt/.

This protocol guarantees adequate storage of information and allows researchers to access the platform through a back office. Then, after a login operation with credentials (user/password) a new REST call is made, that validates information of the admin. If the procedure runs correctly, the system generates a token that gives access, during a limited period of time, to the back office. To guarantee the sample control, essential in social sciences [11], every application must be analysed by members of the research team.

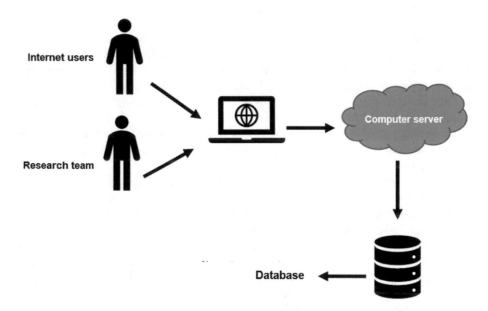

Fig. 1. LLMCP's information system

After login into the system, it is possible for the admin to transfer stored information through an .csv output file, containing data resulting directly from real time monitoring of web navigation actions performed through Google Chrome.

The system also allows access to the participant list and to the creation of subsets by gender, age, civil status or other available variables. This list is located in the back office, developed in HTML, JavaScript and CSS (Fig. 2) and also available for download, through an .xls output file.

In case of approval, the candidate receives an email with a welcome text in Portuguese (*"Obrigado pelo vosso registo. Foi aceite para participar do painel de utilizadores da Internet. Por favor, instale a extensão para Google Chrome, autentifique o email e password para fazer parte da nossa investigação. Em caso de dúvida, não hesite em contactar-nos"*) and a link to download the extension, from the Chrome Web Store.

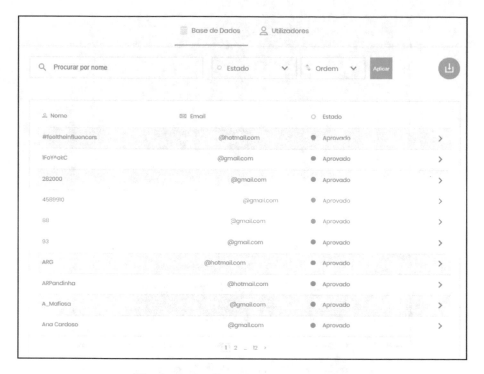

Fig. 2. Back-office (user management area)

In case of rejection, the research team will prepare a personal email, explaining the reasons leading to the decision.

The user is required to login, through the Google Chrome extension, with the chosen username and password, after installing the extension. When login is conducted successfully, as in the above-mentioned case of the admin, the system produces a token that will work until the user logouts.

Whenever active, the extension will connect with the API (Application Programming Interface) of Google Chrome. From this point, with a JavaScript communication, it will track, extract and save information regarding web navigation actions in the database. Otherwise, if deactivated, there is no way to collect data (Fig. 3). Web tracking can be suspended by the panel member at any moment, with a simple click in the user app interface. The system, including the extension and the application form, comply with the General Data Protection Regulation (GDPR), in operation since May 25, 2018, in the European Union.

The data transfer service sends a REST request to the server, that performs a search in the MySQL database. This procedure, as mentioned above, creates an output file in. csv (Comma Separated Values). Every archive stores 50.000 web navigation actions, in chronological order, with approximately 15 mb.

Fig. 3. Web tracking suspension user interface

In the database, every navigation action represents a website, with an URL, tab number, page activity, content date and time of access. To reduce communication time with the server and minimize timeout errors, usually present in large database extractions [12], files are zipped and sent by email. Under normal connection to broadband Internet, the extraction process takes no more than 5 min.

By the end of September 2019, the panel had 165 registered users, the majority being students from ESCS. From these registered users, 90 have been tracked and monitored at least once. The main database, storing 780.000 web navigation actions, occupies approximately 212 mb, in .csv file format.

4   Conclusions

The project team developed an information system to support the deployment and operation of an online panel of internet users, presented here. This effort brings methods from panel research, in the market research industry, to social science research and has the major goal to contribute to the creation of native digital methods.

Considering the accumulated know-how, experience and knowledge, as a consequence of the development of the information system presented here, the team considers there is space and potential to develop new methods to further explore and understand our online behaviour.

Our knowledge about internet use doesn't need to be reduced to information collected exclusively through self-report methods, on one side, or based only in global aggregate data, on the other side. There is room to explore new methods, providing the scientific community with information that can be used to analyse effective online behaviour and web navigation actions performed by individuals. And these methods can, obviously, be combined with self-report methods, in order to further explore representations, opinions or individual motivations.

As next steps, the team expects to explore the available data and to develop the existing information system, improving some of the existing limitations and shortcomings.

References

1. Savage, M., Burrows, R.: The coming crisis of empirical sociology. Sociology **41**(5), 885–899 (2007)
2. Savage, M., Burrows, R.: Some further reflections on the coming crisis of empirical sociology. Sociology **43**(4), 762–772 (2009)
3. Kumar, R., Tomkins, A.: A characterization of online browsing behavior. In: The 19th International Conference on World Wide Web, pp. 561–570. ACM, Raleigh (2010)
4. Jones, S.G. (ed.): Doing Internet Research Critical Issues and Methods for Examining the Net. Sage, London (1999)
5. Rogers, R.: Digital Methods. MIT Press, Cambridge (2013)
6. Rogers, R.: Doing Digital Methods. SAGE, London (2019)
7. Kent, R. (ed.): Measuring Media Audience. Routledge, New York (1994)
8. Montargil, F., Miranda, S., Di Fátima, B.: Medir a sociedade de informação: sistema para um painel online de utilizadores da Internet. In: Iberian Conference on Information Systems and Technologies (CISTI), pp. 1–4. IEEE Xplore Digital Library, Coimbra (2019)
9. Statcounter: Top desktop, tablet, and console browsers per country. Global Stats Tool, New York (2018)
10. Montargil, F., Rodrigues, V., Di Fatima, B.: Sistema de informação para um método digital em ciências sociais. In: Lisboa, 19.ª Conferência da Associação Portuguesa de Sistemas de Informação –CAPSI'2019 (2019)
11. Ander-Egg, E.: Introducción a las técnicas de investigación social. Editorial Humanitas. Buenos Aires (1978)
12. Dai, T., et al.: Understanding real-world timeout problems in cloud server systems. In: Conference on Cloud Engineering (IC2E). IEEE Xplore Digital Library, Orlando, pp. 1–11 (2018)

Health Informatics

Construction of a WBGT Index Meter Using Low Cost Devices

José Varela-Aldás$^{(\boxtimes)}$ ⓘ, Esteban M. Fuentes ⓘ, Belén Ruales ⓘ, and Christian Ichina ⓘ

SISAu Research Group, Universidad Indoamérica, 180103 Ambato, Ecuador
{josevarela, belenruales}@uti.edu.ec,
tebanfuentes@gmail.com, ichinachristian@gmail.com

Abstract. Appropriate working conditions can improve the working performance of an employee on a company, one of the main concerns nowadays is the thermal stress, due to the affection not only to the productivity of an employee but also for the risk of affections to his health, in this sense, WBGT (Wet-Bulb Globe Temperature) index is widely proved to define thermal discomfort and thermal stress but in some cases is out of reach due to its cost or availability on a certain region of the world, so this work explains the construction of a low cost device WBGT meter built by an Arduino pro mini, three lm35 temperature sensors and an LCD to install the electronic system; Globe temperature and wet bulb temperature was achieved using a matte black sphere and moistened fabric, respectively. After testing the dispositive, the collected information was sent to a computer through serial communication for further analysis which showed acceptable errors on the measurement of the WBGT index taken indoors and outdoors thru the device and compared to the data obtained by a commercial WBGT meter.

Keywords: WBGT index · Thermal stress · Sensors · Arduino

1 Introduction

Human body has an exceptional adaptability capacity to its surrounding, especially to the climatic conditions. It is capable to function without any problem at different temperatures, as a prove of this, it is capable to survive extreme temperatures all over the world, according to the seasons or to their weather conditions where they live such as the jungle or the desert, where extreme hot and humid conditions can appear. Body temperature regulator handles temperature variation through the increasing on the consumption of calories, accelerating the metabolism in the case of cold weather or by sweating in the case of extreme warm conditions to low the body temperature perception. However, this capacity of adaptation can be taken in count only during short periods; this does not mean that the human being can develop working activities during the whole working day without any affection to the health caused by the exposition to extreme climatic conditions [1–3].

Terms such as thermal comfort or thermal stress appear based on the sensations produced by the temperature on the human body. Thermal Comfort is related with the

© Springer Nature Switzerland AG 2020
Á. Rocha et al. (Eds.): ICITS 2020, AISC 1137, pp. 459–468, 2020.
https://doi.org/10.1007/978-3-030-40690-5_45

wellness of the temperature feeling, which goes between 19 and 24 °C, and the thermal stress, which is related to the discomfort caused by the temperatures over or under the ones corresponding to the thermal comfort [4, 5]. There are some specific situations where a person can be exposed to high temperature mainly according to a job position, for example by the operation of a machinery which emits a lot of heat, insufficient ventilation, or simply owing to the geographical location where the operator performs his activities is subject to complex conditions of heat and humidity.

Thermal stress by high temperatures can cause dehydration, dizziness, and even skin troubles. The main area of interest related to this risk focuses in Work injuries and accidents due to hot conditions, these sort of issues at work can be produced by physical discomfort, fatigue, loss of psycho-motor performance, concentration and reduced alertness getting dangerous for the health of workers, even putting them at risk of death [6–9].

There are a lot of indexes that allow to determine the environmental or working conditions, most of them not only related to temperature, but also to RH (relative humidity) or metabolism, such as environmental stress index (ESI) [10], outdoor environmental heat index (OEHI) [11], Universal Thermal Climate Index (UTCI), wet-bulb dry temperature (WBDT), wet-bulb globe temperature (WBGT), Tropical Summer Index (TSI), and even some Physiological parameters [12].

Even when exist lots of indexes, WBGT is the most used index to assess 'heat stress' in the management of occupational and sports activities in warm and hot conditions [13, 14]. WBGT represents the temperatures of dry bulb, moist bulb and radiant bulb, the metabolic rate in kilocalories per hour and also the work regime depending on whether it is continuous or allows pauses or intermittent, and in this way allows us to determine if the obtained value is within the permissible limits [15].

Finally, the main objective of this paper is to construct a WBGT low cost device, with parts that can be easy to get [16]. Later the obtained result collected through this device were compared with data obtained by a commercial device and through this ensure that the built device is totally capable to help in the determination of thermal stress by the WBGT index.

This paper consist of an introduction of the study based on WBGT index, WBGT meter, effects of heat over human being and also a short story of the WBGT, later Methods and materials are exposed to explain how the device was developed showing a scheme of the proposal, the connections of the WBGT index meter and the way how the WBGT index meter operates, at the results section the comparison between the commercial device and the low cost device is presented with the obtained data, and finally the conclusions are presented based on the collected data y the two devices.

2 State of the Art

There are just a few studies which falls near this topic, the closest to the mentioned in such as Thermal environment sensor array by Ramirez and others [17], where a thermal environment (TE) monitoring and control device is proposed in 2018, further there is not too much literature boarding the construction of WBGT dispositive.

High cost of WBGT devices and sometimes the low accessibility to these have led certain research groups to develop their own equipment such as the Yantek study and collaborators where the National Institute for Occupational Safety and Health (NIOSH) developed a device to measure the WBGT during refugee alternatives heat/humidity tests for evaluating the thermal environment of underground coal mine refuge alternatives where the black globe temperature sensor was constructed by installing to compression-fitting-mounted, tube-encapsulated Class A RTD inside a matte-black-painted, 152-mm-diameter (6-diameter) copper toilet float. The natural wet-bulb temperature sensor was constructed using a distilled-water-filled glass jar and a tube-encapsulated Class A RTD with a cotton wick covering the RTD and dipping into the water. Another tube-encapsulated RTD was used as the dry-bulb temperature sensor. During data collection, the RTDs for the WBGT device were coed to the data acquisition system for later calculation of the WBGT [18].

3   Methods and Materials

3.1   Formulation of the Problem

Based on the bibliographic analysis of the previous sections, the requirements of the WBGT meter are known, similar work evidences the need for the develop of low cost devices to identify thermal stress. Figure 1 shows the components of the proposal device, the meter requires a microcontroller that performs information processing, an LCD screen to display the information and a DC power supply for its operation.

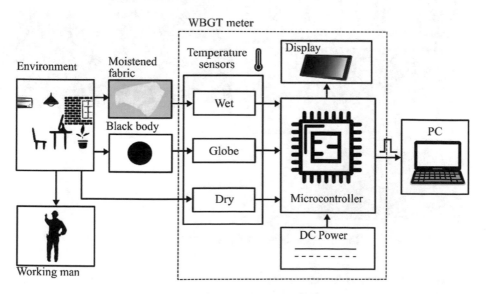

Fig. 1. Scheme of the proposal.

To determine the WBGT index three temperature sensors were needed and the conditioning to measure the temperature of humidity and radiation. The wet bulb temperature was measured using a moist fabric and natural ventilation, and the globe temperature was achieved by a matte black copper coated sphere which absorbs the radiation. In addition, it is important to send the data from the meter to an external computer for thermal stress recollection and analysis.

3.2   Meter Components

Components and electrical connections needed to implement the meter are illustrated on the Fig. 2, an Arduino pro mini as the main microcontroller, the power was provided by a 9 V battery and a voltage regulator at 5 V to stabilize the voltage. Six digital pins of the Arduino perform the activation and control of the LCD screen, and three analog inputs read the temperature sensors, on the other hand, the lm35 are the devices which allow temperature transduction to analog voltage signals.

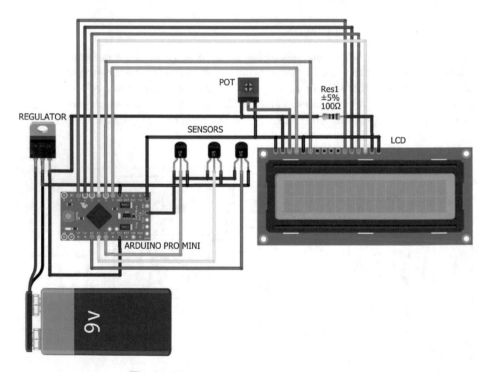

Fig. 2. Connections of the WBGT index meter.

To send the data to the external computer an external USB-TTL serial communicator was used (the Arduino pro mini does not have its own serial communication) to send the information to the PC using serial communication through USB connection.

3.3 System Operation

The program entered in the microcontroller, works according to the flow chart in Fig. 3. When the operation starts, it performs the measurement of the three temperatures (T1 = Globe temperature, T2 = Natural humid temperature and T3 = Bulb temperature dry or air) and the selection of the environment (indoor or outdoor), then it proceed to use the WBGT index formula [12] according to the selection of the environment, in the case of external environments the three temperatures are used and for indoors the air temperature is not used, and the result is shown on the LCD screen.

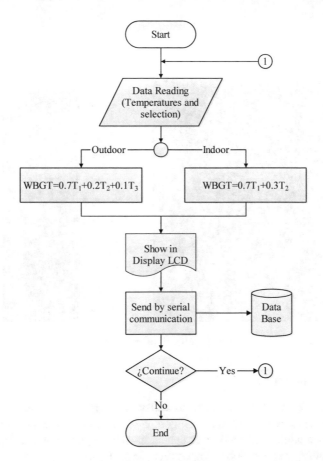

Fig. 3. Operation of the WBGT index meter.

Next, the information is continuously sent to the serial port to be read by a computer and stored in a database for further analysis. The sequence repeats continuously as required, otherwise the meter turns off.

4   Results

Once the WBGT meter is implemented, the electronic circuit was installed in a yellow box, for a better presentation, Fig. 4a show the complete assembled equipment which was employed to develop the measurements, these were taken at the installations of the Universidad Indoamérica which is located in the city of Ambato in Ecuador. The measured data were the temperatures and the WBGT index at the indoors and outdoors, to evaluate the performance of the meter, in addition, the data was checked by measuring through a commercial meter model HT30 of the brand EXTECH Instruments (see Fig. 4b).

(a) Meter implemented   (b) Commercial meter

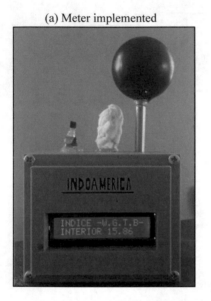

Fig. 4. WBGT index meters.

4.1   Experimental Tests

The three temperatures were measured to verify the correct functioning of the sensors. These values have been calibrated using the local temperature and afterwards the moistened tissue (cotton) and the black sphere have been placed, to measure the natural moist temperature and the globe temperature, respectively.

Figure 5 shows the measurements of each temperature taken outside, noting that the globe temperature has the highest values due to radiation, and that the natural temperature of the humidity had the lowest values, denoting some humidity in the ambient. The mean dry bulb temperature is 16.52 [°C], the average bulb temperature can be 11.55 [°C] and the globe temperature was 19.34 [°C].

Next, Fig. 6 shows the results of the WBGT index in outdoors and indoors, taken during 2 h and 15 min during working hours. Both cases show thermal stress due to

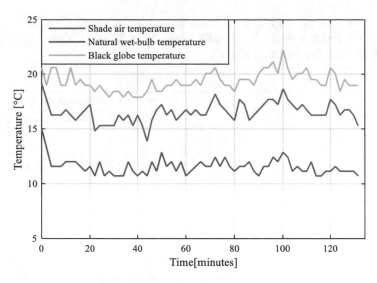

Fig. 5. Data of each temperature.

very low working temperatures, especially in the outdoor. The average WBGT outdoor index is 13.61 [°C] and indoors 14.79 [°C].

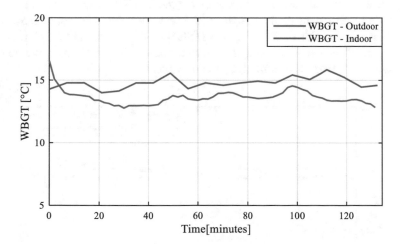

Fig. 6. Data of the WBGT index.

4.2 Comparison

To validate the built WBGT meter device, the measurements were compared with the commercial WBGT meter with certification number 85887. Figure 7 presents the data of both meters for the WBGT index on outdoors, it is important to remark that the implemented meter has values slightly lower than the commercial meter. Analytically, the Mean Absolute Error MAE is 0.5736 [°C], the Mean Square Error MSE is 0.4562 [°C], and the Mean Relative Error MRE obtained is 3.26%, which is an acceptable range of errors in the measurement.

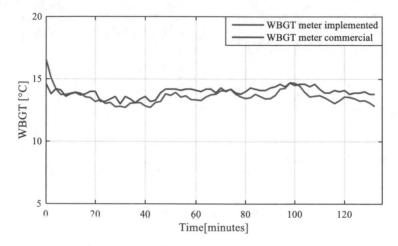

Fig. 7. Comparison of the WBGT index outdoors.

Figure 8 shows the data of both meters for the WBGT index in indoors, noticing that the implemented meter has slightly higher values than the commercial meter. Analytically, the MAE is 0.5290 [°C], the MSE is 0.4224 [°C], and the MRE obtained is 2.64%, which is a range of errors even more acceptable in the measurement, and there were few variations in time due to confinement.

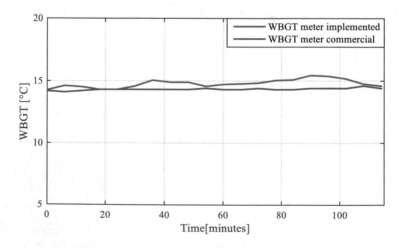

Fig. 8. Comparison of the WBGT index indoors.

5   Conclusions

The WBGT index allows the analysis of thermal stress in work, social and sports environments, but due to the costs of commercial meters, its use can be limited to the use of just few economically stable enterprises or productive sectors, then others with fewer economic resources are relegated to suffer with the damages caused by these situations. To implement low-level or low cost devices, there is a great variety of options for microcontrollers and commercial sensors, so the construction of economic meters is achievable. This work details the implementation of a WBGT meter using an Arduino, an LCD screen and three temperature sensors LM35, to condition the variables wet cotton and a black sphere were used to obtain the natural wet temperature and the temperature of balloon, respectively. The programming code inserted in the Arduino pro mini allows selecting the type of environment to be measured and sending the information to a computer through serial communication.

The results showed the three measured temperatures observing values that correspond to the reality and the WBGT indexes acquired in outdoors and indoors, which shows thermal stress in the work environment due to low temperatures, in both cases less than 15 [°C]. In addition, the comparison of the measurements between the constructed instrument and certified commercial meter is made, obtaining ranges of tolerable relative errors in outdoors and indoors, the constructed instrument can be characterized with an error of 3%.

The main advantage of this proposal is the use of a low-cost device to solve a problem of secular importance, which generally has high cost. There are few works in the literature that implement similar systems, focusing more on fitting the characteristics of thermal meters than on obtaining products for immediate use. In future works, some other can be used several equipment built with this proposal to analyze thermal stress at different levels of height.

References

1. Agüero, M.R., Bethencourt, J.B., Ramírez, R., García, Y.M.: Caracterización del ambiente térmico laboral y su relación con la salud de los trabajadores expuestos. Rev. Cuba. Salud y Trab. **16**, 3–9 (2015)
2. Cheng, Y.T., Lung, S.C.C., Hwang, J.S.: New approach to identifying proper thresholds for a heat warning system using health risk increments. Environ. Res. **170**, 282–292 (2019). https://doi.org/10.1016/j.envres.2018.12.059
3. Deschenes, O.: Temperature, human health, and adaptation: a review of the empirical literature. Energy Econ. **46**, 606–619 (2014). https://doi.org/10.1016/j.eneco.2013.10.013
4. Lan, L., Wargocki, P., Wyon, D.P., Lian, Z.: Effects of thermal discomfort in an office on perceived air quality, SBS symptoms, physiological responses, and human performance. Indoor Air **21**, 376–390 (2011). https://doi.org/10.1111/j.1600-0668.2011.00714.x
5. Castilla, M.M., Álvarez, J.D., Berenguel, M., Pérez, M., Rodríguez, F., Guzmán, J.L.: Técnicas de Control del Confort en Edificios. Rev. Iberoam. automática e informática Ind. **7**, 5–24 (2010). https://doi.org/10.4995/RIAI.2010.03.01
6. Jackson, L.L., Rosenberg, H.R.: Preventing heat-related illness among agricultural workers. J. Agromedicine **15**, 200–215 (2010). https://doi.org/10.1080/1059924X.2010.487021

7. Arias Gallegos, W.L.: Estrés laboral en trabajadores desde el enfoque de los sucesos vitales Occupational stress of workers analyzed from an approach to vital events. Rev. Cuba. Salud Pública. **38**, 525–535 (2012)
8. Jay, O., Brotherhood, J.R.: Occupational heat stress in Australian workplaces. Temperature **3**, 394–411 (2016). https://doi.org/10.1080/23328940.2016.1216256
9. Spector, J.T., Krenz, J., Blank, K.N.: Risk factors for heat-related illness in washington crop workers. J. Agromedicine. **20**, 349–359 (2015). https://doi.org/10.1080/1059924X.2015.1047107
10. Moran, D.S., Pandolf, K.B., Shapiro, Y., Heled, Y., Shani, Y., Mathew, W.T., Gonzalez, R. R.: An environmental stress index (ESI) as a substitute for the wet bulb globe temperature (WBGT). J. Therm. Biol. **26**, 427–431 (2001). https://doi.org/10.1016/S0306-4565(01)00055-9
11. Golbabaei, F., Heidari, H., Shamsipour, A., Forushani, A.R., Gaeini, A.: A new outdoor environmental heat index (OEHI) as a simple and applicable heat stress index for evaluation of outdoor workers. Urban Clim. **29**, 100479 (2019). https://doi.org/10.1016/j.uclim.2019.100479
12. Zare, S., Shirvan, H.E., Hemmatjo, R., Nadri, F., Jahani, Y., Jamshidzadeh, K., Paydar, P.: A comparison of the correlation between heat stress indices (UTCI, WBGT, WBDT, TSI) and physiological parameters of workers in Iran. Weather Clim. Extremes **26**, 100213 (2019). https://doi.org/10.1016/j.wace.2019.100213
13. Brotherhood, J.: What does the WBGT Index tell us: is it a useful index of environmental heat stress? J. Sci. Med. Sport **18**, e60 (2014). https://doi.org/10.1016/j.jsams.2014.11.281
14. D'Ambrosio Alfano, F.R., Malchaire, J., Palella, B.I., Riccio, G.: WBGT index revisited after 60 years of use. Ann. Occup. Hyg. **58**, 955–970 (2014). https://doi.org/10.1093/annhyg/meu050
15. Budd, G.M.: Wet-bulb globe temperature (WBGT)-its history and its limitations. J. Sci. Med. Sport **11**, 20–32 (2008). https://doi.org/10.1016/j.jsams.2007.07.003
16. Saá, F., Varela-Aldás, J., Latorre, F., Ruales, B.: Automation of the feeding system for washing vehicles using low cost devices. In: Advances in Intelligent Systems and Computing, pp. 131–141 (2020). https://doi.org/10.1007/978-3-030-32033-1_13
17. Ramirez, B.C., Gao, Y., Hoff, S.J., Harmon, J.D.: Thermal environment sensor array: part 1 development and field performance assessment. Biosyst. Eng. **174**, 329–340 (2018). https://doi.org/10.1016/j.biosystemseng.2018.08.002
18. Yantek, D.S., Yan, L., Damiano, N.W., Reyes, M.A., Srednicki, J.R.: A test method for evaluating the thermal environment of underground coal mine refuge alternatives. Int. J. Min. Sci. Technol. **29**, 343–355 (2019). https://doi.org/10.1016/j.ijmst.2019.01.004

Software Engineering Issues
for the Development of mHealth Apps

Lornel Rivas[1]([⊠]) ⓘ, Cristhian Ganvini[1] ⓘ, and Luis E. Mendoza[2] ⓘ

[1] Facultad de Ingeniería y Arquitectura, Departamento de Ingeniería
de Sistemas, Cusco Andean University, Universidad Andina del Cusco,
Urbanización Ingeniería Larapa Grande A-7, San Jerónimo, Cusco, Peru
{lrivas, cganvini}@uandina.edu.pe
[2] Escuela Superior Politécnica del Litoral, ESPOL, Facultad de Ingeniería
en Electricidad y Computación, ESPOL Polytechnic University,
Campus Gustavo Galindo Km. 30.5 Via Perimetral,
P.O. Box 09-01-5863, Guayaquil, Ecuador
lemendoza@espol.edu.ec

Abstract. The mobile health (mHealth) can facilitate access to health services in Latin America. Therefore, deepen the process of software development in this domain area has become a key issue for the success of mHealth software development projects. This paper defines a set of issues of interest for mHealth software development teams, organized in three dimensions: Processes, Products, and Actors. The issues are based on Software Engineering knowledge areas and are derived from best practices in mHealth apps and projects, with the aim of providing scalability and sustainability of mHealth apps. The issues were inspected through a project of a public health institution in the Andean Region, in which doctors and engineers interacted positively in a mHealth app development context. The software engineers and doctors confirmed the importance and relevance of the issues and raised the convenience of measuring each of them. In this way, the proposed issues can be used to evaluate the performance of mHealth projects throughout their execution.

Keywords: mHealth · Software engineering issues · eHealth · Healthcare services · Apps

1 Introduction

Technological development in telecommunications and mobile devices facilitates the implementation of mHealth software [1]. Currently, there are difficulties that negatively affect the success of mHealth software development projects. One of the main problems identified is that there are no team guidelines for mHealth software development. On a global scale, the need to support the mHealth apps development process has been recognized, as evidenced in [2–5]. We need guides that involve, among other things, topics such as: (1) joint and permanent participation of health workers and software engineers during project execution, (2) incorporation of policies and priorities established by health systems, and (3) adequate management of medical information.

© Springer Nature Switzerland AG 2020
Á. Rocha et al. (Eds.): ICITS 2020, AISC 1137, pp. 469–479, 2020.
https://doi.org/10.1007/978-3-030-40690-5_46

In order to generate software products with a favorable impact on health services, it is necessary to recognize and incorporate aspects of the institutional, territorial and technological context that characterize the health field and its mHealth services in the software development process. With this, it will be possible to improve services and citizen access to them, in terms of care time, number of patients attended, doctor's workload and service costs. The literature review highlights, for its particular relevance, the need for models, approaches, frameworks and practices for software development, focused on the context of mHealth.

This paper identifies a set of 83 issues of interest to mHealth software development teams, which, when addressed by project stakeholders, ensures scalability and sustainability of mHealth applications. The issues are based on best practices and concepts of Software Engineering (SE), organized in three general dimensions: Processes, Products and Actors, in order to ensure optimal alignment of the project with the health services environment. The set of issues were examined in the context of a real project in a public health institution in the Andean Region. This review allowed the development team to align the project process and outputs to the relevant institutional environment in mHealth. In addition, meetings with doctors and engineers confirmed the importance and relevance of the set of issues. It is expected that the set of issues will constitute a support guide for the software team regarding the aspects that must be taken into account during the whole process of development of mHealth apps, in articulation with health specialists, patients and other actors involved.

The work is structured as follows, in addition to this introduction. In Sect. 2, the framework for the mHealth software development study is established. Section 3 describes the set of issues and their structure. Section 4 presents the results of its review. Section 5 closes with conclusions and future work.

2 Framework for mHealth

The multiple applications of Information Technology (IT) for the health conform the concept of eHealth (electronic health), which involves electronic medical records, telemedicine, healthcare web portals, and hospital management [6, 7]. As part of eHealth, mHealth has been developed, which covers the usage of wireless mobile technologies for public health [8].

The development of mHealth applications constitutes a growing field which provides opportunities to improve healthcare services, giving new possibilities to patients and healthcare workers. Among these advantages outstand [1, 7, 9, 10]: (1) Remote collection of medical data, (2) Remote access to medical information, (3) Information resources designed for mHealth users, (4) Effective communication procedures between doctors and patients, (5) Procedures for emergency responses, (6) Support to decisions in the patient care area, (7) Healthcare extension services oriented to communities, (9) Remote support to diagnosis and treatment for illnesses, and others.

The healthcare is a topic of special relevance in Latin America, considering the current conditions of the sanitary systems. In mHealth, there are essential issues, which can find support in the SE best practices. Estrim and Sim identify the need to delve deeper into software architectures for mHealth [11] and they established that, in order to manage successfully the processes of iterative development, is important: (1) An adequate balance of agility and discipline in the specification of requirements in uncertainty environments [12] and (2) To improve the communication lines among health professionals, research assistants and software developers.

Matters as the responsibility about the information that is managed in mHealth, big amounts of involved data, the access and usage from different places and the necessity of growing at locally and nationally scales, can find support in disciplines such as the risk management, considering the SE approach, same as the software testing [13]. The mHealth environment demands the emphasis in the design with the user; this is because the design, as a knowledge area of the SE, is considered very important.

Additionally, there are global scale initiatives that influence in the scope of mHealth programs and projects. This is the case of the 9 principles for the digital development, promoted by UNESCO, which have implications for the development of applications. These principles are [14]: (1) Designing with the user, (2) Understanding the ecosystem, (3) Designing for the scale, (4) Building for the sustainability, (5) Focused on data, (6) Open standards, open data, open code, innovation, (7) Reuse and improvement, (8) Guarantee the privacy and security, and (9) Be collaborative. These principles can be backed up from the SE. The links of these principles with the practices on mHealth projects served as a base for the establishment of the set of proposed issues.

3   Issues Definition

The issues defined in this work constitute a reference for mHealth software development team and represent aspects that are necessary to recognize and consider for the success of the development process in this domain area. The issues are derived from best organizational practices in programs and mHealth projects, they are also connected with expectations of interest in this area and are backed up in SE best practices. The proposal consists of 83 issues, according to the following general dimensions of analysis, in the outline of the development of mHealth software:

- **Processes:** Encompasses the technical processes, which the work team must follow during the execution of a development project.
- **Products:** Presents the intermediate and final products of the technical processes, which the work team must elaborate during the development of an application.
- **Actors:** Identifies the necessary roles, and the responsibilities that the work team must performance in order to execute the technical processes, in the development process of the application.

The organization of the issues is presented in Table 1. The 'Processes' dimension, is distinguished among Core and Support processes, according to the SE best practices. The set of issues constitutes a checklist for the leader of the project. It is used as support or reference in different stages of the development of mHealth apps. It proposes alerts about issues of special interest for mHealth.

Table 1. Organization of issues.

Dimension		Practice	Focus	Issues
Processes	Core	Planning	*Project context*	01–04
			Agile approaches practices	06–10
		Requirements	*Functional*	11–21
			Non-Functional	22–26
		Architectonic Design		27–29
		Versions development		30–38
		Follow-up and assessment		39–46
	Support	Configuration Management		47–48
		Communication and stakeholder engagement		49–51
		Project Management		52–53
		Validation & Verification		54–55
Products		Developer's Guide		56–59
		Technical manual		60–64
		Scheduling		65–67
Actors		Project team		68–70
		Healthcare Stakeholders		71–76
		Roles		77–83

The affinity of the issues in respect to the mHealth domain, it is observed in the representation of the common needs of software projects and it is focused on relevant attributes of mHealth projects. The set of issues has a systemic approach, and consider the following aspects:

- **Applicability.** The value of the issues depends on the context and the project scope which is studied. Two alternatives can be adopted. "Applicable" or "Not applicable" considering the wide range of opportunities that provides mHealth.
- **Type.** The importance of the issues for the project which is studied depends on its context. The type is established according to the values: "Mandatory" (M), when the issue is essential for mHealth, or "Not Mandatory" (NM), when it can be present or not, according to the project. The set of issues contains a subset of 65 issues of Mandatory type (Table 2).
- **Sources.** The issues require ways of verification to determine their existence in the project which is studied. The sources of this verification can be, agreement records, statistics of the institution, responses to questions made during interviews to involved ones, and others.
- **Stakeholders:** According to the scope of the project under study, the issues are checked by the involved ones in the project, these ones can be, patient, doctor, medical specialists, project leader, director of the institution, researcher and others.

Table 2. Issues for Processes dimension: Core processes.

Practice/*Focus*	ID	Name	Type*
Planning: *Project context*	1	Context analysis	M
	2	Relevance	M
	3	Target audience analysis	M
	4	Work team	M
Planning: *Agile approaches practices*	5	Iterative development	M
	6	Collaborative work	M
	7	Regular meetings	M
	8	Briefings	M
	9	Stakeholders engagement	M
	10	User engagement	M
Requirements: *Functionals*	11	Patient monitoring requirements	M
	12	Visualization requirements for health data	M
	13	Treatment reminders requirements	M
	14	Patient remote visit registration requirements	NM
	15	Data registration requirements in personal file	M
	16	Response requirements for health assessment questionnaires	NM
	17	System feedback requirements	NM
	18	Information requirements for clinical decision support	NM
	19	Automated reporting requirements	NM
	20	Access requirements for medical product information	NM
	21	Emergency response requirements	M
Requirements: *Non functionals*	22	Security requirements	M
	23	Interoperability requirements	M
	24	Scalability requirements	M
	25	Usability requirements	M
	26	Data validation requirements	M
Architectonic Design	27	Hardware and software technologies	M
	28	Contents proofs	M
	29	Prototypes for product usability	M
Versions development	30	Time plans for the project	NM
	31	Relevance of needs over time	NM
	32	Similar initiatives in the project context	NM
	33	Alignment with national priorities	M
	34	Compatibility with existing systems	M
	35	Actors involved in content formulation	NM
	36	Potential impacts	M
	37	Improvement opportunities	NM
	38	Opportunities for integration with other systems	M

(continued)

Table 2. (*continued*)

Practice/*Focus*	ID	Name	Type[*]
Follow-up and assessment	39	Product usability monitoring and evaluation	M
	40	Project integration monitoring and evaluation	M
	41	project sustainability monitoring and evaluation	M
	42	Project scalability monitoring and evaluation	M
	43	Service quality monitoring	M
	44	Information uses and education monitoring at local context	NM
	45	Adoption rate monitoring	M
	46	Implementation cost recording and tracking	M

Notes: [*]**Type:** M = Mandatory; NM = Not Mandatory.

Additionally, it is proposed a scale to determine the grade in which the project is aligned with the SE best practices of mHealth. In Table 3, the ranks that are used to establish the grade of alignment of a project with the issues for mHealth, are presented.

Table 3. Alignment ranks for the mHealth project

Rank	Alignment with the project with mHealth
90%–100%	Optimal alignment
80%–89%	Presents alignment
70%–79%	Doesn't present alignment, subject to recommendations
Less than 70%	Revision of the project plan

In this way, the development team has a support to identify the issues and appreciate the alignment of the project to the SE best practices of mHealth. This facilitates the identification of alerts or improvement opportunities along the process of development of mHealth software product.

4   Issues Inspection

In order to prove the effectivity of the issues, these were checked through a real project for the development of a mHealth software product, in a hospital of regional scope and with institutional cooperation and support.

4.1 Inspection Results

The project, named Chinpuy (Measuring or Marking, in native Quechua Language) corresponds to the development of a mobile application, oriented to the monitoring of the treatment in patients with a diagnosis of Arterial Hypertension; an urgent attention topic [15] according to the policies and healthcare statistics in the region of Cusco, Peru. From an institutional perspective, this hospital institution is an operational, technical and decentralized body, responsible for providing services of comprehensive care of specialized healthcare in the region. It is also in charge of organizing, managing and providing secondary prevention services, recovery and palliative care, healthcare specialized. In addition, it also develops activities of teaching and research. The project expects to reduce the workload in situations in which the technology can take over them. The involved actors in the project development team were:

- Cusco Regional Administration of Health. Establishes policies, priorities and informs about health sector statistics.
- Cusco Regional Hospital (CRH). Guarantees the research-works and provides statistical information about patients care and information about medical specialists.
- Patients. They contribute with the identification of needs, considering the perspective of a cared person. They help with return information about prototypes.
- Medical Specialists. Provide information about pathologies and their treatment. Advise about measures, value ranks, treatments and considerations.
- Technical Coordinator. In charge of managing the iterative development; informs about the progress of the project in coordination with the stakeholders. Assumes technical responsibilities of software development.
- Advisor. Handles conceptual topics about mHealth and about the project context. Assembles with the technical coordinator to get the alignment of the development considering mHealth and the project requirements.
- Project leader. Constitutes the core of the communication between the CRH, medical specialists, patients, technical coordinator and advisor. Assures the vision and recognizes the project context.

The issues were used as a framework to support the development team in the selected project, during its execution. As an example, it is presented in Table 4 the results of Processes dimension: Core Processes – Planning; Focus: Project context. The results of the other issues are available on request by emailing the corresponding author.

Table 4. Processes dimension: Core processes – Planning; Focus: Project context and Agile approaches practices.

Issue			Applicability[#]	Source on which the presence or not of the issue is based	Stakeholders
ID	Type[*]	Name			
Focus: *Project context*					
1	M	Context analysis	A	CRH Statistics	Doctor, Project leader
2	M	Relevance	A	Data collection: Is attention to arterial hypertension a priority for the hospital?	Doctor, Project leader
3	M	Target audience analysis	A	Data collection: Characteristics of the population requiring attention. CRH Statistics	Doctor, Project leader
4	M	Work team	A	Data collection: Does the hospital have a team of technical and medical specialists for Hypertension pathology?	Doctor, Project leader
Focus: *Agile approaches practices*					
5	M	Iterative development	A	Work approach following Scrum	Project leader, Development team
6	M	Collaborative work	A	Availability of physical workplace	Project leader, Development team
7	M	Regular meetings	A	Agenda of daily internal follow-up meetings and weekly stakeholder meetings	Project leader, Development team, Doctor
8	M	Briefings	A	Weekly broadcast in meetings with stakeholders	Project leader Development team, Doctor
9	M	Stakeholder engagement	A	Scheduling meetings with stakeholders	Project leader, Development team, Doctor
10	M	User engagement	A	Scheduling meetings with patients	Project leader, Development team, Patient

Notes: [*]**Type:** M = Mandatory; NM = Not Mandatory.
[#]**Applicability:** A = Applicable; NA = Not Applicable.

In Table 5 it is represented, as an example, the Issues records of which applicability is: Not applicable. In the case of Issue 28, the content proofs are specially directed to applications of which design includes the handling of Short Message Service (SMS). Therefore, content specialists are required to review and validate the information in the messages to be disseminated to users via SMS messaging. In the case of the Chinpuy project, the use of SMS messaging was not required.

Table 5. Processes dimension: Core processes; Focus: Architectonic Design.

Issue			Applicability[#]	Source on which the presence or not of the issue is based	Stakeholders
ID	Type[*]	Name			
27	M	Hardware and software technologies	A	Project planning	Project leader, Development team
28	M	Content proofs	NA		
29	M	Prototypes for product usability	A	Project development strategy	Project leader, Development team, Doctor

Notes: [*]**Type:** M = Mandatory; NM = Not Mandatory.
[#]**Applicability:** A = Applicable; NA = Not Applicable.

4.2 Discussion of Results

After identifying the applicability, the information sources and the involved actors for each issue, it was decided that 64 issues out of the 65 issues considered obligatory for a mHealth project are present; i.e., 98% of the compulsory issues were fulfilled. So, it is possible to establish that the project has an optimal alignment, this way it was 100% achieved the development of the product called Chinpuy.

The issues allow to establish aspects that, at first, were not correctly established. Consequently, they were identified: (a) Statistics and studies about pathology (Issue 2, Relevance), (b) Records about patient care in the Hospital, through enquiry to institutional authorities (Issue 3, Target audience analysis), (c) Patients that contributed from the beginning of the project (Issue 73, Non-professional users), and (d) Hospital doctors that provide service of specialized health to users (Issue 74, Professional users).

The issues facilitated the identification of some functional requirements that are essential for the application: (a) Visualization requirements for health data (Issue 12), (b) Treatment reminders requirements (Issue 13), and (c) Emergency response requirements (Issue 21).

On the other hand, the issues helped with the definition of non-applicable topics according to the project scope: (a) Patient remote visit registration requirements (Issue 14), since the software is not directed to health-worker visits to in situ attention, (b) Access requirements for product information (Issue 20), and (c) Content proofs (Issue 28), which are directed principally to systems based on text messaging SMS.

Through the doctor–patient–developer interaction, it was achieved the following: (a) Functionalities that allow patients get records of their arterial tension values in their smartphones, take control of their medication doses according to medical indications and get reminders, alerts, and even, support in possible dangerous situations; (b) Doctors, from a web interface, observe the behavior of the values registered by their patients, with facilities to add and represent information through histogram and make adjustments on treatments; and (c) It is valued the demand about giving a safe, reliable mechanism and which is backed-up by the knowledge of the medical specialty, attached to institutional lineament, and the expectation of getting a product that pays attention to a regional priority, sustainable in time, able to evolve in its services and architecture.

As for the project team, in a feedback and learning exercise through closing meetings, the following was obtained: (a) The project leader, the technical coordinator and the advisor, gave a positive point of view about the usage of the model as a support for the development of Chinpuy; (b) The medical specialist valued positively the developed product, in consideration of the typical aspects of the medical specialty and the opportunities that can be offered to patients; (c) There is a shared vision among stakeholders about opportunities for improvement, in terms of other diseases that can be treated, and other health service centers; and (d) At the time of preparation of this publication, articulation actions are being coordinated with regional health system agencies, with a view to opportunities for geographical scalability and the scope of priority pathologies that may be covered.

As a final comment, we can confirm that the synergy among doctors, patients and developers was a success which was facilitated by the usage of the issues. The coordinated visualization of the stakeholders about automation opportunities and the opportune identification of material that are relevant to the development process, were also facilitated by the presence of the issues. In case of unfulfillment of obligatory issues, it is possible, through best practices of iterative development, to obtain satisfaction, increasing the project alignment with mHealth. The unfulfillment of Optional issues promotes the discussion about the project focus considering its scope and context.

5 Conclusions and Future Work

mHealth can facilitate citizens the access to information or services of urgent necessity. There is an important potential for the implementation of mHealth services, considering the geographic, social and financial resources gaps. This work proposes a set of 83 issues of reference for mHealth software development teams, which serve as a guide for the leaders of mHealth projects. Although it was applied in a local project, it is generated based on shared aspects by Latin American countries.

Software engineers and doctors, who work in the local environment, confirmed the importance and pertinence of the matter. Opportunities are identified for future work aimed at delving deeper into the model, and operationally defining each of the issues, so that it can also be used to measure the performance of mHealth projects, both in their execution and in their outcome. All of this belongs to strategies of effective implementation in the priority health areas of the countries of the region.

References

1. Vital Wave Consulting: mHealth for Development: The Opportunity of Mobile Technology for Healthcare in the Developing World, Washington, D.C. and Berkshire, UK (2009)
2. Republic of Kenia: Kenya Standards and Guidelines for mHealth Systems. Ministry of Health Republic of Kenya, Kenya (2017)

3. Hannover Medical School: Chances and Risks of Mobile Health Apps (CHARISMHA). Albrecht, U.V. Hannover (2016)
4. Fava, P.: Mobile Health Technology: Key Practices for DRR Implementers, 1st edn. Cooperazione Internazionale, Milano (2014)
5. Broens, T., Van Halteren, A., Van Sinderen, M., Wac, K.: Towards an application framework for context-aware m-health applications. Int. J. Internet Protoc. Technol. 2(2), 109 (2007). https://doi.org/10.1504/IJIPT.2007.012374
6. Oviedo, E., Fernández, A.: Tecnologías de la información y la comunicación en el sector salud: oportunidades y desafíos para reducir inequidades en América Latina y el Caribe, Serie Políticas Sociales, vol. 165, p. 53. CEPAL, NU (2010)
7. World Health Organization: mHealth New Horizons for Health Through Mobile Technologies. Global Observatory for eHealth series, vol. 3, Switzerland (2011)
8. World Health Organization: mHealth Use of appropriate digital technologies for public health. WHO Seventy-First World Health Assembly (2018). https://apps.who.int/gb/ebwha/pdf_files/WHA71/A71_20-en.pdf
9. Ruiz, E., Proaño, A., Ponce, O., Curioso, W.: Tecnologías móviles para la salud pública en el Perú: lecciones aprendidas. Rev. Peru. Med. Exp. Salud Publica. 32(2), 364 (2015). https://doi.org/10.17843/rpmesp.2015.322.1634
10. Källander, E.: Mobile health (mHealth) approaches and lessons for increased performance and retention of community health workers in low- and middle-income countries: a review. J. Med. Internet Res. 15(1), 17 (2013). https://doi.org/10.2196/jmir.2130
11. Estrin, D., Sim, I.: Open mHealth architecture: an engine for health care innovation 330(5) (2010). https://doi.org/10.1126/science.1196187
12. Shawkat, S., Nasrullah, M., Islam, R.: Design and prototypical implementation of a mobile healthcare application: health express. Department of Computer Science and Engineering, BRAC University, Bangladesh
13. Bourque, P., Fairley, R.: Guide to the Software Engineering Body of Knowledge (SWEBOK), 3rd edn. IEEE Computer Society Press, Washington DC (2014)
14. Waugaman, A.: From Principle to Practice: Implementing the Principles for Digital Development. The Principles for Digital Development Working Group, Washington, DC (2016)
15. INEI: Perú: enfermedades no transmisibles y transmisibles. Instituto Nacional de Estadística e Informática del Perú, Lima (2017)

Exploring Breast Cancer Prediction for Cuban Women

José Manuel Valencia-Moreno[1]([⊠]), Everardo Gutiérrez López[1],
José Felipe Ramírez Pérez[2,3], Juan Pedro Febles Rodríguez[3],
and Omar Álvarez Xochihua[1]

[1] Universidad Autónoma de Baja California, Ensenada, BC, Mexico
{jova,everardo.gutierrez,aomar}@uabc.edu.mx
[2] Centro de Informática Médica, La Habana, Cuba
[3] Universidad de las Ciencias Informáticas, La Habana, Cuba
{jframirez,febles}@uci.cu

Abstract. The importance of early detection of breast cancer in the healthcare field has led to the generation of various models for estimating the risk of suffering from it. This paper analyzes the effectiveness of the official model used in the United States for this purpose, known as the Gail model, on a set of cases of Cuban native women. Despite the fact that the version of the model used considers the estimation of risk for Hispanic women born in the United States, certain limitations were found in the results, so the use of computational models based on machine learning applied to the same set of cases is explored as an alternative. The results show that, for the analyzed cases, better results are obtained using some machine learning algorithms, which gives rise to a greater exploration of these as an alternative to traditional models.

Keywords: Breast cancer · Prediction model · Hispanic women

1 Introduction

Breast cancer ranks first in new cases and deaths of the female population worldwide among all types of cancer [1]. However, the incidence and mortality of breast cancer differs from country to country. While in the USA the incidence is 84.9% and the mortality rate is 12.7%, in Latin America and the Caribbean it is 51.9% and 13%, respectively [2].

The difference in the incidence and mortality rates between countries can be explained by the stage at which the cancer is detected and the treatment usually followed. The importance of an early diagnosis is that it drastically reduces the mortality rate and allows more treatment options [3]. The high mortality rate in Latin America in relation to its incidence rate, can be explained by the limited access to health services for early detection, which in turn, causes a high percentage of patients to be diagnosed with breast cancer in advanced stages. Therefore, early detection is the main aspect in order to prevent mortality from breast cancer, especially in developing countries [4–6].

© Springer Nature Switzerland AG 2020
Á. Rocha et al. (Eds.): ICITS 2020, AISC 1137, pp. 480–489, 2020.
https://doi.org/10.1007/978-3-030-40690-5_47

Different models have been developed for the early detection of breast cancer such as risk factor-oriented models [7–12], genetic test-oriented models for the calculation of susceptibility and genetic risk [13–17], and models that are a combination of both previous types [13, 16, 17]. In the case of Latin America genetic tests are economically accessible only to a small part of the population, those who have access to private hospitals; and are mostly unavailable for the general population which makes genetic test-oriented models not accessible for most of the population [18]. The main viable option in order to prevent and measure the risk of breast cancer on time for a proper treatment for Latin American women population is through risk estimation tools based on breast cancer risk factors.

Most estimation models on this type of cancer, based on risk factors [7–11], have been proposed from the examination of a white female population native to the United States. This women population has a different structure and hereditary background than Latin American women, since the last one has an ancestral European component which has been related to an increased risk of breast cancer [19, 20]. In addition, previous studies have reported that the incidence and mortality of breast cancer is due in part to environmental factors that exist in each specific geographic region [21, 22]. This seems to imply that, the models proposed for Hispanic populations, based on Hispanic women born in the United States, should also be extensively tested for cases from populations native to Hispanic countries in order to have an adequate validation of their effectiveness in risk estimation [11].

Section 2 describes some of the main estimation models for breast cancer. In Sect. 3 it is detailed the case study data set from Cuban women. Section 4 states the classification metrics to validate the results, and presents two experiments and their results. First, the Gail model is applied on a set of data from Latin American population born outside the U.S., with the purpose of analyzing the model's behavior. Second, the same data set is provided to machine learning algorithms, with the purpose of validate the estimation of breast cancer risk with this type of computational tools. Finally, Sect. 5 gives some conclusions about these results.

2   Risk Estimation Models

2.1   Risk Factors

Among the best known risk factors for breast cancer are: age, body mass index, alcohol and cigarette consumption, hormonal and reproductive factors (age at menarche, age of first childbirth, age of menopause, use of hormone replacement therapy, use of contraceptive pills, breastfeeding), cancer-related personal history (breast biopsies, atypical ductal hyperplasia, lobular carcinoma in situ, breast density), number of first-degree relatives with breast cancer or ovarian cancer [21, 23]. Other factors to consider are lifestyle, socioeconomic status, and cultural status. The higher the socioeconomic and cultural level, the higher the incidence of breast cancer [22].

2.2   The Gail Model and Its Derivatives

Published in 1989, the Gail model [7] uses five risk factors: patient age, age at menarche, age of first childbirth, number of previous biopsies, and number of first-degree relatives with breast cancer. To create the model, Gail et al. [7] used data from the Breast Cancer Detection Demonstration project. This model has been adopted by the U.S. National Cancer Institute and implemented on the Institute's Web site for public use. It is one of the most widely used and validated models with positive estimation risk results for North American and European white patients [24–26], while it has been shown to give less accurate results for other populations [23–27]. This would motivate the generation of models that followed the same methodology, applied over different women population, like the following.

The Gail Model 2. In 1999, the estimation of breast cancer risk for black population was incorporated into the original model [7], under Gail's supervision. The study included 99 black women with no history of lobular carcinoma in situ, from the Breast Cancer Prevention Trial (BCPT) program. Of this female population, only one developed invasive breast cancer. Therefore, it was not possible to thoroughly evaluate the model's estimation risk accuracy [8]. Although Gail's model was based on American white women data, it has been adapted by incorporating data from other female populations such as: black [8]; African-American [9]; Asian-American and Islanders [10]; and Hispanic Americans [11].

African American Model. In 2007, Gail et al. [9] developed a model to project the absolute risk of invasive breast cancer in a population of African-American women [7]. The data were obtained from the Women's Contraceptive and Reproductive Experiences (CARE) Study.

Asian American Model. Another adaptation to the original Gail model is the projection model of absolute breast cancer risk in Asian American and American Islander women, created by Matsuno et al. in 2011 [10]. They used data from the Asian American Breast Cancer Study. Thus, an Asian American population is incorporated into the Gail model [7].

Hispanic American Model. The most recent adaptation to the Gail model of 1989 [7], was the incorporation of the Hispanic-American female population. In 2017, the Hispanic model for estimating breast cancer risk was created by Banegas et al. [11]. The model is based on data from The San Francisco Bay Area Breast Cancer Study (SFBCS) and the California Cancer Registry (CCR). Because this model was created with data from the SFBCS and the CCR, it is suitable for women in those regions of the United States, specifically in the western states of the United States.

In the specific case of Hispanic women, additional studies are needed to assess its validity, as mentioned by Banegas et al. [11] at the conclusion of his study, while adding the Hispanic Native Americans population to the Gail model. It is important to remember that this Hispanic model was proposed and tuned by using data from Hispanic women born only in western states from USA and not from any other country with Hispanic native population. This represents a problem, since previous results [22] have reported that the age at which breast cancer occurs in the European and U.S.

female population is between 51 and 63 years respectively, while in countries like Mexico, it is from 20 years, increasing its frequency between the years of 40 and 54. That gives a ten years of difference between one region and the other, and it might be as well as between one race and the other.

2.3 Genetic and Mixed Models

Claus Model. In 1991, Elizabeth Claus developed the model that bears her surname [13]. For the development of the Claus model, only cases of white women along with their mothers and sisters were considered. The model took into account the patient's family records, pregnancy and menstruation records, history of benign breast disease, alcohol and cigarette consumption, breast surgery, sociodemographic variables, and oral contraceptive use. Among its main disadvantages it only works for patients who have at least one female relative diagnosed with breast cancer [26].

BRCAPRO Model. It was created by Giovanni Parmigiani and Don Berry in 1998 [14]. The model its available as a software program, and its algorithm is a Bayes rules based model, which evaluates the probability that a person is a carrier of BRCA1 and BRCA2 genes mutations. It uses data from the history of relatives with breast cancer and ovarian cancer, as well as from relatives not affected by cancer. It's main limitation is it does not incorporate data from other genes that cause breast cancer.

Jonker Model. This genetic model is an extension of the Claus model from 2003. The model is based on the family history of breast or ovarian cancer, as published by Jonker [15]. It uses data from the BRCA1, BRCA2, and hypothetical BRCAu genes, but does not include data from breast cancer risk factors. The model was generated by a linear regression of the independent variables and considers estimates of bilateral risk for breast cancer, ovarian cancer, and three or more affected family members.

Tyrer-Cuzick Model. This model is also known as IBIS and it was released by its creators in 2004 [16]. The model used a data set from the International Breast Cancer Intervention Study (IBIS) and other epidemiological data. For its breast cancer risk analysis, it incorporates data from both, genetic family factors and some risk factors, as well as endogenous hormonal factors.

BOADICEA Model. General model that estimates the risk of suffering breast cancer and the probability that a patient is a carrier of this type of cancer, starting from a family history of breast or ovarian type. It was developed by Antoniou et al. in 2004 [17]. The BRCA1 and BRCA2 genes, along with a polygenic component, explain the genetic susceptibility and the risk of breast cancer in a woman.

2.4 Other Risk Factor-Based Models

Korean Model. The model used the Seoul Breast Cancer Study (SeBCS). Its estimations were based on calculations over a data set of 3,789 Korean women. The model considered age, family history of breast cancer, age at menarche, age of successful first birth, menopause status and age, breastfeeding duration, use of oral contraceptives, body mass index, and exercise as major risk factors. The authors concluded that the Korean model is a better tool for estimating breast cancer risk in Korean women than the Gail model [12].

3 Hispanic Women Estimation Risk: A Case Study from Cuba

The National Cancer Institute, has stated that the tool BCRAT, based on the Gail model, performs well but may underestimate the risk in Hispanic women born outside the U.S. Therefore, the model needs validation for that one and other subgroups.[1]

3.1 Tool for Applying the Hispanic American Model

One of the main elements of the experiment is the Gail model updated in December 2017, which was obtained from the official website of the U.S. National Cancer Institute, in its Division of Cancer Epidemiology and Genetics. The BCRAT breast cancer risk assessment tool version 2.0 was chosen, in particular the R-coded version for Windows, which includes the Hispanic model.

3.2 Case Study in Cuban Women

The other important element is the dataset used to validate the BCRTA results for Hispanic woman, in this study a Cuban native group. For this purpose, a set of 1,159 cases of women diagnosed with breast cancer and 404 cases of women not diagnosed with breast cancer were obtained. This set of data was kindly provided by the Medical Informatics Center of the University of Informatics Sciences in Havana, Cuba. The data belong to the hospital records of Villa Clara and Cienfuegos provinces. All women were born in these municipalities and their ages vary between 25 and 90 years. The data set consists of 20 variables. None of the variables contains the name or any information about the patient's identity. For the purposes of this study, only the variables required by the Gail algorithm were considered: age; number of biopsies and whether she has had atypical hyperplasia; age at menarche; age of first birth; number of first-degree family history of breast cancer; and race.

4 Experimental Results

4.1 Classification Results Measurement

To measure the effectiveness of applying the models to the case study data, BCRAT was used to estimate the risk of developing breast cancer in five years for all the records. The risk factor was used to calculate the number of cases correctly and

---

[1] https://bcrisktool.cancer.gov/about.html.

incorrectly classified for women with cancer (positive, P) and cancer-free (negative, N). Then, the True Positive (TP), True Negative (TN), False Positive (FP) and False Negative (FN) totals were counted, and the following metrics were calculated [28]:

$$Sensitivity = TP/P \tag{1}$$

$$Specificity = TN/N \tag{2}$$

$$Precision = TP/(TP + FP) \tag{3}$$

$$Accuracy = (TP+TN)/(P + N) \tag{4}$$

$$F1 = 2^* ((Precision * Sensitivity)/(Precision + Sensitivity)) \tag{5}$$

4.2 Gail Model Applied to Cuban Women's Cases

A first experiment was carried out in order to know the effectiveness of the Gail updated model on a data set of Hispanic population of Cuban women. Two scenarios were considered to apply BCRAT tool to the case study data. In the first scenario (GE1), the value "3" for race was indicated for all records, the model considers all women as Hispanic American regardless of the race indicated in the medical record. In the second scenario (GE2), the race indicated in the medical record was taken into account to code the race as specified by the BCRAT software. In the latter (GE2), Hispanic women are divided between those born in the United States and outside.

Table 1 shows the results of applying the Gail model software to the case study data for both scenarios. Notice that, for the second scenario, an improvement in the correct detection of positive cases while the detection of negative cases remains almost invariant. These results lead to a correct classification rate of 77.41% for GE1, and 86.60% for GE2. Even though the second case shows an improvement, there is still an error rate of 13.4%, mostly as a result of positive cases that were misclassified as negative (FN). These results are analyzed and further described by using the metrics results reported in Table 2, in particular Sensitivity and Accuracy values. This motivates the search for alternative solutions to diminish the error in general, and in particular the False Negatives that represent the greatest hazard for these cases where an early detection is of crucial importance.

4.3 Machine Learning Models Applied to Cuban Women's Cases

Based on the limitations shown in the results of the first experiment, another experiment was carried out applying basic machine learning models to the same data set. This, in order to explore the validation problem of estimating breast cancer risk and exploring the relevance of dealing with that by machine learning computational tools. In this experiment, machine learning algorithms were used as computational tools. The algorithms used were: Naive Bayes (NB), Generalized Linear Model (GLM), Logistic Regression (LR), Deep Learning (DL), Decision Tree (DT), Random Forest (RF), and Gradient Boosted Trees (GBT). The algorithms were executed through RapidMiner©

Table 1. Results of applying the Gail model to both scenarios.

	Scenario 1 (GE1)	Scenario 2 (GE2)
True Positives (TP)	833	976
True Negatives (TN)	379	380
False Positives (FP)	25	24
False Negatives (FN)	326	183

Table 2. Metrics for the Gail results in both scenarios.

	Scenario 1 (GE1)	Scenario 2 (GE2)
Sensitivity	71.87%	84.21%
Specificity	93.81%	94.06%
Precision	97.09%	97.60%
Accuracy	77.54%	86.76%
F1	82.60%	90.41%

Table 3. Metrics for results of applying Machine learning methods.

	Sensitivity	Specificity	Precision	Accuracy	F1
Naive Bayes (NB)	100%	93.8%	97.9%	98.4%	98.9%
Generalized Linear Model (GLM)	100%	93.8%	97.9%	98.4%	98.9%
Logistic Regression (LR)	100%	93.8%	97.9%	98.4%	98.9%
Deep Learning (DL)	98.70%	93.8%	97.9%	97.4%	98.3%
Decision Tree (DT)	100%	93.8%	97.9%	98.4%	98.9%
Random Forest (RF)	100%	93.8%	97.9%	98.4%	98.9%
Gradient Boosted Tree (GBT)	100%	93.8%	97.9%	98.4%	98.9%

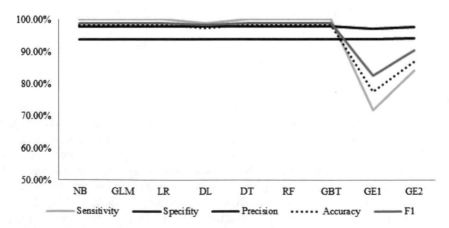

Fig. 1. Metrics comparison among the Gail model and machine learning methods.

Studio version 8.2, on a Windows 64 bits platform. The task to be performed by the algorithms was to predict the breast cancer variable, being "breast cancer = yes" the class of greatest interest.

The same data set from the first experiment was used, with the preprocessing required by the BCRAT tool. Table 3 shows the results obtained using the previously defined metrics. Notice that, for Specificity and Precision, the machine learning methods obtain values equivalent to the Gail model. However, for the remaining three metrics all the methods obtain better results than the Gail model. It should be noted that, the correct identification of positive cases (with cancer) plays a fundamental role in these last three metrics. This seems to indicate that the main limitation shown by the Gail model results in the previous section, could be overcome using machine learning models over the same input data, since their results show a better identification of positive cases, at least for the analyzed cases, minimizing the False Negatives that represent the greatest danger in the strategies of prevention and early detection of breast cancer.

Figure 1 illustrates the comparison of the values obtained for the metrics for both Gail model and machine learning methods. The difference in the Sensitivity, Accuracy and F1 metrics is noticeable.

5   Conclusions

This paper explored the effectiveness of applying a risk factor-based model, the Gail model in its latest version which contemplates the Hispanic American model, to a population of women born in Cuba. The results showed the lack of effectiveness of this model in particular to correctly detect positive cases, with cancer, which represents additional evidence of the limitations of its application to Hispanic women born outside the United States. Such limitations are aggravated by the relevance and negative impact that might have this lack of detection of positive cases, since that represents one of the main challenges from the prevention point of view and an early detection and proper treatment of the disease. Such results lead to emphasize the need for alternative models for estimating the risk of breast cancer. The results of applying an automatic model generation tool based on machine learning to the data are also presented. These results are as good as those obtained by the Gail model for all proposed metrics, and clearly better in three of the five metrics for the analyzed cases. This seems to suggest that machine learning techniques are a really promising alternative in the creation of risk factor-based models alternatives, to obtain a better prediction of breast cancer based on the same data from risk factors. However, a broader consideration of the estimation scenarios is required, in addition to validate the creation of this type of models with a greater variety of cases. As future work, it is considered the application of this type of models to a greater number and diversity of cases, as well as the examination of alternatives for the estimation of breast cancer risk under a multidisciplinary scheme.

Acknowledgments. Special thanks to RapidMiner© team for providing an Educational Licence Program version of their software for this work.

References

1. Ferlay, J., Colombet, M., Soerjomataram, I., Mathers, C., Parkin, D.M., Piñeros, M., Znaor, A., Bray, F.: Estimating the global cancer incidence and mortality in 2018: GLOBOCAN sources and methods. Int. J. Cancer **144**(8), 1941–1953 (2019). https://doi.org/10.1002/ijc. 31937. PMID: 30350310
2. Bray, F., Ferlay, J., Soerjomataram, I., Siegel, R.L., Torre, L.A., Jemal, A.: Global cancer statistics 2018: GLOBOCAN estimates of incidence and mortality worldwide for 36 cancers in 185 countries. CA Cancer J. Clin. **68**(6), 394–424 (2018). https://doi.org/10.3322/caac. 21492. PMID: 30207593
3. American Cancer Society: Cancer Facts & Figures 2019. American Cancer Society, Atlanta (2019)
4. Agarwal, G., Ramakant, P., Forgach, E.R., Rendón, J.C., Chaparro, J.M., Basurto, C.S., Margaritoni, M.: Breast cancer care in developing countries. World J. Surg. 2069–2076 (2009). https://doi.org/10.1007/s00268-009-0150-z
5. Anderson, B.O., Braun, S., Lim, S., Smith, R.A., Taplin, S., Thomas, D.B.: Early detection of breast cancer in countries with limited resources. Breast J. **9**(Suppl 2), S51–S59 (2003). Global Summit Early Detection Panel
6. Smith, R.A., Caleffi, M., Albert, U.S., Chen, T.H.H., Duffy, S.W., Franceschi, D., Nyström, L.: Breast cancer in limited-resource countries: early detection and access to care. Breast J. **12**, S16–S26 (2006). https://doi.org/10.1111/j.1075-122X.2006.00200.x
7. Gail, M., Brinton, L., Byar, D., Corle, D., Green, S., Shairer, C., Mulvihill, J.: Projecting individualized probabilities of developing breast cancer for white females who are being examined annually. J. Natl. Cancer Inst. **81**(24), 1879–1886 (1989). https://doi.org/10.1093/ jnci/81.24.1879
8. Costantino, J.P., Gail, M.H., Pee, D., Anderson, S., Redmond, C.K., Benichou, J., Samuel Wieand, H.: Validation studies for models projecting the risk of invasive and total breast cancer incidence. JNCI: J. Natl. Cancer Inst. **91**(18), 1541–1548 (1999). https://doi.org/10. 1093/jnci/91.18.1541
9. Gail, M., Costantino, J., Pee, D., Bondy, M., Newman, L., Selvan, M., Anderson, G., Malone, K., Marchbanks, P., McCaskill-Stevens, W., Norman, S., Simon, M., Spirtas, R., Ursin, G., Bernstein, L.: Projecting individualized absolute invasive breast cancer risk in African American women. J. Natl. Cancer Inst. **99**(23), 1782–1792 (2007). https://doi.org/ 10.1093/jnci/djm223
10. Matsuno, R., Costantino, J., Ziegler, R., Anderson, G., Li, H., Pee, D., Gail, M.: Projecting individualized absolute invasive breast cancer risk in Asian and Pacific Islander American women. J. Natl. Cancer Inst. **103**(12), 951–961 (2011). https://doi.org/10.1093/jnci/djr154
11. Banegas, M., John, E., Slattery, M., Gomez, S., Yu, M., LaCroix, A., Rowan, D., Hines, C., Thompson, C., Gail, M.: Projecting individualized absolute invasive breast cancer risk in US Hispanic women. J. Natl. Cancer Inst. **109**(2) (2017). https://doi.org/10.1093/jnci/djw215
12. Park, B., Ma, S.H., Shin, A., Chang, M.-C., Choi, J.-Y., et al.: Korean risk assessment model for breast cancer risk prediction. PLoS ONE **8**(10), e76736 (2013). https://doi.org/10.1371/ journal.pone.0076736
13. Claus, E.B., Risch, N., Douglas Thompson, W.: Genetic analysis of breast cancer in the cancer and steroid hormone study. Am. J. Hum. Genet. **48**(2), 232–242 (1991). PMID: 1990835
14. Parmigiani, G., Berry, D., Aguilar, O.: Determining carrier probabilities for breast cancer-susceptibility genes BRCA1 and BRCA2. Am. J. Hum. Genet. **62**, 145–158 (1998). https:// doi.org/10.1086/301670

15. Jonker, M.A., Jacobi, C.E., Hoogendoorn, W.E., Nagelkerke, N.J.D., de Bock, G.H., van Houwelingen, J.C.: Modeling familial clustered breast cancer using published data. Cancer Epidemiol. Biomark. Prev. **12**(12), 1479–1485 (2003)
16. Tyrer, J., Duffy, S., Cuzick, J.: A breast cancer prediction model incorporating familial and personal risk factors. Stat. Med. **23**(7), 1111–1130 (2004). https://doi.org/10.1002/sim.1668
17. Antoniou, A., Pharoah, P., Smith, P., Easton, D.: The BOADICEA model of genetic susceptibility to breast and ovarian cancer. Br. J. Cancer **91**(8), 1580–1590 (2004). https://doi.org/10.1038/sj.bjc.6602175
18. Narod, S.A.: Screening for BRCA1 and BRCA2 mutations in breast cancer patients from Mexico: the public health perspective. Salud Publica de México **51**(supp 2), s191–s196 (2009). PMID: 19967274
19. Hidalgo-Miranda, A., Jiménez-Sánchez, G.: Bases genómicas del cáncer de mama: avances hacia la medicina personalizada. Salud Pública de México **51**(supp 2), s197–s207 (2009)
20. Fejerman, L., John, E.M., Huntsman, S., Beckman, K., Choudhry, S., Perez-Stable, E., Burchard, E.G., Ziv, E.: Genetic ancestry and risk of breast cancer among U.S. Latinas. Cancer Res. **68**(23), 9723–9728 (2008). PMID: 19047150
21. McPherson, K., Steel, C.M., Dixon, J.M.: ABC of breast diseases: breast cancer—epidemiology, risk factors, and genetics **321**, 624 (2000). https://doi.org/10.1136/bmj.321.7261.624
22. Rodríguez Cuevas, S.A., García, M.C.: Epidemiología del cáncer de mama. Ginecología y Obstetricia de México **74**(11), 585–593 (2006)
23. Amir, E., Freedman, O.C., Seruga, B., Evans, D.G.: Assessing women at high risk of breast cancer: a review of risk assessment models. J. Natl. Cancer Inst. **102**(10), 680–691 (2010). https://doi.org/10.1093/jnci/djq088
24. Wang, X., Huang, Y., Li, L., Dai, H., Song, F., Chen, K.: Assessment of performance of the Gail model for predicting breast cancer risk: a systematic review and meta-analysis with trial sequential analysis. Breast Cancer Res. **20**(1), 18 (2018). https://doi.org/10.1186/s13058-018-0947-5
25. Spiegelman, D., Colditz, G.A., Hunter, D., Hertzmark, E.: Validation of the Gail et al. model for predicting individual breast cancer risk. J. Natl. Cancer Inst. **86**(8), 60060–60067 (1994)
26. Eva Singletary, S.: Rating the risk factors for breast cancer. Ann. Surg. **237**(4), 474–482 (2003). PMC1514477
27. Climente, I.P.P., Morales-Suárez-Varela, M.M., González, A.L., Magraner-Gil, J.F.: Aplicación del método de Gail de cálculo de riesgo de cáncer de mama a la población valenciana. Clin. Transl. Oncol. **7**(8), 336–343 (2004)
28. Fawcett, T.: An introduction to ROC analysis. Pattern Recogn. Lett. **27**(8), 861–874 (2006). https://doi.org/10.1016/j.patrec.2005.10.010

Tele-Dentistry Platform for Cavities Diagnosis

Diana Lancheros-Cuesta[⊠], Jose Luis Ramirez,
Herlinto Alveiro Tupaz Erira, Pedro David Hidalgo Caicedo,
Fernando Dvila, Andres Salas, Julin Eduardo Mora, Stella Herrera,
Luz M. Arango Betancourt, and Beatriz Echeverri

Facultad de Ingenieria – Odontologia, Universidad Cooperativa de Colombia,
Bogota, Colombia
diana.lancheros@campusucc.edu.co

Abstract. The influence of technology in the health-care sector has led to new approaches and services. Several services have been modified, improving: (i) primary healthcare, (ii) reaction time, and (iii) information management. Nevertheless, Colombian patients, who are aware of technological advances in health-care, are dissatisfied regarding the available and the offered telemedicines services. Regarding dentistry, several technological tools are available for diagnosis and treatment of oral diseases, which in most of the cases are preventable, i.e. cavities. Nevertheless, technological developments are uncommon and less frequent than in other health-care sectors. Furthermore, the innovation with respect to techniques for proper cavities diagnosis and current technological tools do not offer suitable solutions, in addition, existing methods and tools need to be merged with necessary promotion and prevention. Therefore, the development of a tele-dentistry platform is crucial. The platform should allow the tracking and diagnosis of dental cavities following the standards of the International Caries Detection and Assessment System (ICDAS). Consequently, this paper presents a new tele-dentistry platform for cavities diagnosis.

Keywords: Telemedicine · Computational platform · Document manager · ICDAS · Dental cavities

1 Introduction

Telemedicine has allowed the access of people who live in underdeveloped countries to health-care services including dentistry, generating prevention and diagnosis contexts for patients. One example is the RAFT project in Africa [2] Bagayoko et al. that consist in telemedicine technologies that allow people living in remote areas to access health-care services and learning courses to prevent diseases. The RAFT project shows to be successful for teleteachings and telecosultations. Concerning healthcare education, the World Health Organization (WHO) states that the combination of learning opportunities facilitate voluntary behavior changes leading up to an improvement in life quality [6].

Similarly, the 7th article of the resolution No. 412 of the Ministry of Health of Colombia, issued in 2002 [9], highlights that for early detection of diseases, it is important to take into account all activities, procedures, and interventions that allow the timely and efficient identification of illness. This early detection can be achieved

Á. Rocha et al. (Eds.): ICITS 2020, AISC 1137, pp. 490–497, 2020.
https://doi.org/10.1007/978-3-030-40690-5_48

through healthcare education as is presented in the technical standard of the Ministry of Health of Colombia regarding prevention for oral care, which takes into account the III national study of oral care ENSAB III issued in 1998 [9], and shows the importance of promotion and prevention in oral care and highlight oral morbidity reduction using as example dental cavities. Teledentistry is a relatively new field that merges two main topics: (i) communication technologies and (ii) dental care procedures. According to state of the art [3] Tele-dentistry is defined as: The practice of oral health-care offering diagnosis, medical consultation, treatment, and education using interactive audio, video, or data communication. In Colombia, the main advantage of these technologies is to offer patients, living in rural or conflict areas where health-care is limited or unsuitable, the access to dentistry services. However, the implementation of these tools requires infrastructure resources, i.e. Internet connectivity or network access, that thanks to the governmental efforts these infrastructure requirements are gradually available where is needed. The developed teledentistry platform must take into account the ICDAS standard, which is currently used for oral care professionals in Colombia. This standard applies seven codes between zero to six according to the severity of the damage. In the case of dental cavities, the pits and fissures of teeth are evaluated to establish the corresponding ICDAS code. The dentists must offer an extensive and detailed description of the damage, proposing a code suitable for the prevention and treatment of the cavities.

Consequently, the present paper presents the development of a platform, following three main stages: (i) design, (ii) analysis, and (iii) validation. It is to note that these three stages are carried out in collaboration with a group of experts in the field of dentistry at the Universidad Cooperativa de Colombia. The paper is composed of four sections as follows: (1) Related research studies: In this part, a summary of state of the art is presented, highlighting the most relevant advances in the field od teledentistry. (2) Methodology: This section introduces the stages followed during the development and implementation process. (3) Platform design: This part presents the software design approach, including a layer-based development. (4) Results: This last section exposes platform validation and assessment.

2   Related Research Studies

There is a growing body of literature that recognizes the importance and the social impact of teledentistry and telemedicine. Consequently, in the following, the most relevant research works, concerning the development of telemedicine or teledentistry platforms, are summarized. [7] Petcu et al. have developed the e-DENT project, whose objective was the service access improvement for people who can not have access to an on-site medical appointment (e.g. senior citizens, disabled individuals or prisoners). The result is a system, that is currently working in France, that allow access to dentistry diagnosis for vulnerable persons. Agnisarmana et al. [1] state that telemedicine starts as an option, for people living in rural areas, to have access to medical care improving live quality. Similarly, the authors describe how this kind of systems provides greater attention and assistance for people who can not have an on-site medical appointment.

The methodology proposed considers an intrasubject experimental design, in which participants (20 in total) are in contact with four telemedicine platforms.

Petruzzi and De Benedittis [8] state that one of the most frequent problems during medical consultations concerns the required time for diagnosis, consultations, forecasts, and treatments, which could even require a second opinion that in some cases must be obtained in a different place. Petruzzi and De Benedittis. carry out a research study in which they contact people using phone numbers acquired during conferences, congresses, and meetings about oral diseases. They send photos and questions, about oral clinical cases using WhatsApp application, which are verified by two persons using two different brands of mobile phones. The phone numbers, the information concerning the sender, and the proposed question are stored. Furthermore, Two experts in oral medicine are in charge of giving a possible diagnosis and classify problems into the following categories: Traumatic, infectious, pre-neoplastic/neoplastic, autoimmune and not diagnosable. Images are verified in a period not longer than 30 min, giving a timeframe of two hours to have back an appropriate answer. After that, patients are invited to take an oral mucosa sample in a clinic the elapsed time (in days) to take samples is measured. An independent pathologist studies injuries biopsies to define if the damage is malignant or benign, if it is pre-cancerous, or if it comes from a chronic sore of unknown reason.

The research shows the importance of telemedicine in the reduction of cost, and in efficiency of medical consultations. Chia-Yung et al. [5] implement a mobile application, in collaboration with: dentists, scientists, and computer scientists of the Taiwan University, that facilitate the communication between patients and dentists. The application, developed using cloud services and a Java algorithm, allows communication patient-dentist communication, send an appointment reminder to patients, and dentists can modify schedule in an easy way, offering consultancy to patients at any time. Twenty-six dentist and thirtytwo patients are part of a study, which shows a significant increment in the quality of medical attention proposed to patients, regarding the appointment reminder and schedule adjuster. On the other hand, it is easier for dentists to deal with patients cases. The application shows very efficient results concerning established relation between patients and dentists.

Zakian et al. [10] use infrared (IR) images trying to detect the cavities employing the thermal changes associated with the dehydration. Cavities have been one of the health problems always present in the society and in the human history a timely and correct detection of these would help it treatment avoiding the total loss of teeth. In the research measurements of the tooths dehydration is carried out using the thermal changes, using a camera that captures infrared images at a rate of 14 pictures per second in a minute, including a period of dried. The capture of information is divided into three periods, the first 10 s for capturing a reference temperature, the following 40 s the teeth dried off with a compressor and finally 10 s for stabilization and to register the final temperature. With the help of a video composed of the sequence of captured infrared images, an analysis of 25 teeth is obtained. Besides, a total of 72 histological cuts of different fissures and different gravity of damage are selected. As a result, authors claims that the identification through infrared images is better than conventional in vitro X-ray photograph. The higher performance of IR method is because in vitro X-ray does not manage to have a complete certainty since it needs more studies to be able to confirm a diagnosis. Fricton and Chen [4] affirm that nowadays, the service of oral health is limited to urban areas. Therefore, it was difficult to access this type of services in a rural zone, and this owes to factors as: (i) population socioeconomic level in the areas where would be implemented this new trend (Tele-dentistry), (ii) geographic location, (iii) lack of health insurance (by patients care-lessness or lack of money), and (iv) lack of oral care specialized establishments located closer to patients place of residence. Bearing in mind the facility to access oral care consultation through telemedicine, not only the unattended population would decrease, but the number of expert consultations per patient would increase. The above state-ments owe to that fact that patients do not have to go up to the point of attention, but they can take consultation it in a virtual way, using any available technological device (Smartphone, Tablet, Portable, among others).

Fricton and Chen [4] develop a research study to promote the prevention of oral diseases in rural zones using teledentistry, avoiding the necessity of a dentistry spe-cialized in vivo consultation. Patients have certainty to be accompanied by qualified personnel, and in the case of critical illnesses can receive accurate information for proper attention. Teledentistry is a project merging internet services and oral care, which seeks to adress the problematics by means of videoconferences, images, audio and patients information. In 2004 the University of Minnesota in collaboration with Hibbings university center, incorporates Teledentistry for consultations and attention to families of scanty resources, aiming to: (i) have presence inside the most isolated social contour or rural zones through teledentistry and (ii) show that it is possible to have an improvement in oral care providing teledentistry access. As a result, it is shown that habitants of the zones mentioned above can be trained to fulfill the requirements of prevention of oral diseases. For instance, in Alaska they were managed to attend 830 patients during 1.200 visits a year, 700 of those visits were preventive, and 500 were for reconstructive attention.

3 Methodology

In the following the methodology used for the platform development and implemen-tation. The process is carried out following six stages: (i) information gathering, (ii) information analysis, (iii) design, (iv) feedback, (v) implementation, and (vi) use and maintenance. These stages are described hereunder. Stage 1: information gathering. This stage studies the definition software requirement regarding: (i) already imple-mented solutions, (ii) recommended processes and methodologies, and (iii) characteri-zation of the documents that will be handled using the software tool. The users participation is critical at this stage since they must define the necessities that the software must fulfill. The developer is in charge of the analysis and treatment of the gathered information. Stage 2: information analysis. This stage corresponds to pre-design, and begins cleaning up information (gathered in the previous stage), to define the way for the development, giving priority to user requirements needs of the client. Stage 3: design. In this stage, the following processes are carried out: (i) development of the database scheme, (ii) design of the graphic User Interphase (GUI) to facilitate user interaction, (iii) definition of the methodologies and required recommendations,

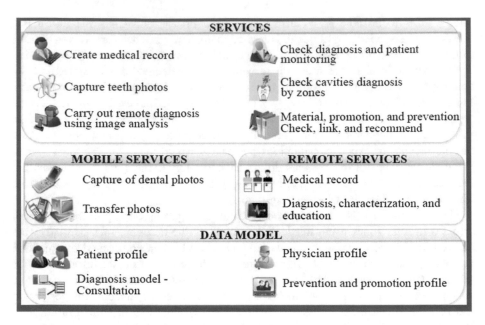

Fig. 1. Telemedicine-dentistry platforms architecture.

and (iv) development of the logic algorithm that interprets users actions. Stage 4: feedback. In this stage, User carries out an evaluation to validate the proper operation of the application, and in the case of inconsistencies, those could be solved in time before the implementation. Stage 5: implementation. The software is deployed and handled information is entered. A production phase is launched aiming to evaluate the software behavior while dealing with the loaded data. So that, process performance and possible difficulties are assessed. Stage 6: use and maintenance. This stage implies constant assistance to the final product, regarding supervision, assessment, and modification of system information, according to end user instructions. This process ensures the improvement of the software, getting closer to an ideal product without fails or errors.

4   Platform Design

The telemedicine platform is developed using a layer based scheme as is shown in Fig. 1. As presented in the figure, from the bottom to the top side, the first layer is composed of the information model, which is constituted by: (i) Patient profile: This module stores main faetures of patients who have cavities, (ii) Diagnosis modelConsultation: This module allows the storage of diagnosis features taking into account ICDAS standard. (iii) Physician profile: This module stores dentists and dental assistants information. (iv) Prevention and promotion profile: this module allows the characterization of required information needed to allow the population to have access to multimedia contents regarding dental cavities prevention. The next layer is composed of the

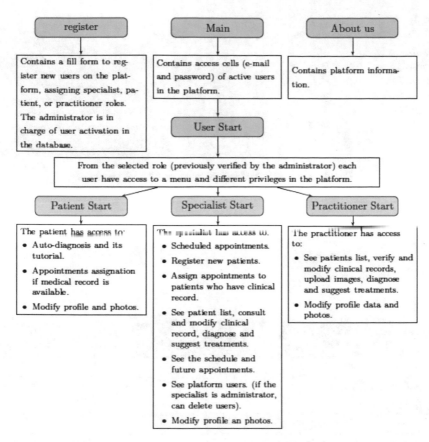

Fig. 2. Telemedicine-dentistry platforms architecture.

platform services in terms of the mobile device that includes the capture of photos of teeth or molars cavities and the communication to transfer the photos to the platform. The remote services allow the consultation of the clinical records, the characterization of the population, the diagnosis and the access to educational materials for promotion and prevention. The third layer addresses the general services offered by the system (creation of clinical records, storage of dental photos, the accomplishment of remote diagnoses, consultation and monitoring of patients, consultation of characterization of the cavities, and recommendation of the material for prevention and promotion). Figure 2 presents the detailed functionalities for users in the platform.

5   Results

The developed platform allows to carry out the diagnosis of dental cavities taking into account the ICDAS standard as is presented in Figs. 3 and 4. The platform facilitates forwarding the information to be checked in different cities where the Universidad Cooperativa de Colombia has dentistry faculties.

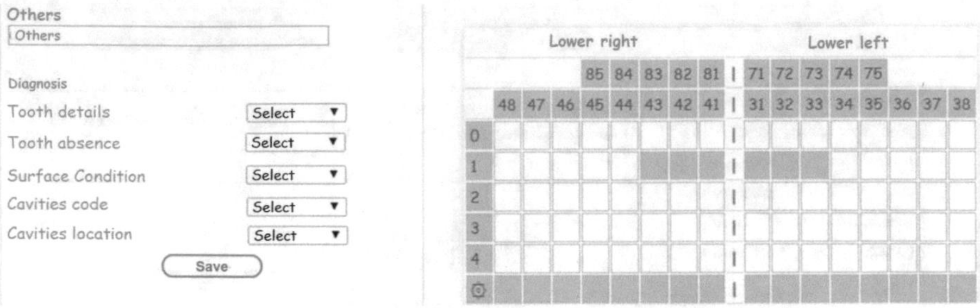

Others

Others

Diagnosis

Tooth details	Select ▼
Tooth absence	Select ▼
Surface Condition	Select ▼
Cavities code	Select ▼
Cavities location	Select ▼

Save

						Lower right										Lower left					
					85	84	83	82	81	I	71	72	73	74	75					
	48	47	46	45	44	43	42	41	I	31	32	33	34	35	36	37	38
0									I								
1									I								
2									I								
3									I								
4									I								
☉									I								

Fig. 3. Evaluation taking into account ICDAS standard.

PATIENT DATA

NAME: CAROLINA MESA

ID: 1019017501

DATE OF BIRTH: 10/8/1992

Symptomatology

Dental surface	State	
	Upper	Lower
1. (K032) Erosion	Not Present	Not Present
2. (K038) Abfracción	Not Present	Not Present
3. (K031) Graze	Not Present	Not Present
4. (Z076) Frontal caries Primary Healthy (CCP)	Not	Not
5. (K020) CCP injury of brown spot	Present	Not Present
6. (K020) CCP injury of white spot	Not Present	Present
7. (K020) CCP micro-cavity	Not Present	Not Present
8. (K020) CCP underlying Shade of dentine	Not Present	Not Present
9. (K029) CCP detectable cavity	Not Present	Not Present
10. (K021) CCP widespread cavity	Not Present	Not Present
11. (K020) Secondary caries not cavitacional	Not Present	Not Present
12. (K021) Secondary caries cavitacional	Not Present	Not Present
13. (K022) Radicular caries not cavitacional	Not Present	Not Present
14. (K022) Radicular caries cavitacional	Not Present	Not Present

Diagnóstico: Caries coronal primaria en 15 y 44

Tratamiento: Amalgama de metal

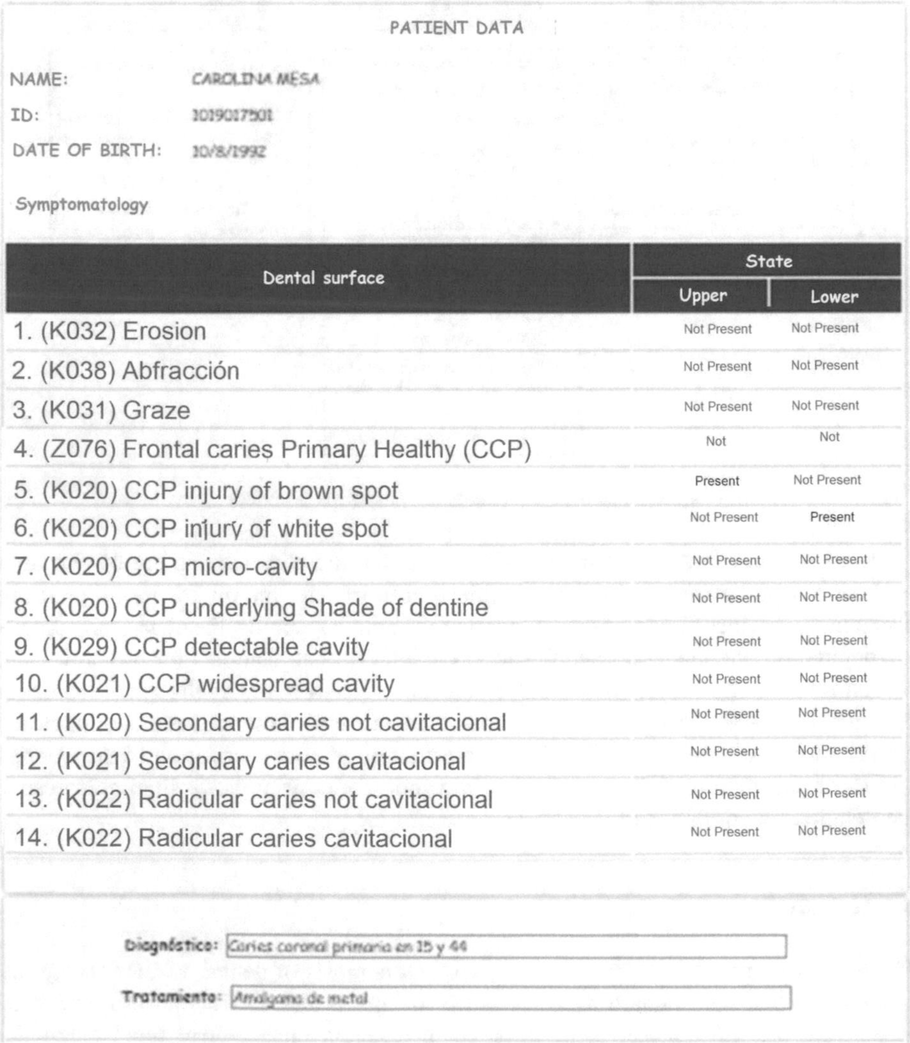

Fig. 4. Diagnosis taking into account ICDAS standard.

6   Conclusions

A platform, for dentistry process management, has been implemented under the concept of telemedicine, for the diagnosis and monitoring of dental cavities using the ICDAS standard. The virtual platform allows connecting a user, previously registered, from any remote location. Furthermore, the platform allows the administrator to control and manage virtual tools from any server, giving an answer to modifications in real time y immediately to users. As perspectives for future work, validation is proposed considering clinical cases in the platform from the different locations of the University.

References

1. Agnisarman, S.O., Madathil, K.C., Smith, K., Ashok, A., Welch, B., McElligott, J.T.: Lessons learned from the usability assessment of home-based telemedicine systems. Appl. Ergon. **58**, 424–434 (2017)
2. Bagayoko, C.O., Müller, H., Geissbuhler, A.: Assessment of internet-based telemedicine in Africa (the raft project). Comput. Med. Imaging Graph. **30**(6), 407–416 (2006)
3. Chen, J.W., Hobdell, M.H., Dunn, K., Johnson, K.A., Zhang, J.: Teledentistry and its use in dental education. J. Am. Dent. Assoc. **134**(3), 342–346 (2003)
4. Fricton, J., Chen, H.: Using teledentistry to improve access to dental care for the underserved. Dent. Clin. North Am. **53**(3), 537–548 (2009)
5. Lin, C.Y., Peng, K.L., Chen, J., Tsai, J.Y., Tseng, Y.C., Yang, J.R., Chen, M.H.: Improvements in dental care using a new mobile app with cloud services. J. Formos. Med. Assoc. **113**(10), 742–749 (2014)
6. World Health Organization: Prevention of Oral Diseases. World Health Organization, Geneva (1987)
7. Petcu, R., Ologeanu-Taddei, R., Bourdon, I., Kimble, C., Giraudeau, N.: Acceptance and organizational aspects of oral tele-consultation: a French study. In: 2016 49th Hawaii International Conference on System Sciences (HICSS), pp. 3124–3132. IEEE (2016)
8. Petruzzi, M., De Benedittis, M.: Whatsapp: a telemedicine platform for facilitating remote oral medicine consultation and improving clinical examinations. Oral Surg. Oral Med. Oral Pathol. Oral Radiol. **121**(3), 248–254 (2016)
9. Ministerio de Salud, C.: actividades, procedimientos e intervenciones de demanda inducida y obligatorio cumplimiento y se adoptan las normas tcnicas y guas de atencin para el desarrollo de las acciones de proteccin especfica y deteccin temprana y la atencin de enfermedades de inters en salud pblica, February 2000
10. Zakian, C., Taylor, A., Ellwood, R., Pretty, I.: Occlusal caries detection by using thermal imaging. J. Dent. **38**(10), 788–795 (2010)

Information Technologies in Education

Technology Integration in a Course for Prospective Mathematics Teachers

Alexander Castrillón-Yepes(⊠) ⓘ,
Jaime Andrés Carmona-Mesa(⊠) ⓘ,
and Jhony Alexander Villa-Ochoa(⊠) ⓘ

Universidad de Antioquia, Medellín 005010, Colombia
{alexander.castrillony,jandres.carmona,
jhony.villa}@udea.edu.co

Abstract. This document presents the results of a study that sought to integrate technologies into a mathematics course for prospective mathematics teachers. Two types of environments were created, one focused on technologies for interaction and communication and the other focused on technologies for the study of mathematics. The results show that the former environment acted as a classroom extension that allowed teachers to recognize students' learning processes and needs, while the later environment allowed the creation of strategies to solve tasks.

Keywords: Digital technology · ICT · Introductory course of linear algebra

1 Introduction

Some investigations point out the importance of training prospective mathematics teachers to effectively integrate technologies in the teaching processes [4, 7]. In this regard, Kay [7] reports that there is a diversity of strategies for initial teacher training, one of them being the integration and use of technologies in a transversal way across all courses. However, the development of this strategy is a complex process given that technology integration depends on the theoretical and empirical nature of each course contents, as well as in the institutional educative purposes and the teacher's knowledge of the relationship between technologies and specific contents [3].

Experiences of integrating technologies in introductory courses of linear algebra are reported in the international literature [1, 13]. For instance, Piñero et al. [13] highlight the value of social networks such as Facebook, YouTube and WhatsApp within the educational field to promote collaborative learning; for these authors, collaboration through these networks can foster the exchange of information and generate fruitful discussions. Furthermore, Aranda and Callejo [1] incorporated technologies in vector-spaces-related teaching topics; the efforts of these researchers focused on improving abstraction processes by using strategies and implementing tasks aiming to strengthen knowledge through different representation systems. On the other hand, Del Pino and Lozano [6] reported the effects of Information Communication Technologies (ICT) on

© Springer Nature Switzerland AG 2020
Á. Rocha et al. (Eds.): ICITS 2020, AISC 1137, pp. 501–510, 2020.
https://doi.org/10.1007/978-3-030-40690-5_49

educative evaluation processes, highlighting the importance of good communication means and skills. They additionally comment:

> The use of technologies in this subject [linear algebra] has a direct impact on the treatment of specific contents because it allows addressing more complex exercises, producing a closer approach to real science and technology problems and displacing the focus of teaching-learning processes towards modeling and discernment, in short, towards the development of general mathematical skills, leaving laborious calculations (from a didactic perspective) to the computer (original text in Spanish, p. 5–6).

According to the aforementioned studies, the need for class environments that integrate technologies can be concluded. On the one hand, technologies provide means that allow students conceptual expansion, production of meaning, experimentation, simulation, decrease in operating load, interconnection and applications in other sciences fields, etc.; on the other hand, technologies are a means for interaction among students, content, and teachers, fostering the communication and dissemination of the generated knowledge.

Linear algebra topics have a conceptual and symbolic complexity that derives, in part, from the abstract nature of those subjects and the ways of understanding and representing them. Part of the difficulties that students have in understanding these topics is also derived from such complexity [14]. In that sense, a study was developed seeking to answer the question: What are the contributions that technologies can offer to prospective teachers in an introductory course of linear algebra?

Guided by our research question, we proceed as follows: In Sect. 2, we present the research methodology, where the analytical tools, data collection, and data analysis are described. Then, based on the constructed categories, the results of the research are presented (Sect. 3). Finally, in Sect. 4, we conclude and discuss the need to transcend a domesticated use of technologies and present some considerations aiming to effectively integrate technology into the training of prospective mathematics teachers.

2   Methodology

Environments with technologies such as a closed Facebook group, GeoGebra and the Casio ClassWiz Fx-991LA X calculator were designed and implemented within a regular course for prospective mathematics teachers at the Universidad de Antioquia. The main objective was to explore the possibilities and limitations that technologies offer in the study of the course mathematical content. This section presents the tools used for empirical data analysis; a description of the course, its principles of design and its structure are also presented. Finally, the data analysis method is presented.

2.1   Analytical Tools

As mentioned earlier, prospective mathematics teachers require first-hand experiences that allow them to recognize the potentials and limitations of technologies in their mathematics learning process. Environments that integrate a diversity of resources and uses of technology for learning should be promoted [10]. A domesticated use of technologies must be transcended [2]. The use of certain technologies should not be limited to the replication of what can be done with other technologies such as pencil and paper. The new possibilities offered for new technologies regarding the construction and production of knowledge should be studied, recognized and promoted.

In the previous section, the opportunities offered by some ICT were presented; on the one hand, they allow to create environments that promote interaction and communication (through social networks, platforms, videoconferences, etc.); on the other hand, digital technologies help to develop activities of the particular discipline (through specific software, calculators, etc.). These two ways of using technology served as a frame to analyze the data gathered during the course.

2.2 Technologies in an Introductory Linear Algebra Course

The introductory Linear Algebra course is based on mathematical argumentation, demonstrations and applications in different contexts. The course also looks for the promotion and establishment of conceptual relationships that encourage the subsequent study of other mandatory courses such as Linear Algebra, Vector Calculus, and some Physics courses.

In this introductory course to linear algebra, some of the vector spaces to be studied require a broad understanding of what "algebra" means, and some of them have a strong visual component (R^2, R^3, E^2, P_n). In this regard, different investigations have shown the importance of incorporating technologies in the study of concepts of algebra and concepts with a broad visual component [8, 14]. Despite this, in a review of the course plan, no evidence was found of any use of digital technologies. To correct this lack, two types of environments were proposed.

The first environment included ICT aiming to configure a virtual space for learning [9]. This space involved a closed Facebook group, the video production of an application project and the dissemination of the video via an academic blog. In the closed group, interactions between students and teachers, development of tasks, dissemination of news, resources and course materials were promoted. Weekly tasks were proposed that involved reflection on key conceptual aspects of the topics covered in the course. As an example, Fig. 1 presents a comic used to generate a discussion on two different concepts having a similar algebraic representation.

The second environment included computer designed tasks using GeoGebra and the calculator Casio ClassWiz Fx-991LA X. Tasks were designed in order to promote a variety of mathematical activities. Prospective teachers were encouraged to solve problems, build strategies, guess, argue, reason, and not just to observe a graph or reduce calculations operating load. As an example, the following task is presented: "Design a strategy to multiply two matrices of order 5×5 using the calculator What would happen if you want to make the product of two 3×3 matrices? Can you use an adapted 5×5 algorithm?".

2.3 Data and Data Analysis

The course took place during the first semester of 2017. It covered four hours per week and lasted 16 weeks. The course had a full professor and a teaching assistant who weekly promoted discussions about the proposed tasks, the means to develop them and the obtained results. Subsequently, new tasks were formulated and interaction was promoted. As a result, 20 documents, 3 audio recordings, 9 videos built by the students and all publications and comments posted on Facebook were collected.

To answer the research question, each of the authors of the chapter made a codification based on the two categories presented in Sect. 2.1 of this chapter. Each author independently encoded the documents produced by prospective teachers, the posts on Facebook and the videos. Once this review was completed, the findings were organized in a list of possible contributions and limitations that, in the opinion of each author, the technology offered to the course. Subsequently, the findings of each author were discussed in the team, seeking to reach consensus; As a result of this debate, a more refined categorization emerged (see Table 1).

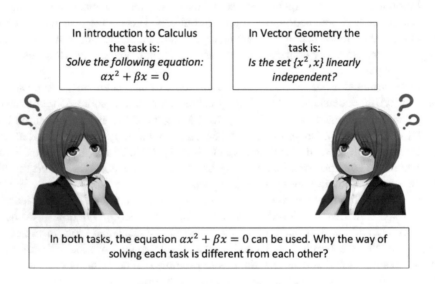

Fig. 1. Comic used on Facebook

Table 1. List of categories, technologies, and codes emerging from data analysis

Categories	Technologies	Codes
ICT as an environment for discussion	Closed Facebook group; Blog	Dialogue, present idea, debate, conceptual expansion
Digital technologies in mathematics classroom	Casio ClassWiz Fx-991LA X calculators, GeoGebra, Spreadsheet	Use of the device, problem-solving, strategy building

3   Results

Results are presented by categories as described in Table 1. Beside each category, conceptual aspects and the uses of technology that promoted it are shown; an episode is also presented to illustrate the findings.

3.1   ICT as an Environment for Discussion

In the course, concepts such as linear independence, linear combination, basis and dimension of a vector space are studied. These concepts are usually presented and defined in the course textbooks; subsequently, examples and demonstrative exercises, expressed in algebraic language, are proposed. For some particular vector spaces, Oropeza and Lezama [12] suggest promoting the use of visualization resources since they "use mathematics in relation to numerical, graphical, algebraic, verbal and also gestural subjects" (Original text in Spanish, p. 18).

To promote diverse kinds of interactions among students, an activity was carried out with GeoGebra in the P_2 space and then a task was proposed in the Closed Group on Facebook promoting analysis on linear dependence or independence of a set of lines (see Fig. 2). In the closed group, students had the opportunity to discuss two different responses and present arguments in favor of each one. This situation allowed the teacher and the teaching assistant to note that a group of students was identified with the P_2 vector space, while the other group was identified with the R^2 vector space. These discussions and arguments were later reconsidered in the classroom to clarify the respective points of view. As an example, the transcription of some posts is presented (See Figs. 3 and 4).

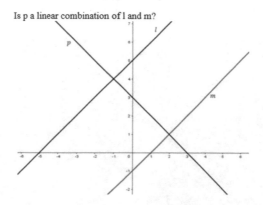

Fig. 2. Linear combination task

In Fig. 3 it can be seen that a student reflects about the possible relationship between the linear combination and the function intercepts with the coordinate axes (points). He/she considered the space R^2 to determine the dependence or linear independence of the lines generated from the vectors: (3, 3), (−5, 5), (1, −1), which he/she constructed, perhaps, based on the graphic representation presented, without recognizing the vector space that underlies the task. In Fig. 4 the argument constructed by the student is presented.

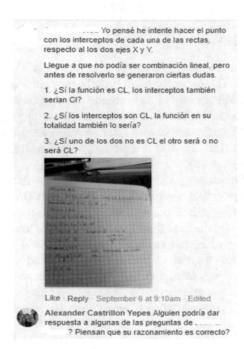

..... Yo pensé he intente hacer el punto con los interceptos de cada una de las rectas, respecto al los dos ejes X y Y.

Llegue a que no podía ser combinación lineal, pero antes de resolverlo se generaron ciertas dudas.

1. ¿Sí la función es CL, los interceptos también serian Cl?

2. ¿Sí los interceptos son CL, la función en su totalidad también lo sería?

3. ¿Sí uno de los dos no es CL el otro será o no será CL?

Like · Reply · September 6 at 9:10am · Edited

Alexander Castrillon Yepes Alguien podría dar respuesta a algunas de las preguntas de .... ? Piensan que su razonamiento es correcto?

Student 1: I thought and tried to construct the point with the intercepts of each of the lines with respect to the two axes X and Y.

I came to the fact that it could not be a linear combination, but before solving it, certain doubts were generated.

1. If the function is LC (Linear combination), would the intercepts be LC?
2. If the intercepts are LC, would the function as a whole also be LC?
3. If one of the two is not LC, will the other be or will not be LC?

Alexander (teaching assistant) replied: Could someone answer some of this student's questions? Do you think his reasoning is correct?

Fig. 3. Student 1 Post about the Homework

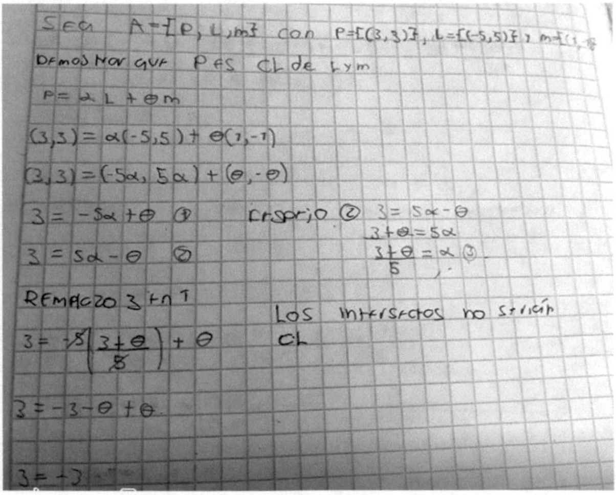

Fig. 4. Student 1 support document

Technologies such as Facebook offered possibilities to 'extend' the classroom and promote complementary actions that would contribute to the production of mathematical knowledge. The main resource is asynchronous participation in academic discussions and debates about mathematical situations or tasks. Other effects of the use of technology were: (i) the possibility of disseminating study material, (ii) the proposition of other tasks (both by the teacher and by the students) and (iii) the possibility to question and generate discussions around the addressed concepts.

3.2   Digital Technologies in Mathematics Classroom

International literature has argued in favor of the need to transcend domesticated uses of technology [2, 11]. For these prospective teachers, the calculator was conceived as a device to streamline operations and calculations; however, the proposed tasks offered opportunities for them to have experiences on other uses of calculators. In the task proposed in Sect. 2.2, students began with an exploration of the menu keys corresponding to matrices and, quickly, they recognized that the calculator only allowed to make calculations with square matrices of order 4×4. Therefore, they realized that to multiply matrices of a higher size (order) they had to create strategies.

Students worked in groups. Gabriel and María (pseudonyms) built the following strategy:

- First of all, they defined two matrices $A, B \in R_{4\times4}$ with entries a_{ij} and b_{jk}, respectively. Then, they defined the matrix $C = AB$ with entries c_{ik}. From the above, they built two matrices $A', B' \in R_{5\times5}$ with entries $a'_{ij} = a_{ij}$ y $b'_{jk} = b_{jk}; \forall i = 1,2,3,4; \forall j = 1,2,3,4, \forall k = 1,2,3,4$ (see Fig. 5), if i = 5 or j = 5 or k = 5 then a'_{ij} and b'_{jk} are any real number. Thus, they defined the product $D = A'B'$ with entries d_{ik}.
- The students identified a relationship between the entries of the matrix $C_{4\times4}$ and the matrix $D_{5\times5}$ (see Fig. 6). From this relationship, they constructed the matrix $E_{4\times4}$ and subsequently defined the matrix $F = C+E$ with entries $f_{ik} = d_{ik}; \forall i = 1,2,3,4; \forall k = 1,2,3,4$.
- Finally, the students stated that the entries in the last row of D are obtained by multiplying the fifth row of A with the matrix B' (they called it G); while the last column of D is obtained by multiplying the matrix A' with the fifth row of B (they called it H). In this way, they build the "general structure of matrix D" presented in Fig. 7.

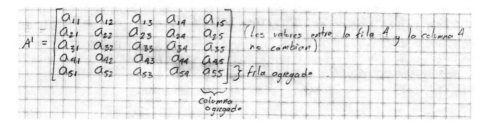

Fig. 5. Matrix A definition

Fig. 6. Relationships among components of two different matrices.

Figure 6 legend reads: "the values of the 4 row and the 4 column of matrix D are equal to matrix C plus the product of the 5 column of A' with the 5 row of B'".

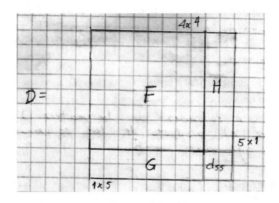

Fig. 7. The general structure of students' reasoning.

A similar reasoning was raised by the students to construct an algorithm that would allow using the calculator for square matrices of order n × n. It is worth noting that this task was extrapolated to make calculations of determinants; and although in neither case was it possible to identify the use of a simpler algebraic language, in both cases a correct reasoning is observed to use a calculator in situations that escape their functionalities. This raises a reflection on how the limitations of different technological means can be used to design tasks that promote mathematical thinking.

Other uses of technology as a tool for creating mathematical problems were also presented in the course. For example, in a given task, the students had to recognize the arithmetic pattern that is generated with the use of the keys EXE e=. From this, with the appropriate mediation of the teacher, a system of equations can be built and solved using the topics of the course and with the use of a calculator.

4   Discussion and Conclusions

The first part of this chapter, examples in favor of the need to transcend domesticated uses of technologies were presented; also, the need to promote an integration of technologies in mathematics classes that allows the use and exploration of devices according to their nature and their possibilities was argued. International research reports that technologies offer possibilities to construct at least two types of environments, (i) ICT as an environment to promote the discussion of tasks and (ii) digital technologies developed for the study of mathematics.

In this respect, this chapter offers two important results. First, it shows that social networks, such as Facebook, offer opportunities to extend the classroom and the time assigned to learning. As other research has reported [5], virtual learning environments (e.g., this closed Facebook group) promote types of student-student and teacher-student interactions that are not present in the conventional classroom. In the case of this course, virtual learning environments allowed us to hold dialogues and present arguments and counter-arguments about class assignments; regarding teachers, such environments allowed them to recognize the points of view and understanding level of students and, based on that information, allowed them to offer feedback via the same space or later in the classroom. The closed Facebook group also offered the opportunity for students to express their concerns, collaborate with their classmates, share material and enable the recognition of aspects of the content that needed to be discussed and expanded in other class spaces. One of the limitations offered by these spaces is the low degree of voluntary participation of some students; this fact affects the richness of the discussion and limits the identification of other topics and paths for class discussion.

A second result has to do with the importance of recognizing the possibilities and limitations of different resources used in class; this recognition can guide the construction of tasks that promote mathematical activity and where technology transcends a domesticated use as a fundamental condition for the construction of knowledge in the classroom [3]. This research shows that it is possible to use the limitations of technologies to promote a mathematical activity. This approach requires more research on the use of ICT and digital technologies in prospective mathematics teachers' training. This process allowed them to recognize that technologies, such as the Casio ClassWiz Fx-991LA X calculator, can be used to pose and solve problems and not only to reduce the operating load. In this sense, the mathematical knowledge that is constructed in mathematics courses may be accompanied by reflection on one's experience with the means with which that knowledge was constructed; in this way, it is expected to provide bases for prospective teachers' pedagogical knowledge.

References

1. Aranda, C., Callejo, M.L.: Construcción del concepto de dependencia lineal en un contexto de geometría dinámica: Un estudio de casos. Revista latinoamericana de investigación en matemática educativa **13**(2), 129–158 (2010)
2. Borba, M.C., Villarreal, M.E.: Humans-with-Media and the Reorganization of Mathematical Thinking, 1st edn. Springer, New York (2005)

3. Carmona-Mesa, J.A., Salazar, J.V.F., Villa-Ochoa, J.A.: Uso de calculadoras simples y videojuegos en un curso de formación de profesores. Uni-pluriversidad **18**(1), 13–24 (2018). https://doi.org/10.17533/udea.unipluri.18.1.02

4. Carmona-Mesa, J.A., Villa-Ochoa, J.A.: Necesidades de formación en futuros profesores para el uso de tecnologías. Resultados de un estudio documental. Revista Paradigma **38**(1), 169–185 (2017)

5. Chugh, R., Ruhi, U.: Social media in higher education: a literature review of Facebook. Educ. Inf. Technol. **23**(2), 605–616 (2018). https://doi.org/10.1007/s10639-017-9621-2

6. Del Pino, P., Lozano, L.: La didáctica del álgebra lineal y su diseño: una visión con el uso de la tecnología. CD de Monografías 2012, Universidad de Matanzas "Camilo Cienfuegos" (2012)

7. Kay, R.H.: Evaluating strategies used to incorporate technology into preservice education. J. Res. Technol. Educ. **38**(4), 385–410 (2006)

8. Kripka, M.R.L., Kripka, M., Pandolfo, P.C.D.N., Pereira, L.H.F., Viali, L., Lahm, R.A.: Aprendizagem de Álgebra Linear: explorando recursos do GeoGebra no cálculo de esforços em estruturas. Acta Scientiae **19**(4), 544–562 (2017)

9. López Dávila, C.: Un modelo de investigación orientado a la implementación de programas estructurados en ambientes virtuales de aprendizaje. Uni-pluriversidad **15**(2), 61–73 (2016)

10. Martínez Aguilar, F.M., Susano García, L.J., Espinosa Delgado, J.M.: Redes sociales y TICS en la cátedra universitaria. Uni-pluriversidad **15**(1), 87–99 (2015)

11. Medina Orellán, J., Ortiz Buitrago, J.: Competencias matemáticas y uso de calculadora gráfica en un contexto de resolución de problemas aplicados. Uni-pluriversidad **13**(3), 14–28 (2013)

12. Oropeza, C., Lezama, J.: Dependencia e independencia lineal: una propuesta de actividades para el aula. Revista Electrónica de Investigación en Educación en Ciencias **2**(1), 23–39 (2007)

13. Piñero, E., Vivas, M., Camacaro, M.: Las TIC como recursos digitales para el estudio del álgebra lineal: Una experiencia con estudiantes de ingeniería en informática de la Universidad Centroccidental Lisandro Alvarado. Revista de Matemática MATUA **3**(2), 47–57 (2016)

14. Sandoval, I., Possani, E.: An analysis of different representations for vectors and planes in R3. Educ. Stud. Math. **92**(1), 109–127 (2016). https://doi.org/10.1007/s10649-015-9675-2

ClassFlow: Performance Indicator Based on Learning Routes

Franklin Chamba[1]([✉]), Susana Arias[2], Ruth Patricia Maldonado[3], and Diego Freire[4]

[1] Facultad de Ciencias Sociales, Universidad Tecnica de Machala, Machala, Ecuador
fchamba@utmachala.edu.ec

[2] Facultad de Ciencias de la Salud, Universidad Tecnica de Ambato, Ambato, Ecuador
Sa.arias@uta.edu.ec

[3] Departamento de Psicología, Universidad Tecnica Particular de Loja, Loja, Ecuador
rpmaldonado@utpl.edu.ec

[4] Universidad Regional Autónoma de los Andes, Turismo, Km 5 1/2 via Baños, Ambato, Ecuador
Ua.diegofreire@uniandes.edu.ec

Abstract. At present, education is subject to a series of changes that generally affects the exact sciences such as physics and mathematics. This makes the methodological or pedagogical process always the same, the same sequence, the same series and some teachers generally accept any process as long as I came to the answer, although in some cases there is resistance from the students. Although most teachers prefer the traditional and reproduce only what he learned during his training, this type of thinking limits learning to the student and makes him a mechanic or a memorist. During the investigation, it has been considered what needs to change the mentality of human behavior in teaching students, transform their method, their thinking in a pedagogical way, using necessary face-to-face resources that facilitate the teacher to make an interactive class, attacking the different styles of learning that exist within a classroom and therefore covering some special educational needs that may or may exist within the classroom. The ClassFlow tool is a teaching support resource that helps and greatly facilitates understanding to foster the ideology that exact sciences can also be practiced online or in real time, inside or outside a laboratory, using the resources or mobile devices that Students have, like their internet data, Class-Flow is a powerful tool that helps control any type of hyperactivity or learning problem within the classroom. Therefore, ClassFlow is being used in several institutions worldwide to improve the student learning process. The Promethean ClassFlow teacher learning platform allows you to plan classes more efficiently and make the classes a more interesting experience for your students.

Keywords: Office skills · MOOC · Human behavior · Software tools · Learning · Virtual environments

Á. Rocha et al. (Eds.): ICITS 2020, AISC 1137, pp. 511–521, 2020.
https://doi.org/10.1007/978-3-030-40690-5_50

1 Introduction

Social human behavior which is not always clearly warned or defined by consciousness, thereby emphasizing the tendency to self-protection of the I or the SAME, of one's own identity and of the participation of consciousness that is generally affected by customs of the culture of each individual, social, economic and family. As human beings behave in the face of the adversity of life and society, it has always raised questions about the motivations that people have, when they provide help to others regarding the relationship between parenting practices and the characteristics of personality, and the regulation of social human behavior (Fernandez 2013), avoids the responsibility that every subject is at the mercy of the unconscious forces emphasizing the behavior that appear in the pathological personality and the conscious and unconscious conflicts that give rise to Affected childhood experiences for your entire life. Currently human behavior manifests actions, reactions, as it develops through the framework, family, social and educational, where ancestral values, knowledge, skills, aptitudes and customs are acquired. However, the technological globalization presented by education and learning methods have not had a significant advance, the traditional is still used even when there are tools to innovate in the teaching process, avoiding the poor performance of learning in students (Sorokin and Sotomayor 2016). In the province of El Oro-Ecuador, due to the complexity of teaching practices and pedagogy, innovation and the search for new ways to energize the teaching chair with teaching resources according to the scope, culture and idiosyncrasies of different societies are necessary. This perspective must be considered, the teaching practice in the development of competences framed in the ability to design permanent and meaningful learning experiences. In which students are the central point of the teaching-learning process through the correct use of ICTs and towards a digital culture to face the new challenges of society around the widespread use of ICTs that affect the educational practice and the social human behavior of the student, of poor academic performance, where teachers must use innovative methodologies that facilitate the improvement in the teaching of students based on learning paths (García 2014). The present work will allow the use of ICTs to help develop new ways of learning, communication and ways of interaction with the sources of knowledge and knowledge as vehicles that promote the circulation, use, access, representation and creation of information that the educational units need. The CEPWOL Altamira Individual Educational Unit of the city of Santa Rosa, El Oro province, with teachers of Basic General Education, shows that there is disinterest in learning in students, presenting the need to innovate in the teaching-learning process in routes of learning that improve academic performance. Faced with the difficulties in treating social human behavior, this research is framed, in learning paths to improve performance, ensuring that the topics that have the greatest difficulty such as dyslexia, dyscalculia, hyperactivity, are taught technologically and achieve meaningful learning; Contribute to the improvement of the teaching-learning process of the students of the CEPWOL Altamira Private Educational Unit. At present, the motivation of the students is key to the change of social human behavior in the learning paths, by sending digital assignments to students and parents. To reinforce the positive learning results, these individual or group power, thanks to an entry code that allows facilitating and recognizing the student entering the class and the type of activity that must be performed, in Fig. 1 you can show the motivation of the students and teachers of the one used by ClassFlow in the classroom.

Fig. 1. ClassFlow

Assignments of activities can be individual or group of a whole class or for specific activities such as educational needs (curricular adaptations), the teacher has a tool that can create badges for valuation of personalized points or select certain students or the whole class, promoting collaboration and educational self-management skills. Class-Flow and its different activities that allow adapting to any need or learning styles that students have, to improve social human behavior in the subject of mathematics such as the recognition of study material, attention and concentration (memory games), organization of information (Visual Learning), alphabet soup that reinforces vocabulary (reading Writer), research (Cause-effect), problem solving by crossword puzzle, definition of objectives, creation (critical and logical criteria), among other benefits that can be found in the cloud (ClassFlow 2018).

2   Art State

Inquiries about social human behavior: performance indicator based on learning paths, during the research process you can write down some topics related to this work at the Technical University of Ambato and other journal articles, like one of the research variables raised. Social skills and child prosocial behavior from positive psychology to the analysis of the results of de (Lacunza 2015), in their article conclude that: "The identification of skills will favor socialization with peers makes it possible to promote healthy social skills and prosocial behaviors, basic resources for the positive development of the child" (p. 6). Social human behavior, are situations that are experienced or referenced in programs or novels that are normally imitated forming a copy copied by the viewer. (Condemaita-Díaz 2017) in his work the reality program "Calle 7-Double Temptation" and its influence on the behavior of the students of Tenth year of the Bolivar Educational Unit states the following: It is concluded that in the classroom the behavior of the students is low and inadequate because there are conflicts between classmates, the norms of institutional coexistence, cultural differences, there is presence of verbal aggressiveness, lack of values and disrespect to the basic norms of behavior for the educational institution (p. 144). (Yépez 2010) in his job the absence and its influence on the Learning path of the first year of basic education at the Alberto Albán Villamarín school in the community of Noetanda states the following: It concludes that the children mostly do not match the tasks received in the learning path when they have failed, in addition these children who are missing (Zambrano-Ríos 2017) classes find it difficult to discuss the topics covered in the route learning (p. 55). In his research report on the Factors in the socio-affective behavior of children aged 3 to 4, he states the following: It concludes that family problems are stressors that influence the development of the child causing complications in the control of sphincters, as well as the image of the child, aspects that alter the child's behavior socially and affectively, also altering family well-being (p. 56). (Jácome Mayorga 2017) in his thesis report Learning routes and communication skills in children in fourth and fifth years of basic education states the following: It concludes that the lack of institutional self-management does not allow teachers to be trained in a continuous and systematic way on these topics of innovative learning strategies, so they are only directed by the guidelines given by the Ministry of Education, but it is also the lack of interest of each of the teachers to innovate, so that the learning outcomes are not ideal, teachers have a profile that is not competitive at the local or provincial level, because they present many gaps throughout their school education (p. 142). The authors (Galarsi et al. 2011) in the article Behavior, history and evolution conclude that: For most of human history man has considered himself as a superior being completely different from animals. But if you consider the contributions of Charles Darwin, who suggested that throughout the evolution we have maintained blood ties with other species. Today, a century later, this relationship is admitted by many thinkers. The information society requires that students learn through the use of ICTs, and in turn be protagonists of their learning, as the human being performs his activities consciously and unconsciously, it is intended that the behavior be related to knowledge, where the student develops his future through a critical paradigm because the educational reality seeks solutions to the problem between learning and social human behavior that is currently being lost, values and personality. The educational processes in the different studies show that the tools offered by ICTs can help in an effective and effective way to find better learning paths for the construction of knowledge, objective experiences, transforming the student's thinking for a better human behavior in the current society. Education today is facing specific and cultural problems, which basically refer to the need for the use of the most modern computer technologies in order to meet quality standards and promote a digital culture that represents the "Information Age" And that has a decisive impact on the specific objectives of current Education.

3 Methodology

Research is part of human behavior in general and therefore knowledge has been defined as a process in which a cognitive subject (who knows) is related to an object of knowledge (that which is known) which results in a product mental new, called knowledge. Thus, the same term designates the process and the result of said process; that is, we call the subjective operation that produces it, as well as the product itself, knowledge. (Rodríguez 2011). It affects social human behavior in education as an indicator of performance based on learning routes for students of Basic General Education of the CEPWOL ALTAMIRA Private Education Unit of the city of Santa Rosa in the province of El Oro. The modality, on the other hand, allows an approach to the problem of study, but with the actors of the educational community, through the collection of information of knowledge, experiences and information that parents, teachers have authorities on the communication strategies executed, evaluating through of quantified indicators, and analysis with the participation of the study group (see Table 1).

Table 1. Independent variable

INDEPENDENT VARIABLE: Social Human Behaviors				
Conceptualización	Category	Indicators	Items	Techniques and instruments
Human behavior has evolved over the years, which is why we can realize that human behavior has different characteristics and values, they have been one of the causes of the deficit that exists in human, social, and social performance. Each environment to which a person belongs can help in their social growth and in turn give attitudes, emotions that allow them to continue in a society that today requires thinking people, to help show change	Human Conduct	There are different types of elements that intervene in human behavior	Are the elements of computational tools necessary for the development of activities necessary?	Survey Questionnaire Resources
	Human Performance	Usability of the tools in human performance	How do you determine the usability of the teaching material?	
	Types of behaviors	Applications development	Do you think that the study of types of behavior is necessary?	
	Evaluation of behavior and beliefs	Use of tools for the study of behavior and its beliefs	Through which medium has the use of tools for the study of behavior developed?	

Table 2. Dependent variable

Dependent variable: Learning routes

Conceptualization	Category	Indicators	Items	Techniques and instruments
Methods used to obtain some type of knowledge, using learning strategies through the technological use in education, the temporary data, which show and record changes through a significant number of relevant dimensions, which determine the efficiency and effectiveness of the skills and knowledge through learning methods, with the different teaching resources, which allow the generation of learning indicators	Learning strategies	Skills Development	Is it necessary for teachers and students to develop skills?	Survey
	Technological use in education	The Tic in the formative process	When is it important to use the Tic in the learning path?	Survey
	Skills and knowledge	Technical methods in the teaching process	How should teachers' techniques be?	Survey
	Learning methods	Applicability of strategies	Is learning environments necessary?	Survey
	Types of didactic re-courses	Development of educational resources	How do you identify teaching resources?	Survey
	Types of indicators	Search for skills	Is it necessary for teachers to look for competencies?	Survey

Research conducted on a sample of subject's representative of a larger population, which is carried out in the context of daily life, using standardized interrogation procedures, in order to obtain quantitative measurements of a wide variety of objective and subjective characteristics of the population. The survey is aimed at students, authorities and teachers of the CEPWOL Altamira Individual Educational Unit, whose instrument is the questionnaire, prepared with closed questions that allow the study variable to be collected. The questionnaire allowed to gather information with open and closed questions established beforehand, they are always posed in the same order and formulated previously preparing and strictly standardized. The research instruments were subject to criteria of validity and reliability. The validity is given through the technique "expert judgments"; while the reliability was carried out through the application of a pilot test to a group of students and teachers with characteristics similar to the established sample, allowing to detect errors in the understanding of the questions and the selection of answers to correct before your application. The internal consistency method based on Cronbach's alpha, to estimate the reliability of the measuring instrument through a set of items that will be measured in the same theoretical dimension by applying the SPSS software (Table 2).

4 Experimentation

To carry out this research, it has been considered that we will work with 3 authorities of the institution, 5 heads of areas, 15 teachers and 100 students from all areas of the CEPWOL Altamira Private Educational Unit of the Province of El Oro. Cronbach's alpha reliability result is 0.886 of the questions raised. Through the development of the research, the ADDIE model was considered because it contains the theories of behaviorism, constructivism, social learning and cognitivism that will help shape and define the result of the materials that ClassFlow possesses for the different activities that were carried out. Figure 2 shows the progress made using the ADDIE methodology, adapting its phases to the needs of the research to achieve the scope of the proposed objective through learning paths to obtain positive social human behavior in the subject of mathematics (Figs. 3, 4, 5, 6 and 7).

4.1 Students

Are computational tools necessary for the development of social activities?

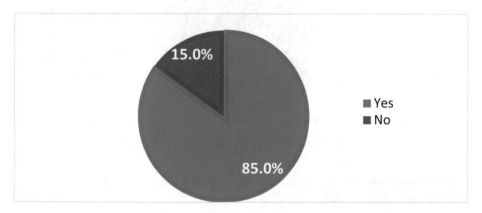

Fig. 2. Computational tools for the development of social activities

Alternative Frequency Percentage
Question When is it important to use the Tic in the learning routes?

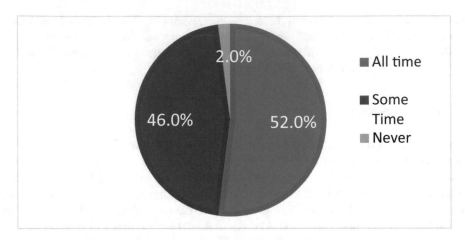

Fig. 3. Importance of ICT in learning routes

Question How should teachers' techniques be for learning processes?

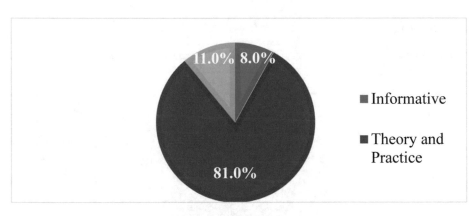

Fig. 4. Teaching techniques for learning processes

Question. Does the teacher handle learning by discovery as the result of good teaching on learning paths?

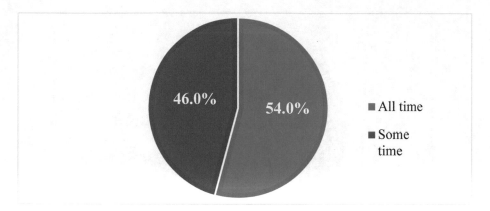

Fig. 5. Learning by discovery on learning paths

4.2 Teachers and Authorities

Question Does the student know that meaningful learning is the result of teaching on learning paths?

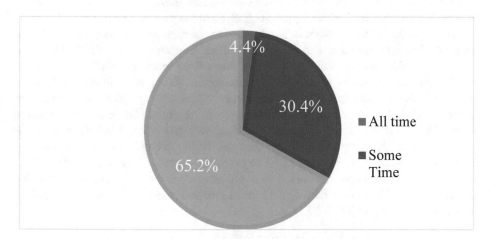

Fig. 6. Meaningful learning results from learning paths

Question. Is it necessary for teachers to develop teaching skills and learning strategies?

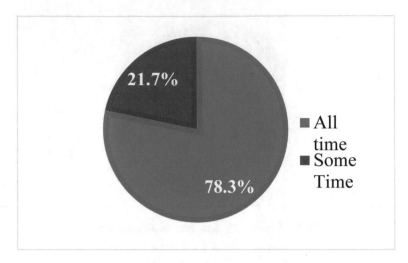

Fig. 7. Skills development and teaching learning strategies

Hypothesis Approach

Ho = Social Human Behavior Does not affect education as a performance indicator based on learning routes for students Basic General Education of the CEPWOL Altamira Private Educational Unit of the city of Santa Rosa in the province of El Oro (Fig. 8).

H1 = Social Human behavior: Yes It affects education as a performance indicator based on learning paths for students Basic General Education of the CEPWOL Altamira Private Educational Unit of the city of Santa Rosa in the province of El Oro.

Normal distribution table the level of confidence of 95% was verified, and the level of significance of 5%, graph 24 illustrates the values corresponding to z and α in graph 25 the standard normal distribution table is presented. The calculated z value ($z = -8,962$) is lower than the level of controlled confidence ($z = 1.96$), therefore, it is located outside the acceptance region, then the null hypothesis is rejected, and the alternate hypothesis is accepted that He says: Social Human Behavior: It affects education as a performance indicator based on learning routes for students of Basic General Education of the CEPWOL Altamira Private Educational Unit of the city of Santa Rosa in the province of El Oro.

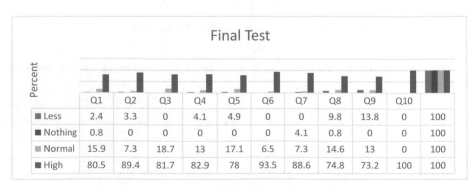

Fig. 8. ClassFlow

5   Conclusions

The evaluation of the survey was conducted empirically to determine if the tool improved Social Human Behavior in the subject of mathematics in General Basic Education students. The highest percentage of students and teachers have experienced changes in the learning process, both in behavior and social human behavior. In the majority of respondents the use of the ClassFlow tool contributes to the learning of collaborative and group work, through the exchange of knowledge between students and teachers (Lacunza 2015). The vast majority of teachers must take into consideration that the students' social human behavior is the basis for better academic performance and performance in any subject. Finally, most consider that the social human behavior of students with the use of ClassFlow, have improved their social behavior, along with their skills and strengths, have shared and worked collaboratively and in a team, which allows them to apprehend significantly, permanently, both inside and outside the institution, improving their social and personal relationships.

References

Fernandez, R.L.: Tendencias Históricas y contemporáneas: Roptura, 12 Noviembre 2013. http://roa.ult.edu.cu/handle/123456789/2114. Último acceso 12 Noviembre 2017

Sorokin, P., Sotomayor, M.A.: Ciencias Sociales Humanas y de Comportamiento: Dificulatades regularias en países latinoamericanos. Revista de la Facultad de Ciencias Humanas de la Universidad Autonoma de Colombia **14**(1), 23 (2016)

Hérnandez, L., Acevedo, J., Martínez, C., Cruz, B.C.: El Uso de las TIC en el aula: un analisis en terminos de efectividad y eficiencia. de Iberoamericano de Ciencia, Tecnología Innovación y Educación, Buenos Aries (2014)

García, D.M.: Enseñando aprendez Más: estrategias de Aprendizaje. CIDUI (2), 9 (2014)

ClassFlow: Herramientas Colaborativa-Nube (2018). https://classflow.com/es/features/

Lacunza, A.B.: Las habilidades sociales y el comportamiento prosocial infantil desde la psicología positiva. Pequén **12**, 20 (2015)

Condemaita-Díaz, C.E.: El programa de Reality "Calle 7- Doble Tentación" y su Influencia en el Comportamiento de los estudiantes de Décimo curso de la Unidad Educativa Bolívar, Ambato (2017)

Yépez, A.A.: La inasistencia y su influencia en la ruta del Aprendizaje del primer año de educación básica en la escuela Alberto Albán Villamarín de la comunidad de noetanda, Ambato (2010)

Zambrano-Ríos, M.F.: Factores en el comportamiento socioafectivo de los niños de 3 a 4 años, Ambato (2017)

Jácome Mayorga, M.E.: Las rutas de Aprendizaje y las habiliades de comunicación en los niños de cuarto y quinto año de educación básica, Ambato (2017)

Galarsi, M.F., Medina, A., Ledezma, C., Zanin, L.: Comportamiento, historia y evolución. Redalyc (24), 36 (2011)

Rodríguez, J.M.: Métodos de investigación cualitativa. Revista de Investigación Silogismo **1**(08) (2011)

Multimedia Systems and Applications

Navigation of Resources from Tangible Object Recognition to Improve Virtual Tours in Botanical Gardens

D. M. Muñoz-Araque[1(✉)], Maycol Hernandez Garcia[1(✉)],
Paulo Gaona Garcia[2(✉)], and Carlos Montenegro[2(✉)]

[1] Universidad Distrital Francisco José de Caldas, Bogotá, Colombia
{dmmunoza, mhernandezg}@correo.udistrital.edu.co
[2] Facultad de Ingeniería, Universidad Distrital Francisco José de Caldas,
Bogotá, Colombia
{pagaonag, cemontenegrom}@udistrital.edu.co

Abstract. In order to create an immersive application that allows access to digital resources through the creation of a navigation structure based on recognition of tangible objects, a study is presented where a group of 12 people interact with the proposed model and evaluate the experience with the immersive model. The results showed a great acceptance in terms of ease of use and comfort surpassing the average usability scale system, the aspects to be taken into account has to do with the functionalities that are expected to implement according to the comments of the participants so it is concluded that the development of the application meets expectations. It is expected to add more features with the motivation of achieving a great reception by the community taking into account the future integrations and learning that makes part of the possibilities that this technology offers both visitors and researchers in the areas addressed by the implementation of this application.

Keywords: Tangible object recognition · Knowledge organization systems · Immersive application · Usability

1 Introduction

Immersive technologies offer people support tools for the processes required in different fields, achieving a change of mentality, transforming traditional techniques and focusing attention on providing personalized assistance, more agile and comfortable [1–3]. We can see them in countries such as Spain or Mexico, where they are encouraging the knowledge of botanical gardens in the community to highlight the sense of belonging to these places. In the Royal Botanic Garden of Madrid, an innovative mobile application has been implemented, being awarded as the best national tourist application. "RJB Museo Vivo" tries to improve the experience of visitors by allowing them to make an interactive visit to the garden [4].

The following study aims to establish methods of interaction based on recognition of tangible objects with digital resources to improve the experience in the tours that are

Á. Rocha et al. (Eds.): ICITS 2020, AISC 1137, pp. 525–534, 2020.
https://doi.org/10.1007/978-3-030-40690-5_51

made on a botanical garden and be the basis to facilitate processes of association of concepts through the integration of taxonomies of plant species. Thus, the following article presents an immersive prototype based on recognition of tangible objects, to carry out virtual tours on some plant species taking as a case study the Botanical Garden of Bogotá "José Celestino Mutis" as one of the research centers in botany and scientific development with the greatest trajectory in Colombia [5] center that contributes to the conservation of the district's flora, the main objective is to consolidate itself as a place for science, with its modernization plans to be recognized in Latin America [6].

One of the motivations of this study is to determine if the technique of Tangible Object Recognition facilitates the process of navigation together with the association of concepts of plant species to facilitate the learning and training process presented to both garden visitors and students.

The following section deals with the background of projects focused on the application of different techniques that provide people with information and learning, later on the methodology section presents the stages of development of the application for the study together with the advances of the proposed model and the tests that are intended to be applied to people and finally the section of conclusions where the evaluated characteristics and the results of the study are presented.

2   Background

The technology is changing the way of visualizing the information, taking into account the different places of cultural interest and the great amount of information that they can provide to their visitors, the interest of the people is focused in new experiences modifying in many occasions the perspective with which they were coming to know different places like museums, classrooms, libraries among others. With emerging technologies such as Virtual Reality, visitors get great advantages from devices with interactive tools. The use of immersive graphical interfaces would allow a better understanding of the information in people, this helps in the process of resource recovery, in the search for specific terminology and when generating systems that last in time, facilitating information to the next generations. Innovation in the development of creative content in projects such as GVAM (Guías Virtuales Accesibles para Museos) has improved the experience of visitors [7].

The use of these technological tools makes it possible to explain concepts and disseminate information that would bring people closer to knowledge and learning in museums, thus solving problems of didactics and communication. This information is present in videos, infographics, books, photographs, animations that dynamize the visit to the museum [8, 9]. In a study conducted at the Shihsanhang Museum in China [10], where different variables are taken into account with the implementation of a virtual reality system by visitors, the results concluded that this implementation is perceived satisfactorily, improving the understanding of the content that the museum provides to users [11]. Taking into account the rise of technologies such as Virtual Reality and its various applications in various environments (such as the field of video games, entertainment and simulations), establishing a change at the educational level is also possible being a field of great importance in society so the use of information organization systems (structures that will allow the association and conceptualization of information on each subject of interest depending on the field of knowledge you want to know) and the navigation structure coupled with immersive interfaces would allow easy and entertaining access to users [12]. It is necessary to establish usable and accessible interfaces to complete the learning process as well as the correct visualization to present the information coherently and clearly when searching by context, how the structure is presented will allow a better understanding in people regarding navigation and access to virtual resources [13].

Many times new challenges are found that immersive tools can satisfy, virtual museums (VM) are being developed with the initiative to promote the cultural heritage of cities with cultural and historical richness. The development of an eco-museum, for example, which is proposed in the city of Tunja to communicate the unknown heritage of this city, encourage appropriation and preservation will allow the construction of a site with information on culture, history, values and traditions through a virtual journey [3].

The realization of inventories is also an interesting application, especially when applying it to fields such as the environment, an example of this is the design of an interactive and multi-touch map of tree survival, which carries out an inventory of trees and their complete study (survival, location, classification, identification of protected species) without the need for a tedious and inefficient physical record, with the option of the multi-touch table, people can access that information and be evaluated with a good acceptance by users [14].

3 Materials and Methods

The analysis of the development for the creation of the graphic interface of the application requires the following steps that allow the implementation of a navigation model from the recognition of tangible objects, taking into account the information on plant species.

3.1 Analysis and Design Approach for Navigation in the Application

The immersive interface approach is generated from 3 main modules: Map, Zones and Garden Routes, in addition to interfaces such as the Main Menu and Application Information (Fig. 1).

Module - Garden Map. The user accesses the garden map from the main menu interface, where the user interaction with the application aims to show general information about the garden, taking into account meeting points, areas and collections that are on display and services offered within the place as a cafeteria, nursing among others (Figs. 2 and 3).

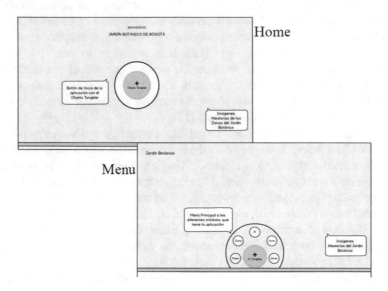

Fig. 1. Diagram of the initial interface "Home" and "Main Menu".

Module - Garden Zones. In the Garden Zones module, there is an approach to each of the collections and exhibitions that the garden has, improving the knowledge of the specimens in the user.

Module - Garden Tours. The routes of the garden, present an interface similar to the module of Map of the Garden, inside this interface the user will find different options of routes inside the place and in the map, the way is visualized to follow.

Finally, the user finds the help and information interface provided by the application for a better understanding of the use.

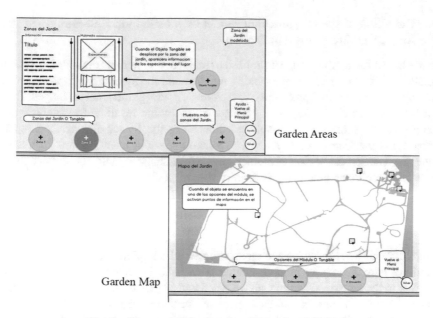

Fig. 2. Diagram of the interface "Areas" and "Map".

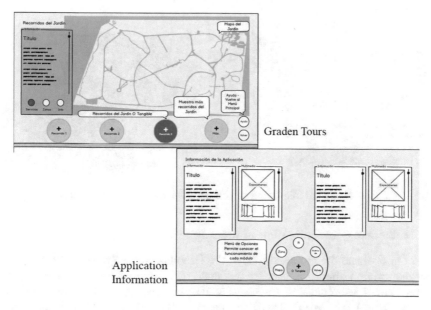

Fig. 3. Diagram of the interface of the "Tours" and "Information" module.

3.2   Definition of the Organization Model of Information Relevant for Navigation in the Application

For the test of the application, we propose to manage a system or model of information organization based on taxonomies taking into account the system that is managed with plant specimens. In the prototype, two zones of the garden are managed for the moment, with which the tests will be made to know the result of the analysis of the usability of the same one. The selected Zones are the Systematic of Cryptogams - Cascade and the Garden of Wetlands. The taxonomy will have 3 levels of hierarchy where the Garden Zones are presented, in turn, associated with the collections and finally the specimens belonging to each of these.

3.3   Development of the Interface Design Using Different Technological Tools to Generate an Immersive Interaction Model

The multitouch table in which the application will be implemented has the possibility of linking with the Unity graphic engine where the interface and the zones corresponding to the Botanical Garden are designed and modeled with the aim of generating scenes of the real zones and of each specimen in the garden, which allows the information presented in the application to be recognized in the path (Fig. 4).

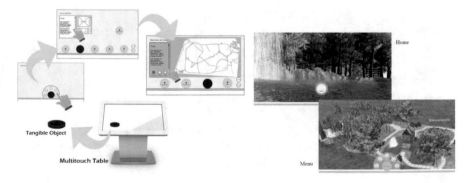

Fig. 4. Visitor interaction as you get immersed in the application.

3.4   Methodology of Evaluation of the Application

In order to evaluate the proposed model, the Usability Scale Systems [15] are used to know the opinion about the users' experience, is a very useful tool in the usability measurement of an application, by means of 10 questions that are qualified from 1 to 5, being the value 1 is "Strongly Disagree" and 5 is "Strongly Agree". The questions are as follows:

(1) I think that I would like to use this system frequently.
(2) I found the system unnecessarily complex.

(3) I thought the system was easy to use.
(4) I think that I would need the support of a technical person to be able to use this system.
(5) I found the various functions in this system were well integrated.
(6) I thought there was too much inconsistency in this system.
(7) I would imagine that most people would learn to use this system very quickly.
(8) I found the system very cumbersome to use.
(9) I felt very confident using the system.
(10) I needed to learn a lot of things before I could get going with this system.

4   Results

The SUS Usability test resulted in an average score of 76.46 which was calculated taking into account the responses of each of the volunteers who performed the usability test. This score sets the test at a 70% percentile which places it in category B indicating an acceptable status [16] for the first version of the program. The study was conducted with 12 people between 18 and 30 years and complement the contributions given in terms of their experience with the application with great acceptance and expectations in its final implementation (Table 1).

Table 1. Results of the System Usability Scale tests

N	SUS	Learnability	Usability
1	60,0	7,5	52,5
2	97,5	17,5	80,0
3	92,5	20	72,5
4	57,5	5	52,5
5	77,5	12,5	65,0
6	92,5	20	72,5
7	82,5	12,5	70,0
8	45,0	7,5	37,5
9	77,5	15	62,5
10	72,5	15	57,5
11	80,0	12,5	67,5
12	82,5	17,5	65,0

Table 2. Collection of SUS test result

#Question	N	Maximum	Minimum	Mean	SD
1	12	5	3	3,81818182	0,63245553
2	12	4	1	2	1,26491106
3	12	5	3	4,27272727	0,78624539
4	12	5	1	2,72727273	1,34839972
5	12	5	2	3,63636364	0,80903983
6	12	3	1	1,45454546	0,687551651
7	12	5	2	4,45454545	1,03572548
8	12	4	1	1,81818182	1,07871978
9	12	5	1	4,18181818	1,25045446
10	12	4	1	2	1

The table above shows the details of the test performed, the SUS that was obtained from the responses of each of the volunteers where learning and usability are measured based on their experience with the application. For this, the question style was taken into account and what topic is covered in terms of the application; Questions 4 and 10 allow us to obtain the dimension of the capacity of learning of the volunteer, as well as question 2, in exchange the other questions manage to establish the capacity of use that can be had of the application [16] (Table 2).

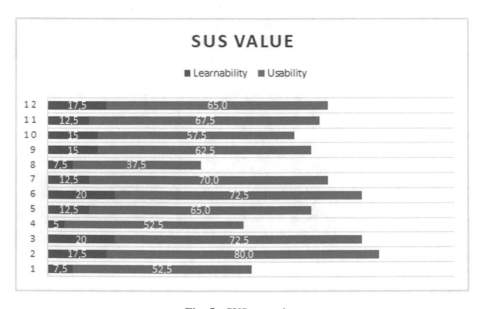

Fig. 5. SUS test values

Taking into account Fig. 5, the way in which the application becomes accessible to users is evidenced, since the relationship between usability and the learning shown by each of the volunteers with their responses indicates the ease at the time of interaction human-machine allowing people to be able to properly handle the application with a slight external help that allows them to take full advantage of the application's capabilities without having to obtain a large amount of knowledge to achieve its use.

5   Conclusions and Future Work

A prototype was made based on techniques of recognition of tangible objects to establish the degree of acceptance of users with the graphical interface according to the case study and the promotion of knowledge that is intended to be done through interaction.

The results were related to the objective of the study to establish the usability of the application. Based on the results obtained, it is possible to corroborate that this type of proposal facilitates the processes of interaction between users and the immersive model where the implementation of the information is intended to be an important part of the objective of the project taking into account the knowledge that the place offers to the community as expressed by the participants of the test, Likewise, there are possibilities to establish strategies that allow to interact with the garden's resources, to carry out training processes through evaluation elements that can be carried out interactively, or in a complementary way with the purpose of strengthening themes that are being approached for different uses by the visitors. This study allows to know the point of view of each volunteer and to establish different situations both positive and of improvement of the application.

Taking into account the contributions of users in the study, it is proposed to address more features that complement the experience. As a future work, the "Tours" module will be developed that will allow visitors to provide a new support tool in the discovery of the garden, informing of the different themes that are exposed, improving their interaction and location in place with the specimens through new technologies such as georeferencing and the implementation of methodologies that allow captivating the public with learning.

References

1. Arredondo, J.: Augmented Reality as a Library Tool in University Libraries. Secretary of Public Education, México (2017)
2. Cervera, A.: The augmented library: an approach for improving users awareness in a campus library. IEEE (2017)
3. Llanos, J.: Is it possible a virtual-eco-museum? Digital Heritage de IEEE (2015)
4. Europa Press: The app 'RJB Museo Vivo', awarded as the best tourist application. https://www.europapress.es/turismo/fitur/noticia-app-rjb-museo-vivo-premiada-mejor-aplicacion-turistica-nacional-cultural-20180119140641.html. Accessed 18 Nov 2018
5. ElTiempo: Archivo. http://www.eltiempo.com/archivo/documento/CMS-16465510. Accessed 18 Nov 2018

6. Diazgranados, M.: The Botanical Garden of Bogotá José Celestino Mutis is anniversary. ResearchGate (2015)
7. Gaona García, D., Martín-Moncunill, P.A., Sánchez-Alonso, S., Fermoso García, A.: A usability study of taxonomy visualisation user interfaces in digital repositories. Emerald Insight **38**(2), 284–304 (2014)
8. Arroyo-Vázquez, N.: Experiencias de realidad aumentada en bibliotecas: estado de la cuestión. Universidad de Barcelona (2016)
9. Gurri, J., Carreras, C.: Realidad virtual en nuestros museos: experiencias de la colaboración entre Dortoka y el grupo Òliba. Internet Interdisciplinary Institute IN3 (2003)
10. Li, P.-P., Chang, P.-L.: A study of virtual reality experience value and learning efficiency of museum using Shihsanhang museum as an example. IEEE (2017)
11. Castellanos, A.: Augmented reality for improved visit to the Telecommunications Museum. Universidad Politécnica de Valencia (2018)
12. Garcia Gaona, P.A., Martin Moncunill, D., Gordillo, K., Gonzalez Crespo, R.: Navigation and visualization of knowledge organization systems using virtual reality glasses. IEEE Latin Am. Trans. **14**(6), 2915–2920 (2016)
13. Martín-Moncunill, D., Garcia Gaona, P.A., Rajabi, E.: Navigating through the Linking Open Data cloud datasets: preliminary ideas of a visual search tool based on its defined domains. IEEE (2015)
14. Quintino, B., Ávila, M., Ávila, F., Bianchetti, M., Gaona, E.: Design of an interactive and multi-touch tree survival map. Tecnológico Nacional de México (2017)
15. Usability.gov: System Usability Scale (SUS). https://www.usability.gov/how-to-and-tools/methods/system-usability-scale.html. Accessed 08 Sept 2019
16. Measuring Usability with the System Usability Scale (SUS): 5 Ways to Interpret a SUS Score. https://measuringu.com/interpret-sus-score/. Accessed 8 Sept 2019

The Potential of Social Media to Enhance Cultural Adaptation: A Study on Iranian Student in the Finnish Context

Zahra Hosseini, Sirkku Kotilainen, and Jussi Okkonen[✉]

Faculty of Information Technology and Communication, Tampere University,
33014 Tampere, Finland
{zahra.hosseini, sirkku.kotilainen,
jussi.okkonen}@tuni.fi

Abstract. This paper aims to find out the adequacy of social media as a trendy technology application to enhance cultural adaptation among Iranian tertiary-level students in Finland. For this purpose, integrative communication theory adapted as the research framework to study the effect of environment and investigate the approach of social and individuals' communication in the process of cultural adaptation. Accordingly, 23 students of the different universities in Finland through a snowball sampling method were interviewed and their opinion, attitudes, feeling and experienced were recorded and thematically analyzed. The results of the study indicated the potential of social media for individual communication among the participants. Iranian students are easily communicating with their family and friends inside and outside of Finland. It assists to reduce the intercultural stress of the participants. Further, social media opens the door for students (or immigrants) to communicate with their compatriots in Finland when they need some information or guidance. However, some participants of this study do not believe the adequacy of the current function of social media for enhancing their cultural adaptation. While Friendship with Finnish is a challenge for Iranian students, social media could not provide a route to facilitate communication between Iranians and Finnish society. Whereas social media is potential to facilitate exchanging the information for a minor community in a host country and providing acting as a bridge to communicate immigrant with local people, it is substantial to find the right approach that social media could enhance cultural adaptation through these functions.

Keywords: Social media · Cultural adaptation · Finnish culture

1 Introduction

To understand the internationalization of higher education in the society, examining the mobility of international students and their cultural integration process are becoming increasingly important. Between 2000 and 2011; the number of international students has more than doubled in the world. In 2016, more than 4.85 million students are enrolled outside their country of citizenship [1]. The largest numbers of international students are from China, India, and Korea. Asian students account for 53% of all

© Springer Nature Switzerland AG 2020
Á. Rocha et al. (Eds.): ICITS 2020, AISC 1137, pp. 535–549, 2020.
https://doi.org/10.1007/978-3-030-40690-5_52

students studying abroad worldwide. OECD [2] estimated an approximate 3.7 million students to have been studying outside their respective home countries in 2011, and the numbers have increased since (p. 320). Many scholars have investigated the mobility of international students and the reasons behind that [3–6]. One study [7] highlighted eleven factors that influence students' decisions to leave their home countries to study abroad. These factors include mobility costs, employment and income in the host country, geographical distance, host country climate, quality of host institutions, financial aid and future career perspective, with factors such as economic status of the home country, language and intercultural training of home institutions, parental influence and personal interest in mobility having been identified as push factors.

Regardless of the reasons for the mobility, international students are facing many challenges during their time abroad, in both academic settings and in their new societal environments. The researchers [8] in particular, looked at students studying in the U.S., China, and France, and revealed these international students to face a myriad of issues in their new host societies, including family problems, financial difficulties, psychological problems and socio-cultural problems. Cultural adjustment appears to be particularly problematic and has been widely studied as a key problem with tertiary-level international students in countries such as Japan [9], Spain [9], the U.S. [11–16] and in African countries [17, 18]. Of the extensive research conducted, certain studies emphasized inadequate language proficiency to be a major issue when examining transnational spaces [14, 19–21]. Lack of language proficiency may affect the international students' communication within academic environments, and may cause them to feel disconnected from their peers or classmates. Further, this may affect students' academic activities and increases interpersonal distress.

Accordingly, studies on the process of cultural adaptation is considered significant and because of the importance of communication in the process of cultural adaptation, the scholars attempt to develop the framework to interpret the dynamism of communication in the process of cultural adaptation. Kim [22] believes that attention to environment and individuals' communication is the missing keys in the literature of cultural adaptation. In this regard, Integrative Communication Theory suggested by Kim offers a "big picture," a systemic insight into what happens over time when someone crosses cultural boundaries and what factors facilitate or impede his or her adaptation to the host culture. In this theory, Cross-cultural adaptation is defined as "*the entirety of the dynamic process by which individuals who, through direct and indirect contact and communication with a new, changing, or changing the environment, strive to establish (or reestablish) and maintain a relatively stable, reciprocal, and functional relationship with the environment*" [22, p. 31].

With the growth of communication facilities in the last two decades, social media have become part of the environment. Therefore, analyzing individual and social communication through social media application in the process of cultural adaptation has received scholars' attention. Many studies have highlighted the potential of the social network to aid cultural adaptation of immigrants and students [23–30]. The researchers [31] found how social network usage increased the cultural adaptation among beginner Chinese students in the United States. In the same context, the other research [32] indicated Facebook and Renren reflected Chinese students' cultural orientation, language proficiency, and length of stay in the United State. These results are supporting the another findings [33] that indicated the students who were using Facebook demonstrated a lower degree of acculturative stress and a higher degree of psychological well-being compared to other groups in the study.

2 Context of the Study

This study conducted on the context of the Finnish-Iranian society and Iranian students are the participants of the study. Although many indices of the Iranian population appear similar to those of developed countries, the rate of immigration to Iran is negative [34]. An average of 300,000 people leaves the country annually, placing Iran in the 15th position among 193 countries in terms of net migration rate. In addition, a large number of Iranians move aboard each year for the purpose of their studies. Based on UNESCO data (2012), Iran has the largest number of students studying abroad among the Middle Eastern countries after Saudi Arabia. An estimate by the Iranian Ministry of Education placed between 350 and 500 thousand Iranians to be studying outside of Iran as of 2014 [35].

Finland as the host country is not a priority destination for International students because of the language barrier and geographic condition [36]. However, the quality of Finland educational system is known over the world, which has raised international interest [37]. It has caused the number of international students after 2005 has increased by over 10% per year as in 2015 more than 20 thousands foreign students enrolled in Finland's universities university of applied science (UAS). *"In total, 76% of international students in Finnish universities came from outside of the EU/EEA countries in 2015. In UAS's, the number is even higher, 80% coming outside of the EU/EEA, whereas in universities the figure is 73%"* [38].

3 Method

The participants of this study were from the different universities in Finland and the snowball sampling method was employed to find Iranian tertiary-level students in Finland. In phenomenography, the size sample of respondents continually increases to achieve the maximum variation regarding the data saturation [39]. The sample size was 23 master and Ph.D. students. The study utilized a semi-structured interview protocol to uncover the interviewees' values, feelings, beliefs, and experiences of the phenomenon. The interview protocol included questions about demographic information, participants' conceptions and experiences of Finnish society and their conceptions and experiences of studying in Finland and how social media was used and in their daily life to learn a new culture.

The participants were from the different universities in Finland. They were included 23 students including 14 male and 9 female 14 of them were studying in master's degree and 9 remained were studying as PhD degree. The demographic information of the participants is shown in Table 1.

Table 1. Some participants' views about the reasons of membership in Finnish society

Participant	Statement
B1	Yes, I know myself as a member of Finnish society. From 1 to 10 I give myself 8. I think more is because of the friends that I have here
L1	I still don't know myself as a member in Finnish society. I don't know the Finnish language so I am still outsider
D1	I give one or 2 from 10 to myself. Because if I am not spending my time with Finn
A1	I see myself as a member of Finnish society. I give myself 6 from 10. Because it is important, that Finland develops
S2	I know myself a member of Finnish society. Because I am working
	If I give myself 4 from 10. If I knew the Finnish language, I was giving myself 8
B2	I don't know their language. Someone who is working or has a Finnish girlfriend is integrated
B3	I give myself 5 from 10. If I learn the Finnish language, it helps me
	If you have a job, you are more in the society
S3	I still don't feel a member of Finnish society. I am a member of the minority here and I hope the size of minority reduces by time. But still, I don't see myself as a member of society. I think I am 4 from 10

The location of interviews based on the selection of participants. Accordingly, the students who are studying in the Tampere University were interviewed face to face in the university campus and other students were interviewed through video skype. The interviews were conducted in the native language of the interviewees (i.e. Farsi) since the native language of the interviewer (one of the authors of this paper) is Iranian. This allowed for control of any culturally determined construct bias. Furthermore, this reduced the risk that the interviewer and interviewee would consider different factors to be of importance, as they may attribute different meanings and interpretations to the events/behavior described by the interviewee [40].

The interview data collected through audio recordings and the data were transcribed verbatim for analysis of the meanings, especially those between the sets of categories. Thematic analysis was employed as the method for analyzing the data. According to Braun and Clarke [41], "Thematic analysis is a method for identifying, analyzing and reporting patterns (themes) within data" (p. 79). This study utilizes the thematic analysis will be conducted following Braun and Clarke's (2014) guidelines. "Thematic analysis is a poorly demarcated, rarely acknowledged, yet widely used qualitative analytic method within psychology" [41, p: 77]. Accordingly, in the first step, the data were transcribed. Subsequently, data coded and categorized.

4 Results

Social media has a place in the daily activities of Iranian students indeed. They are communicating with Iranians, international and Finnish people through social media. They use the different applications for the different purposes. The theme of analysed data determined the characteristics of participants' communication in Finnish-Iranian society concerning social media usage. Further, the different social media applications for individual and social purposes in original and host countries are illustrated.

4.1 Communication in Finnish-Iranian Society

Regardless of the used application type for communicating, Iranian students who live in the Finnish-Iranian society face some challenges in either Finnish or Iranian society in Finland. In this section, Iranian students' feeling, ideas, thoughts, and experiences of communication with Finnish and Iranian people in Finland have studied. The participants' stress and adaptation in Finnish society is categorized as effortless, challenging and hard in communication and the role of social media in this process is highlighted.

4.1.1 Communicating with Finnish People

The result of verbatim analysis regarding the nationality of friends indicated that most of the participants of the study feel communication and making friends with Finnish people is not easy. Nine of 23 participants have more Iranian than other nationalities' friend and 8 of remained participants have more international friends than Finnish. In fact, only two of 23 participants have Finnish friends more than other nationalities. The notable results show that the years of living in Finland necessarily has not increased having Finnish friends. For example, "B1" and "B2" are almost at the same age, level of study, position in the university and both are in Finland for about 7 years. However, "B1" has many Finnish friends while "B2" does not have any Finnish friend and feel very far from Finnish students or colleagues. Findings of the study demonstrated three levels of conceptions about communication in Finnish society as follows:

(a) *Communication is possible and effortless:* Some participants are communicating and building a friendship with the Finnish people. They feel communication with Finnish people is easy and does not need much effort. Although the language is the most important tool for communicating but the level of their knowledge of the Finnish language is different. While, "B1" knows the Finnish language well, "L1" and "F1" still are the beginner in the Finnish language.

"B1" is a doctoral researcher. He has citizenship in Finland, knows himself as a member of Finnish society, and describes his experiences in communication in Finnish society well. He claims *"Most of my friends are foreigners. I liked to have more friends but I didn't have so much time. .. Now I am more with Finnish friends. ... When I came to open communication with Finnish, I lost that sense that these people are very different from us."* Concerning the role of social media to develop his cultural adaptation, he believes that social media did not have much effect in his adaptation and communication in Finland. He believes *"In fact, the culture of virtual communication is less in Finnish people than Iranians."* He trusts What's app application more than others although he is online on Facebook messenger for his friends to him Facebook is more useful to organize face-to-face meetings.

Similarly, "L1" enjoys contacting with Finnish people and feel learning from them, although after being 6 years in Finland still does not know the Finnish language. She prefers having Finnish friends than Iranian friends. She uses social media for communicating with her family in Iran, however, she dislikes the style of using social media among Iranians: "I *do not like to send some jokes or wasting time. I just share something that has meaning or communicate with others*". She believes social media did not help her to integrate with society as much as her Finnish friends did.

Likewise, "F1" who is playing Iranian music instrument in one Finnish band, appreciates Finnish culture and people and believes:" *They (Finn) are very open mind. I think they have learned that truth is not only what we think and we know. The human can have different thoughts.*" He uses social media to communicate with his family but believes social media cannot do much in the process of cultural adaptation. "F1" used to use social media particularly, Facebook more but he thinks the way of using social media in the first two years after entering Finland had made him more connected to the home country and resisting to be integrated into the new culture.

"A3" is working full-time during her master thesis knows herself as part of Finnish society and feel comfortable with Finnish people. She understands Finnish language but still not feeling so comfortable to talk in Finnish. She believes having good friends helped her more to integrate into the new culture than any software or application. She uses Facebook to ask her question from other students or Iranians but not forgetting knowledge or communicating in the new culture.

These participants found Finnish people close to them and feel comfortable spending their time with them. They do not feel a mental wall between themselves and their Finnish friends and sometimes they find themselves more fit with them than some of their compatriots (L1 and F1). They not only did not find social media for helping them to integrate the new culture but they think it could keep them more introverted and far from learning and experiencing the new culture.

(b) *Communication is challenging but valuable:* Some of the Iranian students would like to communicate with Finnish society but building friendship with Finnish people is challenging for them. "Y1", and "D1", are interested in communication with Finnish although do not have any Finnish friend.

"D1" does not know himself as a member of the Finnish society although he feels he is not an Iranian anymore. He believes "*There are many things that I accept in Iranian culture. I appreciate the culture of respect in Iranian society and also the culture and art that Iranian have are very powerful.*" He prefers to communicate and friendship with Finnish but more of his friends are international students. He has a negative thought about social media and feels in the new world the media is controlling people but still, he uses that to know about Finnish culture. For that, he has participated in a Finnish page on Facebook: "*I am in the group that has 20 thousand Finn. I see how they are communicating. That helped me. I use google translate. I am one of the few members who are not Finnish. I don't suggest that link to others. Because I am patient but others may leave the page. ... It helps me to integrate. It has cheap content but it is active.*"

Likewise, "Y1" studies Ph.D. program in the second year and does not have any Finnish friends yet while she has a good image about them and their friendship. More of her experiences about Finnish people and culture are based on what she has heard "*I found it is difficult to get a friend with them. They are friendly but making a connection with them is hard. .. I have a challenge. They are not keeping up with you. You have to ask them a lot. Maybe you should ask them many times. But you are the main one for pushing them. If you ask, they are very helpful... They don't get involved with you. You should start the talk, they don't get involved with you.... I really like to become friend with them. I don't know how to make a connection with them.*"

D1 is getting help from the translation link on Facebook. It shows how participant's Interest toward the new culture is important to get cultural knowledge even if the barrier of language exist. However, he is still learning Finnish culture as an outsider and not integrating because there is not an available route for some students to communicate in Finnish society.

(c) *Communication is hard because we are different:* The third group of the participants are not interested to communicate with Finnish people because they do not feel to have anything to share with them. They feel Iranians life and condition is too far from Finnish culture and life style. Therefore, they may not understand each other well even if they can talk or spend time with each other it would be limited in a few cases.

For example, "H2" doctoral students who lives with his family in Finland does not feel any attachment with Finnish society. He is keeping his communication with his family and friends in other countries through Facebook. He is comfortable living in Finland and although has received the citizenship of Finland but is not interested to make a communication or Friendship with Finnish. Similarly, "B2", a doctoral student who lives in Finland for 7 years, still finds himself as a stranger in Finland and believes "It *is not easy to get a close friend with them (Finn people). Their culture is very different from ours.... In fact, now I am not so interested to have Finnish friends because I see them very far from us.*" He does not have any Finnish friend and more communicating with Iranians in Finland. He is watching two Farsi channels of satellite and get information about events in Finland through Facebook or YouTube. Although he is always available on social media for his personal purposes, does not know himself active in social media because does not share content.

The other student, "H1" believes "*there are too many differences between our culture and Finn ('culture)... when they come in the relationship, they do not much effort to keep it and they just leave it*". He knows social media important for finding other Iranians in Finland and getting some important information about Finland and Finnish culture from them. However, most of his friends are international especially Asian students. "H1" feels Finnish students are cold to make a friendship. Similarly, "S1" and "S2" feel there is not much talk with Finnish people because Iranians and Finns are different. "S2" describes, "*I feel we don't have any talk. Talking becomes very casual after sometimes. I like to communicate with people who are in my own language and culture. I like to spend time with Iranians*". He is following Iranian news by social media and likes to watch football from the Iranian TV channel with Iranian friends. Likewise, "F2", he claims that he more was active in social media but later he realized to be active in social media has disconnected him from real communication. He thinks although some Iranians are trying to behave like Finnish people, in the end, they are not accepted in the society like native people.

Some of the participants are disappointed for making the friendship in Finland as they have had in Iran or other countries. "A1", "A3", "E1", "M1" are frustrated to communicate with Finnish. While "S3" has accepted this problem in Finnish environment, "M1" is feeling irritable living in Finland although he has lived in U.K for four years and has a Finnish wife. He believes that the coldness in Finnish people behaviors has affected on other nationalities and made it as a norm in Finland environment. He states: "*Here the commination become fades. You think the norm is this and you go with the wave. Unfortunately, I couldn't make close communication with anyone here... maybe change in me or change in environment*". He is available in What's app and telegram for contacting his family and getting news from Iran every day. He has Facebook but does not do much activity in that because he does not want to keep time on that. He also receives Iranian news of Iran from Persian channels.

Likewise, "A1" dislike communication rules in Finland. He has very few friends and believes communication in Finnish environment is very limited and the reason is that "*Finnish students don't like to communicate with international students*". He seems to keep a tight communicating with his parents and may contact with them two times a day. He uses What's pp and Facebook for communication with friends and uses Facebook for getting information from events. Social media helped "A1" to accommodate in Finland by getting help from Iranian. However, he claims: "*Social media didn't help me for being adapted in Finnish culture. Most social media are used by international.*"

While "M1", "A1" and "E1" are frustrated about communicating with Finnish people in Finland, "A2" and "S3" are accepting the Finnish culture and feel comfortable living in Finland. "S3" claims that in the first year after coming to Finland, he was living alone and he had social problem but now he does not expect much in his social life in Finland. He has used social media for getting information to come to Finland and even now, he is the member of telegram channel and Facebook pages of Iranians. He uses Persian sites, channels and groups on social media to get the information or help but he believes that assuming social media to make a community for mixing Finnish and Iranian nationality and culture is a wrong expectation. He has an optimistic view of living Iranian and Finnish in the same society although he believes they are very different in background and culture.

Overall, the number of participants who believe Finnish culture and Iranian culture are too far is more than the number of those participants who feel comfortable to communicate with Finnish people. Notwithstanding they agree in the disparity between two cultures, their reflection in the deal with that is different.

4.1.2 Communicating with Iranians People in Finland

Most of the participants are the members of the different Facebook pages that belong to students or Iranian living in Finland. These pages connect Iranians to each other. Facebook pages are meant for asking a question and receiving the answer that may happen once or twice per month. Further, there are two groups in telegram app with 114 and 175 members that their membership may cover students and some other Iranian immigrants in Finland. The content of these groups relates to personal questions about the university, visa, laws, and general rules in Finland. The contents are useful for the group members but usually do not include any cultural and communication issues in Finland. The conversations are usually short and who wants to get more information or discussion, asks permission for a private chat. Therefore, these portals become the individual tool than a social portal when more details are required. A few of students who were interviewed for this study were the members of these telegram groups. But the majority of them are always available for their family and their friends in Iran via two basic applications, What's app and telegram for calling and texting. Iranian TV channels sites (usually available on YouTube), Facebook and telegram channels and groups are the source of news from Iran.

Social life still seems difficult for many of the Iranian students in Finland particularly single students. Many participants are still more attached to Iranian friends although they are not satisfied in their friendship with Iranian. "B2" has a few Iranian friends and sees even the culture of Iranians in Finland is very different. Similarly, "S1" does not see himself matched with the type of Iranians who are living in abroad and feels problem in his social life. "D1" although appreciates Iranian culture and art, prefers to have Finnish friends. "F1" the master student who has left his flight engineering job in Iran and has experienced being a cleaner in Finland does not feel to have any Iranian friend in Finland and thinks Iranians are treating the same way as inside Iran with each other (exploitation and colonialism). "L1" who has a small family in Finland says, *"I have more problem with Iranian friends. For example, I don't want for some gathering but they invite me and then I have to invite them. But for Finn friends, we go to café or we invite for a coffee. The relationship is clear"*. Likewise, "F1" and "S3" are married students who live with their families do not feel problem in their social life but feel challenging to communicate with Iranians in Finland and beliefs. "S3" states: *"I have more challenges with Iranian. Maybe the same guard that we have as Iranian and maybe it is my own problem and I am trying to make it less. In a small city like Oulu, the number of Iranians are less, people know each other as a village, and I think I had more disappointment in my social life. Maybe it is because we have more expectation from our same language people. It is not because of their behaviors."*. "H1", also, likes to be far from Iranians because he sees Iran as a multicultural society itself. He says, *"Inside our own country (Iran) we have more cultural conflicts. Because when I go to one state in the east of Iran they talk in their language that we can't understand and they support themselves but not me because I went there from Tehran."*

4.2 Membership in Finnish Society

Although Communication is the main door to enter the new culture, the member of local friends may not make the immigrant feel being a member of society. While 17, of 23 participants who were interviewed plan to reside in Finland, few of them feeling to be a member of society because of the different reasons (Table 1).

While many participants assume the number of Finnish friends as the necessity for their cultural integration, for plenty of them, having a job and talking with the Finnish language means the membership in Finnish society. As "S3" claims "*I think if you find a job and earn money and pay tax it means being integrated into society, it means you attend in the society. It is right way of citizenship when you accept responsibility.. but if you have 50 Finnish friends and be close to them, I don't call it as integrating. It is a little integration*". To "S3", the meaning of integration in the new culture is a broader conception. He thinks having a deserving job and place in Finland is required for integrating or being a part of Finnish society and the number of Finnish friends is not a symbol of integration in the new culture. He states that social media can be useful if it just provides a source of information through a community for the minority in the society (particularly Iranians) to solve their problem and share their experiences in the portal. He plans to reside in Finland although believes Iranians and Finnish are different and do not have many Finnish friends. Similarly, "S2" considers the difference between Iranian and Finnish culture as a barrier for communication and believes social media has accomplished its responsibility for assisting cultural adaptation: "*After I came, social media helped me. When I have a problem. When I wanted to open a bank account, when I want to extend the visa, I was asking. Sometimes I got help from Finnish like this too.*"

4.3 Social Media Applications

The participants of this study utilize social media for both individual and social communicating purposes. The functions and tools of social media are different toward the original and new culture. They are keeping daily contact with their family and friends (individual purpose) mostly through What's app and telegram chat and groups. Facebook, YouTube channel TV, and telegram channels and groups update them for social and political news (social purpose). The participants profit by What's app, Facebook messenger, telegram texting as the communication tools to call or text to their friends in Finland (individual purpose) and Facebook is usually used for receiving information, organizing meeting or events in Finland (social purpose).

Telegram is the frequent application for texting among Iranians and What's app is usually used for making audio and video call. Facebook messenger is usually used for texts and call outside of Iran because it is filtered in Iran. Although many Iranians are getting news from two Farsi channels of the satellite through YouTube link, telegram groups and channels are the easier portals for updating them about their family, friends, and news of Iran. (Fig. 1).

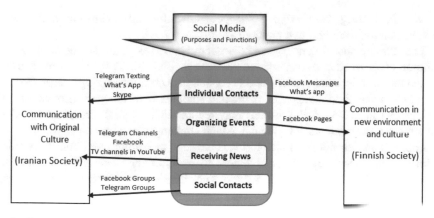

Fig. 1. Purpose and functions of utilizing social media application for Iranian students in Finland

5 Discussion and Conclusion

Communication is considered the main key to cultural adaptation [22]. Entering a new culture and making communication is a challenging and stressful process for every foreigner. On the other hand, nowadays, social media becomes a daily activity, which provides audio and video communication, facilitates finding and sharing information and helps the process of socializing. The social media seems more beneficial for immigrants and international students due to reducing loneliness and rootlessness by making a connection between them and their family and friends in their home country.

Results of the study demonstrated how Iranian tertiary-level students experience stress in their communication with Finnish and Iranians in Finland. Following the non-linear growth in the process of cultural integration, social media is assumed as the communication facilitator for connecting the participants to their family and friends in their home culture and acts as the bridge to attach them to their old culture.

Although all the Iranian students who interviewed are using the different kinds of social media in their daily activities and many of them have used the forums like Applyabroad, Facebook or telegram groups to get help for admission and accommodation process, very few of them claim that social media have been useful for their cultural adaptation. They believe face-to-face communication and friendship has helped them more in the process of cultural integration. Their claims are inconsistent with the results of previous studies. Many studies introduce social media as a beneficial application for enhancing cultural adaptation. Facebook is reported as the potential tool for reducing acculturation stress among East Asian students in the United States [43] and it is shown beneficial for enhancing cultural adaptation among Chinese students in the United States [31]. Further, another researcher [32] demonstrated that Facebook and Renren influenced Chinese cultural orientation language proficiency, and length of stay in the U.S. In other studies, social media has been acknowledged to reduce the stress of immigrant in the face with the cultural shock or some other issues, which relates to studying abroad like Fear of Missing Out [44] and decreases their stress [26]. Further, Facebook has been identified as helpful social media for the student's adaptation through facilitating knowledge exchange, alleviating apprehension, and enabling socialization and building community [33, 45].

The current study demonstrated Facebook and Telegram as the accepted and common social media among Iranian students in Finland. They are using Facebook messenger for individual contacts and Facebook pages for organizing events. There are few posts in the Facebook pages of Iranian students than telegram groups, but only participating in these existing groups may reduce their cultural stress and loneliness. Nevertheless, the students who overcome their challenges in communication in the new environment and do not feel much distance between themselves and Finnish people do believe in the adequacy of social media to develop their cultural adaptation and competence. They believe social media did not help them in their adaptation as their face-to-face friendship had help. This statement shows participants' image and expectation of social media for communication with Finnish is more than what they have experienced. It clarifies that social media does not satisfy their expectations concerning to be a portal for connecting between Finnish and Iranians or the required cultural content.

Obviously, social media is limited to make a ground to communicate the different cultures because of the language barrier. It has reported as the serious issue of international students for cultural adaptation in the previous studies [14, 19–21]. Hence, regarding the language barrier, utilizing social media for being the route for communicating international students or immigrant with the local people is a further step in the use of technology for cultural adaptation [46, 47].

The current study did not show the language barrier as the impediment for communicating with Finnish people. This assertion may limit to the context of the study. According to Official Statistics of Finland (OSF) more than 90% of Finnish residents know at least one foreign language and at least 82% of Finnish people are well in English skills [48]. On the other hand, all participants are studying their education in English language. Yet, lack of language skills and job are identified as two important requirements for linking the participants to Finnish society. In fact, the findings of this study inconsistent with previous studies [32, 33] indicated social media did not help the participant to develop their language skills due to existing literature contexts as the United States and United Kingdom that are English speaking countries. The ability of Finnish people to understand English reduces necessity of learning Finnish language for the foreigners, particularly, international students who are capable in speaking English and making them to communicate with Finnish people. It somehow makes learning Finnish more difficult for English speakers. However, ability in communication with Finnish people does not make Iranian students feel as being member of Finnish society.

The results of the present research along with the previous studies have shown the efficiency of social media as the communication tool. It facilitates individual communication among students. Further, it prepares the appropriate path for students in the same culture to share their knowledge and assist their cultural adaptation process. The context of this study is Iranian tertiary-level students living in Finland, therefore this finding is transferable to similar contexts such as students or immigrants from other nationalities in Finland. Further, the findings may be applicable to other Nordic countries because of the similarity in socio- cultural conditions of the context.

However, the current functions of social media in the present study neither increased students' knowledge toward Finnish culture nor provided the environment for communicating between Finnish and Iranian students. Accordingly, the more studies in the different contexts are suggested due to finding new ways for utilizing social media for cultural adaptation.

Social media is potential for making an environment to connect Finnish and international students particularly Iranian students. Further, social media is talented to provide cultural information for immigrant students living in Finland. It is capable for sharing knowledge, experiences and attitudes toward the new culture and develop cultural adaptation of consumers. This study trusts that social media yet, is not employed and utilized for its full potential to enhance cultural adaptation. Accordingly, the implication of these results suggests designing the particular social platforms with the purpose of developing cultural adaptation.

References

1. UNESCO: http://data.uis.unesco.org/. Accessed 20 July 2018
2. OECD: Education at a Glance 2011. Paris: OECD (2011). https://www.oecd.org/education/skills-beyond-school/48631582.pdf. Accessed 20 July 2018
3. Kondakci, Y.: Student mobility reviewed attraction and satisfaction of international students in Turkey. High. Educ. 62(5), 573–592 (2011). https://doi.org/10.1007/s10734-011-9406-2
4. Cantwell, B., Luca, S.G., Lee, J.J.: Exploring the orientations of international students in Mexico: differences by region of origin. High. Educ. 57, 335–354 (2009)
5. Park, E.: Analysis of Korean Students' international mobility by 2-D model: driving force factor and directional factor. Higher Educ. 57, 741–55 (2009). http://link.springer.com/article/ https://doi.org/10.1007/s10734-008-9173x
6. Altbach, P.G., Knight, J.: The internationalization of higher education: motivations and realities. J. Stud. Int. Educ. 11(3/4), 290–305 (2007)
7. Cao, C., Zhu, C., Meng, Q.: A survey of the influencing factors for international academic mobility of Chinese university students. High. Educ. Quart. 70, 200–220 (2016). https://doi.org/10.1111/hequ.12084
8. Sanchez, C.M., Fornerino, M., Zhang, M.: Motivations and the intent to study abroad among U.S. French and Chinese students. J. Teach. Int. Bus. 1(18), 27–52 (2006)
9. Peltokorpi, V.: Cross-cultural adjustment of expatriates in Japan. Int. J. Hum. Res. Manag. 19(9), 1588–1606 (2008)
10. Zhu, L., Liu, M., Fink, E.L.: The role of person-culture fit in Chinese students' cultural adjustment in the United States: a galileo mental model approach. Hum. Commun. Res. 42(3), 485–505 (2016)
11. Donin, I.B.: The relationships among acculturation, coping, and stress in Puerto Rican and Korean college students. Diss. Abstr. Int. 55(11-A), 3451 (1995)
12. Sodowsky, G.R., Lai, E.M.W.: Asian immigrant variables and structural models of cross-cultural distress. In: Booth, A., Crouter, A.C., Landale, N. (eds.) Immigration and the Family: Research and Policy on U.S. Immigrants, pp. 211–234. Lawrence Erlbaum, Mahwah (1997)
13. Tung, M.L.: The roots of the challenge: undergraduate Chinese students adjusting to American college life. Int. J. High. Educ. 5(3), 121 (2016)

14. Yao, C.W.: Unfulfilled expectations: influence of Chinese international students' Roommate relationships on sense of belonging. J. Int. Stud. **6**(3), 762–778
15. Vang, K.C.: The expectations and perceptions of studying abroad of college students in Yunnan Province. Master thesis. International Studies in the School of Adult and Professional Studies Concordia. University. Irvine, California, 13 December 2016 (2016)
16. Smith, R.A., Khawaja, N.G.: A review of the acculturation experiences of international students. Int. J. Intercult. Relat. **35**, 699–713 (2011)
17. Okpara, J.O.: Cross-cultural adjustment of expatriates: exploring factors influencing adjustment of expatriates in Nigeria. Int. J. Cross Cult. Manag. **16**(3), 259–280 (2016)
18. Maundeni, T., Malinga, T., Kgwatalala, D., Kasule, I.: Cultural adjustment of international students at an African university. J. Psychol. Afr. **20**(1), 79–84 (2010)
19. Lin, J.C.G., Yi, J.K.: Asian international students' adjustment: issues and program suggestions. Coll. Stud. J. **31**, 473–479 (1997)
20. Pineda, P., Moreno, V., Belvis, E.: The mobility of university students in Europe and Spain. Eur. Educ. Res. J. **7**(3), 273–288 (2008)
21. Klahr, S., Ratti, U.: Increasing engineering student participation in study abroad: a study of U.S. and European programs. J. Stud. Int. Educ. **1**(4), 79–102 (2000)
22. Kim, Y.Y.: Becoming Intercultural: An Integrative Theory of Communication and Cross-Cultural Adaptation. Sage Publications, Thousand Oaks (2001)
23. Croucher, S.M.: Looking Beyond the Hijab. Hampton Press, Cresskill (2008)
24. Trebbe, J.: Types of immigration, acculturation strategies and media use of young Turks in Germany. Communications **32**, 171–191 (2007). https://doi.org/10.1515/COMMUN.2007.011
25. Tsai, J.H.: Use of computer technology to enhance immigrant families' adaptation. J. Nurs. Scholarsh. **38**, 87–93 (2006). https://doi.org/10.1111/j.1547-5069
26. Ye, J.: An examination of acculturative stress, interpersonal social support, and use of online ethnic social groups among Chinese international students. Howard J. Commun. **17**, 120 (2006). https://doi.org/10.1080/10646170500487764
27. Wang, W., Huang, T., Huang, S., Wang, L.: Internet use, group identity, and political participation among Taiwanese Americans. China Media Res. **5**(4), 4762 (2009)
28. Hwang, B., He, Z.: Media uses and acculturation among Chinese immigrants in the USA: a uses and gratification approach. Gazette **61**, 522 (1999)
29. Raman, P., Harwood, J.: Acculturation of Asian Indian sojourners in America: application of the cultivation framework. South. Commun. J. **73**, 295311 (2008)
30. Yanagihara, H.: Relationship between media use and cultural adjustment: a study on international students at Marshall University. Master thesis in Marshall University (2017)
31. Forbush, E., Foucault-Welles, B.: Social media use and adaptation among Chinese students beginning to study in the United States. Int. J. Intercult. Relat. **50**, 1–12 (2016)
32. Li, C.: A tale of two social networking sites: how the use of Facebook and Renren influences Chinese consumers' attitudes toward product packages with different cultural symbols. Comput. Hum. Behav. **32**, 162–170 (2014)
33. Ryan, S.D., Magro, M.J., Sharp, J.H.: Exploring educational and cultural adaptation through social networking sites. J. Inf. Technol. Educat. **10**, 1–16 (2011)
34. https://www.worldometers.info. Accessed 09 Sept 2019
35. www.payvand.com/news/14/sep/1002.html. Accessed 09 Sept 2019
36. Choudaha, R., Chang, L.: Trends in International Student Mobility. World Educ. News Rev. **25**(2) (2012)
37. Balbutskaya, E.: Demand for Finnish Education Export in Russia. Case: Saimaa University of Applied Sciences Master thesis (2015). https://www.theseus.fi/bitstream/handle/10024/92047/Balbutskaya_Evgenia.pdf?sequence=1. Accessed Dec 2018

38. CIMO: International mobility in Finnish higher education in 2015: degree students. Centre for International Mobility. Accessed 09 Sept 2019
39. Trotter II, R.T.: Qualitative research sample design and sample size: resolving and unresolved issues and inferential imperatives. Prev. Med. **55**(5), 398–400 (2012)
40. Neyer, A.K., Harzing, A.W.: The impact of culture on interactions: five lessons learned from the European commission. Eur. Manag. J. **26**(5), 325–334 (2008)
41. Braun, V., Clarke, V.: Using thematic analysis in psychology. Qual. Res. Psychol. **3**(2), 77–101 (2006)
42. Krefting, L.: Rigor in qualitative research: the assessment of trustworthiness. Am. J. Occup. Ther. **45**(3), 214–222 (1991)
43. Park, N., Song, H., Lee, K.M.: Social networking sites and other media use, acculturation stress, and psychological well-being among East Asian college students in the United States. Comput. Hum. Behav. **36**, 138–146 (2014)
44. Hetz, P.R., Dawson, C.L., Cullen, T.A.: Social media use and the fear of missing out (FoMO) while studying abroad. J. Res. Technol. Educ. **47**(4), 259–272 (2015). https://doi.org/10.1080/15391523.2015.1080585
45. Adikari, S., Adu, E.K.: Usage of online social networks in cultural adaptation. In: PACIS (2015)
46. Shao, Y., Crook, C.: The potential of a mobile group blog to support cultural learning among overseas students. J. Stud. Int. Educ. **19**(5), 399–422 (2015)
47. Ernst, S., Janson, A., Söllner, M., Leimeister, J.M.: It's about understanding each other's culture – improving the outcomes of mobile learning by avoiding culture conflicts. In: International Conference on Information Systems (ICIS), Dublin, Ireland (2016)
48. Official Statistics of Finland (OSF). https://www.stat.fi/til/aku/2006/03/aku_2006_03_2008-06-03_kat_001_fi.html. Accessed 09 Sept 2019

Cybersecurity and Cyber-Defense

How to Develop a National Cybersecurity Strategy for Developing Countries. Ecuador Case

Mario Ron[1,2(✉)], Geovanni Ninahualpa[1], David Molina[1], and Javier Díaz[2]

[1] Universidad de las Fuerzas Armadas ESPE, Av. General Ruminahui, Sangolquí, Ecuador
{mbron, gninahualpa, damolinall}@espe.edu.ec
[2] Universidad Nacional de la Plata, Avenida 7 877, 1900, La Plata, Buenos Aires, Argentina
jdiaz@unlp.edu.ar

Abstract. The increase use of Information and Communication Technologies entails associated risks, which must be systematically managed by those responsible for state security. This need has led to the development of National Cybersecurity Strategies, which in one way or another have been created and formalized, without applying a specific model previously analyzed in reference to the good practices that have been published by international organizations working in. In these matters, the lessons learned in similar exercises carried out by states that have developed their national strategies in various ways, often with foreign consultants far from local reality, have not been considered. This work aims to cover this need and provide a systematically analyzed methodology based on the good practices developed so far, adapted to Latin American reality with an example applied in Ecuador, so that it can be used by other countries in the region or similar conditions as a methodological guide to develop your strategies.

Keywords: National Cybersecurity Strategy · Methodology · Good practices · Methodological guide

1 Introduction

The strategic solutions to the problem of insecurity in the use of Information and Communication Technologies (ICT) have been driven mainly by countries that make up the group of those developed, both economically and technologically, who have published, managed and updated them in a permanent way, with assessments that have served to feed them back and establish a life cycle.

Some developing countries have also done the same, trying to emulate the great powers, taking into account that technological development in their countries has reached levels not previously anticipated and that it has become necessary for their internal and international economy bringing, as expected, the growth of illicit activities and potential risks that affect their computer systems. In the Latin American and

© Springer Nature Switzerland AG 2020
Á. Rocha et al. (Eds.): ICITS 2020, AISC 1137, pp. 553–563, 2020.
https://doi.org/10.1007/978-3-030-40690-5_53

Caribbean region there are already 11 countries that have developed their national cybersecurity strategies (NCS) [1].

NCS are documents that allow the establishment of specific principles, guidelines, objectives and measures to mitigate the risks associated with cyber security at the national level, following a downward approach from the high level, to subsequent operational actions. The Information Network Services Directive (NIS) [1], adopted by the European Union (EU), requires that its member states develop and adopt an NCS, if necessary, they can ask the European Agency for Information Network Security (ENISA) to support them in the drafting of such a document [2].

The Political Constitution of a state [3], establishes the rights of citizens that the State must guarantee, among them are those related to access to Information and Communication Technologies (ICT), so safe and reliable, responsibility that must be achieved in a planned and structured manner; However, in developing countries, such as Ecuador, NCS have not been developed, certain regulations and laws related to cybersecurity have been issued, but without a systematic and organized process as should be done. For this reason, the responsibility inherent in an organizational structure appropriate to the solution must be formulated based on an NCS consistent with the national reality.

The elaboration of an NCS requires a strategic reflection process on cybersecurity and should have as reference a useful, flexible and easy-to-use framework that allows considering the social-economic context of the country and its current situation in cybersecurity, so it should also help governments to build and manage the strategy taking into account the country's own situation and it's cultural and social conditions, to create an information society that is safe and resilient.

A methodology is presented as a guide or methodological framework, which constitutes a strategic resource used to reach agreements between those interested in national cybersecurity, that is: government agencies, social, productive organizations, technology providers, educators and citizens in general.

The methodologies constituted in practical guides lead an efficient and effective process in the construction, implementation, management and continuous improvement of the cybersecurity conditions of the country; they may be related to national planning policy procedures with which they should agree so that the result, that is the strategy, is executed with the support of all actors, including and mainly the government. They will therefore have to address governance, political, operational, technical and legal aspects, organize and prioritize these aspects using models, general principles and good practices that should be considered when developing and managing an NCS, focusing on the protection of aspects civilians of cyberspace, therefore does not cover the development of defensive or offensive cyber capabilities by the military forces, defense forces or intelligence agencies of a country whose process requires special treatment.

In the present work, Sect. 2 describes several reference methodologies to elaborate NCS, in order to obtain from them later their comparable main characteristics. In Sect. 3 we worked on a comparative analysis of the methodologies, using comparative parameters of the characteristics of each one, so that the components of the eclectic method that is the reason for this research can be defined with them. In Sect. 4 the method is presented with its components and characteristics in a summarized way, a more detailed guide will be carried out as an additional contribution and finally the conclusions of the study and future work are detailed in Sect. 5.

The questions that lead to this research are: are there methodological and logical antecedents to elaborate NCS? Do they exist and what are the differences between the methodologies to elaborate NCS? Which of these differences can be used to construct an eclectic model? These questions are made up of the following hypothesis: "There is an eclectic method with which a NCS can be developed considering the social-economic and political characteristics of the developing countries of South America".

2 Methodologies Which to Construct an NCS

Methodologies Chosen and Selection Criteria. An extensive search of documents related to information security was carried out, several of them were found that during the last decade, governments, organizations and individuals have developed, the available material is considerable [4]. Among the documents we have: Guidelines, National Strategies, Work frames, Codes of good practice, Detailed technical regulations and like these. An initial list was prepared, but when analyzing it's content some documents were omitted because they related to the specific content of the NCS, but not to the procedure to build them, as is the reason for this work. Those issued in the last ten years were considered, with a criterion of global diversity, including as expected, the national regulations of Ecuador. The following is a summary description of the main characteristics of the documents found.

Recommendation of the Organization for Economic Cooperation and Development (OECD) for Digital Security Risk Management for Economic and Social Prosperity [5]. It is a high-level regulatory framework for the development of a National Policy and International Cooperation for the Protection of Critical Information Infrastructures (PCII), develops a concept of Critical Information Infrastructures (CII) and how to identify them; promotes the adoption of regulatory frameworks with safety guidelines defined by the OECD.

The Recommendation specifies how the national government should demonstrate leadership and commitment, manage risks and work together with the private sector and international cooperation to share knowledge, information, experiences and promote a common understanding of the problem. It is the result of a comparative analysis of the way in which the policies of seven OECD member countries have been developed, the aspects that include, risk management practices and methods, the strategies used to mitigate vulnerabilities and monitor threats, roles and responsibilities, cooperation, operation and exchange of information internationally.

CIRCULAR NO. A-130 [6]. It establishes the policy for the management of information resources of the agencies of the Executive Branch of the Federal Government and the information classified for national security purposes in the United States of America (USA), includes appendices to implement the policy, is part of the Paperwork Reduction Act (PRA), which is a policy so that information resource management activities are carried out efficiently, effectively and economically, among other aspects determines that organizations must establish a Business Architecture, include a Technical Reference Model and a Profile of IT Standards, incorporate security in the information architecture in support of the business, prioritize key systems and apply state policies and include security in ICT investments and budget programming.

ENISA Good Practice Guide - Design and Implementation of National Cyber Security Strategies [2]**.** It was developed by the European Network and Information Security Agency (ENISA), so that the member states of the European Union (EU) build and review their NCS, as official documents that establish strategic principles, guidelines, objectives and measures specific to mitigate cybersecurity risks, through a top-level downward approach. This guide updates the original 2012 guide and analyzes the state of the NCS in the EU and the European Free Trade Association (EFTA) area, is aimed at government officials responsible for policy development and stakeholders in the cycle of life of the strategy to define important related areas, help the development, management, evaluation and updating of NCS, identify challenges, good practices and lessons learned from NCS already developed, provide recommendations for governments and policy makers and contribute to integrated cyber security strategy in Europe.

It's a practical and systematic guide to create or align an NCS with the requirements of the current NIS Directive of the European Community and improve, complement, maintain and verify the effectiveness of an NCS, through a life cycle that contains the main steps in its development, tasks and examples for each step and objective, key performance indicators (KPI) and compliance indicators for the objectives of the strategy, gaps and challenges faced by several countries and recommendations for the future.

It establishes six phases for design and development and fifteen standardized objectives for a NCS. In describing the current status of NCS in the EU Member States, it identifies five challenges and gaps, in addition to nine recommendations that should be considered for the development and maintenance of NCS.

Recommendations for the Development of the National Cybersecurity Strategy OAS-Mexico [7]**.** At the request of the government and through a commission of international experts, the OAS Inter-American Committee against Terrorism (CICTE) worked in Mexico, using round tables with the Government, national organizations and international experts to understand the current situation and build a National Cyber Security Framework (MNSC). In the discussions five themes were identified, which were discussed in detail at each table, facilitated by a moderator and a secretary, both international consultants. The topics were: Research and Development, Culture, Education and Prevention, Cooperation and Coordination, Standards, Technical Criteria and Regulation, and Legal Framework. In this process, valuable recommendations for the construction of an MNSC were obtained, which were used to prepare the NCS that was promulgated in 2019 using this procedure.

ITU Guide for the Development of a National Cybersecurity Strategy [8]**.** It presents several principles and good practices to elaborate, establish and apply a NCS, designed as a enlistment to provide resilience, confidence and security in ICT. It was developed by the ITU, public and private sector partners, academia and civil society, with the aim of supporting national stakeholders, government, legislators and regulators in preparing a defensive response to cyber threats.

The guide distinguishes the process of the life cycle of an NCS and the content of the document, does not contemplate the development of cyber capabilities by military forces, defense or intelligence agencies. It presents a vision of the essential elements that a country requires to be cyberprepared and that it should consider when developing its strategies and action plans.

It offers existing approaches and applications, references to complementary resources that serve as the basis for the preparation, drafting and management of the NCS. It begins with a description of the guide and several definitions, then specifies the stages of strategy development and management in it's life cycle, then describes the General Principles for developing the Strategy; It also details the priority areas and good practices finally the reference materials and acronyms.

A NCS translates the government's vision into actions that allow it to achieve it's objectives, the measures, programs and initiatives, the necessary resources and how to use them, also identifies metrics to evaluate the results achieved.

The life cycle defined by the guide includes 5 phases: Initiation, Inventory and analysis, Development of the NCS, Execution and Supervision and evaluation. Nine intersectoral principles applicable in the NCS elaboration process are presented below. It also details a set of good practices in seven priority areas, with several elements each, which serve to focus the objectives of an NCS.

Guide for the Formulation of Sectorial Public Policies [9]. Prepared by the National Secretariat of Planning (SENPLADES) with the intention of harmonizing criteria for formulating public policies, with instruments that allow visualizing the common good and responding to modern requirements.

The Constitution of Ecuador creates the National Decentralized System of Participatory Planning of Ecuador, where the National Development Plan is the highest instrument of national planning and public policy and provides that the Ministers of State exercise the leadership of the public policies of the area in charge. Public policy together with jurisdictional and regulatory guarantees, are mechanisms that guarantee and enforce the rights of citizens, it's conceived as a series of strategies and decisions adopted by an authority in order to solve complex public problems [10], it is a general guideline that shows the political will of the government to act in a given situation. It's an instrument that the State uses to link short-term social needs with medium and long-term political vision, eliminating inequities and future risks. It is a deliberately designed and planned process, with objectives, strategies and guidelines that demand resources and interaction between political and social actors. It is defined as a strategy that institutionalizes public intervention in response to a social problem identified as a priority.

Three stages have been defined to fulfill the life cycle of public policies, with a previous preparatory stage, then the formulation, then the execution or implementation and finally the monitoring and evaluation. The evaluation includes information that serves to control, update, reformulate, terminate or delete the policy. The process is flexible, continuous and allows a new cycle to be restarted periodically. Each of the stages also has activities to accomplish.

In the preparatory and diagnostic stage, the political and technical teams are integrated, the need is identified, spaces for participation of social actors are formed, information and recommendations of international, regional, national and civil society organizations are collected, the situation is analyzed and the problem is characterized, then action alternatives are formulated with the corresponding cost and financing, these alternatives are ranked and finally the viable alternatives are selected.

In the formulation stage, public policy is defined, citizen participation spaces are created with whom an analysis is carried out, guidelines, goals, indicators, programs and projects are detailed and the document is prepared, considering the approximate costs of its implementation. To approve the policy and incorporate it into the system, adjustments are made and the proposal is validated with the technical team and citizen participation, then it is adjusted and validated by the political team. The proposal for revision of the SENPLADES is presented and then it goes on to review and approval of the respective Sector Council and incorporation into the system, to finally materialize in the corresponding institutional planning. Once this process has been completed, the policy, programs and projects are disseminated to the population and all interested parties.

In the monitoring and evaluation stage, several aspects related to the purpose of covering its entire cycle are detailed, including accountability as provided by the Constitution of Ecuador. It also presents the structure of the document as it should be written, with a description of the minimum content of each section and has at the end several work annexes with matrices that are used during the process.

3   Comparative Analysis of the Referenced Methodologies for Preparing NCS

Comparison of the General Characteristics of the Documents. Table 1 details the general characteristics of the selected documents in a comparative way, that is: type of document, objective, the public to whom it is addressed and the approach of the management participation method. As you can see, three of the documents are guides to good practices, to which a high-level regulatory framework of a similar nature, an experimental recommendation of the OAS and a high-level policy are added, based on the one issued the NCS of the USA.

As for the objectives, four of them are oriented to the construction of the NCS, another to the creation of a policy in general and one towards the management of information resources of the state with a higher level than an NCS. As for the target audience, the ITU, ENISA and OECD guidelines are oriented to governments and stakeholders in several countries, the OAS recommendations similarly, the US circular only for the federal government and that of Ecuador towards the government with social participation. In the participation approach, everyone agrees in a direction of the central government with the participation of all the interested parties, in addition the OAS-Mexico includes to direct the effort international consultants hired. This analysis is important to decide the characteristics of the model proposed below.

Table 1. General characteristics of the documents.

Document	Document type	Objective	Addressed to	Participation and direction approach
OECD Recommendation	High Level Regulatory Framework	Development of a National CII Protection Policy	Government and CII operators	Directed by the central government with participation of CII operators
Circular A-130	National Policy	Information resource management	Government Organizations (USA)	Directed by the federal government with participation of CII operators
ENISA Guide 2016	Good practice guide	Design, development, adaptation, updating and evaluation of an NCS through a top-level descending approach	States that make up the European Union (EU)	Directed by the central government with participation of CII operators
OAS-Mexico Recommendation	High level recommendation	Understand the current situation and build an NCS	International experts, government agencies, work tables	Directed by international experts, the central government and working groups
ITU Guide 2018	Good practice guide	Harmonizes principles and good practices to develop, establish and apply NCS, with participatory methodology	Stakeholders, government, national legislators and regulators	Directed by the central government with stakeholder participation
SENPLADES Guide	Good practice guide	Harmonize methodologies and criteria to formulate public policies	Government agencies with citizen participation	Directed by the central government with stakeholder participation

Comparison of the Specific Characteristics of the Documents. Each of the documents describes a life cycle, some do not say it formally, but they make up a method with systematic sequences, some with a greater number of steps, perhaps in more detail than others. They also present general principles of application in the same way as in the case of life cycles. A summary of each document regarding the life cycle is presented below in Table 2.

4 Definition of an Eclectic Method for Developing NCS in Developing Countries

Regional Considerations. Developing an NCS is essential for all states, the threats are the same in both developed and developing countries, the difference is in the ability and capacity available to face these threats [11].

Latin American politics has reached a crisis, the economy weakens, governments lose support from the population and poverty increases. Neoliberal reforms failed to build solid state institutions and the old elites reserve exclusive powers and rights that prevent collective participation [12].

A methodology for developing NCS is dependent on factors related to the economy and society, which must be considered to make it effective and strengthen it in the most crucial points, it must be considered that in the developing countries it has not been possible to create a public services system high quality and accessible to the majority, there is a central administration that greatly favors clientelism and corruption where democratic controls and demands are annoying, especially to its old allied elites and economic oligarchies that maintain consortia with heavily monopolized private media, so that the state is seen as political booty and not as administrator of public affairs, which uses western economic recipes and technocratic recommendations that are really doomed to fail, because there has been a lack of political courage to carry out structural reforms, within an international cooperation not has become effective [12].

Life Cycle of an NCS. With these considerations, a summary of the life cycle developed as an iterative model of construction and improvement of an NCS is presented in Fig. 1, in which the well-known Deming cycle and the documents compared and described above have been taken into account. It is based on general principles of application and includes good practices for each of the six stages as an additional document.

Preparation. It includes the initial activities of the creation of work teams, context analysis, definition of the initiative, political consultation, getting participation of all stakeholders, formalizing the proposal, preparing the project and approving it by the government, this is the main difference between this model and those for developed countries.

Evaluation of the Current Situation. If it is the first time that the exercise is carried out, the work team is conformed to whom the responsibility is assigned to collect the information based on parameters and components involved in the evaluation of the current situation and the model with the metrics respective designed for the effect. In the event that an NCS is already available, the evaluation model established in the formulation stage of the NCS will be used, the organization and responsibilities specified therein, consists of a formal and accountability evaluation.

Risk Analysis. Based on the information collected, a risk analysis model is established that is carried out by defined sectors and is consolidated at the national level by the commission appointed for this purpose, validated with the interested parties and

Table 2. Comparison of the life cycle of the methodology.

OECD Recommendation	Circular NO. A-130 USA	ENISA guide	OAS-Mexico Recommendations	ITU Guide	SENPLADES Guide
CII definition	Establish a Business Architecture	Define vision, scope, objectives and priorities	Hiring Consultants	Define processes, deadlines and identify participants	Diagnosis, conformation of the technical, political team and other participants
CII Identification	Define a Technical Reference Model	Evaluate the risks	Conformation of the work tables	Inventory and analysis of risks, deficiencies and problems	Collect information in a participatory manner and characterize the problem
Adoption of OECD regulatory frameworks	Incorporate the SI into the information architecture to support the business	Define existing policies, regulations and capabilities	Identification of topics of interest	Write the text of the NCS with the participation of the public, private and civil society sectors	Formulate, hierarchize and select action alternatives
	Prioritize key systems and implement OMB and NIST policies	Establish the governance structure	Discussion in detail at each table, facilitated by consultants	Approach in a structured way its implementation, with an Action Plan	Validate by teams: political technician and SENPLADES with citizen participation
	Schedule and budget SI investments in ICT	Identity and engage stakeholders	Development of the NCS, by international consultants	Supervision, evaluation and accountability	Approval of the Sector Council, incorporation into the system and dissemination to stakeholders
		Exchange of information			Execution or implementation
					Monitoring, evaluation and accountability
					Update, reformulate or delete the policy

for-mally approved. In the case of having an NCS already, the analysis is made again with the changes that have occurred in the current situation.

Formulation/Reformulation. A preliminary document of the NCS is structured and written through the participation of all stakeholders from each sector, consolidated at the national level and coordinated by the National Cybersecurity Authority (NCA), taking into account good practices and established principles. Then the preliminary document is disseminated and recommendations are received from the interested

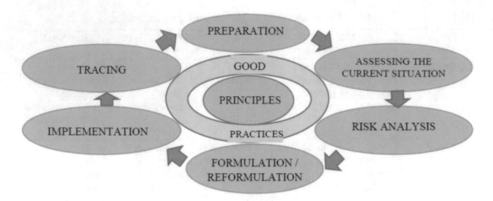

Fig. 1. NCS life cycle

parties and the general public, the recommendations are analyzed and the pertinent ones are incorporated, to then proceed to the official approval, incorporation into the National Planning System and dissemination.

Implementation. Specific programs and projects are prepared by those responsible for each objective, according to national regulations, projects and budgets are approved and resources are allocated for the execution of the actions provided for in the general plan of action that is consolidated by the NCA created as part of the organizational structure designed in the NCS.

Tracing. It includes the activities of permanent and continuous verification of the execution of each program and project, through the use of key metrics and indicators of compliance and closure of the initially approved projects.

5   Conclusions and Future Work

Several guides of good practices have been reviewed, to obtain elements that allow to configure a life cycle of an NCS, oriented to the Latin American reality, especially as part of the developing region. The social-economic vision, the ability and capacity available to face these threats of the region is important to choose and put emphasis on the components of the life cycle and its principles, but they really materialize in the set of good practices that link all the components of the designed iterative model. It should be taken into account that the model is used to create, maintain and update a NCS. As it is an eclectic model, several components have been taken from the models analyzed. For the purposes of this document, due to its extension, the presentation of the principles and other designed elements has been restricted.

As future work are pending guides and manuals of good practices to operationalize the model, considering international standards and good practices. Also the implementation and evaluation in situ of the model in Ecuador.

References

1. Parlamento Europeo: Directiva (UE) 2016/1148 del Parlamento Europeo y del Consejo, vol. 2014, no. 2, p. 30 (2016)
2. ENISA: NCSS good practice guide: designing and implementing national cyber security strategies, November 2016
3. Asamblea Constituyente: Constitución de la República del Ecuador, p. 218 (2008)
4. ICC: Guía de seguridad ICC para los negocios (2015)
5. OECD: Digital Security Risk Management for Economic and Social Prosperity. OECD Publishing (2015)
6. OMB: CIRCULAR NO. A-130 OMB Managing Information as a Strategic Resource, p. 13, USA (2016)
7. OEA: Recomendaciones para el Desarrollo de la Estrategia Nacional de Ciberseguridad-México, Mexico (2017)
8. ITU: Guide to developing a national cybersecurity strategy – strategic engagement in cybersecurity, 1st ed. NATO (2018)
9. SENPLADES: Guía para la formulación de políticas públicas sectoriales, p. 64, Ecuador (2011)
10. Villanueva, L.F.A., Porrua, M.A.: La implementación de las políticas públicas, p. 24, Mexico (1993)
11. Ellefsen, I.: The development of a cyber security policy in developing regions and the impact on stakeholders. In: Ist-Africa 2014 Conference Proceedings, p. 10 (2014)
12. Burchardt, H.-J.: La crisis actual de América Latina: causas y soluciones. Nueva Soc. **267**, 114 (2017)

Predicting Personal Susceptibility to Phishing

Ingvar Tjostheim[1](✉) and John A. Waterworth[2] (iD)

[1] Norwegian Computing Center, P.O. Box 114 Blindern, 0314 Oslo, Norway
Ingvar.Tjostheim@nr.no
[2] Umeå University, Main Campus, 901 87 Umeå, Sweden

Abstract. Phishing is a confidence trick with damaging impacts on both individuals and society as a whole. In this paper, we examine the possible role of thinking styles, as assessed by the Cognitive Reflection Test (CRT), and other factors to predict personal susceptibility to phishes. We report the results of two large-scale national studies conducted on cross-sectional populations in Norway. Using a binary logistic regression method, we analyzed the relationship between CRT scores, willingness to share data and demographic variables, to susceptibility to comply with phishes. Our main finding was that both an intuitive thinking style, as operationalized by the CRT scores, and willingness to share personal, significantly predict the probability of falling for phishing. As these results are based on two large-scale studies of national populations, they can be expected to have greater validity than earlier studies. The finding that CRT scores and other personal characteristics can predict the likelihood of falling for phishing suggests methods of pre-emptive testing of individuals as part of private and organizational strategies for encouraging improved resistance to phishing and other forms of personal data theft.

Keywords: Susceptibility to phishing · Cognitive Reflection Test · National samples · Data-disclosure · Privacy

1 Introduction

Phishing is a confidence trick aimed at getting unsuspecting people to give away personal details on the internet, so that the perpetrator can make fraudulent use of their credentials [1, 2]. Often, phishing attacks are very sophisticated so that even well-educated and cautious internet users are liable to be fooled into revealing key personal data. Many internet users have a tendency towards privacy-compromising behavior, revealing a divergence between their attitudes to privacy and their actual behavior [3–5]. According to Nicholson et al. [6] phishing is a prime example of where users are overconfident. Other factors are inattention, optimism biases, lack of rational behavior, limited mental resources and other "biases and heuristics - well known to behavioral researchers" (Acquisti et al. [5]: 32).

Many research findings are based on student samples, but the phenomenon is not only a problem for students and results with student samples are not necessarily representative of the general population. We carried out two studies targeting the national population of Norway, by recruiting large numbers of participants through two professional market research companies.

© Springer Nature Switzerland AG 2020
Á. Rocha et al. (Eds.): ICITS 2020, AISC 1137, pp. 564–575, 2020.
https://doi.org/10.1007/978-3-030-40690-5_54

According to Toplak et al. [8], the Cognitive Reflection Test (CRT) [7] has the capacity to function as a unique predictor of performance on a number of heuristics-and-biases tasks, as a measure of individuals' ability to suppress intuitive and spontaneous wrong answers in favor of correct answers requiring greater reflection. Our studies tested whether CRT scores are predictive of the tendency to fall for phishes and included questions about phishing and misuse of personal data as well as a choice experiment on sharing of personal data.

2 Related Work and Motivation for the Research

Ferreira and Vieira-Marques [9] provide an overview of ten years of phishing research based on 605 scientific journal abstracts. They conclude that there is no single solution to the phishing threat and, for future research, call for a *"focus on socio-technical and integrated solutions that can reflect a comprehensive understanding of both human computer interaction and **user unique characteristics**"* (our emphasis). Addressing this need to assess user unique characteristics was the main motivation for our research.

According to Volkamer et al. [10] and the APWG Internet Policy Committee Global Phishing Survey it takes, on average, 28.75 h to detect new phishing websites. Users are mostly unprotected from phishing until malicious websites are identified and blocked [11]. To avoid phishing during this period, users have to reflect on whether to go along with what they are being asked to do (for a phish to work), rather than simply complying. This motivates our research into intuitiveness (automatic decision-making behavior) versus reflective problem-solving styles in relation to the tendency to fall for phishing and willingness to share personal data, and why we chose to include a version of the CRT.

The CRT is often thought of as measuring "people's tendency to answer questions with the first idea that comes to their mind without checking it" (Kahneman [12]:65). This has been attributed to a tendency towards "miserly" information processing, to impulsively accept the solution to a problem that involves expending a minimum of cognitive effort [8, 13]. To score highly on the CRT, the respondent needs to reflect on and question their initial intuitive responses [14–16] and this involves cognitive effort. High CRT scoring corresponds to a personal tendency not to rely on intuition (which is fast), rather than analytical reasoning (which takes longer).

Bialek and Pennycook [16] discuss whether or not the cognitive reflection test is robust to multiple exposures. They suggest that it is and write that "...participants who do poorly on the CRT massively overestimate their performance (i.e., they do not realize they are doing poorly; Pennycook et al. 2017), indicating that intuitive individuals may have a metacognitive disadvantage (see also Mata et al.)." [17].

It could be argued that low scores on the CRT simply reflect low mathematical skill or general cognitive ability. But while these factors may influence their scores somewhat, they do not explain them completely [8, 13, 18–20]. The CRT aims to cue intuitions that are common across people and lead to potential responses from nearly all test-takers. Differences in scores can then be taken to reflect an individual's tendency towards reflective versus intuitive thinking. We suggest that the CRT is relevant for phishing research, since in a phishing context a fast and intuitive response style might be expected to correlate with higher vulnerability.

Several studies have used the CRT in relation to phishing susceptibility, though not with national populations. In a study with 42 students in a lab experiment, Kumaraguru et al. [21] found that the low CRT scoring group had a higher probability of clicking on the phishing-no-account e-mails than those in the high CRT scoring group, 0.39 versus 0.04, respectively. In their study with the classic three-items CRT, a CRT score of 0–1 (all wrong or one correct) was coded as the "low CRT group" and 2–3 (two or all correct) as the "high CRT group."

Butavicius et al. [22] performed a phishing study with 121 students. These researchers found a significant negative correlation between CRT scores and link safety judgments for spear-phishing ($\rho = -.23$, $p = .014$, $N = 112$) and phishing ($\rho = -.3$, $p = .001$, $N = 112$) emails, but no significant correlation between performance on the CRT and link safety judgments on genuine emails ($\rho = -.01$, $p = .973$, $N = 114$). Petraityte et al. [23] recruited 100 participants consisting of university students, lecturers and staff, and asked them to assess QR-codes. They found that less impulsive people (those with a higher CRT score) who did not know what the purpose of the test was responded better. Participants with higher CRT scores were less likely to click on the URL held inside the fake QR code. Cognitive impulsivity did not reveal any significant difference for the participants who were informed what the study was about. Finally, in a study by Jones et al. [24] with 224 university students and staff, the participants were asked to examine 36 emails (18 legitimate and 18 phishing emails). Although the analysis of the data primarily indicated that participants who demonstrated higher sensation seeking were poor at discriminating between phishing and legitimate stimuli, the authors write that "Performance on the CRT also predicted susceptibility.".

A further motivation for our work was the tendency that many have of sharing of personal data when they do not have to. In the digital economy, we pay with our data [25–27]. For many applications, we have to give consent to sharing, but not always. All Internet-users can be targeted by phishing and by requests to share personal data generally. We therefore chose to use national population samples rather than convenience samples such as a sample with students only.

3   Hypotheses

As mentioned above, there are many studies documenting that it is hard to detect fraud and phishing [9]. A low score on the CRT indicates a tendency towards intuitive decision-making [7, 8, 13]. Jones et al. [24, 30], in their phishing study, found that performance on CRT predicted susceptibility to phishing.

Based on these earlier studies, we argue that a reflective decision-making style measured with the CRT is one, but not necessarily the only, factor that will predict falling for phishing.

We expected that both level of education and a personal style of thinking would be predictors of an individual's likelihood of falling for phishes, with higher levels of education and more analytical styles of thinking being associated with lower susceptibility. Based on this, our two main hypotheses were as follows:

Hypothesis 1: Education is a predictor of susceptibility to phishing. In comparison to those with low education, those with high education are less susceptible.

Hypothesis 2: CRT scores are a predictor of susceptibility to phishing. In comparison to those with a low score on the CRT, those with high score on the CRT are less susceptible to phishing.

Previous research has also shown that females generally score lower on the CRT [7, 8, 18] and so we expected them to be more susceptible to phishing than males. However, studies on susceptibility to phishing have not found an effect of gender [2, 31]; and on the other hand, some studies have indicated that in some situations, men take more risks than women, e.g., [32]. Our third hypothesis concerns the possibility of gender differences in susceptibility to phishing, as follows:

Hypothesis 3: Gender is a predictor of susceptibility to phishing.

In their responses to the market research company, our participants were asked to provide access to their answers to previous surveys and their Facebook profiles. Our fourth hypothesis was based on the *a priori* assumption that people who were willing to share their personal data are more likely to fall for phishing, as follows:

Hypothesis 4: Willingness to share personal data is a predictor of susceptibility to phishing.

4   Method

We carried out two studies in cooperation with two different market research companies, to achieve our aim of national studies on an issue affecting a broad section of the population. The two surveys included questions from the Eurostat-survey about credit cards and misuse of data [28]. The formulation of these questions was discussed with the national bureau of statistics in Norway. This meant that the findings in our studies, the demographical profile and the number that reported falling for phish can be compared to statistical data published by the national bureau of statistics.

The Cognitive Reflection Test was used to assess participants' thinking styles, i.e. intuitive versus analytical. While in some countries many in the general public know the correct answers to the CRT [29] the CRT has not, to the best of our knowledge, been used in a national survey in Norway before. We also designed a behavioral measure concerning disclosure of personal data and demographics. We asked the participants for consent, to give us access to all the data about the participant that the market research company already had. Since the market research company was the data-processor, and we did not actually receive the data, we did not need ethical approval for the studies.

For the sample sizes we used the Eurostat-stat cybersecurity 2017 survey [28] as a guide. In this survey, 8% answered that they had experienced identity theft, that is someone stealing personal data and impersonating the person. On the basis of this we set a target of at least 100 respondents in each study who had experienced phishing in the past.

4.1   Participants and Survey Format

The participants were recruited from two panels, citizens that were 18 years to 79 years in study 1 and 16 to 69 years in study 2. In total, study 1 had 1340 respondents 18 to 79 years old, and 1148 with the age 18 to 69. In study 2 there were in total 1405 individuals aged 16 to 69 years. We excluded the 70 plus age group from study 1 in order to have a more similar age-profile for the two studies (Table 1).

Table 1. Age profile of the participants.

Age	18–19	20–29	30–39	40–49	50–59	60–69	70–79
Study1 (n = 1340)	2%	15%	19%	18%	21%	14%	14%
Study2 (n = 1405)	8%	21%	16%	18%	20%	18%	

In both studies 49% were male and 51% female. Persons of ages above 19 years were uniformly represented in our sample, with about 15–20% of participants in each age group (Table 1). Table 2 shown the education profiles of participants in the two studies, (not including those aged 70–79 in Study 1).

Table 2. Participants' educational profile.

	Primary	Secondary	1st degree	Higher degree
Study1 (n = 1148)	7%	35%	38%	20%
Study2 (n = 1405)	18%	36%	30%	17%

Participants received an email and answered the web-based survey on a PC or smart-phone, which took them about 10–15 min. The CRT-questions were in the last section. For the CRT, we used the open format in study 1. In study 2, 50% received the open format, and 50% the multiple-choice format for the three CRT-items.

4.2   Measures

The three measures used in the studies were the Cognitive Reflection Test, a self-reported measure on phishing, and a behavioral measure on disclosure of personal data and demographics. The open format, where the respondents fill in the answers them-selves, is the standard CRT format, but recently a multiple-choice format has been developed. The motivation for using a multiple-choice format, is to save time for the respondents [35]. In study 1 we used the open format whereas in study 2 we used both the open format and the 3-item multiple choice format, so that the results could be compared.

The context for our experiment on disclosure of personal data was that the participants in both studies had taken part in surveys before as panel members. The market research company has the answers to these surveys in their database but will normally not share this information with other clients. However, it is possible to link data and build a very detailed profile of each respondent based on answers to previous surveys. This was the context for our experiment on disclosure of personal data. We asked, in cooperation with the market research company, if we could have access to their answers to previous surveys and their Facebook profiles and with all these data build new profiles of them. The market research company, the data-processor, did not the share the personal data with us as client.

Both studies used two questions from the Eurostat-survey about credit cards and misuse of data. The formulation of these questions was discussed with the national bureau of statistics. In these questions, the word phishing is not used. Phishing is a term known in technological contexts, but its meaning is not known to all citizens.

5 Results

When asked, around 10% of participants reported that they had experienced misuse of personal data, which is similar to the numbers reported in the Eurostat-surveys. For the three CRT-questions, the mean time used for the open format was 186 s vs. 108 s for the multiple-choice format.

To test our hypotheses, we chose binary logistic regression with a dichotomous variable, 'has fallen for phish (yes/no)', as the dependent variable. One of the purposes was to investigate the question: is CRT score a good predictor of falling for phishing when we include the other factors gender, age, education and disclosure of data as variables?

Binary logistic regression is a form of regression used when the dependent variable is a dichotomy and the independent variables are of any type. Binary logistic regression can be used to predict a categorical dependent variable on the basis of continuous and/or categorical independent variables, in our case whether or not someone reports that s/he has fallen for phishing in the past. Table 3 presents descriptive statistics of the data in our analytical model.

Table 3. Descriptive statistics of data in our model.

	Min	Max	Mean	SD	Skewness	Kurtosis
Age:Study1 (1148)	18	69	43.9	14.1	0.01	−1.05
Age:Study2 (1405)	16	69	42.0	15.7	−0.02	−1.32
CRT:Study1	0	3	1.25	1.76	0.30	−1.42
CRT:Study2	0	3	0.97	0.72	−0.01	−1.54

Cases with probabilities above a given numerical cut-off are accepted. We chose 0.12 and 0.15 based on the percentages for falling for phishes in the two datasets. The binary logistic, with the chosen cut-offs 1 is categorised as success whereas cases lower than this cut off value are classified as 0 (failure). This method is used to test the null hypothesis that a linear relationship does not exist between the predictor variables and the log odds of the criterion variable.

Table 4. Education and CRT scores

	Education level	All wrong	One correct	Two correct	All three correct
Study 1	Primary education (84)	56%	16%	16%	13%
	Secondary education (398)	48%	22%	17%	13%
	University/college, lower level (432)	32%	23%	23%	22%
	University/college, higher level (234)	24%	16%	20%	40%
Study 2	Primary education (246)	48%	31%	15%	7%
	Secondary education (498)	49%	28%	16%	8%
	University/college, lower level (413)	40%	31%	15%	14%
	University/college higher level (244)	32%	27%	18%	23%

We tested two models, with SPSS version 25, the first with study 1 data and the second with study 2 data. Of our two models the first had a p-value smaller than 0.05 and the second a p-value larger than 0.05 (Table 5). The Nagelkerke R2 is a pseudo R-square and it is impacted by how lopsided the split of dependent variables is. It is often used to assess model adequacy [33]. The Nagelkerke R2 was 13.3% for study 1 and 19.6% for study 2. Misuse of credit-card and ID theft were coded as <u>one</u> binary variable. Skewness represents the extent to which scores have a tendency toward the upper or lower end of a distribution, while kurtosis indicates the extent to which a distribution of scores is relatively flat or relatively peaked. If the result is greater than ±2.0, the variable has a skewness problem. This was not the case for our two studies.

In the binary logistic models, we included gender, age, education, the CRT scores and the behavioral measure of data disclosure as variables. Table 4 shows that it was those with the longest education that performed best on the CRT-test. Since it has been shown that those with good mathematical skills or cognitive abilities often perform better on the CRT, we included an interaction effect of CRT and education in the model. The R squares indicated that the variables in the equation contributed to predicting the dependent variable falling for phishing.

In logistic regression models, the Hosmer-Lemeshow test [33, 34] is often used as a goodness of fit test. Our analysis found that the second model has a good fit, but not the first model. The first model has lower classification accuracy and not a good enough discrimination power according to this test. We report the data from both models in Table 5.

Table 5. The Binary Logistics Regression Model - Variables in the final equations

	Independent variables	Beta estimates	S.E.	Wald	P (Sig)
Study 1. Results of the individual predictors in step 3 (backward model)	Constant	−0.250	0.35	0.52	0.47
	Male/female	0.407	0.19	4.47	0.03
	Age	−0.340	0.07	24.05	0.00
	Disclosure no/yes	0.709	0.19	13.39	0.00
	CRT	−0.386	0.09	18.65	0.00
Study 2. Results of the individual predictors in step 3 (backward model)	Constant	0.446	0.28	2.63	0.11
	Male/female	0.385	0.17	5.42	0.00
	Age	−0.21	0.01	15.47	0.00
	Disclosure yes/no	−1.406	0,18	64.70	0.00
	CRT	−0.383	0.09	10.99	0.00
	- 2 Log likelihood	Cox and Snell R2	Nagelkerke R2	Hosmer and Lemeshow Test Chi-square (Sig.)	
Study 1. Results of the overall model, Classification accuracy: 66%	774.294	0.063	0.123	23.536 (0.003)	
Study 2. Results of the overall model Classification accuracy: 74%	1039.80	0.089	0.157	8.443 (0.391)	

We used the Wald statistic to identify the significant variables in the model. The Wald statistic is the square of the t-statistic and gives equivalent results for a single parameter and can be used to test the significance of particular predictors in a statistical model. As the method for selecting how independent variables are entered into the analysis, we choice backward Wald. The method analyzes the predictor variables and picks the one that predicts the most on the dependent measure. In the backward method, all the predictor variables chosen are added into the model. Then, the variables that do not (significantly) predict anything on the dependent measure are removed from the model one by one. The backward method is generally the preferred method, because the forward method might produce so-called suppressor effects. These suppressor effects occur when predictors are only significant when another predictor is held constant.

The Wald statistic estimates indicated that data disclosure behaviour, CRT scores and age were predictors of falling for phishing. The conclusions from our statistical models were as follows:

Hypothesis 1, that education is a predictor of susceptibility to phishing, was rejected. Education was not a predictor of falling for phish.

Hypothesis 2, that CRT scores are a predictor of susceptibility to phishing was supported. In comparison to those with a low score on the CRT, those with high score on the CRT were less susceptible to phishing.

Hypothesis 3, that gender is a predictor of susceptibility to phishing, was also supported. In both studies the Wald estimates indicated a gender difference, with men being *more* susceptible to falling for phish than women.

Hypothesis 4, that, willingness to share personal data is a predictor of susceptibility to phishing was supported. The respondents that gave consent were more susceptibility to phishing than those that did not.

6 Discussion and Future Work

Our results confirmed the potential of using the CRT as a test for the likelihood of a person's susceptibility to phishing. The CRT provides a useful tool for identifying one of the characteristics of people who would benefit from advice or tuition to help avoid falling for these damaging confidence tricks. Willingness to share data was also, unsurprisingly, associated with susceptibility to phishing. That females were less susceptible to phishing than males was unexpected, given earlier findings. This may reflect that we used a broad cross-section of the population rather than a convenience sample.

CRT has been developed and validated with student samples and very few studies have used the CRT with ordinary citizens, as we did in the present study. When a convenience sample is used, it may not be representative of the population at large so that the results are of limited generalizability. Those with less education and other groups such as the elderly might behave differently from students. National studies can serve as a reference for other studies. This is also why we cooperated with the national bureau of statistics on the wording of the questionnaire. However, it is harder to design experiments with national samples, since the participants are not in a controlled environment. There may also be ethical issues that are more challenging in uncontrolled environments, such as eliciting informed consent. Survey participants will have given consent, but are probably expecting to answer standard questions, and are less used to doing tasks such as the CRT problems.

In the USA and some other English-speaking countries, the CRT is quite well known. If a respondent knows the correct answer in advance, the CRT cannot be used as intended. This is one of the reasons why alternatives to the standard CRT have been developed, tested and used in some recent studies [35]. In non-English speaking countries, such as the country of this study, Norway, it has rarely been used, so that it is unlikely that respondents will know the answers already. However, if someone performs an online search, he or she will be able to find the correct answers easily.

The present results demonstrate that those with more education perform significantly better than others on the CRT. One of the strengths of using the CRT is that it is not a self-reported measurement but rather, assuming that the respondent does not search for the answer (or know the answer in advance), tells us about the respondent's individual behavior and characteristics. Our study indicates that individual citizens can perform well on the CRT without higher education. In our two logistics models that included demographics, a measure on data-sharing and the CRT, it was the data-sharing behavior, the CRT and gender that contributed significantly to predicting susceptibility to phishing, not education. However, we cannot know for certain that the respondents reported honestly when they answered questions and cannot be absolutely sure whether they have actually fallen for phishing or not.

Sirota and Juanchich [36] argue that the standard open format should be replaced by a multiple-choice format because it is less likely that someone will perform a search to find the answer. Another possible solution would be to use a timer and score slow answers as wrong. This approach was used by Da Silva et al. [37] and should be considered for future studies with the CRT.

We speculate that some participants who spent long times on the CRT questions may have searched for the answers online. In a controlled laboratory setting this is less likely to happen. Also, when a respondent is answering a survey on his or her PC or smartphone, they may be distracted and/or not really care much about the questions or the answers given [38]. In a more controlled environment this is less of a problem. Another drawback is the costs of recruiting many respondents for a large-scale study.

Nevertheless, since these results are based on large-scale studies of national populations, in contrast to relying on mostly student populations, we believe that they can be expected to have greater validity and reliability than results from earlier studies. Students by definition have relatively high levels of education, and the characteristics, attitudes and experiences of technology of groups of students can be expected to vary according to their specific study areas.

We have confirmed that the CRT is a useful predictor of which online users are likely to fall for phishing, but it is not the only measure to be recommended. We suggest that in future studies the CRT should be used in semi-natural phishing experiments, together with other measurements of risk propensity such as inattention, optimism bias or overconfidence [39]. This would help narrow down the personal characteristics underlying subjective susceptibility to phishes.

Our finding that CRT scores and other personal characteristics can predict the likelihood of falling for phishing suggests methods of pre-emptively testing and training individuals as part of private and organizational strategies for encouraging improved resistance to phishing and other forms of personal data theft. In forthcoming research, we will investigate the use of serious games on this topic, both as a way of detecting susceptibility to phishing and as a means for training adaptive responses to attempted phishes and other forms of identity theft.

Acknowledgements. This research was supported by Research Council Norway under the grant 270969, the research programme IKTpluss.

References

1. Jagatic, T., Johnson, N., Jakobsson, M., Menczer, F.: Social phishing. Commun. ACM **5** (10), 94–100 (2007)

2. Dhamija, R., Tygar, J.D., Hearst, M.: Why phishing works. In: Grinter, R., Rodden, T., Aoki, P., Cutrell, E., Jeffries, R., Olson, G. (eds.) Proceedings of the SIGCHI Conference on Human Factors in Computing Systems, CHI 2006, Montréal, Québec, Canada, 22–27 April 2006, pp. 581–590. ACM Press, New York (2006)
3. Acquisti, A.: Privacy in electronic commerce and the economics of immediate gratification. In: EC 2004 Proceedings of the 5th ACM Conference on Electronic Commerce, USA, pp. 21–29 (2004)
4. Barnes, S.B.: A privacy paradox: social networking in the United States. First Monday 11(9) (2006). http://firstmonday.org/article/view/1394/1312
5. Acquisti, A., Adjerid, I., Balebako, R., Brandimarte, L., Cranor, L., Komanduri, S., Leon, P., Sadeh, N., Schaub, F., Sleeper, M., Wang, Y., Wilson, S.: Nudges for privacy and security: understanding and assisting users choices online. ACM Comput. Surv. 50(3), 44 (2017). Article 44
6. Nicholson, J., Coventry, L., Briggs, P.: Can we fight social engineering attacks by social means? Assessing social salience as a means to improve phishing detection. In: Proceedings of the Thirteenth Symposium on Usable Privacy and Security, SOUPS 2017. USENIX, Santa Clara (2017)
7. Frederick, S.: Cognitive reflection and decision making. J. Econ. Perspect. 19(4), 25–42 (2005)
8. Toplak, M.E., West, R.F., Stanovich, K.E.: The Cognitive Reflection Test as a predictor of performance on heuristics and biases tasks. Memory Cogn. 39, 1275–1289 (2011)
9. Ferreira, A., Vieira-Marques, P.: Phishing through time: a ten year story based on abstracts. In: Proceedings of the 4th International Conference on Information Systems Security and Privacy, vol. 1, pp. 225–232 (2018)
10. Volkamer, M., Renaud, K., Reinheimer, B., Kunz, A.: User experiences of TORPEDO: tooltip-powered phishing email detection Comput. Secur. 71, 100–113 (2017)
11. Stockhardt, S., Reinheimer, B., Volkamer, M., Mayer, P., Kunz, A., Rack, P., Lehmann, D.: Teaching phishing-security: which way is best? In: 31st IFIP TC 11 International Conference on Systems Security and Privacy Protection, SEC 2016, vol. 471, pp. 135–149. Springer, New York (2016)
12. Kahneman, D.: Thinking. Fast and Slow. Macmillan, New York (2011)
13. Toplak, M.V., West, R.F., Stanovich, K.E.: Assessing miserly information processing: an expansion of the Cognitive Reflection Test. Think. Reason. 20, 147–168 (2014)
14. Pennycook, G., Cheyne, J.A., Koehler, D.J., Fugelsang, J.A.: Is the cognitive reflection test a measure of both reflection and intuition? Behav. Res. Methods 48(1), 341–348 (2016)
15. Pennycook, G., Rand, D.: Lazy, Not biased: susceptibility to partisan fake news is better explained by lack of reasoning than by motivated reasoning. Cognition 188, 39–50 (2018)
16. Bialek, M., Pennycook, G.: The Cognitive Reflection Test is robust to multiple exposures. Behav. Res. Methods. 50, 1953–1959 (2018)
17. Mata, A., Ferreira, M.B., Sherman, S.J.: The metacognitive advantage of deliberative thinkers: a dual-process perspective on overconfidence. J. Pers. Soc. Psychol. 105, 353–373 (2013)
18. Campitelli, G., Gerrans, P.: Does the cognitive reflection test measure cognitive reflection? A mathematical modeling approach. Memory Cogn. 42(3), 434–447 (2014)
19. Cokely, E.T., Kelley, C.M.: Cognitive abilities and superior decision making under risk: a protocol analysis and process model evaluation. Judgm. Decis. Making 4, 20–33 (2009)
20. Liberali, J.M., Reyna, V.F., Furlan, S., Stein, L.M., Pardo, S.T.: Individual differences in numeracy and cognitive reflection, with implications for biases and fallacies in probability judgment. J. Behav. Decis. Making 25, 361–381 (2012)

21. Kumaraguru, P., Rhee, Y., Sheng, S., et al.: Getting users to pay attention to anti-phishing education: evaluation of retention and transfer. In: Proceedings of the Anti-Phishing Working Group's Second Annual eCrime Researchers (2017)
22. Butavicius, M., Parsons, K., Pattinson, M., McCormac, A.: Breaching the Human Firewall: Social engineering in Phishing and Spear-Phishing Emails, May 2016
23. Petraityte, M., Dehghantanha, A., Epiphaniou, G.: Mobile phone forensics: an investigative framework based on user impulsivity and secure collaboration errors (Chap. 6). In: Contemporary Digital Forensic Investigations of Cloud and Mobile Applications, pp. 79–89. Syngress (2017)
24. Jones, H.S., Towse, J.N., Race, N., Harrison, T.: Email fraud: the search for psychological predictors of susceptibility. PLoS One **14**(1), e0209684 (2019)
25. Elvy, S.A.: Paying for privacy and the personal data economy. Columbia Law Rev. **117**(6), 1369–1459 (2017)
26. Hacker, P., Petkova, B.: Reining in the big promise of big data: transparency, inequality, and new regulatory frontiers. Northwest. J, Technol. Intellect. Prop. **16**, 1 42 (2017)
27. Greengard, S.: Weighing the impact of GDPR. Commun. ACM **61**(11), 16–18 (2018)
28. European Union 2017. 5661. Special Eurobarometer 464a "European attitudes towards cyber security", September 2017
29. McCall, R.: Can you pass the world's shortest IQ test? It's just three questions long, but few can get them all right (2017). http://www.iflscience.com
30. Jones, H.: What makes people click: assessing individual differences in susceptibility to email fraud (2016). eprints.lancs.ac.uk
31. Parsons, K., McCormac, A. Pattinson, M., Butavicius, M., Jerram, C.: Phishing for the truth: a scenario-based study of users' behavioural response to emails. In: IFIP International Information Security Conference, pp. 366–378. Springer, Berlin (2013)
32. Charness, G., Gneezy, U.: Strong evidence for gender differences in risk-taking. J. Econ. Behav. Organ. **83**, 50–58 (2012)
33. Hosmer, W., Lemeshow, S.: Applied Logistic Regression. Wiley, New York (1989)
34. Archer, K.J., Lemeshow, S., Hosmer, D.W.: Goodness of fit tests for logistic regression models when data are collected using a complex sampling design. Comput. Stat. Data Anal. **51**, 4450–4464 (2007)
35. Primi, C., Morsanyi, K., Chiesi, F., Donati, M.A., Hamilton, J.: The development and testing of a new version of the cognitive reflection test applying item response theory (IRT). J. Behav. Decis. Making **29**, 453–469 (2016)
36. Sirota, M., Juanchich, M.: Effect of response format on cognitive reflection: validating a two- and four-option multiple choice question version of the Cognitive Reflection Test. Behav. Res. Methods (2018). https://doi.org/10.3758/s13428-018-1029-4
37. Da Silva, S., Da Costa Jr., N., Matsushita, R., Vieira, C., Correa, A., De Faveri, D.: Debt of high-income consumers may reflect leverage rather than poor cognitive reflection. Rev. Behav. Finance **10**, 42–52 (2017)
38. MacKenzie, S.B., Podsakoff, P.M.: Common method bias in marketing: causes, mechanisms, and procedural remedies. J. Retail. **88**, 542–555 (2012)
39. Lejuez, C.W., Read, J.P., Kahler, C.W., Richards, J.B., Ramsey, S.E., Stuart, G.L., Strong, D.R., Brown, R.A.: Evaluation of a behavioral measure of risk taking: the Balloon Analogue Risk Task (BART). J. Exp. Psychol. Appl. **8**(2), 75–84 (2002)

Computer Networks, Mobility and Pervasive Systems

R-IoT: An Architecture Based on Recoding RLNC for IoT Wireless Network with Erase Channel

Yair Rivera Julio$^{(\boxtimes)}$ ⓘ, Ismael Gutiérrez Garcia ⓘ, and José Marquez ⓘ

Universidad del Norte, Barranquilla, Colombia
{yairr, isgutier, jmarquez}@uninorte.du.co

Abstract. A wireless IoT communication architecture is proposed that takes advantage of RLNC recording properties on natives packets of Core network. The platform aims to cover the shortcomings of the TCP/UDP/IP internet protocol for a traffic streaming intrasession sensitive to the delay in wireless networks since the current systems in native mode do not differentiate between transmission failures due to bottlenecks caused by congestion or by the characteristics of the random medium. The system establishes a mapping of the population of packets to be transmitted over an RLNC coding system based on Finite fields (Galois field), which allows forwarding nodes to eliminate redundant information on the network to decrease communication times and increase throughput. Moreover, thanks to its configuration, the model takes advantage of a Cross-Layer communication, to define a dynamic trade-off between the parameters that define the performance of the system, which translates into a coding density in a reduction of the delay in each of the transmissions. The communication platform is presented under a simulation scheme supported by (KODO and NS3) and is compatible with the Multicast, Unicast, and Broadcast transmission types established under a channel with erasure effect.

Keywords: Random Lineal Network Coding (RLNC) · Recoding · Internet of Thing (IoT) · Multicast · Unicast · Broadcast · Finite fields · TCP/UDP/IP · KODO and NS3

1 Introduction

According to the latest trends and updates, the population of IoT devices that connect to the web wirelessly to access high-performance streaming has increased exponentially [1]. Current systems and network devices allow this type of connection through the traditional paradigm of packet storage and forwarding; however, under this perspective, efficiency in using the spectrum efficiently is increasingly relevant. On the other hand, there is a waste of resources and coding at the network level, since the TCP/IP protocol does not contemplate a multicast transmission to a native mode in wireless networks [2]. In response to the above, progress has been made based on code theory, which is intended to break traditional schemes and cover the shortcomings of commercial protocols for the transfer of digital data wirelessly [3].

© Springer Nature Switzerland AG 2020
Á. Rocha et al. (Eds.): ICITS 2020, AISC 1137, pp. 579–588, 2020.
https://doi.org/10.1007/978-3-030-40690-5_55

This article is structured as follows: Sect. 2 presents the basics concepts necessary to understand Network Coding(NC), especially that which refers to RLNC coding and recoding, as well as the characterization of the system; Sect. 3 describes how RLNC it complements with the OSI model protocol stack. It is detailed how to establish the coding model taking into account a channel with erasure effect, as well as the definition of all transition of states through a Markovian system for events; Sect. 4 describes the screen format for RLNC proposed by Muriel and how to establish a trade-off between the different parameters of the coding with respect to the variables that directly influence the performance of the system; Sect. 5 describes the architecture of the coding model for a multicast IoT wireless network with erasure effect; Finally Sect. 6 describes the efficiency of the model through developed simulations..

2   State of the Art and Coding Model

The following Table 1 gives a series of terms to be used throughout the document.

Table 1. Terms to be used.

Heading level	Example
Native Packet	Uncoded Packet
Encoded Packet	A random combination of original Packets
Innovator Vector	Linearly independent vector to those received
Coding Sliding Windows	Number of packets of which the generation consists
Degrees of freedom	Number of packets involved in a codified packet
Generation	Native packet blocks g, selected for coding
GF(q)	Is the finite fields of q elements, where q is a prime power
Relay	Forwarding terminal
Source	Source terminal
Sink	Sink terminal
Degrees of freedom	Number of packets involved in a codified packet

Thanks to the development of Network Coding and the recoding of data through intermediate nodes, it is possible to shorten data transfer times [4]. NC allows data packets to be mapped to a system of finite fields GF(q), where q explicitly states the number of symbols handled by the system or the dimension in which the data packets and the coding matrix are formed, that is to say, $q = 2^m$, where m is the number of bits per symbol over the finite field [5]. This adaptation allows us to work a data packet transmission as a subset of vector subspaces. In addition to performing packet recoding without prior decoding; in other words, it is possible to work on a pre-processing of packets to exploit the algebraic properties of the system. This in order to establish a coding system to shorten the transmission times for sensitive streaming with respect to the delay [6]. One way to simplify the architecture and make it efficient is to adopt a data transfer with an intrasession flow, which means establishing linear combinations of packets on a data flow of the same nature [3]. Through the recoding in intermediate nodes, it is possible to eliminate those packets represented as linearly dependent vectors. This ultimately results in the elimination of duplicate information that is not needed in the transmission [7]. RLNC coding establishes a linear combination of randomly selected packets or symbols within the same generation or within the same packet population [8]. Each packet segment or data vector is associated with a C_i coding vector generated in the GF(q) space, which represents each of the entries in the coding matrix C. Each system symbol is encoded through formalized operations in the vector space GF(q) over the binary field F_2. Due to IoT devices contain a finite energy source for their operation, the *bit-by-bit* Xor (denoted by $\oplus$) operation is determined as the coding operator; Xor exerts minimal operational complexity in its execution. The model for starting the coding process first selects a random set of native packets P. The number of selected packets is defined by the slider or coding window; each symbol of the packet segment is associated with a coefficient c_i randomly generated by a uniform distribution(see Eq. 1) [9].

$$P' = \oplus_{j=1}^{g} c_{i,j} \cdot P_j, \forall_i \in [1, g] \tag{1}$$

Where g is the rank of the coding matrix, then each encoded packet P'_i is sent by the source node to a sink node or relay node, depending on the type of communication, either multiple-hop or one-hop. At the other end of the communication must complete a coding block or vector made up of a set of linearly independent vectors $P^T = \left[P'_1 \ldots P'_g \right]$. Each original packet of information sent is defined as the vector product $P' = C \times P$. For this reason, it is necessary to find C^{-1} to get all native packets P [10]. The decoding process is carried out thanks to a Gaussian elimination [11]. Finally, it is possible to optimize transmission times through relay nodes [12]. For this, a recoding without decoding is established, that is to say, a new coding on the encoded vectors, directly on linear coefficients, which allows eliminating the linearly dependent packets that are not necessary to form a complete rank matrix. The data transfer is modeled taking into account a channel under the effect of erasure, due to it is necessary to establish in the receiver a mechanism that allows us to obtain the information with a high probability of decoding. Specialized algorithms such as Belief Propagation (BP) allow optimal performance in packet decoding since its basic structure is based on Tanner Graph [13]. This allows us to take advantage of the redundancy of the packets generated linearly independent within the same generation of data, and even encoded with different degrees of freedom, to finally obtain a reproduction of the originally sent information [14, 15].

3 State of the Art and Coding Model

The core of the encoding system is located on the transport layer of the internet protocol. This sublayer facilitates compatibility with the services of the upper layers in communication. The identification of parameters and the exchange of information takes place through a Cross-Layer process, thus establishing a collaborative system with feedback support for the control of errors (see Fig. 1).

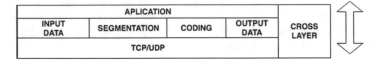

Fig. 1. Communication RLNC Cross-Layer

The process could be repeated without any problem in each Relay node, guaranteeing the elimination of retardation errors when creating new packets or linear combinations in the communication path. A significant advantage of this process through RLNC is the ability to generate this new recoding without adding computational complexity, only with the linear coefficients (see Eq. 2).

$$P'' = \oplus_{j=1}^{g} c_{i,j} \cdot P'_j, \forall_i \in [1, g].$$ (2)

As the linearly dependent vectors are eliminated, there is a higher probability of obtaining an innovative vector to increase the rank of the decoding matrix. On the other hand, it is essential to consider that you want to transmit a file sensitive to the delay and changes of the wireless medium. Therefore, it is known that the quality of the channel between the origin and the destinations is defined by the Frame Error Rate (FER) parameter [16]. Under this scenario, the probability that the packet will reach all destinations (R), is $(1\text{-FER})^R$. Therefore, it can be deduced that the lost probability is: Pro = $1-(1-FER)^R$ [17]. Based on the binomial distribution, it is possible to obtain the probability of X packets lost from a total of L packets to transmit from a generation, and a total of T selected native packets for the generation of encoding. (see Eq. 3).

$$P_{Perdidos}(X, T) = \binom{L}{X} \left((1 - FER)^R\right)^X \times (1 - (1 - FER))^{T-X}$$ (3)

Thanks to RLNC recoding, the likelihood of packets reaching all destinations (R) is (1-FER). This is because each forwarding node exerts its linear combination of vectors, accordingly is being generated a new set of packets; whereby, the probability of loss is lower. This scenario Taking into account the propagation of the error could be modeled through a probability transition scheme characterized by Markov absorbent chains for discrete events [14]. This model allows mapping the system to a binomial probability model with transient states, where each state represents the probability of receiving a packet linearly independent of the previous one, that is, increasing the rank of the matrix through innovative vectors. In the Markovian model, each state of the string is represented by the relation(r, e), where r corresponds to the rank of the matrix of the decoder nodes and e the number of non-null columns of the decoding matrix, in consequence, $r \leq e$ [18]. On the other hand, there is the state (k, k), which is the only absorbent state of the chain, since the different ranks both in the intermediate node and in the final node have no chance of decreasing. Any node that reaches this state is because it has a full rank matrix; as a consequence, it can finally decode and get the original information sent (see Fig. 2) [19].

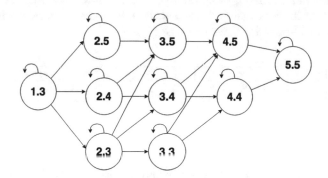

Fig. 2. Markovian model, with k = 5 and w = 3 packet segments.

Generically, to pass from one state (r, e) to another (r + i, e + j), where (i, j) ∈[0,1] it is necessary to define a total of 4 transitions of probabilities T(i, j) (see Eq. 4) [20].

$$
T_{ij}\begin{cases} \beta_1\beta_2 + \beta_1(1-\beta_2)\left(\frac{Q^e}{Q^r}\right)\beta_2 + \beta_1(1-\beta_1)\left(\frac{Q^r}{Q^k}\right)(1-\beta_2)\left(\frac{Q^e}{Q^r}\right) & ,i=0, j=0 \\ \beta_2(1-\beta_1)\left(\frac{Q^r}{Q^k}\right) + \left(\frac{Q^e}{Q^{r+1}}\right)(1-\beta_2)(1-\beta_1)\left(\frac{Q^r}{Q^k}\right) & ,i=1, j=0 \\ (1-\beta_1)\left(1-\frac{Q^r}{Q^k}\right)\left(\frac{Q^e}{Q^{r+1}}\right)(1-\beta_2) & ,i=1, j=1 \\ \beta_1(1-\beta_2)\left(1-\frac{Q^e}{Q^r}\right) + (1-\beta_1)\left(\frac{Q^r}{Q^k}\right)(1-\beta_2)\left(1-\frac{Q^e}{Q^r}\right) & ,i=0, j=1 \end{cases}
\tag{4}
$$

Where β is the probability of getting a vector linearly dependent on those already generated, therefore 1−β is the probability of obtaining a vector linearly independent. For each state (r, e), if it receives a packet with innovative coefficients, the rank of the decoding matrix is always increased. Hence there is a probability of generating an innovative vector [21]. Now, with an emphasis on coding and computational complexity, each linear coefficient generated in the Galois field finally determines the computational complexity of the coding system. This implies that the design of the system must contemplate a trade-off between complexity and Overhead since as the dimensions of the GF(q) system increase, the density of coding also increases. The increase in the dimensions of the generating system leads to a rise in the Throughput; however, it is also necessary to have higher hardware requirements for processing. On the contrary, with fewer dimensions the complexity of the system and the density of coding decreases, although it implies a higher probability of generating more linearly dependent vectors [22]. High complexity in coding also implies a high latency brought to the processing, if there is that the power consumption increases, therefore, the requirements of IoT hardware must be more robust [23].

4 Coding-Recoding Frame Format and Trade off of Parameters

Thanks to the frame format, designed by Medar Muriel, the RLNC-based coding system, it is possible to dynamically adjust the GF(q) coding density within a single transmission for intrasession streaming (see Fig. 3).

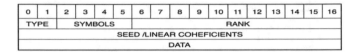

0	1	2	3	4	5	6	7	8	9	10	11	12	13	14	15	16
TYPE		SYMBOLS				RANK										
SEED /LINEAR COHEFICIENTS																
DATA																

Fig. 3. Package structure for RLNC encoded [24].

This parameter update can be established jointly between the RLNC coding and the other flow and congestion control mechanisms, this to improve the data transmissions under a channel with erasure effect. Each of these parameters is described below; TYPE: Indicates whether the coding is systematic, encoded or recodified. SYMBOLS: Indicates the number of symbols or base of the GF(q) system, DATA: segmented native information, RANK: Represents the current rank of the encoder, LINEAL COHEFICIENTS: encoding vector generated in the Galois field GF(q) [8]. Because the model was designed to be compatible with a communication model developed by layers, it is necessary to use the Cross-Layer technique, this to improve the efficiency and performance of the system, in addition to adapting transmissions to the wireless channel status. This mechanism allows decreasing the error in the transmissions through an optimal adjustment of parameters from the application layer [25]. To rearrange the dimensions of the fields to generate the codification in the segments of the fields, are properties that commercial protocols such as TCP/IP do not have. These handle a fixed size Maximum Transmission Unit (MTU) frame format, the only size of which depends on the communication standard, without packet segmentation. For this reason, RLNC facilitates the distribution of operations under a channel with erasure effect [26]. Depending on the frame format, codifications coefficients can also be generated by implementing a pseudo-random generator number; for this, only the generating seed is stored in the data frame. The design of the coding model contemplates what each parameter is configured from the application layer, which makes it easier to adjust these parameters from the beginning of the data transfer. A particular case would be the choice of symbol size, which depends on direct specifications, for example, the transmission systems establish a maximum MTU transmission unit for information fragmentation. Therefore, the selection of the symbol must be linked to the MTU of the system, in such a way that an equilibrium can be established concerning the GF(q) symbol generating system, i.e., a flexible symbol representation adjustable to the size of the MTU can be defined [27]. Advanced solutions introduce independent coding between the transport layer and the network layer of the protocol stack, allowing flow control through the sliding window. While this mechanism is compatible in a recoding environment, it nevertheless increases the coding delay by increasing the interaction at the transport and network layer level with the coding processes. Modifications of this operation establish a trade-off between the sliding window and the computational complexity [28]. A larger coding window increases the probability of obtaining an innovative vector, thus amortizing the retransmission or jacks generated by the Forward Error Correction (FEC), even though it also increases the computational complexity and resources of the system [29]. The generation size or sliding window could be configured through a session exchange, which establishes at the beginning of the communication.

5 Service Architecture and Coding

The problem with Unicast and Multicast transmissions is that they provide a definite utilization of network resources, A scenario that is re evaluated by NC through the use of coding in Relay nodes, so that there can be communication between sink nodes and relays, whit the finally to generate innovative vectors for increasing the rank of the decoding matrix. NC allows the introduction of new scenarios in data transmissions by performing a single transmission of encoded packets that benefits one or many receivers. The sender node establishes linear combinations of packets; therefore, it has the information potentially necessary to correct errors in different terminals and even in forwarding nodes. RLNC's efficiency in the forwarding process focuses on its ability to create new innovative vectors through the recoding process on these relay nodes [30]. This recoding process allows minimizing the signaling between two communication devices since the random combinations provide an implicit solution for coordination, which is evidenced by optimizing the FEC traffic for error control (see Fig. 4) [31].

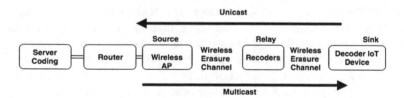

Fig. 4. Flow recoding architecture - streaming intrasession for IoT.

As a significant advantage, the decoder nodes have a memory buffer, which allows sharing the loads caused by the information forwarding of lost packets [32].

6 Simulation and Design of Experiments

Several scenarios will be presented, all of them based on the scheme established by Fig. 4, represented through the ns-3 simulator and a coding system based on the Kodo libraries for coding and recoding based on RLNC. An important parameter to vary is the FER of the link for the same data transmission with different configurations in the ground field: GF (2), GF (2^4) and GF (2^8) (see Fig. 5)

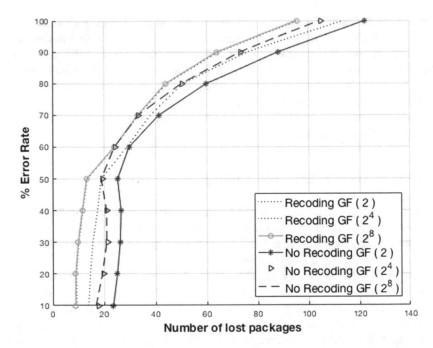

Fig. 5. Sending the same information flow with two Relays nodes, the channel modeled under erasure effect, a packet size of 1000 Bytes per packet and a generation Size of 3.

Thanks to the recoding, fewer packets are needed to send the same information, especially for extension fields GF (2), actually GF (2^4) and GF (2^8). This relation is because, as the size of the field increases, the probability of generating linearly dependent vectors decreases [33]. Recoding involves eliminating redundant and unnecessary information through processing at forwarding nodes. Property that is retained even with a high error rate in the transmission channel, as evidenced in Fig. 5. Finally, the process shortens transmission times by establishing a minimum latency in the delivery of data packets. All this complement of mechanisms establishes high end-to-end communication reliability for delay-sensitive streaming traffic.

7   Conclusion

By meeting the high demand for next-generation IoT services over wireless networks, low-latency and highly reliable communications must be guaranteed. Accordingly, it is essential to design flexible connection architectures that allow exploiting the coding of data packets through the new coding schemes. The proposed method aims to identify the parameters necessary to establish a dynamic trade-off between the variables involved in the coding concerns, in such a way as to offer a smaller number of transmissions at a higher throughput. Although some proposals have already been submitted under the RLNC coding system, the proposed scheme intends to be the starting point to characterize the behavior of the recoding schemes in the network to eliminate unnecessary information in an IoT network. This scheme allows generating stable proposals, especially in wireless networks, new architectures that enable you to differentiate between data congestion created by bottlenecks and random changes to the environment. All to increase the reliability and stability of the system.

References

1. Banafa, A.: Internet of Things: opportunities and challenges. OpenMind, 16–27 (2015). https://doi.org/10.13140/rg.2.1.1113.8409
2. Siasi, N., Sulieman, N.I., Gitlin, R.D.: Ultra-reliable NFV-based 5G networks using diversity and network coding. In: 2018 IEEE 19th Wireless Microwave Technology Conference WAMICON 2018, pp. 1–4 (2018). https://doi.org/10.1109/wamicon.2018.8363900
3. Stolpmann, D., Petersen, C., Eichhorn, V., Timm-Giel, A.: Extending on-the-fly network coding by interleaving for avionic satellite links. In: 2018 IEEE 88th Vehicular Technology Conference, pp. 1–5. IEEE (2018)
4. Sun, Q.T., Li, S.Y.R., Li, Z.: On base field of linear network coding. IEEE Trans. Inf. Theory 62, 7272–7282 (2016). https://doi.org/10.1109/TIT.2016.2613988
5. Márquez Díaz, J., Gutiérrez García, I., Valle Herrera, S., Falco Pastrana, M.: Packet output and input configuration in a multicasting session using network coding. KSII Trans. Internet Inf. Syst. 13, 686–710 (2018). https://doi.org/10.3837/tiis.2019.02.012
6. Garrido, P., Gomez, D., Lanza, J., Serrat, J., Aguero, R.: UPCommons providing reliable services over wireless networks using a low overhead random linear coding scheme, 1–11 (2016). https://doi.org/10.1007/s11036-016-0731-7
7. Calvo, R.A.: Intra-flow sparse network coding : modeling and open issues (2018)
8. Ning, H., Ling, C.: Network coding. Heterog. Cell Netw. Theory Simul. Deploy. (2011). https://doi.org/10.1017/CBO9781139149709.015
9. Medard, M., Fouli, K., Heide, J., Shi, S.: Random Linear Network Coding (RLNC)-based symbol representation draft-heide-nwcrg-rlnc-02, 1–13 (2020)
10. Soon, Y.O., Lee, E.K., Gerla, M.: Adaptive forwarding rate control for network coding in tactical manets. In: Proceedings - IEEE Military Communication Conference MILCOM, pp. 1381–1386 (2010). https://doi.org/10.1109/milcom.2010.5680142
11. Ho, T., Viswanathan, H.: Dynamic algorithms for multicast with intra-session network coding. IEEE Trans. Inf. Theory 55, 797–815 (2009). https://doi.org/10.1109/TIT.2008.2009809
12. Pahlevani, P., Hundebøll, M., Pedersen, M.V., et al.: Novel concepts for device-to-device communication using network coding. IEEE Commun. Mag. 52, 32–39 (2014). https://doi.org/10.1109/MCOM.2014.6807944
13. Chen, T., Vakilinia, K., Member, S., Divsalar, D.: Protograph-based raptor-like LDPC codes, pp. 1–12 (2012)
14. Garrido, P., Lucani, D.E., Agüero, R.: Markov chain model for the decoding probability of sparse network coding. IEEE Trans. Commun. 65, 1675–1685 (2017). https://doi.org/10.1109/TCOMM.2017.2657621
15. Marquez, J., Gutierrez, I.: Coding and decoding packet in a multicast network: programing test. IEEE Lat. Am. Trans. 16, 598–603 (2018). https://doi.org/10.1109/TLA.2018.8327418
16. Garrido, P., Lucani, D.E., Agüero, R.: Role of intermediate nodes in sparse network coding: characterization and practical recoding. In: European Wireless 2017 - 23rd European Wireless Conference, pp. 314–320 (2017)

17. Garrido, P., Agüero, R.: Caracterización experimental del comportamiento de Network Coding para comunicaciones multicast, 288–293 (2017). https://doi.org/10.4995/jitel2017. 2017.6601
18. Chatzigeorgiou, I., Tassi, A.: Decoding delay performance of random linear network coding for broadcast, 1–10 (2017). https://doi.org/10.1109/tvt.2017.2670178
19. Jones, A.L., Chatzigeorgiou, I., Tassi, A.: Binary systematic network coding for progressive packet decoding. In: IEEE International Conference Communications 2015-September, pp. 4499–4504 (2015). https://doi.org/10.1109/icc.2015.7249031
20. Zhao, X.: Notes on "exact decoding probability under random linear network coding". IEEE Commun. Lett. **16**, 720–721 (2012). https://doi.org/10.1109/LCOMM.2012.041112.112564
21. Skevakis, E., Lambadaris, I.: Optimal control for network coding broadcast, 1–6 (2016). https://doi.org/10.1109/glocom.2016.7842079
22. Garrido, P., Sorensen, C.W., Lucani, D.E., Aguero, R.: Performance and complexity of tunable sparse network coding with gradual growing tuning functions over wireless networks. In: IEEE International Symposium Personal, Indoor, and Mobile Radio Communications PIMRC (2016). https://doi.org/10.1109/PIMRC.2016.7794915
23. Lucani, D.E., Pedersen, M.V., Ruano, D., et al.: Fulcrum network codes: a code for fluid allocation of complexity (2014). https://doi.org/10.1109/INFCOM.2009.5061931
24. Shi, S., Fouli, K.: Random Linear Network Coding (RLNC)-based symbol representation draft-heide-nwcrg-rlnc-background-00 draft-heide-nwcrg-rlnc-01, 1–17 (2019)
25. Yang, Y., Member, S., Chen, W., et al.: Truncated-ARQ aided adaptive network coding for cooperative two-way relaying networks : cross-layer design and analysis, 1–15 (2016). https://doi.org/10.1109/access.2016.2604323
26. Sundararajan, J.K., Shah, D., Médard, M., et al.: Network coding meets TCP: Theory and implementation. Proc. IEEE **99**, 490–512 (2011). https://doi.org/10.1109/JPROC.2010. 2093850
27. Pedersen, M.V., Lucani, D.E., Fitzek, F.H.P., et al.: Network coding designs suited for the real world: what works, what doesn't, what's promising. In: 2013 IEEE Information Theory Workshop ITW (2013). https://doi.org/10.1109/itw.2013.6691231
28. Mohandespour, M., Govindarasu, M., Wang, Z.: Rate, energy, and delay tradeoffs in wireless multicast: network coding versus routing. IEEE Trans. Mob. Comput. **15**, 952–963 (2016). https://doi.org/10.1109/TMC.2015.2439258
29. Huang, W., Li, X., Jiang, Y., et al.: Reliable hybrid systematic network coding for multicast services in 5G networks. In: IEEE International Conference on Communications (2018). https://doi.org/10.1109/ICC.2018.8422494
30. Feizi, S., Lucani, D.E., Sørensen, C.W., et al.: Tunable sparse network coding for multicast networks. In: 2014 International Symposium on Network Coding, NetCod 2014 - Conference Proceedings (2014). https://doi.org/10.1109/netcod.2014.6892129
31. Szabó, D., Csoma, A., Megyesi, P., et al.: Network coding as a service. Infocommun. J. **7**, 2–11 (2015)
32. Cai, S., Zhang, S., Wu, G.: Minimum cost opportunistic routing with intra- session network coding. In: 2014 IEEE International Conference on Communications, pp. 502–507 (2014)
33. Farzamnia, A., Zhen, L.H., Fan, L.C., Islam, N.: Investigation on decoding failure probability in erasure network coded channels, 4–5 (2017)

Evaluation of TEEN and APTEEN Hybrid Routing Protocols for Wireless Sensor Network Using NS-3

Orlando Philco Asqui[1]([⊠]), Luis Armando Marrone[2],
and Emily Estupiñan Chaw[1]

[1] Faculty of Technical Education for Development,
Catholic University Santiago de Guayaquil, Guayaquil 090101, Ecuador
luis.philco@cu.ucsg.edu.ec,
emily.estupinan.chaw@gmail.com
[2] National University of La Plata, B1900 La Plata, Argentina
lmarrone@linti.unlp.edu.ar

Abstract. Wireless sensor networks (WSN) with hundreds or thousands of sensor nodes can collect information from a location and transmit the data to a particular user, depending on the application. These sensor nodes have some limitations due to their limited power, storage capacity and computing power. Data is routed from one node to another using different routing protocols. The position of the nodes can significantly affect the useful life of the network. The distribution of a non-uniform node (random location) in a certain area can cause many collisions between packets and therefore unbalanced traffic. Two routing protocols are evaluated; TEEN and APTEEN. As a methodology, the empirical method is used to simulate scenarios with high data traffic, as well as the analytical method to design the connection or topology that converges in energy efficiency. The result suggests that the routing protocol APTEEN in hierarchical cluster topology is the best efficiency in energy consumption.

Keywords: WSN · Energy · Ns-3 · TEEN · APTEEN · QoS · Efficiency

1 Introduction

A wireless sensor network (WSN), are embedded in personal area networks and form a group of spatially dispersed and dedicated sensors to monitor and record the physical conditions of the environment and organize the data collected in a central location (coordinating node) one type WSN network is the Zigbee network, which is characterized by short distance wireless communication and low power consumption [1]. But in urban outdoor environments the signal shared by the sensors tends to be lost and energy efficiency arises as problematic when they communicate with each other and the link is failed [2]. A WSN achieves or operates through heterogeneous nodes autonomously, without dependence on any infrastructure [3]. To access the information present in sensors that form a network, these devices with internal processing capacity must have a process called routing. This technical aspect is a difficulty of the WSN when causes such as high traffic or data loss limit the maximization of the sensor's

Á. Rocha et al. (Eds.): ICITS 2020, AISC 1137, pp. 589–598, 2020.
https://doi.org/10.1007/978-3-030-40690-5_56

network life by continuously routing the collected data (information) to the base station [4]. The routing problem is used to determine a set of different routes with the maximum lifetime added while limiting the life of each sensor for its initial battery life [5]. In WSN, energy efficiency is required to send collected information to the central base station (coordinating node or gateway). At the base station, the received data is further processed, specifically, the maximum number of energy efficiency routes is plotted so that each route is a set of selected sensors (instead of all sensors). Based on the routing protocol, WSN can be granted acceptable performances or performances, as frequent repetitions of failed communications cause the battery to discharge, and energy consumption can be affected by failed link scenarios [6].

The rest of the research work is divided as follows: in Sect. 2, work related to routing protocols is presented, in Sect. 3 the simulations are presented in NS-3, in Sect. 4 the results are presented and in Sect. 5 shows the conclusions.

2   Related Works

Results of investigations with WSN deployed networks with critical data in fire detection application in forest or cities, identify speed problems or data transfer in a WSN when monitoring and data processing must be in real time [7]. Works related to hierarchical routing protocols are detailed [8–16] (Table 1):

Table 1. Hierarchical routing protocols

Routing protocol	Characteristic	Scientific support with indexed publications
Low-Energy Adaptive Clustering Hierarchy (LEACH)	Minimizes power dissipation with random selection of nodes as group heads	[8–10]
Power-Efficient Gathering in Sensor Information System (PEGASIS)	It allows the nodes to communicate exclusively with their closest neighbors. Use a turn-based strategy to communicate with the base station	[11, 12]
Threshold-sensitive Energy Efficient protocol (TEEN)	The nodes are in constant study of the medium, however, the data is transmitted less frequently	[13, 14]
Adaptive Periodic TEEN (APTEEN)	Change the period or limit values used by the TEEN protocol according to the needs of the user and the type of application	[15, 16]

The authors [17, 18] consider the LEACH protocol as one of the basic and simple two-layer cluster routing techniques in WSN, in which single-hop communication is used between the base station and its cluster head (Cluster Head, CH). However, the authors [19] presented the Threshold-sensitive Energy Efficient Network (TEEN) protocol, which is a data-centric protocol, was designed for critical time applications. Later the same authors [20] make an improvement and design Adaptive Threshold TEEN or only APTEEN. Today both protocols are classified as hybrids. To obtain better data in the process of communication between nodes, interference in high-density deployments of the nodes should be reduced [21]. Hierarchical type protocols apply a structure in the network to obtain energy efficiency, stability and scalability.

In this class of protocols, network nodes are organized into groups in which a node with greater residual energy, for example, assumes the role of a cluster head (Cluster Heads, CH). The group leader is responsible for coordinating activities within the group and forwarding information between groups. Clustering has the potential to reduce energy consumption and extend the life of the network, it also uses a data-centric approach to disseminate interest within the network [22]. APTEEN's architecture is the same as that of the TEEN protocol, which uses the concept of hierarchical energy and efficient communication nodes. The APTEEN protocol mainly supports three types of consultation [23]:

1. One-time consultation
2. Historical consultation
3. Persistent query.

When a sensor node detects an event, the data can be transmitted at that specific moment. In most conditions, the quality of the sensory data is not as important as the conserved energy and the network coverage. To reduce the energy intake at the nodes, the network requires reducing the quality of the results.

3 Analysis in the NS-3 Simulator

The simulator NS-3, a free simulator for educational and research use, free of license and available for any platform such as Linux, Windows, Mac OS, is used [24]. This research work is used in the Linux platform (Ubuntu) using the NS-3.29 simulator.

By using the NS3 simulator version 3.29, plus NetAnim xml file analysis tool. that contain traffic flow data generated in the simulation. The data in the network is generated by random variables with an exponential continuous probability distribution, like the behavior of a network with critical data or persistent query.

The WSN network is designed in two topologies that can support nodes with random installation, these topologies are; Star and mesh with 100, 120, 160 and 200 nodes. The Whireshark tool is used to analyze throughput between nodes.

4 Results

This part presents the results obtained from each scenario and checks the metrics obtained when executing the simulations. The TEEN and APTEEN protocols are evaluated in star and mesh networks with 100, 120, 160 and 200 nodes.

4.1   Delay Time

When 160 and 200 nodes are simulated the delay maintains an average of 0.1 s with TEEN, while the delay with APTEEN is between 0.057 and 0.06 s (Fig. 1).

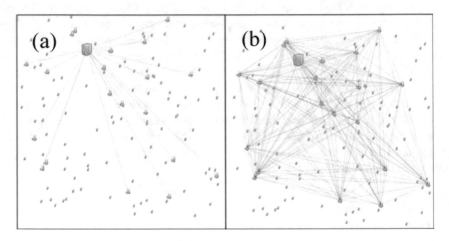

Fig. 1. Communication between nodes (a) Topology star. (b) Topology mesh.

The TEEN and APTEEN protocols are evaluated, in networks with star and mesh topology with different scenarios of 100, 120, 160 and 200 nodes with an area of 300 m × 300 m, network performance with 300, 400, 700 and 1000 nodes is also evaluated using the APTEEN protocol with an area of 700 m × 700 m. Some metrics for the behavior of the WSN network are analyzed, such as (Fig. 2):

1. Delay Time
2. Energy Consumption
3. Throughput
4. QoS

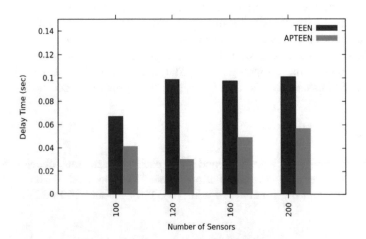

Fig. 2. Delay result with hybrid protocols.

4.2   Energy Consumption

Regarding energy consumption, Fig. 3 shows that the TEEN protocol provides a lower energy dissipation value while increasing the number of nodes. The nodes are randomly placed in the network. All nodes begin with an initial energy of 22 J. However, its radio model is modified to include power dissipation in idle time and detection power dissipation (equal to 10%).

It is indicated that this result is given because TEEN only transmits critical time data while continuously detecting the environment. To overcome the inconveniences of TEEN, the transmission of periodic data is incorporated to form APTEEN. Therefore, as simulated, the TEEN protocol is the one that produces less energy consumption and as stated above, the more nodes it has the more energy savings it provides.

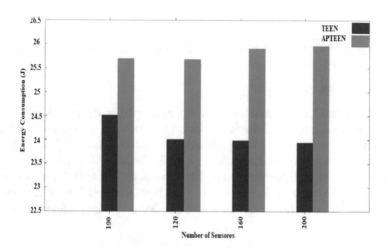

Fig. 3. Energy Consumption of the network.

4.3   Throughput

The performance or performance of the network quantify the transfer of information (Mbps) within the communication between nodes, for better analysis is based on terms of dead nodes and the number of routes of the sensor nodes. Using the NS-3 simulator, the traffic generated by each node sensor can be sent through multiple routes, in the place of a single route, and the appropriate protocol for an important energy conservation. The result of Fig. 4 indicates that the performance in 100 nodes reaches approximately 0.06 Mbps with TEEN and 0.1 Mbps with APTEEN. With 200 sensor nodes the performance is 0.04 Mbps with the TEEN protocol and 0.066 Mbps with APTEEN. Sensor energy is used very efficiently by reducing the number of non-critical data transmissions. The user can change the periodicity, the threshold values and the parameter that will be detected in different zones. This network can emulate the proactive or reactive network by properly changing the periodicity or threshold values. Therefore, this network can be used in any type of application by properly configuring the various parameters.

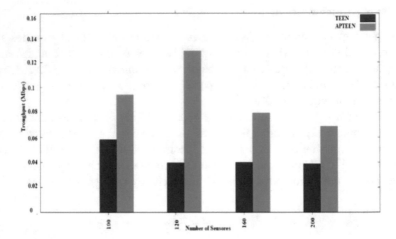

Fig. 4. Network Performance with TEEN and APTEEN Hybrid protocols.

4.4   QoS (Quality of Service)

Figure 5 shows the QoS with an assessment to provide quality of service (QoS) in WSN in medium access control (MAC) and network layer (x axis; minimum to maximum) according to the number of nodes that communicate within the network. It can be seen in Fig. 6 that better QoS is obtained with APTEEN, both in 100, 160 and 200 nodes. However, it is indicated that the standard studied is not suitable for high-speed data applications, such as multimedia transmission. On the other hand, other standards such as Bluetooth and Wifi have higher data rates and use QoS mechanisms for communication.

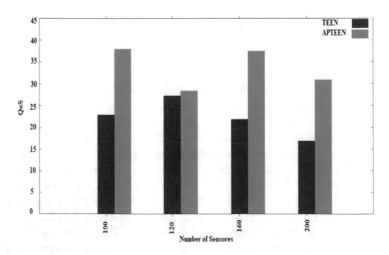

Fig. 5. QoS between 100 to 200 nodes.

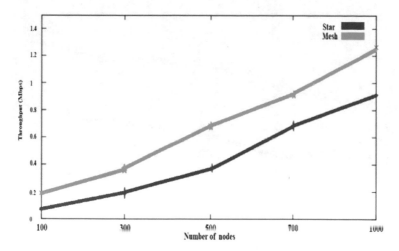

Fig. 6. Throughput protocol APTEEN.

Next, the network performance with 300, 400, 700 and 1000 nodes are evaluated using APTEEN. In the performance of the network using two topologies, the scenario has been implemented in terms of dead nodes and the number of routes of the sensor nodes. The "x" axis was represented by the number of nodes and the "y" axis will represent the number of the Throughput. Figure 7 shows the delay time (delay) of the APTEEN protocol in all tests. The mesh topology obtains less delay (Δ 0.07 s.) Up to 500 nodes, from 700 nodes onwards the mesh topology obtains acceptable delay compared to the star topology (Δ 0.34 s.) With significant delays. Both mesh topology with APTEEN communications between nodes are caused by queries or events, when a user sends a query, it is stored on demand. Figure 8 shows results of energy consumed in Joules when APTEEN is used with the two topologies mentioned above.

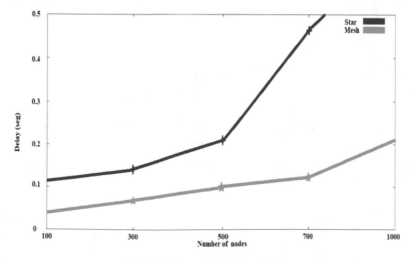

Fig. 7. Delay APTEEN protocol.

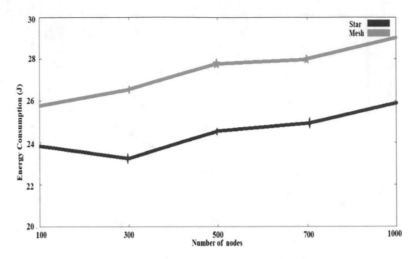

Fig. 8. APTEEN protocol energy consumption up to 1000 nodes.

The APTEEN protocol in the star topology (single jump) provides a lower energy dissipation value while increasing the number of nodes. The nodes are randomly placed in the network. All nodes begin with an initial energy of 22 J. In mesh topology, better levels of QoS are obtained (Fig. 9).

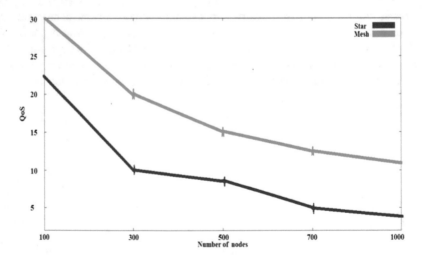

Fig. 9. QoS in APTEEN protocol.

5   Conclusion

In conclusion we have that the performance in 100 nodes reaches approximately 0.06 Mbps with TEEN and 0.1 Mbps with APTEEN. With 200 sensor nodes the performance is 0.04 Mbps with the TEEN protocol and 0.066 Mbps with APTEEN. It is noted that at lower throughput, optimal energy levels can be achieved within the network.

The APTEEN protocol has a satisfactory performance in the network. This network can emulate the proactive or reactive network by properly changing the periodicity or threshold values. Therefore, this network can be used in any type of application by properly configuring the various parameters. The simulation with the mesh topology evaluating APTEEN for 100, 300, 500, 700 and 1,000 nodes facilitates acceptable performance, energy saving and QoS in the communication of critical data.

For future work, the performance of the WSN networks in the cluster tree topology will be evaluated when critical data is processed, and the quality of the services analyzed in a grouping topology of three or more layers, will also be carried out with an intelligent algorithm in routing protocols.

References

1. Patel, N., Kathiriya, H.J., Bavarva, A.: Wireless Sensor Network Using ZigBee (2013)
2. Buratti, C., Conti, A., Dardari, D., Verdone, R.: An overview on wireless sensor networks technology and evolution. Sensors (Basel) 9(9), 6869–6896 (2009)
3. ITU: Y.2060: Visión general de la Internet de las cosas (2012). https://www.itu.int/rec/T-REC-Y.2060-201206-I/es. Accessed 18 Sept 2019
4. Rathi, N., Saraswat, J., Pratim, P.: A review on routing protocols for application in wireless sensor networks. Int. J. Distrib. Parallel Syst. 3(5), 39–58 (2012)
5. Ketshabetswe, L.K., Zungeru, A.M., Mangwala, M., Chuma, J.M., Sigweni, B.: Communication protocols for wireless sensor networks: a survey and comparison. Heliyon 5(5), e01591 (2019)
6. Bansal, S., Sandhya, D.J., Mukherjee, S.: An analysis of real time routing protocols for wireless sensor networks. Int. J. Eng. Sci. Technol. 3 (2011)
7. Mehmood, A., Song, H.: (PDF) Smart Energy Efficient Hierarchical Data Gathering Protocols for Wireless Sensor Networks. https://www.researchgate.net/publication/283768187_Smart_Energy_Efficient_Hierarchical_Data_Gathering_Protocols_for_Wireless_Sensor_Networks. Accessed 18 Sept 2019
8. MSB Technology, Jayakumari, R.B., Senthilkumar, V.J.: Congestion Control in Wireless Sensor Networks using Prioritized Interface Queue (2012)
9. Anand, S., Chandel, S.: Comparison of routing protocols in wireless sensor networks: a detailed survey, p. 8 (2014)
10. Almazroi, A.A., Ngadi, M.: A review on wireless sensor networks routing protocol: challenges in multipath techniques, vol. 59, p. 41 (2005)
11. Loscri, V., Maskooki, A., Mitton, N., Vegni, A.M.: Wireless cognitive networks technologies and protocols, p. 38 (2015)
12. Mehndiratta, N., Bedi, H.: Design issues for routing protocols in WSNs based on classification, vol. 2, no. 3, p. 9 (2013)
13. Lee, S., Noh, Y., Kim, K.: Key schemes for security enhanced TEEN routing protocol in wireless sensor networks. Int. J. Distrib. Sens. Netw. 9(6), 391986 (2013)

14. Javaid, N., Mohammad, S.N., Latif, K., Qasim, U., Khan, Z.A., Khan, M.A.: HEER: hybrid energy efficient reactive protocol for wireless sensor networks. In: 2013 Saudi International Electronics, Communications and Photonics Conference, pp. 1–4 (2013)
15. Ajmeri, M., Upadhyay, K.: Energy efficient hierarchical routing protocols for wireless sensor networks: a survey. Int. J. Innov. Res. Dev. 3(2) (2014)
16. Parmar, B., Munjani, J.: Analysis and improvement of routing protocol LEACH using TEEN, APTEEN and adaptive threshold in WSN. Int. J. Comput. Appl. 95(22), 5–9 (2014)
17. Heinzelman, W.B., Chandrakasan, A.P., Balakrishnan, H.: An application-specific protocol architecture for wireless microsensor networks. IEEE Trans. Wirel. Commun. 1(4), 660–670 (2002)
18. Xu, J., Jin, N., Lou, X., Peng, T., Zhou, Q., Chen, Y.: Improvement of LEACH protocol for WSN. In: 2012 9th International Conference on Fuzzy Systems and Knowledge Discovery, pp. 2174–2177 (2012)
19. Manjeshwar, A., Agrawal, D.P.: TEEN: a routing protocol for enhanced efficiency in wireless sensor networks. In: Proceedings 15th International Parallel and Distributed Processing Symposium. IPDPS 2001, pp. 2009–2015 (2001)
20. Manjeshwar, A., Agrawal, D.: APTEEN: A hybrid protocol for efficient routing and comprehensive information retrieval in wireless sensor networks (2002)
21. Barceló, M.: Wireless sensor networks in the future Internet of Things: density, mobility, heterogeneity and integration, p. 184 (2015)
22. Ullah, I., Safi, A., Arif, M., Azim, N., Ahmad, S.: (PDF) wireless sensor network applications for healthcare ResearchGate (2017). https://www.researchgate.net/publication/317338965_Wireless_Sensor_Network_Applications_for_Healthcare. Accessed 16 Feb 2019
23. Lotf, J.J., Hosseinzadeh, M., Alguliev, R.M.: Hierarchical routing in wireless sensor networks: a survey. In: 2010 2nd International Conference on Computer Engineering and Technology, vol. 3, pp. V3-650–V3-654 (2010)
24. NS-3: NS-3 network simulator Tutorial (2013)

Antenna Systems and Technologies

Miniaturized Wearable Minkowski Planar Inverted-F Antenna

Sandra Costanzo[✉] and Adil Masoud Qureshi

Università della Calabria, 87036 Rende, CS, Italy
costanzo@dimes.unical.it

Abstract. A miniaturized wearable Planar Inverted-F antenna is presented. Miniaturization is achieved by adopting a Minkowski fractal-shaped radiating element instead of the usual polygonal shape. The designed antenna is resonant at a frequency equal to 2.4 GHz and suitable for wearable applications. Simulation results show excellent return loss (~40 dB) and over 18% impedance bandwidth.

Keywords: PIFA · Fractal · Minkowski

1 Introduction

The Planar Inverted-F Antenna (PIFA) is one of the most ubiquitous antennas in modern telecommunication applications. Almost every cellular mobile phone on the market today includes more than one variant of PIFA [1]. The popularity of PIFA stems from its compact form-factor and relatively lower radiation towards the user body. These characteristics also make PIFA a prime candidate for the emerging wearable market [2, 3].

Wearable antennas are a basic component in a body-centric communication system. Applications that make use of body-centric communication include health care and safety monitoring, security and law enforcement as well as search and rescue [4]. In most wearable systems, the antenna is usually placed on the upper torso (shoulders, chest, helmet etc.) in order to maximize coverage [5]. As a result, wearable antennas must be small enough, usually less than 10 cm, to be easily and unobtrusively integrated. A considerable volume of research has been published on the topic of PIFA miniaturization. The most straightforward method is to use a substrate with a higher dielectric permittivity, which results in a smaller guided wavelength, allowing the antenna to resonate at a lower frequency. However, this usually leads to lower impedance bandwidth (>6% [6]) and, in some cases, lower gain. Capacitive loading [7] has also been used to counteract the effect of shortening the radiating element. Unfortunately, this method has the same drawbacks [8] as higher dielectric permittivity materials. Attempts have been made to reduce the size of the PIFA by replacing the shorting post/plate with a lumped resistance [9]. While this technique maintains the bandwidth of the antenna, the miniaturization is achieved at the cost of radiation efficiency [10].

© Springer Nature Switzerland AG 2020
Á. Rocha et al. (Eds.): ICITS 2020, AISC 1137, pp. 601–606, 2020.
https://doi.org/10.1007/978-3-030-40690-5_57

The authors propose the use of fractal geometry, specifically the Minkowski fractal, for PIFA miniaturization. Earlier work by the authors, has shown that it is an effective method for reducing the size of planar radiators [11]. The resonant frequency of a PIFA is related to the length of the current path along the periphery of the patch [12]. Increasing the perimeter of a rectangular or square PIFA, invariably yields a bigger antenna. By using a fractal shape instead, it is possible to increase this path length without increasing the overall size of the antenna. Thus, miniaturization is achieved by virtue of a lower resonant frequency in the same physical envelope.

In this contribution, a compact Minkowski fractal shaped PIFA is presented, which is able to operate within the ISM (Industrial, Scientific, Medical) band (2.4 GHz). The license free band is widely used in consumer devices for communication using popular standards such as Bluetooth and WiFi. The fractal geometry reduces the resonant frequency of the antenna, in comparison with its square counterpart, which is equivalent to miniaturization. A parametric analysis is also presented, to demonstrate the behavior of the Minkowski fractal configuration. Finally, a simple ground plane modification method is presented for enhancing the bandwidth of the miniaturized antenna.

2   Antenna Configuration

The design of the Minkowski Fractal PIFA originates from a simple PIFA, similar to the earliest PIFA designs used for portable communication [13]. A square shaped radiating element is used (Fig. 1) as it is the usual generator shape for the Minkowski fractal. In order to transform the square shape into a Minkowski fractal, identical rectangular slots, having length 'L' and width 'W', are introduced at the centre of each side of the square (Fig. 1). The antenna is excited using a coaxial probe feed, placed in the top right quadrant of the radiating element, close to the shorting plate. The ground plane and the radiating element are separated by air, acting as substrate. The height of the air substrate is equal to 4 mm, while the width of the shorting plate is equal to 6 mm, as shown in Fig. 1.

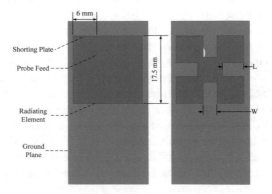

Fig. 1. Square PIFA and Minkowski PIFA (1st Iteration).

3 Simulation Results and Parametric Analysis

As a first step, the square PIFA shown in Fig. 1 is simulated to form a baseline for comparison. The resonant frequency of the square PIFA depends on the feed location and the length of the shorting strip. The feed location is optimized, via simulation, to achieve the lowest possible resonant frequency with a good impedance matching.

Figure 2 shows the return loss of the final square PIFA design (in red), resonant at a frequency approximately equal to 2.82 GHz.

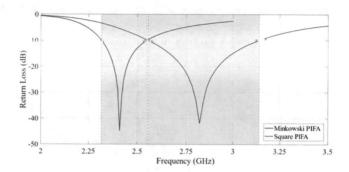

Fig. 2. Return loss for the original PIFA and the 1st stage Minkowski PIFA.

In the next step, the square radiating element is replaced with a Minkowski fractal shaped element [14]. The slot size and the feed location are optimized for operation in the ISM band at 2.4 GHz (Fig. 2). Translation of the resonant frequency from 2.82 GHz to 2.41 GHz without changing the overall size of the antenna is equivalent to a nearly 15% reduction in size [15]. However, the miniaturization is achieved at the cost of the impedance bandwidth. The square PIFA provides a fractional bandwidth of just over 20% at 2.82 GHz. The fractional bandwidth of the 1st stage Minkowski PIFA at 2.4 GHz in comparison is under 10%. Nevertheless, the radiation pattern of the antenna is almost un-affected by the miniaturization (see Fig. 3).

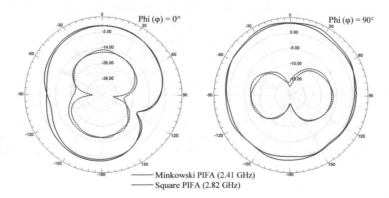

Fig. 3. Co-polar (solid) and Cross-polar (dashed) radiation pattern: comparison between square and Minkowski PIFAs.

The fractal structure increases the perimeter of the radiating element without increasing the area. The resonant frequency decreases as a direct result of this increase in perimeter. Therefore, the extent of the reduction in the resonant frequency depends on the size of the slots. The size of the slot is determined by the length 'L' and the width 'W'. To perform an accurate analysis, each of these parameters is varied one by one, while maintaining fixed the other. Figure 4 shows the simulated return loss for different variations of the Minkowski PIFA. It is clear that the resonant frequency is very sensitive to the length 'L' of the slots, whereas the effect of varying the width 'W' is much less significant. This is to be expected, as changes in the length 'L' are directly translated into the perimeter of the structure. On the other hand, the width 'W' does not change the perimeter of the structure at all. The small variations in resonance due to change in width may be attributed to changes in the reactive behavior of the slots.

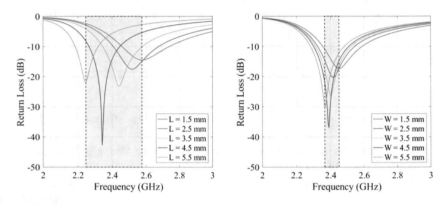

Fig. 4. Parametric analysis of the effects of slot size on the resonant frequency.

4   Bandwidth Enhancement

Wearable antennas are expected to operate in the vicinity of the human body, which is a lossy, high-permittivity medium. Therefore, wearable antennas often suffer de-tuning due to the body being inside the nearfield region of the antenna. The problem can be further worsened due to flexing, moisture, foreign substances, as well as metallic objects carried by the user. As a result, it is desirable for wearable antennas to have a wide bandwidth, in order to ensure reliable operation under all scenarios.

The impedance bandwidth of the Minkowski PIFA covers the ISM band (2.4–2.5 GHz), with a > 50 MHz margin on either side. Nevertheless, the technique introduced in [16], which is based on the adoption of a T-shaped ground plane, can be employed to increase the impedance bandwidth of the Minkowski PIFA antenna up to 18%. This is successfully performed, with the configuration and results shown in Fig. 5.

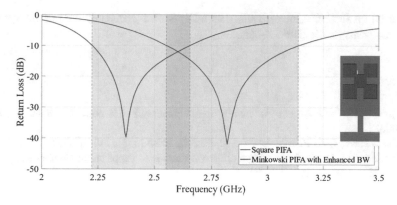

Fig. 5. Return loss for the square PIFA and the Minkowski PIFA with T-shaped ground plane.

The radiation performance of the final design for the Minkowski PIFA with T-shaped ground plane is also similar to the square PIFA (see Fig. 6). The antenna is linearly polarized, and it has a peak gain equal to 3.2 dBi.

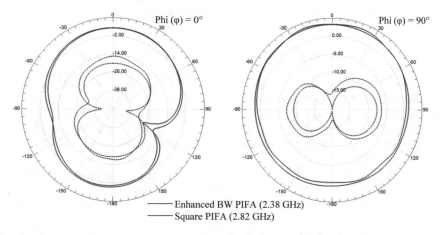

Fig. 6. Co-polar (solid) and Cross-polar (dashed) radiation pattern: comparison between square PIFA and Minkowski PIFA with T-shaped ground plane.

5   Conclusion

A viable method for the miniaturization of wearable PIFA has been proposed. Simulation results have revealed very good performances of the proposed configuration within the ISM band (2.4–2.5 GHz). The design is easy to manufacture and can be produced at low cost. The radiation pattern and impedance bandwidth performances are comparable to the standard PIFA. There is only a negligible loss of the fractional bandwidth (about 2%), in comparison to the miniaturization ($\sim$ 15%). Future work will be focused on further miniaturization, as well as on the investigation of the effects of textile materials as dielectric substrates.

References

1. Fujimoto, K. (ed.): Mobile Antenna Systems Handbook, 3rd edn. Artech House, Boston (2008)
2. Rais, N.H.M., Soh, P.J., Malek, F., Ahmad, S., Hashim, N.B.M., Hall, P.S.: A review of wearable antenna. In: 2009 Loughborough Antennas & Propagation Conference, pp. 225–228. IEEE, Loughborough (2009)
3. Rogier, H.: Textile antenna systems: design, fabrication, and characterization. In: Tao, X. (ed.) Handbook of Smart Textiles, pp. 1–21. Springer, Singapore (2015)
4. Young, P.R., Aanandan, C.K., Mathew, T., Krishna, D.D.: Wearable antennas and systems. Int. J. Antennas Propag. **2012**, 1–2 (2012)
5. Nepa, P., Rogier, H.: Wearable antennas for off-body radio links at VHF and UHF bands: challenges, the state of the art, and future trends below 1 GHz. IEEE Antennas Propag. Mag. **57**(5), 30–52 (2015)
6. Lo, T.K., Hwang, Y.: Bandwidth enhancement of PIFA loaded with very high permittivity material using FDTD. In: IEEE Antennas and Propagation Society International Symposium. 1998 Digest. Antennas: Gateways to the Global Network. Held in conjunction with: USNC/URSI National Radio Science Meeting, vol. 2, pp. 798–801 (1998). (Cat. No.98CH36, vol. 2)
7. Waterhouse, R.B. (ed.): Printed Antennas for Wireless Communications. Wiley, Hoboken (2007)
8. Rowell, C.R., Murch, R.D.: A capacitively loaded PIFA for compact mobile telephone handsets. IEEE Trans. Antennas Propag. **45**(5), 837–842 (1997)
9. Wong, K.L., Lin, F.L.: Small broadband rectangular microstrip antenna with chip-resistor loading. Electron. Lett. **33**, 1593–1594 (1997). https://doi.org/10.1049/el:19971111
10. Godara, L.C. (ed.): Handbook of Antennas in Wireless Communications. CRC Press, Boca Raton (2002)
11. Costanzo, S., Venneri, F.: Miniaturized fractal reflectarray element using fixed-size patch. IEEE Antennas Wireless Propag. Lett. **13**, 1437–1440 (2014)
12. PIFA - Planar Inverted-F Antennas. http://www.antenna-theory.com/antennas/patches/pifa.php
13. Taga, T., Tsunekawa, K.: Performance analysis of a built-in planar inverted-F antenna for 800 MHz band portable radio units. IEEE J. Sel. Areas Commun. **5**, 921–929 (1987). https://doi.org/10.1109/JSAC.1987.1146593
14. Costanzo, S., Venneri, F., Di Massa, G.: Modified Minkowski fractal unit cell for reflectarrays with low sensitivity to mutual coupling effects. Int. J. Antennas Propag. (2009)
15. Costanzo, S., Venneri, F.: Fractal shaped reflectarray element for wide angle scanning capabilities. In: 2013 IEEE Antennas and Propagation Society International Symposium (APSURSI), pp. 1554–1555. IEEE APS, Orlando, FL (2013)
16. Wang, F., Du, Z., Wang, Q., Gong, K.: Enhanced-bandwidth PIFA with T-shaped ground plane. Electron. Lett. **40**(23), 1504–1505 (2004)

Preliminary SAR Analysis of Textile Antenna Sensor for Non-invasive Blood-Glucose Monitoring

Sandra Costanzo[✉] and Vincenzo Cioffi

Department of Computer Engineering, Modelling, Electronics, and Systems
Science (DIMES), University of Calabria, 87036 Rende, CS, Italy
costanzo@dimes.unical.it

Abstract. This work presents the design and the specific absorption rate analysis of a 2.4 GHz wearable patch antenna, realized by using fabrics for the substrate and the conductive parts. This antenna can be used to monitor physiological parameters such as glucose concentration in diabetes patients, thus the real risk of its use for humans should be carefully evaluated. For this reason, simulation tests are performed to evaluate the SAR levels, by adopting a realistic stratified phantom including dry skin, wet skin, fat, blood and muscle. Some preliminary numerical results are discussed to prove the sensor efficiency.

Keywords: Non-invasive monitoring · Specific absorption rate · Wearable antenna

1 Introduction

Nowadays, there is a rapid development in mobile communications systems, such as Body Area Network (BANs) and Personal Area Networks (PANs), that leads to requiring ever more efficient antennas. There are new types of antenna fixed on the cloths, directly attached to the human body or even implanted inside it. Another potential application of wearable antennas is in the framework of Body-Centric Wireless Communications (BCWCs) which is used for health monitoring, military purposes, and security applications [1, 2]. Wearable systems can be adopted for remote patient monitoring, which helps the doctors in diagnoses and control of the disease while working from home or clinic.

The requirement for the wearable antenna is to be flexible and conformable to human body, without affecting his daily life. As a matter of fact, most of the wearable antennas are realized using fabrics for the substrate and also for the conductive parts [3].

Wearable antennas operate in the close proximity of human body, which has high dielectric constant and conductivity. This kind of antennas work with the radiating element facing the human arm, so the radiation gets penetrated and absorbed in the body tissues. The absorption of power per unit mass of human body is evaluated by a term known as the Specific Absorption Rate (*SAR*), a parameter giving a measure of the electromagnetic wave penetration in human body tissues. SAR is a number that measures the speed at which energy is absorbed by the human body when it is exposed

© Springer Nature Switzerland AG 2020
Á. Rocha et al. (Eds.): ICITS 2020, AISC 1137, pp. 607–612, 2020.
https://doi.org/10.1007/978-3-030-40690-5_58

to electromagnetic fields with a carrier frequency between 100 kHz and 10 GHz. In mobile telephony, the SAR establishes the energy absorbed by a particular mass of human tissue within a certain period of time; it is calculated in units of power per mass (W/kg), with a well known expression given as:

$$SAR = \frac{\sigma |E|^2}{\rho} \left[\frac{W}{kg} \right] \tag{1}$$

In the above equation, σ, is the conductivity of the tissue (S/m), ρ gives the mass density of tissue (kg/m³), and E is the root mean square of the electric field strength (V/m). SAR can be also computed as the rate of temperature rise at a given point, such as used in some basic research [4].

This contribution is organized as follows. Section 2 presents the design of a textile rectangular patch antenna for ISM (Industrial, Scientific and Medical) Band, with a detailed description of the used fabrics and their characteristics. Section 3 shows some preliminary results relative to SAR simulations, as well as discussing about the technique adopted to maintain the SAR value below the prescribed limit. Conclusions are finally drawn in Sect. 4.

2 Standard Coaxial Probe-Fed Patch Antenna on Textile Substrate

A very simple sensor configuration is adopted in the present contribution, namely a rectangular coaxial probe-fed patch antenna (see Fig. 1(a)), able to work within the ISM Band.

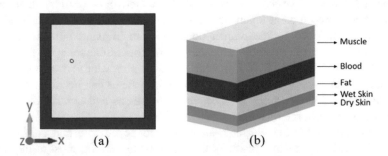

Fig. 1. (a) Wearable sensor geometry and (b) stratified radiation medium.

The textile material adopted as substrate is given by denim, which is durable, inelastic, low-thickness, comfortable for user and low-cost. Its dielectric constant and tangent loss factor are equal to $\varepsilon_r = 1.6$ and $\tan\delta = 0.01$, respectively, at 2.4 GHz.

In the present design, a 2 mm thickness is considered, while PCPTF (Pure Copper Polyester Taffeta Fabric) is adopted for the patch and the ground plane. The optimized sensor has a final size approximately equal to 4 × 4 cm, thus being very small and easily wearable and integrable.

The wearable antenna is optimized on CST, by considering a stratified medium (see Fig. 1(b)) as superstrate, whose characteristics are reported in Table 1.

Table 1. Characteristics of the stratified superstrate

Medium	Thickness [mm]	ε_r	tanδ	σ [S/m]	Density [kg/m^3]
Dry skin	0.015	38.06	0.2835	1.5	1000
Wet skin	0.985	42.92	0.2727	1.62	1000
Fat	0.5	10.84	0.1808	0.27	850
Blood (BGC = 100 mg/dl)	2.5	58.53	0.1743	1.41	1060
Muscle	3	52.34	0.1893	1.37	1050

The standard Cole-Cole model [5] is adopted to obtain the dielectric parameters for the media in the superstrate, while the enhanced dielectric model proposed by the authors in [6] is used for blood. For the accurate design of the wearable sensor, a set of numerical validations are conducted by simulating the return loss at different values for the Blood Glucose Concentration (BGC), with the complex permittivity of blood derived from the enhanced dielectric model [6].

For the SAR analysis, just a single BGC value equal to 100 mg/dl is considered, whose related return loss, revealing a very good matching, is illustrated in Fig. 2.

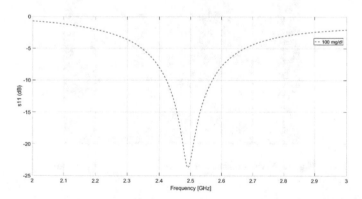

Fig. 2. Return loss of adopted wearable sensor - BGC = 100 mg/dl (blood dielectric parameters [6]: εr = 58.53 and tanδ = 0.1743).

3 Preliminary Specific Absorption Rate (SAR) Analysis

In order to evaluate the real risk for the human body, some numerical simulations are conducted to measure the SAR levels.

The SAR analysis is carried out using CST Microwave Studio, with the standard safe levels reported below:

- **FCC [7]:**

- Body, trunk, head: 1 g-SAR with limit 1.6 W/kg
- Limbs: 10 g-SAR with limit 4 W/kg

- **CE [8]:**

- Body, trunk, head: 10 g-SAR with limit 2 W/kg
- Limbs: 10 g-SAR with limit 4 W/kg

Since the sensor should work on the human arm, a safe limit equal to 4 W/kg for any 10 g of tissue is assumed.

The result in terms of SAR distribution for the configuration described in Sect. 2 is reported in Fig. 3, revealing that the considered geometry exceeds the permitted safety limit, with a maximum value equal to 5.16 W/kg.

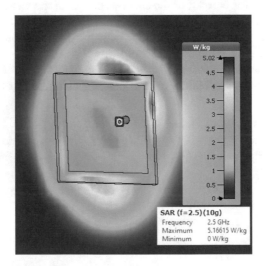

Fig. 3. SAR analysis for the first sensor configuration.

The problem raised in the present application is strictly related to the radiation medium. As a matter of fact, SAR is directly proportional to the body conductivity and inversely proportional to the body permittivity [9]. Human body tissue with higher water content is more susceptible to absorbing radiation; moreover, the body tissue has a greater conductivity, which increases SAR values.

There are some different techniques to reduce the SAR, such as the adoption of metamaterials to restricts the propagation of surface waves within a specific frequency band, and therefore to reduce the level of unwanted radiations towards the human body [10]. An alternative method is based on a proper choice of the dielectric substrate parameters. To this end, a parametric analysis is specifically conducted in the present work by varying the dielectric thickness of the adopted denim material. In particular, it is proved that a value equal to 4 mm is able to satisfy the prescribed SAR limitations, as reported in Fig. 4.

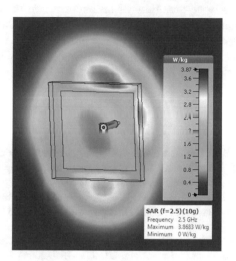

Fig. 4. SAR analysis for the sensor configuration with a 4 mm substrate thickness.

In the above case, the SAR value reveals a sharp decrease, by reaching the value of 3.86 W/kg, which is perfectly below the safe limits for humans.

A complete description of SAR values obtained with the above simulation campaign are reported in Table 2.

Table 2. SAR values at different substrate thickness.

Thickness [mm]	SAR [W/kg]
1.5	5.81
2	5.16
4	3.86
5	3.58

An increase in thickness causes a decrease in SAR values, so it is possible to reduce the SAR level by simply varying the substrate thickness, without any change in the dielectric material.

4 Conclusions

In the present contribution, the design and preliminary SAR analysis of a 2.4 GHz wearable antenna has been discussed. The proposed sensor, made on a textile substrate, is a suitable candidate to monitor the blood glucose levels in a non-invasive way. Due to the operation in the close proximity of human arm, an accurate SAR analysis is conducted. In particular, it has been demonstrated that SAR values also depend on the substrate thickness. Parametric simulations have been performed for different dielectric height, and a thickness value equal to 4 mm has revealed to provide a SAR value equal to 3.86 W/kg, below the safety limit of 4 W/kg.

References

1. Grilo, M., Correra, F.S.: Parametric study of rectangular patch antenna using denim textile material. In: 2013 SBMO/IEEE MTT-S International Microwave & Optoelectronics Conference (IMOC), Rio de Janeiro, Brazil (2013)
2. Mantash, M., Tarot, A.C., Collardey, S.: Investigation of flexible textile antennas and AMC reflectors. Int. J. Antennas Propag. (2012). https://doi.org/10.1155/2012/236505. Article ID 236505, 10 pages
3. Locher, I., Klemm, M., Kirstein, T., Tröster, G.: Design and characterization of purely textile patch antennas. IEEE Trans. Adv. Packag. **29**, 777–788 (2006)
4. Islam, T.M., Abidin, Z.H., Faruque, I.R.M., Misran, N.: Analysis of materials effects on radio frequency electromagnetic fields in human head. Prog. Electromagnet. Res. **128**, 121–136 (2012)
5. Gabriel, S., Lau, R.W., Gabriel, C.: The dielectric properties of biological tissues: III. Parametric models for the dielectric spectrum of tissues. Phys. Med. Biol. **41**, 2271–2293 (1996)
6. Costanzo, S., Cioffi, V., Raffo, A.: Complex permittivity effect on the performances of non-invasive microwave blood glucose sensing: enhanced model and preliminary results. In: WorldCIST 2018, Naples, Italy (2018)
7. https://www.fcc.gov/search/#q=specific%20absorption%20rate&t=web
8. Official Journal of the European Communities, L 199/59, 2 July 1999
9. Husni, N.A., Islam, M.T., Faruque, M.R.I., Misran, N.: Effects of electromagnetic absorption towards human head due to variation of its dielectric properties at 900, 1800 and 1900 MHz with different antenna substrates. Prog. Electromagnet. Res. **138**, 367–388 (2013)
10. Ali, U., et al.: Design and SAR analysis of wearable antenna on various parts of human body using conventional and artificial ground planes. J. Electr. Eng. Technol. **12**(1), 317–328 (2017)

Information and Knowledge
in the Internet of Things

Facial Recognition: Traditional Methods vs. Methods Based on Deep Learning

Shendry Rosero Vasquez[1,2](✉)

[1] Universidad P. Santa Elena, Facsistel, La Libertad, Ecuador
srosero@upse.edu.ec
[2] Escuela Superior Politécnica del Litoral, Guayaquil, Ecuador
shrosero@espol.edu.ec

Abstract. The present study shows the quantitative analysis of the evaluation of traditional methods of facial recognition versus the evaluation of methods based on deep learning, the contribution of this study lies in the conditions of the dataset to be used, which unlike the regular paradigms, it contains a limited size of images to train what the researcher has called a hostile training environment, which will serve as a reference to determine what technique and/or algorithm can work to recognize faces in which you do not have a lot of information to be processed and with a poor quality of the face image to be recognized, factors that in recent years are recurring in the search requests of missing persons or in the absence of the quality of certain security cameras.

Keywords: Facial recognition · Eigenfaces · Fisherfaces · Histograms · Tensorflow · Keras · Facenet · Openface

1 Introduction

In recent years, the number of missing people has increased significantly, as well as the number of requests for help on social networks to find the location of those people. A recurring element of these requests lies in the fact that generally there is not a large number of images for processing, either by traditional techniques or even worse by using techniques based on deep learning, not to mention that in many cases few images that are counted do not have the appropriate resolution to subject them to a facial recognition process. Parallel to this, in many institutions there is the possibility of implementing biometric systems for access and security; However, they do not have high resolution cameras that allow easy identification of unwanted access or that guarantee access to authorized personnel for this purpose. Given this, it is proposed to evaluate a select group of traditional facial recognition techniques, as well as another group of techniques based on deep learning. The contribution of this study[1] will be to provide an analysis of the number of successes that each technique presents to these conditions called hostile training conditions, so that the researcher can in future take these results as a baseline for the development of subsequent algorithms.

---

[1] A set of more detailed results can be found in a parallel study conducted by the same author as part of his studies in computer science.

© Springer Nature Switzerland AG 2020
Á. Rocha et al. (Eds.): ICITS 2020, AISC 1137, pp. 615–625, 2020.
https://doi.org/10.1007/978-3-030-40690-5_59

The organization of this article presents in Sect. 2, a brief evolution of facial recognition techniques, as well as a quick definition of the algorithms to analyze, in Sect. 3 the methodology of the experiment carried out, after that, the Sect. 4 shows the results of the experiments performed, Sect. 5 the conclusions and finally offers an idea of future work based on the results obtained.

2 Evolution of Facial Recognition Techniques

One of the computer vision techniques that has been gaining significant importance in recent years is undoubtedly the development of applications based on facial recognition (FR), then we must distinguish between the detection of the face itself, of the techniques they try give an identification to the face, which is parameterized based on a comparison with a set of previously established data, or from which the subjects of analysis are fully identified. This leads us to define FR today as a technique of searching for patterns in the human face through electronic means.

Facial recognition is not a new technique but it has managed to adapt to the biometric needs of the time, proof of this is the first studies of Woodrow Bledsoe [1], who in the 60 s tried to define the problem of facial recognition as "a problem hindered by the variability of states and inclination of the head, lighting conditions, facial expressions and aging". Bledsoe's studies were basically oriented to look for parameters such as distances between coordinates of objects, for example: eyes, mouth, nose, ears, measurements that were calculated and processed in order to obtain patterns of at least 40 images and that could finally identify to previously recognized test subjects. The limitations of this study were related to the low computational capacity of the equipment of the time, which limited the number of images to be processed. Subsequently, the Stanford Research Institute by Peter Hart performed similar procedures, reaching a processing of around 2000 photographs, and allowing computers for the first time to systematically defeat the human in recognition tasks [2, 3]. It is based on this, that there is a widespread notion of attributing better recognition results to those processes that base their analysis on large amounts of data [4].

From the mentioned studies, diverse techniques are born that base its execution, either in analysis of mathematical models resulting from the obtained patterns, or in the adaptation of modern techniques of processing objects in images such as neural networks, specifically convolutional networks based on learning deep. This study bases its aim to determine how well the facial recognition in these two conditions mentioned groups called hostile training conditions. This is the analysis of results with extremely small datasets to achieve a result that is independent of the quality of images and that allows to define which technique conforms to real recognition application conditions.

2.1 Simplified Facial Recognition Process

Three specific stages in facial recognition can be distinguished: (i) Face detection, (ii) Feature Extraction and (iii) Face recognition. According to Yang, Kriegman, and Ahuja [5], the detection of a face consists in determining the existence or not of a human face, if it exists, the exact location of the face in the image must be extracted.

The extraction of characteristics is a process that is carried out through the analysis of areas of the previously detected face and which is limited by the conditions of the image, posture of the face, and objects other than the face that may appear in reference images, of there, the best results lie in the use of a large number of photographs and the quality of these when they are captured. Finally, the recognition of the face consists of any process that is oriented to transform the parts of the human face into fixed-sized arrangements that allow a facial representation. This step requires that both the images and the features extracted from those images form a database, the following is to capture the image of the face, extract its characteristics and compare them with the previously established database, it is called facial recognition. The difference between FR methods lies in the different techniques and algorithms that are used for each of the previously described steps.

2.2 Traditional Techniques

This section describes the basic facial recognition techniques that are included in open libraries such as OpenCV, a library released under a BSD license (Berkeley Software Distribution), which is one of the reasons to include it in this analysis due to its ease of use and implementation, an aspect that allows many researchers to consider the inclusion of OpenCV in image analysis. Among the techniques analyzed are: (i) Eigenfaces, (ii) FisherFaces, and (iii) LBPH (Local Binary Patterns Histograms).

The detailed and individual analysis of these techniques is beyond the scope of this study, however it is necessary to mention that their bases rest in the studies of Takeo Kanade [6], studies that tried to use positional markers of characteristic objects such as eyes, nose, mouth, to build a vector of features with elements such as distance between points, angles between them, etc., Despite being a robust method against lighting conditions, its technique is limited by the complexity in the registration of marking points even with algorithms current. Likewise, other works based on facial geometric were presented by Poggio and Brunelli [7] in 1992, who used a 22-dimensional feature vector, despite working with large amounts of data it was shown that geometric characterization is not sufficient for facial recognition.

The foregoing generates the birth of techniques such as eigenfaces [8], a holistic facial recognition proposal, whose basic idea is a mathematical model that minimizes variance within a class and maximizes variance between classes. The problem with this is the generation of models with high dimensionality which can generally be treated using techniques such as PCA (Principal Components Analysis).

From this point of view, although high dimensionality can be treated, a new problem is generated with the ability to differentiate between classes, so proposals such as Linear Discriminant Analysis could solve the problem of dimensionality and class differentiation at the same This proposal was developed by the statistician Sir RA Fisher, and published in 1936 (The use of multiple measurements in taxonomic problems), obtaining a technique that maximizes the proportion of dispersion between classes and within classes, instead of Maximize global dispersion. This means that the same classes must be grouped closely, while the different classes move away from each other in the representation of the lower dimension, calling this technique fisherfaces. Finally, these two proposals, give a vector treatment to the data that are technically reduced in their dimensionality to keep only that important information, the general idea with these methods is that the greater the amount of information the index of recognition should improve. However, from the literature investigated to obtain good recognition ranges, at least 8 images with ± 1 image are needed for each person, in this case fisherfaces fails. Details of measurements like these can be found in [6]. Although, from the above, the "faces" techniques remain a compendium of important features of the face that will eventually represent a low dimensional image, which makes them vulnerable to lighting and position problems. An alternative to these methods is to summarize the local structure in an image, comparing each pixel with its neighborhood so that a pixel is taken as the center and threshold against its neighbors. Its intensity is examined and if the intensity of the central pixel is greater than its neighbor, it will be specified with 1 and 0 if not. The result will be a binary number for each pixel. Therefore, with 8 surrounding pixels, you will get 2 ^ 8 possible combinations, what are called Local Binary Patterns (LBP) or, sometimes, simply called LBP codes, which are the basis of the binary local pattern histograms.

2.3 Deep Learning Techniques

An Artificial Neural Network (ANN), is a structure oriented to the simulation of the information processing performed by the human brain, based on inputs, weights, synapses, activation functions and outputs [9], an example of this can be visualized in Fig. 1.

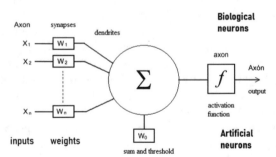

Fig. 1. The image shows the mathematical model of an artificial neuron, presented by psychiatrist and neuroanatomist Warren McCulloch and mathematician Walter Pitts, image taken from: http://www.cs.us.es/ ~ fsancho/?e=72.

Parallel to this, the concept of convolution must also be established, convolution is the operation between two functions that would allow transformations oriented to obtain a third resulting function, from this the concept of Convolutionary Neural Networks in which convolutions allow to detect structures within of an image and from that formulating more complex functions [10], it is necessary to indicate that the strength of a CNN (convolutional neural networks) is image processing although they are also used in other fields such as text processing to cite another example.

A CNN works similarly to an ANN, that is, they have an input layer that is responsible for receiving the pixels of the images, this data is passed to what will be called hidden layers that maintain two fundamental operations: (i) the pooling or grouping, operation in charge of reducing the size of the images, so once processed, the delivery to the following layers will be a set of images of reduced size, and (ii) convolutions, which is the filtering of images in order to find important patterns within them [11]. The advantage of a CNN over an ANN is that a couple of important aspects, the first: an artificial neural network for the case of images, generates a large amount of data to process, amount that increases as the number of images to be processed increases, this in turn generates a second problem related to the way of learning the network, and with the efficiency when working on and tasks such as object detection.

Considering that recognizing a face through computer vision implies detecting the face, extracting characteristics to generate a comparative model and recognizing it, is possible to adapt modern techniques that would allow its implementation. One approach that is currently gaining strength is that of techniques based on deep learning [12]. These techniques mainly make use of convolutional neural networks, which are certain types of artificial neural networks with a variation of the concept of multilayer perceptron and which are based on two-dimensional matrices, this concept facilitates its use effectively in classification and segmentation of images. Convolutional neural networks overcome these problems by using layers that perform convolutions, convolutions are operations that are responsible for processing part of the image instead of the complete image, each convolution sweeps the entire image and generates a new output image with a smaller height and length but increased in depth, to this must be added the pooling that will be responsible for reducing the dimensions of the image even more, maintaining the depth. Regarding convolutions, three parameters stand out: the size of the filter that allows the image process; the depth of the convolutional layer, which is the amount of filters that will be applied to the images (each filter allows to find patterns so the use of different filters allows obtaining different images, this is what is called depth of the image) and finally the step or stride, which is the distance that the filter must travel through the image, the greater the step, the greater the reduction of the image. Regarding pooling, it should be noted that in addition to reducing the size of images, it allows avoiding the over-adjustment of the model, there are two main types of grouping: max-pooling, which is responsible for filtering the images under a concept of maximum pixel and the average-pooling will instead obtain the average pixel value of the window or filter, since the images have a numerical matrix treatment, the idea is to obtain a smaller matrix with respect to the input image. A convolutional model that is considered as a good image classifier should then include the use of convolution layer followed by pooling, convolution layer and again a pooling layer, and so on. The difference that can be found between convolutional neural networks lies in this organization

2.4 Considerations of the Study and Problem to Be Solved

A separate aspect that should be mentioned under this concept is that of the number of images that would allow the training of the network, in ANN the number is conditioned to the growth of the network and is subject to the available technical resource, as an alternative to this is CNN, which tends to think that the greater the amount of input information, the better the training, so there are questions about how deep a deep network should be [13] and what is the appropriate size of the dataset for its training, questions that are outside the scope of the present study, but that allow to introduce the problems to be solved: can a system based on what has been called a hostile training system (very few images), train a traditional system and a deep learning system for facial recognition?, and what techniques present better results when training a system with a limited number of pictures?. Under this context, the researcher maintains two well-defined options: (i) generate an architecture that allows to properly train a network for face recognition or (ii) use a pre-trained architecture for the extraction of features and that these can be used As an input in the development of applications for facial recognition, the researcher can choose the option that best suits their existing hardware resources. In the present study and due to the practical conditions, it is proposed to determine which technique performs a better recognition while maintaining a small dataset and without the standardization of images beyond those necessary by the algorithms themselves, the contribution will be to provide a quantitative analysis of the technique that yield better results under hostile training conditions and outside the quantity paradigm.

3 Methodology

Unlike traditional forms of training such as in [14] or the number of photographs used in both [4] and [15], the evaluation of 6 facial recognition techniques distributed in two well-defined groups is proposed: on the one hand traditional methods with eigenfaces, fisherfaces and histograms of local binary patterns and on the other hand the evaluation of modern deep learning techniques based on the TensorFlow API for object recognition, the Keras API for object recognition and Facenet a tool based in openface [4]. The evaluation was carried out on a dataset composed of 41 classes corresponding to 41 people, mostly from the Faculty of Systems and Telecommunications of the State University Peninsula de Santa Elena, each class was composed with an average 40 images, obtaining a total of 857 images after the pre-processing required by the working conditions of eigenfaces and the respective APIs mentioned above. This dataset in general was called "Group A", and contains the set of photographs that allowed the training of the models to be evaluated with a 90% separation of the dataset for training and about 10% remaining for tests. To this must be added a set of 41 additional images, which was later called "Group B", this dataset consisted of images completely outside Group A, and were used as a control element.

Once the respective datasets were formed, from group A, 10 photographs were randomly taken in 5 different executions with a resolution of 160x160 per image. For group B, the 10 randomly selected photographs of the previous group with better resolution (240 × 300) were taken as control element For the evaluation of each group two parameters were defined: (i) SUCCESSES as the validity of the recognition expressed in that the photo corresponds to the subject labeled; and, (ii) FAILURES of any expression of the implementation in which the face does not correspond to the subject labeled.

Regarding the hardware used, as in the proposed objectives, the nature of the research was aimed at working on average equipment, and although this is a variable that did not influence the results, it was decided to place it in this section as a reference for future work, so it was worked on a computer with Core I7 processor of 8 Gb of RAM and with NVIDIA GEFORCE graphics processor with which basic training tests resulted in an average training of 5 h on average; Due to the nature of the proposal, no tests were planned on equipment with higher capacities, however tests were carried out on equipment with GPU 1080 ti, 32 GB of RAM, and i7 as a processor, in which averages of 1 h of training were achieved.

4 Results

According to the methodology proposed in the previous section, disaggregated results were obtained in what was called Success/Failures, which are shown in Table 1, which shows a better performance of the algorithm based on histogram of local patterns on the tests performed in group A.

Table 1. It shows the global successes and failures for traditional methods in code execution

Model	Gr	Code execution									
		1st		2nd		3rd		4th		5th	
		Successes	Fails	Successes	Fails	Successes	Fails	Successes	Fails	Successes	Fails
Eigenfaces	A	0	10	0	10	0	10	0	10	0	10
	B	0	10	0	10	0	10	0	10	0	10
Fisherfaces	A	0	10	0	10	0	10	0	10	0	10
	B	0	10	0	10	0	10	0	10	0	10
LBPH	A	6	4	5	5	7	3	6	4	5	5
	B	1	9	0	10	2	8	1	9	1	9

In the case of the results of deep learning techniques, the results are shown in Tables 2, 3 and 4.

Table 2. Results of the Tensorflow Object Detection Api.

Model	Gr	Code execution									
		1st		2nd		3rd		4th		5th	
		Successes	Fails	Successes	Fails	Successes	Fails	Successes	Fails	Successes	Fails
TfApi	A	10	0	10	0	10	0	10	0	10	0
	B	0	10	0	10	0	10	0	10	0	10

Table 3. Results of Keras code execution.

Gr		1st		2nd		3rd		4th		5th	
		Successes	Fails	Successes	Fails	Successes	Fails	Successes	Fails	Successes	Fails
KERAS	A	1	9	2	8	1	9	1	9	1	9
	B	3	7	2	8	1	9	1	9	1	9

Table 4. Results of facenet/openface code execution.

Gr		1st		2nd		3rd		4th		5th	
		Successes	Fails	Successes	Fails	Successes	Fails	Successes	Fails	Successes	Fails
Facenet	A	7	3	6	4	4	6	8	2	7	3
	B	3	7	1	9	1	9	2	8	2	8

According to what is observed in the preceding tables, it could be thought that the Tensorflow object detection API could be considered as the best method in terms of number of hits, but it should also be considered that when executing images that are outside the training dataset, the API fails in most executions, but the algorithm based on *openface* that shows better performance with group B, which can be evidenced in Table 5.

Table 5. Global results from all models.

	Group A		Group B	
	Successes	Fails	Successes	Fails
EigenFaces	0	10	0	10
Fisherfaces	0	10	0	10
LBPH	5.8	4.2	1	9
TfApi	10	0	0	10
Keras	1.2	8.8	1.6	8.4
Facenet	6.4	3.6	1.8	8.2

Beyond the tests performed, the effectiveness of each algorithm was evaluated and related to what the literature corresponding to each method offered as performance, results that can be evaluated in Table 6, in which the two grouping methods are observed, TR for traditional and DL for methods based on deep learning.

Table 6. Comparison of evaluations.

		Success rate according to analyzed literature	Reference	Success rate according to evaluations carried out
TR	Eigenfaces	92–100%	[16]	0%
	Fisherfaces	85.75%	[17]	0%
	Histograms	88.73%	[18]	60%
DL	TensorFlow O. API	45% prom.	[19]	100%
	Keras O. API	40.8% prom.	[20]	10%
	Facenet/openface	99.63%	[21]	60%

In a graphical analysis as evidenced in the preceding tables, it can be seen in Fig. 2 that the tensorflow API is imposed on the other models for group A. On X axes.

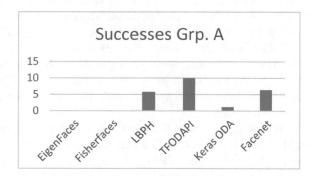

Fig. 2. Comparison of successes found in the executions of group A algorithms, the different evaluated methods can be observed on the X axis, while the percentages of successes reached on the vertical axis.

The same execution, but this time on group B, show a slight advantage of Facenet/openface over the other models, values shown in Fig. 3.

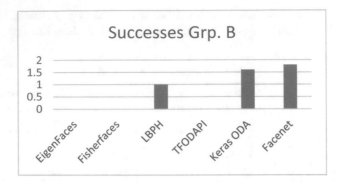

Fig. 3. Comparison of successes found in the executions of group B algorithms, the different evaluated methods can be observed on the X axis, while the percentages of successes reached on the vertical axis

5   Conclusions

This article has presented an evaluation of two groups of facial recognition techniques based on both traditional techniques and deep learning techniques, evaluation that was carried out on particular training conditions. The contribution that is delivered rests on the fact of being able to determine, which technique would specifically allow to work on conditions in which the image to be evaluated does not enjoy the great conditions of resolution, size and standardization of form, contrary to the traditional paradigms in which training datasets have large volumes of photographs and lighting conditions, this in the long run allows us to consider for example Facenet as an algorithm that would work very well in real-time systems and systems with low performance in both image quality and technological resources.

After the evaluation of the proposed techniques, the need to generate experiments that can demonstrate how would improve the accuracy of the methods that achieved poor results, increasing the size of the dataset training as raises the analysis of the two algorithms with better results based on real-time tests, and through the use of security camera systems for biometric application implementations.

References

1. Tucker, J.: How facial recognition technology came to be (2014)
2. Munson, J.H., Duda, R.O., Hart, P.E.: Experiments with Highleyman's data. IEEE Trans. Comput. **C–17**(4), 399–401 (1968)
3. Munson, J.H.: Experiments in the recognition of hand-printed text, Part I. In: Proceedings of the December 9–11, 1968, Fall Joint Computer Conference, Part II on - AFIPS 1968 (Fall, Part II), p. 1125 (1968)
4. Schroff, F., Kalenichenko, D., Philbin, J.: FaceNet: a unified embedding for face recognition and clustering, March 2015
5. Yang, M.H., Kriegman, D., Ahuja, N.: Detecting faces in images: a survey. IEEE Trans. Pattern Anal. Mach. Intell. **24**, 34–58 (2002)

6. Kanade, T.: Picture processing system by computer complex and recognition of human faces, Dep. Inf. Sci. Kyoto Univ. (1973)
7. Brunelli, R., Poggio, T.: Face Recognition Through Geometrical Features, pp. 792–800. Springer, Heidelberg (1992)
8. Kawulok, M., Celebi, M.E., Smolka, B.: Advances in Face Detection and Facial Image Analysis. Springer, Cham (2016)
9. Humphrys, M.: Single-layer Neural Networks (Perceptrons). https://www.computing.dcu.ie/~humphrys/Notes/Neural/single.neural.html. Accessed 10 Nov 2018
10. Wu, S., Zhong, S., Liu, Y.: Deep residual learning for image recon. Multimed. Tools Appl. (2017)
11. Rosebrock, A.: Deep Learning for Computer Vision with Python Starter Bundle, 1st edn. (1.2.2), PyImage Search, New York (2017)
12. Goodfellow, I., Bengio, Y., Courville, A.: Deep learning. Nature **521**(7553), 800 (2016)
13. Ba, L.J., Caruana, R.: Do Deep Nets Really Need to be Deep?, pp. 1–9
14. Liu, T., Fang, S., Zhao, Y., Wang, P., Zhang, J.: Implementation of Training Convolutional
15. Menshawy, A.: Deep Learning by Example: A Hands-on Guide to Implementing Advanced Machine Learning Algorithms and Neural Networks. Packt, Birmingham (2018)
16. Çarıkçı, M., Özen, F.: A face recognition system based on eigenfaces method. Procedia Technol. **1**, 118–123 (2012)
17. Zhang, C.-Y., Ruan, Q.-Q.: Short paper: face recognition using L-Fisherfaces*. J. Inf. Sci. Eng. **26**(4), 1525–1537 (2010)
18. Huang, D., Shan, C., Ardebilian, M., Wang, Y., Chen, L.: Local binary patterns and its application to facial image analysis: a survey. IEEE Trans. Syst. Man Cybern. Part C (Appl. Rev.) **41**(6), 765–781 (2009)
19. Huang, J., et al.: Speed/accuracy trade-offs for modern convolutional object detectors, November 2016
20. Lin, T.-Y., Goyal, P., Girshick, R., He, K., Dollár, P.: Focal loss for dense object detection, August 2017

Multimodal Smartphone-Based System for Long-Term Monitoring of Patients with Parkinson's Disease

Tetiana Biloborodova[1](✉), Inna Skarga-Bandurova[1,2],
Oleksandr Berezhnyi[1], Maksym Nesterov[1],
and Illia Skarha-Bandurov[3]

[1] Volodymyr Dahl East Ukrainian National University, Severodonetsk, Ukraine
biloborodova@snu.edu.ua
[2] Oxford Brookes University, Oxford, UK
[3] Luhansk State Medical University, Rubizhne, Ukraine

Abstract. The paper presents a smartphone-based system for long-term self-monitoring patients with Parkinson's disease. Particularly promising multimodal functionality includes collecting data from different internal measurement units in two modes, evaluating the performance of specific tasks and special non-obtrusive passive sensing features based on day-basis activities. Information about physical activities and symptoms is processed and displayed in the form of a diary. The general system architecture and functional architecture are introduced. We outlined the main design principles and provisions on data completeness and engagement of participants during the first two study months. Some aspects of data analysis are also discussed. The classification accuracy with one data processing method for tremor data is 89,7%. Insofar as MeCo system uses different tremor components, the enhancement of the classification accuracy may be achieved by combination of criteria from different techniques. The combination of criteria provides up to 10% more accurate results in comparison with a single analysis.

Keywords: Parkinson's disease · Smartphone · Application · Long-term monitoring

1 Introduction

Changes in movement control are usually the first noticeable symptom seen in patients with Parkinson's disease (PD). Motor fluctuations and degradation in motor function are reflected in the patient's motor behavioral patterns, e.g., subtle motor movements, speed of reflex moves, and intermittent tremor. Diagnosis of PD is made by a physician who assesses the patient's condition using standardized scales, such as the Unified Parkinson's Disease Rating Scale (UPRDS) [1]. The MDS-sponsored revision of the UPDRS (MDS-UPDRS) [2] includes four parts to measure the range of PD symptomatology. An adequately prescribed dose of the drug enables to return the correct control of movement within a few hours. However, as the disease progresses, the motor complications develop. Obtaining accurate information about the long-term evolution

© Springer Nature Switzerland AG 2020
Á. Rocha et al. (Eds.): ICITS 2020, AISC 1137, pp. 626–636, 2020.
https://doi.org/10.1007/978-3-030-40690-5_60

of these motor symptoms such as tremors and their short-term fluctuations is crucial to ensure optimal treatment for patients with PD and to measure the results of clinical trials correctly. On the other hand, people with PD usually visit a specialist no more than once or twice a year that makes this problem very complex since it becomes unavailable to track the disease progression in detail and means that both efficiency and side effects of medication can be unnoticed. In this context, electronic patients' diaries and self-reports based on the MDS-UPDRS could be the abundant source of information to assess the severity of the PD symptoms. Monitoring patients with PD at home and automatic data analysis can help healthcare providers to understand the impact of PD on daily habits better, even in the early stages, assess the patient's response to drug therapy, level of motor symptoms, and compare their effectiveness with the prescribed treatment. As mentioned in [3], digital health solutions penetrating real-life faced with a challenging task to capture useful disease indicators and achieve long-term statistics.

The goal is to develop an affordable and easy-to-use solution for remote monitoring of symptoms of Parkinson's disease at home, a kind of electronic patient diary for a continuous long-term assessment of the motor conditions. The present work is in line with the efforts toward predictive healthcare data analytics, in capturing early signs of PD as well as assessing the development of Parkinsonian symptoms over time. The idea of this approach involves a combination of well-known methods for testing patients with PD, namely specially designed tests built-in as an application in a smartphone and enable to perform analysis of data in a different time and population scales.

2 Background

Accomplish these ends, we analyzed recent smartphone applications mPower [4], PANDA [5], PVI [6], SmT [7], cloudUPDRS [8] and techniques used for smartphone-based assessment of different PD symptoms. A comprehensive review of methods for assessment of motor malfunctions in Parkinson's disease is presented in [9]. Many previous studies have proposed to use different internal measurement units of smartphones for measuring personal daily activities such as walking, standing and running, etc. Most recent studies of smartphone-based assessments are focused on assessing motor symptoms in Parkinson's disease (PD) such as tremor [10–14], speech [15], gait [16, 17], and bradykinesia [7, 18]. For this study, we add dysarthria and dysphonia to core features selected by Mantri and Morley [19] for prodromal and early Parkinson's disease diagnosis as bradykinesia, rigidity, tremor, postural instability, which comply with MDS-UPDRS [2]. Table 1 summarizes information about internal measurement units (IMU) and their current operational capability to perform both task-specific testing and daily activity monitoring.

- **Accelerometer:** The smartphone-based tools for assessing Parkinsonian tremor are the most common, these tools use signals from the phone's accelerometer and gyroscope to compute a set of metrics to quantify a patient's tremor symptoms [10, 20]. These applications are now ubiquitous due to advanced onboard sensors and wireless connectivity. They are also easy to use compared with most wearable

Table 1. The leveraging IMU of smartphones for PD monitoring

IMU	Task/Test	Time	Target symptom	Type of monitoring
Accelerometer [10, 12]	Hold the phone flat in the hand	4 min	Tremor	Task-specific
Accelerometer [1]	Walk 5 m in a straight line and back with the phone in the pocket	2 min	Postural instability	Task-specific
Camera [14]	Hand opening and closing, finger tapping, Pronation-supination	4 s	Bradykinesia	Task-specific
Camera [14]	Open and close the contralateral hand (Froment's maneuver)	30 s	Rigidity	Task-specific
Camera [14]	Hand and arm extension to the front and to the side	30 s	Tremor	Task-specific
Microphone [6, 22]	Voice-based tests (verbal diadochokinesis)	30 s	Dysphonia	Task-specific
Microphone [21, 23]	Telephone conversation	10 min	Dysarthria	Daily activity monitoring
Microphone [15]	Movement of the hand by extracting the phase of sound waves reflected by the hand	30 s	Tremor	Task-specific
Sensor screen [24]	Keyboard-tapping, Picture-tapping tasks	30 s	Tremor	Task-specific
Sensor screen [18]	Keyboard-tapping tasks, Picture-tapping tasks	20 s	Bradykinesia	Task-specific
Sensor screen [18]	Routine typing activities using virtual keyboard	5 min	Tremor	Daily activity monitoring

solutions previously developed for this purpose. In [20], the smartphone's accelerometer was used for differential diagnosis between Parkinson's disease and essential tremor.

- **Camera:** Tremor detection using camera discussed in [15]. Moreover, Williams et al. [14] have recently used a camera to assess they applied Eulerian magnification to amplify changes in video pixels over time. This made it possible to check the small motions and detect tremors at the early stage.
- **Microphone:** Gillivan-Murphy et al. [21] developed technology for evaluation voice tremors in people with PD using acoustic analysis; it is also enabled to examine correlations with voice disability and disease variables. Nonlinear speech analysis for quantification of average Parkinson's disease symptom severity. In [15], Wang et al. proposed to use speakers and microphones of smartphones to perform tracking of a hand and detect tremors. Smartphone-based acoustic sensing based on low-latency acoustic phase methodology deployed on existing mobile devices as software.

- **Sensor screen:** study [18] confirm that a quantitative assessment of bradykinesia in PD using finger tapping tasks.

Data fusion technique for different PD symptoms captured via smartphone-based tests is discussed in [16], and machine learning approaches has been explored in [17].

The level of everyday interaction people with their smartphones can be transformed into particular behavioral indices, built in a dynamic and personalized way across the short sessions of interaction. Talking about designing monitoring tool, it should be mentioned that in some conditions using active sessions is quite difficult for patients, especially in the elderly groups; in this case, special non-obtrusive and passive sensing features could be useful. In view of this, app working sessions can be based both on task-specific tasks, when user open application and perform tests and daily activity monitoring, when application collect data (voice jitter, specific sound waves, tracking accuracy, speed and pressure on virtual keyboard, etc.) directly from everyday routines and analyze it in long term dynamics. Motivated by those mentioned above, the current study steps further by developing a multimodal smartphone-based system for long-term monitoring of patients with Parkinson's disease.

3 Base Architectures

3.1 General System Architecture

The MeCo system includes the following components (Fig. 1):

(1) A smartphone app for iOS and Android that enables patients to carry out different motor performance tests complete a self-assessment questionnaire and securely submit data to the cloud service.
(2) A web application to collect data from wearable sensors.

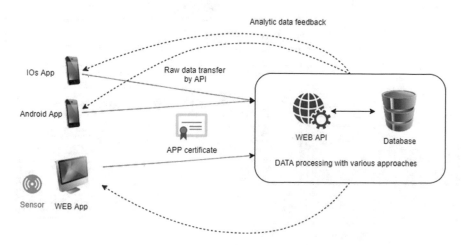

Fig. 1. General system architecture.

(3) Scalable cloud-based data collection services that collect data from patients' smartphones; it also ensures secure data management, and applies the data processing pipeline (see Fig. 2).

(4) A webserver to host the site and store the signals.

(5) Data-mining package for health data analytics incorporating quantitative and semi-structured data, and longitudinal analyses, clustering and classification; and a clinical user interface including short term and long term data visualization.

The MeCo implements a comprehensive workflow that provides audio, video, and textual media to guide patients and their carers to conduct the tests at home and in their homes unsupervised by a clinician. The MeCo service is designed to deliver scalable performance through the use of micro services architecture [25]. This approach differs from traditional monolithic web applications and aims to maximize vertical decomposition and scaling capabilities, which are critical to high performance in data-intensive environments.

3.2   Functional Architecture

The basic structure of our functional architecture is shown in Fig. 2. Raw data from all sensors are processed and then used to perceive the current state. This process is mainly composed of three steps: collection, processing and analysis. These steps are explained below.

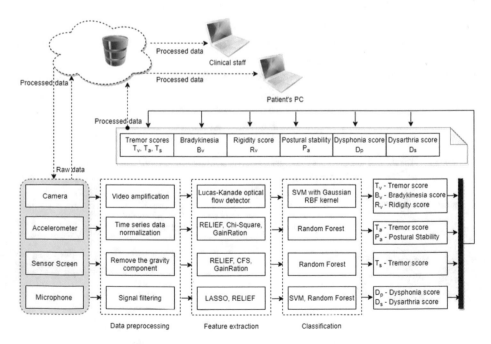

Fig. 2. Functional architecture of MeCo system

Data Collection

The data is collected by the MeCo 1.0 application remotely. Subjects from Ukraine provide multimodal pseudo-anonymized data (e.g., voice, movement sessions) via downloading the mobile app from the Google Play store and registering for the study. Written informed consent should be obtained from all subjects before their participation in research and procedures conducted under institutional and international guidelines on adult-led research. Subjects can recall the process at any time using the available option in the application and even request the removal of the collected data.

Data Processing

The data processing steps are similar to classification works and include the preprocessing, feature extraction, and classification stages. The main difference in this work is that we use various data preprocessing techniques to prepare data. As can be seen from Fig. 2, MeCo can receive and process at least three different parameters for tremor form camera, accelerometer, and sensor screen.

It enables us to proceed with these data both independently and form integrated score for this parameter.

(1) Pre-Processing: At the end of the data collection session, all data gathered from the smartphone is extracted and passed through the data processing pipeline. A technique for data processing stands on the task and type of IMU used for the study. Thus, data from the camera should be amplified applying apply Eulerian magnification technique mentioned in [14] to enhance the quality of output video sequence. For the data obtained from the accelerometer, we remove the gravity component and perform data normalization. Sensor data corresponding to keypress is extracted and framed as well. One frame is considered as the entire duration between a keypress and a key-release.

(2) Feature Extraction: For each frame, we provide features extracted for each frame depending on the task and type of source data. Each feature can either be a sensor-specific feature or a temporal one [26]. All the features are extracted individually and computed either for the three-axis individually or combination of all axis.

Data Analysis

At this stage, tasks roughly divided into four categories: detection, classification, tracking, and motion estimation. To this time, detection and classification are completely realized.

(3) Classification: The activity recognition problem is considered as a supervised classification task where the subsequences of datasets are fed into a machine learning classifier. Thus, for touchscreen data, once the features for every keypress of every user are extracted, we combine the data of all the users and leverage 10-fold cross-validation. In the classification stage, we employed several machine learning classifiers, i.e., Random Forest, k-Nearest Neighbor (kNN), and Support Vector Machine (SVM). We evaluated various classifiers and found that the accuracy obtained using a Random Forest classifier is the highest. For simplest tasks, as described in the example below, Random Tree classifier can be also efficient.

4   An Example of Data Processing Form Accelerometer

4.1   Data Collection

The signal T_s used for analysis is obtained from routine procedure assessments employing a smartphone placed in the hand. The set of exercises includes 3 tasks based on finger touches. These tasks were intended to reflect the most common types of interaction with touchscreen interfaces.

1. Basic tapping: a person touches a spiral, which appeared at a random location on the screen.
2. Sequential tapping: a person taps the numbers appearing on-screen using the virtual keypad.
3. Double-tapping: a person switches off an alarm by tapping twice on a 15-mm circle.

   User actions during the first test comprises from the sequence of the following steps. Participant touches the Start button. At first they needs to follow the spiral line on the screen by the index finger. When line is ended, press the Finish button. Then second round starts. After clicking the Finish button, the spiral starts to disappear and appear, and participant need to run their finger along the spiral and click the Finish button.

   As a result of the tremor test, the time stamp and accelerometer data are obtained in three axes (Table 2).

Table 2. The fragment of accelerometer data

Timestamp	X	Y	Z
924802121	0.009	8.293	4.539
924802355	−0.086	8.619	4.146
924802362	−0.057	8.466	4.079
924802369	−0.114	8.581	4.52
924802378	−0.086	8.676	4.52
924802387	−0.105	8.715	4.462

4.2   Data Processing

Data processing is carried out on a sliding window. Each window is partially overlapped with the acceleration sampling procedure and is processed separately. For each acceleration component the following statistical features are calculated:

- Mean (Mean X, Mean Y, Mean Z)
- Standard deviation (St.dev X, St.dev Y, St.dev Z)
- Energy of the sequence (Energy X, Energy Y, Energy Z)
- Pearson's correlation between each pair of acceleration components (P.correl X–Y, P.correl X–Z, P.correl Y–Z).

After this stage, we obtain 13 variables for each sample window. A fragment of the data set is presented in Table 3.

Table 3. Generated accelerometer data set

Time stamp	7240555	7240565	7240575	7240585	7240595	7240605
Mean X	−0,05905	−0,05838	−0,05752	−0,05599	−0,05504	−0,05409
Mean Y	1,15957	1,17021	1,18065	1,19138	1,20229	1,21599
Mean Z	9,84823	9,85293	9,85465	9,85312	9,85082	9,85025
St.dev X	0,07469	0,07432	0,07427	0,07425	0,07436	0,07481
St.dev Y	0,33684	0,34404	0,35034	0,35858	0,36557	0,37565
St.dev Z	0,17359	0,18000	0,18093	0,18166	0,18259	0,18316
Energy X	0,00901	0,00888	0,00877	0,00859	0,00850	0,00847
Energy Y	1,45693	1,48657	1,51545	1,54668	1,57781	1,61833
Energy Z	97,0175	97,1123	97,1465	97,1166	97,0717	97,0606
P.correl Y–Z	0,12086	0,17299	0,19069	0,16497	0,14077	0,09851
P.correl X–Z	−0,12700	−0,12588	−0,11962	−0,12886	−0,13462	−0,14734
P.correl X–Y	0,39714	0,37929	0,37799	0,38472	0,38819	0,40199
Class	H	H	H	H	H	H

The classification is carried out using labeled data with four output variable: N (no tremor), S (slight tremor), M (moderate tremor), and S (severe tremor).

The original and generated tremor test data set was classified using 10-fold cross-validation method. Classification carried out using the Random Tree algorithm.

4.3 Evaluation

The evaluation of the classification was carried out using the following parameters: accuracy, sensitivity, specificity. To calculate these parameters, we used classification assessment by confusion matrix. Based on the confusion matrix the sensitivity, the specificity, and the accuracy are calculated as follows (1)–(3) using true positive ς_{00}, false negative ς_{10}, true negative ς_{11} and false positive ς_{01} results of n observations.

$$Sensitivity = \frac{\varsigma_{00}}{\varsigma_{00} + \varsigma_{10}} * 100\%, \tag{1}$$

$$Specificity = \frac{\varsigma_{11}}{\varsigma_{11} + \varsigma_{01}} * 100\%, \tag{2}$$

$$Accuracy = \frac{\varsigma_{00} + \varsigma_{11}}{n} * 100\%, \tag{3}$$

The classification comparison results are presented in Table 4.

Table 4. Classification results

Model	Sensitivity, %	Specificity, %	Accuracy, %
Accelerometer test data	100	64,9	89,7

The classification accuracy with this data processing method is 89,7%. As MeCo system uses different tremor components, the enhancement of the classification accuracy may be achieved by combination of criteria from different techniques. The combination of tremor criteria provides 8%–10% more accurate results in comparison with single analysis.

5   Conclusion and Future Work

This solution is designed to help people monitor the disease's progression in time, and provide insights into the understanding as to how Parkinson's arises as well as lifestyle factors that may affect their symptoms. MeCo enables us to acquire large-scale data streams from users' everyday patterns is human-mobile interaction, both patient-reported outcomes, and physical activity data. To this moment, we have preliminary results on data completeness and engagement of participants during the first two study months (from August 10, 2019, to October 10, 2019). Updated information will be presented at the conference. During the first study month, we collected 302 GB of sensor data and 2407 questionnaires. The connection settings developed for the system proved to be efficient when sensor data transmitted from the smartphone to database storage. The time required to transfer data to the database is less than one second. Also, future work will involve developing and implementing additional tests related to speech recognition for PD symptoms detection. In the future, we plan to expand analytical features of the system not only objectively asses Parkinson's symptoms but also to combine the different symptoms with results from smell testing, and keyboard-tapping tasks to divide individuals into high-risk, middle-risk, and low-risk groups.

References

1. Fahn, S.E.: The UPDRS development committee, the unified Parkinson's disease rating scale. Recent Dev. Parkinson's Dis. **2**(153–163), 293–304 (1987)
2. Goetz, C.G., et al.: Movement disorder society-sponsored revision of the unified Parkinson's disease rating scale (MDS-UPDRS): scale presentation and clinimetric testing results. Mov. Disord. **23**(15), 2129–2170 (2008)
3. Iakovakis, D., Hadjidimitriou, S., Charisis, V., et al.: Motor impairment estimates via touchscreen typing dynamics toward Parkinson's disease detection from data harvested in-the-wild. Front. ICT (2018). https://www.frontiersin.org/articles/10.3389/fict.2018.00028/full

4. Parkinson mPower mobile application. https://parkinsonmpower.org
5. Parkinson's Digital Assessment (PANDA). http://cenvigo.com/en/parkinsons_pda_application/
6. PVI: Parkinson's voice initiative. http://www.parkinsonsvoice.org/
7. SmT: Smartphone tapper. https://sites.google.com/site/neurorehabict/downloads/ftapp
8. Stamate, C., Magoulas, G.D., Kueppers, S., Nomikou, E., Daskalopoulos, I., Luchini, M.U., Moussouri, T., Roussos, G.: Deep learning Parkinson's from smartphone data. In: Proceedings IEEE International Conference Pervasive Computing and Communications (PerCom), Kona, HI, USA, pp. 31–40 (2017)
9. Oung, Q.W., Muthusamy, H., Lee, H.L., Basah, S.N., Yaacob, S., Sarillee, M., Lee, C.H.: Technologies for assessment of motor disorders in Parkinson's disease: a review. Sensors **15** (9), 21710–21745 (2015). https://doi.org/10.3390/s150921710
10. Kostikis, N., Hristu-Varsakelis, D., Arnaoutoglou, M., Kotsavasiloglou, C.: A smartphone-based tool for assessing Parkinsonian hand tremor. IEEE J. Biomed. Health Inf. **19**(6), 1835–1842 (2015). https://doi.org/10.1109/jbhi.2015.2471093
11. Sen, S., Grover, K., Subbaraju, V., Misra, A.: Inferring smartphone keypress via smartwatch inertial sensing. In: 2017 IEEE International Conference on Pervasive Computing and Communication Workshops (PerCom Workshops), Kona, HI, 13–17 March 2017, pp. 685–690. https://ink.library.smu.edu.sg/sis_research/3583
12. Carignan, B., Daneault, J.F., Duval, C.: Measuring tremor with a smartphone. Methods Mol. Biol. **1256**, 359–374 (2015). https://doi.org/10.1007/978-1-4939-2172-0_24. PMID: 25626551
13. Wang, W., Liu, A.X., Sun, K.: Device-free gesture tracking using acoustic signals. In: MobiCom 2016 Proceedings of the 22nd Annual International Conference on Mobile Computing and Networking, NY, USA, 3–7 October 2016, pp. 82–94 (2016)
14. Williams, S., Fang, H., Alty, J., Qahwaji, R., Patel, P., Graham, C.D.: A smartphone camera reveals an 'invisible' Parkinsonian tremor: a potential pre-motor biomarker? [Letter to the Editors]. J. Neurol. **265**, 3017–3018 (2018)
15. Wang, W., Wang, X., Xie, L.: Tremor detection using smartphone-based acoustic sensing. In: UBICOMP/ISWC 2017 ADJUNCT, Maui, Hawaii, USA, 11–15 September 2017, pp. 309–312 (2017)
16. Arora, S., Venkataraman, V., Zhan, A., Donohue, S., Biglan, K.M., Dorsey, E.R., et al.: Detecting and monitoring the symptoms of Parkinson's disease using smartphones: a pilot study. Parkinsonism Relat. Disord. **21**, 650–653 (2015). https://doi.org/10.1016/j.parkreldis.2015.02.026
17. Cuzzolin, F., Sapienza, M., Esser, P., Dawes, H., Saha, S., Collet, J., Franssen, M.: Metric learning for Parkinsonian identification from IMU gait 3 measurements. Gait Posture **54**, 127–132 (2017)
18. Lee, C.Y., Kang, S.J., Hong, S.K., Ma, H.I., Lee, U., et al.: A validation study of a smartphone-based finger tapping application for quantitative assessment of bradykinesia in Parkinson's disease. PLoS ONE **11**(7), e0158852 (2016)
19. Mantri, S., Morley, J.F.: Prodromal and early Parkinson's disease diagnosis. Pract. Neurol. **35**, 28–31 (2018). https://practicalneurology.com/articles/2018-may/prodromal-and-early-parkinsons-disease-diagnosis
20. Barrantes, S., Sánchez, E.A.J., González, R.H.A., Martí, M.J., Compta, Y., Valldeoriola, F., et al.: Differential diagnosis between Parkinson's disease and essential tremor using the smartphone's accelerometer. PLoS ONE **12**(8), e0183843 (2017). https://doi.org/10.1371/journal.pone.0183843
21. Gillivan-Murphy, P., Miller, N., Carding, P.: Voice tremor in Parkinson's disease: an acoustic study. J. Voice **33**(4), 526–535 (2019)

22. Yang, C.-C., Chung, Y.-M., Chi, L.-Y., Chen, H.-H., Wang, Y.-T.: Analysis of verbal diadochokinesis in normal speech using the diadochokinetic rate analysis program. J. Dent. Sci. **6**, 221–226 (2011)
23. Tsanas, A., Little, M.A., McSharry, P.E., Spielman, J., Ramig, L.O.: Novel speech signal processing algorithms for high-accuracy classification of Parkinson's disease. IEEE Trans. Biomed. Eng. **59**(5), 1264–1271 (2012)
24. Taylor Tavares, A.L., Jefferis, G.S.X., Koop, M., Hill, B.C., Hastie, T., Heit, G., Bronte-Stewart, H.M.: Quantitative measurements of alternating finger tapping in Parkinson's disease correlate with UPDRS motor disability and reveal the improvement in fine motor control from medication and deep brain stimulation. Mov. Disord. **20**(10), 1286–1298 (2005). https://doi.org/10.1002/mds.20556
25. Newman, S.: Building Microservices: Designing Fine-Grained Systems. O'Reilly Media, Sebastopol (2015)
26. Erdaş, Ç.B., Atasoy, I., Açıcı, K., Oğul, H.: Integrating features for accelerometer-based activity recognition. In: The 3rd International Symposium on Emerging Information, Communication and Networks (EICN 2016), vol. 98, pp. 522–527 (2016). Procedia Computer Science

Digital Competences Desirable in University Students

Paola Cortez[✉], Verónica Benavides, Félix Rosales,
and Lilibeth Orrala

Universidad Estatal Península de Santa Elena,
240350 La Libertad, Ecuador
{pcortez,vbenavides,frosales,lorralas}@upse.edu.ec

Abstract. Currently, Information and Communication Technologies (ICT) are present in all areas of human development and activities. Banks and financial activities, products and labor activity, the government, the media, the health area, among many others, use digital tools and media to carry out their functions. Education does not escape that reality of being part of a digital society in continuous expansion. Knowledge is produced, disseminated, shared and discussed through digital media. It constitutes the most valuable asset of any organization in the in-formation society. This article is derived from a mixed research based on phenomenology and hermeneutics, whose intent was to generate a theoretical approach to desirable digital competencies in university students. The results indicate that the use of ICT focuses mainly on the use of email and the tools of Microsoft Office PowerPoint and Excel. To a lesser extent, the in-corporation of the use of some social networks is observed. As for the digital ones evaluated, the study demonstrates its mastery only by a third of the interviewees, which implies that two thirds of the students require training and education courses to acquire them, being encouraging that half of the sample understands the importance of this type of skills and the need to achieve them.

Keywords: ICT tools, digital skills · Desirable competencies

1 Introduction

Digital competences as priority skills in the management of information and own technological resources to develop the events, situations and action schemes of communication, are indispensable in the appropriation of the criteria aspects addressed in the construction of the formative knowledge to be deployed. As an intelligible formative platform congruent with the daily social practices of global reality. At this orientation, affirm (Spante, Hashemi, Lundin, Algiers and Wang 2018) that are aspects compatible with the communications perspectives necessary to take seriously, as a cross reference of social investigations that arise from the legitimacy of the policies focused on this field in the different universities. In this same order of ideas, the competitive advantages in the facilitation of convergent mobile networks used in the knowledge society, in a particularly attractive way for personal and professional

© Springer Nature Switzerland AG 2020
Á. Rocha et al. (Eds.): ICITS 2020, AISC 1137, pp. 637–644, 2020.
https://doi.org/10.1007/978-3-030-40690-5_61

communication, in addition to its implementation as a third community space, opens up the intelligible possibilities of the field University training, which was called by (McDougall, Readman, and Wilkinson 2018), shows interested in determining the progress in which students are mobilized given digital skills, by adopting their strength in terms of functional literacy in his cognition, in specific national and international environments that are deeply located within the technological, social and economic philosophy, immersed in the progressive changes of university education.

2   Literature Review

Various international and national investigations have been presented in academic fields, related to the digital competences of university students, among which the following are presented, in descending order on their completion dates. To this end, Esteve (2015), titled his research: The Digital Teaching Competition. Analysis of Self-Perception and Evaluation of the Performance of University Education Students through a 3D Environment, presented at the Rovira I Virgili University, Spain.

The work mentioned, methodologically, dealt with an educational design raised as a systematic process of design, cyclical, development and evaluation of educational intervention in relation to technology as a solution to a complex problem, in addition to increasing knowledge generated a series of design principles applicable to other realities. Among the results it was found that students feel the ability to design and develop learning experiences that incorporate ICT tools and resources, thus 38.9% and 55% of them are considered quite or very capable of using the potential in that area to design didactic activities with digital resources that adapt to diversity.

It is concluded that the capacity of future university professionals must continue to improve in professional practice through the use of ICT given their recognized digital capabilities in the training and implementation of research experiences and learning. Technological usability is recommended to freely explore 3D environments, the valuation of ideas and more flexible structured strategies in university practice.

The contribution of this previous study highlights the practicality and usefulness of virtual environments to develop new pedagogical perceptions in potential end users who adapt the components of communication, teamwork and understandability of technological coding to assess the construction of knowledge.

In this same orientation, (Umbleja 2017), titled his research: Competence Based Learning-Framework, Implementation, Analysis and Management of Learning Process, presented at the Technical University of Tallinn, Estonia, Tallinn, with the aim of formulating a learning framework Full focus on the acquisition of new skills and knowledge, by transforming the classic learning environment based on topics proposed by effective learning in the use of technology.

The methodology used was based on the review of the literature to identify the common key aspects of different authors and interwoven approaches to the study framework in the possibility of verifying the effectiveness on the new schemes proposed in digital competencies, their deficiencies, principles and implementation in the formative reality. Data analysis and hypothesis tests were analyzed in an exploratory manner.

Among the findings of the aforementioned research, it was found that the proposed digital construction model worked on the prediction of the satisfactory final grade with more than 80% of the students who worked harder on this technological characteristic during the semester considered. The changes in the study groups were favor-able to the activity in the information search preference.

He concluded that when showing the predictions to the students, the different sections of the computer system developed with the implementation of technological and statistical tools were incorporated into the main page, although the incorporation of predictions on the main page did not allow to measure the frequency with which the students They look for the information. He recommended the prediction model created to develop digital competence in the student as a specific and viable characteristic when assuming changes in the implementation of online courses.

The contribution of this previous study focuses on the multiple possibilities offered by the online courses for mass education open to new educational realities that transcend the classical classroom, the real time of learning, the spaces and environment to build knowledge, in the management of the skills acquired inside the university.

Similarly, (Muller 2017), in his doctoral thesis studied the teaching of digital skills in the 1st year of service at a South American university. In this way, he examined the barriers, the enablers and their impact on the students' successful commitment to the use of ICT as a positive influence on self-efficacy and complementary in digitally literate research techniques.

Methodologically, a quantitative investigation implemented the computer attitude scale (CAS) to determine the digital self-efficiency in the current levels of student's beliefs about skills in terms of self-efficiency in software management and various computer programs a unique application of the theory of the network of actors was used in the analysis of the data. The results found indicate that 43% of students failed the baseline digital proficiency test to enter the higher education institutions of Spain (IES), which implies that the ability to operate digital devices and their relevant applications in the search for student learning in a self-sufficient way as functional ability.

The relationship between previous studies and current research is linked to the possibility of improving the learning, productivity and performance of university students. Defining the digital competences involved in the development of new knowledge as training support, from the strength in technological skills and attitudes towards changes in the effective and self-sufficient use of the media.

In the same order of ideas, (Mendoza 2017) was located, entitled: Analysis of Internet access for high school students in Ecuador, presented at the University of Huelva, Spain. The growing activity on the Internet was addressed as a reflection of the economic, social and cultural activities and relationships that exist off-line, including inequalities, whose access to the skills, use and benefits of ICT represents a path for development and advance in learning.

In these interconnection schemes, it refers to the context of Ecuador with respect to government information sites with the updated information offered in all institutional lines and services, while the government through the Ministry of Education generated a state policy in the year 2014 with plans and initiatives in projects related to the technological platform backed by the need for a permanent national relationship and training for teachers.

The descriptive nature of the research with empirical support defined the variables that could be measured and statistically analyzed., Supported by the foundation of positivism, whose collection of data through the questionnaire, personal interviews with principals and educational authorities allowed interaction and Discussion with student and teachers. Among the results it can be seen that Internet access in the University has a predominance of low intensity. There are central classrooms accessible to the university community favorable option in 84% of cases.

The investigation concluded that the level of active participation with Web 2.0 is favored from the academic point of view in courses, events, communication to work in teams, videoconferences, content creation through Wiki, blogs, multimedia, etc., which benefit the system of improvement in access and right to communication and information. It is suggested that the development of policies that act at different levels of literacy and Internet access be used in schools in order to generate greater autonomy, training, operational capabilities and information management such as skills development and knowledge of academic use.

The relationship between this study and current research is evidenced by the need for progress in the use of the Internet in Ecuador for academic achievement, in order to overcome the digital divide in access to materials and devices that allow taking advantage of the technological possibilities in the different productive fields in accordance with the global information that facilitates the Internet connection. Therefore, the State must establish mechanisms to reduce the cost of the Internet service and regulate the intervention of the private company in this context; in turn improve the quality of the service given the idea of massification of the fiber optic connection.

3   Educational Meetings from Virtuality: Institutional Reality

The valuation of university professors in terms of the technical, technological and attitude implications towards innovation in the performance of their functions, attracts the evolution of new situations of effective action that must reflect on the basis of skills, abilities, competencies and ethical aspects, which specifically exalt the dominant university service within the framework of the necessary adjustments to activate the benefits of information and communication technologies.

However, the expression of the competencies demarcates the direction in the organizational decision making, which, according to (Alles 2005), establishes the conditions applicable to all the members of the organization, regardless of their level, it is also the manifestation of the "competences specific for senior management and for all levels of supervision, which apply to the respective scenarios of the organization, no matter the area to which each position belongs" (p. 69). The author also indicates the value of the specific competences of the area related to planning and organization capacity, collaboration, pressure tolerance, analytical thinking, perseverance and innovation.

Specifically, the global dynamism invites those who develop in the university educational areas to constantly adapt to the demands of the market, which is why they are called to optimize those digital competences that have been assigned to them, for the same characterization of their teaching realities, as the strength of their training in technology-conjugated skills.

4 Digital Competition

By understanding the significance of digital competencies in the proper management of communications, new ways of accessing the world of another's life become evident, seeking network participation in topics of interest within the condition of success in the relationship system personal and professional from scenarios that enhance knowledge and the communication boom on the basis of the interactive experiential of reciprocal information that promote productivity, beyond the profitable actions that are envisioned rather, in the sense of integral, social and collective human behavior of educational attributions in the reference of technological approaches with the other.

Faced with this vision, the United Nations Educational, Scientific and Cultural Organization (UNESCO 2013), invites educational actors primarily to the teacher, to promote a new approach to education in terms of digitized skills aimed at training, for the development of technological skills and attitudes that allow participation in academic, cultural and research meetings, not only in the condition of content teaching, but rather, towards the field of transcendence whose tendency is to open spaces in the communicational relations system with its globalized information environment.

This implies the need to reflect from the point of view of the contextualization of the curriculum, the training in digital competencies that address novel learning in the face of the commitment of university institutions towards the focus of the development of the institutional platform that accommodates this type of strategies, guidelines and forms of connectivity raised at the level of globalization in the attention to the social phenomenon related to the competences open to network communication.

As announced by (Tejada 2007), the integration of knowledge is a digital competence available within the context of the teaching function to accommodate the training processes due to their own characteristics of global implications and information updating in the technological platform.

In fact, the perspective of the integration of digital knowledge in the training of communication students expands the possibility of generating knowledge from education efforts towards the search for cognitive unity in academic, collaborative and educational processes. formation of groups in the information network, as an influence of an exercise of holistic appropriation, which implies that this integration in training must be manifested by pointing towards new teachers' competencies in the instrumental multi-referentiality of the management of teams, knowledge networks and technology appropriation strategies.

5   Student Digital Competition

The need to be on the agenda with innovation and the importance of digital communication, focuses on the formative sense of the students of the new entry in the career of social communication. Therefore, it highlights the ideas expressed by Condemarín, (García and Gutiérrez 2011), in relation to the attitudes, technological expressions and rapprochement of teachers with students in this same innovative dynamic, as an attitude of being able to communicate in a positive environment of learning, values transmitted in thinking, acting and feeling of new practices, in accordance with the new realities of communication.

This advantage of knowing how to use and take advantage of the benefits of technologies in order to achieve educational objectives and offer greater confidence in the participatory atmosphere of the university environment, is exposed in the school focus of adaptations, stimuli and reinforcements of digital skills in a system of collaborative relationships, technological sensitivity, creativity and strengthening of learning. Therefore, the announced situation implies that digital competences are a global and growing phenomenon, which allows communication between people, representing ease and versatility in aspects such as the management of social information and the relationship with others.

This is where these conditions stand out because they shorten distances, promote institutional development and modernize the condition of transformation of the education sector, compared to the benefits of digital skills as available requirements in the condition of multi-functional characteristics, as this represents events that cannot be delayed of adaptation in the global access of the information through all the massive channels, given the formative interest from the university, for the exercise of greater contacts that are equated with the previous consultations and the information requirements in the experiences of teamwork, motivation and networking.

6   Management of Tools and Academic Services Supported by ICT

When exploring among the students about the management of some tasks that require the use of ICTs, it was found that more than 80% reported finding all the tasks explored very easy or easy except for participation in virtual forums where this assessment fell to a 61.3%, the task being perceived with greater difficulty.

On the other hand, the results about the knowledge and use of the technological tools that are offered in the universities for the student activity. It is observed that 57% reported having used the Virtual Library database and 59.8% knowing the repository, however, only 14.95% were registered or had information loaded in this repository. Regarding the use of the moodle platform, we found that a large part of them (60.8%) reported fully understanding the operation of the platform and the tools it offers and its use in its teaching activity (53.6%); However, a high percentage (45.9%) said they found technical problems in the use of it. As we see in these results, there is still a significant percentage in training, which does not know or use these technological tools.

It can be evidenced that despite the technological advances that currently exist and the fact that access to various tools is provided, emails and the tools of Microsoft Office PowerPoint and Excel are still the most used. There is also the incorporation of the use of some social networks such as Facebook and Twitter, as well as the use of Wikis (such as Wikipedia) and Blogs. To a lesser extent, online courses, virtual forums and MOODLE (Modular Object-Oriented Dynamic Learning Environment). It is striking that participants do not report the use of the latter, which could indicate the lack of knowledge of the software tool.

It is observed that around 83% of the students considered themselves capable of understanding the communicative codes of the digital contexts and using them effi ciently to communicate on the network. It was also found that 37.6% understood how meaning is produced through multimedia and how culture is produced through the Internet and social networks. Only 2% indicated that they believed that communication on the network occurs in the same way as in non-digital contexts and that digital information follows the same codes of meaning as non-digital; while for 14.43% the information and communication in the network if they have their own codes, but do not know how to interpret them.

On the other hand, 72.7% of respondents were perceived as having competences to participate in virtual communities, 27% actively participating in several of them. However, a similar number (26%) self-assessed as aware of the importance of participating in virtual communities, but do not know how to do it or what tools they can use, suggesting that this could be an important area at the time of their formation. Regarding the ability to filter and classify the information on the web according to interests, most said they know and use tools to filter, store and classify the information they are interested in (27%) or occasionally (45%). Almost all of those who indicated they did not know them, if they are aware of the importance of filtering and organizing the information, so they may be interested in learning the tools available for it.

On the other hand, about 61% of the students identified themselves as capable of creating and editing digital content using the different tools for their habitual or occasional creation; while 77.3% were considered competent to share on the network. A significant percentage of students despite understanding the importance of creating quality digital content, do not know how to use the necessary tools for it, nor know the most appropriate channels according to the type of resources (37%) or have a positive attitude towards sharing resources in the network, but they don't think they know how to do it correctly (20.6%). Of great concern is that only slightly more than half of respondents are perceived as having the capacity to understand the legal and ethical aspects related to the use of ICT, as well as aspects related to the management of privacy and security. That is, a high percentage is vulnerable when using these technologies because they do not know the protection tools or how to correctly handle the legal and ethical aspects of their use.

The majority of students self-assessed with the ability to create and manage a digital identity and to use collaborative work tools (61.4% and 58.8% respectively). This last competence is very important both in education and in research and in general, for the collective construction of knowledge; however, although they said they knew the tools of collaborative work, only 16% reported that they use them beyond occasionally. Likewise, only 21.1% use technology daily to improve the quality of work and express themselves creatively with ICT. It is observed that students, although they seem to know the ICT tools and their potential, do not yet incorporate them in their usual way, largely because they do not know how to do it. Given that 71.13% of teachers said they had the ability to learn from (and with) digital technologies, it is feasible to propose models of teacher training using their own ICTs, under distance or mixed modalities.

7   Conclusions

As a reflection, it can be said that approximately one third of teachers in training are competent in the aspects explored, which implies that two out of three teachers require training and education courses for the acquisition of digital skills required by professionals of the 21st century. Since half of the sample understands the importance of this type of skills and the need to achieve it, universities must guarantee the training of their teachers in the use of ICT so that they can incorporate them into the teaching and learning process, therefore, also favor the development of the digital competence of its students.

References

Alles, M.: Gestión por competencias. El diccionario, 2ª edn. Granica, Buenos Aires (2005)

Arias, F.: El proyecto de investigación e introducción a la metodología científica. Epísteme, Caracas (2006)

Esteve, F.: La Competencia Digital Docente. Análisis de la Autopercepción y Evaluación del Desempeño de los Estudiantes Universitarios de Educación por medio de un Entorno 3D. Tesis doctoral no publicada. Universitat Rovira I Virgili, España (2015)

Bisquerra, R. (Coord.).: Metodología de la investigación educativa. (2ª ed.).La Muralla, Madrid (2009)

Bonnefoy, J., Armijo, M.: Indicadores de desempeño en el sector público. Naciones Unidas, Santiago de Chile (2005)

Bunge, M.: La investigación científica. Buenos Aires: Ariel. (2001)

Bunge, M.: Crisis y reconstrucción de la filosofía. Gedisa, Barcelona (2002)

Condemarín, E., García, C., Gutiérrez, M.: Con amor se enseña mejor. Propuestas para docentes de hoy, 2ª edn. Universidad Alberto Hurtadom, Santiago de Chile (2011)

Figueroa, A.: Conociendo a los grandes filósofos, 5ª edn. Universitaria, Santiago de Chile (2002)

Hernández, R., Fernández, C., Baptista, P.: Metodología de la investigación, 4ª edn. McGraw Hill Interamericana, México (2006)

Kuhn, T.: The structure of sientific revolucions, 2ª edn. University of Chicago Press, Chicago (1970)

Mahner, M., Bunge, M.: Fundamentos de biofilosofía. Siglo XXI, México (2001)

McDougall, J., Readman, M., Wilkinson, P.: The uses of (digital) literacy. J. Learn. Media Tehnol. 43(3), 263–279 (2018). https://doi.org/10.1080/17439884.2018.1462206

Accounting Transparency
of Non – Governmental Organizations:
A Bibliometric Analysis

Mingming Zhang[1], Manuela Cañizares Espada[2(✉)],
Raquel Pérez Estébanez[1], and Elena Urquía-Grande[1]

[1] Facultad de Ciencias Económicas y Empresariales Campus de Somosaguas,
Universidad Complutense de Madrid (UCM), Madrid, Spain
{mingminz, raperez, eurquiag}@ucm.es
[2] Facultad de Ciencias Económicas y Empresariales,
Universidad a Distancia de Madrid (UDIMA), Collado Villalba, Madrid, Spain
manuela.canizares@udima.es

Abstract. As non-governmental organizations (NGOs) become increasingly involved in international affairs, they face a growing deficit of confidence due to opacity of information environment. This opacity not only affects NGO financing, but also decreases public confidence in NGOs, affecting negatively their sustainability. Due to the crucial role of transparency in NGOs, the present study performs a bibliometric analysis of the research on accounting transparency. The bibliometric, conceptual and normative analysis performed shows that accounting transparency of NGOs is a broad concept. Study of the accounting transparency of NGOs is thus neither well-developed nor very saturated.

Keywords: Transparency · NGOs · Bibliometric analysis

1 Introduction

Non-governmental organizations (NGOs) are organizations that originate neither in the governmental sphere nor as for-profit organizations. They usually serve the public welfare and obtain their financing from the governments of the countries with which they collaborate. NGOs are becoming increasingly involved in international affairs, and their role and influence within and outside the United Nations system is constantly increasing and gaining recognition in various fields. However, NGOs currently face a serious problem: interest groups growing deficit of confidence due to scandals, which negatively influences NGOs public image [14]. For example, in China, the Red Cross Society of China (RCSC) suffered a crisis of confidence in 2011. The media published news of its financial failure, generating great turmoil concerning its internal management and ineffective supervision. The result was great distrust in the general population. From that moment, the RCSC suffered a drastic fall in donations. According to China's 2011 charitable donation report, the RCSC system at all levels received social

© Springer Nature Switzerland AG 2020
Á. Rocha et al. (Eds.): ICITS 2020, AISC 1137, pp. 645–659, 2020.
https://doi.org/10.1007/978-3-030-40690-5_62

donations of approximately 2,867 million yuan (3.4% of total donations in China), a decrease of 59.39% over the previous year. This event increased attention to the transparency of information in charitable organizations and led to publication of the Charity Law of the People's Republic of China, which regulates various aspects of charitable organizations, including information disclosure and supervision.

This situation and similar ones in other NGOs render transparency fundamental, since opacity of information hinders fundraising by undermining trust in NGOs and could lead to mismanagement in some NGOs, causing financial deficits that could even destroy them. This situation not only affects NGOs' financing, but also decreases public confidence in these organisms, influencing their sustainability. Transparency in NGOs is crucial.

According to the authors [1], the exponential increase in funding of NGOs relative to the services provided is associated with both national and foreign donors' interest in greater transparency to make NGOs' effective and honest use of contributions visible. They argue that transparency would improve future performance and make future funds safer.

Given the importance of this topic, the objective of this paper is to study the factors and influence of accounting transparency in NGOs through bibliometric analysis of the topic.

Considering the foregoing arguments, and the fact that NGOs' transparency becomes a tool to guarantee their credibility, this study focuses especially on accounting transparency by answering the following research question: RQ1: Is accounting transparency in NGOs an impact issue?

This study has several parts. The first is an introduction, which presents the objective, motivation and research question. The next section develops the theoretical framework—the bibliometric analysis through which to develop the concept of accounting transparency and assessment of it. Finally, the study ends with conclusions, suggestions and limitations of the research.

2   Theoretical Framework

To focus on the accounting transparency of NGOs, we use the search terms "non-governmental organization transparency" for the bibliometric analysis. A total of 122 related documents on a recent topic were found in the "Web of Science" database.

Table 1. Number of documents indexed with search term "non-governmental organization transparency" located in the Web of Science.

Findings	122
Total times cited	898
Average citations per item	7.36
h-index	16

As Table 1 shows, the total number of citations of all elements found is 898, which represents the total number of documents cited from 1997 to 2018. The average number of citations per item is 7.36, calculated as the total number of times cited divided by the results obtained. The h-index of 16 is based on a list of publications classified in descending order by number of times cited. The index means that we find 16 documents that have been cited at least 16 times. This information is useful because it reduces the disproportionate weighting of documents cited very frequently and of articles that have not yet been cited. The greater the h-index is, the greater the document's influence. The h-index is not high for this topic.

Literature indexed under the keyword "non-governmental organization transparency" began to be published in 1997 (see Fig. 1). This starting date may be due to the United Nations Economic Council's 1996 adoption of Resolution 1996/31 on the consultative relationship between the United Nations and non-governmental organizations which recognized NGOs operating in countries and regions. Within these twenty-one years, papers on transparency of NGOs were intermittent in the first 8, and the field has had a constant and fluctuating development since 2006. The years 2008 and 2016 stand out as having significant contributions. Research in 2008 grew rapidly, due to the participation of NGOs in the legislative process of the European Union and the importance of the Internet to the dissemination of financial and social information, and 2008 shows a strong increase in research in the area of social science. Another important increase in research begins in 2016 and is currently peaking due to the publication of freedom of information laws, protection of ecosystems and the Extractive Industry Transparency Initiative (EITI). In general, NGO transparency research is receiving increasing attention, showing this topic as a growing trend.

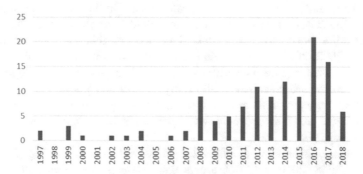

Fig. 1. Contributions per year.

The number of citations has also grown significantly. In 2008, citations also began to increase significantly due to a rapid increase in studies. The year 2012 is important, as the rate of growth of dating times increases. According to the previous graph, the number of documents in 2012 almost equals the academic production of the two previous years. Although the number of published documents decreases, the general trend is growing. Further, after fifteen years of development, the number of cited references can only increase.

As many documents in the first search are not related to the accounting transparency of NGOs, the first set of documents found is subjected to a filtering process. We keep only documents that belong to the Web of Science categories: economics, business, management, public administration and business finance. Since our study includes the regulatory dimension, we also keep documents on legislation. After filtering, 38 documents remain (see Table 2). The total number of times these documents have been cited is 311 for all years (1997–2018), and the average number of citations per item is 8.18, higher than the number obtained without the filter. We can thus assume that most citations come from these documents. The h-index of 11 for the filtered documents is lower, however, than that for the unfiltered search. In the filtered sample, 11 documents have been cited at least 11 times. Comparison with the h-index obtained in the initial sample (h-index = 16) reveals that the majority of the impact stems from these 311 publications. These documents thus have a great influence on the transparency of NGOs.

Table 2. Documents and citations of "non-governmental organization transparency" obtained after the filter process.

Findings	38
Total times cited	311
Average citations per item	8.18
h-index	11

The filtered documents on "non-governmental organization transparency" also started in 1997 and show intermittent development. Unlike the previous graphic of "elements published each year," however, the number of these documents has been growing steadily since 2010, if we take into account that the filter of documents published in 2002, 2003 and 2009 eliminated documents unrelated to the accounting transparency of NGOs (related instead to interdisciplinary social science topics, such as informatics, medicine, chemistry, philosophy and engineering). Further, the same pattern occurs: the years 2008 and 2016 produced many relevant documents, peaking in 2016, and the trend toward published documents in accounting transparency of NGOs increases.

The number of citations of the filtered documents fluctuates (see Fig. 2). The years 2001, 2004, 2008, 2010 and 2016 are peaks; between these years, the number of published documents increases over the tendency. The year 2001 is a special case; but the fact that no articles were published during this year does not mean that documents published in previous years were not cited. The years 2003, 2005, 2009 and 2012 have valleys because there are no publications in these years (except 2012), and citations decrease in 2012 due to the restriction of research areas. As of 2012, significant growth begins. Based on the graphs, we can assume that scholarly interest the accounting transparency of NGOs began to increase in 2012.

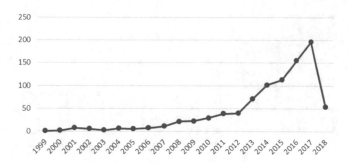

Fig. 2. Number of cites per year.

In the Fig. 3 shows the distribution of the 38 documents across the 15 (already filtered) research areas. The distribution shows that the subject arouses interest in several disciplines. Documents in the field of business economics constitute the highest proportion, followed by those in the field of public administration, with less than half the number in business economics. The field of law also shows a significant number of publications, as do other areas in the social sciences. The documents in accounting transparency are the most numerous, and researchers in this field have been more prolific than in other areas of the social sciences. Each author published only one document on this topic. We could assume that this means the topic has interested many authors but led to few publications. The figures could also indicate, however, that this is a novel subject with insufficient development and in-depth study to date.

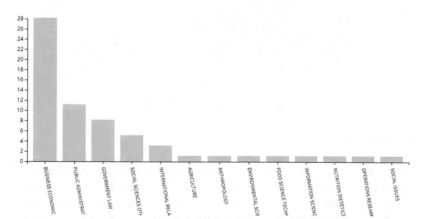

Fig. 3. Documents published by research area.

In the Fig. 4 presents the 10 sources that have published the largest number of documents. The most important sources are *Journal of Business Ethics*, with an impact factor of 2.354 in 2016 and a quartile of Q2 in business scope; *Public Administration and Development*, with an impact factor of 0.860 and a rank of Q3 in public administration; and *Accounting Auditing & Accountability Journal*, with an impact factor of 2.732 and a rank of Q1 in business and finance. The sources of published documents are not limited to journals but also come from international congresses.

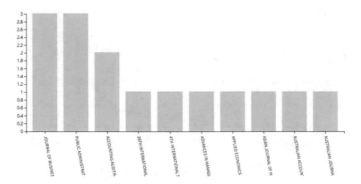

Fig. 4. Ranking of journals.

Finally, the Fig. 5 shows the top 10 countries by number of documents published. These are Western countries. Within the top 10, most documents come from the United States, England and Australia. We can assume that the Western countries research the transparency of the NGOs more, especially in the three countries mentioned.

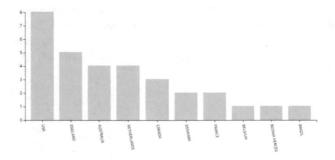

Fig. 5. Ranking of countries.

3   Methodology

The bibliometric analysis performed enables us to determine the general traits and trend of the research on accounting transparency of NGOs. Since the literature focuses on accounting transparency of NGOs, we take "non-governmental organization transparency" as the keyword for the bibliometric analysis and consult the results in the "Web of Science" database. Since many documents are not related to accounting transparency of NGOs, we apply a filter and obtain documents in the categories of economics, business, management, public administration and business finance. The final bibliometric analysis includes explanations of elements published each year, number of times cited per year, research areas, authors, sources, countries or regions, co-occurrence of terms and keywords, bibliographic coupling of documents and cited references. This analysis is performed in the Web of Science database with the VOSviewer tool, which can create the network visualization maps to illustrate super-position and density of the elements.

4   Results

We performed a bibliometric analysis to find more information through the creation, visualization and exploration of maps based on any type of network data located by the dating process between some authors. This bibliometric analysis located co-occurrence of terms and keywords, bibliographic coupling of documents and co-citation of cited references among the 38 filtered documents. Network visualization was based on the data obtained through the search and subsequent filtering. The minimum number of occurrences of a term is 5. As there are only 38 documents on "non-governmental organization transparency," it is difficult to create a map. We thus reduce the minimum number of term occurrences to represent the relationship between the terms more clearly. Under this condition, we obtain 36 terms that reach the threshold. These terms have a total of 369 occurrences with an average of 10 occurrences per term and a fluctuation of 5 30. Table 3 shows the terms by occurrence and relevance. The results indicate that the majority belong to cluster 1. As it is shown, we obtain four main terms: "organization," "non-governmental organization," "process" and "development" regarding occurrence and relevance.

Table 3. Terms, occurrences and relevance.

Terms	Occurrences	Relevance
CLUSTER 1		
transparency	30	0.41
country	18	0.44
organization	16	0.83
ngo	16	0.36
non-governmental organization	15	0.57
process	11	1.12
development	11	1.12
article	10	1.35
problem	9	1.02
accountability	9	0.6
corruption	7	1.17
case	6	1.65
number	6	1.17
term	5	2.6
business	5	1.77

(*continued*)

Table 3. (*continued*)

Terms	Occurrences	Relevance
CLUSTER 2		
paper	16	0.49
study	11	0.71
stakeholder	11	0.27
practice	10	0.77
governance	10	0.53
impact	10	0.47
attention	8	0.8
standard	8	0.42
issue	7	1.08
company	7	0.66
corporate social responsibility	5	2.98
csr	5	2.98
CLUSTER 3		
analysis	13	0.67
level	12	0.41
non-governmental organization	11	1.16
research	11	0.7
activity	11	0.43
participation	10	0.6
order	7	0.64
need	6	2.29
lack	6	0.75

The Fig. 6 represents the three clusters using different colors, and Table 3 displays all terms with their occurrences and relevance in the clusters. Although all clusters are related to each other, three clear lines of research emerge from the clusters.

Cluster 1 is represented by the color red, which includes 15 terms—among them, "transparency," "country," "ngo," "process," "development," etc. This group is large and contains many terms describing aspects of NGO transparency [8, 10, 30] and [16]; accountability [27]; and corruption [34] and [12] at the corporate, industrial, national and international levels. The documents in this cluster usually discuss problems [32], current traits and future trends, or prospects for future development [34] and [8].

Cluster 2 has 12 terms shown in green: "paper," "study," "stakeholder," "practice," etc. Because many studies in this group were based on real cases, they are empirical investigations. Research in this cluster can be divided into three subgroups according to content: corporate social responsibility [2, 18, 19]; interest groups [6, 31] and [4]; and governance [28, 33] and [17]; with relationships among the subgroups.

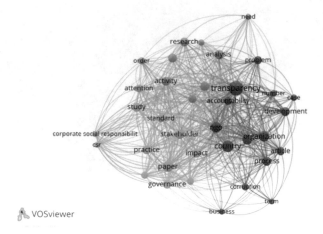

Fig. 6. Network visualization - main terms in 38 documents of "non-governmental organization transparency".

Finally, in blue, is cluster 3. This line of research contains 9 terms, such as "analysis," "level," "non-governmental organization," "research," that represent theoretical investigations. The documents in this cluster typically focus on analysis of and research on transparency at national and international level—for example, corporate anti-corruption disclosure and press freedom at national level [3], participation of NGOs in the legislative process of the European Union [7], and transparency of the World Trade Organization [22]. Quite a few of the studies were motivated by lack of transparency or regulatory governance.

Network visualization of the terms (Fig. 7) represents the elements as circles. The greater the weight of an element is, the larger its circle. The closer two elements are to each other, the stronger their relationship. According to circle size, the term "transparency" has the most weight, followed by "paper", "ngo", "organization" and "country". These terms have heavy weights in the visualization and a large number of occurrences in the real textual data. Most of the heavyweight terms are located in cluster 1, indicating that this information is more important and more relevant than that in the other documents in the set. The term "transparency" is close to "accountability," or the rendering of accounts; that is, accountability has the closest relationship to transparency, as we saw in the previous analysis. Accountability is thus an important aspect of the studies in this cluster. We can see that the main terms correspond to the colors of different years. Here, research has generally evolved over the years analyzed, from the development of NGOs [21] and [5] to their transparency, accountability and corruption, considered at national or international level [12] and [27]. At the end of the sequence, in the most recent years, the studies pay more attention to interest groups, corporate social responsibility, activities and governance, etc...., [18] and [25]. The data thus show gradual progress toward in-depth research.

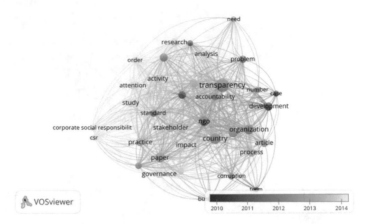

Fig. 7. Overlay visualization - main terms in 38 "non-governmental organization transparency" documents.

The elements in the element density display (Fig. 8) are represented by label in a way similar to the network and overlay displays. The color of each point in the element density display indicates the density of elements at that point. By default, colors very from blue to green to yellow. The more intense the yellow is, the greater the density of documents around that term. Conversely, the more intense is the blue, the lower the density of documents around that term. In this paper, the terms in the accounting transparency of NGOs take the colors of blue to green to yellow but not red, indicating that this topic is not sufficiently consolidated, as the terms do not have sufficiently close relationships to each other.

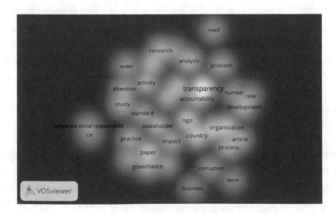

Fig. 8. Density display: density of elements - main terms in 38 "non-governmental organization transparency" documents.

In this image, the "transparency" point is bright yellow, but there is only one surrounding element, "accountability." The elements "transparency" and "account-ability" have the greatest weight and the closest relation to each other. The term "accountability" is usually accompanied by the term "transparency" because transparency and accountability are close in meaning: accountability is a means of accessing transparency. We also find two other concentrations of yellow: (1) the area linking "ngo," "country" and "organization;" and (2) the area linking "paper" and "governance." These terms also have large weights.

The Fig. 9 shows the web display of keyword co-occurrence. By default, the minimum occurrence number of a keyword is 5, but due to the small number of documents, we have reduced the minimum number of occurrences to 4. Under this condition, 7 key words reach the threshold. In this map, "transparency" and "accountability" produce larger circles and yellow color, indicating the highest numbers of occurrence and the greatest relevance. This result corresponds to the results reported above, that rendering of accounts is very closely related to transparency. According to [13], the transparency of NGOs involves the degree of information and the attitude with which they face accountability. For [20], the strength of NGO legitimacy depends on the rendering of NGOs' accounts. Based on the options of two previous authors, [26] concludes that accountability can serve as a sign by which to distinguish more transparent NGOs. In addition, [9] believes that the notion of transparency can be understood as an activity within the general system of accountability, and assumes the action of publishing organizations' information, while the term of transparency also involves issues of ethics, clarity, public morals, honesty, exposure and information, among others. If transparency is not guaranteed or no norms and mechanisms exist to require total and clear rendering of public accounts, then there is a risk of corruption. The term transparency is thus directly and indirectly related to the concept of corruption, and the term transparency in NGOs implies accountability that grants the entity legitimacy to self-govern and protects it from corrupt behavior. As the image shows, the keyword "transparency" is closer to "accountability," "ngo," "governance" and "corruption;" and the keyword "accountability" is closer to "transparency," "politics" and "legitimacy."

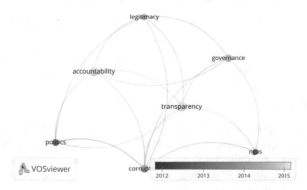

Fig. 9. Overlay visualization - keywords in 38 "non-governmental organization transparency" documents.

The image of the bibliographic coupling of the documents (Fig. 10) shows that the 38 documents can be divided into small groups through correlations of sets that are similar to each other. The largest set consists of 24 connected (similar) documents.

Calculating the total strength of bibliographic links with other documents shows that [18, 22–24] and [27] have the greatest total link strength. And among the main documents, according to the size of circles reflecting the weights of elements, [27] has less weight than the other documents, and [18] have many neighboring documents compared with other main documents.

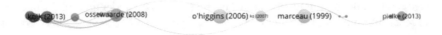

Fig. 10. Network visualization - bibliographic coupling of 38 "non-governmental organization transparency" documents.

The author [18] examines issues concerning multinational companies and conflict, including interactions with NGOs, and presents the potential contributions of multinational companies to peaceful societies, emphasizing innovative collaborations of multinational companies and non-multinational partners in Central Africa in the context of corporate social responsibility policies. They [45] analyze the dynamics of the legitimacy of international NGOs, taking humanitarian intervention after the tsunami (2004/2005) as an example. He [23] explores the systemic role of corruption and its relation to low human development and identifies NGO watchmen with Transparency International's data. They [22] examine the relationship of the World Trade Organization to NGOs, and the demands from civil society for greater transparency and public participation. He [27] explores how the International Federation of Football Associations can be held accountable. This article has broad implications for understanding accountability mechanisms in general in international NGOs.

The documents are divided into 5 clusters with different colors. Some documents in one cluster are related to some documents in other clusters:

- Some documents in cluster 1 (red) are related to cluster 2 (violet) and cluster 3 (green).
- Some documents in cluster 2 are related to clusters 1, 3 and 4 (yellow).
- Some documents in cluster 3 are related to clusters 1, 2 and 4.
- Some documents in cluster 4 are related to clusters 2, 3 and 5.
- Some documents in cluster 5 are related only to cluster 4.

Finally, with respect to co-citation (Fig. 11) of cited references, the minimum number of citations of a quoted reference is 3 by default. Calculating the total strength of the citation links as well as other cited references shows that the largest set of connected elements consists of 3 elements with greater total link strength: [11, 15, 29]. From the network visualization, the 3 references are divided into 2 groups: [11] and [15] belong to one group and [29] to another. Although we cannot know which of the documents cite the 3 previous articles, we can deduce the relationships between the 3 articles. [15] and [29] have governance in common and the studies by [15] and [11] address corporate social responsibility and transparency. As we see in Fig. 11, [15] is related to [29] and [11], but there is no relationship between [29] and [11].

Fig. 11. Network visualization - co-citation of references cited in 38 "non-governmental organization transparency" documents.

5 Conclusions

The bibliometric, conceptual and normative analysis performed shows that accounting transparency of NGOs is a broad concept. Although more than twenty years have passed since the concept was proposed, it is difficult to reach agreement on a definition. This concept can contain many elements, such as quality of information, disclosure of information, voluntary disclosure, etc. Due to the concept's fuzzy definition, no standard method has been established to evaluate it. The same occurs for accounting transparency in the field of NGOs. Further, although several authors have published documents in this field, each author has published only one article, and the total number of documents is insufficient. Study of the accounting transparency of NGOs is thus neither well-developed nor very saturated. As there is a tendency for the number of documents in this field to grow in both number of published works and citations, accounting transparency in NGOs should receive more attention and have an important future projection.

References

1. Ahmed, Z., Hopper, T.: Politics, development and NGO accountability. In: Performance Management in Nonprofit Organizations: Global Perspectives, pp. 17–42. Routledge, London (2014)
2. Berkowitz, H., Bucheli, M., Dumez, H.: Collectively designing CSR through meta organizations: a case study of the oil and gas industry. J. Bus. Ethics **143**(4), 753–769 (2017)
3. Blanc, R., Islam, M.A., Patten, D.M., Branco, M.C.: Corporate anti-corruption disclosure: an examination of the impact of media exposure and country-level press freedom. Account. Audit. Account. J. **30**(8), 1746–1770 (2017)
4. Burchell, J., Cook, J.: Sleeping with the enemy? Strategic transformations in business–NGO relationships through stakeholder dialogue. J. Bus. Ethics **113**(3), 505–518 (2013)
5. Bushman, R.M., Piotroski, J.D., Smith, A.J.: What determines corporate transparency? J. Account. Res. **42**(2), 207–252 (2004)
6. Camilleri, M.A.: Valuing stakeholder engagement and sustainability reporting. Corp. Reput. Rev. **18**(3), 210–222 (2015)
7. De Jesús Butler, I.: Non-governmental organisation participation in the EU law-making process: the example of social non-governmental organisations at the commission, parliament and council. Eur. Law J. **14**(5), 558–582 (2008)
8. Deighton Smith, R.: Regulatory transparency in OECD countries: overview, trends and challenges. Aust. J. Public Adm. **63**(1), 66–73 (2004)
9. Del Castillo, A.: Medición de la corrupción: Un indicador de la rendición de cuentas. Auditoria Superior de la República (2003)

10. Delimatsis, P.: Transparency in the WTO's decision-making. Leiden J. Int. Law **27**(3), 701–726 (2014)
11. Den Hond, F., De Bakker, F.G.: Ideologically motivated activism: how activist groups influence corporate social change activities. Acad. Manag. Rev. **32**(3), 901–924 (2007)
12. Gebel, A.C.: Human nature and morality in the anti-corruption discourse of transparency international. Public Adm. Dev. **32**(1), 109–128 (2012)
13. Grabulosa, J.L.: El debate sobre la transparencia y la rendición de cuentas, 25 October 2007. http://www.observatoritercersector.org/pdf/publicacions/articulo_boletin_l8.pdf.   Accessed 25 June 2009
14. Gugerty, M.K., Prakash, A. (eds.): Voluntary Regulation of NGOs and Nonprofits: An Accountability Club Framework. Cambridge University Press, Cambridge (2010)
15. Haufler, V.: Disclosure as governance: the extractive industries transparency initiative and resource management in the developing world. Glob. Environ. Polit. **10**(3), 53–73 (2010)
16. Jedrzejka, D.: Towards greater transparency – development of corporate disclosures: evidence from the Polish stock market. In: Conference: 26th International Scientific Conference on Economic and Social Development - Building Resilient Society, Zagreb, Croatia (2017)
17. Ko, C.Y.E.: Policy entrepreneur in global governance: the rise and advocacy of transparency international. In: Conference: 3rd International Conference on Public Administration Location: University Electrical Science & Technology, School Political Science & Public Administration, Chengdu, Peoples of China (2007)
18. Kolk, A., Lenfant, F.: Multinationals, CSR and partnerships in Central African conflict countries. Corp. Soc. Responsib. Environ. Manag. **20**(1), 43–54 (2013)
19. Lauwo, S.G., Otusanya, O.J., Bakre, O.: Corporate social responsibility reporting in the mining sector of Tanzania: (Lack of) government regulatory controls and NGO activism. Account. Audit. Account. J. **29**(6), 1038–1074 (2016)
20. Lister, L.: NGO legitimacy: technical issue or social construct? Crit. Anthropol. **23**(2), 175–192 (2003)
21. Liu, X.B., Deng, S.L., Wen, J.X.: The evaluation of China's provincial fiscal transparency in 2010. J. Shanghai Univ. Financ. Econ. **3** (2010)
22. Marceau, G., Pedersen, P.N.: Is the WTO open and transparent? J. World Trade **33**, 5 (1999)
23. O'Higgins, E.R.: Corruption, underdevelopment, and extractive resource industries: addressing the vicious cycle. Bus. Ethics Q. **16**(2), 235–254 (2006)
24. Ossewaarde, R., Nijhof, A., Heyse, L.: Dynamics of NGO legitimacy: how organising betrays core missions of INGOs. Public Adm. Dev.: Int. J. Manag. Res. Pract. **28**(1), 42–53 (2008)
25. Pavlovic, J., Lalic, D., Djuraskovic, D.: Communication of non-governmental organizations via facebook social network. Eng. Econ. **25**(2), 186–193 (2014)
26. Pérez, C., del Carmen, M., Gálvez Rodríguez, M.D.M., López Godoy, M.: Una oportunidad de mayor legitimidad de las ONG a través de la transparencia on-line: experiencia de las ONG españolas y mexicanas. Contaduría y Administración **234**, 55–77 (2011)
27. Pielke Jr., R.: How can FIFA be held accountable? Sport Manag. Rev. **16**(3), 255–267 (2013)
28. Sovacool, B.K., Walter, G., Van de Graaf, T., Andrews, N.: Energy governance, transnational rules, and the resource curse: exploring the effectiveness of the extractive industries transparency initiative (EITI). World Dev. **83**, 179–192 (2016)
29. Suchman, M.C.: Managing legitimacy: strategic and institutional approaches. Acad. Manag. Rev. **20**(3), 571–610 (1995)
30. Tadele, H., Roberts, H., Whiting, R.H.: Microfinance institutions' transparency in Sub-Saharan Africa. Appl. Econ. **50**(14), 1601–1616 (2018)

31. Uldam, J., Hansen, H.K.: Corporate responses to stakeholder activism: partnerships and surveillance. Crit. Perspect. Int. Bus. **13**(2), 151–165 (2017)
32. Vadlamannati, K.C., Cooray, A.: Do freedom of information laws improve bureaucratic efficiency? An empirical investigation. Oxford Econ. Pap. **68**(4), 968–993 (2016)
33. Van der Ven, H.: Correlates of rigorous and credible transnational governance: a cross sectoral analysis of best practice compliance in eco-labeling. Regul. Gov. **9**(3), 276–293 (2015)
34. Vian, T., Brinkerhoff, D.W., Feeley, F.G., Salomon, M., Vien, N.T.K.: Confronting corruption in the health sector in Vietnam: patterns and prospects. Pub. Adm. Dev. **32**(1), 49–63 (2012)

Plagiarism Detection in the Classroom: Honesty and Trust Through the Urkund and Turnitin Software

Simone Belli[1]([✉]) [iD], Cristian López Raventós[2] [iD],
and Teresa Guarda[3,4,5] [iD]

[1] Complutense University of Madrid, 28040 Madrid, Spain
sbelli@ucm.es
[2] Universidad Autónoma Nacional de México, Morelia, Mexico
clopezr@enesmorelia.unam.mx
[3] CIST – Centro de Investigacion en Sistemas y Telecomunicaciones, UPSE,
La Libertad, Ecuador
tguarda@gmail.com
[4] Universidad Estatal Peninsula de Santa Elena, UPSE,
La Libertad, Ecuador
[5] Algoritmi Centre, Minho University, Guimarães, Portugal

Abstract. This text is the result of a research process that has been carried out during the last five years on the realities and problems of plagiarism in contexts of higher education in different countries. We have observed eight groups of university students of different degrees in Engineering in a mandatory course of "Oral and written expression" in two universities in different countries (Ecuador and Spain). This course aims to provide tools to develop skills in oral and written presentations on academic assignments.

Through practical exercises, manuals and presentations, students develop different strategies to generate tools to expose their knowledge as a "way of doing". The development of student tasks follows specific guidelines discussed with the use of technology to detect plagiarism, Turnitin and Urkund tools. The results of our research indicate how the average percentage of plagiarism in student work is reduced with the introduction of this software.

Keywords: Plagiarism · Intellectual integrity · Urkund · Turnitin · Trust

1 Introduction

Many university students tend to minimize the importance of plagiarism in academic work [1, 2]. We have realized an ethnographic work in these classrooms to observe what are the argumentative and ethical strategies in the presence of plagiarism in the students' written works. We have used Turnitin and Urkund software to detect plagiarism in their writings.

We have observed eight groups of the mandatory course "Oral and Written Expression" (ExOEsa), four in a Spanish university and four in an Ecuadorian university. Six groups received a class in Spanish, the other two in English. The main

© Springer Nature Switzerland AG 2020
Á. Rocha et al. (Eds.): ICITS 2020, AISC 1137, pp. 660–668, 2020.
https://doi.org/10.1007/978-3-030-40690-5_63

objective of this course is to offer tools to develop oral and written presentation skills in academic papers throughout Engineering careers.

Previous research [3, 4] has identified the many possible reasons for students to plagiarize. For example, students do not have confidence in their writing skills, do not devote enough time to the elaboration of their tasks, have a positive attitude towards deception, or simply ignore how to quote properly [5]. The cost of plagiarism is very high because it transforms teachers into police officers, dedicating time and efforts in something that does not benefit the learning environment [6]. It is also worth mentioning that plagiarism has to do with intercultural aspects. Numerous studies have shown that plagiarism and the concept of authorship are understood differently depending on the country the student comes from [7–11].

2 What Does the Plagiarism in the Course of Oral and Written Expression?

Plagiarism has always been present in the classrooms [12], but in recent years the phenomenon has acquired a new dimension, also thanks to the Internet [13]. For this reason, teachers have a mission when it comes to detecting and holding students accountable for this illegal practice, and thanks to the digital tools it is possible to have an ally when it comes to enforcing this mission.

In the ExOEsa program, it is mentioned the importance of the originality of students' written works presented during the semester. Each text has to meet this originality criterion, otherwise, there are penalties for students who commit plagiarism.

Plagiarism in the ExOEsa course is punished with a negative grade. In the extreme case of plagiarism, the university can expel the student from the institution. In this way, students are encouraged to assimilate the educational standards of their respective institutions, regardless of origin.

On the first page of the program, it is mentioned: "The written works produced by the students must be original. The plagiarism of other texts or parts of them, will be penalized with the suspension of the subject, in addition to any other action that the university could undertake If you want to reproduce any fragment of a text that has been read, you must transcribe it in quotation marks and you must indicate the author and the written works from which it has been extracted. Similarly, when the text fragments come from the Internet, they must be submitted in quotation marks, with the author's name, the title of the entry and the address of the website where it is located" [14].

In the same way, in the ExOEsa professor's book, some strategies against plagiarism are shown: "You can find the mechanisms to avoid plagiarism in the students' written works. It may be by the subject's own choice (in search of "difficult to plagiarize" problems), the students who ask wrote the works written in the classroom, on the last day (although they need to bring the necessary materials from home), or by any other method" [14].

3   Technologies to Detect Plagiarism

In recent years several computer programs have progressed on the market to detect plagiarism. The most common in Europe and Latin America are Turnitin and Urkund, although there are other automatic plagiarism detectors. Thanks to these programs, teachers have a powerful tool to assess the level of honesty of students.

These software aim to fulfill three objectives:

1. Ensure the originality of each work: Check the students' written works by comparing them with the world's largest comparison database;
2. Smarter rankings: Give students timely feedback and save teachers time for evaluation;
3. Streamline peer review processes.

In our research, these software helped teachers in objectifying plagiarism [15]. Thanks to this tool, the teacher can easily justify a bad grade that shows the percentage of plagiarism in the work presented by the student. At the same time, it saves time spent reviewing a text that is not evaluable due to its illegitimate origin.

Another significant feature is the creation of a shared library with all academic papers analyzed by the software. In this way, it is easier to find plagiarism in the works written by other students in previous courses of the same university or other universities.

These software compare the works of the students with an extensive and indexed file of online documents including works of other students previously analyzed through the software [16]. The database where these software searches for information includes three main sources:

1. The Network, where more than 45 billion web pages are consulted.
2. The works of the students analyzed by the software, more than 337 million documents.
3. Scientific journals and books, where software have partnered with publishers, including databases of libraries, publishers, reference collections, digital publications, etc. (more than 130 million documents).

But one of the main problems caused by the use of this technology is ethical [17]. Many professors working in the institutions that use this software think that this tool is against the principle of presumption of innocence, where everyone is considered innocent until proven guilty, an expression coined by the lawyer William Garrow in the nineteenth century. Instead, these software to detect plagiarism are intended just the opposite, everyone is guilty until proven otherwise. For the teacher who uses this type of technology, each student is guilty of plagiarism until the software proves its innocence. For this reason, it has been observed how many professors, throughout the research, opposed the use of this software in the evaluation of their students' work. Just to demonstrate how ethics enters into decisions made regarding the use or not of this technology.

In the two institutions where student work has been analyzed, university policy imposes the use of this tool to qualify students. But this mandatory requirement often has to be negotiated with teachers. Many of the teachers do not always agree with this use. Many times teachers do not have time to devote to learning this tool. It has been observed how in the training courses to use these software, teachers often do not participate.

On the other hand, students know that the institution's policy requires the use of this tool in a mandatory way, but we also know that in many courses it is not used by the teacher's decision. Students know when the teacher uses this tool because he mentions it in the first classes of the course that will be used when evaluating.

4 Methodology

The purpose of this article is to show how the use of software to detect plagiarism in the classroom can improve and sensitize students about the importance of properly citing sources and not plagiarizing their written work. To fulfill this purpose, an experimental model was designed where 8 groups of students were studied who delivered 2,390 papers throughout the subject. Half of the groups knew about the use of computer tools for plagiarism detection and the other groups did not.

All students of the eight groups are studying different degrees in Engineering. For group 1, 2, 3 and 4, students were previously informed about the use of the software during the course. The teachers of these groups agree with the use of this software and have created an activity on the Moodle (in Spain) and D2L (in Ecuador) educational platform to present the written works.

For groups 5, 6, 7 and 8, students were not previously informed about the use of the software during the course, although they know that the use of this software is required in their institution. The teachers of these groups do not agree with the use of this software for the ethical reasons that we have explained in the previous section.

The 2,390 students' written works have been analyzed with software to detect plagiarism (Turnitin in Spain and Urkund in Ecuador) and we have classified plagiarism into four categories according to the percentage that these software elaborates:

- Legitimate work. From 0% to 24% plagiarized material.
- Small scale plagiarism. From 25% to 49% plagiarized material.
- Plagiarism at medium scale. From 50% to 74% plagiarized material.
- Large scale plagiarism. From 75% to 100% plagiarized material.

In this way, the works presented by the students have been divided into these four categories, based on the methodology proposed by Batane [18] and [19] in their experimental studies to reduce plagiarism through the use of specific software. Of course, we agree with Mphahlele and McKenna [20] that it is irresponsible to arrive at conclusions based on similarity percentages only, as this will lead to false accusations of plagiarism [21].

5 Data Analysis

In Table 1, we have arranged the eight groups and the number of written papers that the students have presented during the course and the total sum of the plagiarism percentages of the students per class. We note that the average rate of plagiarism per class is significantly different between groups 1–4 and 5–8.

Table 1. Percentage of plagiarism in the eight groups

Group	The total sum of student plagiarism per group	Total number of texts delivered per group	Plagiarism average per group
Group 1	214	34	6.30%
Group 2	310	35	8.85%
Group 3	324	36	9.00%
Group 4	230	34	6.76%
Group 5	234	24	9.75%
Group 6	423	32	13.21%
Grupo 7	284	14	20.28%
Group 8	371	19	19.52%
Total	2390	228	10.48%

The average total of plagiarism of the eight groups is 10.48%, but the average percentage by group 7 and group 8 is very significant. This means that in group 7 of each written work, 20.28% can be found of plagiarized material.

If we compare this average percentage of plagiarism with the four categories of plagiarism, the plagiarism is not so high, and in most cases, it drops under the category of "legitimate work." But if we break down the type of plagiarism for each group, we can obtain more details about the difference between the eight groups, as shown in Table 2. These percentages have been calculated by identifying the papers of each category by the total of the papers delivered by each group.

Table 2. Groups and categories of plagiarism.

Groups	Large scale plagiarism	Plagiarism at medium scale	Small scale plagiarism	Legitimate work
Group 1	0	0	0	100.00%
Group 2	0	2.90%	5.70%	91.40%
Group 3	0	0	0	100.00%
Group 4	0	0	0	100.00%
Group 5	21.30%	12.30%	31.40%	35.00%
Group 6	15.30%	0	41.70%	43.00%
Group 7	28.60%	0	7.10%	64.30%
Group 8	0	5.90%	23.50%	70.60%
Total	65.2	21.1	–	–
Average	**8.20%**	**2.6%**	**13.70%**	**75.50%**

In groups 1, 2, 3, and 4, we have almost all of the work delivered in the category of legitimate jobs, and only a few small-scale cases and medium-scale plagiarism. Groups

5, 6, 7, and 8 are where most cases of plagiarism are found—many cases of large-scale plagiarism, and many instances of small-scale plagiarism (30.6%). We can see how teachers who do not use this software receive jobs that are likely to contain plagiarism.

On the other hand, when teachers agree with the use of the tool, the average plagiarism is very low, and there is no medium and large-scale plagiarism. The objective of the subject is to improve students' writing skills and learn to cite the reference material without committing plagiarism correctly. On the opposite side, the teachers of groups 5, 6, 7 and 8, did not agree with the tool, the average plagiarism is high, and there are many cases of small, medium and large scale plagiarism. Teachers do not meet the objective of this subject.

6 Discussion

With this brief analysis of how and why software is used to detect plagiarism, it seems necessary to stress that plagiarism is not only a quantitative but also a qualitative issue. The role of the teacher is essential to analyze the results offered by the software. It is crucial that the teacher, according to the use of the software, accompany these results with a qualitative analysis of plagiarism according to the context in which it occurs. It is essential not to limit yourself exclusively to the percentage of plagiarism, but to observe case by case to decide whether it is real plagiarism or not. Several times the software marks a statement as plagiarism because it does not recognize the quotation marks. It is also significant that teachers understand the different use of sources by students since it is not the same to paraphrase a Wikipedia article than to paraphrase a scientific article. It is essential to explain to students it is necessary to check papers through software and to discuss what these shows us about improving our writing practices [20].

Many misunderstandings are due to plagiarism being a complex phenomenon that requires a multifaceted approach to combat it. The teachers of the ExOEsa subject, apart from using this powerful technological tool, also need a vast knowledge of the writing mechanism. For example, the software does not recognize plagiarism as the translation of texts from other languages. The teacher can recognize the authorship of a text and find out that the thinking of an author who writes in another language is not being used. It has been observed in the research on how the students' style often changes throughout their work at the moment when they begin to translate the work into another language of an author. This form of plagiarism is the most difficult to detect and requires the experience of the teacher to recognize it. The software provides us with this data where it mainly finds the original works that have been plagiarized (Fig. 1), 53% of the students' work, 44% on the Web and only 3% in the scientific publications. Medium and large-scale plagiarisms come from the work of students from previous courses and/or from other universities.

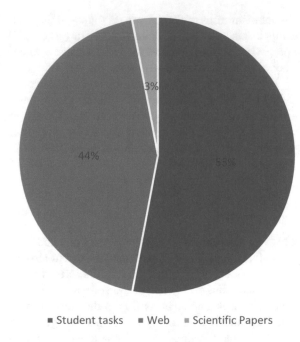

■ Student tasks   ■ Web   ■ Scientific Papers

Fig. 1. Sources of plagiarism

In the first practices of the students, the Web represents the most popular source of searching for information. Only a small percentage use scientific documents. In our investigation, a student plagiarized the complete work of another student of a British university. When the teacher asked the student why he did it, the student reported that his cousin was a student at this British university. He thought the software would not detect jobs from other international universities. The creation of this database of all students' university works represents a great resource where software can detect plagiarism from a horizontal dynamic: Students who copy other students. This is a lack of awareness among students [22], and it is essential for making students and faculty aware of these tensions and issues.

In our eight groups, students plagiarized mainly from the Web, 92% of the information came from their pages, and Wikipedia in many cases was the first source of plagiarism.

7   Conclusions

The results of our research indicate how the average percentage of plagiarism in student work is reduced with the introduction of these software. We can compare our result with other studies that demonstrate the effectiveness of using these tools to detect and stop plagiarism [18, 19, 23]. In our study, we have observed the effectiveness of software to avoid plagiarism in the classroom. We think that these software meet the objective in four points:

– The first point is empowerment. Plagiarism does not allow self-confidence in writing by students. When students know that the teacher uses the software, they gain the confidence to write a text, because there is no doubt about the detection of plagiarism. The software and the teacher will detect each type of plagiarism.
– The second point is effectiveness. The time and energy to evaluate the work of the students will be directed towards the quality of the text, without thinking whether the document contains plagiarism or not. When the teacher must review the students' work, the software will help them to gain time and energy. In a few minutes, the works will be analyzed by the tool to detect cases of plagiarism.
  The third point indicates that the use of the Instrument in the course allows students to learn not to plagiarize.
– The fourth point is that although plagiarism is always objective, honesty and trust is subjective. This is built in the classroom, in the relationship between teacher and students.

Thanks to these technologies to detect plagiarism, students' ability to plagiarize successfully is reduced. Applying Whitley's work [4], in which the ability to cheat and the risk of detecting determines the student's willingness to cheat when specific feedback is provided to the student through the summary of originality that these software offers, the plagiarism in students [24]. When students see their work marked and underlined in red in each paragraph that has been plagiarized and the percentage of plagiarized work the software prepares, this will discourage future plagiarism on the part of the student.

In conclusion, plagiarism detection software is one of the tools to promote honesty and trust in students. Although we have conducted this research on a restricted group of students, our significant results suggest that future research should focus on intercultural differences in student behaviors regarding plagiarism.

References

1. Breen, L., Maassen, M.: Reducing the incidence of plagiarism in an undergraduate course: the role of education. Issues Educ. Res. **15**(1), 1–16 (2005)
2. Thomson Maddox, T.: Plagiarism and the community college. In: Roberts, T.S. (ed.) Student Plagiarism in an Online World. Information Science Reference, Hershey (2008)
3. Council of Writing Program Administrators: Defining and avoiding plagiarism: The Council of Writing Program Administrators' Statement on Best Practices (2003). http://www.wpacouncil.org/node/
4. Whitley, B.: Factors associated with cheating among college students: a review. Res. High. Educ. **39**(3), 235–274 (1998)
5. Martin, D., Rao, A., Sloan, L.: Ethnicity, acculturation, and plagiarism: a criterion study of unethical academic conduct. Hum. Organ. **70**(1), 88–96 (2011)
6. Hannabuss, S.: Contexted texts: issues of plagiarism. Libr. Manag. **22**(6/7), 311–318 (2001)

7. Chapman, K., Lupton, R.: Academic dishonesty in a global educational market: a comparison of Hong Kong and American university business students. Int. J. Educ. Manag. **18**(6/7), 425–435 (2004)
8. Introna, L., Hayes, N., Blair, L., Wood, E.: Cultural attitudes towards plagiarism: developing a better understanding of the needs of students from diverse cultural backgrounds relating to issues of plagiarism (2003)
9. Salter, S., Guffey, D., McMillan, J.: Truth, consequences, and culture: a comparative examination of cheating and attitudes about cheating among United States and United Kingdom students. J. Bus. Ethics **31**(1), 37–50 (2001)
10. Song-Turner, H.: Plagiarism: academic dishonesty or "blind spot" of multicultural education? Aust. Univ. Rev. **50**(2), 39–50 (2008)
11. Wheeler, G.: Plagiarism in the Japanese universities: truly a cultural matter? J. Second Lang. Writ. **18**, 17–29 (2009)
12. Sureda-Negre, J., Comas-Forgas, R., Oliver-Trobat, M.: Plagio académico entre alumnado de secundaria y bachillerato: diferencias en cuanto al género y la procrastinación. Comunicar **44**, 1–13 (2015)
13. Comas, R., Sureda, J.: Academic plagiarism: explanatory factors from students' perspective. J. Acad. Ethics **8**(3), 217–232 (2010)
14. Garcés, P., Pavón, M., Pérez, E.: Técnicas de Expresión Oral y Escrita. Curso 2009/2010. Libro del Alumno. Universidad Carlos III de Madrid, Madrid (2009)
15. Christodoulou, I.: The impact factor of plagiarism. Int. J. Health Sci. **1**(4), 113–114 (2008)
16. Maurer, H., Kappe, F., Zaka, B.: Plagiarism - a survey. J. Univers. Comput. Sci. **1**, 1050–1084 (2006)
17. Zimmerman, T.A.: McLean students file suit against turnitin.com: useful tool or instrument of tyranny? In: The CCCC-IP Annual: Top Intellectual Property Developments of 2007. Intellectual Property Caucus of the Conference on College Composition and Communication, New York, pp. 2–11 (2008)
18. Batane, T.: Turning to turnitin to fight plagiarism among university students. Educ. Technol. Soc. **13**(2), 1–12 (2010)
19. Kose, O., Arikan, A.: Reducing plagiarism by using online software: an experimental study. Contemp. Online Lang. Educ. J. **1**, 122–129 (2011)
20. Mphahlele, A., McKenna, S.: The use of turnitin in the higher education sector: decoding the myth. Assess. Eval. High. Educ. **44**(7), 1079–1089 (2019)
21. Weber-Wulff, D.: Plagiarism detection software: promises, pitfalls, and practices. In: Handbook of Academic Integrity, pp. 625–638. Springer, Singapore (2016)
22. Wrigley, S.: Avoiding 'de-plagiarism': exploring the affordances of handwriting in the essay-writing process. Act. Learn. High. Educ. (2017). https://doi.org/10.1177/1469787417735611
23. Weinstein, J., Dobkin, C.: Plagiarism in US higher education: estimating Internet plagiarism rates and testing a means deterrence (2002)
24. Martin, D.: Plagiarism and technology: a tool for coping with plagiarism. J. Educ. Bus. **80**(3), 149–152 (2005)

Media, Applied Technology and Communication

Public Service Media's Funding Crisis in the Face of the Digital Challenge

Francisco Campos-Freire[1], Marta Rodríguez-Castro[1(✉)], and Olga Blasco-Blasco[2]

[1] Department of Communication Sciences,
Universidade de Santiago de Compostela, 15782 Santiago de Compostela, Spain
francisco.campos@usc.com, m.rodriguez.castro@usc.es
[2] Department of Applied Economics,
Universitat de València, 46022 Valencia, Spain
olga.blasco@uv.es

Abstract. The funding crisis of Public Service Media (PSM) is also a crisis affecting its legitimacy, business model, audience, innovation and transformation required to adapt to the current digital ecosystem, which is dominated by the changes in the access and consumption ways available to citizens, as well as by the new telecommunications global players. Eleven European countries have cut back the budgets of their public service media organizations during the past five years, and those which haven't done it yet are facing adjustment plans until 2020. Besides financial pressures, European PSM is also facing increasing opposition from private operators, populist parties and the constant appetite for manipulation of the governments in power. By applying quantitative and qualitative methods to approach the evolution of the funding of European PSM, this paper describes and contrasts the debate around the future and sustainability of this media model. To this effect, we consider the manifestations and strategic visions of three intertwined parts: (a) the current policies of governments and political parties in the main European countries; (b) the strategic plans and the company reports of European PSM organizations; and (c) a review of the academic work that is being developed around this topic.

Keywords: Public Service Media · Funding · Competition · Media policy · Licence fee

1 Introduction

European public service broadcasting (PSB) was born under the administration and control of the States during the first decades of the 20th century and has remained like that just until the second half of this period, when the liberalization of broadcasting breaks with its monopolistic conception - in the United Kingdom, through the 1954's Television Act [1] which enabled the entry of ITV's third channel in the British system. Some time afterwards, the European Court of Justice decided on the Sacchi Case [2] and consolidated the sentence against RAI's the commercial monopoly, thus opening the Italian commercial market.

© Springer Nature Switzerland AG 2020
Á. Rocha et al. (Eds.): ICITS 2020, AISC 1137, pp. 671–680, 2020.
https://doi.org/10.1007/978-3-030-40690-5_64

Along the same lines, the policies of the European Commission also opened the very concept of public service to the notion of services of general economic interest [3], in order to consolidate the dual model of public and commercial broadcasting. Such coexistence and competition duality was confronted and disputed in additional complaints before the European Court of Justice in relation to the legitimacy of the competition spaces, the funding systems and the broadcasting rights over sports and programs of high commercial value, as well as the fight for high audience numbers. The Protocol to the Amsterdam Treaty on public service broadcasting [4], as well as the 2001 [5] and 2009 [6] Broadcasting Communications legitimized the funding system of public service broadcasters under three conditions: to comply with a sociocultural mission, to monitor and control its democratic pluralism and to avoid acts of unfair competition to the private sector.

In the transition between the 20th and the 21st century, with the arrival of the Internet and the following disruptive innovation [7] triggered by digital platforms and affecting the models of legacy media, the scramble for the market and the media hegemony was no longer an issue affecting two but three types of operators. Mass media of the 20th century evolved into a mass of media in the 21st century [8], leading to a journalistic explosion and to the irruption of self-communication systems [9]. Such outbreak prompted another legitimacy crisis and called upon the need to update the PSM system in the face of the changing uses, consuming habits and behaviours of their audiences in the digital society [10, 11].

This update or reinvention developed in the face of the challenges introduced by the digital society [12–14] set the evolution of the PSB concept [15] towards the one of public service media (PSM) [16–18] and public service Internet [19], which is already present in the BBC's strategy [20]. The direction that PSM should follow to get to the third decade of the 21st century should go through the Internet and cloud distribution, so the public service mission must be reoriented in order to fit within the demands of the Internet Society and the fourth industrial revolution.

The current challenge for PSM is to be able to adapt to the new ecosystem dominated by Internet networks and platforms [21] without losing impact on the traditional means of distribution - digital terrestrial television, cable and satellite – which find themselves stagnated or in recession due to the changes in the access and consumption ways preferred by some types of audiences, like young people. This adaptation requires additional investments and strategies to reinforce the social legitimacy and reputation of PSM through the reincarnation of its core, foundational principles, - the Reithian triad based on informing, educating and entertaining - as well as the application of other new, differentiating values linked to a quality service, such as the ones established by the European Broadcasting Union (EBU): universality, excellence, independence, diversity, innovation and accountability [22].

To the intrinsic challenges that PSM is undergoing, some other contextual demands must be considered. PSM must adapt to regulation changes, both at the European and at a national level, such as the entry into force of the Directive 2018/1808 amending the Audiovisual Media Services Directive of 2010 and its transposition to national legislations, which shall be completed by the end of 2020. Moreover, PSM risks losing penetration during the migration of the digital terrestrial television diffusion from the 700 MHz band to the 470–694 MHz band, in order to enable the implementation of 5G communications [23]. The funding models of PSM are being reconsidered due to the questioning of the licence fee, the decrease in commercial revenues and the public funding budget cuts; and they are also being affected by the polarization and the pressure exerted by political parties that aim either at controlling PSM from the government or at neutralizing its effect from the opposition. This funding crisis comes at a time when competition is growing, namely because of the emergence of new operators and the changes in consumption habits, and thus forces PSM organizations to offer more services with less resources.

All these circumstances increase the pressure on PSM's governance and management, which must be updated in order to overcome an evolution that until now was more vegetative than strategic in approaching the new challenges and values brought by the Internet digital revolution. The evolution of PSM's diverse funding models (based on the licence fee and commercial revenues in some countries, and in public subsidies and advertisement, in others) proves this situation.

2   Funding Results

The pressures exerted on the funding of PSM are translated into the freezing of licence fee rates or even the suppression of such fees, into the increasing losses of advertising revenues or in the direct subsidies and the allocation of public funds to PSM. The EBU estimates that between 2011 and 2015 the funding received by PSM organizations of the 56 countries that are members of this association experienced a 40% cutback.

The total income of PSM organizations in the European Union between 2010 and 2016 decreased more than 5%, reaching a global amount close to 500 million Euros, mainly due to a drop in commercial revenues, i.e. those coming from advertising activities. Two different periods can be identified: between 2010 and 2013, and between 2013–2016. The first period, 2010–2013, reflects the impact of the economic crisis, while the second one shows some signs of light recovery that restores the global PSM funding to the levels preceding the recession experienced by European economies. Even though there is no data available for the period 2016–2020 for all countries, there is a growing trend to keep with the adjustment and restrictions of most of European PSM's budgets. The data collected in Table 1 proves the stagnation of revenues in thirteen countries during the first half of the second decade of the century.

The strongest cutbacks were registered in those European countries that had been intervened due to the recession, such as Portugal, with a budget reduction of a 47%; Spain, −24%; Romania, −18% and Poland, −15%. During the same period, the only PSM organizations that substantially increase their funding are those from Belgium, 8%; Germany, 4%; Denmark, 9%; Finland, 9%; the United Kingdom, 5%; Hungary, 20%; the Netherlands, 4%; Sweden, 11% and Slovenia, 7%.

Table 1. Income of PSM broken down by country, 2010–2016. Source: processed by the authors with data from the EAO [24] (in million €).

Country	2010	2011	2012	2013	2014	2015	2016	%Δ 16/10
Germany	9,034.5	8,787.2	8,591.2	8,823.5	9,445.9	9,212.4	9,177.7	1,6
Austria	971.2	991.6	1,001.9	1,008.1	984.3	978.1	968.7	−0,3
Belgium	716.1	726.5	765.7	775.9	795.9	782.6	787.3	9,9
Bulgaria	66.2	63.9	65.3	62.1	62.1	62.5	61.0	−7,9
Cyprus	45.0	36.8	37.7	30.8	31.7	30.8	33.2	−26,2
Croatia	202.1	190.1	188.9	183.1	179.7	180.0	179.7	−11,1
Denmark	806.7	829.1	861.8	888.9	923.4	893.3	916.0	13,5
Slovakia	101.9	114.5	97.5	100.5	104.1	112.9	115.7	13,5
Slovenia	133.5	109.8	118.9	116.8	117.9	118.3	118.6	−11,2
Spain	2,794.2	2,572.1	2,165.2	1,953.0	1,810.0	1,937.4	2,035.6	−27,1
Estonia	28.4	28.3	28.5	30.5	29.4	33.6	40.3	41,9
Finland	420.0	432.5	456.1	469.1	475.6	472.0	473.2	12,7
France	4,378.5	4,568.5	4,658.7	4,490.7	4,497.8	4,479.6	4,576.0	4,5
Greece	386.0	328.8	n.a.	n.a.	175.6	182.0	190.3	−50,7
Hungary	177.6	238.4	302.0	300.5	279.6	289.2	299.7	68,8
Ireland	408.5	388.1	374.0	364.4	364.3	369.9	374.3	−8,4
Italy	2,821.0	2,825.0	2,625.0	2,562.0	2,594.8	2,493.1	2,809.5	−0,4
Latvia	21.7	21.5	22.5	24.0	26.9	26.6	26.5	22,1
Lithuania	17.3	19.1	20.7	20.9	22.2	25.1	32.9	90,2
Luxembourg	n.a.	n.a.	4.6	5.0	5.3	5.6	5.9	n.a.
Malta	8.2	9.1	10.1	9.6	10.3	12.2	11.7	42,7
The Netherlands	857.2	855.9	864.5	834.0	855.0	848.4	888.2	3,6
Poland	521.5	471.4	383.4	427.5	428.7	460.7	410.2	−21,3
Portugal	309.0	317.0	259.0	234.0	213.5	211.5	215.2	−30,4
United Kingdom	7,076.1	7,053.4	7,539.6	7,135.4	7,230.7	8,115.7	7,386.9	4,4
Czech Republic	359.6	340.6	345.0	337.2	330.9	326.5	323.5	−10,0
Romania	230.6	230.1	223.8	214.0	215.1	205.5	194.4	−15,7
Sweden	726.2	772.5	833.3	830.9	828.5	833.7	867.4	19,4
EU	33,618.8	33,321.7	32,845.0	32,232.6	33,039.0	33,699.1	33,519.5	−0,3

PSM's funding crisis has its origins in the deterioration of its business model, based on three main sources of income: revenues from a licence fee paid by citizens for the access to PSM content in their homes, businesses and/or at an individual level, depending on the country; the provision of State funds coming from the general budgets; and the revenues obtained through the advertising sales or the commercialization of rights.

Changes to the elements of the very concept of the business model [25] explain the meaning of the crisis. The business models based on the allocation of public funds by the States depend on the government administration and parliamentary approval for annual or multiannual periods. When PSM's come from general budgets and have to be annually approved by the parliaments, both the government and the opposition parties tend to use their majorities or minorities in their own political interest, instead of doing so for the benefit of PSM. This political conditioning weakens the reputation, credibility and independence of PSM's value proposal, which must consider all citizens as objective clients, and not just those who identify themselves with the interests of the dominant political force. Against this situation, PSM's budgets must be sustainable, independent and foreseeable, i.e., planned on a multiannual base.

Commercial revenues, on the other hand, represent around 22% of PSM funding - and around 12% corresponds to advertising activities - and tend to diminish and to migrate towards Internet-based services, as it is happening in all kinds of media organizations. The European advertising market, according to data from the EAO [26], grew in 2016 up to 101,297 million Euros, against the 88,818 million Euros from 2011, with an increase of 2,5%.

3 The Licence Fee Crisis

Funding through a tax or licence fee applicable to each household is a system that dates back to the origins of PSB. The United Kingdom applied it to the BBC in 1927 and not only is it still in effect nowadays, but it was also ratified by the Royal Charter approved in 2016 until 2027. This licence fee will be annually updated according to the cost of living index. France and Italy developed their own licence fee systems in 1933 and 1938, respectively, and even Spain introduced a similar funding model up until 2006, although it was never really applied. Most of the EU Member States have or have had a funding model based on the licence fee, but this source of income has been progressively modified or even suppressed. In 2018, the licence fee was working in 18 Member States, with fares ranging from 330 and 283 Euros in Denmark and Austria, respectively, to 33.7 Euros in Portugal and 17.4 in Romania, according to the EAO.

Changes on this direct tax have multiplied during the past years as a consequence of new ways to access content, now not limited to the ownership of a television set, but broadened to online distribution through multiple devices. The structure of the households has also varied, and the concept of the licence fee was replaced by new taxes on the access availability of audiovisual services. However, in countries such as Ireland, the licence fee is still charged by household, a situation which is concerning its PSM, as the figures point that, in 2018, 10,6% of the Irish households stated that they did not own any television set [27].

In some countries, such as Finland since 2013, the licence fee was reformed in order to stop being a tax on households to become a tax on companies and natural persons (up to a maximum of 150 Euros) depending on their income level. Unemployed citizens and pensioners were exempt from paying this tax, as it happens in the United Kingdom. Denmark has also transformed its licence fee into a tax, starting in 2019. France and Spain applied in 2010 a new type of technological fee on the use of the radio spectrum by telecommunication operators, insofar they are part of the converging audiovisual industry. Moreover, due to the 2018 Audiovisual Media Services Directive, digital platforms will also be subjected to the payment of a spectrum fee in some European countries.

The development of new access and consumption ways increased the rejection, evasion or abolition demands of the licence fee. In 2018, this debate was especially intense in Switzerland, Germany, Denmark, Ireland, Poland and Sweden. In both Poland and Ireland, the high evasion rates added to the shortfall of the licence fee to sustain PSM, while in Germany the fee was highly contested due to its unpopularity, in Denmark it was suppressed and in Switzerland it became the object of a referendum celebrated on March 4th 2018.

A growing group of citizens are reluctant to the direct payment for PSM, as they barely consume its content and prefer the services of online global digital platforms. This reluctance reached its most visible expression in Switzerland, where the youth associations of the two main right-wing political parties, supported by small and medium-sized enterprises, prompted a referendum on the suppression of the licence fee to fund the Swiss PSM, consisting in 329€ per household and 519€ for enterprises. SRR-SRG, the Swiss PSM organization, has a particular multicultural and multilingual model, broadcasting in four languages (German, French, Italian and Romansh) to their respective communities. Its annual budget accounts for 1,200 million € for the funding of 21 radio stations and 13 local televisions, most of them of private propriety.

This plebiscite triggered an intense debate in favour and against "No Billag". 54% of the Swiss went to the polls and the referendum concluded, with 71.6% of the votes against the suppression of the licence fee, that the SSR should maintain this income stream, although lowered down to 317€ from 2010 on. The opinion contrary to the licence fee lost in all 26 cantons and it only won in 6 of the 2,250 Swiss municipalities.

The campaign that preceded the referendum was articulated, on the side of the detractors of the fee, on the fight for advertising between public and private media, the lack of austerity of Swiss PSM, the need for greater monitoring of SSR, its lack of trust and the control of media power, instead of focusing on the usefulness of local, public journalism [28]. These arguments were however used by the managers of SSR-SRG to better connect with the citizens, through direct meetings in all the cantos as well as through the partnership with akin collectives.

The "No Billag" referendum stabilized the validity of the licence fee for the next 15 years, although the Swiss corporation has announced budget cutbacks in order to adapt to the reduction of the fee, as well as to the implementation of other reforms to prompt the digital transition, more online services and increased collaboration with private companies. In 2019 the collection system of the licence fee, whose concession had been granted to Billag AG, subsidiary of Swiscom. The Swiss controversy was replicated in Germany, both by the populist politicians and by the association of private broadcasters [29], as a strategy to push for budget cutbacks and competition overview within the German public service broadcasters.

In the case of Denmark, the EU Member State with the highest licence fee rates, this funding method will be replaced in 2019 for a public tax, a change that will also reduce in a 15 to 20% the funding available for PSM. The Danish PSM has its own, particular model, where the State is the owner of the operator receiving public funding

(DR) and also of a competing commercial company (TV2) sustained only by commercial revenues. Public funds for Danish PSM is allocated as follows: 67% goes to DR, 20%, to foster audiovisual production, 9% to TV2's regional broadcasters, and 1% to local radio, while the remaining percentage goes to the general functioning of the sectoral regulator.

Both German PSM organizations, the nationwide broadcaster ZDF and the regional media outlets associated in the ARD, face budget reduction plans leading to a new licence fee model that will entry into force in 2021. These changes have developed in the middle of several lawsuits against users refusing to pay the licence fee and the constant attacks of political parties - such as *Alternative für Deutschland*, with a 13% representation - in favour of suppressing such fee. According to a survey conducted by Yougov, 44% of the Germans consider that the licence fee is too high, while another 13% do not wish to pay it.

The German licence fee will remain at 17.5 Euros per month until 2020. By now, the ZDF has announced a savings plan of 270 million €, while the ARD plans a budget cutback of up to 1,000 million € until 2028, at a rate of 100 million per year, facing a restructuring of its organization through increased cooperation with both the ZDF and the regional public broadcasters. At the same time, German PSM keeps preparing to strengthen its public service legitimacy in the horizon of the fourth audiovisual generation, cloud tv [30].

In Ireland, the public broadcaster RTÉ is claiming a reform of its funding system in order to compensate the 22% budget cutbacks developed between 2008 and 2016 due to the drop on licence fee revenues, caused by the high evasion rate (15%) and to the fact that young people no longer register in its service, by the decrease of advertising income as a result of the strong Internet competition, and by the need to adapt the licence fee to new access ways to media services [31]. According to RTÉ's Director General, Dee Forbes, "Irish public service media is providing a service superior to the funds it is receiving and it is facing urgent challenges".

The BBC has not been an exception in the trend towards budget savings. The British corporation has experienced a 20% budget reduction between 2012 and 2016 (around 300 million Euros) due to the licence fee freeze in 2010 and the obligation to provide its service for free to retirees from 2016 on. Thus, according to the BBC's Chairman, David Clementi, and its Director-General, Tony Hall, the efficiency level of the BBC increased by 25%, as it not only did not reduce its services, but actually increased them, a statement that they ratify by invoking the assessment of the experienced, perceived and estimated value that is developed by the British corporation [20].

France Télévisions, which has also been suffering budget adjustments since 2016, is pending on a reform affecting the French PSM system, composed of this corporation, Radio France, INA and France Médias Monde. This reform's pillars are the integration of all the companies of the public media sector within one same group, the redefinition of the public service remit of the different media outlets, the preservation of its independence from the political power, its funding and its adaptation to the digital scenery, as well as to the strategic role that PSM should play in digital platforms, by supporting French creative industries.

France Télévisions has also developed a digital transformation plan for the period 2016–2020, based on opening the corporation to the society, implementing account-ability measures and improving its efficiency communication through the endorsement of an assessment system of its aggregated value. Between 2016 and 2017, its management team developed a set of 26 public meetings in 22 cities to communicate its social value and to discuss its programming and the identity of its media services.

In Belgium, the visions of the Flemish VRT's for the period 2016–2020 and of the Walloon RTBF's for 2017–2020 highlight the opening up of both corporations to their respective communities, as well as their digital transformations. The new RTBF's structure is organized around two main big multimedia and convergent areas: production and diffusion-distribution. This is the new step towards organizational convergence, which was preceded by a first phase of integration a second one oriented at the planning of the structures around the integrated production and its distribution through all platforms available.

A committee within the Swedish Parliament, chaired by the expert Sture Nordh, studied the future of SVT's funding in order to draft a law that will come into force in 2020. The underlying objective of this change is to replace the current licence fee for an individual tax aimed at the contribution to PSM, in line with the Finnish model. The recommendation issued by the parliamentary committee including reducing the licence bee and setting an annual personal tax of 132 Euros for all citizens over the age of 18, except for students and unemployed. The Swedish Tax Agency would be in charge of the collection of the fee.

The direct pressure exerted on PSM's organizations in Austria, Poland, Hungary or the Czech Republic is not only of a financial nature, but also political and with media repercussions. Heinz-Christian Strache, Austrian vice-chancellor from the FPÖ (Freedom Party of Austria) was reported by a popular ORF anchor, Armin Wofl, and the public broadcaster itself because of his constant attacks against the ORF, claiming that it was adopting a "left-wing bias" and offering manipulated information against him. Similarly, the president of the Czech Republic, Milos Zeman, has led confrontations against the press and PSM journalists, and this situation was repeated in Poland, where several controversies have affected the governing party.

4   Discussion and Conclusions

Even though the crisis that European PSM is undergoing has several diverse and complex causes, the economic and financial ones are at the centre of the problem. In the Internet society, the value proposal of PSM remains necessary, protecting the values of independence, diversity, quality, universality, innovation and social responsibility. In order to protect certain space for PSM in a context characterised by dominance of the new global digital platforms, some countries have already started rethinking their media laws. In 2019, for instance, Ofcom published some guidelines aimed at reinforcing the EPG prominence of British public service broadcasters, as well as to guarantee their prominence in the online environment, by ensuring that PSB content is discoverable in major viewing platforms, such as Smart TVs [32].

However, for PSM to work efficiently towards the public interest and to contribute to society, the contents they offer, their relationships with their audiences and the permanent innovation processes are key. In order to meet these three challenges, PSM organizations need to be well funded. The funding crisis that PSM organizations are undergoing has heavily conditioned them, and sometimes prevented them, to face the digital transformation challenge and to adapt their model to the Internet society, that is to say, to adjust their public service remits and value proposals to the digital demands of both citizens as individuals and societies as a whole.

Relations, partnerships and alliances with stakeholders [33], as well as income streams funding models, need to be rethought in order to enable their adaptation to the changes imposed by society's evolution. This challenge, however, asks for the refunding and restoration of PSM's legitimacy, drawing from the premise that this public service is as necessary nowadays as it was in its origins, and bearing in mind that PSM must serve both foundational and new values linked to universality, quality, independence, diversity, innovation, while also contributing to the consolidation of new values emerging under the digital societies.

References

1. Parliament of the United Kingdom: Television Act 1954 (1954)
2. European Court of Justice: Judgement of the Court of Justice, Saachi, Case 155/73, 155-73 (1974). https://bit.ly/2lSIket
3. European Commission: Communication from the Commission – Services of general interest in Europe (2001). https://bit.ly/2kP00aD
4. Protocol to the Amsterdam Treaty on public service broadcasting (1997). http://data.europa. eu/eli/treaty/ams/pro_9/sign
5. European Commission: Communication from the Commission on the application of State aid rules to public service (2001). https://bit.ly/2kLwiDl
6. European Commission: Communication from the Commission on the application of State aid rules to public service (2009). https://bit.ly/2lSC3zg
7. Christensen, C.M., Raynor, M.E., McDonald, R.: What Is Disruptive Innovation? Harvard Business Review, December 2005. https://bit.ly/2kePVmW
8. Ramonet, I.: La explosión del periodismo. De los medios de masas a la masa de medios. Clave Intelectual, Madrid (2011)
9. Castells, M.: Comunicación y poder. Alianza Editorial, Madrid (2009)
10. Blaug, R., Horner, L., Lekhi, R.: Public Value, Politics and Public Management. The Work Foundation (2006). https://bit.ly/2kmqOij
11. Arriaza Ibarra, K., Nowak, E., Kuhn, R.: Public Service Media in Europe: A Comparative Approach. Routledge, London (2015)
12. Bardoel, J., d'Haenens, L.: Reinventing public service broadcasting in Europe: prospects, promises and problems. Media Cult. Soc. **30**(3), 337–355 (2008)
13. Iosifidis, P.: Retos y estrategias. Servicio público de televisión en Europa. Infoamérica **3–4**, 7–21 (2010)
14. Tremblay, G.: Public service media in the age of digital networks. Can. J. Commun. **41**, 191–206 (2016). https://bit.ly/2mjNvUV
15. Mendel, T.: Public service broadcasting: a comparative legal survey. UNESCO, Paris (2011). https://bit.ly/2mdPnON

16. Donders, K.: Public Service Media and Policy in Europe. Palgrave Macmillan, London (2012)
17. Horowitz, M.A., Car, V.: The future of Public Service Media. Medijske Studije/Media Stud. **6**(12), 2–9 (2015). https://bit.ly/2lTs9xl
18. Tambini, D.: Five theses on public media and digitization: from a 56-country study. Int. J. Commun. **9**, 1400–1424 (2015). https://bit.ly/2kBWG2F
19. Fuchs, Ch.: The Online Advertising Tax as the Foundation of a Public Service Internet. University of Westminster Press, London (2018)
20. BBC: BBC Annual Plan 2018/19 (2018). https://bbc.in/2pJMPam
21. Lowe, G.F., Van den Bulck, H., Donders, K.: Public Service Media in the Networked Society. Nordicom, Gothemburg (2018). https://bit.ly/2kMnVYc
22. EBU: Public Service Values. Editorial Principles and Guidelines (2014). https://bit.ly/2lWhMZy
23. Karppinen, K., Moe, H.: What we talk about when talk about "media independence". Javnost – Public **23**(2), 105–119 (2016). https://bit.ly/2kOfXO6
24. Albújar Villarubia, M.: Don't touch my megahertz! Planning the Second Digital Dividend in Spain amid the battle between telecoms, OTTs and DTT players. Int. J. Digit. Telev. **9**(3), 251–269 (2018)
25. Osterwalder, A., Pigneur, Y., Tucci, C.L.: Clarifying business models: origins, present, and future of the concept. Commun. Assoc. Inf. Syst. **16**, 1–25 (2005). https://aisel.aisnet.org/cgi/viewcontent.cgi?article=3016&context=cais
26. EAO: Yearbook 2017/2018, Key Trends. Television, Cinema, Video and On-Demand Audiovisual Services - The Pan-European Picture (2018). https://bit.ly/2HHnKaR
27. Sheahan, E.: Fewer TVs threaten RTE licence income. The Times, 30 June 2019. https://bit.ly/2lR7QRc
28. Saner, M.: Beobachtung der Wahlberinchtertattung des Regionaljournals Zürich Schalfhausen (2018)
29. Vaunet: Länder starten zweite Konsultation zum Medienstaatsvertrag, 3 July 2019. https://bit.ly/2lOGknn
30. Dörr, D., Holznagel, B., Picot, A.: Legitimation und Auftrag des öffentlich-rechtlichen Fersehens in Zeiten der Cloud, Peter Lang, Bern
31. Ramsey, P.: Public Service Media funding in Ireland faces continuing challenges. LSE's Media Policy Project Blog, 6 September 2017. https://bit.ly/2mdRuCd
32. Ofcom: A new future for public service broadcasting (2019). https://bit.ly/2NXuKny
33. Donders, K., Van den Bulck, H., Raats, T.: The politics of pleasing: a critical analysis of multistakeholderism in Public Service Media policies in Flanders. Media Cult. Soc. **41**(3), 347–366 (2019)

Digital Citizenship and Participation Through Twitter: The Case of Provincial Capital Municipalities in Ecuador (2009–2019)

Narcisa Jessenia Medranda Morales[1]([⊠]),
Victoria Dalila Palacios Mieles[2], Daniel Barredo Ibáñez[3],
and Camila Pérez Lagos[4]

[1] Carrera de Comunicación Social, Universidad Politécnica Salesiana,
170517 Quito, Ecuador
nmedranda@ups.edu.ec
[2] Departamento de Pastoral Universitaria, Pontificia Universidad Católica
del Ecuador, 17012184 Quito, Ecuador
[3] Escuela de Ciencias Humanas, Universidad del Rosario,
110821 Bogotá, Colombia
[4] Centre Censier 13, Sorbonne Nouvelle Paris, rue de Santeuil-Paris Cedex 05,
Paris, France

Abstract. Within the framework of citizen empowerment generated by technological development, social media turn out to be a powerful tool of communication and socialization. That is the reason why governmental entities have turned their look towards the unlimited possibilities that some virtual platforms and applications offer. In that sense, this article provides a general vision regarding participation routines of Ecuadorian users through Twitter. To this goal, a monitoring process has been carried out over eight main municipalities in this Andean republic. The results indicate that users actively interact on Twitter accounts belonging to public municipal organizations, spreading information through retweet. Thus, this social network has become an interesting citizen discussion space promoting a larger democratic evolution.

Keywords: Twitter · Digital citizenship · Content analysis · Social networks · Politics

1 Introduction

Please note that the first paragraph of a section or subsection is not indented. The first paragraphs that follows a table, figure, equation etc. does not have an indent, either.

Subsequent paragraphs, however, are indented.

Web 2.0 is the starting point for developing collective interaction spaces where people exchange information about entertainment, education and politics (Lozares 1996). These new environments foster a greater citizen involvement and they provide a greater knowledge about state agencies policies.

By digital citizenship we will refer to a group of people who make up a community, based on relationships and communication (Cabañez 2010). In addition, they own

© Springer Nature Switzerland AG 2020
Á. Rocha et al. (Eds.): ICITS 2020, AISC 1137, pp. 681–692, 2020.
https://doi.org/10.1007/978-3-030-40690-5_65

online participation skills (Menéndez 2016). This concept is related to a new type of citizenship defined by frequent use of Internet (Mossberger, Tolbert and McNeal 2008). Then, it is clear the digital sphere has transformed conditions for democracy through more participation and social welfare (Mossberger 2010).

Digital citizenship has found a niche in virtual media because it is not limited to geographical and political boundaries, which produces more interaction between users who are looking to move without restrictions. Being a digital citizen implies equality in transmitting and receiving knowledge (Cabañez 2010), regardless of factors that might cause exclusion such as gender, age, social and economic status.

From 2009, politicians worldwide have adopted social media to involve their electors (Chi 2010). They have understood Information and Communication Technologies (ICT) are able to promote changes in society (Calvo 2018; Sádaba 2012). In addition, they have favored a transformation in the communication paradigm (Calvo 2018). It has even been said virtual environments are perceived as a political battlefield (Menéndez 2016). Thus, they are considered as "The power of actors and organizations included in the networks that make up the core of the global network society over groups or people that are not integrated into it" (Castells 2009: 72–73).

Consequently, strategic use of social media is quite important in political campaigns (Campos 2017). It is inconceivable to think, for example, of an electoral process without them (Rendueles 2016); the widespread use of these platforms entails also other benefits, such as stimulating a democratic boost in authoritarian societies (Kruikemeier 2014). According to Achache (2012), political marketing works as a communication pervasive model, adding elements such as propaganda, advertising and persuasion for mobilizing emotions and actions. On the other hand, Komorowska (2016) points out that political discourse establishes the relationship between language, ideology and proposals. The dialogue that occurs through online spaces enhances public debates through a political discourse redefinition (Weller, Bruns, Burgess, Mahrt and Puschmann 2013) and a greater elector's activation.

In that sense, Twitter is one of the most important political communication tools (Andrés and Uceda 2011; Campos 2017) and a powerful media influencer (Enli 2017; Shapiro and Hempfil 2017). It eases interaction and experiences development among users that make up Social Media (Kuz, Falco and Giandini 2016). In addition, it adopts the asymmetric relationships model (Kruikemeier 2014).

The consolidation of Twitter as a political dissemination means (Arceneaux and Weiss 2010) has occurred along with the trend towards professionalization regarding global communication management (Kreiss and Janinski 2012), as it has been shown by the statements on this social network or hashtags displaying topics about outstanding news (English 2016; Heravi and Harrower 2016).

However, this tool does not remain subject to a purely informative-collaborative use (Nabel 2013; Angulo, Estrella and López 2018). In the political context it is understood as a debate and creative space (Angulo, Estrella and López 2018) as well as a tool used for persuasion and propaganda (Mancera and Helfrich 2014; Carrasco, Villar and Tejedor 2018).

This platform enables dynamics, produced in contemporary society, to be transformed, while at the same time, it allows to segment and group citizens participating in political debates by ideological affinities (Congosto, Fernández and Moro 2011). As a result, digital citizens stop being mere information consumers to become active individuals capable of discovering and sharing topics they are interested in within the virtual environment (Kirilenko and Stepchenkova 2014; Hong and Nadler 2012). Therefore, digital citizens' opinions can influence the evolution of government policies, given that digital social networks, such as Twitter, have become a platform for political debate (López and Ulloa 2011).

Shortly after its birth in 2006, Twitter already got more than 328 million users (Toledano 2017) and it has been used in a wide range of political contexts. This has caused a substantial variation in its scope and nature, which has allowed giving life to the so-called cyber politics or techno politics (Gutiérrez-Rubí 2014; Carrasco et al. 2018). These terms are related to microblogging, to the kind of content going around it (Angulo 2018), but above all, to the active and strategic use of this platform to make changes in political agendas.

Most of the research work on Twitter and politics has focused on the use made by politicians and how they communicate and express themselves through this network (Solop 2009; Golbeck, Grimes and Rogers 2010; Lassen and Brown 2011; Olof and Moe 2013; Lafuente and Verón 2013; Zamora and Zurutuza 2014; García and Zugasti 2014). The way citizens use Twitter in the political realm has also been studied (Tumasjan, Sprenger, Sandner and Welpe 2010; Hawthorne, Houston and McKinney 2013; Bekafigo and McBride 2013).

In Ecuador, around four out of every ten households, in 2018, counted on internet broadband (El Universo 2018), although that same year about six out of ten Ecuadorians had internet services on their smartphones (Telecommunications Regulation and Control Agency 2018). In a country with a high emigrant population percentage, the use of ICTs has a large impact on socialization routines and political debates. It is often cited that Rafael Correa's victory, in 2006, was largely caused by an efficient social media management (Rivera 2014), namely Twitter, an enclave commonly used by this Ecuadorian former president because it helps establish an immediate interaction with voters (Freidenberg 2013). Barredo et al. (2015), when studying the possible relationship between social networks and the vote decision in Ecuador, found that exposure to online media tends to encourage higher political participation levels, especially in younger groups.

The fundamental goal of study is to reflect on exposed political contents, analyzing discursive strategies and establishing the kind of response given by citizens. We have followed the studies made by Angulo et al. (2017) who contrasted discursive strategies on Twitter during the Ecuadorian elections in 2017. This essay also follows Calvo (2017) and Carrasco et al. (2018) who addressed the role of Twitter as a second screen in political debates on television.

2   Methodology

In this descriptive article, a content analysis has been used as a research technique. This is a "systematic and objective" quantitative method (Bardin 2002, p. 25) that studies latent or evident contents in a message (Aigeneren 2009, p. 4). In this study, we are interested in visualizing government management actions taken by Ecuadorian municipalities, in addition to monitoring their Twitter accounts to interpret online interactions as well as the way communication occurs with users. The ultimate goal is to measure and quantify relationships. Twitter accounts that belong to the 24 provincial capitals' municipalities of Ecuador have been found. Out of them, those generating the greatest interaction will be identified and analyzed.

Therefore, each account was considered from their creation date until July 2019. For example, the accounts belonging to the Guayaquil, Ibarra and Loja municipalities were monitored from 2009 to 2019. While the Nueva Loja municipality created its account in 2010 and Carchi did the same in 2011, the remaining ones began their activities on Twitter from 2012. In 2015, Riobamba was the last one.

Once the universe for this study was located, data was collected using the following computer tools: Foller.me and TweetReach, from the beginning of each account's activities. For examining trends, an analysis sheet was previously drawn, taking into account the results of some previous studies focused on Twitter analysis and its relationship with political debates (Angulo et al. 2017; Calvo 2017; Carrasco et al. 2018; Justel et al. 2018).

The analysis sheet was set up on a two main factors basis:

1. Scope of publications on Twitter: topics analysis, number of followers, steady traffic data.
2. Interaction with users: tweets and retweets accounting to determine users' interaction in news spaces; "likes" as a social interaction strategy.

3   Results

First, data about followers was checked as well as tweets on monitored accounts:

Table 1. Number of followers, tweets and retweets on monitored Twitter accounts (2019).

Analyzed accounts	Followers	Tweets	Retweets (%)
@MunicipioCuenca	9091	30869	35%
@GADGuaranda	3382	7197	5%
@Mazogues	1458	1450	5%
@tulcanmunicipio	2828	5466	3%
@MuniRiobamba	3778	5597	7%
@gadlatacunga	1439	1482	6%
@AlcaldiaMachala	14095	7787	10%
@AlcaldiaEsme	8207	7949	3%
@gadsantacruz	2466	3791	5%
@alcaldiagye	642147	17537	20%
@AlcaldiaIbarra	8817	14634	34%
@MunicipioDeLoja	18777	26938	13%
@Mbabahoyo	2413	2559	5%

(continued)

Table 1. (*continued*)

Analyzed accounts	Followers	Tweets	Retweets (%)
@PortoviejoGAD	8022	3147	3%
@MunicipioMorona	3128	3027	4%
@alcaldiatena	1406	1767	11%
@GADMFOrellana	644	841	6%
@GADMPastaza	1675	7222	5%
@MunicipioQuito	271808	40951	10%
@GadSantaElena	–	–	–
@GADsantodomingo	3644	3283	1%
@gadmlagoagrio	1006	6073	1%
@gadambato	6503	4153	5%
@Alcaldia_Zamora	3767	4689	4%

Results on Table 1 show not directly proportional relationships between the number of followers, tweets and retweets. The @MunicipioQuito account shows that the number of its tweets is much lower than the number of followers. Out of every 100 tweets published, only 20% were shared by followers, which shows that there are many passive followers within the network: a large number of people are likely devoted to reading news, but they do not intervene on the account, they are limited to consuming contents. This phenomenon is similar on @alcaldiagye (Guayaquil) (Fig. 1).

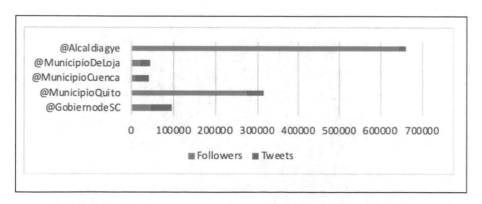

Fig. 1. Accounts with the highest number of tweets among Ecuadorian municipalities (2019).

There is a contrast with the accounts of @MunicipioDeLoja, @MunicipioCuenca and @AlcaldiaIbarra, where there are more retweets the followers, which implies interaction dynamics between participants. In addition, this also shows some variables like mediators and moderators promoting a greater visibility in these municipalities.

In all accounts, satisfaction regarding them is somehow shown through "likes", which refers to a positive appreciation about the information they are reading.

The number of tweets reported on these accounts is higher in @MunicipioQuito, @MunicipioCuenca, @MunicipioDeLoja, @alcaldiagye and @AlcaldiaIbarra, which correspond to municipalities with more followers, except @MunicipioCuenca. Regarding the events happening on Twitter accounts, some details can be noticed from the data retrieved on Foller.me (Fig. 2). That infographic shows the accounts with more tweets and user interactions. The Twitter account belonging to Guayaquil (@alcaldiagye) has the largest number of followers in the country: the estimated range is 34.7% out of the total information issued by the official municipal sites.

Meanwhile, the municipality of Quito's account (@MunicipioQuito) exceeds the others in number of tweets. However, its scope is 21.7%. Likewise, the municipality of Cuenca (@MunicipioCuenca) counts on a significant number of followers, below Guayaquil and Quito. These are cities with more inhabitants than Cuenca. Nevertheless, the scope of its publications is 8, 4%. In @MunicipioDeLoja, its number of followers places this Tweeter portal site in the third place among the accounts with the largest number of followers in the country, with an estimated reach of 6.4%. On the other hand, the municipality of Ibarra (@AlcaldiaIbarra) is the account with the smallest reach, barely exceeding 1% in an account that has totally issued 14634 tweets since its start (Fig. 3).

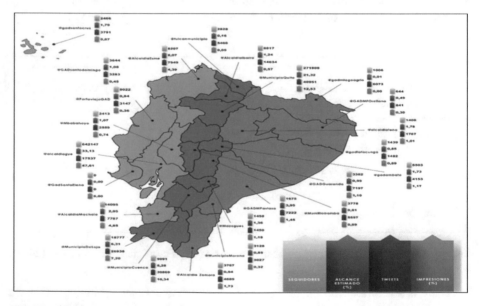

Fig. 2. Tweets scope and impressions from the Ecuadorian monitored municipalities (2019).

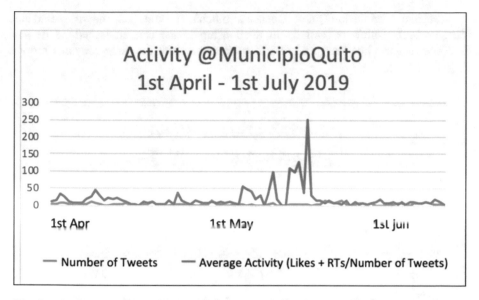

Fig. 3. Municipal accounts with the largest citizen participation - @MunicipioQuito (2019).

Quito has one of the largest population in Ecuador. In addition, it is capital city, which causes more interactions related to problems and activities issued on its Twitter official account. Although during the first two weeks of monitoring -which matches with Jorge Yunda's assumption of office-, users interaction was growing up; from the first days of June 2019, it has been decreasing towards a similar trend, and even lower, to the one observed during April and May of the same year (Fig. 4).

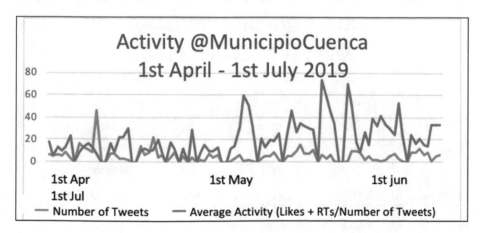

Fig. 4. Municipal accounts with the largest citizen participation - @MunicipioCuenca (2019).

Just like @MunicipioQuito, Cuenca's official account has shown remarkable changes in its followers' activity in relationship to the content issued on its site. A 200% increase has been recorded in comparison with the three previous months (Fig. 5).

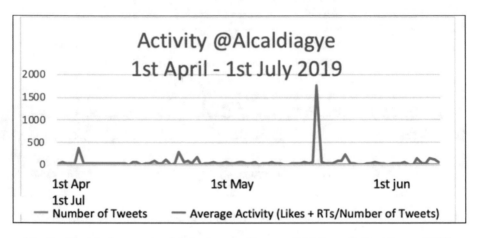

Fig. 5. Municipal accounts with the largest citizen participation - @Alcaldiagye (2019).

For their part, @alcaldiagye keeps a similar trend during the three analyzed months, except for some milestones that caused a higher reaction on the audience. This is about the video published on June 26 2019 about the raid carried out by the National Police to catch criminals (Fig. 6).

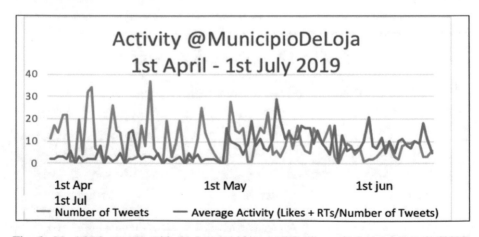

Fig. 6. Municipal accounts with the largest citizen participation - @MunicipioDeLoja (2019).

The Municipality of Loja's Tweeter account displays a tweets decrease, but an increase regarding followers' activity. This trend is still going on so far, until the last analyzed date for this article. Results above point out that interactions on the @MunicipioQuito, @alcaldiagye, @MunicipioCuenca, @MunicipioDeLoja and @AlcaldiaIbarra accounts are more dynamic and have reached a higher scope throughout the analyzed time. Information issued on these accounts has been studied. As a result, it has been set up the most valued hashtags as shown on Chart 4 (Table 2):

Table 2. More frequent hashtags on Tweeter accounts belonging to Ecuadorian municipalities that have the largest number of tweets and interaction (2019)

Account	Hashtags
@MunicipioQuito	#QuitoPeludoOtraVez
	#QuitoIncluyenteOtraVez
	#QuitoVive
@alcaldiagye	
@MunicipioDeLoja	#Loja
	#LojaParaTodos
	#NadaPorLaFuerzaTodoPorLaLey
@MunicipioCuenca	–
@AlcaldiaIbarra	#Ibarra

The most frequently used hashtags are on @MunicipioQuito, @MunicipioDeLoja and @AlcaldiaIbarra. They are about these institutions and the information shared with other accounts (retweets) which is a primary information outreach mechanism. The @alcaldiagye and @MunicipioCuenca accounts do not use hashtags but instead they post a direct tag headed to the account related with the discussed topic.

4   Conclusions

On the accounts @MunicipioQuito, @alcaldiagye and @MunicipioCuenca, a greater activity was observed, which is partially reciprocated by users interactions. Secondly, it is interesting to notice that the size of municipalities is not proportional to the size of the communities they can handle through Twitter. An effective management, in that sense, should be able to appeal a greater number of users. This is something that might indeed cause a positive impact on several aspects such as touristic promotion in favor of the city itself.

Interactions on social networks, and particularly on Twitter, boost new habits for users interested in getting fast and accurate information about public institutions in charge of setting up policies and rules in their communities. From this approach, this social network contributes to identify what problems are arising in the environment as well as the possible solutions proposed, among others, by public institutions in order to face them. By this way, it is configured a digital citizenship which, through approaching between electors and elected ones (authorities), deepens and improves the sense of democratic participation. Then, from online enclaves, public organizations have the chance to foster a greater cohesion among users consolidating a community.

While analyzing topics issued by these Ecuadorian municipal governments as well as the users' responses, it can be noticed a greater interest around public policies topics. People demand to solve this kind of issues as soon as possible. However, citizens in Quito have shown a greater interest over topics regarding animals, even more than the interest for those about the city. For their part, citizens in Guayaquil are more appealed by crime report, which is constantly published on the municipal Twitter account. Finally, account followers in Loja are interested in cultural topics.

In a more comprehensive analysis, it is possible to say that interactions aroused by state organizations bear fruit throughout the path of individuals because their goal is to actively convey information hoping users receive it and expecting their feedback. These users are not necessarily the final ones, as it was clearly established on this study; they are the ones who multiply contents retrieved on Twitter.

References

Achache, G.: El marketing político. In: Mercier, A. (ed.) La Comunicación Política. La Crujía Ediciones, Buenos Aires (2012)

Agencia de Regulación y Control de las Telecomunicaciones: Servicio Móvil Avanzado. Densidad de Abonados y Líneas Activas (2018). http://www.arcotel.gob.ec/servicio-movil-avanzado-sma/. Accessed 02 Mar 2019

Aigeneren, M.: Análisis de contenido: una introducción. U de Antioquia, Medellón (2009)

Andrés, R.R., Uceda, D.U.: Diez razones para el uso de Twitter como herramienta en la comunicación política y electoral. Comunicación y Pluralismo (10), 89–116 (2011)

Angulo, N., Estrella, A, López, M.: La política en Twitter. Un estudio comparativo de las estrategias discursivas de los candidatos finalistas a la Presidencia de Ecuador en 2017. adComunica. Revista Científica de Estrategias, Tendencias e Innovación en Comunicación **16**, 25–44 (2018). https://doi.org/10.6035/2174-0992.2018.16.3

Arceneaux, N., Schmitz, A.: Seems stupid until you try it: press coverage of Twitter, 2006-9. New Media Soc. **12**(8), 1262–1279 (2010)

Bardin, L.: El análisis de contenido. Akal, Madrid (2002)

Bekafigo, M.A., McBride, A.: Who tweets about politics? Political participation of Twitter users during the 2011 gubernatorial elections. Soc. Sci. Comput. Rev. **31**(5), 625–643 (2013)

Broersma, M., Graham, T.: Twitter as a news source: how Dutch and British newspapers used tweets in their news coverage, 2007–2011. J. Pract. **7**(4), 446–464 (2013). https://doi.org/10.1080/17512786.2013.802481

Cabañez, E.: Hacia ciudadanía digital: Una carrera de obstáculos. Universidad de Santiago de Compostela, Compostela (2010)

Calvo, L.M.: Twitter como segunda pantalla en los debates políticos en televisión. Análisis de la etiqueta #L6Nrajoyrivera. In: ICONO 14, vol. 16, no. 1, pp. 160–184 (2018). https://doi.org/10.7195/ri14.v16i1.1138

Campos, E.: Twitter y la comunicación política. El profesional de la información **26**(5), 785–793 (2017). https://doi.org/10.3145/epi.2017.sep.01

Carrasco, R., Villar, E., Tejedor, L.: Twitter como herramienta de comunicación política en el contexto del referéndum independentista catalán: asociaciones ciudadanas frente a instituciones públicas. In: ICONO 14, vol. 16, no. 1, pp. 64–85 (2018). https://doi.org/10.7195/ri14.v16i1.1134

Chi, F., Yang, N.: Twitter in Congress: Outreach vs Transparency. MPRA, Toronto (2010)

Congosto, M.L., Fernández, M., Moro, E.: Twitter y política: información, opinión y ¿predicción? Cuadernos de Comunicación Evoca **4**, 11–16 (2011)

Castells, M.: Comunicación y Poder, pp. 72–73. Alianza Editorial, Madrid (2009)

El Universo: El 41,2% de los hogares de Ecuador cuenta con internet fijo, 17 May 2018. https://www.eluniverso.com/noticias/2018/05/17/nota/6764079/412-hogares-ecuador-cuenta-internet-fijo. Accessed 02 Mar 2019

Enli, G.: Twitter as arena for the authentic outsider: exploring the social media campaigns of Trump and Clinton in the 2016 US presidential election. Eur. J. Commun. **32**(1), 50–61 (2017)

Freidenberg, F.: Ecuador 2013: las claves del éxito de la Revolución Ciudadana. OPEX, Madrid (2013)

García, C., Zugasti, A.: La campaña virtual en Twitter: las cuentas de Rajoy y Rubalcaba en las elecciones generales de 2011. Historia y Comunicación Soc. **19**, 299–311 (2014)

Golbeck, J., Grimes, J.M., Rogers, A.: Twitter use by the U.S. Congress. J. Am. Soc. Inf. Sci. Technol. **61**(8), 1612–1621 (2010)

Gutiérrez-Rubí, A.: Tecnopolítica: Uso y la concepción de las nuevas herramientas tecnológicas para la comunicación, la organización y la acción política colectivas, España (2014)

Hawthorne, J., Houston, J.B., McKinney, M.S.: Live-tweeting a presidential primary debate: exploring new political conversations. Soc. Sci. Comput. Rev. **31**(5), 552–562 (2013)

Heravi, B., Harrower, N.: Twitter journalism in Ireland: sourcing and trust in the age of social media. Inf. Commun. Soc. **19**(9), 1194–1213 (2016). https://doi.org/10.1080/1369118X.2016.1187649

Hong, S., Nadler, D.: Which candidates do the public discuss online in an election campaign? The use of social media by 2012 presidential candidates and its impact in candidate salience. Gov. Inf. Q. **29**, 455–461 (2012)

INEC: Tecnologías de la Información y Comunicaciones (TIC'S) 2016. INEC (2016). http://www.ecuadorencifras.gob.ec/documentos/web-inec/Estadisticas_Sociales/TIC/2016/170125.Presentacion_Tics_2016.pdf. Accessed 02 Mar 2019

Kirilenko, A., Stepchenkova, S.: Public microblogging on climate change: one year of Twitter worldwide. Glob. Environ. Change **26**(171), 171–182 (2014)

Komorowska, A.: Introducción Pragmática del discurso electoral y el uso de nosotros. In: Gornikiewicz, J., Marczuk, B., Piechnik, I. (eds.) Études sur le texte dédiées à Halina Grzmil-Tylutki. Universidad Jaguelonica, Cracovia (2016)

Kreiss, D., Jasinski, C.: The tech industry meets presidential politics: explaining the democratic party's technological advantage in electoral campaigning, 2004–2012. Polit. Commun. **33**(4), 1–19 (2016). https://doi.org/10.1080/10584609.2015.1121941

Kruikemeier, S.: How political candidates use Twitter and the impact on votes. Comput. Hum. Behav. **34**, 131–139 (2014). https://doi.org/10.1016/j.chb.2014.01.025

Kuz, A., Falco, M., Giandini, R.: Análisis de redes sociales: un caso práctico Red de Revistas Científicas de América Latina, el Caribe. España y Portugal. Computación y Sistemas **20**(1), 89–106 (2016). https://doi.org/10.13053/CyS-20-1-2321

Lafuente, P., Verón, J.J.: El uso de Twitter por los líderes de las organizaciones políticas minoritarias en la campaña electoral de las generales de 2011. In: Crespo, I. (ed.) Partidos, medios y electores en procesos de cambio. Las elecciones generales españolas de 2011, pp. 541–565. Tirant Lo Blanc, Valencia (2013)

Larsson, A.O., Moe, H.: Representation or participation? Twitter use during the 2011 Danish election campaign. Javn.- Public **20**(1), 71–88 (2013)

Lassen, D., Brown, A.: Twitter: the electoral connection? Soc. Sci. Comput. Rev. **29**(4), 419–436 (2011). https://doi.org/10.1177/0894439310382749

López, S., Ulloa, P.: Twitter usado como herramienta de debate político. Universidad de Especialidades Espíritu Santo, Guayaquil (2011)

Lozares, C.: La teoría de redes sociales. Rev. REDES **48**, 103–126 (1996)

Mancera, A., Helfrich, U.: La crisis en 140 caracteres: el discurso propagandístico en la red social Twitter. Cult. Lenguaje Y Representación **12**, 59–86 (2014). https://doi.org/10.6035/CLR.2014.12.4

Mossberger, K.: Digital Citizenship: The Internet, Society, and Participation. MIT Press, Cambridge (2008)

Mossberger, K.: Toward digital citizenship. In: Howard, P.N. (ed.) Routledge Handbook of Internet Politics, pp. 173–185. Taylor & Francis, New York (2010)

Morozov, E.: La locura del solucionismo tecnológico. Clave Intelectual, Madrid (2015)

Nabel, L.C.T.: Los mecanismos de lo político en las redes sociales de Internet. Aposta. Revista de Ciencias Sociales (58), 1–27 (2013)

Olof, A., Moe, H.: Representation or participation? Twitter use during the 2011 Danish election campaign. Javn.-Public **20**(1), 71–88 (2013)

Paulussen, S., Harder, R.A.: Social media references in newspapers: Facebook, Twitter and YouTube as sources in newspaper journalism. J. Pract. **8**(5), 542–551 (2014). https://doi.org/10.1080/17512786.2014.894327

Rendueles, C.: La ciudadanía digital. ¿Ágora aumentada o individualismo postmaterialista? Revista Latinoamericana de Tecnología Educativa **15**(2), 15–24 (2016). https://doi.org/10.17398/1695288X.15.2.15

Rivera, J.: Rafael Correa y las elecciones 2006. Inicios del Marketing y Comunicación política digital en Ecuador. Chasqui. Rev. Latinoamericana de Comunicación **126**, 116–123 (2014)

Sádaba, I.: Introducción a la investigación social online. In: Sádaba, I., Arroyo, M. (eds.) Metodología de la investigación social. Técnicas innovadoras y sus aplicaciones. Síntesis, Madrid (2012)

Shapiro, M.A., Hemphill, L.: Politicians and the policy agenda: does use of Twitter by the US congress direct New York Times content? Policy Internet **9**(1), 109–132 (2017). https://doi.org/10.1002/poi3.120

Solop, F.: RT@ BarackObama we just made history. Twitter and the 2008 presidential election. In: Hendricks, J.A., Denton, R.E. (eds.) Communicator-in-Chief. A Look at How Barack Obama used New Media Technology to Win the White House, pp. 37–50. Lexington Books, Lanham (2009)

Toledano, B.: El número de usuarios que ha sumado Twitter en el último trimestre asciende a cero. El Mundo. Recuperado de, 27 July 2017. https://goo.gl/qu8SBj

Weller, K., Bruns, A., Burgess, J., Mahrt, M., Puschmann, C. (eds.): Twitter and Society. Peter Lang, New York (2013)

Zamora, R., Zurutuza, C.: Campaigning on Twitter: towards the 'personal style' campaign to activate the political engagement during the 2011 Spanish general elections. Comunicación y Sociedad **27**(1), 83–106 (2014)

Latin American Indigenous Post-drama Architectures from an Andean Perspective Revisited

Miguel A. Orosa(✉) and Viviana Galarza-Ligña

Pontificia Universidad Católica del Ecuador, Ibarra, Ecuador
{maorosal, vngalarza}@pucesi.edu.ec

Abstract. Throughout this paper we will analyze the deep organizations of the Latin American indigenous post-drama from an Andean point of view and not from a Western standpoint. **Theoretical framework**. To better understand these, we will make comparisons with the dramatic and post-dramatic architectural dispositions of the West, but only in order to highlight and make more understandable those of the indigenous post-drama. **Discussion and results**. Next we are going to analyze and characterize a type of (post)dramatic organization very different from the Medieval and especially from the Greek Western one; that is to say, a (post-)theatrical culture we can describe as millenary, native and autochthonous typical of (one of) the Latin American indigenisms that has never been academically studied (theatrically speaking) until the present. In order to do so, we have carefully observed for a long time the plays with the greatest indigenous influence in countries such as Ecuador, Bolivia, Peru and Colombia. We have characterized their patterns in the light of studies related to this civilization and we will hereunder reflect doctrine and praxis through a canonical play of this culture: La flor de la Chukirawa, by Patricio Vallejo Aristizábal.

Keywords: Latin American indigenous post-drama · Western drama · Western post-drama · Post-dramatic culture in the Spanish language

1 Introduction

Throughout this paper we will analyze the deep organizations of the Latin American indigenous post-drama from an Andean point of view and not from a Western standpoint. To better understand these, we will make comparisons with the dramatic and post-dramatic architectural dispositions of the West, but only in order to highlight and produce an understanding of those in the indigenous post-drama.

This is a subject of maximum significance in the world we live nowadays, even more so when up to the moment (and as far as we know) nobody has written about this particular matter: the architectures and organizational foundations of the contemporary post-theatre of indigenous influence staged in Latin America.

We are going to analyze and characterize a type of (post)dramatic organization very different from the Medieval and especially from the Greek one; that is to say, a theatrical culture that we can describe as millenary, original and autochthonous typical

© Springer Nature Switzerland AG 2020
Á. Rocha et al. (Eds.): ICITS 2020, AISC 1137, pp. 693–702, 2020.
https://doi.org/10.1007/978-3-030-40690-5_66

of the Latin American indigenism. It is an organizational structure that opens the doors to a post-theatrical world that is thoroughly ancient and therefore new and unknown.

In order to do so, we have carefully observed for a long time the plays with the greatest indigenous influence in countries such as Ecuador, Bolivia, Peru and Colombia. We have characterized their patterns in the light of studies related to this civilization and we will now reflect doctrine and praxis through a canonical play of this culture: La flor de la Chukirawa, by Patricio Vallejo Aristizábal.

2 Theoretical Framework: Western (Post-)theatre

2.1 The Dramatic Period

In Western culture, and throughout the entire dramatic period [1–6] (that was born with Greece and ended in the last thirty years of the twentieth century), there is an obsessive idea in the art world and consequently the theater also participates: the notion of unity, composition or the aspiration to harmonize all the artistic work around a single idea. This ruling idea or "theme" of the work of art would be the driving or the organizing principle of the whole; all the parts of the work (even the smallest element in it) would reflect and respond to that thematic idea: unity, universe, synthesis, meaning, composition. This is the fundamental principle of the dramatic period.

And this fundamental principle crystallizes in different realities and variables that can be seen in Table 1 (West dramatic panel) that we show below.

Table 1. West dramatic panel

	Category	Component
1	Plot	Plot: turning points. Emotional projections Selection of facts (story): not all the life or legends of the hero, but only the events necessary to tell the theme of the work
2	Dramatic organization	External: blocks or acts Middle: themes, movements (exposition, conflict, resolution) Unity of action (a protagonist who seeks an objective)
3	Dramatic tension	Conflict or confrontation (obstacles that separate the protagonist from his/her objective) Activation: style, spectacular speech, intellectual discourse, characters, music
4	Narrative organization within drama Type of narrative organization	Narrative organizations involved in the field of drama (arrangements/organization of the "novel" that appears within the drama)
5	Technical-formal organization	Situation, formal drawing, intensity, duration, colors and textures, tempo, story

Source: Orosa y López-López [7]

All the categories of Table 1 above, from 1 to 5 contribute to the unity of the dramatic play. See in this sense Orosa [8, 9].

From segment 1 (Plot) we want to highlight "the selection of the facts of the internal story or plot". This element is especially significant to explain the idea of unity in the field of story construction. When developing a storyline, we do not relate all the legends or events of the protagonist's life, but the author selects those that are relevant to the unitary plot he wants to tell.

With regard to segment 2 (Dramatic organization), just to highlight the concepts of "unity of action" and "dramatic contexts". As is well known, unity of action has to do with the fact that one character, and only one, seeks the fulfillment of its own objective within a dramatic play (this rule prevents the change of character in search of his own goal throughout the entire play). As for the dramatic contexts (setup, confrontation and resolution), they arise from a unitary and teleological vision of existence. Namely, a person is born, develops his preferences and personality, struggles to overcome the obstacles that keep him away from his objective and, at last, achieves his goals, or not.

Segment 3 (Dramatic tension). Confrontation puts the character to the test and, for that very reason, it gives birth to it, it creates it. Simultaneously, this aforementioned character has to overcome, or not, the corresponding conflict that engendered him in search of his dramatic goal, and by the same token (that confrontation) gives unity to the (dramatic) organization of the play in its entirety. Confrontation is also very important from the perspective of dramatic tension. By hindering the conflict that the protagonist achieves his goals, it becomes a source of attention for the viewer from which emanates the interest of the play in terms of tension.

The conflict in Western drama, therefore, is a source of dramatic unity (knot/crux and liaison between the setup and the resolution/denouement); also enjoys a definitive role in the creation of the character and is the main energy from which the tension of the play arises. There is no life or dramatic style without confrontation.

In the age of drama, the term "confrontation" has an exclusionary sense of opposition. The term itself, conflict, enjoys a certain destructive sense, if two forces confront, at least one of them must lose in order for the drama to end. The confrontation must be overcome if the protagonist wants to reach a happy ending, but for that purpose something or someone must fall on the way. We are talking about two or more forces that necessarily end up destroying each other.

Segment 4 (narrative organizations that run within the dramatic framework). We are talking about dispositions/structures from the novel inserted in the dramatic organization for some constructive reasons and which contribute to give "meaning", that is, "unity" to the work of theatrical architecture.

Segment 5 (the formal techniques within the dramatic realm). It is the most abstract foundation of drama and one of the most classical factors, used since antiquity to achieve expressiveness or beauty itself. In short, we are referring to the composition of drama, to unity, to the meaning of the dramatic play: to art in a pure and classical sense. See in this sense Orosa y Pérez [9, 10].

2.2   The Post-dramatic Period

Just as the obsession of the dramatic stage was the unity of the artistic work, the constant preoccupation of the post-dramatic culture is the chaos (of order), the tendency to escape, the rupture of unity, the breakdown of sense, the juxtaposition (parataxis) of different ideas against or versus the composition (synthesis) [11–16]. And is not this post-dramatic trend, again, an obsession with the idea of unity that is tried to be denied in each text?

This fact or tendency (unity or its rejection) is once more emphasized in the works of certain authors of post-drama in which it is a question of combining this tendency to chaos with the insertion of a variable dose of order in their dramatic plays. Particularly, we do not know or remember a total rupture with the idea of sense or unity in the post-drama, which usually shows up (at least) in the construction of the narrative-dramatic style.

However, this notion of unity or sense is weaker in the post-drama than in the previous dramatic stage and strongly shines many times more by its sickly absence than by its obvious presence.

Table 2. West post-dramatic panel

	Category	Component
1	Time and reality	Absence of chronological or linear time Fragmentation of characters, of architectures… "Kairos" or present time Disregard or controversy over the concept of "reality" Positions in conditional or go to potential ones
2	Theme and text	Textual autarchy (the predominance of the text is denied) Visuality and multi-disciplinarity of the show besides the text Thematic fragmentation or dramatic fragmentation of actions on stage Juxtaposition of ideas against organic unity Apparent irrationality, self-referentiality Collage Multiplicity and multi-perspectivism Repeating the opposite of what it looked like before
3	Characters	Constitution of characters tends to fragmentation or puppets in the hands of the author
4	Scenes	Inconsistency of disciplines and events Non-hierarchical (horizontality) of topics or speeches Sensory communication with the viewer Relations between text and scenes are autonomous and noninterpretive Reiterations of scenes Micro-scenes Daily routines and micro-stories
5	Dialogues and language	Organizations and non-logical provisions The truth and the exclusivity of the word are questioned No causal logic
6	Participation and result	Sensitive and emotional communication with the viewer Reflection on the world

Source: Orosa y López-López [7]

As for the vital idea of conflict, the notion that feeds it in the post-drama is one of excluding opposition, already present in the previous stage; and another of nonsense, of anguish: precisely the one that provokes the rejection of unity, of composition, of synthesis. But, again, despite the various plots or dramatic actions that may develop into a post-dramatic play, the theme remains one, one alone, and the conflict tends to highlight it, to give it prominence and importance with its presence and development.

As we can see in Table 2 (West post-dramatic panel), the main patterns of the post-drama show a hesitant tendency to deny, up to a certain extent, the idea of dramatic composition. See Orosa [17, 18].

Category 1: the absence of chronological or linear times or the rejection of the notion of "reality" as something sensitive and close to ourselves (which is also expressed in the acceptance of conditional or different future denouements) does not necessarily attack the idea of unity proper to Western dramatic art. Rather, it seems to us that it introduces new components that give variety to the concept of dramatic art in our time.

Category 2: many of the elements in this category have to do with conflict, with its post-dramatic use, and they do not especially infringe upon the idea of unity, with the exception of two of them: the denial of the autarchy of the text which is placed on an equal footing as the rest of the (post-)dramatic disciplines, and the substitution of organic unity by the idea of juxtaposition of scenes. In addition, the intense use of self-referentiality can also turn the post-dramatic play into an attack on the unity and meaning proper of the classical epoch.

Denying the preponderance of the text and placing it on the same level as other (post-)dramatic disciplines can become an attack on classical unity if the director does not have a prior plan or idea of what he wishes to do with the performance of the play. Concerning the substitution of organic unity by the idea of juxtaposition of scenes, it can be a problem, or not, for the unit of the piece depending on the doses of order that the author uses in the field of parataxis (juxtaposition).

The abusive use of self-referentiality can be a problem too for the unity of the work from the point of view of the meaning of the plot (if development is very abstruse), although not necessarily for the aesthetic unity of the post-dramatic play. The aesthetic-normative approach can respond to certain compositional criteria and leave the understanding of meanings to the choice of each one.

Category 3 to 6: they are comments already made or similar to those we have previously given in this regard.

3   Latin American Indigenous Post-drama Architectures Seen from an Andean Perspective. Discussion and Results

Next we are going to analyze and characterize a type of (post)dramatic organization very different from the Medieval and especially from the Greek Western one; that is to say, a (post-)theatrical culture we can describe as native and typical of (one of) the Latin American indigenisms that has never been academically studied (theatrically speaking) until now. In order to do so, we have carefully observed for a long time the plays with the greatest indigenous influence in countries such as Ecuador, Bolivia, Peru and Colombia [19–24]. We have characterized their patterns in the light of studies related to this civilization and we will hereunder reflect doctrine and praxis through a canonical play of this culture: La flor de la Chukirawa, by Patricio Vallejo Aristizábal.

One might wonder why we have selected La flor de la Chukirawa (The flower of Chukirawa), by Patricio Vallejo Aristizábal. We have chosen this play to deal with the problem of Latin American indigenous post-drama because we have considered that it meets perfectly all the patterns that characterize this millennial artistic trend (as we will see below) and therefore it could be considered a canonical play of this native movement and completely representative of its inspiration and culture.

The patterns that we refer to and which we will discuss below could be the following (among others) [25–30] (Table 3):

Table 3. Latin American indigenous post-dramatic panel

	Category	Component
1	Systemic or organic vision Atmospheres Environements	Post-dramatic pariverse or duoverse (or multiverse) versus western artistic universe (unity) Truth as a dramatic parity. Complementary dramatic duality "Greek" (Native)-Andean heavy liturgies Refusal of homocentrism Harmonies and structures proper to the entire Nature
2	Conflict Complementary forces Post-dramatic organization Denouement	Conflict understood as vital tension, not in a destructive way It is not necessary for one dramatic force to confront and defeat another Unsolved conflict Complementary pairs versus opposite and excluding pairs
3	Time Post-dramatic tension	There is no vision of dramatic tension understood as attention or interest without cracks in the dramatic play Dramatic tension understood as the intensity of the present, regardless of the past or the future Timelessness or non-existence of time
4	Type of story Narrative and Post-drama	Narrative-dramatic organization linked to the narrative cosmovision It is in the interest of the "author" to tell a story (not a plot) within a conceptual space proper to the cultural cosmovision
5	Pariverse Technical-formal organization	Repetitive ties Parity and cosmic bonding Formal organizations do not pursue the same aims as in the West: the thematic and unitary focus of the play

Source: Orosa

The basic idea of the Latin American indigenous post-drama revolves around the concept of Pariverse or Duoverse (as opposed to the western idea of Universe, system, unity, composition). And from this notion (Duoverse) arise two essential concepts in relation to the aforementioned indigenous post-drama, namely: dual organization (bicephalic system) and complementary pairs (which affects the idea of conflict and post-dramatic organization).

This is clearly pointed out by Lajo [30] when he observes that "para el hombre andino todo objeto real o conceptual tiene su par, siendo así que el paradigma principal del hombre andino es que 'todo' o todos hemos sido paridos, es decir, que el origen cosmogónico primigenio NO es la unidad como en occidente, sino la 'paridad'"[1].

From this principle, and as we have already pointed out, two fundamental columns arise in relation to the Latin American indigenous post drama: dual organization and complementary pairs (which affects the idea of conflict and post-dramatic organization). Concerning dual organization principle, Oviedo [27] makes known, as well as Lajo, that "no existe algo impar porque cada ser existe por la proporcionalidad, compensación y equilibrio con su par complementario (paridad)"[2]. This has its reflection and impact on the architectural organizations of the post-dramatic plays of the Andean indigenism. These tend to be planned and structured around two worlds, two guiding principles instead of one as in the western world. This is apparent in Patricio Vallejo Aristizábal's La flor de la Chukirawa (The Flower of Chukirawa), both in its text and in its staging.

One of the guiding or organizing principles of this play would be the world of the Mother (the Ecuadorian Andean peasant), which generates both thematically and structurally speaking a clear post-dramatic block. The other (block) would correspond to the world that is not within the reach of our closest senses, the one that we cannot easily touch: the Angel of the dead son (Emigrant) and the Angel messenger (Television reporter). These two great masses organize the play as founding principles of creative nature.

This organization is completely surprising for a western spectator, transgressor, and generates new aesthetic architectures worthy of study and special attention.

With regard to the law of complementary pairs, Oviedo [27] points out that "la vida se reproduce... por la existencia y participación de dos fuerzas (madre-padre), por lo que no intenta anular o eliminar a una de ellas, al contrario, incentiva y fomenta la oposición..., que es lo que dinamiza y embellece la vida... La vida no se prolongaría si no coexistiera la contradicción de fuerzas, las cuales no están una contra la otra sino que se interrelacionan en forma armónica y equilibrada. Cada mitad sostiene a la otra y se intervinculan en forma complementaria..."[3].

---

[1] "...for the Andean man every real or conceptual object has its pair, being so that the main paradigm of the Andean man is that 'all' or all of us have been born (Lajo plays with the terms 'pair' and 'give birth' = par-paridad-parir), that is to say, that the original cosmogonic origin is NOT the unity as in the West, but the 'parity'."

[2] "...something odd does not exist because each being exists by proportionality, compensation and balance with its complementary pair (parity)".

[3] "...life reproduces itself... by the existence and participation of two forces (mother-father), so it does not try to annul or eliminate one of them, on the contrary, stimulates and encourages opposition ..., which is what energizes and beautifies life ... Life would not be prolonged if the contradiction of forces did not coexist, which are not one against the other but are interrelated in a harmonious and balanced way. Each half supports the other and intervenes in a complementary way..."

As we said above, within western world the term "confrontation" has an exclusionary sense of opposition. We are talking about two or more forces that necessarily end up destroying each other.

This is not how things happen in the indigenous world, where the conflict does not need the destruction or annulment of the other party. The effect of this philosophy is that we eliminate the chance of a possible outcome or denouement in the play. In the western world the ending is the result of a contradictory plot where one of the parties beats the other, where a protagonist overcomes all the difficulties (conflict) to reach his/her objectives. Within the indigenous world, as we face a complementary opposition, the forces at play are not eliminated and, therefore, there will be no solution (denouement), but simply a vital continuation.

This can be observed in La Flor de la Chukirawa, where the conflict of the Mother with herself (because of the death of her son) or those which she sustains with the society that surrounds her do not evolve anywhere. The same happens with the confrontation that this Mother maintains with the Angel messenger (television reporter), it does not change, it does not evolve or become anything other than a mere exposition of a conflictive and tragic situation. This is typical of the indigenous post-drama, so distant (and therefore so interesting) from Western dramatic culture.

We will make now some comments on different matters. First of all, cosmic parity and repetitive ties: as Lajo points out [30] "para el hombre andino lo masculino-femenino es la manera de hacerse en el ser humano la PARIDAD CÓSMICA que es el paradigma de su pensamiento y la clave de la VINCULARIDAD que es su relación obligatoria entre ellos y con el cosmos"[4]. That is to say, we are talking about obligatory relations between the pairs, and between the pairs and the superior (architectural) organization. These pairs can be, besides the already mentioned, major-minor, up-down, active-passive, boss-subordinate, brother-sister, parent-children... These kinds of pairs and bonds are often found in the formal-aesthetic and the storyline organization of La flor de la Chukirawa, namely: mother-child, sensitive world-suprasensible world, nature and man...

Refusal of homocentrism, harmonies and structures proper to the whole of Nature, "Greek"(Native)-Andean heavy liturgies: in this sense we have to point out following Lajo and Oviedo [27, 30] how the human being (unlike Western dramatic culture) is not the center of controversy and construction of the play, but that Nature has a very important contribution to make in the whole of the post-theatrical show. Concerning "Greek"(Native)-Andean liturgies it should be noted that these are quite heavy and tragic, their atmosphere is strong and unbreathable, reminiscent in a sense of the environments of the Greek tragedy.

Dramatic tension understood as the intensity of the present (regardless of the past or the future), timelessness: particularly it could be said that living (not only studying from an intellectual point of view) within an indigenous sphere of influence could give us the sensation of the non-existence of time or at least an experience of a circular sense

---

[4] "For the Andean man the masculine-feminine is the way to materialize in the human being the COSMIC PARITY that is the paradigm of his thought and the key of the VINCULARITY that is his obligatory relation between them and with the cosmos".

of time, closely linked to the times of Nature. The fundamental difference with Greek time, which also uses a present time, is that in the indigenous world the future and the past do not exist at all in the experience and, furthermore, the feeling of "tension" understood as attention or interest in what is happening is not attended to.

Telling a story (not a plot) happens within a conceptual space proper to the cultural cosmovision: within the indigenous space there is no plot, because the pairs are complementary and not excluding (there is no contradiction or confrontation); there is a story that enjoys an interest, in our opinion, similar to the mythical and is told within the indigenous patterns to which we have been making reference.

4 Conclusions

The Latin American indigenous post-drama belongs to a unique, native, autochthonous civilization (totally different from that of the western world itself) which has some characteristics and patterns that we have been stated and commented on in previous sections (especially its dual sense of organization and its innovative concept of dramatic confrontation).

Finding a post-dramatic culture with these characteristics and so ancient is, in our opinion, a tremendous discovery for the world of contemporary theatre, especially bearing in mind that even though it still exists in the real world (at least their remains or influences), nobody has analyzed (as far as we know) anything on this matter.

Sometimes we believe that the Greek theatre or the European medieval theatre (and the latter, timidly) are the fundamental manifestations of the world theatre and, suddenly, we encounter this type of enriching findings for the West itself and for the whole of Humanity.

References

1. Aristóteles: Poética, ed. trilingüe de Valentín García Yebra, Gredos, Madrid (1974)
2. Blacker, I.R.: Guía del escritor de cine y televisión. Eunsa, Navarra (1993)
3. Cabal, F.: Historia y evolución de la estructura dramática. Primer Acto **215**, 92–99 (1986)
4. Chatman, S.: Story and Discourse. Narrative Structure in Fiction and Film. Cornell University Press, New York (1986)
5. Field, S.: El libro del guion. Plot, Madrid (1994)
6. Field, S.: El manual del guionista. Plot, Madrid (2001)
7. Orosa, M.A., Lòpez-Lòpez, P.C.: Postdrama culture in Ecuador and Spain: methodological framework and comparative study. Comunicar **57**(XXVI), 39–47 (2018)
8. Orosa, M.A.: El cambio dramático en el modelo teleserial norteamericano. Publicia, Saarbrücken (2012)
9. Orosa, M.Á., López-Golán, M., Márquez-Domínguez, C., Ramos-Gil, Y.T.: El posdrama teleserial norteamericano: poética y composición/(Cómo entender el guion de las mejores series escritas para la televisión en los Estados Unidos)/USA TV series postdrama: poetics and composition. Revista Latina de Comunicación Social (72), 500 (2017). http://www.revistalatinacs.org/072paper/1176/26es.html. https://doi.org/10.4185/rlcs-2017-1176

10. Pérez, E.: Aproximación al espacio formal. Un ensayo de epistemología estética, tesis doctoral, Universidad de Valladolid, Valladolid (1995)
11. Alcántara, J.R.: Del teatro posdramático al drama posteatral. Investigación teatral **4**, 7–8 (2015)
12. Carlson, M.: Postdramatic Theatre and Postdramatic Performance. Revista Brasileira de Estudos da Presença **5**, 3 (2015)
13. Cornago, Ó.: Teatro postdramático: Las resistencias de la representación. Artea. Investigación y creación escénica [en línea, sin paginación]. www.arte-a.org. Accessed 15 Nove 2014
14. López, J.G.: La escena del siglo XXI. Asociacion de Directores de Escena, Madrid (2016)
15. López, J.G.: Tendencias en el teatro europeo Actual. Teatrología, 184–201 (2010)
16. Sarrazac, J.P.: El drama no será representado. Teatrología, 92–103 (2010)
17. Orosa, M.A., Galarza-Ligña, V., Culqui, A.: Postdrama communication in Spanish language: Ecuador and Spain. Origins and current scene. The indigenous postdrama. (La flor de la Chukirawa, by Patricio Vallejo Aristizábal, and Gólgota Picnic, by Rodrigo García.). RISTI, N.º E20, 05/2019, pp. 350–363 (2019)
18. Orosa M.Á., Romero-Ortega A.: Ecuador, the Non-communication: Postdrama or Performance?. In: Rocha, Á., Guarda, T. (eds.) Proceedings of the International CONFERENCE ON INFORMATION TECHNOLOGY & SYSTEMS (ICITS 2018), vol. 721. Springer, Cham (2018)
19. Vallejo, P.: La flor de la Chukirawa. In: Antología de teatro ecuatoriano contemporáneo, pp. 71–88 (2007)
20. Vallejo, P.: La flor de la Chukirawa 1ª parte (2007). https://www.youtube.com/watch?v=ftzNL4mXtkA
21. Vallejo, P.: La flor de la Chukirawa 2ª parte (2007). https://www.youtube.com/watch?v=ZPY6ZuhbQ10
22. Francés, C.: La fanesca. Teatro Malayerba, Quito (2009)
23. Gallegos, C.: Barrio Caleidoscopio. Fundación Teatro Nacional Sucre, Quito (2010)
24. García, R.: Gólgota picnic. Centro Dramático Nacional, Madrid (2011)
25. De la Torre, L.M., Peralta, C.S.: La reciprocidad en el mundo andino: el caso del pueblo de Otavalo. Runapura makipurarinamanta, otavalokunapak kawsaymanta. Editorial Abya Yala, Quito (2004)
26. Estermann, J., Peña, A.: Filosofía andina. IECTA-CIDSA, La Paz (1997)
27. Oviedo, A.: Buen Vivir vs. Sumak Kawsay: reforma capitalista y revolución alternativa. Ediciones CICCUS, Buenos Aires (2013)
28. Trentini, G.: Identidad cultural contemporánea de los otavalos: tradición, modernidad y cambios sociales. Sarance **29**, 42–57 (2013)
29. Trentini, G.: Identidad cultural contemporánea de los otavalos: tradición, modernidad y cambios sociales. Parte segunda: análisis de los datos iy comentarios conclusivos. Sarance **30**, 38–58 (2013)
30. Lajo, J., Ñan, K.: la ruta inka de sabiduría. Amaro Runa Ediciones, Lima (2005)

Assessment of the Transparency of Spanish Local Public Administrations: Methodology and Results

Pedro Molina Rodríguez-Navas[1]([⊠]) [iD]
and Vanessa Rodríguez Breijo[2] [iD]

[1] Universidad Autónoma de Barcelona, Cerdanyola del Vallès,
08193 Barcelona, Spain
pedro.molina@uab.cat
[2] Universidad de La Laguna, 38200 San Cristobal de La Laguna,
Santa Cruz de Tenerife, Spain
vrbreijo@ull.edu.es

Abstract. The availability of information about public management is a key factor in the democratic participation of citizens, as it enables public management to be assessed and makes contributions to the decision-making processes regarding public affairs. In this article, we present some results of the Infoparticipa Project, whose objective was to determine to what extent local public institutions meet the minimal requirements of transparency and whether there is a relationship between the behaviour of municipal governments and the number of inhabitants, the governing political party and the mayor's gender. In order to do this, in a first phase, the information published by the websites of the local public administrations was analysed through 41 indicators. In a second phase, the geolocalized outcomes were published and disseminated through the media and a consultancy procedure was offered to the administrations' politicians and policymakers. The results of the evaluation of municipalities with more than 20,000 inhabitants in six Spanish Autonomous Communities indicate that the information published by the councils is still very scarce, especially in those with fewer inhabitants. No clear relationships could be established between the political party governing in each municipality or the mayor's gender and the level of transparency on their websites. However, we have been able to confirm that the full application of the Project, including the consultancy phase, has led to an improvement in the information published by the municipalities in which it was applied.

Keywords: Transparency · Public communication · Local governments · E-government · Democracy · Citizenship

1 Transparency and Government

The increasing demand for transparency does not only affect the governmental sector. Consumers are demanding greater transparency when obtaining mortgages, loans and carrying out financial transactions and company stakeholders want to know in greater detail the internal functioning and the decisions taken at the heart of corporations.

© Springer Nature Switzerland AG 2020
Á. Rocha et al. (Eds.): ICITS 2020, AISC 1137, pp. 703–715, 2020.
https://doi.org/10.1007/978-3-030-40690-5_67

Internationally, pro-transparency activists are emerging and even the creators of open code software are positioning themselves in favour of sharing information and against the defence of its ownership [1]. In the context of governments, transparency is considered to be a democratic value that should be present so that governments are reliable, effective and responsible [2]. Thus, parliaments and municipal administrations are encouraged to be more transparent, despite their public audiences, public access to records and written declarations.

Transparency is developed through two different dimensions: as a facilitating element of democracy that regulates inadequate behaviours by enabling the reduction of corruption, bribery and other forms of government misbehaviours, and as a vehicle for the monitoring of governors by their governed individuals, which is an essential aspect of democracy, as it makes public control possible. From this perspective, the availability of information makes it easier to make decisions in democracy, which is a relevant component of governability [1].

Some authors maintain that further research is needed regarding the relationship between the process of citizen participation and its impact on the building of public trust in government and its legitimacy [3]. Nevertheless, in the study carried out by Kim and Lee [4], a positive association was verified between satisfaction of citizens with participation applications and the capacity they had to assess government transparency. Likewise, it was demonstrated that citizens were more satisfied when they perceived they had the capacity to influence directly the decision-making process. Other studies [4–7] have shown that when governments make efforts to interact with citizens and allow the evaluation of public management and participation in decision-making, they develop a greater trust towards the government offering them these opportunities. Understood from this perspective, transparency is not only openness and the availability of information, but also includes operative processes and the way they are organized [8]. Transparency exists to the extent that an organization provides complete information about all its attributes freely and universally, and also maintains timely communication directly with all key public audiences [9].

For Schauer [1], the most important virtue of transparency is accountability due to the power it grants to citizens. When talking about transparency, there is reference to citizens controlling government and public powers as the basis of representative democracy [10]. When citizens have access to information about governmental management, they can evaluate this management and try to prevent those politicians who did not perform adequately from having access to power and, on the contrary, vote for those who have done their job correctly [11]. However, without governmental transparency, it is difficult to assign responsibilities to elected and appointed official through their actions. Therefore, publication of information promotes democratic accountability [12]. In this sense, use of ICTs is one of the most efficient ways of spreading information about public management, communicating with citizens and creating a more transparent government in the digital era [8, 13]. E-government and institutional websites facilitate the dissemination of public information to citizens, who can access them without visiting public entity offices and without any spatial or time limitations. In addition, technology considerably reduces the costs of collecting, distributing and accessing government information [14].

Citizen participation in the decisions of public administrations also takes advantage of Internet-based applications, which facilitate the development of community, bidirectional communication with governors and the offering of online services [4]. This possibility of electronic participation, which allows voters to directly intervene in public decisions, has opened "unprecedented avenues in the field of immediate democracy" [15]. The dissemination of financial information through the Internet enables public administrators to be publicly held accountable and to promote dialogue regarding the use of public financial resources [16]. E-government is a tool to make the relationships between governors and citizens closer and more open and to involve individuals in public interest affairs, to have access to services and to interact [17].

Hence, the official websites of public administrations have become the corporative image, the first and most significant, of these administrations before the citizens. They are a powerful instrument to justify and maintain legitimacy and an important technical support on which to make concepts such as responsibility and accountability operational [18]. However, on many occasions government structure, bureaucracy and partisanship interfere in the implementation of e-government [19]. Governors often use digital media as a bulletin board or as party political noticeboard [20]. Governments are often reluctant to provide information because the disclosure of this information entails transferring power in the favour of citizens [8].

In order to elucidate the indicators and the degree of transparency of governments, and to determine whether these issues are directly related with specific features such as the size of municipalities, political affiliation of the governors or the institutional structure itself, different studies have been carried out. One of these includes the study of West [21], which focused on recognizing particular traits with regards to transparency presented by the different classes of government. West did not find significant differences in the executive, legislative or judicial branches, but he did find that urban and rural local governments do not offer the same access to information on their websites, nor do they have the same resources available to provide online information.

The research of Armstrong [19] focused on the websites of counties and school committees in the State of Florida, which has one of the most open laws of public records of the United States. The results demonstrated that the objectives of a local government website play an important role in determining the availability of public records within that community. The smallest communities, which are more focused on guaranteeing a greater access to information, were found to be more transparent than large communities with different priorities. Websites of school committees in Florida provide more public information than counties. This is partly due to the fact that they work in a more autonomous manner before State and Federal governments than counties, which depend much more on changes in State and Federal laws and suffer greater inequalities in financing. Finally, the results indicated that counties with a higher proportion of Republicans had a greater level of transparency.

Sanders, Crespo and Holtz-Bacha [22] evaluated the transparency of governments by analysing the degree of professionalization of the communication of central governments in Germany, Spain and the United Kingdom. The results of this research show that, although formal rules have been drawn up that distinguish between partisan political communication and the management of public information, and that mechanisms to enhance the communication process are being implemented, practices in the three countries analysed are still far from systematic. Foremost in the case of Spain is the lack of professionalization of official spokespersons along with the presence of press offices in ministerial departments that are almost always headed by journalists with a questionable degree of independence. The coordination and participation of public officials in the strategic planning of communication continues to be one of the main challenges. Likewise, improvisation and the lack of evaluation of public perception are also problems to be solved. Other studies confirmed the endemic nature of these problems, since in Spain, from the start of democracy at the end of the 1970s, the use of public communication to convey partisan or personalistic messages to the benefit of individuals in the governing administration has been usual [23].

The work of Rivero, Mora and Flores [18] focused on evaluating the websites of large Spanish local governments with more than 300,000 inhabitants in order to determine the amount of information related to accountability offered by local municipal entities. It showed that 86% of them provided budgetary and financial information but that it was not enough, since although they published the annual budget, very few detailed their chapters or informed about their efficient settlement. Similarly, although the purposes of the entity were made explicit through government plans or municipal action plans, information related to the monitoring of the compliance of those objectives was not published. Moreover, although information regarding the organizational structure of the local governments was present in all the entities and that 95% of them published the minutes of the plenary sessions, the salaries of government positions were only found in 7 of the 22 local governments analysed.

Likewise, a project conducted by the University of Vigo measured budgetary transparency in Galician municipalities using a questionnaire of fifteen questions based on the second, third and fourth pillar of the revised Code of Good Practices on Fiscal Transparency of the International Monetary Fund (IMF) dated 2007. Officials in charge of transactions and/or the pre-audit accounting of 40 municipalities answered the questionnaire and the results showed that political polarization increased transparency, although they also showed that a more polarized system of government could hinder a coherent reform policy and become an obstacle for transparency. It was also demonstrated that governments are more likely to increase transparency when they inherit a high level of debt [24].

2 Infoparticipa Project

2.1 Objectives

The current research is part of the project Infoparticipa (www.mapainfoparticipa.com), which started before the approval of the Law of Transparency, which has been in effect in Spain since December 2015.

The purpose of the project was to get public administrations to improve their quality of information and transparency, as well as:

- To demonstrate the relevance of the Internet as a facilitating tool for participation and for monitoring and assessing the actions of political leaders and public policies by citizens.
- To determine to what extent the public institutions meet the minimal requirements of transparency and whether there are relationships between the behaviour of municipal governments and some of its basic characteristics such as the number of inhabitants and the gender of mayor.

This was carried out through the analysis of information published by public local administrations on their websites through the study of 41 indicators and the posterior application of a communication and consultancy procedure for political leaders and policymakers of administrations.

2.2 Methodology

In a first phase, the information that local governments published on their websites was analysed. For the analyses, a relationship of 41 indicators were defined and divided in four groups of issues [23].

The evaluation process was conceived as a civic audit. Evaluators searched for information on the website as any other citizen would do and answer to the question affirmatively or negatively without intermediate responses. The results obtained in each analysis, both the compliance percentage as a whole as well as the indicator-by-indicator results, were published on the website www.mapainfoparticipa.com. In order for any person to compare the results with their own findings, the same page included the guideline used by the evaluators to clarify the indicators and the application criteria. In this way, both the political leaders and policymakers and any other citizen could know how it was assessed and manifest their disagreement in any way.

These results were also presented in a geolocalized manner, that is, on a map where populations could be selected individually or by groups of municipalities, applying criteria such as territorial areas (district, region, etc.), number of inhabitants, political party of the mayor or mayoress, and others. These functions allowed the evaluation of the results by comparing them with nearby localities or localities of similar characteristics or using any other criteria that could be of interest for the visitor of the website.

Following the analysis and publication on the project's platform, the municipality leaders were informed of the availability of this evaluation on the website in order that they could review it, have the opportunity to improve or to dispute with the assessment team any incidence or discrepancy regarding the results presented and that could be publicly consulted on the digital platform of the Infoparticipa Map. Reports of the results were then also published and sent to the media for publication. This communication strategy ensured that local governments became interested in the procedure and used the indicators as a guideline to the good practices they needed to comply with.

When this was the case, the project team responded to the requests, advising on compliance with each indicator. Similarly, they received news regarding any improvements made and, once their suitability had been confirmed, the project team modified the results of the first assessment and the information published on the platform.

The procedure was first applied in the Autonomous Community of Catalonia in a first wave between 2012 and 2013. Also in Catalonia two more waves took place, in 2013 and in 2014. From the second wave onwards, the Infoparticipa Seal was created as an award to those local governments that best met the criteria. In the Autonomous Community of Aragon, two assessment waves were carried out and in the last one (2014) Seals were also awarded. In contrast, just a single assessment wave was carried out in the communities of Andalusia, Canary Islands, Galicia and Madrid.

2.3   Sample

The sample was comprised of the websites of 34 municipalities of more than 20,000 inhabitants from the 6 Spanish Autonomous Communities. The distribution by Autonomous Communities is as follows: Andalusia, 83 municipalities; Aragon, 4; Canary Islands, 27; Catalonia, 64; Galicia, 22; and Madrid, 34.

All the results correspond to the ones obtained on April 7, 2015 but, as mentioned before, some Autonomous Communities were assessed in one occasion and others in two or three occasions: Catalonia, 3 assessment waves; Aragon, 2 waves; and Andalusia, Canary Islands, Galicia and Madrid, 1 wave.[1]

The four Communities where one assessment wave was already carried out totalled 166 municipalities of more than 20,000 inhabitants, 65 of them with more than 50,000 inhabitants and 29 of more than 100,000 inhabitants.

The two communities where the two or three waves were conducted have 68 municipalities of more than 20,000 inhabitants (although only 4 of these belong to Aragon and the rest to Catalonia). Of these, 25 have more than 50,000 inhabitants and 11, more than 100,000 inhabitants.

---

[1] Data of the first wave re-elaborated from the reports with the results of Aragon of November 7, 2013, "Report about information published on websites of local governments of the municipalities of Aragon with more than 10,000 inhabitants", and Catalonia, of October 25, 2013, "Report about information published on the websites of local governments of municipalities of Catalonia with more than 50,000 inhabitants". Data of the last wave in all Communities, April 7, 2015.

3   Results

3.1   Results of the Assessment

Table 1 presents the results of the analysis of all the municipalities in the sample. From the total studied (234), 54.7% exceeded 50% of compliance. With regard to the 90 municipalities with more than 50,000 inhabitants, compliance was exceeded by 78.88%, while this figure was 87.5% for the 40 municipalities with more than 100,000 inhabitants. It can also be seen that no municipality of more than 100,000 inhabitants obtained lower than 30%.

Table 1. Results of the assessments in the 6 Autonomous Communities. Source: Authors

Percentage compliance	Municipalities with more than 20,000 inhabitants (234)		Municipalities with more than 50,000 inhabitants (90)		Municipalities with more than 100,000 inhabitants (40)	
	Cases	%	Cases	%	Cases	%
100	19	8.12	13	14.44	7	17.50
Between 90 and 99.99	19	8.12	11	12.22	6	15.00
Between 80 and 89.99	10	4.27	3	3.33	0	0
Between 70 and 79.99	22	9.40	14	15.56	8	20.00
Between 60 and 69.99	28	11.97	17	18.89	9	22.50
Between 50 and 59.99	30	12.82	13	14.44	5	12.50
Subtotal >50%	128	54.7	57	78.88	35	87.5
Between 40 and 49.99	36	15.38	6	6.67	2	5.00
Between 30 and 39.99	37	15.81	9	10.00	3	7.50
Between 20 and 29.99	27	11.54	3	3.33	0	0
Between 10 and 10.99	4	1.71	0	0.00	0	0
Between 0 and 9.99	2	0.85	1	1.11	0	0
Subtotal <50%	106	45.29	19	21.11	5	12.5

Therefore, a clear direct relationship can be seen between the number of inhabitants of a municipality and the percentage compliance; therefore, the greater the number of inhabitants, the higher the percentage compliance.

However, although the percentage of municipalities achieving 100% compliance was greater in the municipalities with more than 100,000 inhabitants compared to the sample total, it was very low in any of the population bands (8.12%–14.44%–17.5%). The same occurs for municipalities surpassing 90%, although the percentage is higher (16.24%–26.66%–32.5%) and the percentage of those achieving between 80 and 89.99% compliance is always extremely low (4.27%–3.33%–0.00%). Lastly, the greatest percentage of cases in the three bands was seen among those obtaining between 50 and 79.99% compliance: 34.19%–48.89%–55.00%. Despite these results, in the two only Spanish cities with more than a million inhabitants, Madrid (3,165,235 inhabitants), the capital, achieved 90.24% compliance, while Barcelona (1,611,822 inhabitants) achieved 100%.

These results reveal that the level of compliance is rather low considering any studied parameter: very few municipalities reached 100% or surpassed 90% in all bands. The highest percentage of municipalities obtained between 50–79% compliance, which is clearly not enough. The percentage of municipalities that did not reach 50% was very high among municipalities of more than 20,000 inhabitants and, although lower, it is intolerable in municipalities that manage the lives of more than 50,000 or more than 100,000 individuals. Finally, it is worth mentioning that Madrid, which should be a model, did not reach 100%.

Table 2. Results of the assessment: Autonomous Communities in which two or more assessment waves were conducted (Aragon and Catalonia: Ar/Cat) and Autonomous Communities in which a single assessment was conducted (Andalusia, Canary Islands, Galicia and Madrid: ACGM). Source: Authors.

Percentage compliance	Municipalities with more than 20,000 inhabitants (234)				Municipalities with more than 50,000 inhabitants (90)				Municipalities with more than 100,000 inhabitants (40)			
	Ar/Cat	%	ACGM	%	Ar/Cat	%	ACGM	%	Ar/Cat	%	ACGM	%
100	18	26.5	1	0.6	12	48.0	1	1.5	6	54.5	1	3.4
90-99.99	18	26.5	1	0.6	10	40.0	1	1.5	5	45.5	1	3.4
80-89.99	7	10.3	3	1.8	1	4.0	2	3.1	0	0.0	0	0.0
70-79.99	7	10.3	15	9.0	1	4.0	13	20.0	0	0.0	8	27.6
60-69.99	4	5.9	24	14.5	0	0.0	17	26.2	0	0.0	9	31.0
50-59.99	8	11.8	22	13.3	1	4.0	12	18.5	0	0.0	5	17.2
Subtotal >50%	62	91.2	66	39.8	25	100	46	70.8	11	100	24	82.8
40-49.99	4	5.9	32	19.3	0	0.0	6	9.2	0	0.0	2	6.9
30-39.99	2	2.9	35	21.1	0	0.0	9	13.8	0	0.0	3	10.3
20-29.99	0	0.0	27	16.3	0	0.0	3	4.6	0	0.0	0	0.0
10-10.99	0	0.0	4	2.4	0	0.0	0	0.0	0	0.0	0	0.0
0-9.99	0	0.0	2	1.2	0	0.0	1	1.5	0	0.0	0	0.0
Subtotal <50%	6	8.8	100	60.2	0	0.0	19	29.2	0	0.0	5	17.2

Table 2 shows the results, differentiating the Autonomous Communities where two (Aragon) or three (Catalonia) waves of assessments were carried out from the Communities where a single wave of assessment was conducted.

It can be seen that more than 90% of the municipalities assessed exceeded more than once half of indicators and that 53% exceeded a 90% positive evaluation. In addition, none of these municipalities achieved an evaluation lower than 30% and only 8.8% obtained less than 50%.

Continuing with the municipalities that were assessed more than once, all with more than 50,000 inhabitants surpassed a 50% positive evaluation, while 88% exceeded 90%. If we reduce the band to those with more than 100,000 inhabitants, all achieved a result superior to 90% and 54.5% reached 100%.

In contrast, municipalities assessed once presented practically inverse results: 60.2% did not reach a 50% positive evaluation and only 3.0% exceeded 80%. When we only consider municipalities of more than 50,000 inhabitants, their results improved, with 70.8% exceeding a 50% positive evaluation, while only 6.1% exceeded 80%. Only 6.8% of municipalities of more than 100,000 inhabitants exceeded 80%, although in this band, 82% achieved a result greater than 50%.

Therefore, it is clear that municipalities assessed more than once obtained results that were significantly superior to those obtained by municipalities that were assessed just once, which indicates that the procedure has been useful. Not only did they obtain better results as a whole, but it was also observed that a high percentage of municipalities with more than 50,000 or 100,000 inhabitants obtained a level of compliance of 100% or very close to it when they were assessed more than once, while very few of those that were assessed just once achieved these compliance figures.

These data gain a greater sense if we compare results of the first assessment of municipalities in Catalonia and Aragon with the later ones in order to know the starting situation and the subsequent evolution.

Table 3. Results of the assessments of the first and last wave in municipalities with more than 50,000 inhabitants in the Autonomous Communities in which two or more waves of assessments were conducted (Aragon and Catalonia: Ar/Cat)[2] and of the Autonomous Communities in which only a single wave was conducted (Andalusia, Canary Islands, Galicia and Madrid: ACGM). Source: Authors

Percentage compliance	Municipalities with more than 50,000 inhabitants. Aragon and Catalonia (25), first wave		Municipalities with more than 50,000 inhabitants. Aragon and Catalonia (25), final wave		Municipalities with more than 50,000 inhabitants. ACGM (65), single wave.	
	Ar/Cat	%	Ar/Cat	%	ACGM	%
100	1	4.0	12	48.0	1	1.5
Between 90 and 99.99	2	8.0	10	40.0	1	1.5
Between 80 and 89.99	4	16.0	1	4.0	2	3.1
Between 70 and 79.99	6	24.0	1	4.0	13	20.0
Between 60 and 69.99	9	36.0	0	0.0	17	26.2
Between 50 and 59.99	2	8.0	1	4.0	12	18.5
Subtotal >50%	25	96.00	25	100.0	46	70.8
Between 40 and 49.99	1	4.0	0	0.0	6	9.2
Between 30 and 39.99	0	0.0	0	0.0	9	13.8
Between 20 and 29.99	0	0.0	0	0.0	3	4.6
Between 10 and 10.99	0	0.0	0	0.0	0	0.0
Between 0 and 9.99	0	0.0	0	0.0	1	1.5
Subtotal <50%	1	4.0	0	0.0	19	29.2

In Table 3, we observe that only 12% of municipalities in Aragon and Catalonia with more than 50,000 inhabitants that were included in the first wave, performed in

2013, exceeded 90% of results, while after the last wave, this figure rose to 88% municipalities. Meanwhile, in the case of municipalities where just a single wave was conducted, only 3% exceeded 90% in the assessment. Therefore, although municipalities of Aragon and Catalonia obtained better results than the results of communities analysed in the first wave, the percentage of those obtaining satisfactory results, that is, over 90%, was also very low.

On the other hand, in the first assessment in Aragon and Catalonia, only 4% of municipalities did not reach 50%, while in the last wave, none achieved below 50%. Meanwhile, in the municipalities where just a single wave was carried out, 29% did not reach 50% of the evaluation and 19.5% of municipalities did not reach 40% of the evaluation. It was also observed that in the first evaluation in Aragon and Catalonia no municipality achieved below 40%.

Hence, we conclude that although the starting situation was somewhat better in the Communities of Aragon and Catalonia, the analysis confirmed that the procedure was effective, since in these Communities the results improved substantially and, especially, 88% of the municipalities exceeded 90% on assessment, whereas in the first wave only 12% exceeded this.

3.2   Results of the Assessment According to the Mayor's Gender

In Table 4 we can see that there are no very significant differences in relation to whether the government was led by a mayor or mayoress. In the case of Communities where more than one wave was conducted, although in the two bands by number of inhabitants slightly better results were obtained when a man governed, the difference in the number of municipalities governed by a mayoress or a mayor was so significant that a different proportion could modify these results.

Table 4. Results of the assessment according to the gender of the mayor in the Communities in which more than one wave (Aragon and Catalonia) or a single wave (Andalusia, Canary Islands, Galicia, Madrid: ACGM) was conducted. Source: Authors.

Mayor/ mayoress	Inhabitants	Aragon-Catalonia		ACGM	
		N° of municipalities	% evaluation	N° of municipalities	% evaluation
Mayor	20,000-100,000	45	79.3	113	44.0
	More than 100,000	9	98.1	23	62.0
Mayoress	20,000-100,000	12	77.8	24	39.1
	More than 100,000	2	97.6	6	68.3

In the case of local governments assessed only once, the municipalities governed by mayors and with less than 100,000 inhabitants obtained better results. On the other hand, municipalities larger than this obtained better results when they were governed by a mayoress, with a similar difference percentage. As in the previous case, the difference in the number of municipalities could be determinant in the results.

We therefore conclude with the following considerations:

- The large difference in the number of municipalities governed by a mayor (190) or a mayoress (44) does not allow the results to be analysed in a conclusive manner.
- The differences observed are very small and a significant variation in the percentage of municipalities governed by a mayoress could produce noticeable variations.
- Considering the results obtained, the man-woman condition in government is not a significant variable for the incorporation of good practices on transparency in local governments.

4 Conclusions

The analysis performed on the 6 Autonomous Communities studied gives us a panoramic view of the situation of local governments in Spain prior to the Transparency Law coming into effect and before new councils were constituted following the municipal elections in May, 2015. These results can be subsequently compared in order to appreciate the possible impact that the application of the Transparency Law may have and the change in behaviour of the new governments.

On the other hand, the results allows a balance to be obtained of the application of the procedure designed by the Infoparticipa Project team when comparing results in different Autonomous Communities where one or more waves were conducted or from different waves.

Firstly, a general lack of information was observed. It was also observed that, although larger municipalities offered more information, in the Communities where the Infoparticipa Project did not have any impact due to just one wave of assessments being conducted, the results are significantly inferior to the ones obtained in Communities where more than one wave was carried out. This indicates there was no clear will from the councils to apply transparency policies and also that they did not have clear criteria regarding how to develop them. The methodology of the project has been useful in offering these criteria and advice on how to apply them, without forgetting that the procedure of communication and granting of certified seals has encouraged compliance.

It is surprising that although political discourses call for transparency and that its application is constantly demanded, very little has been done to put it into practice. The indicators used in the Infoparticipa project refer to elemental practices and it has been confirmed that even in the Community municipalities where the methodology was applied, no general compliance occurred prior to the electoral process in May 2015 in which citizens could evaluate local governments, granting them a new mandate or preferring a change of political party. It is evident there has been a lack of will with regard to transparency and lack of criteria and examples to follow that would hinder the immediate application of the Law. The results also show that where the Infoparticipa methodology was fully applied, the situation improved and that political leaders and policymakers were better prepared.

The territorial analysis also shows how among the Communities in which only a single wave was conducted, there are some noticeable differences, highlighting the low results in Andalusia especially in and Galicia, although in the Canary Islands and Madrid, although better, they are still very low.

Lastly, we observed that there was no clear influence on the results if a mayor or mayoress governed. This work also reveals, in an indirect but clear way, the huge disproportion still existing in this area, which is unfavourable for women.

The Infoparticipa Project differentiates from others because it does not only intend to diagnose reality, but also to transform it, contributing to the improvement of municipality websites through a methodology of communication and consultancy. Therefore, its continuity is required to keep guiding the work of political leaders and policymakers of the public administrations.

The new platform Infoparticipa Map has been adapted to the demands of the Transparency Law in Spain and Catalonia in order to continue the work of improving the information and transparency of local governments in order to guarantee the rights of citizens to information and participation. This adaptation obliges an assessment with new and greater number of indicators, considering the obligations of law, which will complicate comparisons between the previous results and results obtained in the future. We cannot overlook the fact that the law must encourage public administrations to improve their transparency. Knowing which improvements are consequences of the application of the law and which are due to the influence of different aspects of the Infoparticipa Project will oblige the results to be analysed incorporating qualitative methodologies that have not been considered up until now.[2]

References

1. Schauer, F.: Transparency in three dimensions. Univ. Illinois Law Rev. **4**, 1339–1358 (2011)
2. Grimmelikhuijsen, S.G., Welch, E.W.: Developing and testing a theoretical framework for computer-mediated transparency of local governments. Public Adm. Rev. **72**(4), 562–571 (2012)
3. De Fine Licht, J.: Do we really want to know? The potentially negative effect of transparency in decision making on perceived legitimacy. Scand. Polit. Stud. **34**(3), 183–201 (2011)
4. Kim, S., Lee, J.: E-participation, transparency, and trust in local government. Public Adm. Rev. **72**(6), 819–828 (2012)
5. West, D.M.: E-government and the transformation of service delivery and citizen attitudes. Public Adm. Rev. **64**(1), 15–26 (2004)
6. Tolbert, C.J., Mossberger, K.: The effects of e-government on trust and confidence in government. Public Adm. Rev. **66**(3), 354–369 (2006)

---

[2] This article is the result of the research project *Methods and Models of Information for the Monitoring of the Actions of Local Government Policy-makers and Accountability* (CSO2015-64568-R), financed by the Secretary of State for Research, Development and Innovation, of the Ministry of Economy and Competitiveness of Spain and the European Regional Development Fund (ERDF), within the State Program for Research, Development and Innovation Oriented at the Challenges of Society. Principal investigators: Dr Amparo Moreno Sardà and Dr Núria Simelio Solà.

7. Kweit, M.G., Kweit, R.W.: Participation, perception of participation, and citizen support. Am. Polit. Res. **35**(3), 407–425 (2007)
8. Eom, S.J.: Improving governmental transparency in Korea: toward institutionalized and ICT-enabled transparency. Korean J. Policy Stud. **29**(1), 69–100 (2014)
9. La Porte, T.M., Demchak, C.C., de Jong, M.: Democracy and bureaucracy in the age of the Web. Adm. Soc. **34**(4), 411–446 (2002)
10. Larach, C.: Transparencia y buen gobierno en España. Comentarios a la Ley 19/2013, de 9 de diciembre, de Transparencia, Acceso a la Información Pública y Buen Gobierno. Revista digital de Derecho Administrativo **13**, 255–268 (2015)
11. Alt, J.E., Lassen, D.D.: Fiscal transparency, political parties, and debt in OECD countries. Eur. Econ. Rev. **50**(6), 1403–1439 (2006)
12. Piotrowski, S.J., Van Ryzin, G.: Citizen attitudes toward transparency in local government. Am. Rev. Public Adm. **37**(3), 306–323 (2007)
13. Bertot, J.C., Jaeger, P.T., Grimes, J.M.: Using ICTs to create a culture of transparency: e-government and social media as openness and anti-corruption tools for societies. Gov. Inf. Q. **27**(3), 264–271 (2010)
14. Roberts, A.: Blacked Out: Government Secrecy in the Information Age. Cambridge University Press, New York (2006)
15. García Costa, F.M.: Participación y democracia electrónicas en el Estado representativo. In: Cotino Hueso, L. (coord.) Democracia, participación y voto a través de las nuevas tecnologías. Comares, Granada, Spain, pp. 3–24 (2007)
16. Alcaide Muñoz, L., Rodríguez Bolívar, M.P., López Hernández, A.M.: Transparency in governments: a meta-analytic review of incentives for digital versus hard-copy public financial disclosures. Am. Rev. Public Adm. 1–34 (2016)
17. Pina, V., Torres, R., Royo, S.: Is e-government leading to more accountable and transparent local governments? An overall view. Financ. Account. Manag. Gov. Public Serv. Charities **26**(1), 3–20 (2010)
18. Rivero Menéndez, J.A., Mora Agudo, L., Flores Ureba, S.: Un estudio de la rendición de cuentas a través del e-gobierno en la administración local española. In: Mercado Idoeta, C. (coord.) Empresa global y mercados locales. Universidad Rey Juan Carlos, Madrid, Spain, pp. 1–16 (2007)
19. Armstrong, C.L.: Providing a clearer view: an examination of transparency on local government websites. Gov. Inf. Q. **28**(1), 11–16 (2011)
20. Del Rey Morató, J.: Comunicación política, Internet y campañas electorales. Tecnos, Madrid (2007)
21. West, D.M.: Fourth annual urban e-government study (2004). http://www.insidepolitics.org/PressRelease04city.html
22. Sanders, K., Canel Crespo, M.J., Holtz-Bacha, Ch.: Governments: a three-country comparison of how governments communicate with citizens. Int. J. Press/Polit. **16**(4), 523–547 (2011)
23. Molina Rodríguez-Navas, P., Simelio Solà, N., Corcoy Rius, M., Aguilar Pérez, A.: Mapa Infoparticipa: cartografía interactiva para la mejora de la calidad y la transparencia de la comunicación pública local. Ar@cne. Revista Electrónica de recursos en Internet sobre Geografía y Ciencias Sociales **202**, 1–31 (2015)
24. Caamaño Alegre, J., Lago Peñas, S., Reyes Santias, F., Santiago Boubeta, A.: Budget transparency in local governments: an empirical analysis. Local Gov. Stud. **39**(2), 1–30 (2011)

Brand Valuation Proposal Model for Family and Non-family Cold Meats Companies. Case Study: Sausages La Italiana and Plumrose

Danny C. Barbery-Montoya[1](✉) (iD), Ashley D. Franco-Lara[2](✉),
and Isaac D. Calle Vargas[2](✉)

[1] Universidad Espíritu Santo, Samborondón 092301, Ecuador
dbarbery@uees.edu.ec
[2] Escuela Superior Politécnica del Litoral, Guayaquil 090604, Ecuador
adfranc@espol.edu.ec, idcallev89@gmail.com

Abstract. The study evaluates the family business of cold meats La Italiana, and the non-family business Plumrose, taking into account the trajectory of both companies in the Ecuadorian market (La Italiana with 30 years and Plumrose from Pronaca, with 40 years) and being the objective, to analyze consumer perception and organizational perspective, in addition to discovering the attributes that generate value in each brand. The methodology used is a descriptive and analytical investigation that determines phenomena and realities of each company, using the inductive and relational method. The quantitative study was the CSI survey with questions about the perception of family and non-family business brands, applied to 384 housewives stratified by socioeconomic level. Likewise, a qualitative study was carried out on the commercial managers of each company through in-depth interviews to determine their relevant characteristics. The results indicate that the family business has as a relevant positive factor the satisfaction of the need of consumers, while in its negative part there is confusion in the recognition of its brands. In contrast, the non-familiar company presents brand recognition and importance of quality in its products. Finally, this study identifies brand elements that define the differences between the two companies, thus knowing every detail that the consumer considers important to achieve brand acceptance, recognition and loyalty.

Keywords: Brand attributes · Perception · Brand value

1 Introduction

Family businesses in Ecuador have great influence in the country's economy, but they are very vulnerable to market conditions and family problems. Currently, a study conducted by the Universidad Espíritu Santo (UEES), indicates that 90.5% of companies in the country and registered in the Superintendence of Companies have family structure, 7.9% are non-family owned and 1.7% of unknown property. On the other hand, in Ecuador there is a lot of diversity between regions and the business branch varies by region and province, since each one has different characteristics in the production of goods and services. Such is so, that these differences make the most important regions in Ecuador Costa and Sierra [1].

Á. Rocha et al. (Eds.): ICITS 2020, AISC 1137, pp. 716–724, 2020.
https://doi.org/10.1007/978-3-030-40690-5_68

On the other hand, brands are an intangible asset and their importance lies within companies because they reflect their personality and their competitive advantage [2]. This asset is reflected in a monetary and non-monetary value [3] that depends consumer's perception [4] which leads to constant measurements to determine whether the brand wins or not, reputation and positioning [5]. Therefore, family businesses must be measured not only in financial terms but with other metrics that serve to define their future [6], in which their history can be a way to seize opportunities [7] as well as the high emotional component they possess their brands [8].

Our questioning focuses on the differences that the consumer may perceive in the brands of products from family and non-family businesses; for this reason, this study seeks to perform the perceptual brand assessment of family and non-family businesses, by comparing two companies: "Embutidos La Italiana" and "Pronaca/Plumrose" to determine the relevant characteristics in each of them and identify their differentiating attributes.

1.1 Brand Value and the CSI Model

Today the brand is a determining and differential factor in any product or service in which the final consumer feels identified, provided that it meets their needs and also provides added value to users [9]. Similarly, brand value is addressed from two fundamental perspectives: the financial point of view where it is analyzed as an intangible asset of a company, and from the point of view of the consumer (perception), where it analyzes the product or service to decide the purchase [10]. On the other hand, being a notorious and reputable brand is a necessary requirement for the creation of brand value since, if consumers do not have a brand in mind they cannot associate the information with that brand [11]. However, the tangible attributes of the brand are basic elements to understand the construction of brand value in the mind of the consumer and influence the attitude towards the brand at any level of involvement [12]. Similarly, brand value is the value achieved by current customers and how it is perceived by potential consumers, where some variables such as the name, symbols and personality of the brand are located. These variables constitute some elements such as knowledge, affection and behavior towards the brand and it is a difficult advantage to imitate that leads to the transfer of value between the client and the brand owner [13].

Within the valuation models we have the Consumer Style Inventory (CSI) model [14], defined by its authors, as the consumer's decision-making style, that is, the mental orientation that characterizes a consumer's approach to make purchasing decisions. In other words, all consumers make their purchases with certain fundamental styles such as rational purchases, brand awareness, price, quality, among others. The CSI model defines eleven decision making dimensions [15]:

- **Perfectionist consumer:** High quality products, purchased in a methodical and orderly manner.
- **Brand consumer:** Brand decision and choice, either because it is expensive or known, considering that the price reflects quality.
- **Fashion consumer:** New and innovative products that generate satisfaction and excitement of new fashion.

- **Consumed for recreation and hedonistic:** Seeks pleasure in the act of shopping and is pleasant and entertaining.
- **Price Conscious Consumer:** They look for reduced prices.
- **Loyal Consumer:** Buy in the same store and the same brands.
- **Impulsive consumer:** Sudden and unplanned purchase.
- **Consumer Confused by Excess Options:** They are blocked in their purchase decision due to the excess of alternatives and not having clear criteria.
- **Conscientious consumer of the ecological:** Responsible consumption implies an ethical, ecological and social consumption.
- **Health conscious consumer:** The consumer is concerned about their health and is informed, through conscious consumption.
- **Purchase intention:** Preference for a brand or product over the rest of the competing options.

1.2   Family Business

Family businesses are very important for the development of a country because they predominant worldwide. Within them, the family system consists of three subsystems that allow us to understand how a family business is formed, called the "three-circle model" and which is made up of the family, property and business [16] where the family is the vision and control mechanism of the company, which contributes to the creation of unique resources and capabilities for property continuity [17].

Thus, when talking about family businesses, they are defined as their strengths and weaknesses, everything that has a significant influence on strategic decisions and the triumph achieved by them. Because of this, to understand the family business it is necessary to consider its history and permanence in time [16, 18], its growth and development [19, 20], professionalism [21], continuity and succession [22], equity [16], profitability and investments [23], socialization [24] and temporality [25, 26]; all these variables, as influential elements in your brand value.

2   Methodology

A descriptive, inductive and correlational research was applied for the development of the study. In an exploratory phase, qualitative data was taken through interviews with commercial managers of the family business La Italiana and the non-family business Plumrose considering the trajectory of both companies in the Ecuadorian market (30 and 40 years respectively), with questions developed with the different topics presented in the literary review to achieve a deeper and more relevant analysis. The issues addressed in the in-depth interviews are related to commercial and marketing decisions within each of the companies, to understand their work in the construction of their brands.

On the other hand, a quantitative study based on an infinite population was applied with a sample of 95% confidence and 5% error of 384 housewives of the city of Guayaquil selected by means of a weighted stratified sampling according to the socioeconomic levels high (A), medium-high (B), typical medium (C+), medium-low

(C−) and low (D). The study is based on the application of the CSI questionnaire [14] that additionally includes a dimension made up of nine questions of own elaboration and suggested by the interviewees of each company, in which the variables are measured: importance of origin, quality, added value, manufacturing practices, prices, availability, savings, sound decisions and brand recognition. These variables are included according to the literature reviewed and are intended to understand brand building according to the type of company. Data was worked through SPSS 25 with contingency tables and Chi-square analysis to determine the dependence of variables. Dependency values were measured by Cramer's V tests and the contingency coefficient to define the degree of association of the variables.

3 Results

According to the results obtained with the elaboration of the questions to the commercial managers of the family business (La Italiana) and the non-family business (Plumrose), six differences are determined: (a) the family business sees innovation in its processes as an objective while the non-family business aims to grow; (b) The values of the family business are based on integrity, solidarity and innovation and those of the non-family business in leadership and commitment; (c) There is less autonomy and greater control in decision making within the family business; (d) The family business does not know the requirements and requests of the clients, while the non-family seeks to be informed on what the client wants; (e) The family business measures its growth according to its financial performance as opposed to the non-family, that does so, through its growth in brand awareness (looking for new distribution channels); and (f) Marketing actions are similar, except for the use of sales promotion by the non-family company (Table 1).

Table 1. Differences in the management of processes and brands in the family and non-family business

Variables	Family business	Non-family business
Objectives	Innovation in processes	Growth
Values	Integrity, solidarity, innovation	Leadership, commitment
Autonomy	Low	High
Control	High	Low
Clients	Under client's knowledge	High customer's knowledge
Growth measurement	Based on financial performance	Based on your brand awareness
Varied marketing actions	Varied	Varied with attention to sales promotion

For the consumer's analysis dimensions, the average of each of the dimensions of the CSI model was first obtained. After obtaining the mean, the deviation index on this was calculated in each of the variables of the study dimension. In this way, the highest indices represent the most significant variable within said dimension as shown in Table 2.

Table 2. Featured variables for each of the study marks, in the CSI analysis

Dimension	Family business – La Italiana		Non-family business - Plumrose	
	Element	Index	Element	Index
Perfectionism	Need satisfaction	1,15	Quality importance	1,16
Brand	Supermarkets offer better products	1,18	Supermarkets offer better products	1,18
Fashion	Buy new product	1,08	Buy new product	1,12
Hedonism	Pleasant purchase	1,07	Pleasant purchase	1,15
Price awareness	Price consideration	1,06	Price consideration	1,05
Impulse	Purchase detail	1,19	Purchase detail	1,18
Confusion	Brand confusion	1,05	Brand confusion	1,03
Loyalty	Favorite Brand	1,10	Favorite Brand	1,14
Ecology	Balance between cold meats and healthy foods	1,08	Balance between cold meats and healthy foods	1,11
Health awareness	Choice of healthy food	1,01	Choice of healthy food	1,02
Purchase intention	Initial purchase of vegetable cold meats	1,04	Initial purchase of vegetable cold meats	1,02
Type of Company	Brand recognition	1,14	Brand recognition	1,17

Despite showing a lot of similarity in the valuation of brands, you can notice the difference in the perfectionist dimension and the confusion dimension. In the case of La Italiana, customers believe that the product must meet their needs, unlike Plumrose in which they consider the importance of quality as a relevant variable, similarly, in the confusion dimension there is a difference since while La Italiana customers show confusion in the brand, Plumrose customers are confused by the product. It should also be considered that the impulsive dimension is the one with the highest score in both Italian and Plumrose with a rating of 1.19 and 1.18 respectively, which indicates that customers carefully look at what they spend when they buy cold meats. On the contrary, the health dimension has the lowest indices in both brands, because clients mostly do not consider important health issues related to the purchase of cold meats.

With the data presented, a statistical analysis was carried out applying the chi-square test in order to understand the dependence between the brand and the dimensions of the CSI model. Together with obtaining asymptotic significance ($p < .05$ demonstrates dependence on variables), Cramer's V tests and the contingency coefficient were applied to determine the degree of association of the variables that are dependent on each other. In this way, the results of positive dependence are shown in Table 3.

According to the asymptotic significance (p value), there is a dependency relationship between five of the twelve dimensions, taking into account that the score of each is less than 0.05. On the other hand, in the analysis of association of V of Cramer it is observed that there is no perfect relationship between any of the brand variables (La Italiana and Plumrose) considering that the average of scores ranges between 0.15 and 0.20, being the highest ratio, the importance of national brands. Similarly, the contingency coefficient demonstrates values similar to those obtained with the Cramer V test, which shows a low association between the study marks and the dimensions of the CSI model and family business.

4 Discussion and Conclusion

Data found around the entire investigation contributed with knowledge to identify the main variables that give value to the brand, since consumption habits were analyzed to know the acceptance, recognition and empowerment of each of them. Each brand is linked to the profile of a segment, where each one grants a complete, dynamic and convenient offer of products to its customers, thus generating notoriety, creating reputation [11] and generating involvement [12] in such a way that its value is created [13].

The CSI model [14, 15] defines that values are very similar between the two brands selected for the study, differing in their confusion and satisfaction of need in the brand of family business, with brand recognition, confusion of the product and the importance of quality in the brand of the unfamiliar company; In addition, the company's own values are considered as an internal vision that also seeks to be part of brand building.

When carrying out the analysis of variable's dependence, we can confirm that the predominant variable is the confusion in the family business and brand recognition in the non-family business, elements included in the dimensions of confusion and the dimension related to the type of company, which intentionally was placed to determine the differentiation and in which it is shown that the product of the family business does not show high recognition as that of the non-family business in the elements described in the literature review, that is, the history and permanence in time of the company [18, 16], its growth and development [19, 20], professionalism [21], continuity and succession [22], heritage [16], profitability and investments [23], socialization [24] and temporality [25, 26]. The above-described evidence that the family business does not yet define characteristic features of their family status in the construction of the brand of their products and differs rather, for their satisfaction of the need of the housewife, that is, in product's functionality.

The methodology described in relation to the CSI model can be applied to brands of all types of companies, creating new lines of research that may be of interest to know and generate their respective valuation. We also consider that the study, despite being defined with a representative sample of housewives selected by their socioeconomic status, may have a better sample selection in the future for a comparative analysis that

Table 3. Dependent variables and degree of association between brand and elements of CSI dimensions

Type of consumer according to CSI (dimension)	Brand elements La Italiana and Plumrose	Value	df	Dependency	Association	
				Bilateral asymptotic significance	Cramer V	Contingency coefficient
Brand consumer (Brand)	Importance of national brands	16,201	4	0,003	0,205	0,201
	Choice of expensive brands	10,409	4	0,034	0,164	0,162
	Higher price, better quality	13,866	4	0,008	0,190	0,186
	Advertised brands/ Excellent purchase option	10,456	4	0,033	0,165	0,163
Recreational and hedonistic consumer (Hedonism)	Time wasting	10,869	4	0,028	0,168	0,166
Price conscious consumer (Price awareness)	Lower priced product	11,654	4	0,020	0,174	0,171
Consumer confused by the excess of options (Brand confusion)	Brand Confusion	9,391	4	0,052	0,156	0,154
Consumer by type of company (Type of company)	Ease of finding products	11,46	4	0,022	0,173	0,170
	Higher savings	11,77	4	0,019	0,175	0,172
	Decisions for customer satisfaction	10,933	4	0,027	0,169	0,166
	Brand recognition	9,614	4	0,047	0,158	0,156

goes beyond two brands, as well as also, you can determine a more rigorous statistical model to determine correlations of variables. In the same way, the study can be applied in recognized family and non-family business for international markets with similar dimensions or scope, to determine if there are other differences, beyond those obtained in this case.

To conclude, we consider that the application of this model will be used to analyze the different variables of two or more competing brands, in order to thoroughly know each detail that the consumer considers of greater relevance in the purchase of its products through quantifiable variables.

References

1. Camino-Mogro, S., Bermúdez-Barrezueta, N.: las empresas familiares en el ecuador: definición y aplicación metodológica. X-Pedientes Económicos **2**(3), 46–72 (2018)
2. Siabato, M.F., Oliva, E.J.: Evolución y caracterización de los modelos de Brand Equity. Suma Negocios **5**(12), 158–168 (2014)
3. Tasci, A.D.A.: Consumer value and brand value: rivals or allies in consumer-based brand equity? Tour. Anal. **21**(5), 481–496 (2016)
4. Mohan, M., Jiménez, F., Brown, B., Cantrell, C.: Brand skill: linking brand functionality with consumer- based brand equity. J. Prod. Brand Manage. Relat. **26**(5), 477–491 (2017)
5. Capriotti, P.: Branding Corporativo. Fundamentos para la gestión estratégica de la identidad corporativa. Colección de Libros de la Empresa, Santiago de Chile (2013)
6. Williams, R.I.: Measuring family business performance: research trends and suggestions. J. Fam. Bus. Manage. **8**(2), 146–168 (2018)
7. Tegtmeier, S., Classen, C.: How do family entrepreneurs recognize opportunities? Three propositions. Review of International Business and Strategy **27**(2), 199–216 (2017)
8. Arenas-Cardona, H.A., Rico-Balvín, D.: La empresa familiar, el protocolo y la sucesión familiar. Estud. Gerenc. **30**(132), 252–258 (2014)
9. Vera-Martínez, J.: Perfil de Valor de Marca y la Medición de sus componentes. Acad. Revista Latinoamericana de Administración **41**, 69–89 (2008)
10. Aaker, D.A.: Measuring brand equity across products and markets. Calif. Manage. Rev. **38**(3), 102–120 (1996)
11. Keller, K.L.: Conceptualization, measuring and managing consumer based brand equity. J. Mark. **57**(1), 1–22 (1993)
12. González, E., Orozco, M., De la Paz, A.: Consumer-Based Brand Equity: La mejora de la medición - Evidencia empírica. Contaduría y Administración **235**, 217–239 (2014)
13. Escobar-Naranjo, S.: La equidad de marca "brand equity" Una Estrategia para Crear y Agregar Valor. Estud. Gerenc. **1**, 35–41 (2000)
14. Sproles, G.B., Kendall, E.: A methodology for profiling consumers' decision-making styles. J, Consum. Aff. **20**(2), 267–279 (1986)
15. Ungaretti, J.: Adaptación del Consumer Style Inventory (CSI) al contexto argentino. Revista de Investigación en Psicología Social **1**(1), 25–34 (2013)
16. Goyzueta, S.: Modelo de gestión para las empresas familiares con perspectivas de crecimiento y sostenibilidad. Departamento de Administración, Economía y Finanzas (31), 87–132 (2013)
17. Tàpies, J.: Empresa familiar: un enfoque multidisciplinar. Universia Business Rev. **32**, 12–25 (2011)
18. Echaiz-Moreno, D.: El Protocolo Familiar. La Contractualización en las Familias Empresarias para la Gestión de las Empresas Familiares. Boletín Mexicano de Derecho Comparado **43**(127), 101–130 (2010)
19. Almaraz, A., Ramírez, L.A.: Familias Empresariales en México. Sucesión generacional y continuidad en el siglo XX. El Colegio de la Frontera Norte, Tijuana (2018)
20. Gallo, M.A.: Tipología de las empresas familiares. Revista Empresa y Humanismo **3**(2), 241–258 (2003)
21. Rueda-Galvis, J.F.: La Profesionalización, Elemento Clave del Éxito de la Empresa Familiar. Revista Científica Visión del Futuro **15**(1) (2011)
22. Omaña-Guerrero, L.M., Briceño-Barrios, M.A.: Gerencia de las empresas familiares y no familiares: análisis comparativo. Estudios Gerenciales **29**, 293–302 (2013)

23. Miralles-Marcelo, J.L., Miralles-Quirós, M.M., Lisboa, I.: Empresa familiar y bolsa: Análisis de rentabilidad y estrategias de inversion. Revista Española de Financiación y Contabilidad **41**(155), 393–416 (2012)
24. Cebotarev, E.A.: Familia, socialización y nueva paternidad. Segunda Sección: estudios e investigaciones, pp. 1–19 (2003)
25. Rodríguez, J., Rodríguez, M.: La singularidad de la empresa familiar: Conceptos básicos para llegar a entenderla. Cátedra PRASA de Empresa Familiar de la Universidad de Córdoba. Córdoba (2004)
26. Molina-Parra, P., Botero-Botero, S., Montoya-Restrepo, A.: Estudios de rendimiento en las empresas de familia. Una nueva perspectiva. Estudios Gerenciales **33**, 76–86 (2017)

What Are You Offering?: An Overview of VODs and Recommender Systems in European Public Service Media

Martín Vaz Álvarez[1]([⊠]), José Miguel Túñez López[1], and María José Ufarte Ruíz[2]

[1] Faculty of Communication Sciences, University of Santiago de Compostela, Av. Castelao s/n, 15782 Santiago, Spain
{martin.vaz.alvarez,miguel.tunez}@usc.es
[2] Faculty of Journalism, University of Castilla-la Mancha, Av. Camilo José Cela 50, 13071 Ciudad Real, Spain
mariajose.ufarte@uclm.es

Abstract. This article aims to give a general overview of the situation of recommender systems within European Public Service Media. As Video On Demand content consumptions increase worldwide, Public Service Media must step up their platforms with technologies that are able to compete with privately-owned giants of the industry and also deliver a technologically efficient solution to fulfill their public service commitments, and establish the standards by which the television of the future will be assessed. Through literature review and an empirical analysis of how VODs are implemented in European PSM, we will gauge the relevance of recommender systems within PSM, determine the number of them currently in use and describe them in relation with the messages on their VOD platforms.

Keywords: PSM · Recommender system · Europe · Public television · VOD · Universality · Diversity · Innovation

1 Introduction: The VOD Scenario

During the last few years we have seen audiences worldwide adapt new forms of content consumption with the bloom of VOD (Video on Demand) services [2] and algorithmic automatizations [17] propelled among other things by the proliferation of SmartTV's [10] Particularly in Europe, we have observed how spectators are increasingly spending more time watching non-linear television (free, subscription video on demand and transaction video on demand), and increasingly less time watching traditional linear TV [4, 10].

Although there has been an overall increase on minutes per day per person spent watching audiovisual contents, in countries such as the United Kingdom, Italy and Spain, VOD services have scratched minutes off of traditional television viewing [2]. This tendency is expected to boost in the coming years, as new competitors like Apple and Disney, among others, are set to enter the market by the end of 2019 and throughout 2020 [4].

© Springer Nature Switzerland AG 2020
Á. Rocha et al. (Eds.): ICITS 2020, AISC 1137, pp. 725–732, 2020.
https://doi.org/10.1007/978-3-030-40690-5_69

This places two challenges for European Public Service Broadcasters: on one hand, the swift changes in the audience's habits will eventually require them to develop increasingly more advanced and VOD-oriented content strategies in order to retain the moving audiences; and on the other, as the fight for the viewer's attention intensifies, with the exponentially-growing subscription platforms like Netflix, HBO or Amazon and the forthcoming Disney + or AppleTV, Public Service Broadcasters will need to dedicate a special attention to the development of their VOD platforms.

> *"The main competitive advantages of these new players come from their global scale, huge financial resources, state-of-the art consumer experience and the fact that they own the customer relationship, which is not the case for most of the traditional players in the European market."* [4].

This prospective scenario has already provoked some reaction on behalf of European Public Broadcasters, as it is the case for Spain, where public network RTVE has partnered with privately-owned Atresmedia and Mediaset to launch a new HbbTV (Hybrid Broadcast Broadband TV) service, which will probably become an OTT ("Over The Top") service in the future, called Loves TV [5, 6, 14]. On UK's side, last year OFCOM head Sharon White called for UK broadcasters BBC, ITV and Channel 4 to collaborate on a common service in order to gain scale and compete with global tech giants [4].

> *"Will the current PSBs be successful in the on-demand world? There will be significant new revenue streams in the digital world, and it is unclear how much of that revenue the traditional commercial broadcasters will be able to capture."* [1].

'Capturing' that revenue is seen here as a way to maintain relevance within the context of the future audiovisual landscape, but along with the production of high quality content, there is too the need to display this in an appealing, user friendly way. Indecision being such a relatable feeling when sitting in front of the TV, PWC put it into numbers in late 2017 with a survey on US viewer's attitudes that showed that 55% of VOD users search for new shows every week, with 62% expressing a struggle to find something to watch [12].

> *"Entertainment and media companies have long competed on two dimensions: content and distribution. To thrive in today's increasingly competitive, crowded, slow-growth marketplace, however, they must focus on a third dimension: user experience."* [12].

According to the same study, 33% of the respondents noted that they would choose a particular VOD platform because it is easy to use, and another 22% says that they'd make this choice if they know they'll always be able to choose something to watch. More importantly, 74% agree that despite there being a lot of choices available, they often struggle to find something to watch [12]. These numbers show the importance of coordinating content plus user experience, as it's become clear by both public and private agents dedicating a significant part of its developing efforts to a tool that combines content and layout: recommender systems.

1.1   Recommender Systems for Public Service

Recommender systems signify the technological response to the ever growing accumulation of contents online, as a tool to retrieve, curate and propose elements to be presented to the spectator according to different parameters [15, 19] with the purpose of keeping the spectator engaged to their platform. These parameters represent the core of the recommender system, it's 'personality' or rationale, and are the basis for artificial intelligence and machine learning developments across all disciplines [8, 9].

Depending on the type of platform in use, parameters may vary in order to conciliate the user's demands with the platform-owner's interest [8]. Personalization parameters are here relevant in helping the user find relevant content [13]. So for example, a profit-oriented SVOD like Netflix will take into account parameters such as ratings for shows, demographic particularities, tagging of contents, or similarity to what other users have watched [15] in order to present the contents that have a higher chance of keeping you engaged. But how does this apply in Public Service Media? Should the rationale of a recommender system within PSM be restricted to engagement? How do we apply public service values in algorithmic recommendation systems?

These questions have been already and object for research internationally, but more intensely within the European context. One of the most common conflicts when assessing the rationale behind recommender systems is the principle of engagement. Whilst engagement, is key to maintain relevance, most academics agree this must not come at any cost.

Based on John Reith's remarks on 'giving the public what it wants even thought it doesn't know it wants it' [7, 16] recommender systems must take into account notions of universality, diversity, transparency and impartiality [7, 16, 18] which are too fundamental to reinforce the PSM's legitimacy [3]. In other words, comply with their mission to fulfill their public commitments, as the failure to do so would compromise its commitment to society and could risk exposing the audience to one-sided or bubble-filtered content based exclusively on their own consumption patterns or pre-defined 'likable' content [11].

1.2   The Values in Recommender Systems

As recommender systems become more advanced (and needed) in Public Service Media's VOD platforms, the need for a more in-detail analysis of its functioning grows too, as these algorithms will be, in time, responsible for the audiovisual consumption of increasingly more VOD users.

But in order to asses these functionalities, we must take as a reference the values over which these systems will be programmed upon, what will be the rules they'd have to follow for public service. The way we interpret these rules will be key to the development of these programs, as some authors express: "The notion that PSM values offer distinct frameworks for recommender systems is underpinning new EBU initiatives to develop distinctly PSM approaches to recommendations". [7].

The European Broadcasting Union and its members have pledged to 6 core values for Public Media Service, which are: universality, independence, excellence, diversity, accountability and innovation. Out of these 6, at least 3 are directly connected to a PSM recommender system's task: universality, diversity and innovation:

- Universality, as it must cater contents for all the population, which has to be balanced with the nature of public service content and the personalization features of the recommender system.

- Diversity, as the recommender system would have to assume the responsibility to balance the audience's inputs with contents that inform and educate the spectator through a wide social, cultural and political scope.
- Innovation, as in the process of developing its own algorithms and perfecting its curation capabilities will fulfill its commitment to creating value through novelty and the application of technologies for an improved public service.

With all, every recommender system must take into account the specifics of their society, as the needs, tastes and media consumption habits of each country may require particular recommendation parameters [7].

2 Results

After analyzing a total of 56 Public Service Media in Europe, we have found that 16 of them include in their VODs at least some type of suggestion or recommendation to visualize content, either it being through a 'we recommend...', or a more specific 'recommended for u...'.

The following table represents the list of the European public broadcasters that currently have an active recommender system or at least recommendations elements in their VOD platforms. There is a total of 16 PSM that currently include recommendation elements in their audiovisual online services, 2 of them in Germany, as in the particular case of the ARD, its VOD is based on aggregated contents from all the other 9 regional television networks (Table 1).

Table 1. List of all European PSM with recommendation elements in their VODs.

Public Service Media in Europe with Recommendation Elements	
Country	Name
Germany	ARD
United Kingdom	BBC
Russia	Channel One
Greece	ERT
Estonia	ETV ERR
France	France TV
Croatia	HRTi
Holland	NPO
Italy	RAI
Montenegro	RTCG
Ireland	RTÉ
Portugal	RTP
Switzerland	RTS
Malta	TVM
Finland	YLE
Germany [2]	ZDF

In total, 28,57% of Public Service Media in Europe with a Video On Demand service have content recommender elements. Although in some cases these recommendations do not respond to an algorithmic logic, or "system", it is notable that an intention to personalize and cater specific content is present in all 16 PSM.

In order to asses more in-depth the level of personalization in each of the VOD systems, a further semantic analysis was made, identifying notable differences when proposing contents.

The following wordcloud tries to visually illustrate the most common words found in the VOD's suggestion titles (more common shows up as bigger). The two most common words are 'recommended' and 'recommend' which, although they may seem similar, they refer to two different perspectives on recommendations: (we) recommend, and recommended (to you), which suggest a contraposition between corporative curation and personalization (Fig. 1).

Fig. 1. Wordcloud with the recommendation elements present in the PSMs' VODs.

The results showed that there is at least 3 types of messages present in these VODs, which describe the level of personalization on each case and allow for a glimpse of the logic behind the recommender system:

- **Direct appeal to the spectator:** 'we advice you watch', 'recommended for you', 'have you seen this?', 'you can't miss this', 'options for you', 'recommendations for you'.
- **On behalf of the PSM:** 'it's worth watching', 'watch this week', 'recommended emissions', 'we recommend', 'our selection'
- **Content oriented:** 'binge from the start', 'popular', 'highlighted', 'more seen', 'daily topic', 'something new to watch', 'more recently seen'.

A direct appeal to the spectator suggest that the reccommendation system has taken into account certain aspects of the user's profile in order to make these suggestions. This direct recommendation is present in 9 out of 16 cases, although a more in-depth, exhaustive use of these platforms would be required in order to state if these messages actually remit to a personalized suggestion or if it's just a birdcall for the spectator to click on it.

On the other hand, messages such as 'we recommend', or 'our selection', which are present in 7 out of 16 PSM's, clearly refer to a pre-made recommender structure and do not suggest any particular personalization of the contents, although it shows and intent to promote certain objects over others.

Lastly, the third type of message suggests a curation in relation to the content itself, this is, referring to the show's own nature either by its format, popularity or novelty. It's present in 6 out of the 16 VODs, with words such as 'popular', 'highlighed' or phrases like 'most recently seen' we understand there is a system behind the recommendations that uses big data to organize the VOD's content.

Again, a further extensive use of the recommender systems would be appropriate to determine whether the viewing history of the user is also computed and used for more precise, purposeful recommendations.

3   Discussion and Conclusions

In this article we have drawn an overview of how VODs are starting to take a leading role in the distribution of contents. As privately-owned SVODs are growing and expanding exponentially, PSM's have already started to probe the terrain for possible alliances that will help keeping their space in the audiovisual sector.

Recommender systems are in this matter one of the keys for the improvement of these platforms, and also present opportunities for PSM to fulfill their public service commitments of universality, diversity and innovation. However, ethical questions have arisen regarding what the parameters for these recommender systems should be, not to lead the spectator into a deceptive or unbalanced 'media diet'.

As far as European PSMs are concerned, recommendation elements are present in almost one third of all PSMs that present a Video On Demand platform in their websites. The current projections of this technology and the already official appearance of new commercial rivals suggests that the number of these may increment in the near future.

When it comes to the messaging by which the recommendations are catered, we have found notable differences. There are at least 3 stages of recommendation that seem to refer to different designs. In 2 of them, the direct appeal and the content oriented design, there are nuances of a big data based algorithmic system working in the background. In the remaining, a non data-driven corporative curation appears to be behind.

With all, more empyrical research through a longitudinal, concious use of these recommender systems is required in order to unravel the nature of its algorithms and thus, its true potential to deliver content that fits within the public service values.

4 Research Acknowledgements

This article has been elaborated within the framework of the I+D Project "New values, governance, financing and public audiovisual services for the Internet society: European and Spanish contrasts", (reference: RTI2018-096065-B-I00) from the I+D+i State Program oriented to the Challenges of Society from the Ministry of Science, Innovation and Universities and the European Fund for Regional Development (FEDER).

The author, Martín Vaz Álvarez is granted with a Training University Lecturers contract (FPU) from the Ministry of Science, Innovation and Universities of Spain (ref. FPU19/06204).

References

1. BBC. Summary of market modelling, https://tinyurl.com/yx4drwav. Accessed 02 Sept 2019
2. Begum, F., Moyser, R.: Cross-Platform Television Viewing Time Report- 2019. IHS Markit (2019)
3. Campos Freire, F., Rodríguez Castro, M., de Mateo Pérez, R.: The trend of assessment indicators for public service media in europe. Communication: Innovation & Quality, pp. 3–19 (2019). https://doi.org/10.1007/978-3-319-91860-0_1
4. Council of Europe. Yearbook 2018/2019. Key Trends. Television, Cinema, Video and On-Demand Audiovisual Services - The Pan-European Picture. European Audiovisual Observatory (2019)
5. Europa Press, RTVE, Atresmedia y Mediaset lanzan hoy 'LOVEStv' con la creación de un servicio OTT como posible objetivo final, https://tinyurl.com/us2da7s. Accessed 03 Sept 2019
6. El Español, Cellnex: el gran ganador del nuevo 'Netflix' de Mediaset, Atresmedia y RTVE, https://tinyurl.com/vff8tmu. Accessed 02 Sept 2019
7. Fields, B., Jones, R., Cowlishaw, T.: The Case for Public Service Recommender Algorithms. BBC London, p. 22–24 (2018)
8. Horváth, T., de Carvalho, A.: Evolutionary computing in recommender systems: a review of recent research. Nat. Comput. 16(3), 441–462 (2017)
9. Kane, F.: Building Recommender Systems With Machine Learning and AI: Help People Discover New Products and Content With Deep Learning, Neural Networks, and Machine Learning Recommendations. Amazon Publishing, Seattle (2018)
10. Ofcom. Media Nations: UK, 2018. Ofcom Annual Media Nations Report (2018)
11. Pariser, E.: The troubling future of internet search. The Futurist: World Trends and Forecasts (2011)
12. PWC, How tech will transform content discovery. PricewaterhouseCoopers Consumer Intelligence Series, https://tinyurl.com/r69zrq5. Accessed 02 Sept 2019
13. Ricci, F., Rokach, L., Shapira, B., Kantor, P.B.: Recommender systems handbook. Springer, Heidelberg (2011)
14. RTVE, RTVE, Atresmedia y Mediaset España presentan la identidad corporativa de LOVEStv, la plataforma conjunta de contenidos con tecnología HbbTV, https://tinyurl.com/y97eefw8. Accessed 03 Sept 2019
15. Sohail, S., Siddiqui, J., Ali, R.: Classifications of recommender systems: a review. J. Eng. Sci. Technol. Rev. 10(4), 132–153 (2017)

16. Sørensen, J., Hutchinson, J.: Algorithms and public service media. In: Public Service Media in the Networked Society RIPE, pp. 91–106 (2018)
17. Túñez López, J.M., Toural-Bran, C., Valdiviezo-Abad, C.: Automatización, bots y algoritmos en la redacción de noticias. Impacto y calidad del periodismo artificial. Revista Latina de Comunicación Social, 74, pp. 1.411– 1.433 (2019)
18. Van den Bulck, H., Moe, H.: Public service media, universality and personalisation through algorithms: mapping strategies and exploring dilemmas. Media Cult. Soc. **40**(6), 875–892 (2017)
19. Zhang, J., Wang, Y., Yuan, Z.: Personalized real-time movie recommendation system: Practical prototype and evaluation. Tsinghua Sci. Technol. **25**(2), 180–191 (2019)

Printed in the United States
By Bookmasters